Third Edition

THE SCIENCE OF WATER

Concepts and Applications

Third Edition

THE SCIENCE OF WATER

Concepts and Applications

Frank R. Spellman

CRC Press is an imprint of the
Taylor & Francis Group, an **informa** business

CRC Press
Taylor & Francis Group
6000 Broken Sound Parkway NW, Suite 300
Boca Raton, FL 33487-2742

Printed on acid-free paper
Version Date: 20140324

International Standard Book Number-13: 978-1-4822-4293-5 (Hardback)

Library of Congress Cataloging-in-Publication Data

Spellman, Frank R.
The science of water : concepts and applications / author, Frank R. Spellman. -- Third edition.
pages cm
Includes bibliographical references and index.
ISBN 978-1-4822-4293-5 (alk. paper)
1. Water. 2. Water--Industrial applications. 3. Water--Pollution. I. Title.

GB665.S64 2014
553.7--dc23 2014011463

Visit the Taylor & Francis Web site at
http://www.taylorandfrancis.com

and the CRC Press Web site at
http://www.crcpress.com

Contents

To the Reader

While reading this text, you will spend some time following a drop of water on its travels. Dip your finger in a basin of water and lift it up again. Out of the water below you will bring with it a small glistening drop. Do you have any idea where this drop has been? What changes it has undergone during all the long ages that water has lain on and under the face of the Earth?

Take life easy; eat, drink and be merry.

Luke 12:19

Preface

Hailed on its first publication as a masterly account for the general reader, the water practitioner, and the student, *The Science of Water: Concepts and Applications* continues to ask the same questions: Water, water everywhere … right? Can the Earth's supply of finite water resources be increased to meet growing demand? Many hold the persistent belief that the Earth's finite water resources can somehow grow. History has demonstrated that consumption and waste increase in response to rising supply, but the fact of the matter is that freshwater is a finite resource that can be increased only slightly through desalinization or some other practices, all at tremendous cost.

In addition to asking the same questions, this text has now been completely revised and expanded for the third edition. The text still deals with the essence of water—that is, what water is all about. Further, although water is one of the simplest and most common chemical compounds on Earth, this text does not ignore the fact that water is one of the most mysterious and awe-inspiring substances we know. Important to this discussion about water and its critical importance to the inhabitants of Earth are humans—their use, abuse, and reuse of wastewater. Furthermore, this text takes the view that because water is the essence of all life on Earth it is precious—too precious to squander. Thus, as you might guess, the common thread woven throughout the fabric of this presentation is water resource utilization and protection. Additionally, it is important to point out that this new edition provides additional information not covered in previous renditions; for example, the chapter dealing with still waters includes an upgrade on waste stabilization ponds. Current updates on pertinent regulations dealing with important aspects of water supply and treatment are provided. Additionally, a new chapter covering the economics of water and another new chapter dealing with economical use of water are included in this updated edition.

The text follows a pattern that is nontraditional; that is, the presentation here is based on real-world experience—not on theoretical gobbledygook. Clearly written and user friendly, this timely revision of *The Science of Water: Concepts and Applications* builds on the remarkable success of the first and second editions. Still written as an information source, it should be pointed out that this text is not limited in its potential for other uses. For example, although this text can be utilized by water/wastewater practitioners to gain valuable insight into the substance they work hard to collect, treat, and supply, it can just as easily provide important information for policymakers who may be tasked with making decisions concerning water resource utilization. Consequently, this book will serve a varied audience—students, lay personnel, regulators, technical experts, attorneys, business leaders, and concerned citizens.

What makes this third edition of *The Science of Water: Concepts and Applications* different from other available science books? Consider the following: The author has worked in and around water/wastewater treatment and taught water/wastewater science, operations, safety, and math for several years at the apprenticeship level and at various college levels for students, in addition to teaching short courses for operators. Additionally, the author has sat at the table of licensure examination preparation boards to review, edit, and write state water licensure exams. This text was written by an author who has personally spent a considerable portion of his adult life wading into the Teton, Yellowstone, Shoshone, Columbia–Snake, and Flathead rivers to study, sample, and assess their condition, as well as into the somewhat warmer waters of the Virgin River, Moose Creek, and Bitch Creek of Idaho.

This step-by-step book provides concise, practical knowledge that practitioners and operators must have to pass licensure and certification tests. The text is user friendly; no matter the difficulty of the problem to be solved, each operation is explained in straightforward, plain English. Moreover, several hundred example math problems are presented throughout to enhance the learning process.

The first and second editions were highly successful and well received, but like any flagship and follow-up editions of any practical manual there is always room for improvement. Thankfully, many users have provided constructive criticism, advice, and numerous suggestions, and these inputs from actual users have been incorporated into this new edition. The bottom line is that the material in this text is presented in manageable chunks to make learning quick and painless through clear explanations that readily provide an understanding of the subject matter; it is packed with worked examples and exercises.

A knowledgeable reader might wonder whether water treatment, wastewater treatment, and other work with water are more of an art than a science, but it should be pointed out that the study of water is a science—a science that is closely related to other scientific disciplines such as biology, microbiology, chemistry, mathematics, hydrology, and others. Thus, to solve the problems and understand the issues related to water, water practitioners need a broad base of scientific information from which to draw. Consider, for example, a thoracic surgeon (thoracic surgery is the Major League of surgery, according to a thoracic surgeon I know) who has a reputation of being an artist with a scalpel. This information might be encouraging to the would-be patient who requires thoracic surgery. However, this same patient might further inquire about the surgeon's education, training, experience, and knowledge of the science of medicine. If I were the patient, I would want my surgeon to understand the science of my heart and other vital organs before taking up a scalpel and performing the surgery. Wouldn't you?

Author

Frank R. Spellman, PhD, is a retired assistant professor of environmental health at Old Dominion University, Norfolk, Virginia, and the author of more than 90 books covering topics ranging from concentrated animal feeding operations (CAFOs) to all areas of environmental science and occupational health. Many of his texts are readily available online, and several have been adopted for classroom use at major universities throughout the United States, Canada, Europe, and Russia; two have been translated into Spanish for South American markets. Dr. Spellman has been cited in more than 450 publications. He serves as a professional expert witness for three law groups and as an incident/accident investigator for the U.S. Department of Justice and a northern Virginia law firm. In addition, he consults on homeland security vulnerability assessments for critical infrastructures, including water/wastewater facilities, and conducts audits for Occupational Safety and Health Administration and Environmental Protection Agency inspections throughout the country. Dr. Spellman receives frequent requests to co-author with well-recognized experts in various scientific fields; for example, he is a contributing author to the prestigious text *The Engineering Handbook*, 2nd ed. Dr. Spellman lectures on sewage treatment, water treatment, and homeland security, as well as on safety topics, throughout the country and teaches water/wastewater operator short courses at Virginia Tech in Blacksburg. He holds a BA in public administration, a BS in business management, an MBA, and both an MS and a PhD in environmental engineering.

1 Introduction

When color photographs of the Earth as it appears from space were first published, it was a revelation: they showed our planet to be astonishingly beautiful. We were taken by surprise. What makes the earth so beautiful is its abundant water. The great expanses of vivid blue ocean with swirling, sunlit clouds above them should not have caused surprise, but the reality exceeded everybody's expectations. The pictures must have brought home to all who saw them the importance of water to our planet.

Pielou (1998)

Fresh water is unique from other commodities in that it has no substitutes.

Postel et al. (1996)

Whether we characterize it as ice, rainbow, steam, frost, dew, soft summer rain, fog, flood, or avalanche, or as something as stimulating as a stream or cascade, water is special—water is strange—water is different.

Water is the most abundant inorganic liquid in the world; moreover, it occurs naturally anywhere on Earth. Literally awash with it, life on this planet depends on it, and yet water is so very different.

Water is different, scientifically. With its rare and distinctive property of being denser as a liquid than as a solid, it is different. Water is different in that it is the only chemical compound found naturally in solid, liquid, gaseous states. Water is sometimes called the universal solvent. This is a fitting name, especially when you consider that water is a powerful reagent capable of dissolving everything on Earth.

Water is different. It is usually associated with all the good things on Earth; for example, water is associated with quenching thirst, with putting out fires, and with irrigating our farms. The question, though, is can we really say, emphatically and definitively, that water is associated with only those things that are good?

Not really! Remember, water is different; nothing, absolutely nothing, is safe from it.

Water is different. This unique substance is odorless, colorless, and tasteless. Water covers 71% of the earth completely. Even the driest dust ball contains 10 to 15% water.

Water and life, life and water—inseparable.

The prosaic becomes wondrous as we perceive the marvels of water. Covering the Earth are 326,000,000 cubic miles of water, but only 3% of this total is fresh, with most of that is locked up in polar ice caps, glaciers, and lakes, or in flows through soil and in river and stream systems that return to an increasingly saltier sea (only 0.027% is available for human consumption).

Water is different. Standing at a dripping tap, water is so palpably wet that one can literally hear the drip–drop–plop.

Water is special, water is strange, water is different. More importantly, water is critical to our survival, yet we sometimes abuse it, discard it, foul it, curse it, damn it, or ignore it. However, because water is special, strange, and different, the dawn of tomorrow is pushing for quite a different treatment of water.

Along with being special, strange, and different, water is also a contradiction, a riddle. Why? Consider the Chinese proverb that "water can both float and sink a boat." Saltwater is different from freshwater. This text deals with freshwater and ignores saltwater because saltwater fails water's most vital duties: to be pure and sweet and serve to nourish us.

The presence of water everywhere feeds these contradictions. Lewis (1996, p. 90) pointed out that "water is the key ingredient of mother's milk and snake venom, honey and tears." Leonardo da Vinci gave us insight into more of water's apparent contradictions:

> Water is sometimes sharp and sometimes strong, sometimes acid and sometimes bitter;
> Water is sometimes sweet and sometimes thick or thin;
> Water sometimes brings hurt or pestilence, sometimes health-giving, sometimes poisonous.
> Water suffers changes into as many natures as are the different places through which it passes.
> Water, as with the mirror that changes with the color of its object, so it alters with the nature of the place, becoming: noisome, laxative, astringent, sulfurous, salt, incarnadined, mournful, raging, angry, red, yellow, green, black, blue, greasy, fat or slim.
> Water sometimes starts a conflagration, sometimes it extinguishes one.
> Water is warm and is cold.
> Water carries away or sets down.
> Water hollows out or builds up.
> Water tears down or establishes.
> Water empties or fills.
> Water raises itself or burrows down.
> Water spreads or is still.
> Water is the cause at times of life or death, or increase of privation, nourishes at times and at others does the contrary.
> Water, at times has a tang, at times it is without savor.
> Water sometimes submerges the valleys with great flood.
> In time and with water, everything changes.

We can sum up water's contradictions by simply stating that, although the globe is awash in it, water is no single thing but an elemental force that shapes our existence. Da Vinci's observation that "in time and with water, everything changes" concerns us most in this text.

Many of Leonardo's water contradictions are apparent to most observers, but with water there are other factors that do not necessarily stand out, that are not always so apparent. This is made clear by the following example—what you see on the surface is not necessarily what lies beneath.

DID YOU KNOW?

There is a lot of salty water on our planet. By some estimates, if the salt in the ocean could be removed and spread evenly over the Earth's land surface it would form a layer more than 500 feet (166 meters) thick, about the height of a 40-story office building. The question is: Where did all this salt come from? Stories in folklore and mythology from almost every culture offer explanations for how the oceans became salty, but the answer is really quite simple. Salt in the ocean comes from rocks on land. Here's how it works: The rain that falls on the land contains some dissolved carbon dioxide from the surrounding air. This causes the rainwater to be slightly acidic due to carbonic acid. The rain physically erodes the rock and the acids chemically break down the rocks; the resulting salts and minerals are carried along in the water in a dissolved state as ions. The ions in the runoff are carried to the streams and rivers and then to the ocean. Many of the dissolved ions are used by organisms in the ocean and are removed from the water. Others are not used up and are left for long periods of time so their concentrations increase over time. The two ions that are present most often in seawater are chloride and sodium. These two make up over 90% of all dissolved ions in seawater (USGS, 2013).

STILL WATER

Consider a river pool, isolated by fluvial processes and time from the main stream flow. We are immediately struck by one overwhelming impression: It appears so still … so very still … still enough to soothe us. The river pool provides a kind of poetic solemnity, if only at the pool's surface. No words of peace, no description of silence or motionless can convey the perfection of this place, in this moment stolen out of time.

We consider that the water is still, but does the term *still* correctly describe what we are viewing? Is there any other term we can use besides *still*—is there any other kind of still? Yes, of course, we know many ways to characterize still. *Still* can mean inaudible, noiseless, quiet, or silent. *Still* can also mean immobile, inert, motionless, or stationary—which is how the pool appears to the casual visitor on the surface. The visitor sees no more than water and rocks.

The rest of the pool? We know very well that a river pool is more than just a surface. How does the rest of the pool (the subsurface, for example) fit the descriptors we tried to use to characterize its surface? Maybe they fit, maybe they don't. In time, we will go beneath the surface, through the liquid mass, to the very bottom of the pool to find out. For now, remember that images retained from first glances are almost always incorrectly perceived, incorrectly discerned, and never fully understood.

On second look, we see that the fundamental characterization of this particular pool's surface is correct enough. Wedged in a lonely riparian corridor—formed by a river bank on one side and sand bar on the other—between a youthful, vigorous river system on its lower end and a glacier- and artesian-fed lake on its headwater end, almost entirely overhung by mossy old Sitka spruce, the surface of the large pool, at least at this particular location, is indeed still. In the proverbial sense, the pool's surface is as still and as flat as a flawless sheet of glass.

The glass image is a good one, because like perfect glass, the pool's surface is clear, crystalline, unclouded, definitely transparent, and yet perceptively deceptive as well. The water's clarity, accentuated by its bone-chilling coldness, is apparent at close range. Further back, we see only the world reflected in the water—the depths are hidden and unknown. Quiet and reflective, the polished surface of the water perfectly reflects in mirror-image reversal the spring greens of the forest at the pond's edge, without the slightest ripple. Up close, looking straight into the depths of the pool we are struck by the water's transparency. In the motionless depths, we do not see a deep, slow-moving reach with the muddy bottom typical of a river or stream pool; instead, we clearly see the warm variegated tapestry of blues, greens, blacks stitched together with threads of fine, warm-colored sand that carpets the bottom, at least 12 feet below. Still waters can run deep.

No sounds emanate from the pool. The motionless, silent water does not, as we might expect, lap against its bank or bubble or gurgle over the gravel at its edge. Here, the river pool, held in temporary bondage, is patient, quiet, waiting, withholding all signs of life from its surface visitor.

Then the reality check: This stillness, like all feelings of calm and serenity, could be fleeting, momentary, temporary, you think. And you would be correct, of course, because there is nothing still about a healthy river pool. At this exact moment, true clarity is present, it just needs to be perceived … and it will be.

We toss a small stone into the river pool and watch the concentric circles ripple outward as the stone drops through the clear depths to the pool bottom. For a brief instant, we are struck by the obvious: The stone sinks to the bottom, following the laws of gravity, just as the river flows according to those same inexorable laws—downhill in its search for the sea. As we watch, the ripples die away, leaving as little mark as the usual human lifespan creates in the waters of the world, then disappear as if they had never been. Now the river water is as before, still. At the pool's edge, we peer down through the depth to the very bottom—the substrate.

We determine that the pool bottom is not flat or smooth but instead is pitted and mounded occasionally with discontinuities. Gravel mounds alongside small corresponding indentations—small, shallow pits—make it apparent to us that gravel was removed from the indentations and piled into slightly higher mounds. From our topside position, as we look down through the cool, quiescent liquid, the exact height of the mounds and the depth of the indentations are difficult for us to judge; our vision is distorted through several feet of water.

However, we can detect near the low gravel mounds (where female salmon have buried their eggs and where their young will grow until they are old enough to fend for themselves), and actually through the gravel mounds, movement—water flow—an upwelling of groundwater. This water movement explains our ability to see the variegated color of pebbles. The mud and silt that would normally cover these pebbles have been washed away by the water's subtle, inescapable movement. Obviously, in the depths, our still water is not as still as it first appeared.

The slow, steady, inexorable flow of water in and out of the pool, along with the upflowing of groundwater through the pool's substrate and through the salmon redds (nests) is only a small part of the activities occurring within the pool, including the air above it, the vegetation surrounding it, and the damp bank and sandbar forming its sides.

Let's get back to the pool itself. If we could look at a cross-sectional slice of the pool, at the water column, the surface of the pool may carry those animals that can literally walk on water. The body of the pool may carry rotifers and protozoa and bacteria—tiny microscopic animals—as well as many fish. Fish will also inhabit hidden areas beneath large rocks and ledges, to escape predators. Going down further in the water column, we come to the pool bed. This is the benthic zone, and certainly the greatest number of creatures live here, including larvae and nymphs of all sorts, worms, leeches, flatworms, clams, crayfish, dace, brook lampreys, sculpins, suckers, and water mites.

We need to go down even farther, down into the pool bed, to see the whole story. How far we have to go and what lives here, beneath the water, depends on whether it is a gravelly bed or a silty or muddy one. Gravel will allow water, with its oxygen and food, to reach organisms that live underneath the pool. Many of the organisms that are found in the benthic zone may also be found underneath, in the hyporheic zone.

But to see the rest of the story we need to look at the pool's outlet, where its flow enters the main river. This is a riffle area—a shallow place where water runs fast and is disturbed by rocks. Only organisms that cling very well, such as net-winged midges, caddisflies, stoneflies, some mayflies, dace, and sculpins can spend much time here, and the plant life is restricted to diatoms and small algae. Riffles are a good place for mayflies, stoneflies, and caddisflies to live because they offer plenty of gravel to hide in.

Earlier, we struggled to find the right words to describe the river pool. Eventually, we settled on *still waters*. We did this because of our initial impression, and because of our lack of understanding and lack of knowledge. Even knowing what we know now, we might still describe the river pool as still waters. However, in reality, we must call the pool what it really is—a dynamic habitat. Each river pool has its own biological community, the members interwoven with each other in complex fashion, all depending on each other. Thus, our river pool habitat is part of a complex, dynamic ecosystem. On reflection, we realize, moreover, that anything dynamic certainly cannot be accurately characterized as still—including our river pool.

Maybe you have not had the opportunity to observe a river pool like the one described above. Maybe such an opportunity does not even interest you. Take a moment out of your hectic schedule, though, to perform an action most people never think about doing. Hold a glass of water (like the one in Figure 1.1) and think about the substance within the glass—about the substance you are getting ready to drink. You are aware that the water inside a drinking glass is not one of those items people usually spend much thought on, unless they are tasked with providing the drinking water—or are dying of thirst.

Earlier we stated that water is special, strange, and different. We find water fascinating—a subject worthy of endless interest because of its unique behavior, endless utility, and ultimate and intimate connection with our existence. Remember, there is no substitute for water. Perhaps you might agree with Robbins (1976, pp. 1–2), whose description of water follows:

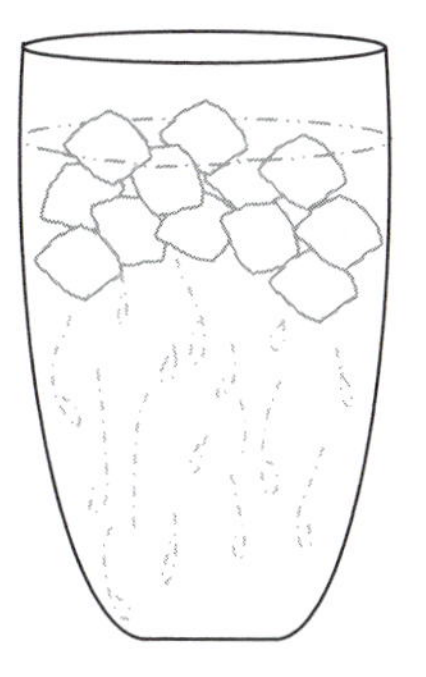

FIGURE 1.1 A glass of drinking water.

Stylishly composed in any situation—solid, gas, or liquid—speaking in penetrating dialects understood by all things—animal, vegetable or mineral—water travels intrepidly through four dimensions, *sustaining* (Kick a lettuce in the field and it will yell "Water!"), *destroying* (The Dutch boy's finger remembered the view from Ararat) and *creating* (It has even been said that human beings were invented by water as a device for transporting itself from one place to another, but that's another story). Always in motion, ever-flowing (whether at stream rate or glacier speed), rhythmic, dynamic, ubiquitous, changing and working its changes, a mathematics turned wrong side out, a philosophy in reverse, the ongoing odyssey of water is irresistible.

As Robbins observed, water is always in motion. The one most essential characteristic of water is that it is dynamic: Water constantly evaporates from seas, lakes, and soil and transpires from foliage. It is transported through the atmosphere and falls to Earth, where it runs across the land and filters down to flow along rock strata into aquifers. Eventually, water finds its way to the sea again—indeed, water never stops moving.

A thought that might not have occurred to most people looking at a glass of water is, "Has someone tasted this same water before us?" Absolutely. Remember, water is almost a finite entity. What we have now is what we have had in the past. We are drinking the same water consumed by Cleopatra, Aristotle, da Vinci, Napoleon, Joan of Arc (and several billion other folks who preceded us)—because water is dynamic (never at rest) and because water constantly cycles and recycles, as we discuss in the next section.

Water never goes away, disappears, or vanishes; it always returns in one form or another. As Dove (1989) noted, "All water has a perfect memory and is forever trying to get back to where it was."

SETTING THE STAGE

The availability of a water supply adequate in terms of both quantity and quality is essential to our very existence. One thing is certain: History has shown that the provision of an adequate quantity of quality potable water has been a matter of major concern since the beginning of civilization.

We know that we need water—especially clean, safe water—to survive. We know a lot about it, but the more we know the more we discover we don't know. Modern technology has allowed us to tap potable water supplies and to design and construct elaborate water distribution systems. Moreover, we have developed technology to treat used water (wastewater)—that is, water we foul, soil, pollute, discard, and flush away. Have you ever wondered where the water goes when you flush the toilet? Probably not.

An entire technology has developed around treating water and wastewater. The technological expertise required ranges from environmental, structural, and civil engineering to environmental science, geology, hydrology, chemistry, and biology, among others.

Along with those who design and construct water and wastewater treatment works, there is a large cadre of specialized technicians who operate the water and wastewater treatment plants. These operators are tasked, obviously, with either providing a water product that is both safe and palatable for consumption or treating a wastestream before it is returned to its receiving body (usually a river or stream). It is important to point out that water practitioners who treat potable and used water streams are not only responsible for ensuring the quality, quantity, and reuse of their product but are also tasked with, because of the events of 9/11, protecting this essential resource from terrorist acts.

The fact that most water practitioners know more about water than the rest of us comes as no surprise. For the average person, our knowledge of water usually extends to knowing no more than whether the water is good or bad: Does it taste good, or does it have a terrible taste? Does it look clean, cool, and sparkling, or is it full of scum, dirt, rust, or chemicals? Is it great for the skin or hair? Does it have some medicinal qualities? Thus, to say the water experts know more about water than the average person is probably an accurate statement.

At this point, the reader is probably asking: What does all this have to do with anything? Good question. What it has to do with water is quite simple. We need to accept the fact that we simply do not know what we do not know about water. As a case in point, consider this: Have you ever tried to find a text that deals exclusively and extensively with the science of water? Such texts are few and far between.

But, why would you want to know anything about water in the first place? Another good question. This text makes an effort to answer this question. To start with, let's talk a little about the way in which we view water. Earlier, brief mention was made about the water contents of a simple drinking water glass. Let's face it—drinking a glass of water is something that normally takes little effort and even less thought. The trouble is, our view of water and its importance is relative.

Consider, for example, a young woman who considered herself an adventurer, an outdoor person. She liked to jump into her four-wheel-drive vehicle and head out for new adventure. One fateful day she decided to drive through Death Valley, California—from one end to another and back again on seldom-used dirt roads. She had done this a few times before and made it back safely, but today she decided to take a side road that seemed to lead to the mountains to her right.

She traveled along this isolated, hardpan road for approximately 50 miles before the motor in her four-wheel-drive vehicle quit. No matter what she did, the vehicle would not start. Eventually, the vehicle's battery died, as she had cranked on it too much.

Realizing that the vehicle was not going to start, she also realized she was alone and deep within a most inhospitable area. What she did not know was that the nearest human being was about 60 miles to the west. She had another problem—a problem more pressing than any other. She did not have a canteen or container of water—an oversight on her part. Obviously, she told herself, this is not a good situation. What an understatement this turned out to be.

Just before noon, on foot, she started back down the same road she had traveled. She reasoned she did not know what was in any other direction other than the one she had just traversed. She also knew the end of this side road intersected the major highway that bisected Death Valley. She could flag down a car or truck or bus and get help, she reasoned.

She walked … and walked … and walked some more. "Gee, if only it weren't so darn hot," she muttered to herself, to sagebrush, to scorpions, to rattlesnakes, and to cacti. And no wonder she was so uncomfortable, as the temperature was about 107°F.

She continued on for hours, but now she was not really walking; instead, she was forcing her body to move along. The heat began playing games with her mind and her reasoning. She recalled the tale about Tantalus, a Greek visitor to the underworld; he had been sent there because of things he had done: cannibalism, human sacrifice, and infanticide. Tantalus' punishment, temptation without satisfaction, became the source of the English word tantalize. He had to stand in a pool of water beneath a fruit tree with low branches. Whenever be bent down to get a drink the water receded before he could get any. This fate cursed him with eternal deprivation of quenching his intolerable thirst. Our adventurer looked for fruit trees with low branches and imagined she was standing in a pool of water. Like Tantalus, she kept reaching down but she was only able to grab handfuls of sand and rocks.

She was thirsty, almost mindless, and in terrible need of nourishment. Each step hurt. She was burning up. She was thirsty. How thirsty was she? Well, right about now just about anything liquid would do, thank you very much! Later that night, after hours of walking through that hostile land, she just couldn't go any farther. Deep down in her heat-stressed mind, she knew she was in serious trouble. Trouble of the life-threatening variety.

Just before passing out, she used her last ounce of energy to issue a dry pathetic scream. This scream of lost hope and imminent death was heard, but only by the sagebrush, the scorpions, the rattlesnakes, and the cacti—and by the vultures that were now circling above her parched, dead remains. They had heard these screams before. They were indifferent; they had all the water they needed, and their food supply wasn't all that bad, either.

This tale perhaps offers a completely different view of water, but a very basic one: We cannot live without it.

HISTORICAL PERSPECTIVE

Early humans, wandering from place to place, hunting and gathering to subsist, probably would have had little difficulty in obtaining drinking water, because they could only survive in areas where drinking water was available with little travail. Little travail, that is, if they did not forget that cave bears, saber-toothed tigers, and wolves drank from the same sources as early humans.

The search for clean, fresh, and palatable water has been a human priority from the very beginning. The author takes no risk in stating that when humans first walked the Earth many of the steps they took were in the direction of water. When early humans were alone or in small numbers, finding drinking water was a constant priority, to be sure, but it is difficult for us to imagine today just how big a priority finding drinking water became as the number of humans proliferated.

Eventually, communities formed, and with their formation came the increasing need to find clean, fresh, and palatable drinking water, as well as a means of delivering it from the source to the point of use. Archeological digs are replete with the remains of ancient water systems. Those digs, spanning the history of the last 20 or more centuries, testify to the importance of maintaining a supply of drinking water. For well over 2000 years, piped water supply systems have been in existence. Whether the pipes were fashioned from logs or clay or carved from stone or other materials is not the point—the point is that these systems were designed to serve a vital purpose: to deliver clean, fresh, and palatable water to where it was needed.

These early systems were not arcane. Today, we readily understand their intended purpose. As we might expect, they could be rather crude, but they were reasonably effective, even though they lacked in two general areas we take for granted today. First, of course, they were not pressurized, but instead relied on gravity flow, as their designers lacked the technology necessary to pressurize the water mains. Even if such pressurized systems were known, they certainly could not have been used to pressurize water delivered via hollowed-out logs and clay pipe.

Second, early civilizations lacked sanitation. Remember, to recognize that a need exists (in this case, the ability to sanitize or disinfect water supplies), the nature of the problem must be defined. Not until the middle of the 1800s, after countless millions of deaths from waterborne disease over the centuries, did people realize that a direct connection between contaminated drinking water and disease existed. At that point, sanitation of water supplies became an issue. When the relationship between waterborne diseases and the consumption of drinking water was established, evolving scientific discoveries led the way toward development of the technology required for processing and disinfection. Drinking water standards were developed by health authorities, scientists, and sanitary engineers.

With the current state of effective technology that we in the United States and the rest of the developed world enjoy today, it would be easy to assume that the water sanitation techniques discovered through the years mean that all is well with us, that problems related to maintaining a source of clean, fresh, palatable drinking water are all in the past. But are they? Have we solved all the problems related to having clean, fresh, palatable drinking water? Is the water delivered to our tap as clean, fresh, and palatable as we think it is? Does anyone really know?

What we do know is that we have made progress. We have come a long way from the days of gravity-flow water delivered via mains made of logs and clay or stone. Many of us on this planet have come a long way from the days of cholera epidemics. However, perhaps we should remember those who were asked to boil their water for weeks on end in Sydney, Australia, in 1998 due to suspected

Cryptosporidium and *Giardia* contamination. Or, better yet, we could speak with those who drank the water in Milwaukee or in Las Vegas in 1993 and survived an onslaught of *Cryptosporidium* from contaminated water coming out of their taps. Or, we should consider a little boy named Robbie who died of acute lymphatic leukemia, the probable cause of which is far less understandable to us: toxic industrial chemicals, unknowingly delivered to him via his local water supply.

If water is so precious, so necessary for sustaining life, then we must ask: (1) Why do we ignore water? and (2) Why do we abuse it? We ignore water because it is so common, accessible, available, and unexceptional (unless you are lost in the desert without a supply of it). We pollute and waste water for many reasons, some of which will be discussed later in this text. You might be asking yourself if water pollution is really that big of a deal. Simply stated, yes, it is. Humans have left their footprints (in the form of pollution) on the environment, including on our water sources. Humans have a bad habit of doing this. What it really comes down to is "out of sight, out of mind" thinking. When we abuse our natural resources, it is tempting to think, "Why worry about it? Let the experts sort it all out."

The science of water is defined as the study of the physical, chemical, and biological properties of water; water's relationship to the biotic and abiotic components of the environment; the occurrence and movement of water on and beneath the surface of the Earth; and the fouling and treatment of water. This text is designed to fill the obvious and unsatisfactory gap in available information about water. Finally, before moving on with the rest of the text, it should be pointed out that the view held throughout this work is that water is special, strange, and different—and very vital. This view is held for several reasons, but the most salient factor driving this view is that on this planet water, without any substitute, is life.

THOUGHT-PROVOKING QUESTIONS

1.1 Do you think we will have enough water for the future? Explain.
1.2 In the future, will drinking water be too unsafe to drink? Explain.
1.3 Do you think the current infrastructure is sufficient for the future? If not, what needs to be done? Explain.
1.4 The growth of the human population inevitably limits the availability of freshwater per person. Do you agree with this statement? Explain.
1.5 Is water scarcity a local or temporary problem? Explain.

REFERENCES AND RECOMMENDED READING

DeZuane, J. (1997). *Handbook of Drinking Water Quality*, 2nd ed. New York: John Wiley & Sons.
Dove, R. (1989). *Grace Notes*. New York: Norton.
Gerba, C.P. (1996). Risk assessment. In *Pollution Science*, Pepper, I.L., Gerba, C.P., and M.L. Brusseau, Eds. San Diego, CA: Academic Press.
Hammer, M.J. and Hammer, Jr., M.J. (1996). *Water and Wastewater Technology*, 3rd ed. Englewood Cliffs, NJ: Prentice Hall.
Harr, J. (1995). *A Civil Action*. New York: Vintage Books.
Lewis, S.A. (1996). *The Sierra Club Guide to Safe Drinking Water*. San Francisco, CA: Sierra Club Books.
Metcalf & Eddy, Inc. (1991). *Wastewater Engineering: Treatment, Disposal, Reuse*, 3rd ed. New York: McGraw-Hill.
Meyer, W.B. (1996). *Human Impact on Earth*. New York: Cambridge University Press.
Nathanson, J.A. (1997). *Basic Environmental Technology: Water Supply, Waste Management, and Pollution Control*. Upper Saddle River, NJ: Prentice Hall.
Pielou, E.C. (1998). *Fresh Water*. Chicago, IL: University of Chicago.
Postel, S., Daily, G.C., and Ehrlich, P.R. (1996). Human appropriation of renewable fresh water. *Science*, 271, 785–788.
Robbins, T. (1976). *Even Cowgirls Get the Blues*. Boston: Houghton Mifflin.
USGS. (2013). *Why Is the Ocean Salty?* Washington, DC: U.S. Geological Survey (http://ga.water.usgs.gov/edu/whyoceansalty.html).

2 All about Water

Water can both float and sink a ship.

—Chinese proverb

A generous person will be enriched, and one who gives water will get water.

—Proverbs 11:25

Unless you are thirsty, in real need of refreshment, when you look upon that glass of water shown earlier in Figure 1.1, you might ask, well, what could be more boring? The curious might wonder what the physical and chemical properties of water are that make it so unique and necessary for living things. Pure water is virtually colorless and has no taste or smell, but the hidden qualities of water make it a most interesting subject.

When the uninitiated become initiated to the wonders of water, one of the first surprises is that the total quantity of water on Earth is much the same now as it was more than three or four billion years ago, when the over 320 million cubic miles of it were first formed. Ever since then, the water reservoir has gone round and round, building up, breaking down, cooling, and then warming. Water is very durable but remains difficult to explain, because it has never been isolated in a completely undefiled state.

Remember, water is special, strange, and different.

HOW SPECIAL, STRANGE, AND DIFFERENT IS WATER?

Have you ever wondered what the nutritive value of water is? Well, the fact is water has no nutritive value. It has none; yet, it is the major ingredient of all living things. Consider yourself, for example. Think of what you need to survive—just to survive. Food? Air? PS-3? MTV? Water? Above all else, you need water, which is the focus of this text. Water is of major importance to all living things; in fact, up to 90% of the body weight of some organisms comes from water. Overall, about 60% of the human body is water; the brain is composed of 70% water, and the lungs are nearly 90% water. About 83% of our blood is water, which helps digest our food, transport waste, and control body temperature. Each day humans must replace 2.4 L of water, some through drinking and the rest from the foods we eat.

There wouldn't be any you, me, or Lucy the dog without the existence of an ample liquid water supply on Earth. The unique qualities and properties of water are what make it so important and basic to life. The cells in our bodies are full of water. The excellent ability of water to dissolve so many substances allows our cells to use valuable nutrients, minerals, and chemicals in biological processes. The "stickiness" of water (due to surface tension) plays a part in our body's ability to transport these materials. The carbohydrates and proteins that out bodies use as food are metabolized and transported by water in the bloodstream. No less important is the ability of water to transport waste material out of our bodies.

Chemically, water is hydrogen oxide; however, upon more advanced analysis, it has been found to be a mixture of more than 30 possible compounds. In addition, all of its physical constants are abnormal. Water is used to fight forest fires; yet, we spray water on coal in a furnace to make it burn better. At a temperature of 2900°C, some substances that contain water cannot be forced to part with it, but others that do not contain water will liberate it when even slightly heated.

When liquid, water is virtually incompressible; as it freezes, it expands by an eleventh of its volume. Water has a high specific heat index; it absorbs a lot of heat before it begins to get hot. This is why water is valuable to industries and in your car's radiator as a coolant. The high specific heat index of water also helps regulate the rate at which air changes temperature, which is why the temperature change between seasons is gradual rather than sudden, especially near the oceans.

Air pressure affects the boiling point of water; thus, it takes longer to boil an egg in Denver, Colorado, than at the beach. The higher the altitude, the lower the air pressure, the lower the boiling point of water, and, thus, the longer time required to hard boil an egg. At sea level, water boils at 212°F (100°C), but at 5000 feet water boils at 202.9°F (94.9°C). The density of water means that sound moves through it for long distances. In seawater at 30°C, sound has a velocity of 1545 meters per second (about 3500 miles per hour).

For these reasons, and many others, we can truly say that water is special, strange, and different.

CHARACTERISTICS OF WATER

Up to this point many things have been said about water; however, it has not been said that water is plain. This is because nowhere in nature is plain water to be found. Here on Earth, with a geologic origin dating back over 3 to 5 billion years, water found in even its purest form is still composed of many constituents. You probably know that the chemical description of water is H_2O—that is, one atom of oxygen bound to two atoms of hydrogen. The hydrogen atoms are "attached" to one side of the oxygen atom, resulting in a water molecule having a positive charge on the side where the hydrogen atoms are and a negative charge on the other side, where the oxygen atom is. Because opposite electrical charges attract, water molecules tend to attract each other, making water "sticky." The positive charge of the hydrogen atoms attracts the negative charge of the oxygen side of a different water molecule.

> ***Note:*** Because these water molecules attract each other, they tend to clump together. This is why water drops are, in fact, drops! If it wasn't for some of Earth's forces, such as gravity, a drop of water would be ball shaped—a perfect sphere. Even if water doesn't form a perfect sphere on Earth, we should be happy that water is sticky.

Along with H_2O molecules, hydrogen (H^+), hydroxyl (OH^-), sodium, potassium, and magnesium, other ions and elements are present. Additionally, water contains dissolved compounds including various carbonates, sulfates, silicates, and chlorides. Rainwater, often assumed to be the equivalent of distilled water, is not immune to contamination as it descends through the atmosphere. The movement of water across land contributes to its contamination, as it takes up dissolved gases, such as carbon dioxide and oxygen, and a multitude of organic substances and minerals leached from the soil. Don't let that crystal clear lake or pond fool you. It is not filled with water alone but instead is composed of a complex mix of chemical ingredients far exceeding the brief list presented here; it is a special medium in which highly specialized life can occur.

How important is water to life? To answer this question all we need to do is to take a look at the common biological cell. It easily demonstrates the importance of water to life. Living cells are comprised of a number of chemicals and organelles within a liquid substance, the cytoplasm, and the cell's survival may be threatened by changes in the proportion of water in the cytoplasm. This change in proportion of water in the cytoplasm can occur due to desiccation (evaporation), oversupply, or the loss of either nutrients or water to the external environment. A cell that is unable to control and maintain homeostasis (i.e., the correct equilibrium/proportion of water) in its cytoplasm may be doomed—it may not survive.

> ***Note:*** Water is called the "universal solvent" because it dissolves more substances than any other liquid. This means that wherever water goes, either through the ground or through our bodies, it takes along valuable chemicals, minerals, and nutrients.

Inflammable Air + Vital Air = Water

In 1783, the brilliant English chemist and physicist Henry Cavendish was investigating electric current. Specifically, Cavendish was passing electric current through a variety of substances to see what resulted. Eventually, he got around to water. He filled a tube with water and sent his current through it. The water vanished. To say that Cavendish was flabbergasted by the results of this experiment would be a mild understatement. "The tube has to have a leak in it," he reasoned. He repeated the experiment again—same result. Then again—same result. The fact is he made the water disappear again and again. Actually, what Cavendish had done was convert the liquid water to its gaseous state—into an invisible gas. When Cavendish analyzed the contents of the tube, he found it contained a mixture of two gases, one of which was *inflammable air* and the other a heavier gas. This heavier gas had only been discovered a few years earlier by his colleague, the English chemist and clergyman Joseph Priestly, who, upon finding that it kept a mouse alive and supported combustion, called it *vital air*.

Just Two H's and One O

Cavendish had been able to separate the two main constituents that make up water. All that remained was for him to put the ingredients back together again. He accomplished this by mixing a measured volume of inflammable air with different volumes of its vital counterpart and setting fire to both. He found that most mixtures burned well enough, but when the proportions were precisely two to one there was an explosion and the walls of his test tubes were covered with liquid droplets. He quickly identified these as water. Cavendish made an announcement: Water is not water. Moreover, water is not just an odorless, colorless, tasteless substance that lies beyond the reach of chemical analysis. Water is not an element in its own right, but a compound of two independent elements, one a supporter of combustion and the other combustible. When united, these two elements become the preeminent quencher of thirst and flames. It is interesting to note that a few year later, the great French genius Antoine Lavoisier tied the compound neatly together by renaming the ingredients *hydrogen* ("water producer") and *oxygen*. In a fitting tribute to his guillotined corpse (he was an innocent victim of the French Revolution), his tombstone came to carry a simple and telling epitaph, a fitting tribute to the father of a new age in chemistry: *just two H's and one O.*

SOMEWHERE BETWEEN 0° AND 105°

We take water for granted now. Every high-school level student knows that water is a chemical compound of two simple and abundant elements. And yet scientists continue to argue the merits of rival theories on the structure of water. The fact is we still know little about water. For example, we don't know how water works. Part of the problem lies with the fact that no one has ever seen a water molecule. It is true that we have theoretical diagrams and equations. We also have a disarmingly simple formula—H_2O. The reality, however, is that water is very complex. X-rays, for example, have shown that the atoms in water are intricately laced. It has been said over and over again that water is special, strange, and different. Water is also almost indestructible. Sure, we know that electrolysis can separate water atoms, but we also know that once they get together again they must be heated up to more than 2900°C to separate again.

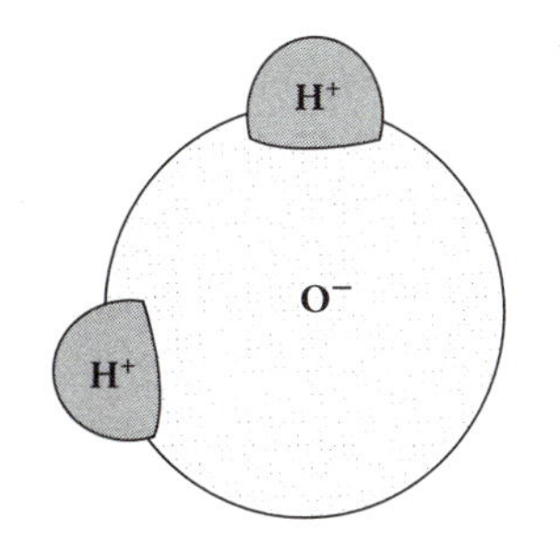

FIGURE 2.1 Molecule of water.

Water is also idiosyncratic. This can be seen in the way in which the two atoms of hydrogen in a water molecule (see Figure 2.1) take up a very precise and strange (different) alignment to each other. Not at all angles of 45°, 60°, or 90°—oh no, not water. Remember, water is different. The two hydrogen atoms always come to rest at an angle approximately 105° from each other, making all diagrams of their attachment to the larger oxygen atom look like Mickey Mouse ears on a very round head (see Figure 2.1; remember that everyone's favorite mouse is mostly water, too). This 105° relationship makes water lopsided, peculiar, and eccentric—it breaks all the rules. You're not surprised are you?

One thing is certain, however; this 105° angle is crucial to all life as we know it. Thus, the answer to why water is special, strange, and different—and vital—lies somewhere between 0° and 105°.

PHYSICAL PROPERTIES OF WATER

Water has several unique physical properties:

- Water is unique in that it is the only natural substance that is found in all three states—liquid, solid (ice), and gas (steam)—at the temperatures normally found on Earth. Earth's water is constantly interacting, changing, and in movement.
- Water freezes at 32°F and boils at 212°F (at sea level, but 186.4°F at 14,000 feet). In fact, water's freezing and boiling points are the baseline with which temperature is measured: 0° on the Celsius scale is the freezing point of water, and 100° is the boiling point of water. Water is unusual in that the solid form, ice, is less dense than the liquid form, which is why ice floats.

Here is a quick rundown of some of water's basic properties:

- Weight:
  62.416 lb/ft^3 (1000 kg/m^3) at 32°F
  61.998 lb/ft^3 at 100°F
  8.33 lb/gal
  0.036 lb/in.^{3}
- Density = 1 g/cm^3 at 39.2°F, 0.95865 g/cm^3 at 212°F
- 1 gal = 4 quarts = 8 pints = 128 ounces = 231 in.^{3}
- 1 L = 0.2642 gal = 1.0568 quarts = 61.02 in.^{3}
- 1 million gal = 3.069 acre-ft = 133,685.64 ft^3

CAPILLARY ACTION

If we were to mention the term *capillary action* to the average man and woman on the street, they might instantly nod their heads and respond that their bodies are full of them—that capillaries are the tiny blood vessels that connect the smallest arteries and the smallest of the veins. This would be true, of course, but in the context of water science the term *capillary* refers to something different than capillary action in the human body.

Even if you've never heard of capillary action, it is still important in your life. Capillary action is important for moving water (and all of the things that are dissolved in it) around. It is defined as the movement of water within the spaces of a porous material due to the forces of adhesion, cohesion, and surface tension. Surface tension is a measure of the strength of the water's surface film. The attraction between the water molecules creates a strong film, which among other common liquids is only surpassed by that of mercury. This surface tension allows water to hold up substances heavier and denser than it. A steel needle carefully placed on the surface of that glass of water will float. Some aquatic insects such as the water strider rely on surface tension to walk on water.

Capillary action occurs because water is sticky, thanks to the forces of cohesion (water molecules like to stay closely together) and adhesion (water molecules are attracted and stick to other substances). So, water tends to stick together, as in a drop, and it sticks to glass, cloth, organic tissues, and soil. Dip a paper towel into a glass of water and the water will climb up the paper towel. In fact, it will keep climbing up the towel until the pull of gravity is too much for it to overcome.

WATER CYCLE

The natural *water cycle* or *hydrologic cycle* is the means by which water in all three forms—solid, liquid, and vapor—circulates through the biosphere. The water cycle is all about describing how water moves above, on, and through the Earth. Much more water, however, is "in storage" for long periods of time than is actually moving through the cycle. The storehouses for the vast majority of all water on Earth are the oceans. It is estimated that of the 332,500,000 cubic miles of the world's water supply, about 321,000,000 cubic miles are stored in oceans. That represents about 96.5%. It is also estimated that the oceans supply about 90% of the evaporated water that goes into the water cycle.

Water—lost from the Earth's surface to the atmosphere either by evaporation from the surface of lakes, rivers, and oceans or through the transpiration of plants—forms clouds that condense to deposit moisture on the land and sea. Evaporation from the oceans is the primary mechanism supporting the surface-to-atmosphere portion of the water cycle. Note, however, that a drop of water may travel thousands of miles between the time it evaporates and the time it falls to Earth again as rain, sleet, or snow. The water that collects on land flows to the ocean in streams and rivers or seeps into the earth, joining groundwater. Even groundwater eventually flows toward the ocean for recycling (see Figure 2.2). The cycle constantly repeats itself—a cycle without end.

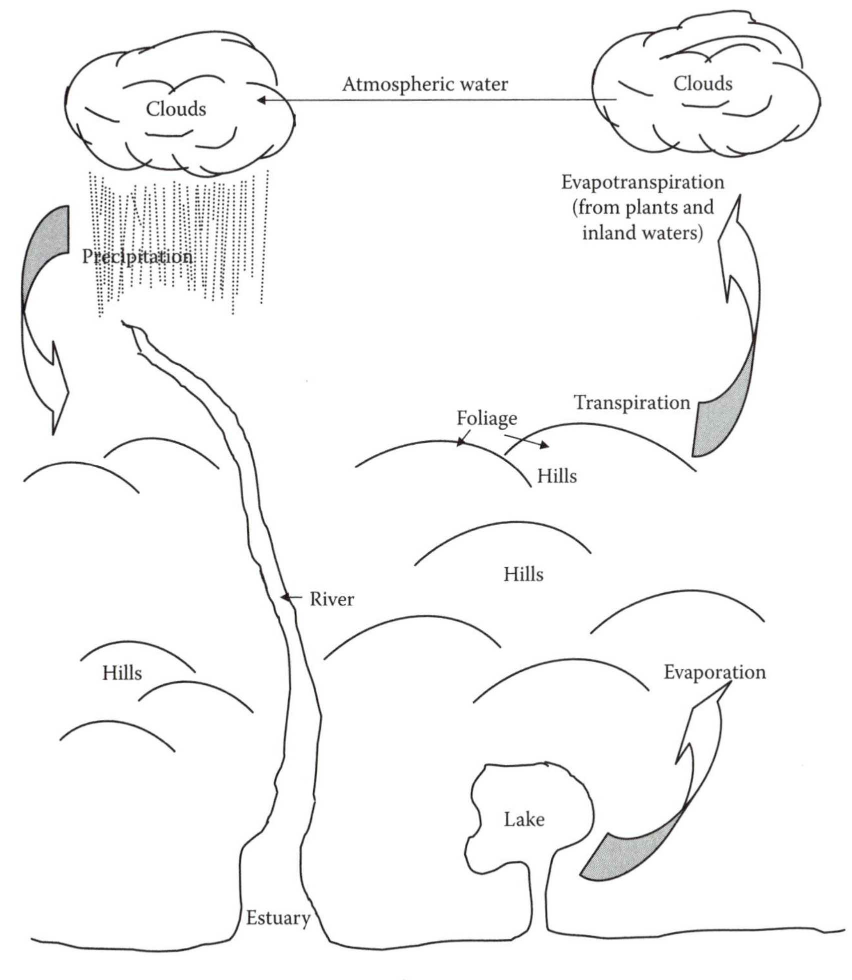

FIGURE 2.2 Water cycle.

Note: How long water that falls from the clouds takes to return to the atmosphere varies tremendously. After a short summer shower, most of the rainfall on land can evaporate into the atmosphere in only a matter of minutes. A drop of rain falling on the ocean may take as long as 37,000 years before it returns to the atmosphere, and some water has been in the ground or caught in glaciers for millions of years.

Note: Only about 2% of the water absorbed into plant roots is used in photosynthesis. Nearly all of it travels through the plant to the leaves, where transpiration to the atmosphere begins the cycle again.

SPECIFIC WATER MOVEMENTS

After having reviewed the water cycle in very simple terms to provide foundational information, it is important to point out that the actual movement of water is much more complex. Three different methods of transport are involved: *evaporation*, *precipitation*, and *runoff*. The evaporation of water is a major factor in hydrologic systems. Evaporation is a function of temperature, wind velocity, and relative humidity. Evaporation (or vaporization), as the name implies, is the formation of vapor. Dissolved constituents such as salts remain behind when water evaporates. Evaporation of the surface water of oceans provides most of the water vapor. It should be pointed out, however, that water can also vaporize through plants, especially from leaf surfaces. This process is called *evapotranspiration*. Plant transpiration is pretty much an invisible process—because the water is evaporating from the leaf surfaces, you cannot just go out and see the leaves breathe. During a growing season, a leaf will transpire many times more water than its own weight. A large oak tree can transpire 40,000 gallons (151,000 liters) per year (USGS, 2006).

The amount of water that plants transpire varies greatly geographically and over time (USGS, 2006). There are a number of factors that determine transpiration rates:

- *Temperature*—Transpiration rates go up as the temperature goes up, especially during the growing season, when the air is warmer due to stronger sunlight and warmer air masses.
- *Relative humidity*—As the relative humidity of the air surrounding the plant rises the transpiration rate falls. It is easier for water to evaporate into drier air than into more saturated air.
- *Wind and air movement*—Increased movement of the air around a plant will result in a higher transpiration rate.

DID YOU KNOW?

Sublimation is the conversion from the solid phase to gaseous phase with no intermediate liquid stage. For those of us interested in the water cycle, sublimation is most often used to describe the process of snow and ice changing into water vapor in the air without first melting into water. The opposite of sublimation is *deposition*, where water vapor changes directly into ice—such as snowflakes and frost. It is not easy to actually see sublimation occurring, at least not with ice. One way to see the results of sublimation is to hang a wet shirt outside on a below-freezing day. Eventually the ice in the shirt will disappear. Actually, the best way to visualize sublimation is not to think of water at all, but instead to consider carbon dioxide in the form of dry ice. Dry ice is actually solid, frozen carbon dioxide, which happens to sublimate, or turn to gas, at a chilly –78.5°C (–109.3°F). The fog you see is actually a mixture of cold carbon dioxide gas and cold, humid air created as the dry ice melts, or sublimates.

DID YOU KNOW?

In order to have precipitation you must have a cloud. A cloud is made up of water in the air (water vapor). Along with this water are tiny particles called *condensation nuclei*—for example, the little pieces of salt leftover after seawater evaporates or a particle of dust for smoke. Condensation occurs when the water vapor wraps itself around the tiny particles. Each particle (surrounded by water) becomes a tiny droplet between 0.0001 and 0.005 centimeter in diameter. These small droplets coalesce and form raindrops.

- *Soil-moisture availability*—When moisture is lacking, plants can begin to senesce (i.e., age prematurely, which can result in leaf loss) and transpire less water.
- *Type of plant*—Plants transpire water at different rates. Some plants, such as cacti and succulents, that grow in arid regions conserve precious water by transpiring less water than other plants.

Note: It may surprise you that ice can vaporize without melting first; however, this *sublimation* process is slower than vaporization of liquid water.

Evaporation rates are measured with evaporation pans. These evaporation pans provide data that indicate the atmospheric evaporative demand of an area and can be used to estimate (1) the rates of evaporation from ponds, lakes, and reservoirs, and (2) evapotranspiration rates. It is important to note that several factors affect the rate of pan evaporation. These factors include the type of pan, type of pan environment, method of operating the pan, exchange of heat between pan and ground, solar radiation, air temperature, wind, and temperature of the water surface (Jones, 1992).

Precipitation includes all of the forms in which atmospheric moisture descends to earth—rain, snow, sleet, and hail. Before precipitation can occur, the water that enters the atmosphere by vaporization must first condense into liquid (clouds and rain) or solid (snow, sleet, and hail) before it can fall. This vaporization process absorbs energy. This energy is released in the form of heat when the water vapor condenses. A good example of this phenomenon is the heat lost when water evaporates from your skin, making you feel cold.

Note: The annual evaporation from ocean and land areas is the same as the annual precipitation.

Runoff is the flow back to the oceans of the precipitation that falls on land. This journey to the oceans is not always unobstructed—flow back may be intercepted by vegetation (from which it later evaporates), a portion is held in depressions, and a portion infiltrates into the ground. A part of the infiltrated water is taken up by plant life and returned to the atmosphere through evapotranspiration, while the remainder either moves through the ground or is held by capillary action. Eventually, water drips, seeps, and flows its way back into the oceans.

Assuming that the water in the oceans and ice caps and glaciers is fairly constant when averaged over a period of years, the Earth's water balance can be expressed by the relationship water lost = water gained (Turk and Turk, 1988).

Q AND Q FACTORS

Whereas drinking water practitioners must have a clear and complete understanding of the natural and manmade water cycles, they must also factor in the two major considerations of *quantity* and *quality*, the Q and Q factors. They are responsible for (1) providing a *quality* potable water supply—one that is clean, wholesome, and safe to drink; and (2) finding a water supply in adequate *quantities* to meet the anticipated demand.

Note: Two central facts important to our discussion of freshwater supplies are that (1) water is very much a local or regional resource, and (2) problems of its shortage or pollution are equally local problems. Human activities affect the quantity of water available at a locale at any time by changing either the total volume that exists there or aspects of quality that restrict or devalue it for a particular use. Thus, the total human impact on water supplies is the sum of the separate human impacts on the various drainage basins and groundwater aquifers. In the global system, the central, critical fact about water is the natural variation in its availability (Meyer, 1997). Simply put, not all lands are watered equally.

To meet the Q and Q requirements of a potential water supply, the drinking water practitioner (whether it be the design engineer, community planner, plant manager, plant administrator, plant engineer, or other responsible person in charge) must determine the answers to a number of questions, including the following:

1. Does a potable water supply exist nearby with the capacity to be distributed in sufficient quantity and pressure at all times?
2. Will constructing a centralized treatment and distribution system for the entire community be best, or would using individual well supplies be better?
3. If a centralized water treatment facility is required, will the storage capacity at the source as well as at intermediate points of the distribution system maintain the water pressure and flow (*quantity*) within the conventional limits, particularly during loss-of-pressure events, such as major water main breaks, rehabilitation of the existing system, or major fires, for example?
4. Is a planned or preventive maintenance program in place (or anticipated) for the distribution system that can be properly planned, implemented, and controlled at the optimum level possible?
5. Is the type of water treatment process selected in compliance with federal and state drinking water standards?

Note: Water from a river or a lake usually requires more extensive treatment than groundwater does to remove bacteria and suspended particles.

Note: The primary concern for the drinking water practitioner involved with securing an appropriate water supply, treatment process, and distribution system must be the protection of public health. Contaminants must be eliminated or reduced to a safe level to minimize menacing waterborne diseases (to prevent another Milwaukee *Cryptosporidium* event) and to avoid long-term or chronic injurious health effects.

6. When the source and treatment processes are selected, has the optimum hydraulic design of the storage, pumping, and distribution network been determined to ensure that sufficient quantities of water can be delivered to consumers at adequate pressures?
7. Have community leaders and the consumer (the general public) received continuing and realistic information about the functioning of the proposed drinking water service?

Note: Drinking water practitioners are wise to direct their attention toward considering point 7, simply because public buy-in for any proposed drinking water project that involves new construction or retrofitting, expansion, or upgrade of an existing facility is essential to ensure that necessary financing is forthcoming. In addition to the finances required for any type of waterworks construction project, public and financial support is also required to ensure the safe operation, maintenance, and control of the entire water supply system. The acronym POTW stands for "*publicly* owned treatment works," and the public foots the bills.

8. Does planning include steps to ensure elimination of waste, leakages, and unauthorized consumption?

Note: Industrywide operational experience has shown that the cost per cubic foot, cubic meter, liter, or gallon of water delivered to the customer has steadily increased because of manpower, automation, laboratory, and treatment costs. To counter these increasing costs, treatment works must meter consumers, measure the water supply flow, and evaluate the entire system annually.

9. Does the water works or proposed water works physical plant include adequate laboratory facilities to ensure proper monitoring of water quality?

Note: Some water works facilities routinely perform laboratory work; however, water pollution control technologists must ensure that the water works laboratory or other laboratory used is approved by the appropriate health authority. Keep in mind that the laboratory selected to test and analyze the water works samples must be able to analyze chemical, microbiological, and radionuclide parameters.

10. Are procedures in place to evaluate specific problems such as the lead content in the distribution systems and at the consumer's faucet or suspected contamination due to cross-connection potentials?
11. Is a cross-connection control program in place to make sure that the distribution system (in particular) is protected from plumbing errors and illegal connections that may lead to injection of nonpotable water into public or private supplies of drinking water?
12. Are waterworks operators and laboratory personnel properly trained and licensed?
13. Are waterworks managers properly trained and licensed?
14. Are proper operating records and budgetary records kept?

SOURCES OF WATER

Approximately 40 million cubic miles of water cover or reside within the planet. The oceans contain about 97% of all water on Earth. The other 3% is freshwater: (1) snow and ice on the surface represent about 2.25% of the freshwater; (2) usable groundwater represents approximately 0.3%; and (3) surface freshwater represents less than 0.5%. In the United States, for example, average rainfall is approximately 2.6 ft (a volume of 5900 km^3). Of this amount, approximately 71% evaporates (about 4200 cm^3), and 29% goes to stream flow (about 1700 km^3). Beneficial freshwater uses include manufacturing, food production, domestic and public needs, recreation, hydroelectric power production, and flood control. Stream flow withdrawn annually is about 7.5% (440 km^3). Irrigation and industry use almost half of this amount (3.4% or 200 km^3/year). Municipalities use only about 0.6% (35 km^3/year) of this amount.

Historically, in the United States, water usage has been increasing (as might be expected); for example, in 1900, 40 billion gallons of freshwater were used. In 1975, usage increased to 455 billion gallons. Projected use in 2002 was about 720 billion gallons. The primary sources of freshwater include the following:

1. Captured and stored rainfall in cisterns and water jars
2. Groundwater from springs, artesian wells, and drilled or dug wells
3. Surface water from lakes, rivers, and streams
4. Desalinized seawater or brackish groundwater
5. Reclaimed wastewater

Current federal drinking water regulations actually define three distinct and separate sources of freshwater: surface water, groundwater, and groundwater under the direct influence of surface water (GUDISW). This last classification is the result of the Surface Water Treatment Rule (SWTR). The definition of the conditions that constitute GUDISW, although specific, is not obvious. This classification is discussed in detail later.

WATERSHED PROTECTION

Watershed protection is one of the barriers in the multiple-barrier approach to protecting source water. In fact, watershed protection is the primary barrier, the first line of defense against contamination of drinking water at its source.

Multiple-Barrier Concept

On August 6, 1996, during the Safe Drinking Water Act Reauthorization signing ceremony, President Bill Clinton stated:

> A fundamental promise we must make to our people is that the food they eat and the water they drink are safe.

No rational person could doubt the importance of the promise made in this statement.

The Safe Drinking Water Act (SDWA), passed in 1974, amended in 1986, and reauthorized in 1996, gives the U.S. Environmental Protection Agency (USEPA) the authority to set drinking water standards. This document is important for many reasons but is even more important because it describes how USEPA establishes these standards.

Drinking water standards are regulations that the USEPA sets to control the level of contaminants in the nation's drinking water. These standards are part of the Safe Drinking Water Act's *multiple-barrier approach* to drinking water protection which includes the following elements:

- *Assessing and protecting drinking water sources*, which means doing everything possible to prevent microbes and other contaminants from entering water supplies. Minimizing human and animal activity around our watersheds is one part of this barrier.
- *Optimizing treatment processes*, which provides a second barrier. This usually means filtering and disinfecting the water. It also means making sure that the people who are responsible for our water are properly trained and certified and knowledgeable of the public health issues involved.
- *Ensuring the integrity of distribution systems*, which consists of maintaining the quality of water as it moves through the system on its way to the customer's tap.
- *Effecting correct cross-connection control procedures*, which is a critical fourth element in the barrier approach. It is critical because the greatest potential hazard in water distribution systems is associated with cross-connections to nonpotable waters. Many connections exist between potable and nonpotable systems (every drain in a hospital constitutes such a connection), but cross-connections are those through which backflow can occur (Angele, 1974).
- *Continuous monitoring and testing of the water before it reaches the tap* are critical elements in the barrier approach and should include specific procedures to follow should potable water ever fail to meet quality standards.

With the involvement of the USEPA, local governments, drinking water utilities, and citizens, these multiple barriers ensure that the tap water in the United States and territories is safe to drink. Simply, in the multiple-barrier concept, we employ a holistic approach to water management that begins at the source and continues with treatment through disinfection and distribution.

The bottom line on the multiple-barrier approach to protecting the watershed can best be summed up as ideally, under the general concept of "quality in, means quality out," a protected watershed ensures that surface runoff and inflow to the source waters occur within a pristine environment (Spellman, 2003).

Watershed Management

Water regulates population growth, influences world health and living conditions, and determines biodiversity. For thousands of years, people have tried to control the flow and quality of water. Water provided the resources and a means of transportation for development in some areas. Even today, the presence or absence of water is critical in determining how we can use land. Yet, despite this long experience in water use and water management, humans often fail to manage water well. Sound water management was pushed aside in favor of rapid, never-ending economic development in many countries. Often, optimism about the applications of technology (e.g., dam building, wastewater treatment, irrigation measures) exceeded concerns for, or even interest in, environmental shortcomings. Pollution was viewed as the inevitable consequence of development, the price that must be paid to achieve economic progress.

Clearly, we now have reached the stage of our development when the need for management of water systems is apparent, beneficial, and absolutely imperative. Land use and activities in the watershed directly impact raw water quality. Effective watershed management improves raw water quality, controls treatment costs, and provides additional health safeguards. Depending on the goals, watershed management can be simple or complex.

This section discusses the need for watershed management on a multiple-barrier basis and provides a brief overview of the range of techniques and approaches that can be used to investigate the biophysical, social, and economic forces affecting water and its use. Water utility directors are charged with providing potable water in a quantity and quality to meet the public's demand. They are also charged with providing effective management of the entire water supply system, and such management responsibility includes proper management of the relevant watershed.

Note: Integrated water management means putting all of the pieces together, including the social, environmental, and technical aspects.

Remarkable consensus exists among worldwide experts over the current issues confronted by waterworks managers and others (Viessman, 1991). These issues include the following:

- *Water availability, requirements, and use*
  Protection of aquatic and wetland habitat
  Management of extreme events (droughts, floods, etc.)
  Excessive extractions from surface and groundwater
  Global climate change
  Safe drinking water supply
  Waterborne commerce
- *Water quality*
  Coastal and ocean water quality
  Lake and reservoir protection and restoration
  Water quality protection, including effective enforcement of legislation
  Management of point- and nonpoint-source pollution
  Impacts on land, water, and air relationships
  Health risks
- *Water management and institutions*
  Coordination and consistency
  Capturing a regional perspective
  Respective roles of federal and state or provincial agencies
  Respective roles of projects and programs
  Economic development philosophy that should guide planning
  Financing and cost sharing

Information and education
Appropriate levels of regulation and deregulation
Water rights and permits
Infrastructure
Population growth

- *Water resources planning*

Consideration of the watershed as an integrated system
Planning as a foundation for, not a reaction to, decision making
Establishment of dynamic planning processes incorporating periodic review and redirection
Sustainability of projects beyond construction and early operation
A more interactive interface between planners and the public
Identification of sources of conflict as an integral part of planning
Fairness, equity, and reciprocity between affected parties

Water Quality Impact

Generally, in a typical river system, water quality is impacted by about 60% nonpoint-source pollution, 21% municipal discharge, 18% industrial discharge, and 1% sewer overflows. Of the nonpoint-source pollution, about 67% is from agriculture, 18% is urban, and 15% comes from other sources. Land use directly impacts water quality. The impact of land use on water quality is clearly evident in Table 2.1. From the point of view of waterworks operators, water quality issues for nutrient contamination can be summarized quite simply:

$$\text{Nutrients} + \text{Algae} = \text{Taste and odor problems}$$

$$\text{Nutrients} + \text{Algae} + \text{Macrophytes} + \text{Decay} = \text{Trihalomethane (THM) precursors}$$

Watershed Protection and Regulations

The Clean Water Act (CWA) and Safe Drinking Water Act (SDWA) address source water protection. Implementation of regulatory compliance requirements (with guidance provided by the U.S. Department of Health) is left up to state and local health department officials. Water protection regulations in force today provide not only guidance and regulations for watershed protection but also additional options for those tasked with managing drinking water utilities.

TABLE 2.1
How Land Use Impacts Water Quality

Type of Land Use	Sediment	Nutrients	Viruses, Bacteria	Trihalomethane (THM)	Fe, Mn
Urban	✓	✓	✓	✓	✓
Agriculture	✓	✓	✓	✓	✓
Logging	✓	✓	✓	✓	
Industrial	✓	✓	✓	✓	
Septic tanks	✓	✓	✓		
Construction	✓	✓			

Source: Spellman, F.R., *The Handbook for Wastewater Operator Certification*, CRC Press, Boca Raton, FL, 2001.

The typical drinking water utility that provides safe drinking water to the consumer has two choices in water pollution control: keep it out or take it out. The "keep it out" part pertains to effective watershed management. In contrast, contaminants that have found their way into the water supply must be removed by treatment, which is the "take it out" part. Obviously, utility directors and waterworks managers are concerned with controlling treatment costs. An effective watershed management program can reduce treatment costs by reducing source water contamination. The "take it out" option is much more expensive and time consuming than keeping it out in the first place. Proper watershed management also works to maintain consumer confidence. If the consumer is aware that the water source from the area's watershed is of the highest quality, then, logically, confidence in the quality of the water is high. High-quality water also works directly to reduce public health risks.

Watershed Protection Plan

Watershed protection begins with planning. The watershed protection plan consists of several elements, which include the need to

- Inventory and characterize water sources.
- Identify pollutant sources.
- Assess vulnerability of intake.
- Establish program goals.
- Develop protection strategies.
- Implement the program.
- Monitor and evaluate program effectiveness.

Reservoir Management Practices

To ensure an adequate and safe supply of drinking water for a municipality, watershed management must utilize proper reservoir management practices. These practices include proper lake aeration, harvesting, dredging, and use of algaecide. Water quality improvements from lake aeration include reduced iron, manganese, phosphorus, ammonia, and sulfide content. Lake aeration also reduces the costs of capital and operation for water supply treatment. Algaecide treatment controls algae which in turn reduces taste and odor problems. The drawback of using algaecides is that they are successful for only a brief period.

POTABLE WATER

Because of huge volume and flow conditions, the quality of natural water cannot be modified significantly within a body of water. Accordingly, humans must augment Nature's natural purification processes with physical, chemical, and biological treatment procedures. Essentially, this quality control approach is directed to the water withdrawn, which is treated, from a source for a specific use.

Potable water is water fit for human consumption and domestic use. It is sanitary and normally free of minerals, organic substances, and toxic agents in excess of reasonable amounts for domestic usage in the area served and is normally adequate in quantity for the minimum health requirements of the persons served.

With regard to a potential potable water supply, the key words, as previously noted, are *quality* and *quantity*. Obviously, if we have a water supply that is unfit for human consumption, we have a quality problem. If we do not have an adequate supply of quality water, we have a quantity problem.

In the following sections, the focus is on surface water and groundwater hydrology and the mechanical components associated with the collection and conveyance of water from its source to the public water supply system for treatment. Well supplies are also discussed.

KEY DEFINITIONS

Annular space—The space between the well casing and the wall of the drilled hole.

Aquifer—A porous, water-bearing geologic formation.

Caisson—Large pipe placed in a vertical position.

Cone of depression—As the water in a well is drawn down, the water near the well drains or flows into it. The water will drain farther back from the top of the water table into the well as drawdown increases.

Confined aquifer—An aquifer that is surrounded by formations of less permeable or impermeable material.

Contamination—The introduction into water of toxic materials, bacteria, or other deleterious agents that make the water unfit for its intended use.

Drainage basin—An area from which surface runoff or groundwater recharge is carried into a single drainage system; it is also called a *catchment area*, *watershed*, or *drainage area*.

Drawdown—The distance or difference between the static level and the pumping level. When the drawdown for any particular capacity well and rate pump bowls is determined, the pumping level is known for that capacity. The pump bowls are located below the pumping level so they will always be under water. When the drawdown is fixed or remains steady, the well is then furnishing the same amount of water as is being pumped.

Groundwater—Subsurface water occupying a saturated geological formation from which wells and springs are fed.

Hydrology—The applied science pertaining to properties, distribution, and behavior of water.

Impermeable—A material or substance through which water will not pass.

Overland flow—The movement of water on and just under the Earth's surface.

Permeable—A material or substance that water can pass through.

Porosity—The ratio of pore space to total volume; a measure of that portion of a cubic foot of soil that is air space and could therefore contain moisture.

Precipitation—The process by which atmospheric moisture is discharged onto the Earth's crust; precipitation takes the form of rain, snow, hail, and sleet.

Pumping level—The level at which the water stands when the pump is operating.

Radius of influence—The distance from the well to the edge of the cone of depression; the radius of a circle around the well from which water flows into the well.

Raw water—The untreated water to be used after treatment for drinking water.

Recharge area—An area from which precipitation flows into underground water sources.

Specific yield—The geologist's method for determining the capacity of a given well and the production of a given water-bearing formation; it is expressed as gallons per minute per foot of drawdown.

Spring—A surface feature where, without the aid of humans, water issues from rock or soil onto the land or into a body of water, the place of issuance being relatively restricted in size.

Static level—The height to which the water will rise in the well when the pump is not operating.

Surface runoff—The amount of rainfall that passes over the surface of the Earth.

Surface water—The water on the Earth's surface as distinguished from water underground (groundwater).

Unconfined aquifer—An aquifer that sits on an impervious layer but is open on the top to local infiltration; the recharge for an unconfined aquifer is local. It is also called a *water table aquifer*.

Water rights—The rights, acquired under the law, to use the water accruing in surface or groundwater for a specified purpose in a given manner and usually within the limits of a given time period.

Water table—The average depth or elevation of the groundwater over a selected area; the upper surface of the zone of saturation, except where that surface is formed by an impermeable body.
Watershed—A drainage basin from which surface water is obtained.

SURFACE WATER

Where do we get our drinking water from? What is the source? Our answer would most likely turn to one of two possibilities: groundwater or surface water. This answer seems simple enough, because these two sources are, indeed, the primary sources of most water supplies. From our earlier discussion of the hydrologic or water cycle, we know that, from whichever source we obtain our drinking water, the source is constantly being replenished (we hope) with a supply of freshwater. This water cycle phenomenon was best summed up by Heraclitus of Ephesus, who said, "You could not step twice into the same rivers, for other waters are ever flowing on to you." In this section, we discuss one of the drinking water practitioner's primary duties—to find and secure a source of potable water for human use.

LOCATION! LOCATION! LOCATION!

In the real estate business, location is everything. We say the same when it comes to sources of water. In fact, the presence of water defines "location" for communities. Although communities differ widely in character and size, all have the common concerns of finding water for industrial, commercial, and residential use. Freshwater sources that can provide stable and plentiful supplies for a community do not always occur where we wish. Simply put, on land, the availability of a regular supply of potable water is the most important factor affecting the presence—or absence—of many life forms. A map of the world immediately shows us that surface waters are not uniformly distributed over the surface of the Earth. American lands hold rivers, lakes, and streams on only about 4% of their surface. The heaviest populations of any life forms, including humans, are found in regions of the United States (and the rest of the world) where potable water is readily available, because lands barren of water simply cannot support large populations.

One thing is certain: If a local supply of potable water is not readily available, the locality affected will seek a source. This is readily apparent (absolutely crystal clear), for example, when one studies the history of water "procurement" for the communities located within the Los Angeles Basin.

Note: The volume of freshwater sources depends on geographic, landscape, and temporal variations, as well as on the impact of human activities.

HOW READILY AVAILABLE IS POTABLE WATER?

Approximately 326 million cubic miles of water comprise Earth's entire water supply. Although it provides us indirectly with freshwater through evaporation from the oceans, only about 3% of this massive amount of water is fresh, and most of that minute percentage of freshwater is locked up in polar ice caps and glaciers. The rest is held in lakes, in flows through soil, and in river and stream systems. Only 0.027% of Earth's freshwater is available for human consumption (see Table 2.2 for the distribution percentages of Earth's water supply). We see from Table 2.2 that the major sources of drinking water are from surface water, groundwater, and from groundwater under the direct influence of surface water (i.e., springs or shallow wells).

Surface waters are not uniformly distributed over the Earth's surface. In the United States, for example, only about 4% of the landmass is covered by rivers, lakes, and streams. The volumes of these freshwater sources depend on geographic, landscape, and temporal variations, and on the impact of human activities.

TABLE 2.2
World Water Distribution

Location	Percent (%) of Total
Land areas	
Freshwater lakes	0.009
Saline lakes and inland seas	0.008
Rivers (average instantaneous volume)	0.0001
Soil moisture	0.005
Groundwater (above depth of 4000 m)	0.61
Ice caps and glaciers	2.14
Total	2.8
Atmosphere (water vapor)	0.001
Oceans	97.3
Total all locations (rounded)	100

Source: Adapted from USGS, *Water Science in Schools*, U.S. Geological Survey, Washington, DC, 2006.

Again, surface water is that water that is open to the atmosphere and results from overland flow (i.e., runoff that has not yet reached a definite stream channel). Put a different way, surface water is the result of surface runoff. For the most part, however, surface water (as used in the context of this text) refers to water flowing in streams and rivers, as well as water stored in natural or artificial lakes or in manmade impoundments such as lakes made by damming a stream or river; springs that are affected by a change in level or quantity; shallow wells that are affected by precipitation; wells drilled next to or in a stream or river; rain catchments; and/or muskeg and tundra ponds.

Specific sources of surface water include

- Rivers
- Streams
- Lakes
- Impoundments (manmade lakes made by damming a river or stream)
- Very shallow wells that receive input via precipitation
- Springs affected by precipitation (flow or quantity directly dependent upon precipitation)
- Rain catchments (drainage basins)
- Tundra ponds or muskegs (peat bogs)

Surface water has advantages as a source of potable water. Surface water sources are usually easy to locate; unlike groundwater, finding surface water does not require a geologist or hydrologist. Normally, surface water is not tainted with minerals precipitated from the Earth's strata. Ease of discovery aside, surface water also presents some disadvantages. Surface water sources are easily contaminated with microorganisms that can cause waterborne diseases and are polluted by chemicals that enter from surrounding runoff and upstream discharges. Water rights can also present problems.

As we have said, most surface water is the result of surface runoff. The amount and flow rate of this surface water are highly variable for two main reasons: (1) human interference, and (2) natural conditions. In some cases, surface water quickly runs off land surfaces. From a water resources standpoint, this is generally undesirable, because quick runoff does not provide enough time for the water to infiltrate into the ground and recharge groundwater aquifers. Also, surface water that quickly runs off land causes erosion and flooding problems. Probably the only good thing that can be said about surface water that runs off quickly is that it usually does not have enough contact time to increase its mineral content. Surface water that drains slowly off land has all the opposite effects.

Drainage basins collect surface water and direct it on its gravitationally influenced path to the ocean. The drainage basin is normally characterized as an area measured in square miles, acres, or sections. Obviously, if a community is drawing water from a surface water source, the size of its drainage basin is an important consideration. Surface water runoff, like the flow of electricity, flows or follows the path of least resistance. Surface water within the drainage basin normally flows toward one primary watercourse (e.g., river, stream, brook, creek), unless some manmade distribution system (canal or pipeline) diverts the flow.

Note: Many people probably have the overly simplified idea that precipitation falls on the land, flows overland, and runs into rivers, which then empty into the oceans. That is overly simplified because rivers also gain and lose water to the ground. Still, it is true that much of the water in rivers comes directly from runoff from the land surface, which is defined as surface runoff.

Surface water runoff from land surfaces depends on several factors:

- *Rainfall duration*—Even a light, gentle rain, if it lasts long enough, can, with time, saturate soil and allow runoff to take place.
- *Rainfall intensity*—With increases in intensity, the surface of the soil quickly becomes saturated. This saturated soil can hold no more water; as more rain falls and water builds up on the surface, surface runoff results.
- *Soil moisture*—The amount of existing moisture in the soil has a definite impact on surface water runoff. Soil already wet or saturated from a previous rain causes surface water runoff to occur sooner than if the soil were dry. Surface water runoff from frozen soil can be as high as 100% of the snow melt or rain runoff because frozen ground is basically impervious.
- *Soil composition*—The composition of the surface soil directly affects the amount of runoff; for example, it is obvious that hard rock surfaces result in 100% runoff. Clay soils have very small void spaces that swell when wet; when these void spaces fill and close, they do not allow infiltration. Coarse sand possesses large void spaces that allow water to flow easily through it, thus producing the opposite effect of clay soil, even in a torrential downpour.
- *Vegetation cover*—Groundcover limits runoff. Roots of vegetation and pine needles, pine cones, leaves, and branches create a porous layer (a sheet of decaying natural organic substances) above the soil. This porous, organic layer readily allows water into the soil. Vegetation and organic waste also act as cover to protect the soil from hard, driving rains, which can compact bare soils, close off void spaces, and increase runoff. Vegetation and groundcover work to maintain the infiltration and water-holding capacity of the soil and also work to reduce soil moisture evaporation.
- *Ground slope*—When rain falls on steeply sloping ground, 80% or more may become surface runoff. Gravity moves the water down the surface more quickly than it can infiltrate the surface. Water flow off flat land is usually slow enough to provide opportunity for a higher percentage of the rainwater to infiltrate the ground.
- *Human influences*—Various human activities have a definite impact on surface water runoff. Most human activities tend to increase the rate of water flow; for example, canals and ditches are usually constructed to provide steady flow, and agricultural activities generally remove groundcover that would work to retard the runoff rate. At the opposite extreme, manmade dams are generally built to retard the flow of runoff.

Paved streets, tarmac, paved parking lots, and buildings are impervious to water infiltration, greatly increasing the amount of stormwater runoff from precipitation events. These manmade surfaces hasten the flow of surface water and often cause flooding, sometimes with devastating consequences. In badly planned areas, even relatively light precipitation can cause local flooding. Impervious surfaces not only present flooding problems but also do not allow water to percolate into the soil to recharge groundwater supplies—a potentially devastating blow to a location's water supply.

Advantages and Discharges of Surface Water

The biggest advantage of using a surface water supply as a water source is that these sources are readily located; finding surface water sources does not demand sophisticated training or equipment. Many surface water sources have been used for decades and even centuries (in the United States, for example), and considerable data are available on the quantity and quality of the existing water supply. Surface water is also generally softer (i.e., not mineral laden), which makes its treatment much simpler.

The most significant disadvantage of using surface water as a water source is *pollution*. Surface waters are easily contaminated with microorganisms that cause waterborne diseases and chemicals that enter the river or stream from surface runoff and upstream discharges. Another problem with many surface water sources is *turbidity*, which fluctuates with the amount of precipitation. Increases in turbidity increase treatment costs and operator time. Surface water temperatures can be a problem because they fluctuate with ambient temperature, making consistent water quality production at a waterworks plant difficult. Drawing water from a surface water supply might also present problems; for example, intake structures may clog or become damaged from winter ice, or the source may be so shallow that it completely freezes in the winter. *Water rights* are another issue, in that removing surface water from a stream, lake, or spring requires a legal right. The lingering, seemingly unanswerable, question is who owns the water?

Using surface water as a source means that the purveyor is obligated to meet the requirements of the Surface Water Treatment Rule and Interim Enhanced Surface Water Treatment Rule (IESWTR), which applies only to large public water systems (PWSs) serving more than 10,000 people. The IESWTR tightened controls on disinfection byproducts and turbidity and regulates *Cryptosporidium*.

Surface Water Hydrology

To properly manage and operate water systems, it is important to have a basic understanding of the movement of water and the factors that affect water quality and quantity—in other words, *hydrology*. A discipline of applied science, hydrology includes several components, such as the physical configuration of the watershed, the geology, soils, vegetation, nutrients, energy, wildlife, and the water itself. Specifically, like the science of water itself, hydrology is the study of the properties of water, the relationship of water to the biotic and abiotic components of the environment, and the occurrence and movement of water on and beneath the surface of the Earth. Hydraulics is different from hydrology in that hydraulics is the science of fluids (liquids and gases). Water hydraulics is the study of water flow and is covered in detail in Chapter 3.

The area from which surface water flows is a *drainage basin* or catchment area. With a surface water source, this drainage basin is most often referred to, in nontechnical terms, as a *watershed* (when dealing with groundwater, we call this area a *recharge area*).

> ***Note:*** The area that directly influences the quantity and quality of surface water is the *drainage basin* or *watershed*.

When we trace on a map the course of a major river from its meager beginnings along its seaward path, it is readily apparent that its flow becomes larger and larger. Every tributary adds to its size, and between tributaries the river grows gradually due to overland flow entering it directly (see Figure 2.3). Not only does the river grow, but its entire watershed or drainage basin, basically the land it drains into, also grows in the sense that it embraces an ever-larger area. The area of the watershed is commonly measured in square miles, sections, or acres. When taking water from a surface water source, knowing the size of the watershed is desirable.

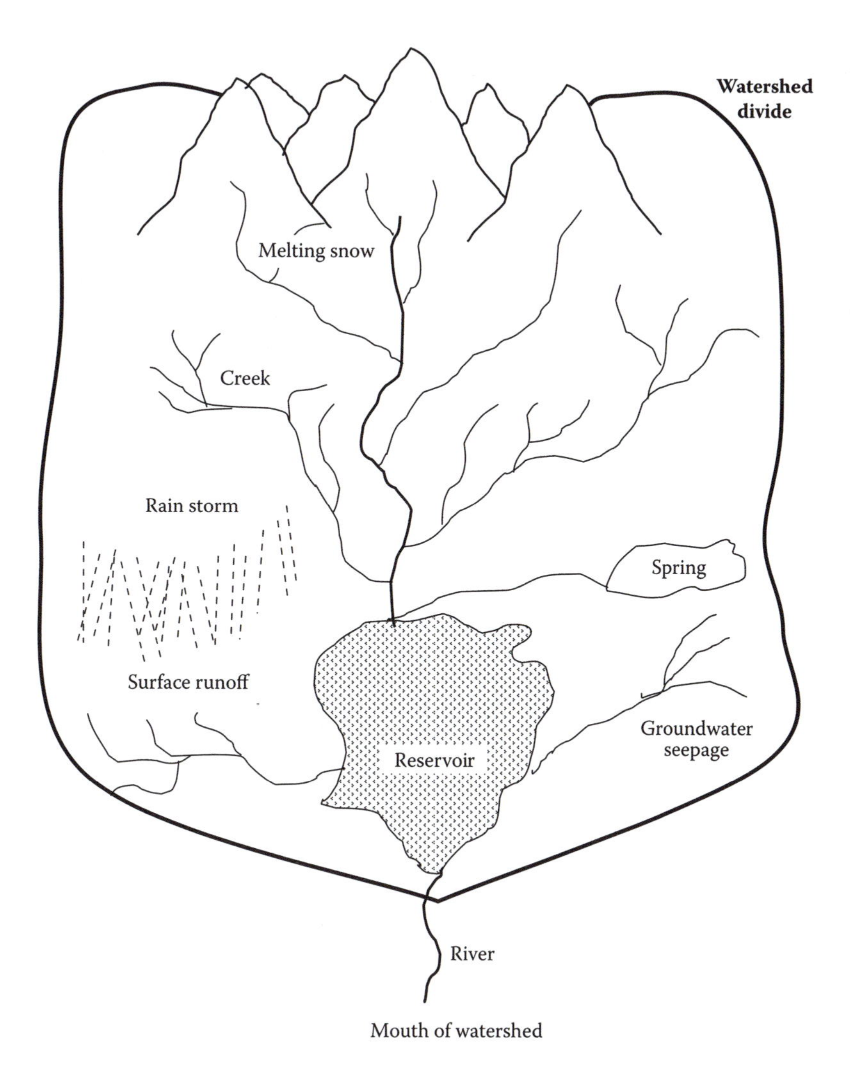

FIGURE 2.3 Watershed.

Raw Water Storage

Raw water (i.e., water that has not been treated) is stored for single or multiple uses, such as navigation, flood control, hydroelectric power, agriculture, water supply, pollution abatement, recreation, and flow augmentation. The primary reason for storing water is to meet peak demands or to store water to meet demands when the flow of the source is below demand. Raw water is stored in natural storage sites (such as lakes, muskegs, and tundra ponds) or in manmade storage areas such as dams. Manmade dams are either masonry or embankment. If embankment dams are used, they are typically constructed of local materials with an impermeable clay core.

Surface Water Intakes

Withdrawing water from a river, lake, or reservoir so it may be conveyed to the first unit process for treatment requires an intake structure. Intakes have no standard design and can range from a simple-pump suction pipe sticking out into the lake or stream to more involved structures costing several thousands of dollars. Typical intakes include submerged intakes, floating intakes, infiltration galleries, spring boxes, and roof catchments. Their primary functions are to supply the highest quality water from the source and to protect piping and pumps from clogging and damage due to wave action, ice formation, flooding, and submerged debris. A poorly conceived or constructed intake can cause many problems. Failure of the intake could result in water system failure.

For a small stream, the most common intake structures are small gravity dams placed across the stream or a submerged intake. In the gravity type of dam, a gravity line or pumps can remove water behind the dam. In the submerged intake type of dam, water is collected in a diversion and carried away by gravity or pumped from a caisson. Another common intake used on small and large streams is an end-suction centrifugal pump or submersible pump placed on a float. The float is secured to the bank, and the water is pumped to a storage area.

Often, the intake structure placed in a stream is an infiltration gallery. The most common infiltration galleries are built by placing well screens or perforated pipe into the streambed. The pipe is covered with clean, graded gravel. When water passes through the gravel, coarse filtration removes a portion of the turbidity and organic material. The water collected by the perforated pipe then flows to a caisson placed next to the stream and is removed from the caisson by gravity or pumping. Intakes used in springs are normally implanted into the water-bearing strata, then covered with clean, washed rock and sealed, usually with clay. The outlet is piped into a spring box. In some locations, a primary source of water is rainwater. Rainwater is collected from the roofs of buildings with a device called a *roof catchment.*

After determining that a water source provides a suitable quality and quantity of raw water, choosing an intake location includes consideration of the following factors:

- Best quality water location
- Dangerous currents
- Sandbar formation
- Wave action
- Ice storms
- Flood factors
- Avoiding navigation channels
- Intake accessibility
- Power availability
- Floating or moving objects that pose a hazard
- Distance from pumping station
- Upstream uses that may affect water quality

Surface Water Screens

Generally, screening devices are installed to protect intake pumps, valves, and piping. A coarse screen of vertical steel bars, with openings of 1 to 3 inches, placed in a near-vertical position excludes large objects. It may be equipped with a trash rack rake to remove accumulated debris. A finer screen, one with 3/8-inch openings, removes leaves, twigs, small fish, and other material passing through the bar rack. Traveling screens consist of wire mesh trays that retain solids as the water passes through them. Drive chains and sprockets raise the trays into a head enclosure, where the debris is removed by water sprays. The screen travel pattern is intermittent and controlled by the amount of accumulated material.

> ***Note:*** When considering what type of screen should be employed, the most important consideration is ensuring that they can be easily maintained.

Surface Water Quality

Surface waters should be of adequate quality to support aquatic life and be aesthetically pleasing; also, waters used as sources of supply should be treatable by conventional processes to provide potable supplies that can meet drinking water standards. Many lakes, reservoirs, and rivers are maintained at a quality suitable for swimming, water skiing, and boating as well as for drinking water. Whether the surface water supply is taken from a river, stream, lake, spring, impoundment, reservoir, or dam, the surface water quality can vary widely, especially in rivers, streams, and small lakes. These water bodies are susceptible not only to waste discharge contamination but also to flash contamination (which can occur almost immediately and not necessarily over time). Lakes are subject to summer/winter stratification (turnover) and to algal blooms. Pollution sources include runoff (agricultural, residential, and urban), spills, municipal and industrial wastewater discharges, and recreational users, as well as natural occurrences. Surface water supplies are difficult to protect from contamination and must always be treated.

Public water systems must comply with applicable federal and state regulations and must provide quantity and quality water supplies including proper treatment (where/when required) and competent/qualified waterworks operators. The USEPA's regulatory requirements insist that all public water systems using any surface water or groundwater under the direct influence of surface water must disinfect and may be required by the state to filter, unless the water source meets certain requirements and site-specific conditions. Treatment technique requirements are established in lieu of maximum contaminant levels (MCLs) for *Giardia*, viruses, heterotrophic plate count bacteria, *Legionella*, and turbidity. Treatment must achieve at least 99.9% removal (3-log removal) and/or inactivation of *Giardia lamblia* cysts and 99.9% removal and/or inactivation of viruses.

GROUNDWATER

Earth possesses an unseen ocean. This ocean, unlike the surface oceans that cover most of the globe, is freshwater, the groundwater that lies contained in aquifers beneath Earth's crust. This gigantic water source forms a reservoir that feeds all the natural fountains and springs of the planet. But how does water travel into the aquifers that lie under Earth's surface? Groundwater sources are replenished from a percentage of the average approximately 3 feet of water that falls to Earth each year on every square foot of land. Water falling to Earth as precipitation follows three courses. Some runs off directly to rivers and streams (roughly 6 inches of that 3-foot yearly total), eventually working its way back to the sea. Evaporation and transpiration through vegetation take up about 2 feet. The remaining 6 inches seep into the ground, entering and filling every interstice, each hollow and cavity. Gravity pulls water toward the center of the Earth. That means that water on the surface will try to seep into the ground below it. Although groundwater comprises only one sixth of the total (1,680,000 miles of water), if we could spread this water out over the land, it would blanket it to a depth of 1000 feet.

Aquifers

Part of the precipitation that falls on land infiltrates the land surface, percolates downward through the soil under the force of gravity, and becomes groundwater. Groundwater, like surface water, is extremely important to the hydrologic cycle and to our water supplies. Almost half of the people in the United States drink public water from groundwater supplies. Overall, more water exists as groundwater than surface water in the United States, including the water in the Great Lakes. But sometimes pumping it to the surface is not economical, and in recent years pollution of groundwater supplies from improper disposal has become a significant problem.

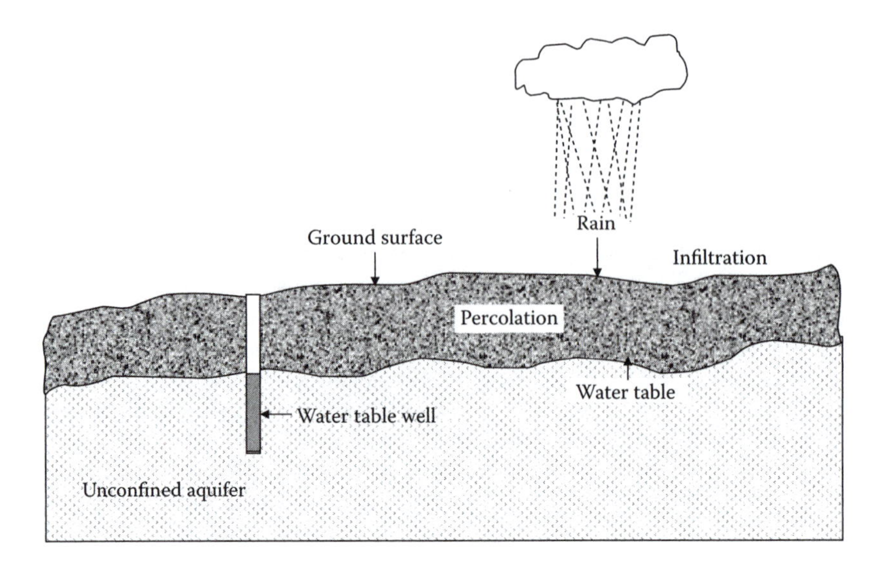

FIGURE 2.4 Unconfined aquifer. (Adapted from Spellman, F.R., *Stream Ecology and Self-Purification: An Introduction for Wastewater and Water Specialists*, CRC Press, Boca Raton, FL, 1996.)

We find groundwater in saturated layers called *aquifers* under the Earth's surface. Three types of aquifers exist: *unconfined*, *confined*, and *springs*. Aquifers are made up of a combination of solid material such as rock and gravel and open spaces called *pores*. Regardless of the type of aquifer, the groundwater in the aquifer is in a constant state of motion. This motion is caused by gravity or by pumping.

The actual amount of water in an aquifer depends on the amount of space available between the various grains of material that make up the aquifer. The amount of space available is called *porosity*. The ease of movement through an aquifer is dependent on how well the pores are connected; for example, clay can hold a lot of water and has high porosity, but the pores are not connected, so water moves through the clay with difficulty. The ability of an aquifer to allow water to infiltrate is referred to as *permeability*.

The unconfined aquifer that lies just under the Earth's surface is called the *zone of saturation* (see Figure 2.4). The top of the zone of saturation is the *water table*. An unconfined aquifer is only contained on the bottom and is dependent on local precipitation for recharge. This type of aquifer is often referred to as a *water table aquifer*. Unconfined aquifers are a primary source of shallow well water (see Figure 2.4). Because these wells are shallow they are not desirable as public drinking water sources. They are subject to local contamination from hazardous and toxic materials, such as fuel and oil, as well as septic tank and agricultural runoff that provides increased levels of nitrates and microorganisms. These wells may be classified as groundwater under the direct influence of surface water and therefore require treatment for control of microorganisms.

A confined aquifer is sandwiched between two impermeable layers that block the flow of water. The water in a confined aquifer is under hydrostatic pressure. It does not have a free water table (see Figure 2.5). Confined aquifers are referred to as *artesian aquifers*. Wells drilled into artesian aquifers are *artesian wells* and commonly yield large quantities of high-quality water. An artesian well is any well where the water in the well casing would rise above the saturated strata. Wells in confined aquifers are normally referred to as *deep wells* and are not generally affected by local hydrological events. A confined aquifer is recharged by rain or snow in the mountains where the aquifer lies close to the surface. Because the recharge area is some distance from areas of possible contamination, the possibility of contamination is usually very low; however, once contaminated, confined aquifers may take centuries to recover.

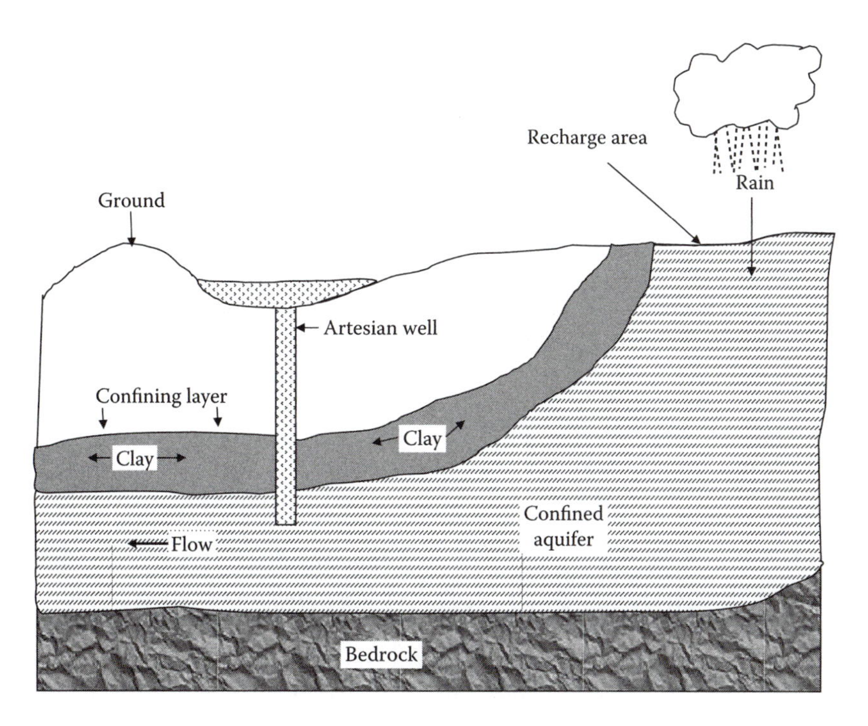

FIGURE 2.5 Confined aquifer. (Adapted from Spellman, F.R., *Stream Ecology and Self-Purification: An Introduction for Wastewater and Water Specialists*, CRC Press, Boca Raton, FL, 1996.)

Groundwater naturally exits the Earth's crust in areas called *springs*. The water in a spring can originate from a water table aquifer or from a confined aquifer. Only water from a confined spring is considered desirable for a public water system.

Groundwater Quality

Generally, groundwater is of high chemical, bacteriological, and physical quality. When pumped from an aquifer composed of a mixture of sand and gravel and when not directly influenced by surface water, groundwater is often used without filtration. It can also be used without disinfection if it has a low coliform count; however, groundwater can become contaminated. To understand how an underground aquifer becomes contaminated, we must understand what occurs when pumping is taking place within the well. When groundwater is removed from its underground source (i.e., from the water-bearing stratum) via a well, water flows toward the center of the well. In a water table aquifer, this movement causes the water table to sag toward the well. This sag is the *cone of depression*. The shape and size of the cone depend on the relationship between the pumping rate and the rate at which water can move toward the well. If the rate is high, the cone is shallow and its growth stabilizes. The area that is included in the cone of depression is the *cone of influence*, and any contamination in this zone will be drawn into the well.

Groundwater under the Direct Influence of Surface Water

Water under the direct influence of surface water (GUDISW) is not classified as a groundwater supply. A supply designated as GUDISW must be treated under the state's surface water rules rather than the groundwater rules. The Surface Water Treatment Rule of the Safe Drinking Water Act requires each site to determine which groundwater supplies are influenced by surface water (e.g., when surface water can infiltrate a groundwater supply and could contaminate it with *Giardia*, viruses, turbidity, and organic material from the surface water source). To determine whether a groundwater supply is under the direct influence of surface water, the USEPA has developed procedures that focus on significant and relatively rapid shifts in water quality characteristics, including turbidity, temperature, and pH. When these shifts can be closely correlated with rainfall or other surface water conditions, or when certain indicator organisms associated with surface water are found, the source is said to be under the direct influence of surface water.

Almost all groundwater is in constant motion through the pores and crevices of the aquifer in which it occurs. The water table is rarely level; it generally follows the shape of the ground surface. Groundwater flows in the downhill direction of the sloping water table. The water table sometimes intersects low points of the ground, where it seeps out into springs, lakes, or streams.

Usual groundwater sources include wells and springs that are not influenced by surface water or local hydrologic events. As a potable water source, groundwater has several advantages over surface water. Unlike surface water, groundwater is not easily contaminated. Groundwater sources are usually lower in bacteriological contamination than surface waters. Groundwater quality and quantity usually remain stable throughout the year. In the United States, groundwater is available in most locations.

As a potable water source, groundwater does present some disadvantages compared to surface water sources. Operating costs are usually higher, because groundwater supplies must be pumped to the surface. Any contamination is often hidden from view, and removing contaminants can be very difficult. Groundwater often possesses high mineral levels and thus an increased level of hardness, because it is in contact longer with minerals. Near coastal areas, groundwater sources may be subject to saltwater intrusion.

Note: Groundwater quality is influenced by the quality of its source. Changes in source waters or degraded quality of source supplies may seriously impair the quality of the groundwater supply.

Prior to moving onto water use it is important to point out that our freshwater supplies are constantly renewed through the hydrologic cycle, but the balance between the normal ratio of freshwater to saltwater is not subject to our ability to change. As our population grows and we move into lands without ready freshwater supplies, we place ecological strains upon those areas and on their ability to support life. Communities that build in areas without an adequate local water supply are at risk in the event of emergency. Proper attention to our surface and groundwater sources, including remediation, pollution control, and water reclamation and reuse, can help to ease the strain, but technology cannot fully replace adequate local freshwater supplies, whether from surface or groundwater sources.

WELL SYSTEMS

The most common method for withdrawing groundwater is to penetrate the aquifer with a vertical well, then pump the water up to the surface. In the past, when someone wanted a well, they simply dug (or hired someone to dig) and hoped that they would find water in a quantity suitable for their needs. Today, in most locations in the United States, for example, developing a well supply usually involves a more complicated step-by-step process. Local, state, and federal requirements specify the actual requirements for development of a well supply in the United States. The standard sequence for developing a well supply generally involves a seven-step process:

1. *Application*—Depending on location, filling out and submitting an application (to the applicable authorities) to develop a well supply is standard procedure.
2. *Well site approval*—Once the application has been made, local authorities check various local geological and other records to ensure that the siting of the proposed well coincides with mandated guidelines for approval.
3. *Well drilling*—The well is drilled.

4. *Preliminary engineering report*—After the well is drilled and the results documented, a preliminary engineering report is made on the suitability of the site to serve as a water source. This procedure involves performing a pump test to determine if the well can supply the required amount of water. The well is generally pumped for at least 6 hours at a rate equal to or greater than the desired yield. A stabilized drawdown should be obtained at that rate and the original static level should be recovered within 24 hours after pumping stops. During this test period, samples are taken and tested for bacteriological and chemical quality.
5. *Submission of documents for review and approval*—The application and test results are submitted to an authorized reviewing authority that determines if the well site meets approval criteria.
6. *Construction permit*—If the site is approved, a construction permit is issued.
7. *Operation permit*—When the well is ready for use, an operation permit is issued.

Well Site Requirements

To protect the groundwater source and provide high-quality safe water, the waterworks industry has developed standards and specifications for wells. The following listing includes industry standards and practices, as well as those items included in example State Department of Environmental Compliance regulations.

Note: Check with your local regulatory authorities to determine well site requirements.

1. Minimum well lot requirements:
    - 50 feet from well to all property lines
    - All-weather access road provided
    - Lot graded to divert surface runoff
    - Recorded well plat and dedication document
2. Minimum well location requirements:
    - At least 50 feet horizontal distance from any actual or potential sources of contamination involving sewage
    - At least 50 feet horizontal distance from any petroleum or chemical storage tank or pipeline or similar source of contamination, except where plastic-type well casing is used the separation distance must be at least 100 feet
3. Vulnerability assessment:
    - Is the wellhead area 1000 ft radius from the well?
    - What is the general land use of the area (residential, industrial, livestock, crops, undeveloped, other)?
    - What are the geologic conditions (sinkholes, surface, subsurface)?

Types of Wells

Water supply wells may be characterized as shallow or deep. In addition, wells are classified as follows:

- Class I, cased and grouted to 100 ft
- Class II A, cased to a minimum of 100 ft and grouted to 20 ft
- Class II B, cased and grouted to 50 ft

Note: During the well development process, mud/silt forced into the aquifer during the drilling process is removed, allowing the well to produce the best-quality water at the highest rate from the aquifer.

Shallow Wells

Shallow wells are those that are less than 100 ft deep. Such wells are not particularly desirable for municipal supplies because the aquifers they tap are likely to fluctuate considerably in depth, making the yield somewhat uncertain. Municipal wells in such aquifers cause a reduction in the water table (or phreatic surface) that affects nearby private wells, which are more likely to utilize shallow strata. Such interference with private wells may result in damage suits against the community. Shallow wells may be dug, bored, or driven:

- *Dug wells*—Dug wells are the oldest type of well and date back many centuries; they are dug by hand or by a variety of unspecialized equipment. They range in size from approximately 4 to 15 ft in diameter and are usually about 20 to 40 ft deep. Such wells are usually lined or cased with concrete or brick. Dug wells are prone to failure from drought or heavy pumpage. They are vulnerable to contamination and are not acceptable as a public water supply in many locations.
- *Driven wells*—Driven wells consist of a pipe casing terminating in a point slightly greater in diameter than the casing. The pointed well screen and the lengths of pipe attached to it are pounded down or driven in the same manner as a pile, usually with a drop hammer, to the water-bearing strata. Driven wells are usually 2 to 3 inches in diameter and are used only in unconsolidated materials. This type of shallow well is not acceptable as a public water supply.
- *Bored wells*—Bored wells range from 1 to 36 inches in diameter and are constructed in unconsolidated materials. The boring is accomplished with augers (either hand or machine driven) that fill with soil and then are drawn to the surface to be emptied. The casing may be placed after the well is completed (in relatively cohesive materials) but must advance with the well in noncohesive strata. Bored wells are not acceptable as a public water supply.

Deep Wells

Deep wells are the usual source of groundwater for municipalities. Deep wells tap thick and extensive aquifers that are not subject to rapid fluctuations in water level (remember that the *piezometric surface* is the height to which water will rise in a tube penetrating a confined aquifer) and that provide a large and uniform yield. Deep wells typically yield water of more consistent quality than shallow wells, although the quality is not necessarily better. Deep wells are constructed by a variety of techniques; we discuss two of these techniques below:

- *Jetted wells*—Jetted well construction commonly employs a jetting pipe with a cutting tool. This type of well cannot be constructed in clay or hardpan or where boulders are present. Jetted wells are not acceptable as a public water supply.
- *Drilled wells*—Drilled wells are usually the only type of well allowed for use in most public water supply systems. Several different methods of drilling are available, all of which are capable of drilling wells of extreme depth and diameter. Drilled wells are constructed using a drilling rig that creates a hole into which the casing is placed. Screens are installed at one or more levels when water-bearing formations are encountered.

Components of a Well

The components that make up a well system include the well itself, the building and the pump, and related piping system. In this section, we focus on the components that make up the well itself. Many of these components are shown in Figure 2.6.

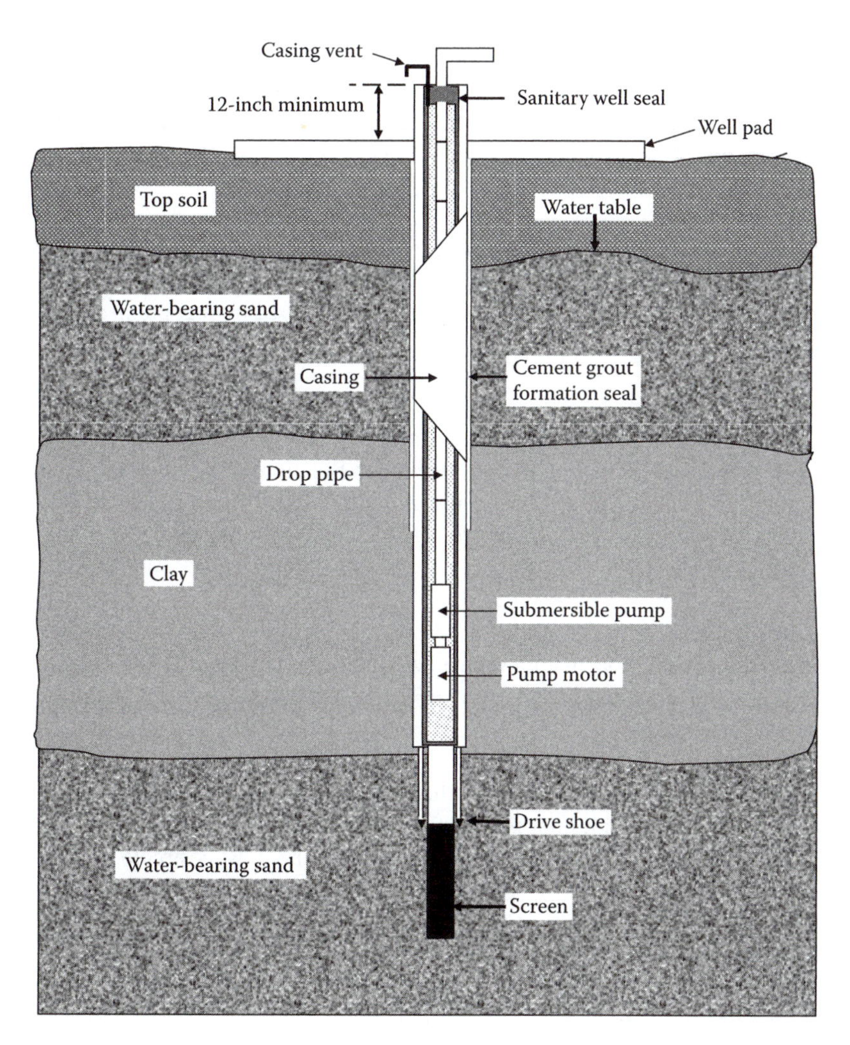

FIGURE 2.6 Components of a well.

Well Casing

A well is a hole in the ground called the *borehole*. To prevent collapse, a casing is placed inside the borehole. The well casing prevents the walls of the hole from collapsing and prevents contaminants (either surface or subsurface) from entering the water source. The casing also provides a column of stored water and housing for the pump mechanisms and pipes. Well casings constructed of steel or plastic material are acceptable. The well casing must extend a minimum of 12 inches above grade.

Grout

To protect the aquifer from contamination, the casing is sealed to the borehole near the surface and near the bottom where it passes into the impermeable layer with grout. This sealing process keeps the well from being polluted by surface water and seals out water from water-bearing strata that have undesirable water quality. Sealing also protects the casing from external corrosion and restrains unstable soil and rock formations. Grout consists of near cement that is pumped into the annular space (it is completed within 48 hours of well construction); it is pumped under continuous pressure starting at the bottom and progressing upward in one continuous operation.

Well Pad

The well pad provides a ground seal around the casing. The pad is constructed of reinforced concrete 6 feet by 6 feet (6 inches thick) with the well head located in the middle. The well pad prevents contaminants from collecting around the well and seeping down into the ground along the casing.

Sanitary Seal

To prevent contamination of the well, a sanitary seal is placed at the top of the casing. The type of seal varies depending on the type of pump used. The sanitary seal contains openings for power and control wires, pump support cables, a drawdown gauge, discharge piping, pump shaft, and air vent, while providing a tight seal around them.

Well Screen

Screens can be installed at the intake points on the end of a well casing or on the end of the inner casing on gravel packed well. These screens perform two functions: (1) support the borehole, and (2) reduce the amount of sand that enters the casing and the pump. They are sized to allow the maximum amount of water while preventing the passage of sand, sediment, and gravel.

Casing Vent

The well casing must have a vent to allow air into the casing as the water level drops. The vent terminates 18 inches above the floor with a return bend pointing downward. The opening of the vent must be screened with No. 24 mesh stainless steel to prevent entry of vermin and dust.

Drop Pipe

The drop pipe or riser is the line leading from the pump to the well head. It provides adequate support so an aboveground pump does not move and so a submersible pump is not lost down the well. This pipe is either steel or polyvinylchloride (PVC). Steel is the most desirable.

Miscellaneous Well Components

Miscellaneous well components include the following:

- *Gauge and air line* measure the water level of the well.
- *Check valve* is located immediately after the well to prevent system water from returning to the well. It must be located above ground and protected from freezing.
- *Flowmeter* is required to monitor the total amount of water withdrawn from the well, including any water blown off.
- *Control switches* control well pump operation.
- *Blowoff valve* is located between the well and storage tank and is used to flush the well of sediment or turbid or super-chlorinated water.
- *Sample taps* include (1) raw water sample taps, which are located before any storage or treatment to permit sampling of the water directly from the well, and (2) entry-point sample taps located after treatment.
- *Control valves* isolate the well for testing or maintenance or are used to control water flow.

Well Evaluation

After a well is developed, conducting a pump test determines if it can supply the required amount of water. The well is generally pumped for at least 6 hours (many states require a 48-hour yield and drawdown test) at a rate equal to or greater than the desired yield. *Yield* is the volume or quantity of water discharged from a well per unit of time (e.g., gpm, ft^3/sec). Regulations usually require that a well produce a minimum of 0.5 gpm per residential connection. *Drawdown* is the difference between the static water level (level of the water in the well when it has not been used for some time and has stabilized) and the pumping water level in a well. Drawdown is measured by using an air line and pressure gauge to monitor the water level during the 48 hours of pumping.

The procedure calls for the air line to be suspended inside the casing down into the water. At the other end are the pressure gauge and a small pump. Air is pumped into the line (displacing the water) until the pressure stops increasing. The highest pressure reading on the gauge is recorded. During the 48 hours of pumping, the yield and drawdown are monitored more frequently during the beginning of the testing period, because the most dramatic changes in flow and water level usually occur then. The original static level should be recovered within 24 hours after pumping stops.

Testing is performed on a bacteriological sample for analysis by the most probable number (MPN) method every half hour during the last 10 hours of testing. The results are used to determine if chlorination is required or if chlorination alone will be sufficient to treat the water. Chemical, physical, and radiological samples are collected for analysis at the end of the test period to determine if treatment other than chlorination may be required.

Note: Recovery from the well should be monitored at the same frequency as during the yield and drawdown testing and for at least the first 8 hours, or until 90% of the observed drawdown is obtained.

Specific capacity (often referred to as the *productivity index*) is a test method for determining the relative adequacy of a well; over a period of time, it is a valuable tool for evaluating well production. Specific capacity is expressed as a measure of well yield per unit of drawdown (yield divided by drawdown). When conducting this test, if possible always run the pump for the same length of time and at the same pump rate.

Well Pumps

Pumps are used to move the water out of the well and deliver it to the storage tank or distribution system. The type of pump chosen should provide optimum performance based on the location and operating conditions, required capacity, and total head. Two types of pumps commonly installed in groundwater systems are *lineshaft turbines* and *submersible turbines*. Whichever type of pump is used, they are rated on the basis of pumping capacity expressed in gpm (e.g., 40 gpm), not on horsepower.

Routine Operation and Recordkeeping Requirements

Ensuring the proper operation of a well requires close monitoring; wells should be visited regularly. During routine monitoring visits, check for any unusual sounds in the pump, line, or valves and for any leaks. In addition, as a routine, cycle valves to ensure good working condition. Check motors to make sure they are not overheating. Check the well pump to guard against short cycling. Collect a water sample for a visual check for sediment. Also, check chlorine residual and treatment equipment. Measure gallons on the installed meter for one minute to obtain the pump rate in gallons per minute (look for gradual trends or big changes). Check water level in the well at least monthly (perhaps more often in summer or during periods of low rainfall). Finally, from recorded meter readings, determine gallons used and compare with water consumed to determine possible distribution system leaks. Along with meter readings, other records must be accurately and consistently maintained for water supply wells. Such recordkeeping is absolutely imperative. The records (an important resource for troubleshooting) can be useful when problems develop or can be helpful in identifying potential problems. A properly operated and managed waterworks facility keeps the following records of well operation.

Well Log

The well log provides documentation of what materials were found in the borehole and at what depth. The well log includes the depths at which water was found, the casing length and type, the depth at which various types of soils were found, testing procedures, well development techniques, and well production. In general, the following items should be included in the well log:

1. Well location
2. Who drilled the well
3. When the well was completed
4. Well class
5. Total depth to bedrock
6. Hole and casing size
7. Casing material and thickness
8. Screen size and locations
9. Grout depth and type
10. Yield and drawdown (test results)
11. Pump information (type, horsepower, capacity, intake depth, and model number)
12. Geology of the hole
13. A record of yield and drawdown data

Well Maintenance

Wells do not have an infinite life, and their output is likely to reduce with time as a result of hydrological and/or mechanical factors. Protecting the well from possible contamination is an important consideration. Potential problems can be minimized if a well is properly located (based on knowledge of the local geological conditions and a vulnerability assessment of the area).

During the initial assessment, ensuring that the well is not located in a sinkhole area is important. Locations where unconsolidated or bedrock aquifers could be subject to contamination must be identified. Several other important determinations must also be made: Is the well located on a floodplain? Is it located next to a drainfield for septic systems or near a landfill? Are petroleum or gasoline storage tanks nearby? Is any pesticide or plastics manufacturing conducted near the well site?

Along with proper well location, proper well design and construction prevent wells from acting as conduits for the vertical migration of contaminants into the groundwater. Basically, the pollution potential of a well equals how well it was constructed. Contamination can occur during the drilling process, and an unsealed or unfinished well is an avenue for contamination. Any opening in the sanitary seal or break in the casing may cause contamination, as can a reversal of water flow. In routine well maintenance operations, corroded casing or screens are sometimes withdrawn and replaced, but this is difficult and not always successful. Simply constructing a new well may be less expensive.

Well Abandonment

In the past, common practice was simply to walk away and forget about a well when it ran dry. Today, while dry or failing wells are still abandoned, we know that they must be abandoned with care (and not completely forgotten). An abandoned well can become a convenient (and dangerous) receptacle for wastes, thus contaminating the aquifer. An improperly abandoned well could also become a haven for vermin or, worse, a hazard for children. A temporarily abandoned well must be sealed with a watertight cap or wellhead seal. The well must be maintained so it does not become a source or channel of contamination during temporary abandonment.

When a well is permanently abandoned, all casing and screen materials may be salvaged. The well should be checked from top to bottom to ensure that no obstructions interfere with plugging and sealing operations. Prior to plugging, the well should be thoroughly chlorinated. Bored wells should be completely filled with cement grout. If the well was constructed in an unconsolidated formation, it should be completely filled with cement grout or clay slurry introduced through a pipe that initially extends to the bottom of the well. As the pipe is raised, it should remain submerged in the top layers of grout as the well is filled.

Wells constructed in consolidated rock or that penetrate zones of consolidated rock can be filled with sand or gravel opposite zones of consolidated rock. The sand or gravel fill is terminated 5 feet below the top of the consolidated rock. The remainder of the well is filled with sand–cement grout.

WATER USE

In the United States, rainfall averages approximately 4250×10^9 gallons a day. About two thirds of this rainfall returns to the atmosphere through evaporation directly from the surface of rivers, streams, and lakes and transpiration from plant foliage. This leaves approximately 1250×10^9 gallons a day to flow across or through the Earth to the sea.

It was estimated that in the United States about 408 billion gallons per day (abbreviated Bgal/d) were withdrawn for all uses in 2000 (USGS, 2004). This total has varied less than 3% since 1985 as withdrawals have stabilized for the two largest uses—thermoelectric power and irrigation. Fresh groundwater withdrawals (83.3 Bgal/d) during 2000 were 14% more than during 1985. Fresh surface water withdrawals for 2000 were 262 Bgal/d, varying less than 2% since 1985.

About 195 Bgal/d, or 8% of all freshwater and saline water withdrawals for 2000, were used for thermoelectric power. Most of this water was derived from surface water and used for once-through cooling at power plants. About 52% of fresh surface water withdrawals and about 96% of saline-water withdrawals were for thermoelectric-power use. Withdrawals for thermoelectric power have been relatively stable since 1985.

Irrigation totaled 137 Bgal/d and represented the largest use of freshwater in the United States in 2000. Since 1950, irrigation has accounted for about 65% of total water withdrawals, excluding those for thermoelectric power. Historically, more surface water than groundwater has been used for irrigation; however, the percentage of total irrigation withdrawals from groundwater has continued to increase, from 23% in 1950 to 42% in 2000. Total irrigation withdrawals were 2% more for 2000 than for 1995, because of a 16% increase in groundwater withdrawals and a small decrease in surface water withdrawals. Irrigated acreage more than doubled between 1950 and 1980, then remained constant before increasing nearly 7% between 1995 and 2000. The number of acres irrigated with sprinkler and microirrigation systems has continued to increase, and they now account for more than one half of the total irrigated acreage.

Public-supply withdrawals were more than 43 Bgal/d for 2000 compared to public-supply withdrawals in 1950 of 14 Bgal/d. During 2000, about 85% of the population in the United States obtained drinking water from public suppliers, compared to 62% during 1950. Surface water provided 63% of the total during 2000, whereas surface water provided 74% during 1950.

Self-supplied industrial withdrawals totaled nearly 20 Bgal/d in 2000, or 12% less than in 1995; compared to 1985, industrial self-supported withdrawals declined by 24%. Estimates of industrial water use in the United State were largest during the years from 1965 to 1980, but usage estimates for 2000 were at the lowest level since reporting began in 1950. Combined withdrawals for self-supplied domestic, livestock, aquaculture, and mining were less than 13 Bgal/d for 2000, and represented about 3% of total withdrawals. California, Texas, and Florida accounted for one fourth of all water withdrawals for 2000. States with the largest surface water withdrawals were California, Texas, and Nebraska, all of which had large withdrawals for irrigation.

In this text, the primary concern with water use is with regard to municipal demand. Later, in Chapter 10, a more in-depth and detailed treatment of water use, or demand, in both micro- and macro-usage is presented. For now, municipal demand is the focus. Demand is usually classified according to the nature of the user:

1. *Domestic*—Domestic water is supplied to houses, schools, hospitals, hotels, restaurants, etc. for culinary, sanitary, and other purposes. Use varies with the economic level of the consumer, the range being 20 to 100 gal/capita/d. It should be pointed out that these figures include water used for watering gardens and lawns and for washing cars.
2. *Commercial and industrial*—Commercial and industrial water is supplied to stores, offices, and factories. The importance of commercial and industrial demand is based, of course, on whether there are large industries that use water supplied from the municipal system. These large industries demand a quantity of water directly related to the number of persons employed, to the actual floor space or area of each establishment, and to the number of units manufactured or produce. Industry in the United States uses an average of 150 Bgal/d of water each day.
3. *Public use*—Public use water is the water furnished to public buildings and used for public services. This includes water for schools, public buildings, fire protection, and flushing streets.
4. *Loss and waste*—Water that is lost or wasted (unaccounted for) can be attributed to leaks in the distribution system, inaccurate meter readings, and unauthorized connections. Loss and waste of water can be expensive. In order to reduce loss and waste a regular program that includes maintenance of the system and replacement and/or recalibration of meters is required (McGhee, 1991).

WATER CONTENT OF CROPS, GOODS, BEVERAGES, ANIMALS, AND INDUSTRIAL PRODUCTS

Table 2.3 shows the water content of some common items. The data contained in the table were derived and adapted from USGS (2014), Gleick (2000), Pacific Institute, (2007), and Hoekstra (2013). The reader is advised that the figures presented in Table 2.3 are approximations or best guesses and are complicated and fraught with problems and uncertainties; the reader should use caution when using the data for simple comparisons.

THOUGHT-PROVOKING QUESTION

2.1 Where does the water cycle begin? Explain.

REFERENCES AND RECOMMENDED READING

Angele, Sr., F.J.. (1974). *Cross Connections and Backflow Protection*, 2nd ed. Denver, CO: American Water Association.

Gleick, P.H. (2000). *Water in Crisis: The World's Water 2000–2001*. Washington, DC: Island Press.

Hoekstra, A.Y. (2013). *The Water Footprint of Modern Consumer Society*. Oxford: Routledge.

Jones, F.E. (1992). *Evaporation of Water*. Chelsea, MI: Lewis Publishers.

Lewis, S.A. (1996). *The Sierra Club Guide to Safe Drinking Water*. San Francisco, CA: Sierra Club Books.

McGhee, T.J. (1991). *Water Supply and Sewerage*, 6th ed. New York: McGraw-Hill.

Meyer, W.B. (1996). *Human Impact on Earth*. New York: Cambridge University Press.

Pacific Institute. (2007). *Bottled Water and Energy Fact Sheet*. Oakland, CA: Pacific Institute (http://pacinst.org/publication/bottled-water-and-energy-a-fact-sheet/).

Peavy, H. S. et al. (1985). *Environmental Engineering*. New York: McGraw-Hill.

TABLE 2.3
Water Content of Selected Items

Item	Liters Water	Item	Liters Water	Notes
		Beverages (per liter)		
Glass of beer	300	Cup of coffee	1120	
Malt beverages (processing)	50	Cup of tea	120	
Glass of water	~1	Glass of wine	960	
Bottle water	3 to 4	Glass of apple juice	950	
Milk	1000	Glass of orange juice	850	
		Assorted produced goods (per kilogram)		
Roasted coffee	21,000	Sheet paper	125	
Tea	9200	Potato chips	925	
Bread	1300	Hamburger	16,000	
Cheese	5000	Leather shoes	16,600	
Cotton textile finished	11,000	Microchip	16,000	
Microchip	16,000			
		Assorted crops (per kilogram)		
Barley	1300	Alfalfa	900 to 2000	To grow; depends on climate and weight of finished crop vs. total yield
Coconut	2500	Sorghum	1000 to 1800	
Corn	900	Corn/maize	1000 to 1800	
Sugar	1500	Rice	1900 to 5000	
Potato	500 to 1500	Soybeans	1100 to 2000	
Wheat	900 to 2000			
		Assorted animals (per kilogram of meat)		
Sheep	6100	Chicken	3500 to 5700	Includes water for feed
Goat	4000	Eggs	3300	
Beef	15,000 to 70,000			
		Assorted industrial products (per kilogram)		
Steel	260	Nitrogenous fertilizer	120	Processing water; there is great variation depending on process
Primary copper	440	Synthetic rubber	460	
Primary aluminum	410	Inorganic pigments	410	
Phosphatic fertilizer	150			

Source: Data from Gleick (2000), Hoekstra (2013), Pacific Institute (2007), USGS (2014).

Pielou, E.C. (1998). *Fresh Water*. Chicago: University of Chicago Press.
Powell, J.W. (1904). *Twenty-Second Annual Report of the Bureau of American Ethnology to the Secretary of the Smithsonian Institution, 1900–1901*. Washington, DC: U.S. Government Printing Office.
Spellman, F.R. (1996). *Stream Ecology and Self-Purification: An Introduction for Wastewater and Water Specialists*. Boca Raton, FL: CRC Press.
Spellman, F.R. (2001). *The Handbook for Wastewater Operator Certification*. Boca Raton, FL: CRC Press.
Spellman, F.R. (2003). *Handbook of Water and Wastewater Treatment Plant Operations*. Boca Raton, FL: Lewis Publishers.
Turk, J. and Turk, A. (1988). *Environmental Science*, 4th ed. Philadelphia, PA: Saunders College Publishing.
USEPA. (2006). *What Is a Watershed?* Washington, DC: U.S. Environmental Protection Agency (http://www.epa.gov/owow/watershed/whatis.html).
USGS. (2004). *Estimated Use of Water in the United States in 2000*. Washington, DC: U.S. Geological Survey.

USGS. (2006). *Water Science in Schools*. Washington, DC: U.S. Geological Survey.
USGS. (2014). *How Much Water Does It Take to Grow a Hamburger?* Washington, DC: U.S. Geological Survey (http://ga2.er.usgs.gov/edu/activity-water-content.cfm).
Viessman, Jr., W. (1991). Water management issues for the nineties. *Water Research Bulletin*, 26(6), 883–981.

3 Water Hydraulics

Anyone who has tasted natural spring water knows that it is different from city water, which is used over and over again, passing from mouth to laboratory and back to mouth again, without ever being allowed to touch the earth. We need to practice such economics these days, but in several thirsty countries, there are now experts in hydrodynamics who are trying to solve the problem by designing flowforms that copy the earth, producing rhythmic and spiral motions in moving water. And these pulsations do seem to vitalize and energize the liquid in some way, changing its experience, making it taste different and produce better crops.

Watson (1988)

Note: The practice and study of water hydraulics is not new. Even in medieval times, water hydraulics was not new, as medieval Europe inherited a highly developed range of Roman hydraulic components (Magnusson, 2001). The basic conveyance technology, based on low-pressure systems of pipe and channels, was already established. In studying "modern" water hydraulics, it is important to remember that, as Magnusson put it, the science of water hydraulics is the direct result of two immediate and enduring problems: "the acquisition of freshwater and access to a continuous strip of land with a suitable gradient between the source and the destination."

TERMINOLOGY

- *Friction head*—The energy needed to overcome friction in the piping system. It is expressed in terms of the added system head required.
- *Head*—The equivalent distance water must be lifted to move from the supply tank or inlet to the discharge. Head can be divided into three components: *static head*, *velocity head*, and *friction head.*
- *Pressure*—The force exerted per square unit of surface area; may be expressed as pounds per square inch.
- *Static head*—The actual vertical distance from the system inlet to the highest discharge point.
- *Total dynamic head*—The total of the static head, friction head, and velocity head.
- *Velocity*—The speed of a liquid moving through a pipe, channel, or tank; may be expressed in feet per second.
- *Velocity head*—The energy needed to keep the liquid moving at a given velocity; it is expressed in terms of the added system head required.

WHAT IS WATER HYDRAULICS?

The word "hydraulics" is derived from the Greek words *hydro* ("water") and *aulis* ("pipe"). Originally, the term referred only to the study of water at rest and in motion (flowing through pipes or channels). Today, it is taken to mean the flow of *any* liquid in a system.

What is a liquid? In terms of hydraulics, a liquid can be either oil or water. In fluid power systems used in modern industrial equipment, the hydraulic liquid of choice is oil. Some common examples of hydraulic fluid power systems include automobile braking and power steering systems, hydraulic elevators, and hydraulic jacks or lifts. Probably the most familiar hydraulic fluid power systems in water/wastewater operations are those in dump trucks, front-end loaders, graders, and earth-moving and excavation equipment. In this text, we are concerned with liquid water.

Many find the study of water hydraulics difficult and puzzling (especially the licensure examination questions), but we know it is not mysterious or incomprehensible. It is the function or output of practical applications of the basic principles of water physics.

WATER BALANCE

In water hydraulics, the water balance or water budget can be used, in equation form, to describe the flow of water in and out of a system. For example, if we are adding 3 grams of water to a container every minute and 1 gram of water is leaking out each minute, then the mass stored with the container is increasing at the rate of 2 grams per minute. Symbolically, we can write this as

$$P = Q + E + \Delta S \tag{3.1}$$

where
- P = Precipitation.
- Q = Runoff.
- E = Evapotranspiration.
- ΔS = Change in storage.

BASIC CONCEPTS

Air pressure (at sea level) = 14.7 pounds per square inch (psi)

This relationship is important because our study of hydraulics begins with air. A blanket of air many miles thick surrounds the Earth. The weight of this blanket on a given square inch of the Earth's surface will vary according to the thickness of the atmospheric blanket above that point. As shown above, at sea level the pressure exerted is 14.7 pounds per square inch (psi). On a mountain top, air pressure decreases because the blanket is not as thick.

$$1 \text{ ft}^3 \text{ H}_2\text{O} = 62.4 \text{ lb}$$

This relationship is also important; note that both cubic feet and pounds are used to describe a volume of water. A defined relationship exists between these two methods of measurement. The specific weight of water is defined relative to a cubic foot. One cubic foot of water weighs 62.4 lb. This relationship is true only at a temperature of 4°C and at a pressure of 1 atmosphere, conditions that are referred to as *standard temperature and pressure* (STP). Note that 1 atmosphere = 14.7 lb/in.^{2} at sea level and 1 ft^3 of water contains 7.48 gal.

The weight varies so little that, for practical purposes, this weight is used for temperatures ranging from 0 to 100°C. One cubic inch of water weighs 0.0362 lb. Water 1 ft deep will exert a pressure of 0.43 lb/in.^{2} on the bottom area (12 in. × 0.0362 lb/in.^{3}). A column of water 2 ft high

DID YOU KNOW?

Some hydrologists believe that a predevelopment water budget for a groundwater system (that is, a water budget for the natural conditions before humans used the water) can be used to calculate the amount of water available for consumption (or safe yield). This concept has been referred to as the "water-budget myth."

exerts 0.86 psi (2 ft × 0.43 psi/ft); one 10 ft high exerts 4.3 psi (10 ft × 0.43 psi/ft); and one 55 ft high exerts 23.65 psi (55 ft × 0.43 psi/ft). A column of water 2.31 feet high will exert 1.0 psi (2.31 ft × 0.43 psi/ft). To produce a pressure of 50 psi requires a 115.5-ft water column (50 psi × 2.31 ft/psi).

Remember the important points being made here:

1. 1 ft^3 H_2O = 62.4 lb (see Figure 3.1).
2. A column of water 2.31 ft high will exert 1.0 psi.

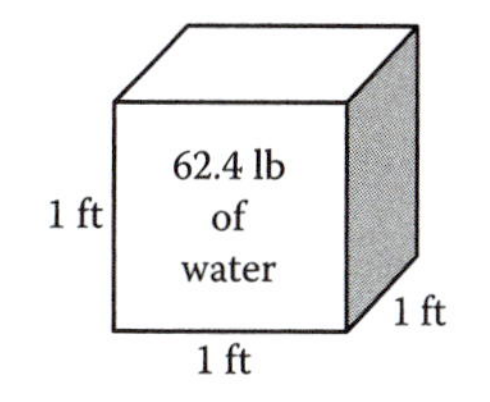

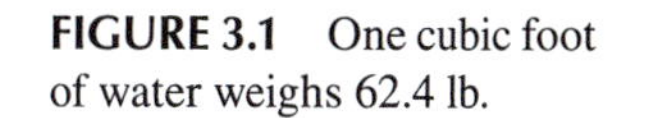
FIGURE 3.1 One cubic foot of water weighs 62.4 lb.

Another relationship is also important:

$$1 \text{ gal } H_2O = 8.34 \text{ lb}$$

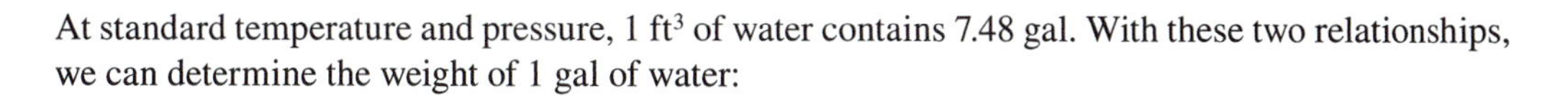
At standard temperature and pressure, 1 ft^3 of water contains 7.48 gal. With these two relationships, we can determine the weight of 1 gal of water:

$$\text{Weight of 1 gal of water} = 62.4 \text{ lb} \div 7.48 \text{ gal} = 8.34 \text{ lb/gal}$$

Thus,

$$1 \text{ gal } H_2O = 8.34 \text{ lb}$$

Note: Further, this information allows cubic feet to be converted to gallons by simply multiplying the number of cubic feet by 7.48 gal/ft^3.

■ Example 3.1

Problem: Find the number of gallons in a reservoir that has a volume of 855.5 ft^3.

Solution:

$$855.5 \text{ ft}^3 \times 7.48 \text{ gal/ft}^3 = 6399 \text{ gal (rounded)}$$

Note: The term *head* is used to designate water pressure in terms of the height of a column of water in feet; for example, a 10-ft column of water exerts 4.3 psi. This can be referred to as 4.3-psi pressure or 10 ft of head.

Stevin's Law

Stevin's law deals with water at rest. Specifically, it states: "The pressure at any point in a fluid at rest depends on the distance measured vertically to the free surface and the density of the fluid." Stated as a formula, this becomes

$$p = w \times h \tag{3.2}$$

where

p = Pressure in pounds per square foot (lb/ft^2 or psf).
w = Density in pounds per cubic foot (lb/ft^3).
h = Vertical distance in feet.

■ **Example 3.2**

Problem: What is the pressure at a point 18 ft below the surface of a reservoir?

Solution: To calculate this, we must know that the density of the water (w) is 62.4 lb/ft³.

$$p = w \times h = 62.4 \text{ lb/ft}^3 \times 18 \text{ ft} = 1123 \text{ lb/ft}^2 \text{ (psf)}$$

Waterworks operators generally measure pressure in pounds per square inch rather than pounds per square foot; to convert, divide by 144 in.²/ft² (12 in. × 12 in. = 144 in.²):

$$p = 1123 \text{ psf} \div 144 \text{ in.}^2/\text{ft}^2 = 7.8 \text{ lb/in.}^2 \text{ (psi)}$$

DENSITY AND SPECIFIC GRAVITY

Table 3.1 shows the relationships among temperature, specific weight, and density of water. When we say that iron is heavier than aluminum, we say that iron has a greater density than aluminum. In practice, what we are really saying is that a given volume of iron is heavier than the same volume of aluminum.

Note: What is density? Density is the *mass per unit volume* of a substance.

Suppose you have a tub of lard and a large box of cold cereal, each having a mass of 600 g. The density of the cereal would be much less than the density of the lard because the cereal occupies a much larger volume than the lard occupies.

The density of an object can be calculated by using the following formula:

$$\text{Density} = \text{Mass} \div \text{Volume} \tag{3.3}$$

In water treatment operations, perhaps the most common measures of density are pounds per cubic foot (lb/ft³) and pounds per gallon (lb/gal):

- 1 ft³ of water weighs 62.4 lb; density = 62.4 lb/ft³.
- 1 gal of water weighs 8.34 lb; density = 8.34 lb/gal.

TABLE 3.1
Water Properties (Temperature, Specific Weight, and Density)

Temperature (°F)	Specific Weight (lb/ft³)	Density (slugs/ft³)	Temperature (°F)	Specific Weight (lb/ft³)	Density (slugs/ft³)
32	62.4	1.94	130	61.5	1.91
40	62.4	1.94	140	61.4	1.91
50	62.4	1.94	150	61.2	1.90
60	62.4	1.94	160	61.0	1.90
70	62.3	1.94	170	60.8	1.89
80	62.2	1.93	180	60.6	1.88
90	62.1	1.93	190	60.4	1.88
100	62.0	1.93	200	60.1	1.87
110	61.9	1.92	210	59.8	1.86
120	61.7	1.92			

The density of a dry material, such as cereal, lime, soda, or sand, is usually expressed in pounds per cubic foot. The density of a liquid, such as liquid alum, liquid chlorine, or water, can be expressed either as pounds per cubic foot or as pounds per gallon. The density of a gas, such as chlorine gas, methane, carbon dioxide, or air, is usually expressed in pounds per cubic foot.

As shown in Table 3.1, the density of a substance like water changes slightly as the temperature of the substance changes. This occurs because substances usually increase in volume (size) by expanding as they become warmer. Because of this expansion with warming, the same weight is spread over a larger volume, so the density is lower when a substance is warm than when it is cold.

> ***Note:*** What is *specific gravity*? Specific gravity is the weight (or density) of a substance compared to the weight (or density) of an equal volume of water. The specific gravity of water is 1.

This relationship is easily seen when a cubic foot of water, which weighs 62.4 lb, is compared to a cubic foot of aluminum, which weighs 178 lb. Aluminum is 2.8 times heavier than water.

It is not that difficult to find the specific gravity of a piece of metal. All you have to do is weigh the metal in air, then weigh it under water. The loss of weight is the weight of an equal volume of water. To find the specific gravity, divide the weight of the metal by its loss of weight in water.

$$\text{Specific gravity} = \text{Weight of substance} \div \text{Weight of equal volume of water} \tag{3.4}$$

■ Example 3.3

Problem: Suppose a piece of metal weighs 150 lb in air and 85 lb under water. What is the specific gravity?

Solution:

$$150\text{ lb} - 85\text{ lb} = 65\text{ lb loss of weight in water}$$

$$\text{Specific gravity} = 150\text{ lb} \div 65\text{ lb} = 2.3$$

> ***Note:*** When calculating specific gravity, it is *essential* that the densities be expressed in the same units.

As stated earlier, the specific gravity of water is 1, which is the standard, the reference against which all other liquid or solid substances are compared. Specifically, any object that has a specific gravity greater than 1 will sink in water (e.g., rocks, steel, iron, grit, floc, sludge). Substances with specific gravities of less than 1 will float (e.g., wood, scum, gasoline). Considering the total weight and volume of a ship, its specific gravity is less than 1; therefore, it can float.

The most common use of specific gravity in water treatment operations is in gallon-to-pound conversions. In many cases, the liquids being handled have a specific gravity of 1 or very nearly 1 (between 0.98 and 1.02), so 1 may be used in the calculations without introducing significant error. For calculations involving a liquid with a specific gravity of less than 0.98 or greater than 1.02, however, the conversions from gallons to pounds must consider specific gravity. The technique is illustrated in the following example.

■ Example 3.4

Problem: A basin contains 1455 gal of a liquid. If the specific gravity of the liquid is 0.94, how many pounds of liquid are in the basin?

Solution: Normally, for a conversion from gallons to pounds, we would use the factor 8.34 lb/gal (the density of water) if the specific gravity of the substance is between 0.98 and 1.02. In this instance, however, the substance has a specific gravity outside this range, so the 8.34 factor must be adjusted by multiplying 8.34 lb/gal by the specific gravity to obtain the adjusted factor:

$$8.34 \text{ lb/gal} \times 0.94 = 7.84 \text{ lb/gal (rounded)}$$

Then convert 1455 gal to pounds using the correction factor:

$$1455 \text{ gal} \times 7.84 \text{ lb/gal} = 11{,}407 \text{ lb (rounded)}$$

FORCE AND PRESSURE

Water exerts force and pressure against the walls of its container, whether it is stored in a tank or flowing in a pipeline. Force and pressure are different, although they are closely related. *Force* is the push or pull influence that causes motion. In the English system, force and weight are often used in the same way. The weight of 1 ft^3 of water is 62.4 lb. The force exerted on the bottom of a 1-ft cube is 62.4 lb (see Figure 3.1). If we stack two 1-ft cubes on top of one another, the force on the bottom will be 124.8 lb. *Pressure* is the force per unit of area. In equation form, this can be expressed as

$$P = F \div A \tag{3.5}$$

where
P = Pressure.
F = Force.
A = Area over which the force is distributed.

Earlier we pointed out that pounds per square inch (lb/in.^{2} or psi) or pounds per square foot (lb/ft^2 or psf) are common expressions of pressure. The pressure on the bottom of the cube is 62.4 lb/ft^2 (see Figure 3.1). It is normal to express pressure in pounds per square inch. This is easily accomplished by determining the weight of 1 in.^{2} of a 1-ft cube. If we have a cube that is 12 in. on each side, the number of square inches on the bottom surface of the cube is 12 × 12 = 144. Dividing the weight by the number of square inches determines the weight on each square inch:

$$62.4 \text{ lb/ft} \div 144 \text{ in.}^2 = 0.433 \text{ psi/ft}$$

This is the weight of a column of water 1 in. square and 1 ft tall. If the column of water were 2 ft tall, the pressure would be 2 ft × 0.433 psi/ft = 0.866.

Note: 1 foot of water = 0.433 psi. To convert feet of head to psi, multiply the feet of head by 0.433 psi/ft.

■ Example 3.5

Problem: A tank is mounted at a height of 90 ft. Find the pressure at the bottom of the tank.

Solution:

$$90 \text{ ft} \times 0.433 \text{ psi/ft} = 39 \text{ psi (rounded)}$$

Note: To convert psi to feet, divide the psi by 0.433 psi/ft.

■ Example 3.6

Problem: Find the height of water in a tank if the pressure at the bottom of the tank is 22 psi.

Solution:

$$\text{Height} = 22 \text{ psi} \div 0.433 \text{ psi/ft} = 51 \text{ ft (rounded)}$$

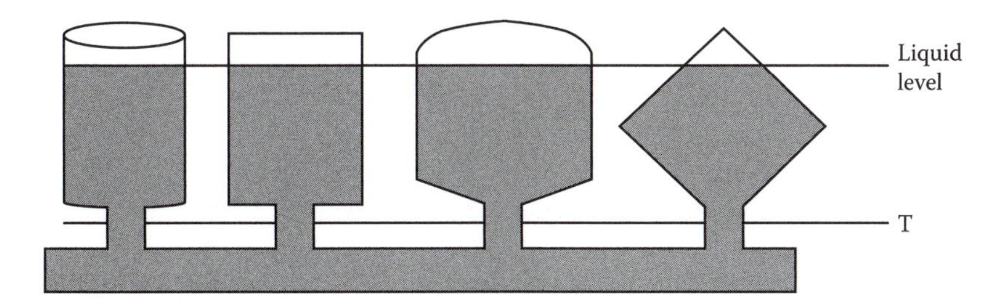

FIGURE 3.2 Hydrostatic pressure.

Note: One of the problems encountered in a hydraulic system is storing the liquid. Unlike air, which is readily compressible and is capable of being stored in large quantities in relatively small containers, a liquid such as water cannot be compressed. It is not possible to store a large amount of water in a small tank, as 62.4 lb of water occupies a volume of 1 ft^3, regardless of the pressure applied to it.

Hydrostatic Pressure

Figure 3.2 shows a number of differently shaped, connected, open containers of water. Note that the water level is the same in each container, regardless of the shape or size of the container. This occurs because pressure is developed within a liquid by the weight of the liquid above. If the water level in any one container is momentarily higher than that in any of the other containers, the higher pressure at the bottom of this container would cause some water to flow into the container having the lower liquid level. In addition, the pressure of the water at any level (such as line T) is the same in each of the containers. Pressure increases because of the weight of the water. The farther down from the surface, the more pressure is created. This illustrates that the weight, not the volume, of water contained in a vessel determines the pressure at the bottom of the vessel.

Some very important principles that always apply for hydrostatic pressure are listed below (Nathanson, 1997):

1. The pressure depends only on the depth of water above the point in question (not on the water surface area).
2. The pressure increases in direct proportion to the depth.
3. The pressure in a continuous volume of water is the same at all points that are at the same depth.
4. The pressure at any point in the water acts in all directions at the same depth.

Effects of Water under Pressure*

Water under pressure and in motion can exert tremendous forces inside a pipeline (Hauser, 1993). One of these forces, called *hydraulic shock* or *water hammer*, is the momentary increase in pressure that occurs due to a sudden change of direction or velocity of the water. When a rapidly closing valve suddenly stops water from flowing in a pipeline, pressure energy is transferred to the valve and pipe wall. Shock waves are set up within the system. Waves of pressure move in a horizontal yo-yo fashion—back and forth—against any solid obstacles in the system. Neither the water nor the pipe will compress to absorb the shock, which may result in damage to pipes and valves and shaking of loose fittings.

* Adapted from Hauser, B.A., *Hydraulics for Operators*, Lewis Publishers, Boca Raton, FL, 1993, pp. 16–18; AWWA, *Basic Science Concepts and Applications: Principles and Practices of Water Supply Operations*, 2nd ed., American Water Works Association, Denver, CO, 1995, pp. 351–353.

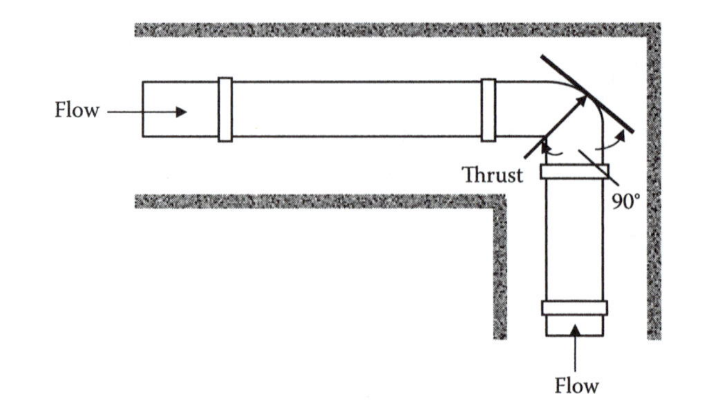

FIGURE 3.3 Direction of thrust in a pipe in a trench (viewed from above).

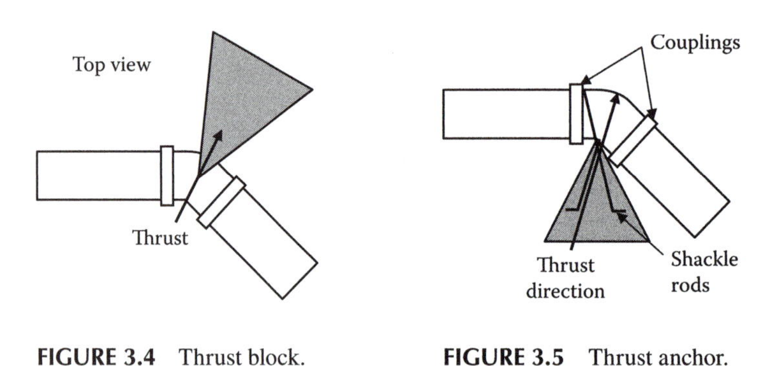

FIGURE 3.4 Thrust block.

FIGURE 3.5 Thrust anchor.

Another effect of water under pressure is called *thrust*, which is the force that water exerts on a pipeline as it rounds a bend. As shown in Figure 3.3, thrust usually acts perpendicular (90°) to the inside surface it pushes against. It affects not only bends in a pipe but also reducers, dead ends, and tees. Uncontrolled, the thrust can cause movement in the fitting or pipeline, which will lead to separation of the pipe coupling away from both sections of pipeline or at some other nearby coupling upstream or downstream of the fitting.

Two types of devices commonly used to control thrust in larger pipelines are thrust blocks and thrust anchors. A *thrust block* is a mass of concrete cast in place onto the pipe and around the outside bend of the turn. An example is shown in Figure 3.4. Thrust blocks are used for pipes with tees or elbows that turn left or right or slant upward. The thrust is transferred to the soil through the larger bearing surface of the block. A *thrust anchor* is a massive block of concrete, often a cube, cast in place below the fitting to be anchored (see Figure 3.5). As shown in Figure 3.5, imbedded steel shackle rods anchor the fitting to the concrete block, effectively resisting upward thrusts. The size and shape of a thrust control device depend on pipe size, type of fitting, water pressure, water hammer, and soil type.

HEAD

Head is defined as the vertical distance water must be lifted from the supply tank to the discharge or as the height a column of water would rise due to the pressure at its base. A perfect vacuum plus atmospheric pressure of 14.7 psi would lift the water 34 ft. When the top of the sealed tube is open to the atmosphere and the reservoir is enclosed, the pressure in the reservoir is increased; the water will rise in the tube. Because atmospheric pressure is essentially universal, we usually ignore the first 14.7 psi of actual pressure measurements and measure only the difference between the water pressure and the atmospheric pressure; we call this *gauge pressure*. Consider water in an open reservoir subjected to 14.7 psi of atmospheric pressure; subtracting this 14.7 psi leaves a gauge pressure of 0 psi, indicating that the water would rise 0 feet above the reservoir surface. If the gauge pressure in a water main were 120 psi, the water would rise in a tube connected to the main:

$$120 \text{ psi} \times 2.31 \text{ ft/psi} = 277 \text{ ft (rounded)}$$

The *total head* includes the vertical distance the liquid must be lifted (static head), the loss to friction (friction head), and the energy required to maintain the desired velocity (velocity head):

$$\text{Total head} = \text{Static head} + \text{Friction head} + \text{Velocity head} \tag{3.6}$$

Static Head

Static head is the actual vertical distance the liquid must be lifted:

$$\text{Static head} = \text{Discharge elevation} - \text{Supply elevation} \tag{3.7}$$

■ Example 3.7

Problem: The supply tank is located at elevation 118 ft. The discharge point is at elevation 215 ft. What is the static head in feet?

Solution:

$$\text{Static head} = 215 \text{ ft} - 118 \text{ ft} = 97 \text{ ft}$$

Friction Head

Friction head is the equivalent distance of the energy that must be supplied to overcome friction. Engineering references include tables showing the equivalent vertical distance for various sizes and types of pipes, fittings, and valves. The total friction head is the sum of the equivalent vertical distances for each component:

$$\text{Friction head} = \text{Energy losses due to friction} \tag{3.8}$$

Velocity Head

Velocity head is the equivalent distance of the energy consumed to achieve and maintain the desired velocity in the system:

$$\text{Velocity head} = \text{Energy losses to maintain velocity} \tag{3.9}$$

Total Dynamic Head (Total System Head)

$$\text{Total head} = \text{Static head} + \text{Friction head} + \text{Velocity head} \tag{3.10}$$

Pressure and Head

The pressure exerted by water or wastewater is directly proportional to its depth or head in the pipe, tank, or channel. If the pressure is known, the equivalent head can be calculated:

$$\text{Head (ft)} = \text{Pressure (psi)} \times 2.31 \text{ (ft/psi)} \tag{3.11}$$

■ Example 3.8

Problem: The pressure gauge on the discharge line from the influent pump reads 72.3 psi. What is the equivalent head in feet?

Solution:

$$\text{Head} = 72.3 \times 2.31 \text{ ft/psi} = 167 \text{ ft}$$

Head and Pressure

If the head is known, the equivalent pressure can be calculated by

$$\text{Pressure (psi)} = \text{Head (ft)} \div 2.31 \text{ ft/psi} \tag{3.12}$$

■ Example 3.9

Problem: A tank is 22 ft deep. What is the pressure in psi at the bottom of the tank when it is filled with water?

Solution:

$$\text{Pressure} = 22 \text{ ft} \div 2.31 \text{ ft/psi} = 9.52 \text{ psi (rounded)}$$

FLOW/DISCHARGE RATE: WATER IN MOTION

The study of fluid flow is much more complicated than that of fluids at rest, but it is important to have an understanding of these principles because the water in a waterworks system is nearly always in motion. *Discharge* (or flow) is the quantity of water passing a given point in a pipe or channel during a given period. Stated another way for open channels, the flow rate through an open channel is directly related to the velocity of the liquid and the cross-sectional area of the liquid in the channel:

$$Q = A \times V \tag{3.13}$$

where

Q = Flow or discharge (cfs).
A = Cross-sectional area of the pipe or channel (ft^2).
V = Water velocity (fps or ft/sec).

■ Example 3.10

Problem: A channel is 6 ft wide and the water depth is 3 ft. The velocity in the channel is 4 fps. What is the discharge or flow rate in cubic feet per second?

Solution:

$$\text{Flow} = 6 \text{ ft} \times 3 \text{ ft} \times 4 \text{ ft/sec} = 72 \text{ cfs}$$

Discharge or flow can be recorded as gal/day (gpd), gal/min (gpm), or cubic feet per second (cfs). Flows treated by many waterworks plants are large and are often referred to in million gallons per day (MGD). The discharge or flow rate can be converted from cubic feet per second to other units such as gpm or MGD by using appropriate conversion factors.

■ Example 3.11

Problem: A 12-in.-diameter pipe has water flowing through it at 10 fps. What is the discharge in (a) cfs, (b) gpm, and (c) MGD?

Solution: Before we can use the basic formula, we must determine the area (A) of the pipe. The formula for the area of a circle is

$$\text{Area } (A) = \pi \times (D^2/4) = \pi \times r^2 \tag{3.14}$$

where

π = Constant value 3.14159, or simply 3.14.
D = Diameter of the circle in feet.
r = Radius of the circle in feet.

Therefore, the area of the pipe is

$$A = \pi \times (D^2/4) = 3.14 \times (1 \text{ ft}^2/4) = 0.785 \text{ ft}^2$$

(a) Now we can determine the discharge in cfs:

$$Q = V \times A = 10 \text{ ft/sec} \times 0.785 \text{ ft}^2 = 7.85 \text{ ft}^3\text{/sec (cfs)}$$

(b) We need to know that 1 cfs is 449 gpm, so 7.85 cfs × 449 gpm/cfs = 3525 gpm (rounded).
(c) 1 million gallons per day is 1.55 cfs, so

$$7.85 \text{ cfs} \div 1.55 \text{ cfs/MGD} = 5.06 \text{ MGD}$$

Note: Flow may be *laminar* (i.e., streamline; see Figure 3.6) or *turbulent* (see Figure 3.7). Laminar flow occurs at extremely low velocities. The water moves in straight parallel lines, called *streamlines* or *laminae,* which slide upon each other as they travel, rather than mixing up. Normal pipe

FIGURE 3.6 Laminar (streamline) flow.

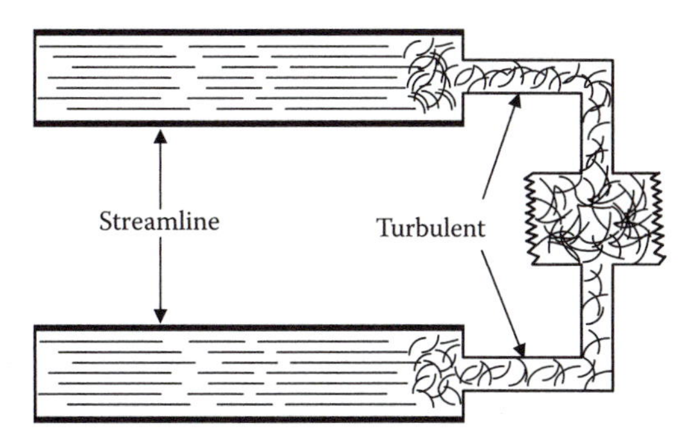

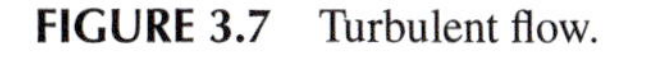
FIGURE 3.7 Turbulent flow.

flow is turbulent flow, which occurs because of friction encountered on the inside of the pipe. The outside layers of flow are thrown into the inner layers; the result is that all of the layers mix and are moving in different directions and at different velocities; however, the direction of flow is forward.

Note: Flow may be steady or unsteady. For our purposes, most of the hydraulic calculations in this manual assume steady-state flow.

Area and Velocity

The *law of continuity* states that the discharge at each point in a pipe or channel is the same as the discharge at any other point (if water does not leave or enter the pipe or channel). That is, under the assumption of steady-state flow, the flow that enters the pipe or channel is the same flow that exits the pipe or channel. In equation form, this becomes

$$Q_1 = Q_2 \quad \text{or} \quad A_1 \times V_1 = A_2 \times V_2 \tag{3.15}$$

Note: With regard to the area/velocity relationship, Equation 3.15 also makes clear that, for a given flow rate, the velocity of the liquid varies indirectly with changes in cross-sectional area of the channel or pipe. This principle provides the basis for many of the flow measurement devices used in open channels (weirs, flumes, and nozzles).

■ Example 3.12

Problem: A pipe 12 inches in diameter is connected to a 6-in.-diameter pipe. The velocity of the water in the 12-in. pipe is 3 fps. What is the velocity in the 6-in. pipe?

Solution: Using the equation $A_1 \times V_1 = A_2 \times V_2$, we need to determine the area of each pipe.

- 12-inch pipe:

$$A = \pi \times (D^2/4) = 3.14 \times (1 \text{ ft}^2/4) = 0.785 \text{ ft}^2$$

- 6-inch pipe:

$$A = \pi \times (D^2/4) = 3.14 \times (0.5 \text{ ft}^2/4) = 0.196 \text{ ft}^2$$

The continuity equation now becomes

$$0.785 \text{ ft}^2 \times 3 \text{ ft/sec} = 0.196 \text{ ft}^2 \times V_2$$

Solving for V_2,

$$V_2 = \frac{0.785 \text{ ft}^2 \times 3 \text{ ft/sec}}{0.196 \text{ ft}^2} = 12 \text{ ft/sec (fps)}$$

Pressure and Velocity

In a closed pipe flowing full (under pressure), the pressure is indirectly related to the velocity of the liquid. This principle, when combined with the principle discussed in the previous section, forms the basis for several flow measurement devices (Venturi meters and rotameters), as well as the injector used for dissolving chlorine into water:

$$\text{Velocity}_1 \times \text{Pressure}_1 = \text{Velocity}_2 \times \text{Pressure}_2 \tag{3.16}$$

or

$$V_1 \times P_1 = V_2 \times P_2$$

PIEZOMETRIC SURFACE AND BERNOULLI'S THEOREM

> They will take your hand and lead you to the pearls of the desert, those secret wells swallowed by oyster crags of wadi, underground caverns that bubble salty rust water you would sell your own mothers to drink.
>
> **Holman (1998)**

To keep the systems in a waterworks operating properly and efficiently, operators must understand the basics of hydraulics—the laws of force, motion, and others. As stated previously, most applications of hydraulics in water treatment systems involve water in motion—in pipes under pressure or in open channels under the force of gravity. The volume of water flowing past any given point in the pipe or channel per unit time is called the *flow rate* or *discharge*—or just *flow*. The *continuity of flow* and the *continuity equation* have already been discussed (see Equation 3.16). Along with the continuity of flow principle and continuity equation, the law of conservation of energy, piezometric surface, and Bernoulli's theorem (or principle) are also important to our study of water hydraulics.

CONSERVATION OF ENERGY

Many of the principles of physics are important to the study of hydraulics. When applied to problems involving the flow of water, few of the principles of physical science are more important and useful to us than the *law of conservation of energy*. Simply, the law of conservation of energy states that energy can be neither created nor destroyed, but it can be converted from one form to another. In a given closed system, the total energy is constant.

ENERGY HEAD

In hydraulic systems, two types of energy (kinetic and potential) and three forms of mechanical energy (potential energy due to elevation, potential energy due to pressure, and kinetic energy due to velocity) exist. Energy is measured in units of foot-pounds (ft-lb). It is convenient to express hydraulic energy in terms of *energy head* in feet of water. This is equivalent to foot-pounds per pound of water (ft-lb/lb = ft).

PIEZOMETRIC SURFACE

We have seen that when a vertical tube, open at the top, is installed onto a vessel of water the water will rise in the tube to the water level in the tank. The water level to which the water rises in a tube is the *piezometric surface*. That is, the piezometric surface is an imaginary surface that coincides with the level of the water to which water in a system would rise in a *piezometer* (an instrument used to measure pressure).

The surface of water that is in contact with the atmosphere is known as *free water surface*. Many important hydraulic measurements are based on the difference in height between the free water surface and some point in the water system. The piezometric surface is used to locate this free water surface in a vessel where it cannot be observed directly. To understand how a piezometer actually measures pressure, consider the following example. If a clear, see-through pipe is connected to the side of a clear glass or plastic vessel, the water will rise in the pipe to indicate the level of the water

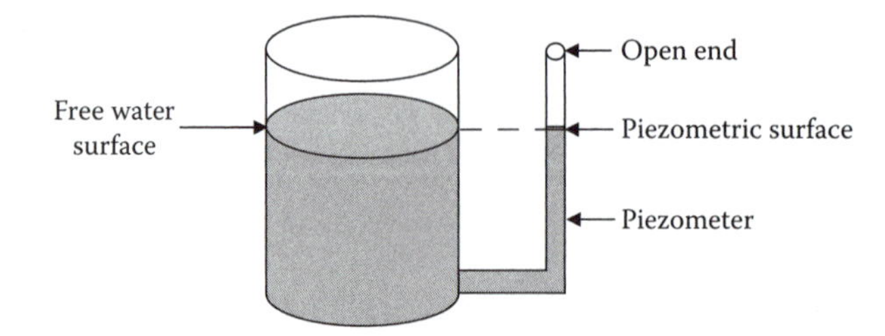

FIGURE 3.8 A container not under pressure; the piezometric surface is the same as the free water surface in the vessel.

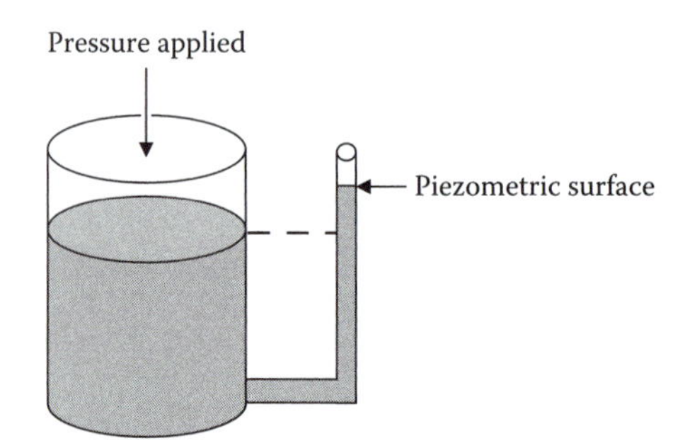

FIGURE 3.9 A container under pressure; the piezometric surface is above the level of the water in the tank.

in the vessel. Such a see-through pipe—a piezometer—allows us to see the level of the top of the water in the pipe; this is the piezometric surface. In practice, a piezometer is connected to the side of a tank or pipeline. If the water-containing vessel is not under pressure (as is the case in Figure 3.8), the piezometric surface will be the same as the free water surface in the vessel, just as when a drinking straw (the piezometer) is left standing in a glass of water.

When pressurized in a tank and pipeline system, as they often are, the pressure will cause the piezometric surface to rise above the level of the water in the tank. The greater the pressure, the higher the piezometric surface (see Figure 3.9). An increased pressure in a water pipeline system is usually obtained by elevating the water tank.

> ***Note:*** In practice, piezometers are not installed on water towers because water towers are hundreds of feet high, or on pipelines. Instead, pressure gauges are used that record pressure in feet of water or in psi.

Water only rises to the water level of the main body of water when it is at rest (static or standing water). The situation is quite different when water is flowing. Consider, for example, an elevated storage tank feeding a distribution system pipeline. When the system is at rest, with all of the valves closed, all of the piezometric surfaces are the same height as the free water surface in storage. On the other hand, when the valves are opened and the water begins to flow, the piezometric surface changes. This is an important point because, as water continues to flow down a pipeline, less and less pressure is exerted. This happens because some pressure is lost (used up) to keep the water moving over the interior surface of the pipe (friction). The pressure that is lost is called *head loss.*

Head Loss

Head loss is best explained by example. Figure 3.10 shows an elevated storage tank feeding a distribution system pipeline. When the valve is closed (Figure 3.10A), all the piezometric surfaces are the same height as the free water surface in storage. When the valve opens and water begins to flow (Figure 3.10B), the piezometric surfaces *drop.* The farther along the pipeline, the lower the

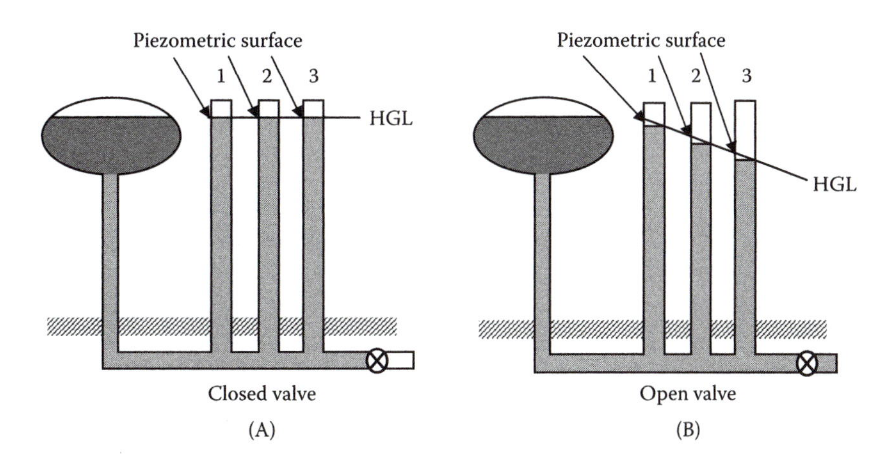

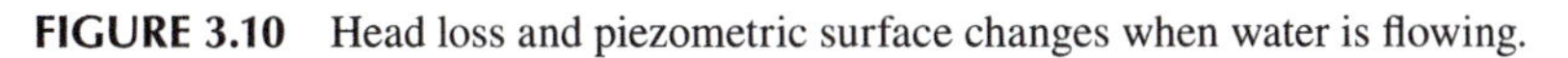
FIGURE 3.10 Head loss and piezometric surface changes when water is flowing.

piezometric surface, because some of the pressure is used up keeping the water moving over the rough interior surface of the pipe. Thus, pressure is lost and is no longer available to push water up in a piezometer; this, again, is the head loss.

Hydraulic Grade Line

When the valve shown in Figure 3.10 is opened, flow begins with a corresponding energy loss due to friction. The pressures along the pipeline can measure this loss. In Figure 3.10B, the difference in pressure heads between sections 1, 2, and 3 can be seen in the piezometer tubes attached to the pipe. A line connecting the water surface in the tank with the water levels at sections 1, 2, and 3 shows the pattern of continuous pressure loss along the pipeline. This is the *hydraulic grade line* (HGL), or *hydraulic gradient*, of the system.

> ***Note:*** It is important to point out that in a static water system the HGL is always horizontal. The HGL is a very useful graphical aid when analyzing pipe flow problems.

> ***Note:*** During the early design phase of a treatment plant, it is important to establish the hydraulic grade line across the plant because both the proper selection of the plant site elevation and the suitability of the site depend on this consideration. Typically, most conventional water treatment plants require 16 to 17 ft of head loss across the plant.

> ***Note:*** Changes in the piezometric surface occur when water is flowing.

Bernoulli's Theorem*

Swiss physicist and mathematician Samuel Bernoulli developed the calculation for the total energy relationship from point to point in a steady-state fluid system in the 1700s. Before discussing Bernoulli's energy equation, it is important to understand the basic principle behind Bernoulli's equation. Water (and any other hydraulic fluid) in a hydraulic system possesses two types of energy—kinetic and potential. *Kinetic energy* is present when the water is in motion. The faster the water moves, the more kinetic energy is used. *Potential energy* is a result of the water pressure. The *total energy* of the water is the sum of the kinetic and potential energy. Bernoulli's principle states that the total energy of the water (fluid) always remains constant; therefore, when the water flow in

* Adapted from Nathanson, J.A., *Basic Environmental Technology: Water Supply, Waste Management, and Pollution Control*, 2nd ed., Prentice Hall, Upper Saddle River, NJ, 1997, p. 29.

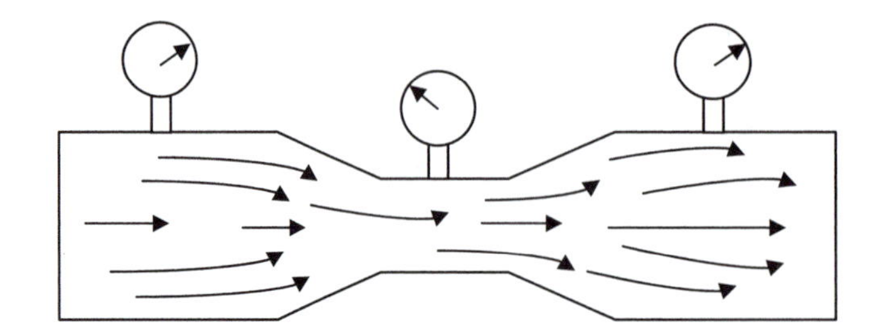

FIGURE 3.11 Demonstration of Bernoulli's principle.

a system increases, the pressure must decrease. When water starts to flow in a hydraulic system, the pressure drops. When the flow stops, the pressure rises again. The pressure gauges shown in Figure 3.11 illustrate this balance more clearly.

Bernoulli's Equation

In a hydraulic system, total energy head is equal to the sum of three individual energy heads. This can be expressed as

Total head = Elevation head + Pressure head + Velocity head

where elevation head is the pressure due to the elevation of the water, pressure head is the height of a column of water that a given hydrostatic pressure in a system could support, and velocity head is the energy present due to the velocity of the water. This can be expressed mathematically as

$$E = z + \frac{p}{w} + \frac{V^2}{2g} \tag{3.17}$$

where

- E = Total energy head.
- z = Height of the water above a reference plane (ft).
- p = Pressure (psi).
- w = Unit weight of water (62.4 lb/ft^3).
- V = Flow velocity (ft/sec).
- g = Acceleration due to gravity (32.2 ft/sec^2).

Consider the constriction in the section of pipe shown in Figure 3.12. We know, based on the law of energy conservation, that the total energy head at section A (E_1) must equal the total energy head at section B (E_2). Using Equation 3.17, we get Bernoulli's equation:

$$z_A + \frac{p_A}{w} + \frac{V_A^2}{2g} = z_B + \frac{p_B}{w} + \frac{V_B^2}{2g} \tag{3.18}$$

The pipeline system shown in Figure 3.12 is horizontal; therefore, we can simplify Bernoulli's equation because $z_A = z_B$. Because they are equal, the elevation heads cancel out from both sides, leaving

$$\frac{p_A}{w} + \frac{V_A^2}{2g} = \frac{p_B}{w} + \frac{V_B^2}{2g} \tag{3.19}$$

As water passes through the constricted section of the pipe (section B), we know from continuity of flow that the velocity at section B must be greater than the velocity at section A, because of the smaller flow area at section B. This means that the velocity head in the system increases as the water flows into the constricted section; however, the total energy must remain constant. For this to occur, the pressure head, and therefore the pressure, must drop. In effect, pressure energy is converted into kinetic energy in the constriction.

The fact that the pressure in the narrower pipe section (constriction) is less than the pressure in the bigger section seems to defy common sense; however, it does follow logically from continuity of flow and conservation of energy. The fact that there is a pressure difference allows measurement of flow rate in the closed pipe.

■ Example 3.13

Problem: In Figure 3.12, the diameter at section A is 8 in., and at section B it is 4 in. The flow rate through the pipe is 3.0 cfs and the pressure at section A is 100 psi. What is the pressure in the constriction at section B?

Solution: Compute the flow area at each section, as follows:

$$A_A = \pi \times (D^2/4) = 3.14 \times (0.666\ \text{ft}^2/4) = 0.349\ \text{ft}^2$$

and

$$A_B = \pi \times (D^2/4) = 3.14 \times (0.333\ \text{ft}^2/4) = 0.087\ \text{ft}^2$$

From $Q = A \times V$ or $V = Q/A$, we get

$$V_A = 3.0\ \text{ft}^3/\text{sec} \div 0.349\ \text{ft}^2 = 8.6\ \text{ft/sec}$$

and

$$V_B = 3.0\ \text{ft}^3/\text{sec} \div 0.087\ \text{ft}^2 = 34.5\ \text{ft/sec}$$

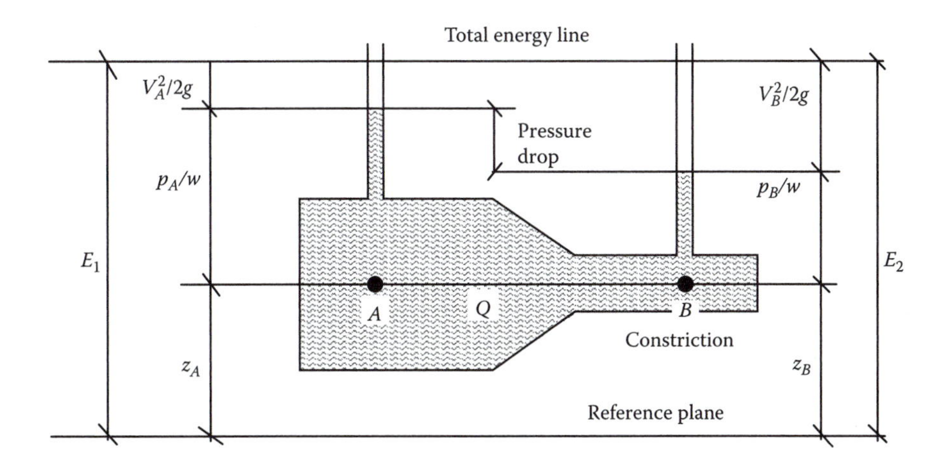

FIGURE 3.12 The result of the law of conservation of energy. Because the velocity and kinetic energy of the water flowing in the constricted section must increase, the potential energy may decrease. This is observed as a pressure drop in the constriction. (Adapted from Nathanson, J.A., *Basic Environmental Technology: Water Supply, Waste Management, and Pollution Control*, 2nd ed., Prentice Hall, Upper Saddle River, NJ, p. 29, 1997.)

Applying Equation 3.19, we get

$$\frac{p_A}{w} + \frac{V_A^2}{2g} = \frac{p_B}{w} + \frac{V_B^2}{2g}$$

$$\frac{100 \times 144}{62.4} + \frac{(8.6)^2}{2 \times 32.2} = \frac{p_B \times 144}{62.4} + \frac{(34.5)^2}{2 \times 32.2}$$

$$231 + 1.15 = 2.3p_B + 18.5$$

$$(231 + 1.15) - 18.5 = 2.3p_B$$

$$p_B = \frac{232.2 - 18.5}{2.3}$$

$$p_B = \frac{213.7}{2.3}$$

$$p_B = 93 \text{ psi}$$

Note: The pressures are multiplied by 144 in.^{2}/ft^2 to convert from psi to lb/ft^2 to be consistent with the units for *w*; the energy head terms are in feet of head.

HYDRAULIC MACHINES (PUMPS)

Conveying water to and from process equipment is an integral part of the water industry that requires energy consumption. The amount of energy required depends on the height to which the water is raised, the length and diameter of the conveying conduits, the rate of flow, and the water's physical properties (in particular, viscosity and density). In some applications, external energy for transferring water is not required. For example, when water flows to a lower elevation under the influence of gravity, a partial transformation of the water's potential energy into kinetic energy occurs. However, to convey water or wastewater through horizontal conduits, especially to higher elevations within a system, mechanical devices such as pumps are employed. Requirements vary from small units used to pump only a few gallons per minute to large units capable of handling several hundred cubic feet per second. Table 3.2 lists pump applications in water/wastewater treatment operations.

Note: When determining the amount of pressure or force a pump must provide to move the water, the term *pump head* was established.

Several methods are available for transporting water, wastewater, and chemicals for treatment between process equipment:

- Centrifugal force inducing fluid motion
- Volumetric displacement of fluids, either mechanically or with other fluids
- Transfer of momentum from another fluid
- Mechanical impulse
- Gravity induced

Depending on the facility and unit processes contained within, all of the methods above may be important to the maintenance operator.

TABLE 3.2
Pump Applications in Water/Wastewater Systems

Application	Function	Pump Type
Low service	To lift water from the source to treatment processes or from storage to filter-backwashing system	Centrifugal
High service	To discharge water under pressure to the distribution system; to pump collected or intercepted wastewater to the treatment facility	Centrifugal
Booster	To increase pressure in the distribution/collection system or to supply elevated storage tanks	Centrifugal
Well	To lift water from shallow or deep wells and discharge it to the treatment plant, storage facility, or distribution system	Centrifugal or jet
Chemical feed	To add chemical solutions at desired dosages for treatment processes	Positive-displacement
Sampling	To pump water/wastewater from sampling points to the laboratory or automatic analyzers	Positive-displacement or centrifugal
Sludge/biosolids	To pump sludge or biosolids from sedimentation facilities to further treatment or disposal	Positive-displacement or centrifugal

Source: Adapted from AWWA, *Water Transmission and Distribution*, 2nd ed., American Water Works Association, Denver, CO, 1996, p. 358.

Pumping Hydraulics

During operation, water enters a pump on the suction side, where the pressure is lower (Arasmith, 1993). Because the function of the pump is to add pressure to the system, discharge pressure will always be higher. An important concept to keep in mind is that, in pump systems, measurements are taken from the point of reference to the centerline of the pump (horizontal line drawn through center of pump). To understand pump operation, or *pumping hydraulics*, we need to be familiar with certain basic terms and then relate these terms to how water is pumped from one point to another (see Figure 3.13):

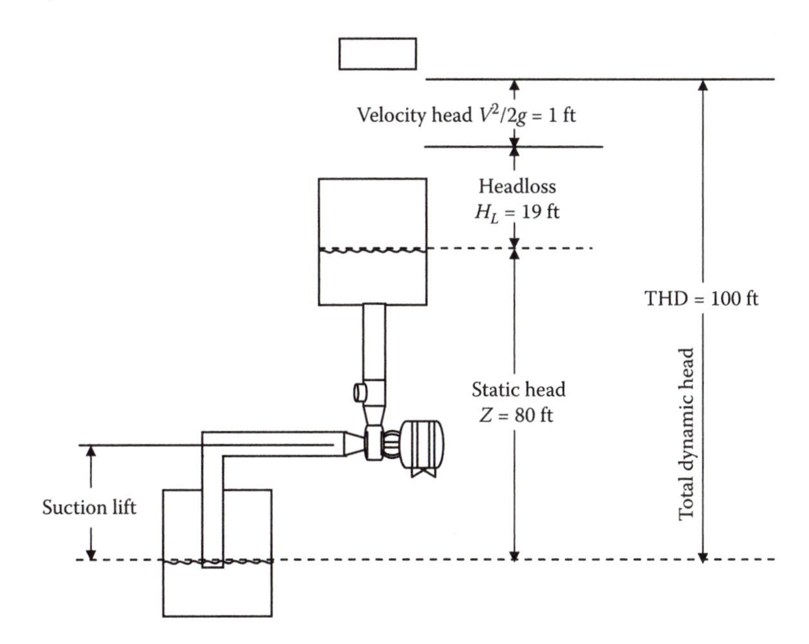

FIGURE 3.13 Components of total dynamic head.

Static head—The distance between the suction and discharge water levels when the pump is shut off. Static head conditions are represented by the letter *Z* (see Figure 3.13).

Suction lift—The distance between the suction water level and the center of the pump impeller. This term is used only when the pump is in a suction lift condition; the pump must have the energy to provide this lift. A pump is said to be in a suction lift condition any time the center (eye) of the impeller is above the water being pumped (see Figure 3.13).

Suction head—A pump is said to be in a suction head condition any time the center (eye) of the impeller is below the water level being pumped. Specifically, suction head is the distance between the suction water level and the center of the pump impeller when the pump is in a suction head condition (see Figure 3.13).

Velocity head—The amount of energy required to bring water or wastewater from standstill to its velocity. For a given quantity of flow, the velocity head will vary indirectly with the pipe diameter. Velocity head is often represented mathematically as $V^2/2g$ (see Figure 3.13).

Total dynamic head—The total energy needed to move water from the centerline of a pump (eye of the first impeller of a lineshaft turbine) to some given elevation or to develop some given pressure. This includes the static head, velocity head, and the head loss due to friction (see Figure 3.13).

Well and Wet Well Hydraulics

When the source of water for a water distribution system is from a groundwater supply, knowledge of well hydraulics is important to the operator. In this section, basic well hydraulics terms are presented and defined, and they are related pictorially (see Figure 3.14). Also discussed are wet wells, which are important in both water and wastewater operations.

Well Hydraulics

- *Static water level*—The water level in a well when no water is being taken from the groundwater source (i.e., the water level when the pump is off; see Figure 3.14). Static water level is normally measured as the distance from the ground surface to the water surface. This is an important parameter because it is used to measure changes in the water table.
- *Pumping water level*—The water level when the pump is operating. When water is pumped out of a well, the water level usually drops below the level in the surrounding aquifer and eventually stabilizes at a lower level; this is the pumping level (see Figure 3.14).
- *Drawdown*—The difference, or the drop, between the static water level and the pumping water level, measured in feet. Simply, it is the distance the water level drops when pumping begins (see Figure 3.14).
- *Cone of depression*—In unconfined aquifers, water flows in the aquifer from all directions toward the well during pumping. The free water surface in the aquifer then takes the shape of an inverted cone or curved funnel line. The curve of the line extends from the pumping water level to the static water level at the outside edge of the zone (or radius) of influence (see Figure 3.14).

Note: The shape and size of the cone of depression are dependent on the relationship between the pumping rate and the rate at which water can move toward the well. If the rate is high, the cone will be shallow and its growth will stabilize. If the rate is low, the cone will be sharp and continue to grow in size.

- *Zone (or radius) of influence*—The distance between the pump shaft and the outermost area affected by drawdown (see Figure 3.14). The distance depends on the porosity of the soil and other factors. This parameter becomes important in well fields with many pumps. If wells are set too closely together, the zones of influence will overlap, increasing the drawdown in all wells. Obviously, pumps should be spaced apart to prevent this from happening.

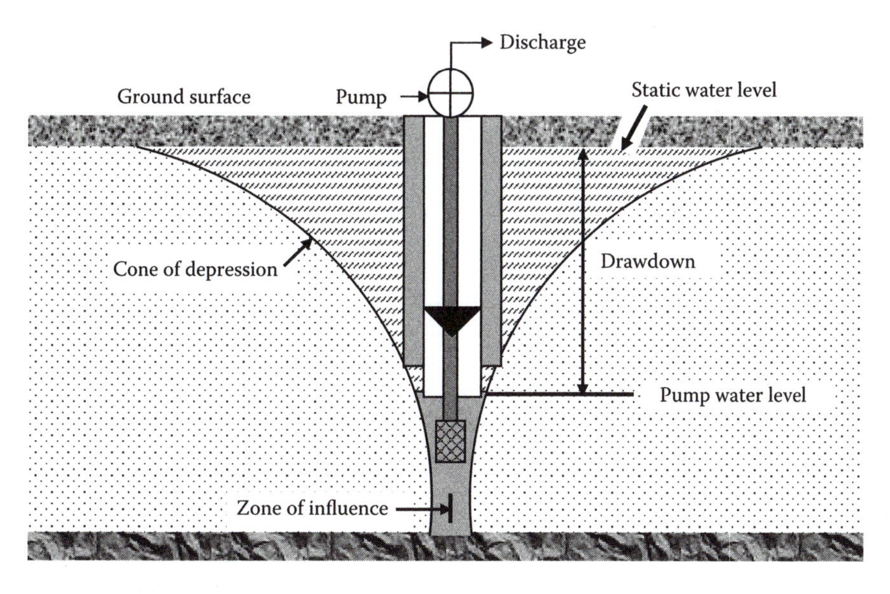

FIGURE 3.14 Hydraulic characteristics of a well.

Two important parameters not shown in Figure 3.14 are well yield and specific capacity:

1. *Well yield* is the rate of water withdrawal that a well can supply over a long period, or, alternatively, the maximum pumping rate that can be achieved without increasing the drawdown. The yield of small wells is usually measured in gallons per minute (liters per minute) or gallons per hour (liters per hour). For large wells, it may be measured in cubic feet per second (cubic meters per second).
2. *Specific capacity* is the pumping rate per foot of drawdown (gpm/ft), or

$$\text{Specific capacity} = \text{Well yield} \div \text{Drawdown} \tag{3.20}$$

■ **Example 3.14**

Problem: If the well yield is 300 gpm and the drawdown is measured to be 20 ft, what is the specific capacity?

Solution:

$$\text{Specific capacity} = 300 \text{ gpm} \div 20 \text{ ft} = 15 \text{ gpm per ft of drawdown}$$

Specific capacity is one of the most important concepts in well operation and testing. The calculation should be made frequently in the monitoring of well operation. A sudden drop in specific capacity indicates problems such as pump malfunction, screen plugging, or other problems that can be serious. Such problems should be identified and corrected as soon as possible.

Wet Well Hydraulics

Water pumped from a wet well by a pump set above the water surface exhibits the same phenomena as the groundwater well. In operation, a slight depression of the water surface forms right at the intake line (drawdown), but in this case it is minimal because there is free water at the pump entrance at all times (at least there should be). The most important consideration in wet well operations is to ensure that the suction line is submerged far enough below the surface so air entrained by the active movement of the water at this section is not able to enter the pump. Because water flow is not always constant or at the same level, variable speed pumps are commonly used in wet well operations, or several pumps are installed for single or combined operation. In many cases, pumping is accomplished in an on/off mode. Control of pump operation is in response to water level in the well. Level control devices such as mercury switches are used to sense high or low levels in the well and to transmit the signal to pumps for action.

FRICTION HEAD LOSS

Materials or substances capable of flowing cannot flow freely. Nothing flows without encountering some type of resistance. Consider electricity, the flow of free electrons in a conductor. Each type of conductor (e.g., copper, aluminum, silver) offers some resistance. In hydraulics, the flow of water is analogous to the flow of electricity. Within a pipe or open channel, for example, flowing water, like electron flow in a conductor, encounters resistance; however, resistance to the flow of water is generally termed *friction loss* (or, more appropriately, head loss).

FLOW IN PIPELINES

The problem of waste flow in pipelines—the prediction of flow rate through pipes of given characteristics, the calculation of energy conversions therein, and so forth—is encountered in many applications of water/wastewater operations and practice. Although the subject of pipe flow embraces only those problems in which pipes flow completely full (as in water lines), we also address pipes that flow partially full (wastewater lines, normally treated as open channels) in this section. Also discussed is the solution of practical pipe flow problems resulting from application of the energy principle, the equation of continuity, and the principle and equation of water resistance. Resistance to flow in pipes occurs due not only to long reaches of pipe but also to pipe fittings, such as bends and valves, which dissipate energy by producing relatively large-scale turbulence.

PIPE AND OPEN FLOW BASICS

To gain an understanding of what friction head loss is all about, it is necessary to review a few terms presented earlier in the text and to introduce some new terms pertinent to the subject (Lindeburg, 1986):

- *Laminar and turbulent flow*—Laminar flow is ideal flow; that is, water particles move along straight, parallel paths in layers or streamlines. Moreover, laminar flow has no turbulence and no friction loss. This is not typical of normal pipe flow because the water velocity is too great, but it is typical of groundwater flow. Turbulent flow (characterized as normal for a typical water system) occurs when water particles move in a haphazard fashion and continually cross each other in all directions, resulting in pressure losses along a length of pipe.
- *Hydraulic grade line (HGL)*—Recall that the hydraulic grade line (shown in Figure 3.15) is a line connecting two points to which the liquid would rise at various places along any pipe or open channel if piezometers were inserted in the liquid. It is a measure of the pressure head available at these various points.

Note: When water flows in an open channel, the hydraulic grade line coincides with the profile of the water surface.

- *Energy grade line*—The total energy of flow in any section with reference to some datum (i.e., a reference line, surface, or point) is the sum of the elevation head (z), the pressure head (y), and the velocity head ($V^2/2g$). Figure 3.15 shows the energy grade line or energy

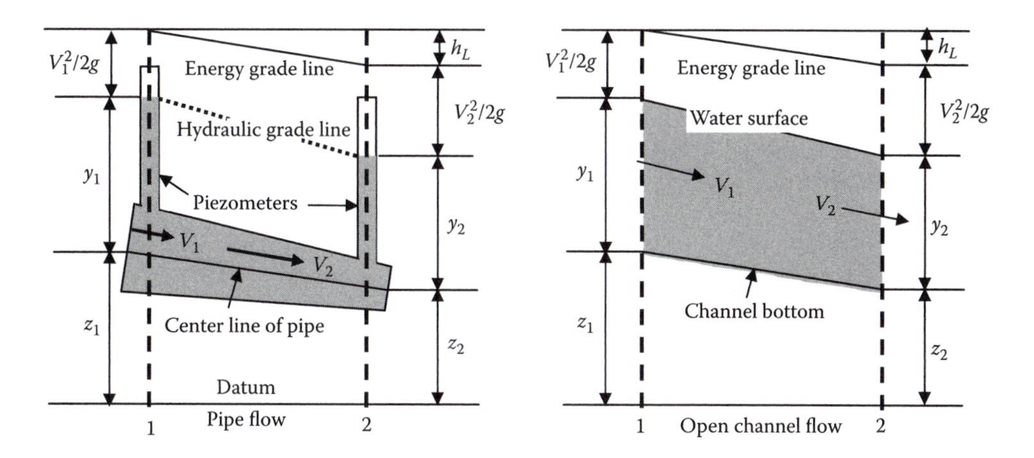

FIGURE 3.15 Comparison of pipe flow and open-channel flow.

gradient, which represents the energy from section to section. In the absence of frictional losses, the energy grade line remains horizontal, although the relative distribution of energy may vary among the elevation, pressure, and velocity heads. In all real systems, however, losses of energy occur because of resistance to flow, and the resulting energy grade line is sloped (i.e., the energy grade line is the slope of the specific energy line).

- *Specific energy (E)*—Sometimes called *specific head*, the specific energy is the sum of the pressure head (y) and the velocity head ($V^2/2g$). The specific energy concept is especially useful in analyzing flow in open channels.
- *Steady flow*—Specific flow occurs when the discharge or rate of flow at any cross-section is constant.
- *Uniform and nonuniform flow*—Uniform flow occurs when the depth, cross-sectional area, and other elements of flow are substantially constant from section to section. Nonuniform flow occurs when the slope, cross-sectional area, and velocity change from section to section. The flow through a Venturi section used for measuring flow is a good example.
- *Varied flow*—Flow in a channel is considered varied if the depth of flow changes along the length of the channel. The flow may be gradually varied or rapidly varied (i.e., when the depth of flow changes abruptly) as shown in Figure 3.16.
- *Slope*—Slope (gradient) is the head loss per foot of channel.

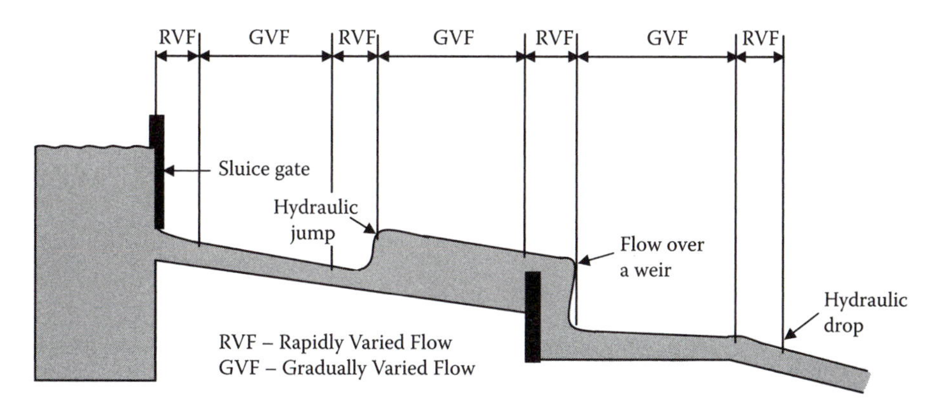

FIGURE 3.16 Varied flow.

Major Head Loss

Major head loss consists of pressure decreases along the length of pipe caused by friction created as water encounters the surfaces of the pipe. It typically accounts for most of the pressure drop in a pressurized or dynamic water system. The components that contribute to major head loss are roughness, length, diameter, and velocity:

- *Roughness*—Even in new pipes, the interior surfaces are rough. The roughness varies, of course, depending on the pipe material, corrosion (tuberculation and pitting), and age. Because normal flow in a water pipe is turbulent, the turbulence increases with pipe roughness, which, in turn, causes pressure to drop over the length of the pipe.
- *Pipe length*—With every foot of pipe length, friction losses occur. The longer the pipe, the greater the head loss. Friction loss because of pipe length must be factored into head loss calculations.
- *Pipe diameter*—Generally, small-diameter pipes have more head loss than large-diameter pipes. In large-diameter pipes, less of the water actually touches the interior surfaces of the pipe (encountering less friction) than in a small-diameter pipe.
- *Water velocity*—Turbulence in a water pipe is directly proportional to the speed (or velocity) of the flow; thus, the velocity head also contributes to head loss.

Note: For pipe with a constant diameter, when flow increases, head loss increases.

Calculating Major Head Loss

Darcy, Weisbach, and others developed the first practical equation used to determine pipe friction in about 1850. The equation or formula now known as the *Darcy–Weisbach* equation for circular pipes, is

$$h_f = f \times \frac{LV^2}{D2g} \tag{3.21}$$

In terms of the flow rate Q, the equation becomes

$$h_f = \frac{8fLQ^2}{\pi^2 g D^5} \tag{3.22}$$

where

h_f = Head loss (ft).
f = Coefficient of friction.
L = Length of pipe (ft).
V = Mean velocity (ft/sec).
D = Diameter of pipe (ft).
g = Acceleration due to gravity (32.2 ft/sec^2).
Q = Flow rate (ft^3/sec).

The Darcy–Weisbach formula was meant to apply to the flow of any fluid, and into this friction factor was incorporated the degree of roughness and an element known as the *Reynold's number*, which is based on the viscosity of the fluid and the degree of turbulence of flow. The Darcy–Weisbach formula is used primarily for head loss calculations in pipes. For open channels, the *Manning* equation was developed during the latter part of the 19th century. Later, this equation was used for both open channels and closed conduits.

TABLE 3.3
C Factors

Type of Pipe	C Factor
Asbestos cement	140
Brass	140
Brick sewer	100
Cast iron	
10 years old	110
20 years old	90
Ductile iron (cement-lined)	140
Concrete or concrete-lined	
Smooth, steel forms	140
Wooden forms	120
Rough	110
Copper	140
Fire hose (rubber-lined)	135
Galvanized iron	120
Glass	140
Lead	130
Masonry conduit	130
Plastic	150
Steel	
Coal-tar-enamel-lined	150
New unlined	140
Riveted	110
Tin	130
Vitrified	120
Wood stave	120

Source: Adapted from Lindeburg, M.R., *Civil Engineering Reference Manual*, 4th ed., Professional Publications, San Carlos, CA, 1986.

In the early 1900s, a more practical equation, the *Hazen–Williams* equation, was developed for use in making calculations related to water pipes and wastewater force mains:

$$Q = 0.433 \times C \times D^{2.63} \times S^{0.54} \tag{3.23}$$

where

Q = Flow rate (ft^3/sec).
C = Coefficient of roughness (C decreases with roughness).
D = Inside pipe diameter (ft).
S = Slope of energy grade line (ft/ft).

C Factor

The *C* factor, as used in the Hazen–Williams formula, designates the coefficient of roughness. *C* does not vary appreciably with velocity, and by comparing pipe types and ages it includes only the concept of roughness, ignoring fluid viscosity and Reynold's number. Based on experience (experimentation), accepted tables of *C* factors have been established for pipe (see Table 3.3). Generally, the *C* factor decreases by one with each year of pipe age. Flow for a newly designed system is often calculated with a *C* factor of 100, based on averaging it over the life of the pipe system.

Note: A high *C* factor means a smooth pipe; a low *C* factor means a rough pipe.

Note: An alternative to calculating the Hazen–Williams formula, called an *alignment chart,* has become quite popular for fieldwork. The alignment chart can be used with reasonable accuracy.

Slope

Slope is defined as the head loss per foot. In open channels, where the water flows by gravity, slope is the amount of incline of the pipe and is calculated as feet of drop per foot of pipe length (ft/ft). Slope is designed to be just enough to overcome frictional losses, so the velocity remains constant, the water keeps flowing, and solids will not settle in the conduit. In piped systems, where pressure loss for every foot of pipe is experienced, slope is not provided by slanting the pipe but instead by adding pressure to overcome friction.

Minor Head Loss

In addition to the head loss caused by friction between the fluid and the pipe wall, losses also are caused by turbulence created by obstructions (i.e., valves and fittings of all types) in the line, changes in direction, and changes in flow area.

Note: In practice, if minor head loss is less than 5% of the total head loss, it is usually ignored.

BASIC PUMPING HYDRAULICS

Water, regardless of the source, is conveyed to the waterworks for treatment and distributed to the users. Conveyance from the source to the point of treatment occurs by aqueducts, pipelines, or open channels, but the treated water is normally distributed in pressurized closed conduits. After use, whatever the purpose, the water becomes wastewater, which must be disposed of somehow but almost always ends up being conveyed back to a treatment facility before being outfalled to some water body, to begin the cycle again. We call this an *urban water cycle*, because it provides a human-generated imitation of the natural water cycle. Unlike the natural water cycle, however, without pipes the cycle would be nonexistent or, at the very least, short circuited.

Piping

For use as water mains in a distribution system, pipes must be strong and durable to resist applied forces and corrosion. The pipe is subjected to internal pressure from the water and to external pressure from the weight of the backfill (soil) and vehicles above it. The pipe may also have to withstand water hammer. Damage due to corrosion or rusting may also occur internally because of the water quality or externally because of the nature of the soil conditions.

Of course, pipes must be constructed to withstand the expected conditions of exposure, and pipe configuration systems for water distribution systems must be properly designed and installed in terms of water hydraulics. Because the water and wastewater operator should have a basic knowledge of water hydraulics related to commonly used standard piping configurations, piping basics are briefly discussed in this section.

Piping Networks

It would be far less costly and make for more efficient operation if municipal water systems were built with individual, single-pipe networks extending from the treatment plant to the user's residence. Unfortunately, this ideal single-pipe scenario is not practical for real-world applications.

Instead of a single piping system, a network of pipes is laid under the streets. Each of these piping networks is composed of different materials that vary (sometimes considerably) in diameter, length, and age. These networks range in complexity to varying degrees, and each of these joined-together pipes contributes energy losses to the system.

Energy Losses in Pipe Networks

Water flow networks may consist of pipes arranged in series, parallel, or some complicated combination. In any case, an evaluation of friction losses for the flows is based on energy conservation principles applied to the flow junction points. Methods of computation depend on the particular piping configuration. In general, however, they involve establishing a sufficient number of simultaneous equations or employing a friction loss formula where the friction coefficient depends only on the roughness of the pipe (e.g., Hazen–Williams equation, Equation 3.23).

> ***Note:*** Demonstrating the procedure for making these complex computations is beyond the scope of this text. Here, we present only an operator's need-to-know aspects of complex or compound piping systems.

Pipes in Series

When two pipes of different sizes or roughnesses are connected in series (see Figure 3.17), head loss for a given discharge, or discharge for a given head loss, may be calculated by applying the appropriate equation between the bonding points, taking into account all losses in the interval; thus, head losses are cumulative. Series pipes may be treated as a single pipe of constant diameter to simplify the calculation of friction losses. The approach involves determining an equivalent length of a constant-diameter pipe that has the same friction loss and discharge characteristics as the actual series pipe system. In addition, application of the continuity equation to the solution allows the head loss to be expressed in terms of only one pipe size.

> ***Note:*** In addition to the head loss caused by friction between the water and the pipe wall, minor losses are caused by obstructions in the line, changes in directions, and changes in flow area. In practice, the method of equivalent length is often used to determine these losses. The method of equivalent length uses a table to convert each valve or fitting into an equivalent length of straight pipe.

When making calculations involving pipes in series, remember these two important basic operational tenets:

1. The same flow passes through all pipes connected in series.
2. The total head loss is the sum of the head losses of all of the component pipes.

In some operations involving series networks where the flow is given and the total head loss is unknown, we can use the Hazen–Williams formula to solve for the slope and the head loss of each pipe as if they were separate pipes. Adding up the head losses to get the total head loss is then a simple matter.

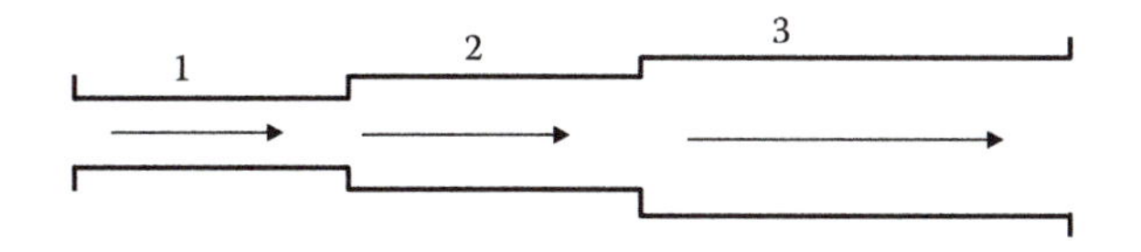

FIGURE 3.17 Pipes in series.

Other series network calculations may not be as simple to solve using the Hazen–Williams equation; for example, one problem we may be faced with is what diameter to use with varying sized pipes connected together in a series combination. Moreover, head loss is applied to both pipes (or other multiples), and it is not known how much loss originates from each one; thus, determining slope would be difficult—but not impossible.

In such cases, the equivalent pipe theory, as mentioned earlier, can be used. Again, one single "equivalent pipe" is created that will carry the correct flow. This is practical because the head loss through it is the same as that in the actual system. The equivalent pipe can have any *C* factor and diameter, just as long as those same dimensions are maintained all the way through to the end. Keep in mind that the equivalent pipe must have the correct length so it will allow the correct flow through that yields the correct head loss (the given head loss) (Lindeburg, 1986).

Pipes in Parallel

Two or more pipes connected (as in Figure 3.18) so flow is first divided among the pipes and then rejoined comprise a parallel pipe system. A parallel pipe system is a common method for increasing the capacity of an existing line. Flows in pipes arranged in parallel are determined by applying energy conservation principles—specifically, energy losses through all pipes connecting common junction points must be equal. Each leg of the parallel network is treated as a series piping system and converted to a single equivalent length pipe. The friction losses through the equivalent length parallel pipes are then considered equal, and the respective flows are determined by proportional distribution.

> ***Note:*** Computations used to determine friction losses in parallel combinations may be accomplished using a simultaneous solution approach for a parallel system that has only two branches; however, if the parallel system has three or more branches, a modified procedure using the Hazen–Williams loss formula is easier.

OPEN-CHANNEL FLOW

> Water is transported over long distances through aqueducts to locations where it is to be used and/or treated. Selection of an aqueduct type rests on such factors as topography, head availability, climate, construction practices, economics, and water quality protection. Along with pipes and tunnels, aqueducts may also include or be solely composed of open channels.
>
> **Viessman and Hammer (1998)**

In this section, we discuss water passage in open channels, which allow part of the water to be exposed to the atmosphere. This type of channel—an open-flow channel—includes natural waterways, canals, culverts, flumes, and pipes flowing under the influence of gravity.

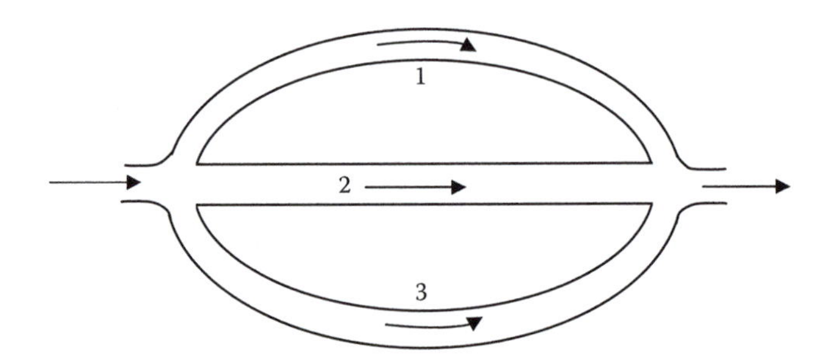

FIGURE 3.18 Pipes in parallel.

Characteristics of Open-Channel Flow

Basic hydraulic principles apply in open-channel flow (with water depth constant) although there is no pressure to act as the driving force (McGhee, 1991). Velocity head is the only natural energy this water possesses, and at normal water velocities it is a small value ($V^2/2g$). Several parameters can be (and often are) used to describe open-channel flow; however, we begin our discussion by addressing several characteristics of open-channel flow, including whether it is laminar or turbulent, uniform or varied, or subcritical, critical, or supercritical.

Laminar and Turbulent Flow

Laminar and *turbulent* flows in open channels are analogous to those in closed pressurized conduits (e.g., pipes). It is important to point out, however, that flow in open channels is usually turbulent. In addition, laminar flow essentially never occurs in open channels in either water or wastewater unit processes or structures.

Uniform and Varied Flow

Flow can be a function of time and location. If the flow quantity is invariant, it is said to be steady. *Uniform* flow is flow in which the depth, width, and velocity remain constant along a channel; that is, if the flow cross-section does not depend on the location along the channel, the flow is said to be uniform. *Varied* or *nonuniform* flow involves a change in these variables, with a change in one producing a change in the others. Most circumstances of open-channel flow in water and wastewater systems involve varied flow. The concept of uniform flow is valuable, however, in that it defines a limit that the varied flow may be considered to be approaching in many cases.

Note: Uniform channel construction does not ensure uniform flow.

Critical Flow

Critical flow (i.e., flow at the critical depth and velocity) defines a state of flow between two flow regimes. Critical flow coincides with minimum specific energy for a given discharge and maximum discharge for a given specific energy. Critical flow occurs in flow measurement devices at or near free discharges and establishes controls in open-channel flow. Critical flow occurs frequently in water/wastewater systems and is very important in their operation and design.

Note: Critical flow minimizes the specific energy and maximizes discharge.

Parameters Used in Open-Channel Flow

The three primary parameters used in open channel flow are *hydraulic radius*, *hydraulic depth*, and *slope* (S).

Hydraulic Radius

The hydraulic radius is the ratio of the cross-sectional area of the flow to its wetted perimeter:

$$r_H = A \div P \tag{3.24}$$

where

r_H = Hydraulic radius.
A = Cross-sectional area of the water.
P = Wetted perimeter.

Why is hydraulic radius important? Good question.

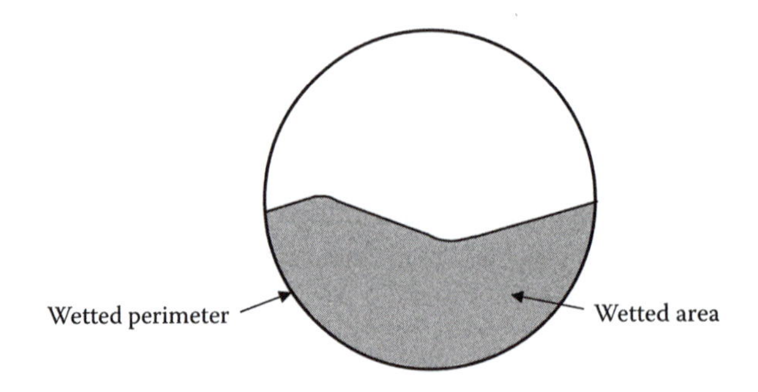

FIGURE 3.19 Hydraulic radius.

Probably the best way to answer this question is by illustration. Consider, for example, that in open channels it is of primary importance to maintain the proper velocity. This is the case, of course, because if velocity is not maintained then flow stops (theoretically). To maintain velocity at a constant level, the channel slope must be adequate to overcome friction losses. As with other flows, calculation of head loss at a given flow is necessary, and the Hazen–Williams equation is useful ($Q = 0.433 \times C \times D^{2.63} \times S^{0.54}$). Keep in mind that the concept of slope has not changed. The difference? We are now measuring, or calculating for, the physical slope of a channel (ft/ft), equivalent to head loss.

The preceding seems logical and makes sense, but there is a problem. The problem is with the diameter. In conduits that are not circular (e.g., grit chambers, contact basins, streams, rivers) or in pipes only partially full (e.g., drains, wastewater gravity mains, sewers), where the cross-sectional area of the water is not circular, there is no diameter. Without a diameter, what do we do? Another good question.

Because we do not have a diameter in situations where the cross-sectional area of the water is not circular, we must use another parameter to designate the size of the cross-section and the amount of it that contacts the sides of the conduit. This is where the hydraulic radius (r_H) comes in. The hydraulic radius is a measure of the efficiency with which the conduit can transmit water. Its value depends on pipe size and amount of fullness. We use the hydraulic radius to measure how much of the water is in contact with the sides of the channel or how much of the water is not in contact with the sides (see Figure 3.19).

Note: For a circular channel flowing either full or half full, the hydraulic radius is *D*/4. Hydraulic radii of other channel shapes are easily calculated from the basic definition.

Hydraulic Depth

The hydraulic depth is the ratio of area in flow to the width of the channel at the fluid surface (note that other names for hydraulic depth are *hydraulic mean depth* and *hydraulic radius*):

$$d_H = A \div w \tag{3.25}$$

where

d_H = Hydraulic depth.
A = Area in flow.
w = Width of the channel at the fluid surface.

Slope

The slope (S) in open-channel equations is the slope of the energy line. If the flow is uniform, the slope of the energy line will parallel the water surface and channel bottom. In general, the slope can be calculated from the Bernoulli equation as the energy loss per unit length of channel:

$$S = d_H \div d_L \tag{3.26}$$

Open-Channel Flow Calculations

The calculation for head loss at a give flow is typically accomplished by using the Hazen–Williams equation. In addition, in open-channel flow problems, although the concept of slope has not changed, a problem again rises with the diameter. In pipes only partially full where the cross-sectional area of the water is not circular, we have no diameter to work with, and the hydraulic radius is used for these noncircular areas. In the original version of the Hazen–Williams equation, the hydraulic radius was incorporated. Moreover, similar versions developed by Chezy (pronounced "shay-zee"), Manning, and others incorporated the hydraulic radius. For use in open channels, Manning's formula has become the most commonly used:

$$Q = \left(\frac{1.5}{n}\right) \times A \times R^{0.66} \times S^{0.5} \tag{3.27}$$

where

Q = Channel discharge capacity (ft^3/sec).
1.5 = Constant.
n = Channel roughness coefficient.
A = Cross-sectional flow area (ft^2).
R = Hydraulic radius of the channel (ft).
S = Slope of the channel bottom (dimensionless).

The hydraulic radius (R) of a channel is defined as the ratio of the flow area to the wetted perimeter (P). In formula form, $R = A/P$. The channel roughness coefficient (n) depends on the material and age for a pipe or lined channel and on topographic features for a natural streambed. It approximates roughness in open channels and can range from a value of 0.01 for a smooth clay pipe to 0.1 for a small natural stream. The value of n commonly assumed for concrete pipes or lined channels is 0.013. The n values decrease as the channels become smoother (see Table 3.4). The following example illustrates the application of Manning's formula for a channel with a rectangular cross-section.

TABLE 3.4
Manning Roughness Coefficient (n)

Type of Conduit	n	Type of Conduit	n
	Pipe		
Cast iron, coated	0.012–0.014	Cast iron, uncoated	0.013–0.015
Wrought iron, galvanized	0.015–0.017	Wrought iron, black	0.012–0.015
Steel, riveted and spiral	0.015–0.017	Corrugated	0.021–0.026
Wood stave	0.012–0.013	Cement surface	0.010–0.013
Concrete	0.012–0.017	Vitrified	0.013–0.015
Clay, drainage tile	0.012–0.014		
	Lined channels		
Metal, smooth semicircular	0.011–0.015	Metal, corrugated	0.023–0.025
Wood, planed	0.010–0.015	Wood, unplaned	0.011–0.015
Cement lined	0.010–0.013	Concrete	0.014–0.016
Cement rubble	0.017–0.030	Grass	0.020
	Unlined channels		
Earth, straight and uniform	0.017–0.025	Earth, dredged	0.025–0.033
Earth, winding	0.023–0.030	Earth, stony	0.025–0.040
Rock, smooth and uniform	0.025–0.035	Rock, jagged and irregular	0.035–0.045

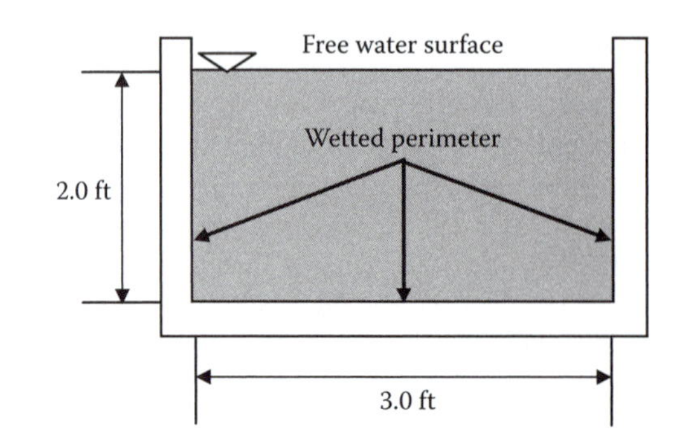

FIGURE 3.20 Illustration for Example 3.15.

■ Example 3.15

Problem: A rectangular drainage channel is 3 ft wide and is lined with concrete, as illustrated in Figure 3.20. The bottom of the channel drops in elevation at a rate of 0.5 per 100 ft. What is the discharge in the channel when the depth of water is 2 ft?

Solution: Assume that $n = 0.013$. Referring to Figure 3.20, we see that the cross-sectional flow area $(A) = 3 \text{ ft} \times 2 \text{ ft} = 6 \text{ ft}^2$, and the wetted perimeter $(P) = 2 \text{ ft} + 3 \text{ ft} + 2 \text{ ft} = 7 \text{ ft}$. The hydraulic radius $(R) = A/P = 6 \text{ ft}^2/7 \text{ ft} = 0.86 \text{ ft}$. The slope $(S) = 0.5/100 = 0.005$. Applying Manning's formula, we get

$$Q = \frac{2.0}{0.013} \times 6 \times 0.86^{0.66} \times 0.005^{0.5} = 59 \text{ cfs}$$

Open-Channel Flow: The Bottom Line

To this point, we have set the stage for explaining (in the simplest possible way) what open-channel flow is—what it is all about. Thus, now that we have explained the necessary foundational material and important concepts, we are ready to explain open-channel flow in a manner in which it can be easily understood.

We stated that, when water flows in a pipe or channel with a free surface exposed to the atmosphere, it is referred to as *open-channel flow*. We also know that gravity provides the motive force, the constant push, while friction resists the motion and causes energy expenditure. River and stream flows are open-channel flows. Flows in sanitary sewers and stormwater drains are open-channel flows, except in force mains where the water is pumped under pressure.

The key to solving routine stormwater and sanitary sewer problems is a condition known as *steady uniform flow*; that is, we assume steady uniform flow. Steady flow, of course, means that the discharge is constant with time. Uniform flow means that the slope of the water surface and the cross-sectional flow area are also constant. It is common practice to call a length of channel, pipeline, or stream that has a relatively constant slope and cross-section a *reach* (Nathanson, 1997).

The slope of the water surface under steady uniform flow conditions is the same as the slope of the channel bottom. The *hydraulic grade line* (HGL) lies along the water surface and, as in pressure flow in pipes, the HGL slopes downward in the direction of flow. Energy loss is evident as the water surface elevation drops. Figure 3.21 illustrates a typical profile view of uniform steady flow. The slope of the water surface represents the rate of energy loss.

> ***Note:*** The rate of energy loss (see Figure 3.21) may be expressed as the ratio of the drop in elevation of the surface in the reach to the length of the reach.

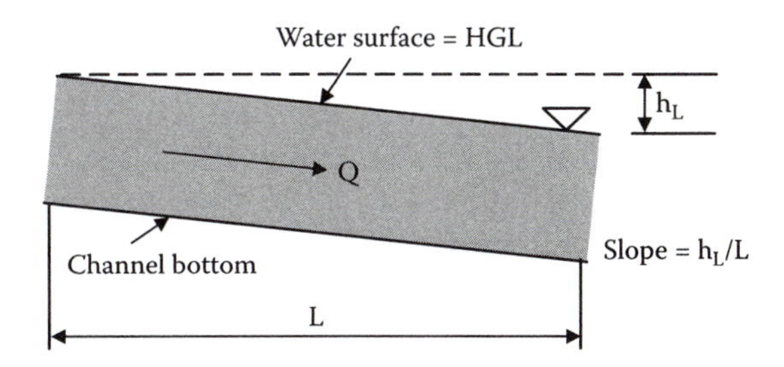

FIGURE 3.21 Steady uniform open-channel flow, where the slope of the water surface (or HGL) is equal to the slope of the channel bottom.

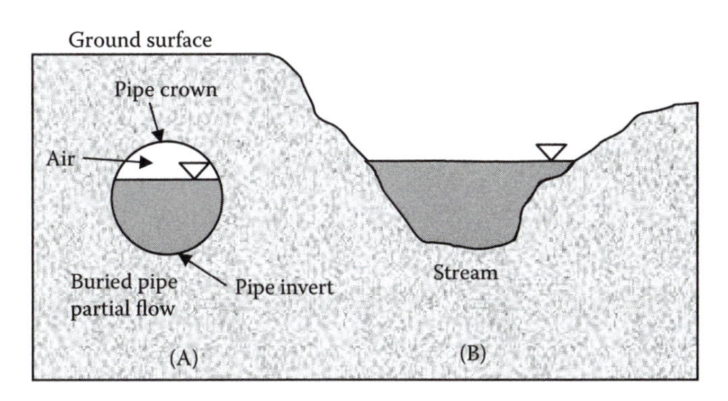

FIGURE 3.22 Open-channel flow in (A) an underground pipe and (B) a surface stream. (Adapted from Nathanson, J.A., *Basic Environmental Technology: Water Supply, Waste Management, and Pollution Control*, Prentice Hall, Upper Saddle River, NJ, 1997, p. 35.)

Figure 3.22 shows typical cross-sections of open-channel flow. In Figure 3.22A, the pipe is only partially filled with water and there is a free surface at atmospheric pressure. This is still open-channel flow, although the pipe is a closed underground conduit. Remember, the important point is that gravity and not a pump is moving the water.

FLOW MEASUREMENT

Although it is clear that maintaining water and wastewater flow is at the heart of any treatment process, clearly it is the measurement of flow that is essential to ensuring the proper operation of a water and wastewater treatment system. Few knowledgeable operators would argue with this statement. Hauser (1996) asked: "Why measure flow?" Then she explained: "The most vital activities in the operation of water and wastewater treatment plants are dependent on a knowledge of how much water is being processed." In this statement, Hauser made it clear that flow measurement is not only important but also routine in water/wastewater operations. Routine, yes, but also the most important variable measured in a treatment plant. Hauser also discussed several reasons for measuring flow in a treatment plant. Additional reasons to measure flow include the following (AWWA, 1995):

- The flow rate through the treatment processes must be controlled so it matches distribution system use.
- It is important to determine the proper feed rate of chemicals added in the processes.
- The detention times through the treatment processes must be calculated. This is particularly applicable to surface water plants that must meet $C \times T$ values required by the Surface Water Treatment Rule.

- Flow measurement allows operators to maintain a record of water furnished to the distribution system for periodic comparison with the total water metered to customers. This provides a measure of water accounted for or, conversely (as pointed out by Hauser), the amount of water wasted, leaked, or otherwise not paid for—that is, lost water.
- Flow measurement allows operators to determine the efficiency of pumps. Pumps that are not delivering their designed flow rate are probably not operating at maximum efficiency, so power is being wasted.
- For well systems, it is very important to maintain records of the volume of water pumped and the hours of operation for each well. The periodic computation of well pumping rates can identify problems such as worn pump impellers and blocked well screens.
- Reports that must be furnished to the state by most water systems must include records of raw and finished water pumpage.
- Wastewater generated by a treatment system must also be measured and recorded.
- Individual meters are often required for the proper operation of individual pieces of equipment; for example, the makeup water to a fluoride saturator is always metered to assist in tracking the fluoride feed rate.

Note: Simply put, measurement of flow is essential for operation, process control, and recordkeeping of water and wastewater treatment plants.

All of the uses just discussed create the need, obviously, for a number of flow measuring devices, often with different capabilities. In this section, we discuss many of the major flow measuring devices currently used in water and wastewater operations.

Flow Measurement the Old-Fashioned Way

An approximate but very simple method to determine open-channel flow has been used for many years. The procedure involves measuring the velocity of a floating object moving in a straight uniform reach of the channel or stream. If the cross-sectional dimensions of the channel are known and the depth of flow is measured, then flow area can be computed. From the relationship $Q = A \times V$, the discharge Q can be estimated. In preliminary fieldwork, this simple procedure is useful in obtaining a ballpark estimate for the flow rate but is not suitable for routine measurements.

■ Example 3.16

Problem: A floating object is placed on the surface of water flowing in a drainage ditch and is observed to travel a distance of 20 m downstream in 30 sec. The ditch is 2 m wide, and the average depth of flow is estimated to be 0.5 m. Estimate the discharge under these conditions.

Solution: The flow velocity is computed as distance over time, or

$$V = D/T = 20 \text{ m}/30 \text{ sec} = 0.67 \text{ m/sec}$$

The channel area is $A = 2 \text{ m} \times 0.5 \text{ m} = 1.0 \text{ m}^2$. The discharge $(Q) = A \times V = 1.0 \text{ m}^2 \times 0.66 \text{ m}^2 = 0.66 \text{ m}^3\text{/sec}$.

Basics of Traditional Flow Measurement

Flow measurement can be based on flow rate or flow amount. *Flow rate* is measured in gallons per minute (gpm), million gallons per day (MGD), or cubic feet per second (cfs). Water/wastewater operations require flow rate meters to determine process variables within the treatment plant, in wastewater collection, and in potable water distribution. Typically, the flow rate meters used are pressure differential meters, magnetic meters, and ultrasonic meters. Flow rate meters are designed for metering flow in closed pipe or open channel flow.

Flow amount is measured in either gallons (gal) or cubic feet (ft^3). Typically, a totalizer, which sums up the gallons or cubic feet that pass through the meter, is used. Most service meters are of this type. They are used in private, commercial, and industrial activities where the total amount of flow measured is used for customer billing. In wastewater treatment, where sampling operations are important, automatic composite sampling units—flow proportioned to grab a sample every so many gallons—are used. Totalizer meters can be of the velocity (propeller or turbine), positive displacement, or compound type. In addition, weirs and flumes are used extensively for measuring flow in wastewater treatment plants because they are not affected (to a degree) by dirty water or floating solids.

Flow Measuring Devices

In recent decades, flow measurement technology has evolved rapidly from the old-fashioned way of measuring flow discussed earlier to the use of simple practical measuring devices to much more sophisticated devices. Physical phenomena discovered centuries ago have been the starting point for many of the viable flowmeter designs used today. Moreover, the recent technology explosion has enabled flowmeters to handle many more applications that could only have been imagined centuries ago. Before selecting a particular type of flow measurement device, Kawamura (2000) recommended considering several questions:

1. Is liquid or gas flow being measured?
2. Is the flow occurring in a pipe or in an open channel?
3. What is the magnitude of the flow rate?
4. What is the range of flow variation?
5. Is the liquid being measured clean, or does it contain suspended solids or air bubbles?
6. What is the accuracy requirement?
7. What is the allowable head loss by the flow meter?
8. Is the flow corrosive?
9. What types of flowmeters are available to the region?
10. What types of post-installation service are available to the area?

Differential Pressure Flowmeters

For many years, differential pressure flowmeters have been the most widely applied flow-measuring device for water flow in pipes that require accurate measurement at reasonable cost (Kawamura, 2000). The differential pressure type of flowmeter makes up the largest segment of the total flow measurement devices currently being used. Differential pressure-producing meters currently on the market include the Venturi, Dall, Hershel Venturi, universal Venturi, and Venturi inserts.

The differential pressure-producing device has a flow restriction in the line that causes a differential pressure, or head, to be developed between the two measurement locations. Differential pressure flowmeters are also known as *head meters*, and, of all the head meters, the orifice flowmeter is the most widely applied device. The advantages of differential pressure flowmeters include the following:

- Simple construction
- Relatively low cost
- No moving parts
- External transmitting instruments
- Low maintenance

- Wide application of flowing fluid, suitable for measuring both gas and liquid flow
- Ease of instrument and range selection
- Extensive product experience and performance database

Disadvantages include the following:

- Flow rate being a nonlinear function of the differential pressure
- Low flow rate range with normal instrumentation

Operating Principle

Differential pressure flowmeters operate on the principle of measuring pressure at two points in the flow, which provides an indication of the rate of flow that is passing by. The difference in pressures between the two measurement locations of the flowmeter is the result of the change in flow velocities. Simply, there is a set relationship between the flow rate and volume, so the meter instrumentation automatically translates the differential pressure into a volume of flow. The volume of flow rate through the cross-sectional area is given by

$$Q = A \times V$$

where

Q = Volumetric flow rate.
A = Flow in the cross-sectional area.
V = Average fluid velocity.

Types of Differential Pressure Flowmeters

Differential pressure flowmeters operate on the principle of developing a differential pressure across a restriction that can be related to the fluid flow rate.

> ***Note:*** Optimum measurement accuracy is maintained when the flowmeter is calibrated, the flowmeter is installed in accordance with standards and codes of practice, and the transmitting instruments are periodically calibrated.

The differential pressure flowmeter types most commonly used in water/wastewater treatment are

1. Orifice
2. Venturi
3. Nozzle
4. Pitot–static tube

Orifice

The most commonly applied orifice is a thin, concentric, and flat metal plate with an opening in the plate (see Figure 3.23), installed perpendicular to the flowing stream in a circular conduit or pipe. Typically, a sharp-edged hole is bored in the center of the orifice plate. As the flowing water passes through the orifice, the restriction causes an increase in velocity. A concurrent decrease in pressure occurs as potential energy (static pressure) is converted into kinetic energy (velocity). As the water leaves the orifice, its velocity decreases and its pressure increases as kinetic energy is converted back into potential energy according to the laws of conservation of energy; however, some permanent pressure loss due to friction always occurs, and the loss is a function of the ratio of the diameter of the orifice bore (d) to the pipe diameter (D).

For dirty water applications (e.g., wastewater), the performance of a concentric orifice plate will eventually be impaired due to dirt buildup at the plate, so eccentric or segmental orifice plates (see Figure 3.24) can be used. Measurements are typically less accurate than those obtained from the concentric orifice plate, so eccentric or segmental orifices are rarely applied in current practice.

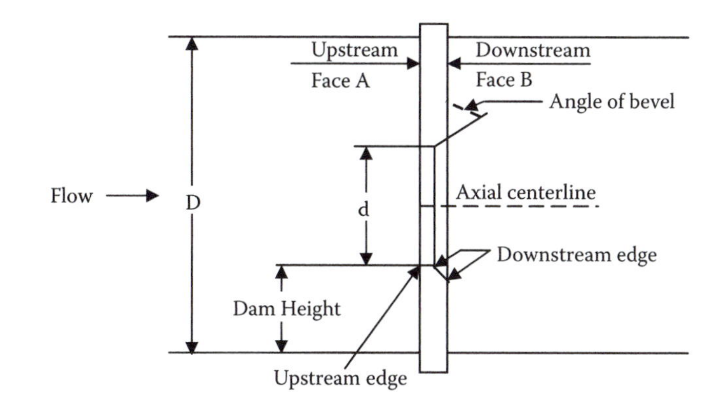

FIGURE 3.23 Orifice plate.

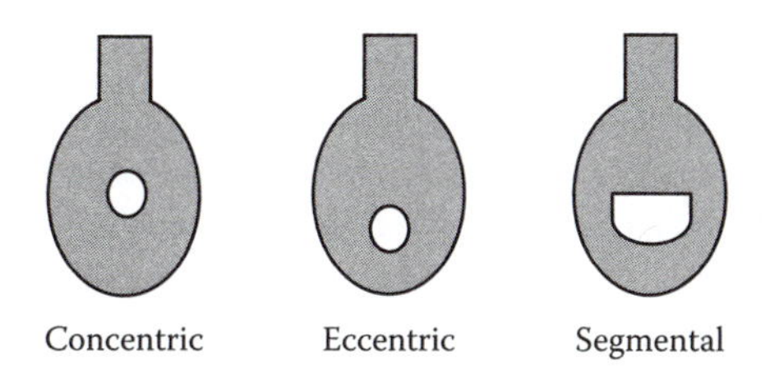

FIGURE 3.24 Types of orifice plate.

The orifice differential pressure flowmeter is the lowest cost differential flowmeter. It is easy to install and has no moving parts; however, it also has high permanent head loss (ranging from 40 to 90%), higher pumping costs, and an accuracy of ±2% for a flow range of 4:1. Also, it is affected by wear or damage.

> ***Note:*** Orifice flowmeters are not recommended for permanent installation to measure wastewater flow; solids in the water easily catch on the orifice, throwing off accuracy. For installation, it is necessary to have ten diameters of straight pipe ahead of the orifice meter to create a smooth flow pattern, and five diameters of straight pipe on the discharge side.

Venturi

A Venturi is a restriction with a relatively long passage with smooth entry and exit (see Figure 3.25). It features long life expectancy, simplicity of construction, and relatively high-pressure recovery (i.e., produces less permanent pressure loss than a similar sized orifice), but it is more expensive, is not linear with flow rate, and is the largest and heaviest differential pressure flowmeter. It is often used in wastewater flows because the smooth entry allows foreign material to be swept through instead of

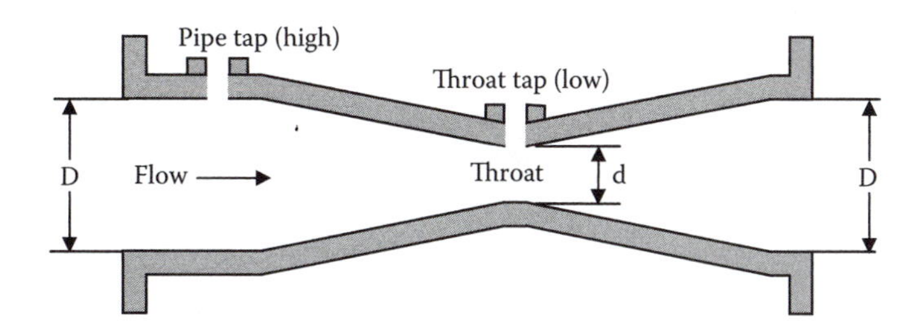

FIGURE 3.25 Venturi tube.

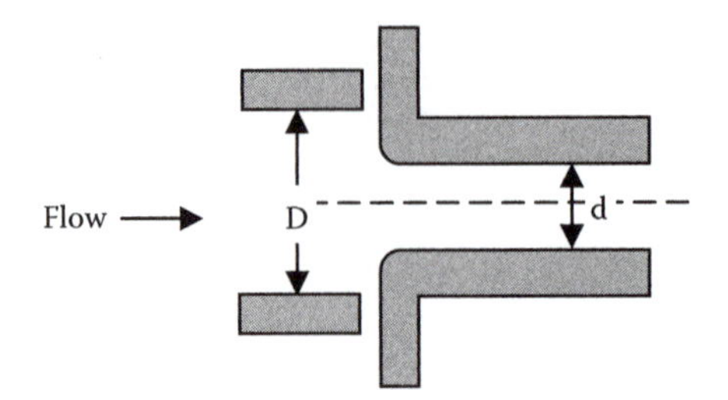

FIGURE 3.26 Long-radius flow nozzle.

building up as would occur in front of an orifice. The accuracy of this type flowmeter is ±1% for a flow range of 10:1. The head loss across a Venturi flowmeter is relatively small, ranging from 3 to 10% of the differential, depending on the ratio of the throat diameter to the inlet diameter (i.e., beta ratio).

Nozzle

Flow nozzles (flow tubes) have a smooth entry and sharp exit (Figure 3.26). For the same differential pressure, the permanent pressure loss of a nozzle is of the same order as that of an orifice, but it can handle wastewater and abrasive fluids better than an orifice can. Note that, for the same line size and flow rate, the differential pressure at the nozzle is lower (head loss ranges from 10 to 20% of the differential) than the differential pressure for an orifice; hence, the total pressure loss is lower than that of an orifice. Nozzles are primarily used in steam service because of their rigidity, which makes them dimensionally more stable at high temperatures and velocities than orifices.

> ***Note:*** A useful characteristic of nozzles it that they reach a critical flow condition—that is, a point at which further reduction in downstream pressure does not produce a greater velocity through the nozzle. When operated in this mode, nozzles are very predictable and repeatable.

Pitot Tube

A Pitot tube is a point velocity-measuring device (see Figure 3.27) with an impact port. As fluid hits the port, the velocity is reduced to zero, and kinetic energy (velocity) is converted to potential energy (pressure head). The pressure at the impact port is the sum of the static pressure and the velocity head. The pressure at the impact port is also known as the *stagnation pressure* or *total pressure*. The pressure difference between the impact pressure and the static pressure measured at the same point is the velocity head. The flow rate is the product of the measured velocity and the cross-sectional area at the point of measurement. Note that the Pitot tube has negligible permanent pressure drop in the line, but the impact port must be located in the pipe where the measured velocity is equal to the average velocity of the flowing water through the cross-section.

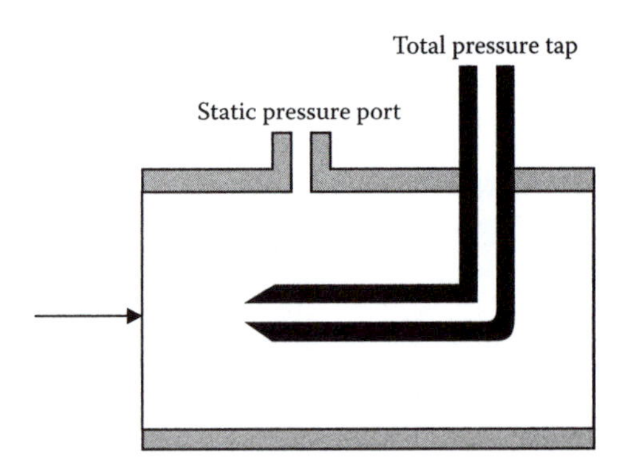

FIGURE 3.27 Pitot tube.

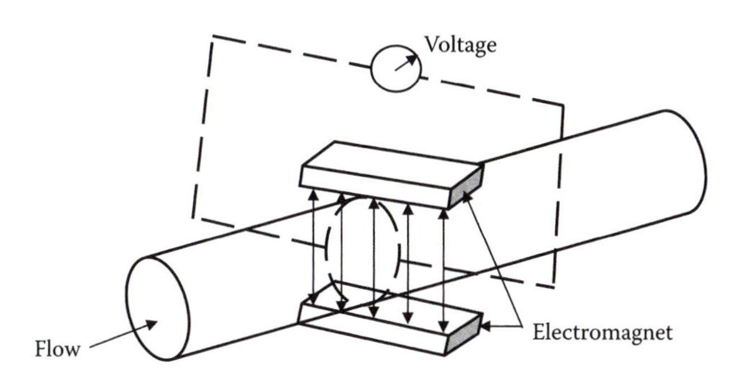

FIGURE 3.28 Magnetic flowmeter.

Magnetic Flowmeters

Magnetic flowmeters are relatively new to the water/wastewater industry. They are volumetric flow devices designed to measure the flow of electrically conductive liquids in a closed pipe. They measure the flow rate based on the voltage created between two electrodes (in accordance with Faraday's law of electromagnetic induction) as the water passes through an electromagnetic field (see Figure 3.28). Induced voltage is proportional to flow rate. Voltage depends on magnetic field strength (constant), distance between electrodes (constant), and velocity of flowing water (variable). Properties of the magnetic flowmeter include: (1) minimal head loss (no obstruction with line size meter); (2) no effect on flow profile; (3) suitability for sizes ranging between 0.1 and 120 inches; (4) accuracy rating of from 0.5 to 2% of flow rate; and (5) an ability to measure forward or reverse flow. The advantages of magnetic flowmeters include the following:

- Obstructionless flow
- Minimal head loss
- Wide range of sizes
- Bidirectional flow measurement
- Negligible effect of variations in density, viscosity, pressure, and temperature
- Suitability for wastewater
- No moving parts

Disadvantages include the following:

- The metered liquid must be conductive (but this type of meter would not be used on clean fluids anyway).
- They are bulky and expensive in smaller sizes.
- They may require periodic calibration to correct drifting of the signal.

The combination of a magnetic flowmeter and transmitter is considered to be a system. A typical system, illustrated in Figure 3.29, has a transmitter mounted remote from the magnetic flowmeter. Some systems are available with transmitters mounted integral to the magnetic flowmeter. Each device is individually calibrated during the manufacturing process, and the accuracy statement of the magnetic flowmeter includes both pieces of equipment. One is not sold or used without the other (WWITA and USEPA, 1991). It is also interesting to note that since 1983 almost every manufacturer now offers the microprocessor-based transmitter.

With regard to a minimum piping, straight-run requirement, magnetic flowmeters are quite forgiving of piping configuration. The downstream side of the magnetic flowmeter is much less critical than the upstream side. Essentially, all that is required of the downstream side is that sufficient backpressure be provided to keep the magnetic flowmeter full of liquid during flow measurement. Two diameters downstream should be acceptable (Mills, 1991).

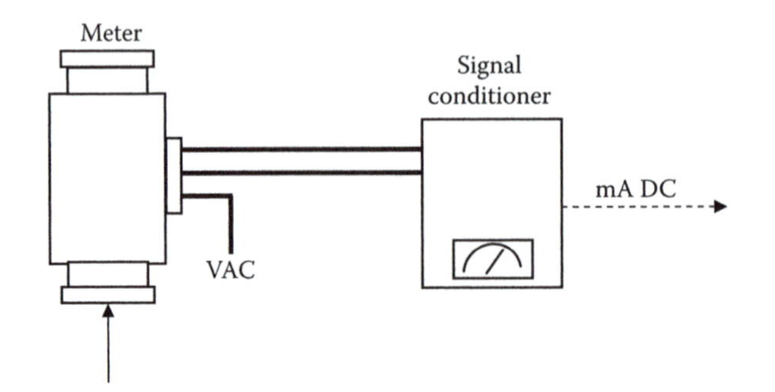

FIGURE 3.29 Magnetic flowmeter system.

Note: Magnetic flowmeters are designed to measure conductive liquids only. If air or gas is mixed with the liquid, the output becomes unpredictable.

Ultrasonic Flowmeters

Ultrasonic flowmeters use an electronic transducer to send a beam of ultrasonic sound waves through the water to another transducer on the opposite side of the unit. The velocity of the sound beam varies with the liquid flow rate, so the beam can be electronically translated to indicate flow volume. The accuracy is ±1% for a flow velocity ranging from 1 to 25 ft/sec, but the meter reading is greatly affected by a change in the fluid composition Two types of ultrasonic flowmeters are in general use for closed-pipe flow measurements. The first (time-of-flight or transit time) usually uses pulse transmission and is intended for use with clean liquids; the second (Doppler) usually utilizes continuous wave transmission and is intended for use with dirty liquids.

Time-of-Flight Ultrasonic Flowmeters

Time-of-flight flowmeters make use of the difference in the time for a sonic pulse to travel a fixed distance, first in the direction of flow and then against the flow. This is accomplished by positioning opposing transceivers on a diagonal path across a meter spool, as shown in Figure 3.30. Each transmits and receives ultrasonic pulses with flow and against flow. The fluid velocity is directly proportional to the time difference of pulse travel. The time-of-flight ultrasonic flowmeter operates with minimal head loss and has an accuracy range of 1 to 2.5% full scale. These flowmeters can be mounted as integral spool piece transducers or as externally mountable clamp-ons. They can measure flow accurately when properly installed and applied (Brown, 1991). The advantages of time-of-flight ultrasonic flowmeters include the following:

- No obstruction or interruption of flow
- Minimal head loss
- Clamp on

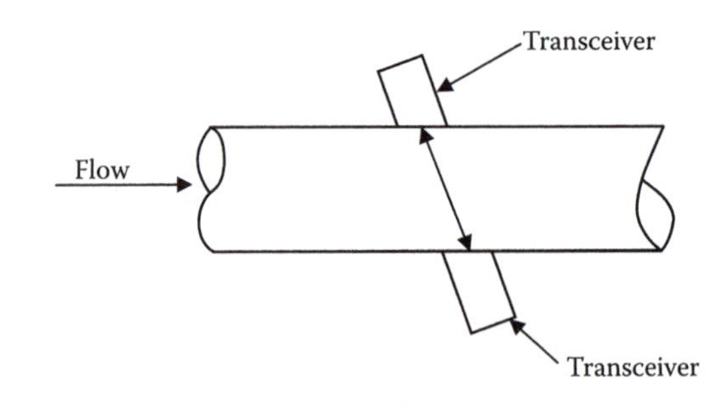

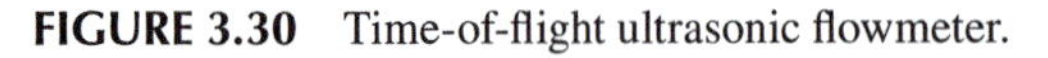

FIGURE 3.30 Time-of-flight ultrasonic flowmeter.

- Can be portable
- No moving parts
- Linear over wide range
- Wide range of pipe sizes
- Bidirectional flow measurement

Disadvantages include the following:

- Sensitivity to solids or bubble content
- Interference with sound pulses
- Sensitivity to flow disturbances
- Critical alignment of transducers
- Requirement for pipe walls to freely pass ultrasonic pulses (clamp-on type)

Doppler Ultrasonic Flowmeters

Doppler ultrasonic flowmeters make use of the Doppler frequency shift caused by sound scattered or reflected from moving particles in the flow path. Doppler meters are not considered to be as accurate as time-of-flight flowmeters; however, they are very convenient to use and are generally more popular and less expensive than time-of-flight flowmeters. In operation, a propagated ultrasonic beam is interrupted by particles in moving fluid and reflected toward a receiver. The difference of propagated and reflected frequencies is directly proportional to fluid flow rate. Ultrasonic Doppler flowmeters feature minimal head loss with an accuracy of 2 to 5% full scale. They are either of the integral spool piece transducer type or externally mountable clamp-ons. The advantages of the Doppler ultrasonic flowmeter include the following:

- No obstruction or interruption of flow
- Minimal head loss
- Clamp on
- Can be portable
- No moving parts
- Linear over wide range
- Wide range of pipe sizes
- Low installation and operating costs
- Bidirectional flow measurement

Disadvantages include the following:

- Minimum concentration and size of solids or bubbles required for reliable operation (see Figure 3.31)
- Minimum speed required to maintain suspension
- Limited to sonically conductive pipe (clamp-on type)

Velocity Flowmeters

Velocity or turbine flowmeters use a propeller (Figure 3.32A) or turbine (Figure 3.32B) to measure the velocity of the flow passing the device. The velocity is then translated into a volumetric amount by the meter register. Sizes exist from a variety of manufacturers to cover the flow range from 0.001 gpm to over 25,000 gpm for liquid service. End connections are available to meet the various piping systems. The flowmeters are typically manufactured of stainless steel but are also available in a wide variety of materials, including plastic. Velocity meters are applicable to all clean fluids. Velocity meters are particularly well suited for measuring intermediate flow rates on clean water (Oliver, 1991). The advantages of the velocity meter include the following:

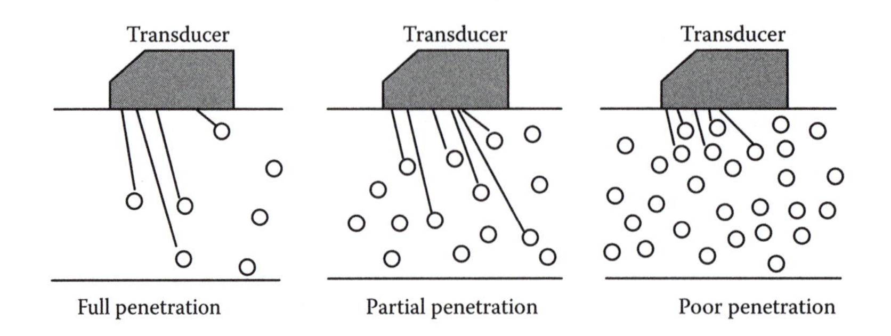

FIGURE 3.31 Particle concentration effect: the greater number of particles, the greater number of errors.

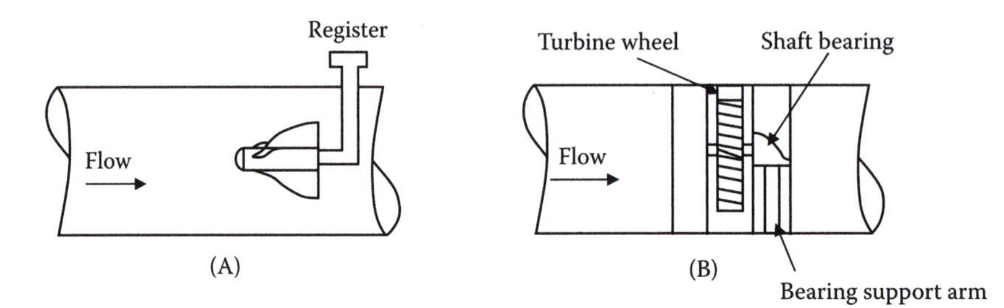

FIGURE 3.32 (A) Propeller meter; (B) turbine flowmeter.

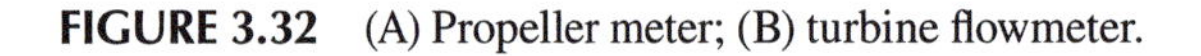

- Accuracy
- Composed of corrosion-resistant materials
- Long-term stability
- Liquid or gas operation
- Wide operating range
- Low pressure drop
- Wide temperature and pressure limits
- High shock capability
- Wide variety of electronics available

As shown in Figure 3.32B, a turbine flowmeter consists of a rotor mounted on a bearing and shaft in a housing. The fluid to be measured is passed through the housing, causing the rotor to spin with a rotational speed proportional to the velocity of the flowing fluid within the meter. A device to measure the speed of the rotor is employed to make the actual flow measurement. The sensor can be a mechanically gear-driven shaft connected to a meter or an electronic sensor that detects the passage of each rotor blade generating a pulse. The rotational speed of the sensor shaft and the frequency of the pulse are proportional to the volumetric flow rate through the meter.

Positive-Displacement Flowmeters

Positive-displacement flowmeters are most commonly used for customer metering and have long been used to measure liquid products. These meters are very reliable and accurate for low flow rates because they measure the exact quantity of water passing through them. Positive-displacement flowmeters are frequently used for measuring small flows in a treatment plant because of their accuracy. Repair or replacement is easy because they are so common in the distribution system.

In essence, a positive-displacement flowmeter is a hydraulic motor with high volumetric efficiency that absorbs a small amount of energy from the flowing stream. This energy is used to overcome internal friction in driving the flowmeter and its accessories and is reflected as a pressure drop across the flowmeter. Pressure drop is regarded as being unavoidable but something that must be minimized.

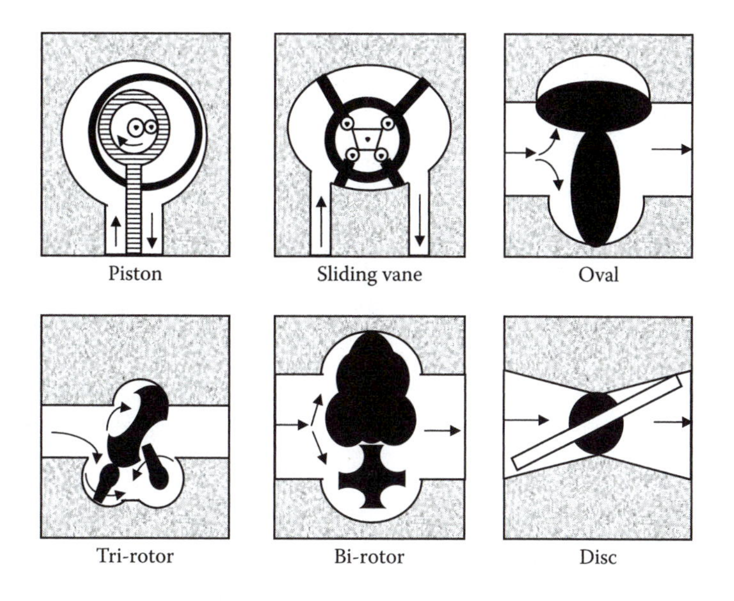

FIGURE 3.33 Six common positive-displacement flowmeter principles.

It is the pressure drop across the internals of a positive-displacement flowmeter that actually creates a hydraulically unbalanced rotor, which causes rotation (Barnes, 1991). A positive-displacement flowmeter continuously divides the flowing stream into known volumetric segments, isolates the segments momentarily, and returns them to the flowing stream while counting the number of displacements.

A positive-displacement flowmeter can be broken down into three basic components: the external housing, the measuring unit, and the counter drive train. The external housing is the pressure vessel that contains the product being measured. The measuring unit is a precision metering element made up of a measuring chamber and a displacement mechanism. The most common displacement mechanisms include the oscillating piston, sliding vane, oval gear, trirotor, birotor, and nutating disc types (see Figure 3.33). The counter drive train is used to transmit the internal motion of the measuring unit into a usable output signal. Many positive-displacement flowmeters use a mechanical gear train that requires a rotary shaft seal or packing gland where the shaft penetrates the external housing.

The positive-displacement flowmeter can offer excellent accuracy, repeatability, and reliability in many applications. It has satisfied many needs in the past and should play a vital role in serving future needs as required.

Open-Channel Flow Measurement Using Hydraulic Structures

The majority of industrial liquid flows are carried in closed conduits that flow completely full and under pressure; however, this is not the case for high-volume flows of liquids in waterworks, sanitary, and stormwater systems that are commonly carried in open channels. Low system heads and high volumetric flow rates characterize flow in open channels. The most commonly used method of measuring the rate of flow in open-channel flow configurations is that of hydraulic structures. In this method, flow in an open channel is measured by inserting a hydraulic structure into the channel, which changes the level of liquid in or near the structure. By selecting the shape and dimensions of the hydraulic structure, the rate of flow through or over the restriction will be related to the liquid level in a known manner. Thus, the flow rate through the open channel can be derived from a single measurement of the liquid level in or near the structure (Grant, 1991). The hydraulic structures used in measuring flow in open channels are known as primary measuring devices and may be divided into two broad categories—weirs and flumes, which are covered in the following subsections.

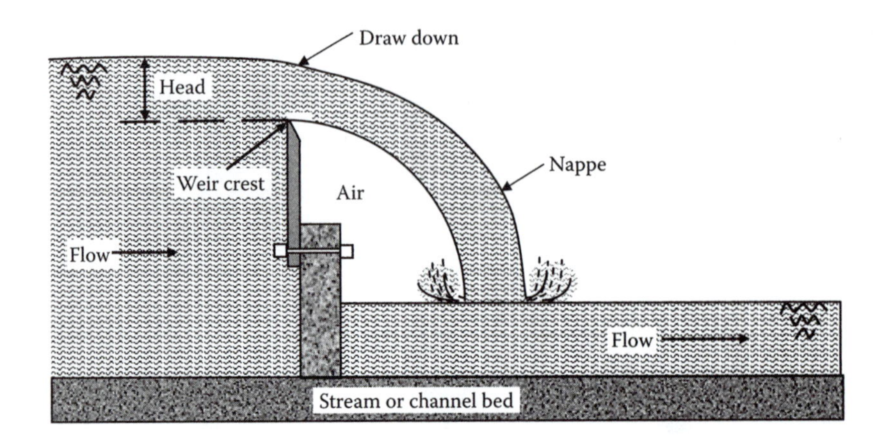

FIGURE 3.34 Side view of a weir.

Weirs

A weir is a widely used device to measure open-channel flow. As can be seen in Figure 3.34, a weir is simply a dam or obstruction placed in the channel so water backs up behind it and then flows over it. The sharp crest or edge allows the water to spring clear of the weir plate and to fall freely in the form of a *nappe*. When the nappe discharges freely into the air, a relationship exists between the height or depth of water flowing over the weir crest and the flow rate (Nathanson, 1997). This height, the vertical distance between the crest and the water surface, is called the *head* on the weir; it can be measured directly with a meter or yardstick or automatically by float-operated recording devices. Two common weirs, rectangular and triangular, are shown in Figure 3.35.

Rectangular weirs are commonly used for large flows (see Figure 3.35A). The formula to compute the rectangular weir is

$$Q = 3.33 \times L \times h^{1.5} \tag{3.28}$$

where

Q = Flow.

L = Width of weir.

h = Head on weir (measured from edge of weir in contact with the water, up to the water surface).

■ Example 3.17

Problem: A weir 4-ft high extends 15 ft across a rectangular channel in which the flow is 80 cfs. What is the depth just upstream from the weir?

Solution:

$$Q = 3.33 \times L \times h^{1.5}$$

$$80 = 3.33 \times 15 \times h^{1.5}$$

$$h = 1.4 \text{ ft (with calculator: 1.6 INV } y^{x1.5} = 1.36\text{, or 1.4)}$$

$$4 \text{ ft height of weir} + 1.4 \text{ ft head of water} = 5.4 \text{ ft depth}$$

Triangular weirs, also called V-notch weirs, can have notch angles ranging from 22.5° to 90°, but right-angle notches are the most common (see Figure 3.35B). The formula used to make 90° V-notch weir calculations is

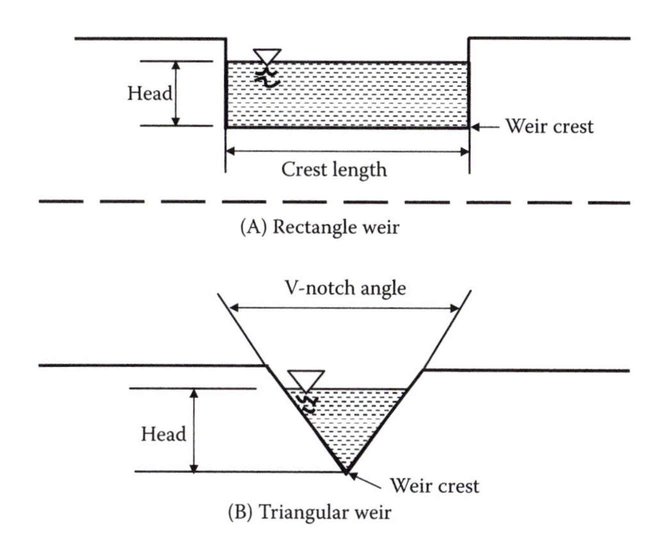

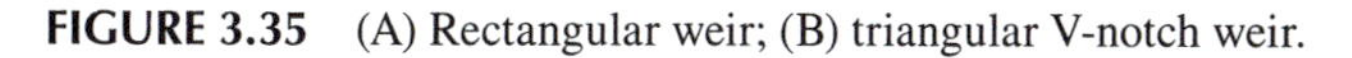

FIGURE 3.35 (A) Rectangular weir; (B) triangular V-notch weir.

$$Q = 2.5 \times h^{2.5} \tag{3.29}$$

where

Q = Flow.

h = Head on weir (measured from bottom of notch to water surface).

■ Example 3.18

Problem: What should be the minimum weir height for measuring a flow of 1200 gpm with a 90° V-notch weir, if the flow is moving at 4 fps in a 2.5-ft-wide rectangular channel?

Solution:

$$\frac{1200 \text{ gpm}}{60 \text{ sec/min} \times 7.48 \text{ gal/ft}^3} = 2.67 \text{ cfs}$$

$$Q = A \times V$$
$$2.67 \text{ cfs} = 2.5 \text{ ft} \times d \times 4 \text{ fps}$$
$$0.27 \text{ ft} = d$$

$$Q = 2.5 \times h^{2.5}$$
$$2.67 \text{ cfs} = 2.5 \times h^{2.5}$$
$$1.03 \text{ ft} = h \quad \text{(with calculator, 1.06 INV } y^{x2.5} = 1.026\text{, or 1.03)}$$

$$0.27 \text{ ft (original depth)} + 1.03 \text{ ft (head on weir)} = 1.3 \text{ ft}$$

It is important to point out that weirs, aside from being operated within their flow limits, must also be operated within the available system head. In addition, the operation of the weir is sensitive to the approach velocity of the water, often necessitating a stilling basin or pond upstream of the weir. Weirs are not suitable for water that carries excessive solid materials or silt, which deposit in the approach channel behind the weir and destroy the conditions required for accurate discharge measurements.

Note: Accurate flow rate measurements with a weir cannot be expected unless the proper conditions and dimensions are maintained.

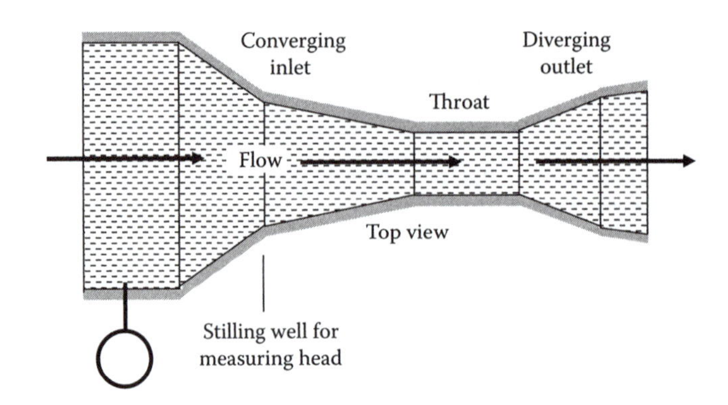

FIGURE 3.36 Parshall flume.

Flumes

A flume is a specially shaped constricted section in an open channel (similar to the Venturi tube in a pressure conduit). The special shape of the flume (see Figure 3.36) restricts the channel area and changes the channel slope, resulting in an increased velocity and a change in the level of the liquid flowing through the flume. The flume restricts the flow and then expands it in a definite fashion. The flow rate through the flume may be determined by measuring the head on the flume at a single point, usually at some distance downstream from the inlet.

Flumes can be categorized as belonging to one of three general families, depending on the state of flow induced—subcritical, critical, or supercritical. Typically, flumes that induce a critical or supercritical state of flow are most commonly used. This is because when critical or supercritical flow occurs in a channel, one head measurement can indicate the discharge rate if it is made far enough upstream so the flow depth is not affected by the drawdown of the water surface as it achieves or passes through a critical state of flow. For critical or supercritical states of flow, a definitive head–discharge relationship can be established and measured, based on a single head reading. Thus, most commonly encountered flumes are designed to pass the flow from subcritical through critical or near the point of measurement.

The most common flume used for a permanent wastewater flowmetering installation is called the *Parshall flume*, shown in Figure 3.36. Formulas for flow through Parshall flumes differ, depending on throat width. The formula below can be used for widths of 1 to 8 ft and applies to a medium range of flows:

$$Q = 4 \times W \times H_a^{1.52} \times W^{0.026} \tag{3.30}$$

where

Q = Flow.
W = Width of throat.
H_a = Depth in stilling well upstream.

Note: Parshall flumes are low-maintenance items.

THOUGHT-PROVOKING QUESTIONS

3.1 In nature, does water inflow always equal water outflow in the water balance or water budget? Explain.
3.2 Is measuring water in a well the same as measuring water in a stream? Explain.
3.3 Why do water levels in wells rise and fall? Explain.

REFERENCES AND RECOMMENDED READING

Arasmith, S. (1993). *Introduction to Small Water Systems*. Albany, OR: ACR Publications, pp. 59–61.

AWWA. (1995a). *Basic Science Concepts and Applications: Principles and Practices of Water Supply Operations Series*, 2nd ed. Denver, CO: American Water Works Association.

AWWA. (1995b). *Water Treatment: Principles and Practices of Water Supply Operations*, 2nd ed. Denver, CO: American Water Works Association, pp. 449–450.

AWWA. (1996). *Water Transmission and Distribution*, 2nd ed. Denver, CO: American Water Works Association, p. 358.

Barnes, R.G. (1991). Positive displacement flowmeters for liquid measurement. In *Flow Measurement*, Spitzer, D.W., Ed. Research Triangle Park, NC: Instrument Society of America.

Brown, A.E. (1991). Ultrasonic flowmeters. In *Flow Measurement*, Spitzer, D.W., Ed. Research Triangle Park, NC: Instrument Society of America.

Cheremisinoff, N.P. and Cheremisinoff, P.N. (1989). *Pumps, Compressors, Fans: Pocket Handbook*. Lancaster, PA: Technomic, p. 3.

Garay, P.N. (1990). *Pump Application Desk Book*. Lilburn, GA: Fairmont Press, p. 10.

Grant, D.M. (1991). Open channel flow measurement. In *Flow Measurement*, Spitzer, D.W., Ed. Research Triangle Park, NC: Instrument Society of America.

Grimes, A.S. (1976). Supervisory and monitoring instrumentation. In *Pump Handbook*, Karassik, I.J. et al., Eds. New York: McGraw-Hill.

Hauser, B.A. (1993). *Hydraulics for Operators*. Boca Raton, FL: Lewis Publishers.

Hauser, B.A. (1996). *Practical Hydraulics Handbook*, 2nd ed. Boca Raton, FL: Lewis Publishers.

Holman, S. (1998). *A Stolen Tongue*. New York: Anchor Press.

Husain, Z.D. and Sergesketter, M.J. (1991). Differential pressure flowmeters. In *Flow Measurement*, Spitzer, D.W., Ed. Research Triangle Park, NC: Instrument Society of America.

Hydraulic Institute. (1990). *Engineering Data Book*, 2nd ed. Cleveland, OH: Hydraulic Institute.

Hydraulic Institute. (1994). *Complete Pump Standards*, 4th ed. Cleveland, OH: Hydraulic Institute.

Kawamura, S. (2000). *Integrated Design and Operation of Water Treatment Facilities*, 2nd ed. New York: John Wiley & Sons.

Krutzsch, W.C. (1976). Introduction and classification of pumps. In *Pump Handbook*, Karassik, I.J. et al., Eds. New York: McGraw-Hill, p. 1-1.

Lindeburg, M.R. (1986). *Civil Engineering Reference Manual*, 4th ed. San Carlos, CA: Professional Publications.

Magnusson, R.J. (2001). *Water Technology in the Middle Ages*. Baltimore, MD: The John Hopkins University Press.

McGhee, T.J. (1991). *Water Supply and Sewerage*, 2nd ed. New York: McGraw-Hill.

Metcalf & Eddy, Inc. (1981). *Wastewater Engineering: Collection and Pumping of Wastewater*. New York: McGraw-Hill.

Nathanson, J.A. (1997). *Basic Environmental Technology: Water Supply Waste Management, and Pollution Control*, 2nd ed. Upper Saddle River, NJ: Prentice Hall.

OCDDS. (1986). *Basic Maintenance Training Course*. North Syracuse, NY: Onondaga County Department of Drainage and Sanitation.

Oliver, P.D. (1991). Turbine flowmeters. In *Flow Measurement*, Spitzer, D.W., Ed. Research Triangle Park, NC: Instrument Society of America.

Spellman, F.R. (2000). *The Handbook for Waterworks Operator Certification*. Vol. 2. *Intermediate Level*. Lancaster, PA: Technomic.

TUA. (1988). *Manual of Water Utility Operations*, 8th ed. Austin: Texas Utilities Association.

Viessman, Jr., W. and Hammer, M.J. (1998). *Water Supply and Pollution Control*, 6th ed. Menlo Park, CA: Addison–Wesley.

Wahren, U. (1997). *Practical Introduction to Pumping Technology*. Houston, TX: Gulf Publishing.

Watson, L. (1988). *The Water Planet: A Celebration of the Wonder of Water*. New York: Crown.

WWITA and USEPA. (1991). *Flow Instrumentation: A Practical Workshop on Making Them Work*. Sacramento, CA: Water and Wastewater Instrumentation Testing Association and U.S. Environmental Protection Agency.

4 Water Chemistry

> Chemical testing can be divided into two types. The first type measures a bulk physical property of the sample, such as volume, temperature, melting point, or mass. These measurements are normally performed with an instrument, and one simply has to calibrate the instrument to perform the test. Most analyses, however, are of the second type, in which a chemical property of the sample is determined that generates information about how much of what is present.
>
> **Smith (1993)**

Although no one has seen a water molecule, we have determined that atoms in water are elaborately meshed. Moreover, although it is true that we do not know as much as we need to know about water—our growing knowledge of water is a work in progress—we have determined many things about water. A large amount of our current knowledge comes from studies of water chemistry.

Water chemistry is important because several factors about water to be treated and then distributed or returned to the environment are determined through simple chemical analysis. Probably the most important determination that the water practitioner makes about water is its hardness.

Why chemistry? "I'm not a chemist," you say. But, when you add chlorine to water to make it safe to drink or safe to discharge into a receiving body (usually a river or lake), you *are* a chemist. Chemistry is the study of substances and the changes they undergo. Water specialists and those interested in the study of water must possess a fundamental knowledge of chemistry. Before beginning our discussion of water chemistry, it is important for the reader to have some basic understanding of chemistry concepts and chemical terms. The following section presents a review of chemistry terms, definitions, and concepts.

CHEMISTRY CONCEPTS AND DEFINITIONS

Chemistry, like the other sciences, has its own language; thus, to understand chemistry, it is necessary to understand the following concepts and key terms.

Concepts

Miscibility and Solubility

Substances that are *miscible* are capable of being mixed in all proportions. Simply, when two or more substances disperse themselves uniformly in all proportions when brought into contact, they are said to be completely soluble in one another, or completely miscible. The precise chemistry definition is a "homogeneous molecular dispersion of two or more substances" (Jost, 1992). Examples include the following observations:

- All gases are completely miscible.
- Water and alcohol are completely miscible.
- Water and mercury (in its liquid form) are immiscible liquids.

Between the two extremes of miscibility is a range of *solubility*; that is, various substances mix with one another up to a certain proportion. In many environmental situations, a rather small amount of a contaminant may be soluble in water in contrast to the complete miscibility of water and alcohol. The amounts are measured in parts per million (ppm).

Suspension, Sediment, Particles, and Solids

Often water carries *solids* or *particles* in *suspension*. These dispersed particles are much larger than molecules and may be comprised of millions of molecules. The particles may be suspended in flowing conditions and initially under quiescent conditions, but eventually gravity causes settling of the particles. The resultant accumulation by settling is referred to as *sediment* or *biosolids* (sludge) or *residual solids* in wastewater treatment vessels. Between this extreme of readily falling out by gravity and permanent dispersal as a solution at the molecular level are intermediate types of dispersion or suspension. Particles can be so finely milled or of such small intrinsic size as to remain in suspension almost indefinitely and in some respects similarly to solutions.

Emulsions

Emulsions represent a special case of a suspension. As you know, oil and water do not mix. Oil and other hydrocarbons derived from petroleum generally float on water with negligible solubility in water. In many instances, oils may be dispersed as fine oil droplets (an emulsion) in water and not readily separated by floating because of size or the addition of dispersal-promoting additives. Oil and, in particular, emulsions can prove detrimental to many treatment technologies and must be treated in the early steps of a multistep treatment train.

Ion

An ion is an electrically charged particle; for example, sodium chloride or table salt forms charged particles on dissolution in water. Sodium is positively charged (a cation), and chloride is negatively charged (an anion). Many salts similarly form cations and anions on dissolution in water.

Mass Concentration

Concentration is often expressed in terms of parts per million (ppm) or mg/L. Sometimes parts per thousand (ppt) and parts per billion (ppb) are also used:

$$\text{ppm} = \text{Mass of substance} \div \text{Mass of solutions} \tag{4.1}$$

Because 1 kg of solution with water as a solvent has a volume of approximately 1 liter,

$$1 \text{ ppm} \approx 1 \text{ mg/L}$$

Definitions

Chemistry—The science that deals with the composition and changes in composition of substances. Water is an example of this composition; it is composed of two gases: hydrogen and oxygen. Water also changes form from liquid to solid to gas but does not necessarily change composition.

Matter—Anything that has weight (mass) and occupies space. Kinds of matter include elements, compounds, and mixtures.

Solids—Substances that maintain definite size and shape. Solids in water fall into one of the following categories:

- *Dissolved solids*—Solids in a single-phase (homogeneous) solution consisting of dissolved components and water. Dissolved solids are the material in water that will pass through a glass fiber filter and remain in an evaporating dish after evaporation of the water.
- *Colloidal solids* (sols)—Solids that are uniformly dispersed in solution. They form a solid phase that is distinct from the water phase.

- *Suspended solids*—Solids in a separate phase from the solution. Some suspended solids are classified as settleable solids. Settleable solids are determined by placing a sample in a cylinder and measuring the amount of solids that have settled after a set amount of time. The size of solids increases moving from dissolved solids to suspended solids. Suspended solids are the material deposited when a quantity of water, sewage, or other liquid is filtered through a glass fiber filter.
- *Total solids*—The solids in water, sewage, or other liquids; include suspended solids (largely removable by a filter) and filterable solids (those that pass through the filter).

Liquids—Having a definite volume but not shape, liquids will fill containers to certain levels and form free level surfaces.

Gases—Having neither definite volume nor shape, gases completely fill any container in which they are placed.

Mixture—A physical, not chemical, intermingling of two of more substances. Sand and salt stirred together form a mixture.

Element—The simplest form of chemical matter. Each element has chemical and physical characteristics different from all other kinds of matter.

Compound—A substance of two or more chemical elements chemically combined. Examples include water (H_2O), which is a compound formed by hydrogen and oxygen. Carbon dioxide (CO_2) is composed of carbon and oxygen.

Molecule—The smallest particle of matter or a compound that possesses the same composition and characteristics as the rest of the substance. A molecule may consist of a single atom, two or more atoms of the same kind, or two or more atoms of different kinds.

Atom—The smallest particle of an element that can unite chemically with other elements. All the atoms of an element are the same in chemical behavior, although they may differ slightly in weight. Most atoms can combine chemically with other atoms to form molecules.

Radical—Two or more atoms that unite in a solution and behave chemically as if a single atom.

Ion—An atom or group of atoms that carries a positive or negative electric charge as a result of having lost or gained one or more electrons.

Ionization—The formation of ions by the splitting of molecules or electrolytes in solution. Water molecules are in continuous motion, even at lower temperatures. When two water molecules collide, a hydrogen ion is transferred from one molecule to the other. The water molecule that loses the hydrogen ion becomes a negatively charged hydroxide ion. The water molecule that gains the hydrogen ion becomes a positively charged hydronium ion. This process is commonly referred to as the *self-ionization of water.*

Cation—A positively charged ion.

Anion—A negatively charged ion.

Organic—Chemical substances of animal or vegetable origin made of carbon structure.

Inorganic—Chemical substances of mineral origin.

Solvent—The component of a solution that does the dissolving.

Solute—The component of a solution that is dissolved by the solvent.

Saturated solution—The physical state in which a solution will no longer dissolve more of the dissolving substance (solute).

Colloidal—Any substance in a certain state of fine division in which the particles are less than 1 micron in diameter.

Turbidity—A condition in water caused by the presence of suspended matter. Turbidity results in the scattering and absorption of light rays.

Precipitate—A solid substance that can be dissolved but is separated from the solution because of a chemical reaction or change in conditions such as pH or temperature.

CHEMISTRY FUNDAMENTALS

Whenever water and wastewater practitioners add a substance to another substance (from adding sugar to a cup of tea to adding chlorine to water to make it safe to drink), they perform chemistry. Water and wastewater operators (as well as many others) are chemists, because they are working with chemical substances, and it is important for operators to know and to understand how those substances react.

MATTER

Going through a day without coming into contact with many kinds of matter is impossible. Paper, coffee, gasoline, chlorine, rocks, animals, plants, water, air—all the materials of which the world is made—are all different forms or kinds of matter. Earlier matter was defined as anything that has mass (weight) and occupies space; matter is distinguishable from empty space by its presence. Thus, obviously, the statement about the impossibility of going through a day without coming into contact with matter is correct; in fact, avoiding some form of matter is virtually impossible. Not all matter is the same, even though we narrowly classify all matter into three groups: solids, liquids, and gases. These three groups are the *physical states of matter* and are distinguishable from one another by means of two general features: shape and volume.

> ***Note:*** *Mass* is closely related to the concept of *weight*. On Earth, the weight of matter is a measure of the force with which it is pulled by gravity toward the center of the Earth. As we leave Earth's surface, the gravitational pull decreases, eventually becoming virtually insignificant, while the weight of matter accordingly reduces to zero. Yet, the matter still possesses the same amount of mass. Hence, the mass and weight of matter are proportional to each other.

> ***Note:*** Because matter occupies space, a given form of matter is also associated with a definite volume. Space should not be confused with air, as air is itself a form of matter. *Volume* refers to the actual amount of space that a given form of matter occupies.

Solids have a definite, rigid shape; their particles are closely packed together and stick firmly to each other. A solid does not change its shape to fit a container. Put a solid on the ground and it will keep its shape and volume—it will never spontaneously assume a different shape. Solids also possess a definite volume at a given temperature and pressure.

Liquids maintain a constant volume but change shape to fit the shape of their container; they do not possess a characteristic shape. The particles of the liquid move freely over one another but still stick together enough to maintain a constant volume. Consider a glass of water. The liquid water takes the shape of the glass up to the level it occupies. If we pour the water into a drinking glass, the water takes the shape of the glass; if we pour it into a bowl, the water takes the shape of the bowl. Thus, if space is available, a liquid assumes whatever shape its container possesses. Like solids, liquids possess a definite volume at a given temperature and pressure, and they tend to maintain this volume when they are exposed to a change in either of these conditions.

Gases have no definite fixed shape, and their volume can be expanded or compressed to fill different sizes of containers. A gas or mixture of gases such as air can be put into a balloon and will take the shape of the balloon. Particles of gases do not stick together at all and move about freely, filling containers of any shape and size. A gas is identified by its lack of a characteristic volume. When confined to a container with nonrigid, flexible walls, for example, the volume that a confined gas occupies depends on its temperature and pressure. When confined to a container with rigid walls, however, the volume of the gas is forced to remain constant.

Internal linkages among its units, including between one atom and another, maintain the constant composition associated with a given substance. These linkages are called *chemical bonds*. When a particular process occurs that involves the making and breaking of these bonds, we say that a *chemical reaction* or a *chemical change* has occurred. Let's take a closer look at both chemical and physical changes of matter.

Chemical changes occur when new substances are formed that have entirely different properties and characteristics. When wood burns or iron rusts, a chemical change has occurred; the linkages—the chemical bonds—are broken. *Physical changes* occur when matter changes its physical properties such as size, shape, and density, as well as when it changes its state (e.g., from gas to liquid to solid). When ice melts or when a glass window breaks into pieces, a physical change has occurred.

Content of Matter: The Elements

Matter is composed of pure basic substances. Earth is made up of the fundamental substances of which all matter is composed. These substances that resist attempts to decompose them into simpler forms of matter are called *elements*. To date, more than 100 elements are known to exist. They range from simple, lightweight elements to very complex, heavyweight elements. Some of these elements exist in nature in pure form; others are combined.

The smallest unit of an element is the *atom*. The simplest atom possible consists of a nucleus having a single proton with a single electron traveling around it. This is an atom of hydrogen, which has an atomic weight of one because of the single proton. The *atomic weight* of an element is equal to the total number of protons and neutrons in the nucleus of an atom of an element.

To gain an understanding of basic atomic structure and related chemical principles it is useful to compare the atom to our solar system. In our solar system, the sun is the center of everything, whereas the *nucleus* is the center in the atom. The sun has several planets orbiting around it, whereas the atom has *electrons* orbiting about the nucleus. It is interesting to note that astrophysicists, who would likely find this analogy overly simplistic, are concerned primarily with activity within the nucleus. This is not the case, however, with chemists, who deal principally with the activity of the planetary electrons. Chemical reactions between atoms or molecules involve only electrons, with no changes in the nuclei.

The nucleus is made up of positive electrically charged *protons* and *neutrons*, which are neutral (no charge). The negatively charged electrons orbiting the nucleus balance the positive charge in the nucleus. An electron has negligible mass (less than 0.02% of the mass of a proton), which makes it practical to consider the weight of the atom as the weight of the nucleus.

Atoms are identified by name, atomic number, and atomic weight. The *atomic number* or *proton number* is the number of protons in the nucleus of an atom. It is equal to the positive charge on the nucleus. In a neutral atom, it is also equal to the number of electrons surrounding the nucleus. The atomic weight of an atom depends on the number of protons and neutrons in the nucleus, as the electrons have negligible mass. Atoms (elements) have received their names and symbols in interesting ways. The discoverer of the element usually proposes a name for it. Some elements get their symbols from languages other than English. The following is a list of common elements with their common names and the names from which the symbol is derived.

Chlorine	Cl
Copper	Cu (Latin *cuprum*)
Hydrogen	H
Iron	Fe (Latin *ferrum*)
Nitrogen	N
Oxygen	O
Phosphorus	P
Sodium	Na (Latin *natrium*)
Sulfur	S

As shown above, a capital letter or a capital letter and a small letter designate each element. These are called *chemical symbols*. As is apparent from the above list, most of the time the symbol is easily recognized as an abbreviation of the atom name, such as O for oxygen.

Typically, we do not find most of the elements as single atoms. They are more often found in combinations of atoms called *molecules*. Basically, a molecule is the least common denominator of what makes a substance what it is. A system of formulas has been devised to show how atoms are combined into molecules. When a chemist writes the symbol for an element, it stands for one atom of the element. A subscript following the symbol indicates the number of atoms in the molecule. O_2 is the chemical formula for an oxygen molecule. It shows that oxygen occurs in molecules consisting of two oxygen atoms. As you know, a molecule of water contains two hydrogen atoms and one oxygen atom, so the formula is H_2O.

Note: H_2O, the chemical formula of the water molecule, was defined in 1860 by the Italian scientist Stanisloa Cannizzarro.

Some elements have similar chemical properties; for example, bromine (atomic number 35) has chemical properties that are similar to the chemical properties of the element chlorine (atomic number 17), with which most water operators are familiar, and iodine (atomic number 53). In 1865, English chemist John Newlands arranged some of the known elements in an increasing order of atomic weights. Newlands arranged the lightest element known at the time at the top of his list and the heaviest element at the bottom. Newlands was surprised when he observed that starting from a given element, every eighth element repeated the properties of the given element.

Later, in 1869, Dmitri Mendeleev, a Russian chemist, published a table of the 63 known elements. In his table, Mendeleev, like Newlands, arranged the elements in increasing order of atomic weight. He also grouped them in eight vertical columns so the elements with similar chemical properties would be found in one column. It is interesting to note that Mendeleev left blanks in his table. He correctly hypothesized that undiscovered elements existed that would fill in the blanks when they were discovered. Because he knew the chemical properties of the elements above and below the blanks in his table, he was able to predict quite accurately the properties of some of the undiscovered elements.

Today our modern form of the periodic table is based on work done by the English scientist Henry Moseley, who was killed during World War I. Following the work of Ernest Rutherford (a New Zealand physicist) and Niels Bohr (a Danish physicist), Moseley used x-ray methods to determine the number of protons in the nucleus of an atom. The atomic number, or number of protons, of an atom is related to its atomic structure. In turn, atomic structure governs chemical properties. The atomic number of an element is more directly related to its chemical properties than is its atomic weight. It is more logical to arrange the periodic table according to atomic numbers than atomic weights. By demonstrating the atomic numbers of elements, Moseley helped chemists to make a better periodic table.

In the periodic table, a horizontal row of boxes is called a *period* or *series*. Hydrogen is all by itself because of its special chemical properties. Helium is the only element in the first period. The second period contains lithium, beryllium, boron, carbon, nitrogen, oxygen, fluorine, and neon. Other elements may be identified by looking at the table. A vertical column is called a *group* or *family*. Elements in a group have similar chemical properties. The periodic table is useful because knowing where an element is located in the table gives us a general idea of its chemical properties.

For convenience, elements have a specific name and symbol but are often identified by chemical symbol only. The symbols of the elements consist of either one or two letters, with the first letter capitalized. Table 4.1 lists some of the elements and indicates those that are important to the water practitioner (about a third of the over 100 known elements).

Compound Substances

If we take a pure substance such as calcium carbonate (limestone) and heat it, the calcium carbonate ultimately crumbles to a white powder; however, careful examination of the process shows that carbon dioxide also evolves from the calcium carbonate. Substances such as calcium carbonate that

TABLE 4.1
Elements and Their Symbols

Element	Symbol	Element	Symbol
Aluminum[a]	Al	Iron[a]	Fe
Arsenic	As	Lead	Pb
Barium	Ba	Magnesium[a]	Mg
Cadmium	Ca	Manganese[a]	Mn
Carbon[a]	C	Mercury	Hg
Calcium	Ca	Nitrogen[a]	N
Chlorine[a]	Cl	Nickel	Ni
Chromium	Cr	Oxygen[a]	O
Cobalt	Co	Phosphorus	P
Copper	Cu	Potassium	K
Fluoride[a]	F	Silver	Ag
Helium	He	Sodium[a]	Na
Hydrogen[a]	H	Sulfur[a]	S
Iodine	I	Zinc	Zn

[a] Elements familiar to water/wastewater treatment operators.

can be broken down into two or more simpler substances are called *compound substances* or simply *compounds*. Heating is a common way of decomposing compounds, but other forms of energy are often used as well. Chemical elements that make up compounds such as calcium carbonate combine with each other in definite proportions. When atoms of two or more elements are bonded together to form a compound, the resulting particle is called a *molecule*.

Note: Only a certain number of atoms or radicals of one element will combine with a certain number of atoms or radicals of a different element to form a chemical compound.

Water (H_2O) is a compound. As stated, compounds are chemical substances made up of two or more elements bonded together. Unlike elements, compounds can be separated into simpler substances by chemical changes. Most forms of matter in nature are composed of combinations of the over 100 pure elements.

If we have a particle of a compound—for example, a crystal of salt (sodium chloride)—and subdivide, subdivide, and subdivide until we get the smallest unit of sodium chloride possible, we would have a molecule. As stated, a molecule (or least common denominator) is the smallest particle of a compound that still has the characteristics of that compound.

Note: Because the weights of atoms and molecules are relative and the units are extremely small, the chemist works with units known as *moles*. A mole (mol) is defined as the amount of a substance that contains as many elementary entities (atoms, molecules, and so on) as there are atoms in 12 g of the isotope carbon-12.

Note: An *isotope* of an element is an atom having the same structure as the element—the same electrons orbiting the nucleus and the same protons in the nucleus—but having more or fewer neutrons.

One mole of an element that exists as single atoms weighs as many grams as its atomic number, so 1 mole of carbon weighs 12 g, and it contains 6.022045×10^{23} atoms, which is the *Avogadro's number*.

Symbols are used to identify elements and are a shorthand method for writing the names of the elements. This shorthand method is also used for writing the names of compounds. Symbols used in this manner show the kinds and numbers of different elements in the compound. These shorthand representations of chemical compounds are called *chemical formulas*; for example, the formula for table salt (sodium chloride) is NaCl. The formula shows that one atom of sodium combines with one atom of chlorine to form sodium chloride. Let's look at a more complex formula for the compound sodium carbonate (soda ash): Na_2CO_3. The formula shows that this compound is made up of three elements: sodium, carbon, and oxygen. In addition, each molecule has two atoms of sodium, one atom of carbon, and three atoms of oxygen.

When depicting chemical reactions, *chemical equations* are used. The following equation shows a chemical reaction with which most water/wastewater operators are familiar—chlorine gas added to water. It shows the molecules that react together and the resulting product molecules:

$$Cl_2 + H_2O \rightarrow HOCl + HCl$$

A chemical equation tells what elements and compounds are present before and after a chemical reaction. Sulfuric acid poured over zinc will cause the release of hydrogen and the formation of zinc sulfate. This is shown by the following equation:

$$Zn + H_2SO_4 \rightarrow ZnSO_4 + H_2$$

One atom (also one molecule) of zinc unites with one molecule of sulfuric acid to give one molecule of zinc sulfate and one molecule (two atoms) of hydrogen. Notice the same number of atoms of each element on each side of the arrow, even though the atoms are combined differently.

Let's look at another example. When hydrogen gas is burned in air, the oxygen from the air unites with the hydrogen and forms water. The water is the product of burning hydrogen. This can be expressed as an equation:

$$2H_2 + O_2 \rightarrow 2H_2O$$

This equation indicates that two molecules of hydrogen unite with one molecule of oxygen to form two molecules of water.

WATER SOLUTIONS

A *solution* is a condition in which one or more substances are uniformly and evenly mixed or dissolved. A solution has two components, a *solvent* and a *solute*. The solvent is the component that does the dissolving. The solute is the component that is dissolved. In water solutions, water is the solvent. Water can dissolve many other substances; in fact, given enough time, not too many solids, liquids, or gases exist that water cannot dissolve. When water dissolves substances, it creates solutions with many impurities. Generally, a solution is usually transparent and not cloudy; however, a solution may have some color when the solute remains uniformly distributed throughout the solution and does not settle with time.

When molecules dissolve in water, the atoms making up the molecules come apart (dissociate) in the water. This dissociation in water is called *ionization*. When the atoms in the molecules come apart, they do so as charged atoms (both negatively and positively charged) called *ions*. The positively charged ions are called *cations* and the negatively charged ions are called *anions*. A good example of the ionization process is when calcium carbonate ionizes:

$$\underset{\text{Calcium carbonate}}{CaCO_3} \leftrightarrow \underset{\text{Calcium ion (cation)}}{Ca^{2+}} + \underset{\text{Carbonate ion (anion)}}{CO_3^{2-}}$$

Another good example is the ionization that occurs when table salt (sodium chloride) dissolves in water:

$NaCl$	$\leftrightarrow$	Na^+	+	Cl^-
Sodium chloride		Sodium ion (cation)		Chloride ion (anion)

Some of the common ions found in water and their symbols are provided below:

Hydrogen	H^+
Sodium	Na^+
Potassium	K^+
Chloride	Cl^-
Bromide	Br^-
Iodide	I^-
Bicarbonate	HCO^{3-}

Water dissolves polar substances better than nonpolar substances. This makes sense when we consider that water is a polar substance. Polar substances such as mineral acids, bases, and salts are easily dissolved in water, while nonpolar substances such as oils, fats, and many organic compounds do not dissolve easily in water. Water dissolves polar substances better than nonpolar substances, but only to a point; for example, only so much solute will dissolve at a given temperature. When that limit is reached, the resulting solution is saturated. When a solution becomes saturated, no more solute can be dissolved. For solids dissolved in water, if the temperature of the solution is increased, the amount of solids (solutes) required to reach saturation increases.

WATER CONSTITUENTS

Natural water can contain a number of substances (what we may call *impurities*) or constituents in water treatment operations. The concentrations of various substances in water in dissolved, colloidal, or suspended form are typically low but vary considerably. A hardness value of up to 400 ppm of calcium carbonate, for example, is sometimes tolerated in public supplies, whereas 1 ppm of dissolved iron would be unacceptable. When a particular constituent can affect the good health of the water user or the environment, it is considered a *contaminant* or *pollutant*. These contaminants, of course, are what the water operator removes from or tries to prevent from entering the water supply. In this section, we discuss some of the more common constituents of water.

SOLIDS

Other than gases, all contaminants of water contribute to the solids content. Natural water carries many dissolved and undissolved solids. The undissolved solids are nonpolar substances and consist of relatively large particles of materials such as silt, that will not dissolve. Classified by their size and state, by their chemical characteristics, and by their size distribution, solids can be dispersed in water in both suspended and dissolved forms.

The sizes of solids in water can be classified as *suspended*, *settleable*, *colloidal*, or *dissolved*. *Total solids* are the suspended and dissolved solids that remain behind when the water is removed by evaporation. Solids are also characterized as being *volatile* or *nonvolatile*.

The distribution of solids is determined by computing the percentage of filterable solids by size range. Solids typically include inorganic solids such as silt and clay from riverbanks and organic matter such as plant fibers and microorganisms from natural or manmade sources.

Note: Though not technically accurate from a chemical point of view because some finely suspended material can actually pass through the filter, *suspended solids* are defined as those that can be filtered out in the suspended solids laboratory test. The material that passes through the filter is defined as *dissolved solids*.

Colloidal solids are extremely fine suspended solids (particles) less than 1 micron in diameter; they are so small (though they still can make water cloudy) that they will not settle even if allowed to sit quietly for days or weeks.

Turbidity

Simply, turbidity refers to how clear the water is. The clarity of water is one of the first characteristics people notice. Turbidity in water is caused by the presence of suspended matter, which results in the scattering and absorption of light rays. The greater the amount of *total suspended solids* (TSS) in the water, the murkier it appears and the higher the measured turbidity. Thus, in plain English, turbidity is a measure of the light-transmitting properties of water. Natural water that is very clear (low turbidity) allows us to see images at considerable depths. High turbidity water, on the other hand, appears cloudy. Keep in mind that water of low turbidity is not necessarily without dissolved solids. Dissolved solids do not cause light to be scattered or absorbed; thus, the water looks clear. High turbidity causes problems for the waterworks operator, as components that cause high turbidity can cause taste and odor problems and will reduce the effectiveness of disinfection.

Color

Color in water can be caused by a number of contaminants such as iron, which changes in the presence of oxygen to yellow or red sediments. The color of water can be deceiving. In the first place, color is considered an aesthetic quality of water with no direct health impact. Second, many of the colors associated with water are not true colors but the result of colloidal suspension and are referred to as the *apparent color*. This apparent color can often be attributed to iron and to dissolved tannin extracted from decaying plant material. *True color* is the result of dissolved chemicals (most often organics) that cannot be seen. True color is distinguished from apparent color by filtering the sample.

Dissolved Oxygen

Although water molecules contain an oxygen atom, this oxygen is not what is needed by aquatic organism living in our natural waters. A small amount of oxygen, up to about ten molecules of oxygen per million molecules of water, is actually dissolved in water. This dissolved oxygen (DO) is breathed by fish and zooplankton and is needed by them to survive. Other gases can also be dissolved in water. In addition to oxygen, carbon dioxide, hydrogen sulfide, and nitrogen are examples of gases that dissolve in water. Gases dissolved in water are important; for example, carbon dioxide is important because of the role it plays in pH and alkalinity. Carbon dioxide is released into the water by microorganisms and consumed by aquatic plants. Dissolved oxygen in water, however, is of the most importance to us here, not only because it is important to most aquatic organisms but also because dissolved oxygen is an important indicator of water quality.

DID YOU KNOW?

If you fill a glass from your faucet the water will look colorless to you; however, the water isn't really colorless. Even pure water is not colorless but has a slight blue tint to it, best seen when looking through a long column of water.

Like terrestrial life, aquatic organisms need oxygen to live. As water moves past their breathing apparatus, microscopic bubbles of oxygen gas in the water—dissolved oxygen—are transferred from the water to their blood. Like any other gas diffusion process, the transfer is efficient only above certain concentrations. In other words, oxygen can be present in the water but at too low a concentration to sustain aquatic life. Oxygen also is needed by virtually all algae and macrophytes and for many chemical reactions that are important to water body functioning.

Rapidly moving water, such as in a mountain stream or large river, tends to contain a lot of dissolved oxygen, while stagnant water contains little. Bacteria in water can consume oxygen as organic matter decays; thus, excess organic material in our lakes and rivers can cause an oxygen-deficient situation to occur. Aquatic life can have a difficult time surviving in stagnant water that has a lot of rotting, organic material in it, especially in summer, when dissolved oxygen levels are at a seasonal low.

Note: Solutions can become saturated with solute. This is the case with water and oxygen. As with other solutes, the amount of oxygen that can be dissolved at saturation depends on the temperature of the water. In the case of oxygen, the effect is just the opposite of other solutes. The higher the temperature, the lower the saturation level; the lower the temperature, the higher the saturation level.

Metals

Metals are elements present in chemical compounds as positive ions or in the form of cations (+ ions) in solution. Metals with a density over 5 kg/dm^3 are known as *heavy metals*. Metals are one of the constituents or impurities often carried by water. Although most of the metals are not harmful at normal levels, a few metals can cause taste and odor problems in drinking water. In addition, some metals may be toxic to humans, animals, and microorganisms. Most metals enter water as part of compounds that ionize to release the metal as positive ions. Table 4.2 lists some metals commonly found in water and their potential health hazards.

Note: Metals may be found in various chemical and physical forms. These forms, or *species*, can be particles or simple organic compounds, organic complexes, or colloids. The dominating form is determined largely by the chemical composition of the water, the matrix, and in particular the pH.

Organic Matter

Organic matter or compounds are those that contain the element carbon and are derived from material that was once alive (i.e., plants and animals). Organic compounds include fats, dyes, soaps, rubber products, plastics, wood, fuels, cotton, proteins, and carbohydrates. Organic compounds in

TABLE 4.2
Common Metals Found in Water

Metal	Health Hazard
Barium	Circulatory system effects and increased blood pressure
Cadmium	Concentration in the liver, kidneys, pancreas, and thyroid
Copper	Nervous system damage and kidney effects; toxic to humans
Lead	Same as copper
Mercury	Central nervous system disorders
Nickel	Central nervous system disorders
Selenium	Central nervous system disorders
Silver	Gray skin
Zinc	Taste effects; not a health hazard

water are usually large, nonpolar molecules that do not dissolve well in water. They often provide large amounts of energy to animals and microorganisms. *Natural organic matter* (NOM) is used to describe the complex mixture of organic material, such as humic and hydrophilic acids present in all drinking water sources. NOM can cause major problems in the treatment of water as it reacts with chlorine to form *disinfection byproducts* (DBPs). Many of the DBPs formed by the reaction of NOM with disinfectants are reported to be toxic and carcinogenic to humans if ingested over an extended period. The removal of NOM and subsequent reduction in DBPs are major goals in the treatment of any water source.

Inorganic Matter

Inorganic matter or compounds are carbon free, not derived from living matter, and easily dissolved in water; inorganic matter is of mineral origin. The inorganics include acids, bases, oxides, and salts. Several inorganic components are important in establishing and controlling water quality. Two important inorganic constituents in water are nitrogen and phosphorus.

Acids

Lemon juice, vinegar, and sour milk are acidic or contain acid. The common acids used in waterworks operations are hydrochloric acid (HCl), sulfuric acid (H_2SO_4), nitric acid (HNO_3), and carbonic acid (H_2CO_3). Note that in each of these acids, hydrogen (H) is one of the elements.

Note: An acid is a substance that produces hydrogen ions (H^+) when dissolved in water. Hydrogen ions are hydrogen atoms stripped of their electrons. A single hydrogen ion is nothing more than the nucleus of a hydrogen atom.

The relative strengths of acids in water (listed in descending order of strength) are shown in Table 4.3.

Note: Acids and bases become solvated and loosely bond to water molecules.

Bases

A base is a substance that produces hydroxide ions (OH^-) when dissolved in water. Lye, or common soap, contains bases. The bases used in waterworks operations are calcium hydroxide, $Ca(OH)_2$; sodium hydroxide, NaOH; and potassium hydroxide, KOH. Note that the hydroxyl group (OH) is

TABLE 4.3
Relative Strengths of Acids in Water

Acid	Formula
Perchloric acid	$HClO_4$
Sulfuric acid	$H2SO_4$
Hydrochloric acid	HCl
Nitric acid	HNO_3
Phosphoric acid	H_3PO_4
Nitrous acid	HNO_2
Hydrofluoric acid	HF
Acetic acid	CH_3COOH
Carbonic acid	H_2CO_3
Hydrocyanic acid	HCN
Boric acid	H_3BO_3

found in all bases. In addition, note that bases contain metallic substances, such as sodium (Na), calcium (Ca), magnesium (Mg), and potassium (K). These bases contain the elements that produce the alkalinity in water.

Salts

When acids and bases chemically interact, they neutralize each other. The compound (other than water) that forms from the neutralization of acids and bases is called a *salt*. Salts constitute, by far, the largest group of inorganic compounds. If water contains significant amounts (referred to as *concentrations*) of dissolved salts, water is known as *saline water*. The most common salt we all know well is sodium chloride (NaCl). In this case, the concentration is the amount (by weight) of salt in water, as expressed in parts per million (ppm). If water has a concentration of 10,000 ppm of dissolved salts, then 1% (10,000 divided by 1,000,000) of the weight of the water comes from dissolved salts. The parameters for saline water are

- Fresh water—Less than 1000 ppm
- Slightly saline water—1000 to 3000 ppm
- Moderately saline water—3000 to 10,000 ppm
- Highly saline water—10,000 to 35,000 ppm
- Ocean water—About 35,000 ppm salt

When we think of saline water it is only natural for us to think of the oceans, But hundreds of miles from the Pacific Ocean the residents of states such as Colorado and Arizona can enjoy a day at the beach just by walking outside their house, for they may very well be in the vicinity of saline water. An extensive amount of very salty water can be found in the ground in the western United States. In New Mexico, approximately 75% of groundwater is too saline for most uses without treatment (Reynolds, 1962). Water in this area may be leftover from ancient times when saline seas occupied the western United States; also, as rainfall infiltrates downward into the ground, it can encounter rocks that contain highly soluble minerals, which turn the water saline. Groundwater can exist and move for thousands of years and thus become as saline as ocean water. A common salt used in waterworks operations, copper sulfate, is utilized to kill algae in water.

pH

pH is a measure of the hydrogen ion (H^+) concentration. Solutions range from very acidic (having a high concentration of H^+ ions) to very basic (having a high concentration of OH^- ions). The pH scale ranges from 0 to 14, with 7 being the neutral value (see Figure 4.1). The pH of water is important

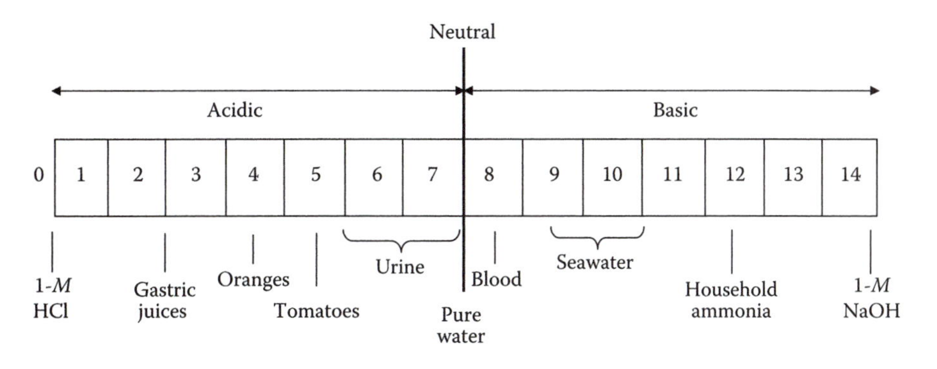

FIGURE 4.1 The pH of selected liquids.

to the chemical reactions that take place within water, and pH values that are too high or low can inhibit the growth of microorganisms. High pH values are considered basic, and low pH values are considered acidic. Stated another way, low pH values indicate a high H^+ concentration, and high pH values indicate a low H^+ concentration. Because of this inverse logarithmic relationship, each pH unit represents a tenfold difference in H^+ concentration.

Natural water varies in pH depending on its source. Pure water has a neutral pH, with equal H^+ and OH^-. Adding an acid to water causes additional positive ions to be released, so the H^+ ion concentration goes up and the pH value goes down:

$$HCl \leftrightarrow H^+ + Cl^-$$

To control water coagulation and corrosion, the waterworks operator must test for the hydrogen ion concentration of the water to determine the pH of the water. In a coagulation test, as more alum (acid) is added, the pH value lowers. If more lime (alkali) is added, the pH value raises. This relationship should be remembered—if a good floc is formed, the pH should then be determined and maintained at that pH value until the raw water changes.

Pollution can change the pH of water, which in turn can harm animals and plants living in the water. Water coming out of an abandoned coal mine, for example, can have a pH of 2, which is very acidic and would definitely affect any fish crazy enough to try to live in that water. By using the logarithm scale, this mine-drainage water would be 100,000 times more acidic than neutral water—so stay out of abandoned mines.

The pH of precipitation and water bodies varies widely across the United States. Natural and human processes determine the pH of water. The National Atmospheric Deposition Program has developed maps of pH patterns (http://nadp.sws.uiuc.edu/ntn/annualmapsbyanalyte.aspx).

> ***Note:*** Seawater is slightly more basic (the pH value is higher) than most natural freshwater. Neutral water (such as distilled water) has a pH of 7, which is in the middle of being acidic and alkaline. Seawater happens to be slightly alkaline (basic), with a pH of about 8. Most natural water has a pH range of 6 to 8, although acid rain can have a pH as low as 4.

OPTICAL PHENOMENA: WATER AND LIGHT

Have you ever looked up into the sky and seen eleven suns? Have you been at sea and witnessed the towering, spectacular Fata Morgana? (Do you even know what Fata Morgana is? If not, hold on; we'll get to it shortly.) How about a "glory"—have you ever seen one? Or how about an albino rainbow—have you seen one lately? Or a regular rainbow or a sun dog—have you searched for that pot of gold at the end of the rainbow, or wondered where all the color in rainbows and sun dogs comes from? Do you know what these are or how they are formed?

Normally, when we look up into the sky, we see what we expect to see: an ever-changing backdrop of color, with dynamic vistas of blue sky; white, puffy clouds; gray storms; or gold and red sunsets. Sometimes, though, when atmospheric conditions are just right, we can look up at the sky or out over the horizon and see strange phenomena such as those mentioned above. What causes these momentary wonders?

Because Earth's atmosphere is composed of gases (air) it is actually a sea of molecules. These molecules of air scatter the shorter blue, indigo, and violet wavelengths of light more than the longer orange and red wavelengths, which is why the sky appears blue. What are wavelengths of light? Simply put, a wavelength of light actually refers to the electromagnetic spectrum. The portion of the spectrum visible to the human eye falls between the infrared and ultraviolet wavelengths. The colors that make up the visible portion of the electromagnetic spectrum are commonly abbreviated by the acronym ROY G BIV (red, orange, yellow, green, blue, indigo, and violet). The word *light* is commonly given to visible electromagnetic radiation, but only the frequency (or wavelength) distinguishes visible electromagnetic radiation from the other portions of the spectrum.

DID YOU KNOW?

Similar to a glory, a *heiligenschein* ("holy light") is a bright halo of light that appears around the shadow of the observer's head. As with a glory, the observer sees only the light surrounding his or her own head, not around those of any companions.

Let's get back to looking up at the sky. Have you ever noticed right after a rain shower how dark a shade of blue the sky appears to be? Have you looked out over the horizon at night or in the morning and noticed that the sun's light gives the sky a red appearance? This phenomenon is caused by sunlight passing through large dust particles, which scatter the longer wavelengths. Have you ever noticed that fog and cloud droplets, with diameters larger than the wavelength of light, scatter all colors equally and make the sky look white? Maybe you have noticed a fleeting greenish light that appears just as the sun sets? It occurs because different wavelengths of light are *refracted* (bent) in the atmosphere by differing amounts. Because green light is refracted more than red light by the atmosphere, green is the last to disappear.

What causes *rainbows*? A rainbow is really nothing more than an airborne prism. When sunlight enters a raindrop, refraction and reflection take place, splitting white light into the spectrum of colors from red to blue and making a rainbow. Because the light is reflected inside the raindrops, rainbows appear on the opposite side of the sky as the sun.

What is a *sun dog*? Sun dogs (i.e., parhelions) are a lot more common than rainbows. They are formed by light refracting through the ice crystals of cirriform clouds (thin, wispy strands). The most common sun dogs are seen about 22° on either side of the sun; they are created by falling ice crystals in the Earth's atmosphere. As water freezes in the atmosphere, small, flat, six-sided ice crystals can form. As these crystals flutter to the ground, much time is spent with their faces flat, parallel to the ground.

Interactions of light waves can produce a *glory*, another optical phenomenon. A glory often appears as an iconic saint's halo (a series of colored rings) about the head of the observer and is produced by a combination of diffraction, reflection, and refraction (backscattered light) toward its source by a cloud of uniformly sized water droplets. For example, if you were standing on a mountain, with the sun to your back, you may cast a shadow on the fog in the valley. Your shadow may appear to be surrounded by colored halos—a glory. The glory is caused by light entering the edges of tiny droplets and being returned in the same direction from which it arrived. These light waves interfere with each other, sometimes canceling out and sometimes adding to each other.

The sighting of well-defined glories at Gausta Mountain, Norway, in 1893 was reported in *Nature* (Wille, 1893): "We mounted to the flagstaff in order to obtain a better view of the scenery, and there we at once observed in the fog, in an easterly direction, a double rainbow forming a complete circle and seeming to be 20 to 30 feet distant from us. In the middle of this we all appeared as black, erect, and nearly life-size silhouettes. The outlines of the silhouettes were so sharp that we could easily recognise the figures of each other, and every movement was reproduced. The head of each individual appeared to occupy the centre of the circle, and each of us seemed to be standing on the inner periphery of the rainbow."

Why do we sometimes see multiple suns? Reflection and refraction of light by ice crystals can create bright halos in the form of arcs, rings, spots, and pillars. Mock suns (sun dogs) may appear as bright spots 22 or 46° to the left or right of the sun. A one-sided mock sun was observed by several passengers aboard the ship *Fairstar* in the Sea of Timor on June 15, 1965. "About 10 minutes before sunset, Mrs. N.S. noticed to her surprise a second sun, much less bright than the real one, somewhat to the left, but at the same height above the horizon. There was nothing at the symmetrical point to the right" (Corliss, 1996). Sun pillars occur when ice crystals act as mirrors, creating a bright column of light extending above the sun. Such a pillar of bright light may be visible even when the sun has set.

What is the *Fata Morgana*? It is an illusion, a mirage, which often fools sailors into seeing mountain ranges floating over the surface of the ocean. Henry Wadsworth Longfellow had his own take on Fata Morgana, which basically explains the essence of the phenomenon:

Fata Morgana

O sweet illusions of song
That tempt me everywhere,
In the lonely fields, and the throng
Of the crowded thoroughfare!
I approach and ye vanish away,
I grasp you, and ye are gone;
But over by night and by day,
The melody soundeth on.
As the weary traveler sees
In desert or prairie vast,
Blue lakes, overhung with trees
That a pleasant shadow cast;
Fair towns with turrets high,
And shining roofs of gold,
That vanish as he draws nigh,
Like mists together rolled—
So I wander and wander along,
And forever before me gleams
The shining city of song,
In the beautiful land of dreams.
But when I would enter the gate
Of that golden atmosphere,
It is gone, and I wonder and wait
For the vision to reappear.

The flagship account of the Fata Morgana comes to us from friar Antonio Minasi who described a Fata Morgana seen across the Straits of Messina in 1773. "In Italian legend, Morgan Le Fay, or, in Italian, Fata Morgan, falls in love with a mortal youth and gives him the gift of eternal life in return for her love; when he becomes restless and bored with captivity, she summons up fairy spectacles for his entertainment" (Minasi, 1773).

An *albino rainbow* (white rainbow or fogbow) is an eerie phenomenon that can only be seen on rare occasions in foggy conditions. It forms when the sun or moon shines on minute droplets of water suspended in the air. The fog droplets are so small that the usual prismatic colors of the rainbow merge together to form a white arc opposite the sun or moon (Corliss, 1996).

COMMON WATER MEASUREMENTS

Water and wastewater practitioners and regulators, such as waterworks operators and the U.S. Environmental Protection Agency (USEPA), along with their scientific counterparts at the U.S. Geological Survey (USGS), have been measuring water for decades. Millions of measurements and analyses have been made. Some measurements are taken almost every time water is sampled and investigated, no matter where in the United States the water is being studied. Even these simple measurements can sometimes reveal something important about the water and the environment around it.

The results of a single measurement of the properties of water are actually less important than looking at how those properties vary over time (USGS, 2006). Suppose we take the pH of the river running through our town and find that it is 5.5. We might say, "Wow, the water is acidic!" But, a pH of 5.5 might be normal for that particular river. It is similar to how an adult's normal body

DID YOU KNOW?

In common water measurement lab operations, to perform accurate test-tube readings the water practitioner must be familiar with the *meniscus*, which is a curve in the surface of a molecular substance (water) when it touches another material. With water, you can think of it as when water sticks to the inside of a glass.

temperature is about 97.5°, but a youngster's normal temperature is *really* normal—right on the 98.6 mark. As with our temperatures, if the pH of a river begins to change, then we might suspect that something is going on somewhere that is affecting the water and possibly the water quality. For this reason, changes in water measurements are more important than the actual measured values.

To this point, the important constituents and parameters of turbidity, dissolved oxygen, pH, and others have been discussed, but there are others we should address. In the following, the parameters of alkalinity, water temperature, specific conductance, and hardness are discussed.

Alkalinity

Alkalinity is defined as the capacity of water to accept protons; it can also be defined as a measure of the ability of water to neutralize an acid. Bicarbonates, carbonates, and hydrogen ions cause alkalinity and create hydrogen compounds in a raw or treated water supply. Bicarbonates are the major components because of carbon dioxide action on basic materials of soil; borates, silicates, and phosphates may be minor components. The alkalinity of raw water may also contain salts formed from organic acids such as humic acids.

Alkalinity in water acts as a buffer that tends to stabilize and prevent fluctuations in pH. In fact, alkalinity is closely related to pH, but the two must not be confused. Total alkalinity is a measure of the amount of alkaline materials in the water. The alkaline materials act as buffers to changes in the pH. If the alkalinity to too low (below 80 ppm), the pH can fluctuate rapidly because of insufficient buffer. High alkalinity (above 200 ppm) results in the water being too buffered. Thus, having significant alkalinity in water is usually beneficial, because it tends to prevent quick changes in pH that interfere with the effectiveness of common water treatment processes. Low alkalinity also contributes to the corrosive tendencies of water.

Note: When alkalinity is below 80 mg/L, it is considered to be low.

Water Temperature

Water temperature is important not only to fisherman but also to industries and even fish and algae. A lot of water is used for cooling purposes in power plants that generate electricity. These plants need to cool the water to begin with and then generally release warmer water back to the environment. The temperature of the released water can affect downstream habitats. Temperature can also affect the ability of water to hold oxygen as well as the ability of organisms to resist certain pollutants.

Specific Conductance

Specific conductance is a measure of the ability of water to conduct an electrical current. It is highly dependent on the amount of dissolved solids (such as salt) in the water. Pure water, such as distilled water, will have a very low specific conductance, and seawater will have a high specific conductance. Rainwater often dissolves airborne gases and airborne dust while it is in the air and thus often has a higher specific conductance than distilled water. Specific conductance is an important water

TABLE 4.4
Water Hardness

Water Hardness Classification	mg/L $CaCo_3$
Soft	0–75
Moderately hard	75–150
Hard	150–300
Very hard	Over 300

quality measurement because it gives a good idea of the amount of dissolved material in the water. When electrical wires are attached to a battery and light bulb and the wires are put into a beaker of distilled water, the light will not light. But, the bulb does light up when the beaker contains saline (saltwater). In saline water, the salt has dissolved and released free electrons, so the water will conduct an electric current.

Hardness

Hardness may be considered a physical or chemical characteristic or parameter of water. It represents the total concentration of calcium and magnesium ions, reported as calcium carbonate. Simply, the amount of dissolved calcium and magnesium in water determines its hardness. Hardness causes soaps and detergents to be less effective and contributes to scale formation in pipes and boilers. Hardness is not considered a health hazard; however, water that contains hardness must often be softened by lime precipitation or ion exchange. Hardwater can even shorten the life of fabrics and clothes. Low hardness contributes to the corrosive tendencies of water. Hardness and alkalinity often occur together, because some compounds can contribute both alkalinity and hardness ions. Hardness is generally classified as shown in Table 4.4.

WATER TREATMENT CHEMICALS

To operate a water treatment process correctly and safely, water operators need to know the types of chemical used in the processes, the purpose of each, and the safety precautions required for the use of each. This section briefly discusses chemicals used in

- Disinfection (also used in wastewater treatment)
- Coagulation
- Taste and odor removal
- Water softening
- Recarbonation
- Ion exchange softening
- Scale and corrosion control

Disinfection

In water practice, disinfection is often accomplished using chemicals. The purpose of disinfection is to selectively destroy disease-causing organisms. Chemicals commonly used in disinfection include chlorine and its compounds (most widely used), ozone, bromide, iodine, and hydrogen peroxide. Many factors must be considered when choosing the type of chemical to be used for disinfection, such as contact time, intensity and nature of the physical agent, temperature, and type and number of organisms.

Coagulation

Chemical coagulation conditions water for further treatment by the removal of

- Turbidity, color, and bacteria
- Iron and manganese
- Tastes, odors, and organic pollutants

In water treatment, normal sedimentation processes do not always settle out particles efficiently. This is especially the case when attempting to remove particles less than 50 μm in diameter.

In some instances, it is possible to agglomerate (to make or form into a rounded mass) particles into masses or groups. These rounded masses are of increased size and therefore increased settling velocities. For colloidal-sized particles, however, agglomeration is difficult because the colloidal particles are difficult to clarify without special treatment. Chemical coagulation is usually accomplished by adding metallic salts such as aluminum sulfate (alum) or ferric chloride. Alum is the most commonly used coagulant in water treatment and is most effective between pH ranges of 5.0 and 7.5. Sometimes polymer is added to alum to help form small floc together for faster settling. Ferric chloride, effective down to a pH of 4.5, is sometimes used.

In addition to pH, a variety of other factors influence the chemical coagulation process:

- Temperature
- Influent quality
- Alkalinity
- Type and amount of coagulant used
- Type and length of flocculation
- Type and length of mixing

Taste and Odor Removal

Although odor can be a problem with wastewater treatment, the taste and odor parameters are primarily associated with potable water. Either organic or inorganic materials may produce tastes and odors in water. The perceptions of taste and odor are closely related and often confused by water practitioners as well as by consumers; thus, it is difficult to precisely measure either one. Experience has shown that a substance that produces an odor in water almost invariably imparts a perception of taste as well; however, taste is generally attributed to mineral substances present in the water, but most of these minerals do not cause odors.

Along with the impact minerals can have on water taste, other substances can affect both water tastes and odors (e.g., metals, salts from the soil, constituents of wastewater, end products generated from biological reactions). When water has a distinct taste but no odor, the taste might be the result of inorganic substances. Anyone who has tasted alkaline water knows its biting bitterness. Salts not only give water a salty taste but also contribute to a bitter taste. In addition to natural causes, water can take on a distinctive color or taste, or both, from human contamination of the water. Organic materials can produce both taste and odor in water. Petroleum-based products are probably the prime contributors to both of these problems in water. Biological degradation or decomposition of organics in surface waters also contributes to both taste and odor problems in water. Algae are another problem. Certain species of algae produce oily substances that may result in both altered taste and an odor. Synergy can also work to produce taste and odor problems in water. Mixing water and chlorine is one example.

With regard to chemically treating water for odor and taste problems, oxidants such as chlorine, chlorine dioxide, ozone, and potassium permanganate can be used. These chemicals are especially effective when water is associated with an earthy or musty odor caused by the nonvolatile metabolic products of actinomycetes and blue–green algae. Tastes and odors associated with dissolved gases and some volatile organic materials are normally removed by oxygen in aeration processes.

Water Softening

The reduction of hardness, or softening, is a process commonly practiced in water treatment. Chemical precipitation and ion exchange are the two softening processes most commonly used. Softening of hardwater is desired (for domestic users) to reduce the amount of soap used, increase the life of water heaters, and reduce encrustation of pipes (cementing together the individual filter media grains).

In chemical precipitation, it is necessary to adjust pH. To precipitate the two ions most commonly associated with hardness in water, calcium (Ca^{2+}) and magnesium (Mg^{2+}), the pH must be raised to about 9.4 for calcium and about 10.6 for magnesium. To raise the pH to the required levels, lime is added. Chemical precipitation is accomplished by converting calcium hardness to calcium carbonate and magnesium hardness to magnesium hydroxide. This is normally accomplished by using the lime–soda ash or caustic soda processes. The lime–soda ash process reduces the total mineral content of the water, removes suspended solids, removes iron and manganese, and reduces color and bacterial numbers. The process, however, has a few disadvantages; for example, the process produces large quantities of sludge, requires careful operation, and, as stated earlier, if the pH is not properly adjusted may create operational problems downstream of the process (McGhee, 1991).

In the caustic soda process, the caustic soda reacts with the alkalinity to produce carbonate ions for reduction with calcium. The process works to precipitate calcium carbonate in a fluidized bed of sand grains, steel grit, marble chips, or some other similar dense material. As particles grow in size by deposition of $CaCO_3$, they migrate to the bottom of the fluidized bed from which they are removed. This process has the advantages of requiring short detention times (about 8 seconds) and producing no sludge.

Recarbonation

Recarbonation (stabilization) is the adjustment of the ionic condition of water so it will neither corrode pipes nor deposit calcium carbonate, which produces an encrusting film. During or after the lime–soda ash softening process, this recarbonation is accomplished through the reintroduction of carbon dioxide into the water. Lime softening of hardwater supersaturates the water with calcium carbonate and may produce a pH of greater than 10. Because of this, pressurized carbon dioxide is bubbled into the water, lowering the pH and removing calcium carbonate. The high pH can also create a bitter taste in drinking water. Recarbonation removes this bitterness.

Ion Exchange Softening

Hardness can be removed by ion exchange. In water softening, ion exchange replaces calcium and magnesium with a nonhardness cation, usually sodium. Calcium and magnesium in solution are removed by interchange with sodium within a solids interface (matrix) through which the flow is passed. Similar to the filter, the ion exchanger contains a bed of granular material, a flow distributor, and an effluent vessel that collects the product. The exchange media include greensand (a sand or sediment given a dark greenish color by grains of glauconite), aluminum silicates, synthetic siliceous gels, bentonite clay, sulfonated coal, and synthetic organic resins and are generally in particle form, usually ranging up to a diameter of 0.5 mm. Modern applications more often employ artificial organic resins. These clear, BB-sized resins are sphere shaped and have the advantage of providing a greater number of exchange sites. Each of these resin spheres contains sodium ions, which are released into the water in exchange for calcium and magnesium. As long as exchange sites are available, the reaction is virtually instantaneous and complete.

When all of the exchange sites have been utilized, hardness begins to appear in the influent, a process known as *breakthrough*. Breakthrough requires the regeneration of the medium by bringing it into contact with a concentrated solution of sodium chloride.

DID YOU KNOW?

Water is essentially incompressible, especially under normal conditions. If you fill a plastic bag with water and put a piece of hose into it, when you squeeze the bag the water won't compress but rather will shoot out the hose. If the water compressed, it would not push back out of the hose. Incompressibility is a common property of liquids, but water is especially incompressible.

Ion exchange used in water softening has both advantages and disadvantages. One of its major advantages is that it produces a softer water than does chemical precipitation. Additionally, ion exchange does not produce the large quantity of sludge encountered in the lime–soda process. One disadvantage is that, although it does not produce sludge, ion exchange does produce concentrated brine. Moreover, the water must be free of turbidity and particulate matter or the resin might function as a filter and become plugged.

Scaling and Corrosion Control

Controlling scale and corrosion is important in water systems. Carbonate and noncarbonate hardness constituents in water cause *scale*, a chalky-white deposit frequently found on teakettle bottoms. When controlled, this scale can be beneficial, forming a protective coating inside tanks and pipelines. A problem arises when scale is not controlled. Excessive scaling reduces the capacity of pipelines and the efficiency of heat transfer in boilers. *Corrosion* is the oxidation of unprotected metal surfaces. Of particular concern, in water treatment, is the corrosion of iron and its alloys (i.e., the formation of rust). Several factors contribute to the corrosion of iron and steel. Alkalinity, pH, DO, and carbon dioxide can all cause corrosion. Along with the corrosion potential of these chemicals, their corrosive tendencies are significantly increased when water temperature and flow are increased.

CHEMICAL DRINKING WATER PARAMETERS

> Water, in any of its forms, also ... [has] scant respect for the laws of chemistry.
>
> Most materials act either as acids or bases, settling on either side of a natural reactive divided. Not water. It is one of the few substances that can behave both as an acid and as a base, so that under certain conditions it is capable of reacting chemically with itself. Or with anything else.
>
> Molecules of water are off balance and hard to satisfy. They reach out to interfere with every other molecule they meet, pushing its atoms apart, surrounding them, and putting them into solution. Water is the ultimate solvent, wetting everything, setting other elements free from the rocks, making them available for life. Nothing is safe. There isn't a container strong enough to hold it.
>
> **Watson (1988)**

Water chemical parameters are categorized into two basic groups: *inorganic* and *organic*. Both groups enter water from natural causes or pollution.

Note: The solvent capabilities of water are directly related to its chemical parameters.

In this section, we do not look at each organic or inorganic chemical individually; instead, we look at general chemical parameter categories such as dissolved oxygen organics (BOD and COD), dissolved oxygen (DO), synthetic organic chemicals (SOCs), volatile organic compounds (VOCs), total dissolved solids (TDS), fluorides, metals, and nutrients—the major chemical parameters of concern.

Organics

Natural organics contain carbon and consist of biodegradable organic matter such as wastes from biological material processing, human sewage, and animal feces. Microbes aerobically break down complex organic molecules into simpler, more stable end products. Microbial degradation end products include carbon dioxide, water, phosphate, and nitrate. Organic particles in water may harbor harmful bacteria and pathogens. Infection by microorganisms may occur if the water is used for primary contact or as a raw drinking water source. Treated drinking water will not present the same health risks. In a potable drinking water plant, all organics should be removed in the water before disinfection.

Organic chemicals also contain carbon; they are substances that come directly from, or are manufactured from, plant or animal matter. Plastics provide a good example of organic chemicals that are made from petroleum, which originally came from plant and animal matter. Some organic chemicals (such as those discussed above) released by decaying vegetation occur naturally and by themselves tend not to pose health problems when they get in our drinking water; however, more serious problems are caused by the more than 100,000 different manufactured or synthetic organic chemicals in commercial use today. These include paints, herbicides, synthetic fertilizers, pesticides, fuels, plastics, dyes, preservatives, flavorings, and pharmaceuticals, to name a few.

Many organic materials are soluble in water and are toxic, and many of them are found in public water supplies. The presence of organic matter in water is troublesome, as organic matter causes (1) color formation, (2) taste and odor problems, (3) oxygen depletion in streams, (4) interference with water treatment processes, and (5) the formation of halogenated compounds when chlorine is added to disinfect water (Tchobanoglous and Schroeder, 1987). Remember, organics in natural water systems may come from natural sources or may result from human activities. Generally, the principle source of organic matter in water is from natural sources including decaying leaves, weeds, and trees; the amount of these materials present in natural waters is usually low. *Anthropogenic* (manmade) sources of organic substances come from pesticides and other synthetic organic compounds.

Again, many organic compounds are soluble in water, and surface waters are more prone to contamination by natural organic compounds than are groundwaters. In water, dissolved organics are usually divided into two categories: *biodegradable* and *nonbiodegradable*. Material that is biodegradable (capable of breaking down) consists of organics that can be used for food (nutrients) by naturally occurring microorganisms within a reasonable length of time. Alcohols, acids, starches, fats, proteins, esters, and aldehydes are the main constituents of biodegradable materials. They may result from domestic or industrial wastewater discharges, or they may be end products of the initial microbial decomposition of plant or animal tissue. Biodegradable organics in surface waters cause problems mainly associated with the effects that result from the action of microorganisms. As the microbes metabolize organic material, they consume oxygen.

When this process occurs in water, the oxygen consumed is dissolved oxygen (DO). If the oxygen is not continually replaced in the water by artificial means, the DO level will decrease as the organics are decomposed by the microbes. This need for oxygen is the *biochemical oxygen demand* (BOD), which is the amount of dissolved oxygen demanded by bacteria to break down the organic materials during the stabilization action of the decomposable organic matter under aerobic conditions over a 5-day incubation period at 20°C (68°F). This bioassay test measures the oxygen consumed by living organisms using the organic matter contained in the sample and dissolved oxygen in the liquid. The organics are broken down into simpler compounds, and the microbes use the energy released for growth and reproduction. A BOD test is not required for monitoring drinking water.

> ***Note:*** The more organic material in the water, the higher the BOD exerted by the microbes will be. Some biodegradable organics can cause color, taste, and odor problems.

Nonbiodegradable organics are resistant to biological degradation. The constituents of woody plants are a good example. These constituents, including tannin and lignic acids, phenols, and cellulose, are found in natural water systems and are considered *refractory* (resistant to biodegradation). Some polysaccharides with exceptionally strong bonds and benzene (associated, for example, with the refining of petroleum) with its ringed structure are essentially nonbiodegradable. Certain nonbiodegradable chemicals can react with oxygen dissolved in water. The *chemical oxygen demand* (COD) is a more complete and accurate measurement of the total depletion of dissolved oxygen in water. *Standard Methods* (Greenberg et al., 1999) defines COD as a test that provides a measure of the oxygen equivalent of that portion of the organic matter in a sample that is susceptible to oxidation by a strong chemical oxidant. The procedure is detailed in *Standard Methods*.

Note: COD is not normally used for monitoring water supplies but is often used for evaluating contaminated raw water.

Synthetic Organic Chemicals

Synthetic organic chemicals (SOCs) are manmade, and because they do not occur naturally in the environment, they are often toxic to humans. More than 50,000 SOCs are in commercial production, including common pesticides, carbon tetrachloride, chloride, dioxin, xylene, phenols, aldicarb, and thousands of other synthetic chemicals. Unfortunately, even though they are so prevalent, few data have been collected on these toxic substances. Determining definitively just how dangerous many of the SOCs are is rather difficult.

Volatile Organic Compounds

Volatile organic compounds (VOCs) are a type of organic chemical that is particularly dangerous. VOCs are absorbed through the skin during contact with water—as in the shower or bath. Hot-water exposure allows these chemicals to evaporate rapidly, and they are harmful if inhaled. VOCs can be in any tap water, regardless of where in the country one lives and the water supply source.

Total Dissolved Solids

Earlier we pointed out that solids in water occur either in solution or in suspension and are distinguished by passing the water sample through a glass-fiber filter. By definition, the suspended solids are retained on top of the filter, and the dissolved solids pass through the filter with the water. When the filtered portion of the water sample is placed in a small dish and then evaporated, the solids in the water remain as residue in the evaporating dish. This material represents the total dissolved solids (TDS). Dissolved solids may be organic or inorganic. Water may come into contact with these substances within the soil, on surfaces, and in the atmosphere. The organic dissolved constituents of water can be the decay products of vegetation, organic chemicals, or organic gases. Removing these dissolved minerals, gases, and organic constituents is desirable, because they may cause physiological effects and produce aesthetically displeasing color, taste, and odors.

Note: In water distribution systems, a high TDS means high conductivity with consequent higher ionization during corrosion control efforts; however, high TDS also suggests a greater likelihood of a protective coating, a positive factor in corrosion control.

Fluorides

Water fluoridation may not be the safe public health measure we have been led to believe. Concerns about uncontrolled dosage, accumulation in the body over time, and effects beyond those on the teeth (e.g., brain as well as bones) have not been resolved for fluoride. The health of citizens requires that all the facts be considered, not just those that are politically expedient (Mullenix, 1997).

Medical authorities might take issue with Dr. Mullenix's view on the efficacy of fluoride in reducing tooth decay, as most seem to hold that a moderate amount of fluoride ions (F^-) in drinking water contributes to good dental health. Fluoride is seldom found in appreciable quantities in surface waters and appears in groundwater in only a few geographical regions, although it is sometimes found in a few types of igneous or sedimentary rocks. Fluoride is toxic to humans in large quantities (the key words are "large quantities" or, in Dr. Mullenix's view, "uncontrolled dosages") and is also toxic to some animals.

Fluoride used in small concentrations (about 1.0 mg/L in drinking water) can be beneficial. Experience has shown that drinking water containing a proper amount of fluoride can reduce tooth decay by 65% in children between the ages 12 and 15. When the concentration of fluorides in untreated natural water supplies is excessive, however, either alternative water supplies must be used or treatment to reduce the fluoride concentration must be applied, because excessive amounts of fluoride can cause mottled or discolored teeth, a condition known as *dental fluorosis*.

Heavy Metals

Heavy metals are elements with atomic weights between 63.5 and 200.5, and a specific gravity greater than 4.0. Living organisms require trace amounts of some heavy metals, including cobalt, copper, iron, manganese, molybdenum, vanadium, strontium, and zinc. Excessive levels of essential metals, however, can be detrimental to the organism. Nonessential heavy metals of particular concern to surface water systems are cadmium, chromium, mercury, lead, arsenic, and antimony.

Heavy metals in water are classified as either *nontoxic* or *toxic*. Only those metals that are harmful in relatively small amounts are labeled toxic; other metals fall into the nontoxic group. In natural waters (other than in groundwaters), sources of metals include dissolution from natural deposits and discharges of domestic, agricultural, or industrial wastes.

All heavy metals exist in surface waters in colloidal, particulate, and dissolved phases, although dissolved concentrations are generally low. The colloidal and particulate metal may be found (1) in hydroxides, oxides, silicates, or sulfides, or (2) adsorbed to clay, silica, or organic matter. The soluble forms are generally ions or unionized organometallic chelates or complexes. The solubility of trace metals in surface waters is predominantly controlled by water pH, the type and concentration of liquids on which the metal could adsorb, the oxidation state of the mineral components, and the redox environment of the system. The behavior of metals in natural waters is a function of the substrate sediment composition, the suspended sediment composition, and the water chemistry. Sediment composed of fine sand and silt will generally have higher levels of adsorbed metal than will quartz, feldspar, and detrital carbonate-rich sediment.

The water chemistry of the system controls the rate of adsorption and desorption of metals to and from sediment. Adsorption removes the metal from the water column and stores the metal in the substrate. Desorption returns the metal to the water column, where recirculation and bioassimilation may take place. Metals may be desorbed from the sediment if the water experiences increases in salinity, decreases in redox potential, or decreases in pH.

Although heavy metals such as iron (Fe) and manganese (Mn) do not cause health problems, they do impart a noticeable bitter taste to drinking water, even at very low concentrations. These metals usually occur in groundwater in solution, and these and others may cause brown or black stains on laundry and on plumbing fixtures.

Nutrients

Elements in water (such as carbon, nitrogen, phosphorus, sulfur, calcium, iron, potassium, manganese, cobalt, and boron—all essential to the growth and reproduction of plants and animals) are called *nutrients* (or biostimulants). The two nutrients that concern us in this text are nitrogen and phosphorus. Nitrogen (N_2), an extremely stable gas, is the primary component of the Earth's atmosphere

(78%). The nitrogen cycle is composed of four processes. Three of the processes—fixation, ammonification, and nitrification—convert gaseous nitrogen into usable chemical forms. Denitrification, the fourth process, converts fixed nitrogen back to the unusable gaseous nitrogen state.

Nitrogen occurs in many forms in the environment and takes part in many biochemical reactions. Major sources of nitrogen include runoff from animal feedlots, fertilizer runoff from agricultural fields, municipal wastewater discharges, and certain bacteria and blue–green algae that obtain nitrogen directly from the atmosphere. Certain forms of acid rain can also contribute nitrogen to surface waters.

Nitrogen in water is commonly found in the form of *nitrate* (NO_3), which indicates that the water may be contaminated with sewage. Nitrates can also enter the groundwater from chemical fertilizers used in agricultural areas. Excessive nitrate concentrations in drinking water pose an immediate health threat to infants, both human and animal, and can cause death. The bacteria commonly found in the intestinal tract of infants can convert nitrate to highly toxic nitrites (NO_2). Nitrite can replace oxygen in the bloodstream, resulting in oxygen starvation, which causes a bluish discoloration of the infant ("blue baby" syndrome).

Note: Lakes and reservoirs usually have less than 2 mg/L of nitrate measured as nitrogen. Higher nitrate levels are found in groundwater ranging up to 20 mg/L, but much higher values are detected in shallow aquifers polluted by sewage or excessive use of fertilizers.

Phosphorus (P) is an essential nutrient that contributes to the growth of algae and the eutrophication of lakes, although its presence in drinking water has little effect on health. In aquatic environments, phosphorus is found in the form of phosphate and is a limiting nutrient. If all phosphorus is used, plant growth ceases, no matter the amount of nitrogen available. Many bodies of freshwater currently experience influxes of nitrogen and phosphorus from outside sources. The increasing concentration of available phosphorus allows plants to assimilate more nitrogen before the phosphorus is depleted. If sufficient phosphorus is available, high concentrations of nitrates will lead to phytoplankton (algae) and macrophyte (aquatic plant) production. Major sources of phosphorus include phosphates in detergents, fertilizer and feedlot runoff, and municipal wastewater discharges. The USEPA 1976 Water Quality Standards, Criteria Summaries for Phosphorus, recommended a criterion of 0.10 µg/L (elemental) phosphorus for marine and estuarine waters but offered no freshwater criterion.

The biological, physical, and chemical condition of our water is of enormous concern to us all, because we must live in such intimate contact with water. When these parameters shift and change, the changes affect us, often in ways science cannot yet define for us. Water pollution is an external element that can and does significantly affect our water. But, what exactly is water pollution? We quickly learn that water pollution does not always go straight from source to water. Controlling what goes into our water is difficult, because the hydrologic cycle carries water (and whatever it picks up along the way) through all of our environment's media, affecting the biological, physical, and chemical condition of the water we must drink to live.

THOUGHT-PROVOKING QUESTION

4.1 Is desalinization of saline water the answer to solving the potable water shortage problem? Explain.

REFERENCES AND RECOMMENDED READING

AMS. (2001). *Glossary*. Geneseo, NY: American Meteor Society (http://www.amsmeteors.org/resources/glossary/).

Anon. (2010). Large swaths of Earth drying up, study suggests. *LiveScience*, October 11 (http://www.livescience.com/8755-large-swaths-earth-drying-study-suggests.html).

Corliss, W.R. (1996). *Handbook of Unusual Natural Phenomena*. New York: Random House.

Greenberg, A.E. et al., Eds. (1999). *Standard Methods for Examination of Water and Wastewater*, 20th ed. Washington, DC: American Public Health Association.

Heinlein, R.A. (1973). *Time Enough for Love*. New York: G.P. Putnam's Sons.

Hesketh, H.E. (1991). *Air Pollution Control: Traditional and Hazardous Pollutants*. Lancaster, PA: Technomic.

Hodanbosi, C. (1996). *Pascal's Principle and Hydraulics*. Washington, DC: National Aeronautics and Space Administration (ww.grc.nasa.gov/WWW/k-12/WindTunnel/Activities/Pascals_principle.html).

Jost, N.J. (1992). Surface and ground water pollution control technology. In *Fundamentals of Environmental Science and Technology*, Knowles, P.-C., Ed. Rockville, MD: Government Institutes.

McGhee, T.J. (1991). *Water Supply and Sewerage*, 6th ed. New York: McGraw-Hill.

Metcalf & Eddy, Inc. (1991). *Wastewater Engineering: Treatment, Disposal, Reuse*, 3rd ed. New York: McGraw-Hill.

Minasi, A. (1773). *Dissertazioni*. Rome.

Mullenix, P.J. (1997). Letter to the Operations and Environmental Committee, City of Calgary, Canada.

NASA. (2005). *What's the Difference Between Weather and Climate?* Washington, DC: National Aeronautics and Space Administration (http://www.nasa.gov/mission_pages/noaa-n/climate/climate_weather.html).

NAS. (1962). *Water Balance in the United States*, Research Council Publ. 100-B. Washington, DC: National Academy of Sciences.

Reynolds, S.E. (1962). *Twenty-Fifth Biennial Report of the State Engineer of New Mexico for the 49th and 50th Fiscal Years, July 1, 1960, to June 30, 1962*. Albuquerque, NM: The Valliant Company.

Shipman, J.T., Adams, J.L., and Wilson, J.D. (1987). *An Introduction to Physical Science*. Lexington, MA: D.C. Heath.

Smith, R.K. (1993). *Water and Wastewater Laboratory Techniques*. Alexandria, VA: Water Environment Federation.

Spellman, F.R. (1997). *Wastewater Biosolids to Compost*. Boca Raton, FL: CRC Press.

Spellman, F.R. (2007). *The Science of Water*, 2nd ed. Boca Raton, FL: CRC Press.

Spellman, F.R. (2009). *The Science of Air*, 2nd ed. Boca Raton FL: CRC Press.

Spellman, F.R. and Whiting, N. (2006). *Environmental Science and Technology: Concepts and Applications*. Rockville, MD: Government Institutes.

Tchobanoglous, G. and Schroeder E.D. (1987). *Water Quality*. Reading, MA: Addison-Wesley.

USEPA. (2014). *SI: 409—Basic Air Pollution Meteorology*. Washington, DC: U.S. Environmental Protection Agency (http://yosemite.epa.gov/oaqps/eogtrain.nsf/DisplayView/SI_409_0-5?OpenDocument).

USGS. (2006). *Water Science for Schools: Water Measurements*. Washington, DC: U.S. Geological Survey.

USGS. (2014). *The Water Cycle: Evapotranspiration*. Washington, DC: U.S. Geological Survey (http://water.usgs.gov/edu/watercycleevaporation.html).

Watson, L. (1988). *The Water Planet: A Celebration of the Wonder of Water*. New York: Crown Publishers.

Wille, A. (1893). A curious optical phenomenon. *Nature*, 48(1243), 391.

5 Water Biology

> Scientists picture the primordial Earth as a planet washed by a hot sea and bathed in an atmosphere containing water vapor, ammonia, methane and hydrogen. Testing this theory, Stanley Miller at the University of Chicago duplicated these conditions in the laboratory. He distilled seawater in a special apparatus, passed the vapor with ammonia, methane and hydrogen through an electrical discharge at frequent intervals, and condensed the "rain" to return to the boiling seawater. Within a week, the seawater had turned red. Analysis showed that it contained amino acids, which are the building blocks of protein substances. Whether this is what really happened early in the Earth's history is not important; the experiment demonstrated that the basic ingredients of life could have been made in some such fashion, setting the stage for life to come into existence in the sea. The saline fluids in most living things may be an inheritance from such early beginnings.
>
> **Kemmer (1979)**

Because microorganisms are significant in water and in disease transmission they are the primary agents of water treatment processes. Water practitioners must have considerable knowledge of the biological characteristics of water. Simply put, waterworks operators and students of the science of water cannot fully comprehend the principles of effective water treatment and water science without knowing the fundamentals concerning microorganisms and their relationships to one another; their effect on the treatment process; and their impact on consumers, animals, and the environment. Water practitioners must know what principal groups of microorganisms are typically found in water supplies (surface and groundwater). They must be able to identify those microorganisms that must be treated (pathogenic organisms) and removed or controlled for biological treatment processes. They must be able to identify the organisms used as indicators of pollution or contamination and know their significance, and they must know the methods used to enumerate the indicator organisms. Finally, water treatment operators must be familiar with those organisms that indicate process conditions to optimize process operation.

Note: In order to have microbiological activity, the body of water or wastewater must possess the appropriate environmental conditions. The majority of wastewater treatment processes, for example, are designed to operate using an aerobic process. The conditions required for aerobic operation are (1) sufficient free, elemental oxygen; (2) sufficient organic matter (food); (3) sufficient water; (4) enough nitrogen and phosphorus (nutrients) to permit oxidation of the available carbon materials; (5) proper pH (6.5 to 9.0); and (6) lack of toxic materials.

BIOLOGY AND MICROBIOLOGY: WHAT ARE THEY?

Biology is generally defined as the study of living organisms (i.e., the study of life). *Microbiology* is a branch of biology that deals with the study of microorganisms so small in size that they must be studied under a microscope. Microorganisms of interest to the water and wastewater operator include bacteria, protozoa, viruses, algae, and others.

Note: The science and study of bacteria are known as *bacteriology.*

The primary concern of waterworks operators is how to control microorganisms that cause waterborne diseases—waterborne pathogens—to protect the consumer (human and animal).

WATER MICROORGANISMS

Microorganisms of interest to water practitioners and students of the science of water include bacteria, protozoa, rotifers, viruses, algae, and fungi. These organisms are the most diverse group of living organisms on Earth and occupy important niches in the ecosystem. Their simplicity and minimal survival requirements allow them to exist in diverse situations. Water practitioners, in particular, are concerned with water supply and water purification through a treatment process. In treating water, the primary concern, of course, is producing potable water that is safe to drink (free of pathogens), with no accompanying offensive characteristics—foul taste and odor. To accomplish this, the drinking water practitioner must possess a wide range of knowledge. In short, to correctly examine raw water for pathogenic microorganisms and to determine the type of treatment necessary to ensure that the quality of the end product—potable water—meets regulatory standards, as well as to accomplish all the other myriad requirements involved in drinking water processing, the water practitioner must be a combination specialist and generalist.

A practitioner whose narrowly focused specialty is not water microbiology must at least have enough knowledge in biological science to enable full comprehension of the fundamental factors concerning microorganisms and their relationships to one another, their effect on the treatment process, and their impact on the environment, human beings, and other organisms. The drinking water practitioner as a generalist must know the importance of microbiological parameters and what they indicate—the potential for waterborne disease. Microbiological contaminants are associated with undesirable tastes and odors and are considered generators of treatment problems in drinking water technology (algae and fungi, for example), and they are important enough to the practitioner that knowledge of them is essential. This chapter provides fundamental knowledge of water biology for the water practitioner (primarily for the generalist).

Because microorganisms are a major health concern, water treatment specialists are mostly concerned with how to control those that cause *waterborne diseases* (e.g., typhoid, tetanus, hepatitis, dysentery, gastroenteritis). Waterborne diseases are carried by *waterborne pathogens* (e.g., bacteria, virus, protozoa). To understand how to minimize or maximize growth of microorganisms and control pathogens one must study the structure and characteristics of the microorganisms. In the sections that follow, the major groups of microorganisms (those important to water practitioners and students of the science of water) are described and discussed in relation to their size, shape, types, nutritional needs, and control.

Note: In a water environment, water is not the medium for the growth of microorganisms but is instead a means of transmission of the pathogen (that is, it serves as a conduit, thus the name *waterborne*) (Koren, 1991). Individuals drink the water carrying the pathogens and thus begins an outbreak of disease. When the topic of waterborne disease is brought up, many might mistakenly assume that waterborne diseases are at home in water. Nothing could be further from the truth. A water-filled environment is not one in which pathogenic organisms would choose to live—that is, if they had such a choice. The point is that microorganisms do not normally grow, reproduce, languish, and thrive in watery surroundings. Pathogenic microorganisms temporarily residing in water are simply biding their time, going with the flow, waiting for their opportunity to meet up with their unsuspecting hosts. To some degree, when the pathogenic microorganism finds a host, it is finally home and may have found its final resting place (Spellman, 1997).

Key Terms

Algae, simple—Plants, many microscopic, containing chlorophyll. Freshwater algae are diverse in shape, color, size, and habitat. They are the basic link in the conversion of inorganic constituents in water into organic constituents.

Algal bloom—Sudden spurts of algal growth which can affect water quality adversely and indicate potentially hazardous changes in local water chemistry.

Anaerobic—Refers to being able to live and grow in the absence of free oxygen.
Autotrophic organisms—Organisms that produce food from inorganic substances.
Bacteria—Single-celled microorganisms that possess rigid cell walls. They may be aerobic, anaerobic, or facultative. They can cause disease, but some are important in pollution control.
Biogeochemical cycle—The chemical interactions among the atmosphere, hydrosphere, and biosphere.
Coliform organism—Microorganisms found in the intestinal tract of humans and animals. Their presence in water indicates fecal pollution and potentially adverse contamination by pathogens.
Denitrification—The anaerobic biological reduction of nitrate to nitrogen gas.
Fungi—Simple plants lacking in ability to produce energy through photosynthesis.
Heterotrophic organism—Organisms that are dependent on organic matter for foods.
Prokaryotic cell—The simple cell type, characterized by a lack of nuclear membrane and the absence of mitochondria.
Virus—The smallest form of microorganisms capable of causing disease.

Microorganisms in General

The microorganisms we are concerned with are tiny organisms made up of a large and diverse group of free-living forms; they exist as single cells, cell bunches, or clusters. Found in abundance almost anywhere on Earth, the vast majority of microorganisms are not harmful. Many microorganisms, or microbes, occur as single cells (unicellular), others are multicellular, and still others—viruses—do not have a true cellular appearance. A single microbial cell, for the most part, exhibits the characteristic features common to other biological systems, such as metabolism, reproduction, and growth.

Classification of Organisms

For centuries, scientists classified the forms of life visible to the naked eye as either animal or plant. Much of the current knowledge about living things was organized by the Swedish naturalist Carolus Linnaeus in 1735. The importance of classifying organisms cannot be overstated, for without a classification scheme establishing criteria for identifying organisms and arranging similar organisms into groups would be difficult. Probably the most important reason for classifying organisms is to make things less confusing (Wistreich and Lechtman, 1980). Linnaeus was innovative in the classification of organisms. One of his innovations still with us today is the *binomial system of nomenclature*. Under the binomial system, all organisms are generally described by a two-word scientific name: *genus* and *species*. Genus and species are groups that are part of a hierarchy of groups of increasing size, based on their nomenclature (taxonomy):

Kingdom
    Phylum
        Class
            Order
                Family
                    Genus
                        Species

Using this system, a fruit fly might be classified as follows:

Animalia
Arthropoda
Insecta
Diptera
Drosophilidae
Drosophila
melanogaster

This means that this organism is the species *melanogaster* in the genus *Drosophila* in the family Drosophilidae in the order Diptera in the class Insecta in the phylum Arthropoda in the kingdom Animalia.

To further illustrate how the hierarchical system is exemplified by the classification system, the standard classification of the mayfly is provided below:

Kingdom—Animalia
Phylum—Arthropoda
Class—Insecta
Order—Ephermeroptera
Family—Ephemeridae
Genus—*Hexagenia*
Species—*limbata*

Utilizing this hierarchy and Linnaeus' binomial system of nomenclature, the scientific name of any organism includes both the generic and specific names. To uniquely name a species, it is necessary to supply both the genus and the species; for our examples, those would be *Drosophila melanogaster* for the fruit fly and *Hexagenia limbota* for the mayfly. The first letter of the generic name is usually capitalized; hence, for example, *E. coli* indicates that *coli* is the species and *Escherichia* (abbreviated as *E.*) is the genus. The largest, most inclusive category is the kingdom. The genus and species names are always in Latin, so they are usually printed in italics. Some organisms also have English common names. Microbe names of particular interest in water/wastewater treatment include the following:

- *Escherichia coli* (a coliform bacterium)
- *Salmonella typhi* (the typhoid bacillus)
- *Giardia lamblia* (a protozoan)
- *Shigella* spp.
- *Vibrio cholerae*
- *Campylobacter*
- *Leptospira* spp.
- *Entamoeba histolytica*
- *Cryptosporidia*

Note: *Escherichia coli* is commonly known as simply *E. coli*, and *Giardia lamblia* is usually referred to by only its genus name, *Giardia*.

Generally, we use a simplified system of microorganism classification in water science by breaking down the classification into the kingdoms of Animal, Plant, and Protista. As a general rule, the Animal and Plant kingdoms contain all of the multicell organisms, and the Protista kingdom includes all single-cell organisms. Along with a microorganism classification based on the Animal, Plant, and Protista kingdoms, microorganisms can be further classified as being *eucaryotic* or *prokaryotic* (see Table 5.1).

TABLE 5.1
Simplified Classification of Microorganisms

Kingdom	Members	Cell Classification
Animal	Rotifers	Eucaryotic
	Crustaceans	
	Worms and larvae	
Plant	Ferns	
	Mosses	
Protista	Protozoa	
	Algae	
	Fungi	
	Bacteria	Prokaryotic
	Lower algae forms	

Note: A eucaryotic organism is characterized by a cellular organization that includes a well-defined nuclear membrane. The prokaryotes have a structural organization that sets them off from all other organisms. They are simple cells characterized by a nucleus *lacking* a limiting membrane, an endoplasmic reticulum, chloroplasts, and mitochondria. They are remarkably adaptable and exist abundantly in the soil, sea, and freshwater.

Differentiation

Differentiation among the higher forms of life is based almost entirely on morphological (form or structure) differences; however, differentiation (even among the higher forms) is not as easily accomplished as we might expect, because normal variations among individuals of the same species occur frequently. Because of this variation, even within a species, securing accurate classifications when dealing with single-celled microscopic forms that present virtually no visible structural differences becomes extremely difficult. Under these circumstances, it is necessary to consider physiological, cultural, and chemical differences, as well as structure and form. Differentiation among the smaller groups of bacteria is based almost entirely on chemical differences.

CELLS

Amoebas at the start
Were not complex;
They tore themselves apart
And started Sex.

—Arthur Guiterman (1871–1943)

Cells are the fundamental units of life. The cell retains a dual existence as a distinct entity and as a building block in the construction of organisms. These conclusions about cells were observed and published in 1838 by Matthias Jakob Schleiden (1804–1881). Later, Rudolph Virchow (1821–1902) offered the powerful dictum, "*Omnis cellula e cellula*" ("All cells come from cells"). This important tenet, along with others, formed the basis of what we call *cell theory*. The modern tenets of the cell theory include the following:

- All known living things are made up of cells.
- The cell is the structural and functional unit of all living things.
- All cells come from pre-existing cells by division.

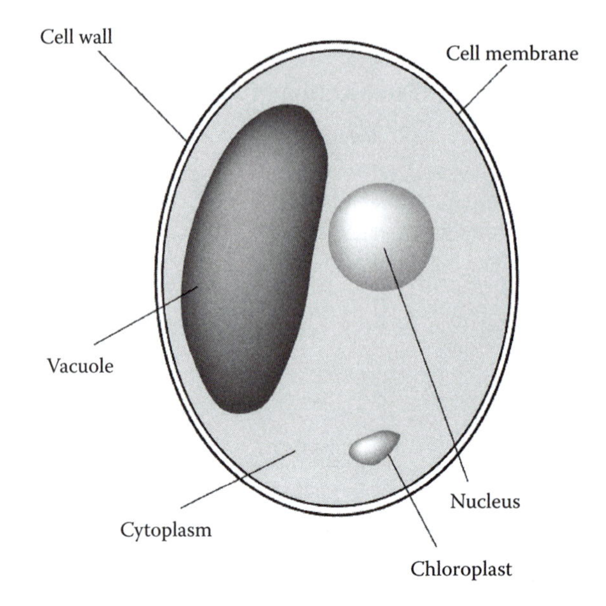

FIGURE 5.1 Plant cell.

- Cells contain hereditary information which is passed from cell to cell during cell division.
- All cells are basically the same in chemical composition.
- All of life's energy flow occurs within cells.

These modern tenets, of course, post-date Robert Hooke's 1663 discovery of cells in a piece of cork, which he examined under his primitive microscope. Hooke drew the cell (actually it was the cell wall he observed) and coined the word "cell," derived from the Latin word *cellula*, which means "small compartment."

Thus, we have known for quite some time that all living things, whether animal or plant, are made up of cells, the fundamental unit of all living things, no matter how complex. A typical cell is an entity isolated from other cells by a membrane or cell wall. The cell membrane contains protoplasm (the living material found within them) and the nucleus.

In a typical mature plant cell (Figure 5.1) the cell wall is rigid and is composed of nonliving material, but in the typical animal cell (Figure 5.2) the wall is an elastic living membrane. Cells exist in a very great variety of sizes and shapes, as well as functions. The cell is the smallest functioning unit of a living thing that still has the characteristics of the whole organism. The size of cells ranges from bacteria too small to be seen with a light microscope to the largest single cell known, the ostrich egg. Microbial cells also have an extensive size range, some being larger than human cells.

> **Note:** The smallest size of a cell is determined by the volume capable of holding the genetic material, proteins, etc., necessary to carry out the basic cell functions and reproduction. The largest size of a cell is limited by metabolism, as a cell must take in adequate amounts of oxygen and nutrients and get rid of wastes.

Types of Cells

Cells are of two fundamental types: *prokaryotic* and *eukaryotic*. Prokaryotic ("before nucleus") cells are simpler in design compared to eukaryotic ("true nucleus") cells, as they possess neither a nucleus nor the organelles (i.e., internal cell structures, each of which has a specific function within the cells) found in the cytoplasm of eukaryotic cells. Because prokaryotes do not have a nucleus,

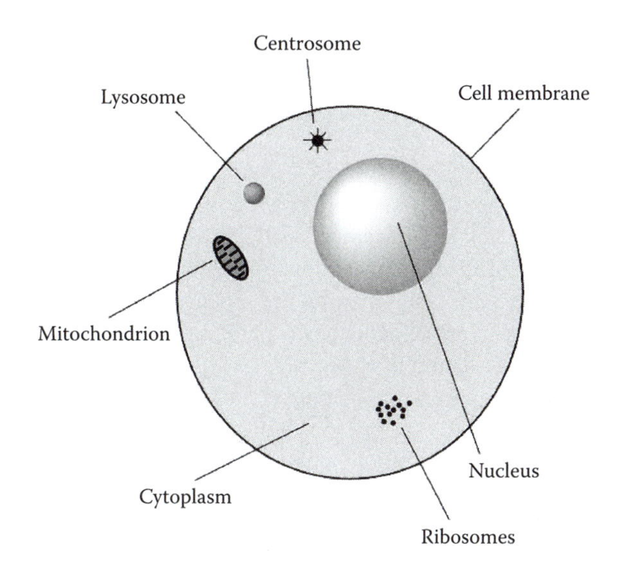

FIGURE 5.2 Animal cell.

the DNA is found within a "nucleiod" region. With the exception of archaebacteria, proteins are not associated with bacterial DNA. Bacteria are the best known and most studied form of prokaryotic organisms (Figure 5.3).

> **Note:** Cells may exist as independent units (e.g., protozoa) or as parts of multicellular organisms, in which the cells may develop specializations and form tissues and organs with specific purposes.

Prokaryotes are unicellular organisms that do not develop or differentiate into multicellular forms. Some bacteria grow in filaments, or masses of cells, but each cell in the colony is identical and capable of independent existence. Prokaryotes are capable of inhabiting almost every place on the planet, from the deep ocean to the edges of hot springs to just about every surface of our bodies.

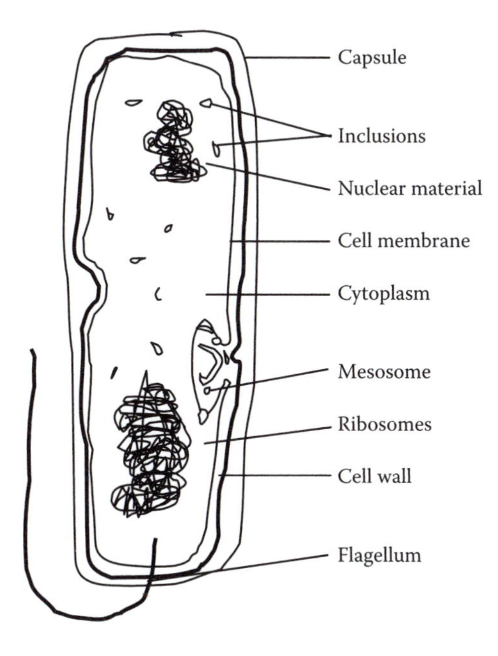

FIGURE 5.3 Bacterial cell.

Note: It is often stated that prokaryotic cells are among the most primitive forms of life on Earth; however, it is important to point out that being primitive does not mean they are outdated in the evolutionary sense, as primitive bacteria seem little changed and thus may be viewed as well adapted.

Prokaryotes are distinguished from eukaryotes on the basis of their nuclear organization, specifically their lack of a nuclear membrane. Also, prokaryotes are smaller and simpler than eukaryotic cells. Again, prokaryotes also lack any of the intracellular organelles and structures that are characteristic of eukaryotic cells. Most of the functions of organelles, such as mitochondria, chloroplasts, and the Golgi apparatus, are taken over by the prokaryotic plasma membrane. Prokaryotic cells have three architectural regions: appendages called *flagella* and *pili* (proteins attached to the cell surface); a *cell envelope* consisting of a *capsule*, a *cell wall*, and a *plasma membrane*; and a *cytoplasmic region*, which contains the *cell genome* (DNA) and ribosomes and various sorts of inclusions.

Eukaryotic cells evolved about 1.5 billion years ago. Protists, fungi, plants, and animals have eukaryotic cells—all plants and animals are eukaryotes. They are larger, as much as 10 times the size of prokaryotic cells, and most of their genetic material is found within a membrane-bound nucleus (a true nucleus), which is generally surrounded by several membrane-bound organelles. The presence of these membrane-bound organelles points to the significant difference between prokaryotes and eukaryotes. Although eukaryotes use the same genetic code and metabolic processes as prokaryotes, their higher level of organizational complexity has permitted the development of truly multicellular organisms.

Note: An enormous gap exists between prokaryote cells and eukaryote type cells: "... prokaryotes and eukaryotes are profoundly different from each other and clearly represent a marked dichotomy in the evolution of life. ... The organizational complexity of the eukaryotes is so much greater than that of prokaryotes that it is difficult to visualize how a eukaryote could have arisen from any known prokaryote" (Hickman, 1997).

Prokaryotic and eukaryotic cells also have their similarities. All cell types are bounded by a plasma membrane that encloses proteins and usually nucleic acids such as DNA and RNA. Table 5.2 shows a comparison of key features of both cell types.

Note: Plant cells can generally be distinguished from animal cells by (1) the presence of cell walls, chloroplasts, and central vacuoles in plants and their absence in animals, and (2) the presence of lysosomes and centrioles in animals and their absence in plants.

BACTERIA

The simplest wholly contained life systems are *bacteria* or *prokaryotes*, which are the most diverse group of microorganisms. They are among the most common microorganisms in water. They are primitive, unicellular (single-celled) organisms possessing no well-defined nucleus and presenting a variety of shapes and nutritional needs. Bacteria contain about 85% water and 15% ash or mineral matter. The ash is largely composed of sulfur, potassium, sodium, calcium, and chlorides, with small amounts of iron, silicon, and magnesium. Bacteria reproduce by binary fission.

Note: *Binary fission* occurs when one organism splits or divides into two or more new organisms.

Bacteria, once considered the smallest living organism (although now it is known that smaller forms of matter exhibit many of the characteristics of life), range in size from 0.5 to 2 μm in diameter and about 1 to 10 μm long.

Note: A *micron* is a metric unit of measurement equal to 1/1000 of a millimeter. To visualize the size of bacteria, consider that about 1000 average bacteria lying side by side would reach across the head of a straight pin.

TABLE 5.2
Comparison of Typical Prokaryotic and Eukaryotic Cells

Characteristic	Prokaryotic	Eukaryotic
Size	1 to 10 μm	10 to 100 μm
Nuclear envelope	Absent	Present
Cell wall	Usually	Present (plants); absent (animals)
Plasma membrane	Present	Present
Nucleolus	Absent	Present
DNA	Present (single loop)	Present
Mitochondria	Absent	Present
Chloroplasts	Absent	Present (plants only)
Endoplasmic reticulum	Absent	Present
Ribosomes	Present	Present
Vacuoles	Absent	Present
Golgi apparatus	Absent	Present
Lysosomes	Absent	Often present
Cytoskeleton	Absent	Present

TABLE 5.3
Forms of Bacteria

	Technical Name		
Form	**Singular**	**Plural**	**Example**
Sphere	Coccus	Cocci	*Streptococcus*
Rod	Bacillus	Bacilli	*Bacillus typhosis*
Curved or spiral	Spirillum	Spirilla	*Spirillum cholera*

Bacteria are categorized into three general groups based on their physical form or shape (although almost every variation has been found; see Table 5.3). The simplest form is the sphere. Spherical-shaped bacteria are called *cocci* (meaning "berries"). They are not necessarily perfectly round but may be somewhat elongated, flattened on one side, or oval. Rod-shaped bacteria are called *bacilli*. Spiral-shaped bacteria called *spirilla* have one or more twists and are never straight (see Figure 5.4). Such formations are usually characteristic of a particular genus or species. Within these three groups are many different arrangements. Some exist as single cells; others as pairs, as packets of four or eight, as chains, or as clumps.

Most bacteria require organic food to survive and multiply. Plant and animal material that gets into the water provides the food source for bacteria. Bacteria convert the food to energy and use the energy to make new cells. Some bacteria can use inorganics (e.g., minerals such as iron) as an energy source and can multiply even when organics (pollution) are not available.

Structure of the Bacterial Cell

The structural form and various components of the bacterial cell are probably best understood by referring to the simplified diagram of the rod-form bacterium shown in Figure 5.3. When studying Figure 5.3, keep in mind that cells of different species may differ greatly, both in structure and chemical composition; for this reason, no typical bacterium exists. Figure 5.3 shows a generalized bacterium and should be referred to for the discussion that follows; however, not all bacteria have all of the features shown in the figure, and some bacteria have structures not shown in the figure.

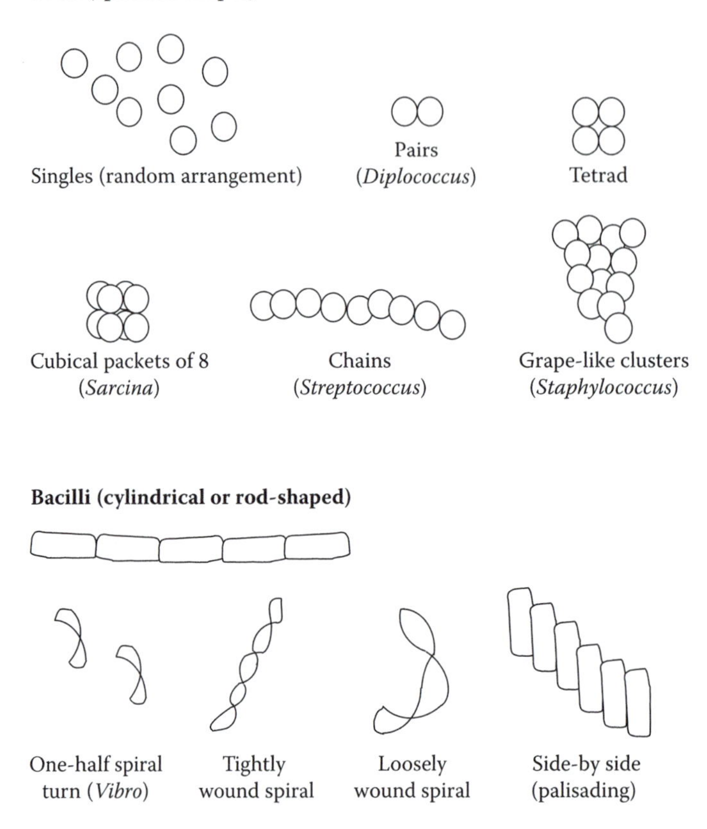

FIGURE 5.4 Bacterial shapes and arrangements.

Capsules

Bacterial capsules are organized accumulations of gelatinous materials on cell walls, in contrast to *slime layers* (a water secretion that adheres loosely to the cell wall and commonly diffuses into the cell), which are unorganized accumulations of similar material. The capsules are usually thick enough to be seen under the ordinary light microscope (macrocapsules), whereas thinner capsules (microcapsules) can be detected only by electron microscopy (Singleton and Sainsbury, 1994). The production of capsules is determined largely by genetics as well as by environmental conditions and depends on the presence or absence of capsule-degrading enzymes and other growth factors. Varying in composition, capsules are mainly composed of water; the organic contents are made up of complex polysaccharides, nitrogen-containing substance, and polypeptides. Capsules confer several advantages when bacteria grow in their normal habitat; for example, they help to: (1) prevent desiccation; (2) resist phagocytosis by host phagocytic cells; (3) prevent infection by bateriophages; and (4) aid bacterial attachment to tissue surfaces in plant and animal hosts or to surfaces of solids objects in aquatic environments. Capsule formation often correlates with pathogenicity.

Flagella

Many bacteria are motile, and this ability to move independently is usually attributed to a special structure, the flagella (singular: flagellum). Depending on species, a cell may have a single flagellum (*monotrichous* bacteria, *trichous* meaning "hair"), one flagellum at each end (*amphitrichous* bacteria, *amphi* meaning "on both sides"), a tuft of flagella at one or both ends (*lophotrichous* bacteria, *lopho* meaning "tuft"), or flagella that arise all over the cell surface (*peritrichous* bacteria, *peri* meaning "around").

A flagellum is a threadlike appendage extending outward from the plasma membrane and cell wall. Flagella are slender, rigid, locomotor structures, about 20 nm across and up to 15 to 20 μm long. Flagellation patterns are very useful in identifying bacteria and can be seen by light microscopy, but only after being stained with special techniques designed to increase their thickness. The detailed structure of flagella can be seen only in the electron microscope.

Bacterial cells benefit from flagella in several ways. They can increase the concentration of nutrients or decrease the concentration of toxic materials near the bacterial surfaces by causing a change in the flow rate of fluids. They can also disperse flagellated organisms to areas where colony formation can take place. The main benefit of flagella to organisms is their increased ability to flee from areas that might be harmful.

Cell Wall

The main structural component of most prokaryotes is the rigid cell wall. Functions of the cell wall include (1) providing protection for the delicate protoplast from osmotic lysis (bursting); (2) determining the shape of a cell; (3) acting as a permeability layer that excludes large molecules and various antibiotics and plays an active role in regulating the intake of ions by the cell; and (4) providing a solid support for flagella. Cell walls of different species may differ greatly in structure, thickness, and composition. The cell wall accounts for about 20 to 40% of the dry weight of a bacterium.

Plasma Membrane (Cytoplasmic Membrane)

Surrounded externally by the cell wall and composed of a lipoprotein complex, the plasma membrane or cell membrane is the critical barrier separating the inside from outside the cell. About 7 to 8 nm thick and comprising 10 to 20% of the dry weight of a bacterium, the plasma membrane controls the passage of all material into and out of the cell. The inner and outer faces of the plasma membrane are embedded with water-loving (hydrophilic) lipids, whereas the interior is hydrophobic. Control of material into the cell is accomplished by screening, as well as by electric charge. The plasma membrane is the site of the surface charge of the bacteria.

In addition to serving as an osmotic barrier that passively regulates the passage of material into and out of the cell, the plasma membrane participates in the entire active transport of various substances into the bacterial cell. Inside the membrane, many highly reactive chemical groups guide the incoming material to the proper points for further reaction. This active transport system provides bacteria with certain advantages, including the ability to maintain a fairly constant intercellular ionic state in the presence of varying external ionic concentrations. In addition to participating in the uptake of nutrients, the cell membrane transport system participates in waste excretion and protein secretions.

Cytoplasm

Within a cell and bounded by the cell membrane is a complicated mixture of substances and structures called the cytoplasm. The cytoplasm is a water-based fluid containing ribosomes, ions, enzymes, nutrients, storage granules (under certain circumstances), waste products, and various molecules involved in synthesis, energy metabolism, and cell maintenance.

Mesosome

A common intracellular structure found in the bacterial cytoplasm is the mesosome. Mesosomes are invaginations of the plasma membrane in the shape of tubules, vesicles, or lamellae. Their exact function is unknown. Currently, many bacteriologists believe that mesosomes are artifacts generated during the fixation of bacteria for electron microscopy.

Nucleoid (Nuclear Body or Region)

The nuclear region of the prokaryotic cell is primitive and a striking contrast to that of the eucaryotic cell. Prokaryotic cells lack a distinct nucleus, the function of the nucleus being carried out by a single, long, double strand of DNA that is efficiently packaged to fit within the nucleoid. The nucleoid is attached to the plasma membrane. A cell can have more than one nucleoid when cell division occurs after the genetic material has been duplicated.

Ribosomes

The bacterial cytoplasm is often packed with ribosomes. Ribosomes are minute, rounded bodies made of RNA and are loosely attached to the plasma membrane. Ribosomes are estimated to account for about 40% of the dry weight of a bacterium; a single cell may have as many as 10,000 ribosomes. Ribosomes serve as sites for protein synthesis and are part of the translation process.

Inclusions

Inclusions (storage granules) are often seen within bacterial cells. Some inclusion bodies are not bound by a membrane and lie free in the cytoplasm. A single-layered membrane 2 to 4 nm thick encloses other inclusion bodies. Many bacteria produce polymers that are stored as granules in the cytoplasm.

Bacterial Growth Factors

Several factors affect the rate at which bacteria grow, including temperature, pH, and oxygen levels. The warmer the environment, the faster the rate of growth. Generally, for each increase of 10°C, the growth rate doubles. Heat can also be used to kill bacteria. Most bacteria grow best at neutral pH. Extreme acidic or basic conditions generally inhibit growth, although some bacteria may require acidic conditions and some alkaline conditions for growth. Bacteria are aerobic, anaerobic, or facultative. If *aerobic*, they require free oxygen in the aquatic environment. *Anaerobic* bacteria exist and multiply in environments that lack dissolved oxygen. *Facultative* bacteria (e.g., iron bacteria) can switch from aerobic to anaerobic growth or can grow in an anaerobic or aerobic environment.

Under optimum conditions, bacteria grow and reproduce very rapidly. Bacteria reproduce by *binary fission*. An important point to consider with regard to bacterial reproduction is the rate at which the process can take place. The total time required for an organism to reproduce and the offspring to reach maturity is the *generation time*. Bacteria growing under optimal conditions can double their number about every 20 to 30 minutes. Obviously, this generation time is very short compared with that of higher plants and animals. Bacteria continue to grow at this rapid rate as long as nutrients hold out—even the smallest contamination can result in a sizable growth in a very short time.

> ***Note:*** Even though wastewater can contain bacteria counts in the millions per milliliter, in wastewater treatment under controlled conditions bacteria can help to destroy and identify pollutants. In such a process, bacteria stabilize organic matter (e.g., activated sludge processes) and thereby assist the treatment process in producing effluent that does not impose an excessive oxygen demand on the receiving body. Coliform bacteria can be used as an indicator of pollution by human or animal wastes.

Destruction of Bacteria

In water and wastewater treatment, the destruction of bacteria is usually called *disinfection*. Disinfection does not mean that all microbial forms are killed. That would be *sterilization*. Instead, disinfection reduces the number of disease-causing organisms to an acceptable number. Growing bacteria are generally easy to control by disinfection; however, some bacteria form survival structures known as *spores*, which are much more difficult to destroy.

> ***Note:*** Inhibiting the growth of microorganisms is termed *antisepsis*, whereas destroying them is called *disinfection*.

Waterborne Bacteria

All surface waters contain bacteria. Waterborne bacteria, as we have said, are responsible for infectious epidemic diseases. Bacterial numbers increase significantly during storm events when streams are high. Heavy rainstorms increase stream contamination by washing material from the ground surface into the stream. After the initial washing occurs, few impurities are left to be washed into the stream, which may then carry relatively "clean" water. A river of good quality shows its highest bacterial numbers during rainy periods; however, a much-polluted stream may show the highest numbers during low flows because of the constant influx of pollutants. Water and wastewater operators are primarily concerned with bacterial pathogens responsible for disease. These pathogens enter potential drinking water supplies through fecal contamination and are ingested by humans if the water is not properly treated and disinfected.

> ***Note:*** Regulations require that the owners of all public water supplies collect water samples and deliver those samples to a certified laboratory for bacteriological examination at least monthly. The number of samples required is usually in accordance with federal standards, which generally require that one sample per month be collected for each 1000 persons served by the waterworks.

Water practitioners determine bacteria counts on prepared media to determine the potential presence of harmful bacteria as well as disease-causing organisms such as viruses and protozoa. Sampling for total coliform, fecal coliform, and *Escherichia coli* bacteria is the generally accepted protocol for determining the presence of harmful biological disease-causing agents in water.

Total Coliform Bacteria

Total coliform bacteria are Gram-negative, aerobic or facultative anaerobic, non-spore-forming rods. These bacteria were originally believed to indicate the presence of fecal contamination; however, total coliforms have been found to be widely distributed in nature and not always associated with the gastrointestinal tract of warm-blooded animals. The number of total coliform bacteria in the environment is still widely used as an indicator for potable water in the United States.

Fecal Coliform Bacteria

Fecal coliform bacteria are a subgroup of coliform bacteria that were used to establish the first microbial water quality criteria. The ability to grow at an elevated temperature (44.5°C) separates this bacteria from the total coliforms and makes it a more accurate indicator of fecal contamination by warm-blooded animals. Fecal coliform bacteria are detected by counting the dark-blue to blue-gray colonies that grow on 0.65-µm filters placed on m-FC agar incubated in a 44.5°C oven for 22 to 24 hours. The presence of fecal coliforms in water indicates that fecal contamination of the water by a warm-blooded animal has occurred; however, recent studies have found no statistical relationship between fecal coliform concentrations and swimmer-associated sickness.

Escherichia coli

Escherichia coli is a rod-shaped bacteria commonly found in the gastrointestinal tract and feces of warm-blooded animals. It is a member of the fecal coliform group of bacteria and distinguished by its inability to break down urease. *E. coli* numbers in freshwater are determined by counting the number of yellow and yellow-brown colonies growing on a 0.45-µm filter placed on m-TEC media and incubated at 35.0°C for 22 to 24 hours. The addition of urea substrate confirms that colonies are *E. coli*. This bacteria is a preferred indicator for freshwater recreation, and its presence provides direct evidence of fecal contamination from warm-blooded animals. Although usually harmless, *E. coli* can cause illnesses such as meningitis, septicemia, and urinary tract and intestinal infections. A recently discovered strain of *E. coli* (*E. coli* 0157:H7) can cause severe disease and may be fatal in small children and the elderly.

PROTOZOA

Protozoa (or "first animals") are a large group of eucaryotic organisms of more than 50,000 known species belonging to the kingdom Protista. They have adapted a form of cell to serve as the entire body; in fact, protozoa are one-celled, animal-like organisms with complex cellular structures. In the microbial world, protozoa are giants, many times larger than bacteria. They range in size from 4 to 500 μm. The largest ones can almost be seen by the naked eye. They can exist as solitary or independent organisms, such as the stalked ciliates (e.g., *Vorticella* spp.) (Figure 5.5), or they can colonize (e.g., the sedentary *Carchesium* spp.). Protozoa get their name because they employ the same type of feeding strategy as animals; that is, they are *heterotrophic*, meaning that they obtain cellular energy from organic substances such as proteins. Most are harmless, but some are parasitic. Some forms have two life stages: *active trophozoites* (capable of feeding) and *dormant cysts*.

The major groups of protozoa are based on their method of locomotion (motility). The *Mastigophora* are motile by means of one or more *flagella*, the whip-like projection that propels the free-swimming organisms (*Giardia lamblia* is a flagellated protozoa). The *Ciliophora* move by means of shortened modified flagella called *cilia*, which are short hair-like structures that beat rapidly and propel them through the water. The *Sarcodina* rely on *amoeboid movement*, which is a streaming or gliding action; the shape of these amoebae change as they stretch and then contract to move from place to place. The *Sporozoa*, in contrast, are nonmotile; they are simply swept along, riding the current of the water.

Protozoa consume organics to survive; their favorite food is bacteria. Protozoa are mostly aerobic or facultative with regard to their oxygen requirements. Toxic materials, pH, and temperature affect protozoan rates of growth in the same way as they affect bacteria.

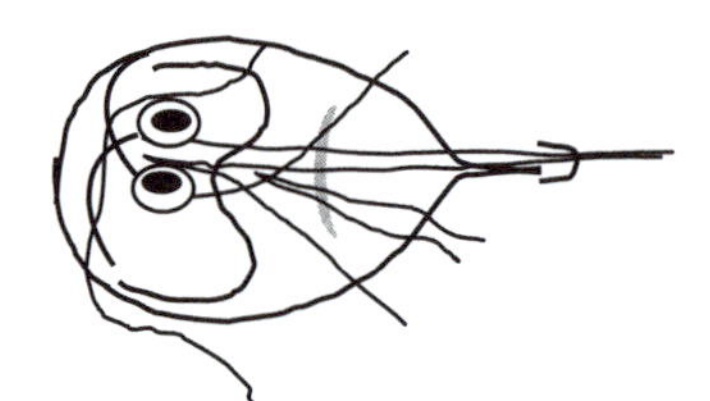

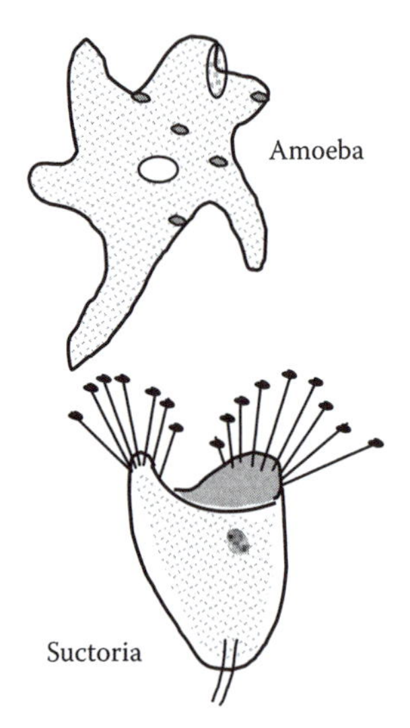

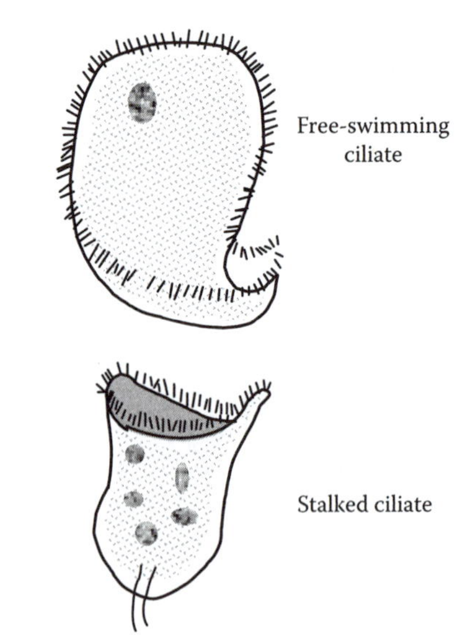

FIGURE 5.5 Protozoa.

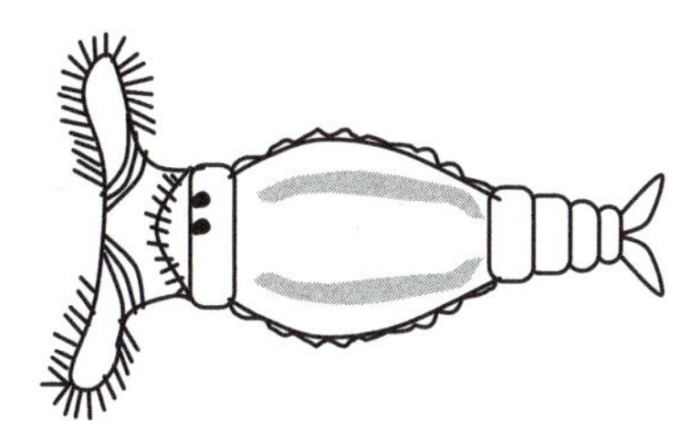

FIGURE 5.6 *Philodina*, a common rotifer.

Most protozoan life cycles alternate between an active growth phase (trophozoites) and a resting stage (cysts). Cysts are extremely resistant structures that protect the organism from destruction when it encounters harsh environmental conditions—including chlorination.

Note: Those protozoa not completely resistant to chlorination require higher disinfectant concentrations and longer contact time for disinfection than normally used in water treatment.

The protozoa and associated waterborne diseases of most concern to waterworks operators include the following:

- *Entamoeba histolytica*—amoebic dysentery
- *Giardia lamblia*—giardiasis
- *Cryptosporidium*—cryptosporidiosis

In wastewater treatment, protozoa are a critical part of the purification process and can be used to indicate the condition of treatment processes. Protozoa normally associated with wastewater include amoebae, flagellates, free-swimming ciliates, and stalked ciliates.

Amoebae are associated with poor wastewater treatment of a young biosolids mass (see Figure 5.5). They move through wastewater by a streaming or gliding motion. Moving the liquids stored within the cell wall effects this movement. They are normally associated with an effluent high in biochemical oxygen demand (BOD) and suspended solids. *Flagellates* (flagellated protozoa) have a single, long, hair-like or whip-like projection (flagellum) that is used to propel the free-swimming organisms through wastewater and to attract food (see Figure 5.5). Flagellated protozoa are normally associated with poor treatment and a young biosolids. When the predominate organism is the flagellated protozoa, the plant effluent will contain large amounts of BODs and suspended solids.

The *free-swimming ciliated protozoan* uses its tiny, hair-like projections (cilia) to move itself through the wastewater and to attract food (see Figure 5.5). This type of protozoan is normally associated with a moderate biosolids age and effluent quality. When the free-swimming ciliated protozoan is the predominate organisms, the plant effluent will normally be turbid and contain a high amount of suspended solids. The *stalked ciliated protozoan* attaches itself to the wastewater solids and uses its cilia to attract food (see Figure 5.5). The stalked ciliated protozoan is normally associated with a plant effluent that is very clear and contains low amounts of both BOD and suspended solids.

Rotifers make up a well-defined group of the smallest, simplest multicellular microorganisms and are found in nearly all aquatic habitats (see Figure 5.6). Rotifers are a higher life form associated with cleaner waters. Normally found in well-operated wastewater treatment plants, they can be used to indicate the performance of certain types of treatment processes.

MICROSCOPIC CRUSTACEANS

Because they are important members of freshwater zooplankton, microscope *crustaceans* are of interest to water and wastewater operators. These microscopic organisms are characterized by a rigid shell structure. They are multicellular animals that are strict aerobes, and as primary producers they feed on bacteria and algae. They are important as a source of food for fish. Additionally, microscopic crustaceans have been used to clarify algae-laden effluents from oxidations ponds.

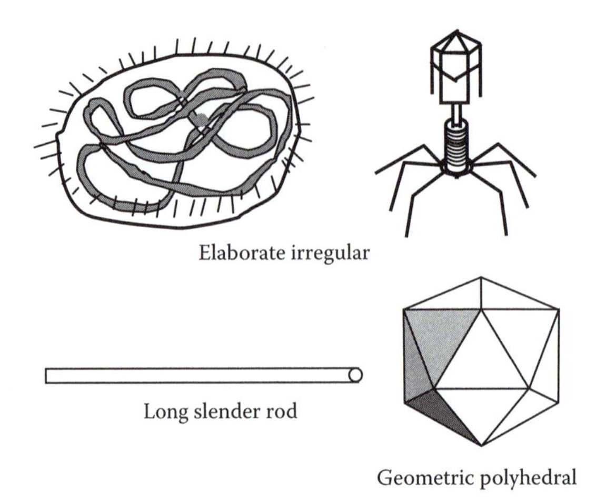

FIGURE 5.7 Viral shapes.

VIRUSES

Viruses are very different from the other microorganisms. Consider their size relationship, for example. Relative to size, if protozoa are the Goliaths of microorganisms, then viruses are the Davids. Stated more specifically and accurately, viruses are intercellular parasitic particles that are the smallest living infectious materials known—the midgets of the microbial world. Viruses are very simple life forms consisting of a central molecule of genetic material surrounded by a protein shell called a *capsid* and sometimes by a second layer called an *envelope.*

Viruses contain no mechanisms by which to obtain energy or reproduce on their own, thus viruses must have a host to survive. After they invade the cells of their specific host (animal, plant, insect, fish, or even bacteria), they take over the cellular machinery of the host and force it to make more viruses. In the process, the host cell is destroyed and hundreds of new viruses are released into the environment. The viruses of most concern to the waterworks operator are the pathogens that cause hepatitis, viral gastroenteritis, and poliomyelitis.

Smaller and different from bacteria, viruses are prevalent in water contaminated with sewage. Detecting viruses in water supplies is a major problem because of the complexity of the procedures involved, although experience has shown that the normal coliform index can be used as a rough guide for viruses as well as for bacteria. More attention must be paid to viruses, however, when surface water supplies have been used for sewage disposal. Viruses occur in many shapes, including long slender rods, elaborate irregular shapes, and geometric polyhedrals (see Figure 5.7). They are difficult to destroy by normal disinfection practices, as they require increased disinfectant concentration and contact time for effective destruction.

> ***Note:*** Viruses that infect bacterial cells cannot infect and replicate within cells of other organisms. It is possible to utilize this specificity to identify bacteria, a procedure called *phage typing.*

ALGAE

You do not have to be a water/wastewater operator to understand that algae can be a nuisance. Many ponds and lakes in the United States are currently undergoing *eutrophication*, which is enrichment of an environment with inorganic substances (e.g., phosphorus and nitrogen), causing excessive algae growth and premature aging of the water body. When eutrophication occurs, especially when filamentous algae such as *Caldophora* breaks loose in a pond or lake and washes ashore, algae makes its stinking, noxious presence known.

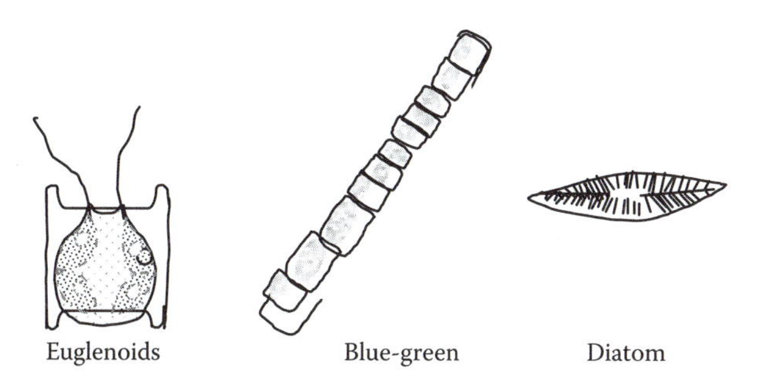

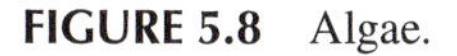
FIGURE 5.8 Algae.

Algae are a form of aquatic plants and are classified by color (e.g., green algae, blue–green algae, golden–brown algae). Algae come in many shapes and sizes (see Figure 5.8). Although they are not pathogenic, algae do cause problems with water/wastewater treatment plant operations. They grow easily on the walls of troughs and basins, and heavy growth can plug intakes and screens. Additionally, some algae release chemicals that give off undesirable tastes and odors. Although algae are usually classified by their color, they are also commonly classified based on their cellular properties or characteristics. Several characteristics are used to classify algae, including (1) cellular organization and cell wall structure; (2) the nature of the chlorophylls; (3) the type of motility, if any; (4) the carbon polymers that are produced and stored; and (5) the reproductive structures and methods.

Many algae (in mass) are easily seen by the naked eye, but others are microscopic. They occur in fresh and polluted water, as well as in saltwater. Because they are plants, they are capable of using energy from the sun in photosynthesis. They usually grow near the surface of the water because light cannot penetrate very far through the water. Algae are controlled in raw waters with chlorine and potassium permanganate. Algal blooms in raw water reservoirs are often controlled with copper sulfate.

Note: By producing oxygen, which is utilized by other organisms, including animals, algae play an important role in the balance of nature.

FUNGI

Fungi are of relatively minor importance in water/wastewater operations (except for biosolids composting, where they are critical). Fungi, like bacteria, are extremely diverse. They are multicellular, autotrophic, photosynthetic protists. They grow as filamentous, mold-like forms or as yeast-like (single-celled) organisms. They feed on organic material.

Note: Aquatic fungi grow as *parasites* on living plants or animals and as *saprophytes* on those that are dead.

MICROBIOLOGICAL PROCESSES

The primary goal in water treatment is to protect the consumer of potable drinking water from disease. Drinking water safety is a worldwide concern. Drinking nontreated or improperly treated water is a major cause of illness in developing countries. Water contains several biological (as well as chemical) contaminants that must be removed efficiently to produce safe drinking water that is also aesthetically pleasing to the consumer. The finished water must be free of microbial pathogens and parasites, turbidity, color, and odor. To achieve this goal, raw surface water or groundwater is subjected to a series of treatment processes that will be described in detail later. Disinfection alone is sufficient if the raw water originates from a protected source. More commonly, several processes are used to treat water; for example, disinfection may be combined with coagulation, flocculation, and filtration.

Several unit processes are used in the water treatment process to produce microbiologically (and chemically) safe drinking water. The extent of treatment depends on the source of raw water; surface waters generally require more treatment than groundwaters. With the exception of disinfection, the other unit processes in the treatment train do not specifically address the destruction or removal of pathogens.

Water treatment unit processes include (1) storage of raw water, (2) prechlorination, (3) coagulation–flocculation, (4) water softening, (5) filtration, and (6) disinfection. Filtration and disinfection are the primary means of removing contaminants and pathogens from drinking water supplies. In each of these unit processes, the reduction or destruction of pathogens is variable and influenced by a number of factors such as sunlight, sedimentation, and temperature.

These water treatment unit processes are important and are described in detail later in this text. For the moment, because of relatively recent events involving pathogenic protozoa causing adverse reactions, including death, among consumers in various locations throughout the United States (and elsewhere), it is important to turn our attention to the pathogenic protozoa. One thing is certain—these pathogenic protozoa have the full attention of water treatment operators everywhere.

Pathogenic Protozoa

As mentioned, certain types of protozoa can cause disease. Of particular interest to the drinking water practitioner are *Entamoeba histolytica* (amoebic dysentery and amoebic hepatitis), *Giardia lamblia* (giardiasis), *Cryptosporidium* (cryptosporidiosis), and the emerging *Cyclospora* (cyclosporasis). Sewage contamination transports eggs, cysts, and oocysts of parasitic protozoa and helminths (tapeworms, hookworms, etc.) into raw water supplies, leaving water treatment (in particular, filtration) and disinfection as the means by which to diminish the danger of contaminated water for the consumer.

To prevent the occurrence of *Giardia* and *Cryptosporidium* in surface water supplies and to address increasing problems with waterborne diseases, the U.S. Environmental Protection Agency (USEPA) implemented its Surface Water Treatment Rule (SWTR) in 1989. The rule requires both filtration and disinfection of all surface water supplies as a means of primarily controlling *Giardia* and enteric viruses. Since implementation of SWTR, the USEPA has also recognized that *Cryptosporidium* species are agents of waterborne disease. In its 1996 series of surface water regulations, the USEPA included *Cryptosporidium*.

To test the need for and effectiveness of the Surface Water Treatment Rule, LeChevallier et al. (1991) conducted a study on the occurrence and distribution of *Giardia* and *Cryptosporidium* organisms in raw water supplies to 66 surface water filter plants. These plants were located in 14 states and a Canadian province. A combined immunofluorescence test indicated that cysts and oocysts were widely dispersed in the aquatic environment. *Giardia* was detected in more than 80% of the samples. *Cryptosporidium* was found in 85% of the sample locations. Taking into account several variables, *Giardia* or *Cryptosporidium* were detected in 97% of the raw water samples. After evaluating their data, the researchers concluded that the Surface Water Treatment Rule might have to be upgraded (subsequently, it has been) to require additional treatment.

Giardia

Giardia lamblia (also known as the hiker's/traveler's scourge or disease) is a microscopic parasite that can infect warm-blooded animals and humans. Although *Giardia* was discovered in the 19th century, not until 1981 did the World Health Organization (WHO) classify *Giardia* as a pathogen. An outer shell called a *cyst* allows *Giardia* to survive outside the body for long periods of time. If viable cysts are ingested, *Giardia* can cause the illness known as *giardiasis*, an intestinal illness that can cause nausea, anorexia, fever, and severe diarrhea.

In the United States, *Giardia* is the most commonly identified pathogen in waterborne disease outbreaks. Contamination of a water supply by *Giardia* can occur in two ways: (1) by the activity of animals in the watershed area of the water supply, or (2) by the introduction of sewage into the water supply. Wild and domestic animals are major contributors to the contamination of water supplies. Studies have also shown that, unlike many other pathogens, *Giardia* is not host specific. In short, *Giardia* cysts excreted by animals can infect and cause illness in humans. Additionally, in several major outbreaks of waterborne diseases, the *Giardia* cyst source was sewage-contaminated water supplies.

Treating the water supply, however, can effectively control waterborne *Giardia*. Chlorine and ozone are examples of two disinfectants known to effectively kill *Giardia* cysts. Filtration of the water can also effectively trap and remove the parasite from the water supply. The combination of disinfection and filtration is the most effective water treatment process available today for prevention of *Giardia* contamination.

In drinking water, *Giardia* is regulated under the Surface Water Treatment Rule. Although the SWTR does not establish a maximum contaminant level (MCL) for *Giardia*, it does specify treatment requirements to achieve at least 99.9% (3-log) removal or inactivation of *Giardia*. This regulation requires that all drinking water systems using surface water or groundwater under the influence of surface water must disinfect and filter the water. The Enhanced Surface Water Treatment Rule (ESWTR), which includes *Cryptosporidium* and further regulates *Giardia*, was established in 1996.

Giardiasis

Giardiasis is recognized as one of the most frequently occurring waterborne diseases in the United States. *Giardia lamblia* cysts have been discovered in places as far apart as Estes Park, Colorado (near the Continental Divide); Missoula, Montana; Wilkes-Barre, Scranton, and Hazleton, Pennsylvania; and Pittsfield and Lawrence, Massachusetts, just to name a few (CDC, 1995).

Giardiasis is characterized by intestinal symptoms that usually last a week or more and may be accompanied by one or more of the following: diarrhea, abdominal cramps, bloating, flatulence, fatigue, and weight loss. Although vomiting and fever are commonly listed as relatively frequent symptoms, people involved in waterborne outbreaks in the United States have not commonly reported them. Although most *Giardia* infections persist only for 1 or 2 months, some people experience a more chronic phase that can follow the acute phase or may become manifest without an antecedent acute illness. Loose stools and increased abdominal gassiness with cramping, flatulence, and burping characterize the chronic phase. Fever is not common, but malaise, fatigue, and depression may ensue; for a small number of people, the persistence of infection is associated with the development of marked malabsorption and weight loss (Weller, 1985). Similarly, lactose (milk) intolerance can be a problem for some people. This can develop coincidentally with the infection or be aggravated by it, causing an increase in intestinal symptoms after ingestion of milk products.

Some people may have several of these symptoms without evidence of diarrhea or have only sporadic episodes of diarrhea every three or four days. Still others may have no symptoms at all. The problem, then, may not be one of determining whether or not someone is infected with the parasite but how harmoniously the host and the parasite can live together. When such harmony does not exist or is lost, it then becomes a problem of how to get rid of the parasite, either spontaneously or by treatment.

> ***Note:*** Three prescription drugs are available in the United States to treat giardiasis: quinacrine, metronidazole, and furazolidone. In a review of drug trials in which the efficacies of these drugs were compared, quinacrine produced a cure in 93% of patients, metronidazole cured 92%, and furazolidone cured about 84% of patients (Davidson, 1984).

Giardiasis occurs worldwide. In the United States, *Giardia* is the parasite most commonly identified in stool specimens submitted to state laboratories for parasitologic examination. During a 3-year period, approximately 4% of 1 million stool specimens submitted to state laboratories tested positive for *Giardia* (CDC, 1979). Other surveys have demonstrated *Giardia* prevalence rates ranging from 1 to 20%, depending on the location and ages of persons studied. Giardiasis ranks among the top 20 infectious diseases causing the greatest morbidity in Africa, Asia, and Latin America; it has been estimated that about 2 million infections occur per year in these regions (Walsh, 1981). People who are at highest risk for acquiring *Giardia* infection in the United States may be placed into five major categories:

1. People in cities whose drinking water originates from streams or rivers and whose water treatment process does not include filtration, or where filtration is ineffective because of malfunctioning equipment
2. Hikers, campers, and those who enjoy the outdoors
3. International travelers
4. Children who attend daycare centers, daycare center staff, and parents and siblings of children infected in daycare centers
5. Homosexual men

People in categories 1, 2, and 3 have in common the same general source of infection; that is, they acquire *Giardia* from fecally contaminated drinking water. The city resident usually becomes infected because the municipal water treatment process does not include the filter necessary to physically remove the parasite from the water. The number of people in the United States at risk (i.e., the number who receive municipal drinking water from unfiltered surface water) is estimated to be 20 million. International travelers may also acquire the parasite from improperly treated municipal waters in cities or villages in other parts of the world, particularly in developing countries. In Eurasia, only travelers to Leningrad appear to be at increased risk. In prospective studies, 88% of U.S. and 35% of Finnish travelers to Leningrad who had negative stool tests for *Giardia* on departure to the Soviet Union developed symptoms of giardiasis and had positive tests for *Giardia* after they returned home (Brodsky et al., 1974). With the exception of visitors to Leningrad, however, *Giardia* has not been implicated as a major cause of traveler's diarrhea, as it has been detected in fewer than 2% of travelers developing diarrhea. Hikers and campers, however, risk infection every time they drink untreated raw water from a stream or river. Persons in categories 4 and 5 become exposed through more direct contact with feces or an infected person by exposure to the soiled diapers of an infected child in cases associated with daycare centers or through direct or indirect anal–oral sexual practices in the case of homosexual men.

Although community waterborne outbreaks of giardiasis have received the greatest publicity in the United States, about half of the *Giardia* cases discussed with staff of the Centers for Disease Control and Prevention (CDC) over a 3-year period had a daycare exposure as the most likely source of infection. Numerous outbreaks of Giardia in daycare centers have been reported in recent years. Infection rates for children in daycare center outbreaks range from 21 to 44% in the United States and from 8 to 27% in Canada (Black et al., 1981; Pickering et al., 1984). The highest infection rates are usually observed in children who wear diapers (1 to 3 years of age).

Local health officials and managers or water utility companies need to realize that sources of *Giardia* infection other than municipal drinking water exist. Armed with this knowledge, they are less likely to make a quick (and sometimes wrong) assumption that a cluster of recently diagnosed cases in a city is related to municipal drinking water. Of course, drinking water must not be ruled out as a source of infection when a larger than expected number of cases is recognized in a community, but the possibility that the cases are associated with a daycare center outbreak, drinking untreated stream water, or international travel should also be entertained.

To understand the finer aspects of *Giardia* transmission and strategies for control, the drinking water practitioner must become familiar with several aspects of the biology of the parasite. Two forms of the parasite exist: a *trophozoite* and a *cyst*, both of which are much larger than bacteria (see Figure 5.9). Trophozoites live in the upper small intestine, where they attach to the intestinal

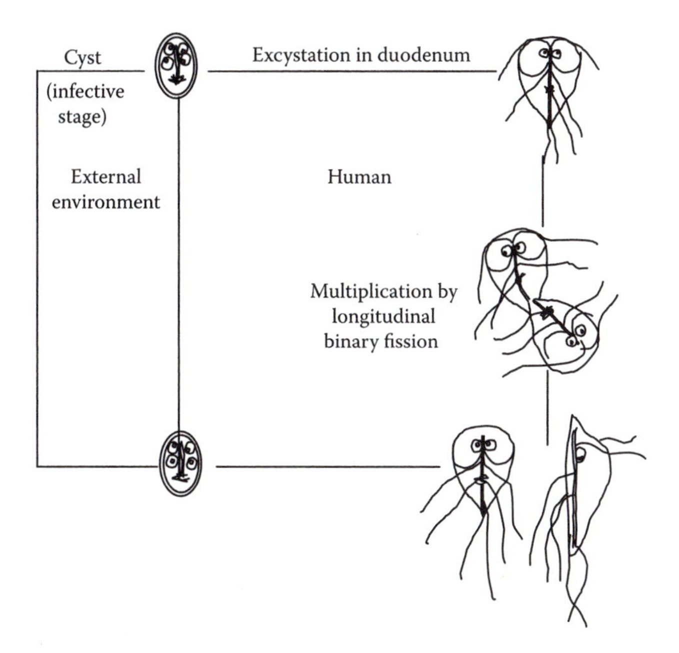

FIGURE 5.9 Life cycle of *Giardia lamblia.*

wall by means of a disc-shaped suction pad on their ventral surface. Trophozoites actively feed and reproduce at this location. At some time during the trophozoite's life, it releases its hold on the bowel wall and floats in the fecal stream through the intestine. As it makes this journey, it undergoes a morphologic transformation into the egg-like cyst. The cyst, about 6 to 9 nm in diameter and 8 to 12 μm in length, has a thick exterior wall that protects the parasite against the harsh elements that it will encounter outside the body. This cyst form of the parasite is infectious to other people or animals. Most people become infected either directly (by hand-to-mouth transfer of cysts from the feces of an infected individual) or indirectly (by drinking feces-contaminated water). Less common modes of transmission included ingestion of fecally contaminated food and hand-to-mouth transfer of cysts after touching a fecally contaminated surface. After the cyst is swallowed, the trophozoite is liberated through the action of stomach acid and digestive enzymes and becomes established in the small intestine.

Although infection after ingestion of only one *Giardia* cyst is theoretically possible, the minimum number of cysts shown to infect a human under experimental conditions is 10 (Rendtorff, 1954). Trophozoites divide by binary fission about every 12 hours. What this means in practical terms is that if a person swallowed only a single cyst then reproduction at this rate would result in more than 1 million parasites 10 days later—1 billion parasites by day 15.

The exact mechanism by which *Giardia* causes illness is not yet well understood, but it apparently is not necessarily related to the number of organisms present. Nearly all of the symptoms, however, are related to dysfunction of the gastrointestinal tract. The parasite rarely invades other parts of the body, such as the gall bladder or pancreatic ducts. Intestinal infection does not result in permanent damage.

Note: *Giardia* has an incubation period of 1 to 8 weeks.

Data reported by the CDC indicate that *Giardia* is the most frequently identified cause of diarrheal outbreaks associated with drinking water in the United States. The remainder of this section is devoted specifically to waterborne transmissions of *Giardia. Giardia* cysts have been detected in 16% of potable water supplies (lakes, reservoirs, rivers, springs, groundwater) in the United States at an average concentration of 3 cysts per 100 L (Rose et al., 1991). Waterborne epidemics of giardiasis are a relatively frequent occurrence. In 1983, for example, *Giardia* was identified as the cause of diarrhea in 68% of waterborne outbreaks in which the causal agent was identified; from 1965 to 1982, more than 50 waterborne outbreaks were reported (CDC, 1984). In 1984, about 250,000 people in Pennsylvania were advised to boil drinking water for 6 months because of *Giardia*-contaminated water.

Many of the municipal waterborne outbreaks of *Giardia* have been subjected to intense study to determine their cause. Several general conclusions can be made from data obtained in those studies. Waterborne transmission of *Giardia* in the United States usually occurs in mountainous regions where community drinking water obtained from clear running streams is chlorinated but not filtered before distribution. Although mountain streams appear to be clean, fecal contamination upstream by human residents or visitors, as well as by *Giardia*-infected animals such as beavers, has been well documented. Water obtained from deep wells is an unlikely source of *Giardia* because of the natural filtration of water as it percolates through the soil to reach underground cisterns. Wells that pose the greatest risk of fecal contamination are poorly constructed or improperly located ones. A few outbreaks have occurred in towns that included filtration in the water treatment process but the filtration was not effective in removing *Giardia* cysts because of defects in filter construction, poor maintenance of the filter media, or inadequate pretreatment of the water before filtration. Occasional outbreaks have also occurred because of accidental cross-connections between water and sewage systems.

From these data, we can conclude that two major ingredients are necessary for waterborne outbreaks: *Giardia* cysts must be present in untreated source water, and the water purification process must fail to either kill or remove *Giardia* cysts from the water.

Although beavers are often blamed for contaminating water with *Giardia* cysts, the fact that they are responsible for introducing the parasite into new areas seems unlikely. Far more likely is that they are also victims: *Giardia* cysts may be carried in untreated human sewage discharged into the water by small-town sewage disposal plants or originate from cabin toilets that drain directly into streams and rivers. Backpackers, campers, and sports enthusiasts may also deposit *Giardia*-contaminated feces in the environment which are subsequently washed into streams by rain. In support of this concept is a growing amount of data indicating a higher *Giardia* infection rate in beavers living downstream from U.S. national forest campgrounds when compared with beavers living in more remote areas that have a near zero rate of infection.

Beavers may be unwitting victims of the *Giardia* story, but they still play an important part in the contamination scheme because they can (and probably do) serve as amplifying hosts. An *amplifying host* is one that is easy to infect, serves as a good habitat for reproduction of the parasite, and, in the case of *Giardia*, returns millions of cysts to the water for every one ingested. Beavers are especially important in this regard, because they tend to defecate in or very near the water, which ensures that most of the *Giardia* cysts excreted are returned to the water.

The microbial quality of water resources and the management of the microbially laden wastes generated by the burgeoning animal agriculture industry are critical local, regional, and national problems. Animal wastes from cattle, hogs, sheep, horses, poultry, and other livestock and commercial animals can contain high concentrations of microorganism, such as *Giardia,* that are pathogenic to humans.

The contribution of other animals to waterborne outbreaks of *Giardia* is less clear. Muskrats (another semiaquatic animal) have been found in several parts of the United States to have high infection rates (30 to 40%) (Frost et al., 1984). Studies have shown that muskrats can be infected with *Giardia* cysts from humans and beavers. Occasional *Giardia* infections have been reported in coyotes, deer, elk, cattle, dogs, and cats (but not in horses and sheep) encountered in mountainous regions of the United States. Naturally occurring *Giardia* infections have not been found in most other wild animals (bear, nutria, rabbit, squirrel, badger, marmot, skunk, ferret, porcupine, mink, raccoon, river otter, bobcat, lynx, moose, and bighorn sheep).

Scientific knowledge about what is required to kill or remove *Giardia* cysts from a contaminated water supply has increased considerably. We know, for example, that cysts can survive in cold water (4°C) for at least 2 months and that they are killed instantaneously by boiling water (100°C) (Frost et al., 1984). We do not know how long the cysts will remain viable at other water temperatures (e.g., 0°C or in a canteen at 15 to 20°C), nor do we know how long the parasite will survive on various environment surfaces, such as under a pine tree, in the sun, on a diaper-changing table, or in carpets in a daycare center.

The effect of chemical disinfection (chlorination, for example) on the viability of *Giardia* cysts is an even more complex issue. The number of waterborne outbreaks of *Giardia* that have occurred in communities where chlorination was employed as a disinfectant process demonstrates that the amount of chlorine used routinely for municipal water treatment is not effective against *Giardia* cysts. These observations have been confirmed in the laboratory under experimental conditions (Jarroll et al., 1979). This does not mean that chlorine does not work at all. It does work under certain favorable conditions. Without getting too technical, gaining some appreciation of the problem can be achieved by understanding a few of the variables that influence the efficacy of chlorine as a disinfectant:

- *Water pH*—At pH values above 7.5, the disinfectant capability of chlorine is greatly reduced.
- *Water temperature*—The warmer the water, the higher the efficacy. Chlorine does not work in ice-cold water from mountain streams.
- *Organic content of the water*—Mud, decayed vegetation, or other suspended organic debris in water chemically combines with chlorine, making it unavailable as a disinfectant.
- *Chlorine contact time*—The longer that *Giardia* cysts are exposed to chlorine, the more likely it is that the chemical will kill them.
- *Chlorine concentration*—The higher the chlorine concentration, the more likely it is that chlorine will kill *Giardia* cysts. Most water treatment facilities try to add enough chlorine to give a free (unbound) chlorine residual at the customer tap of 0.5 mg per liter of water.

These five variables are so closely interrelated that improving one can often compensate for another; for example, if chlorine efficacy is expected to be low because water is obtained from an icy stream, the chlorine contact time or chlorine concentration, or both, could be increased. In the case of *Giardia*-contaminated water, producing safe drinking water with a chlorine concentration of 1 mg per liter and contact time as short as 10 minutes might be possible if all the other variables are optimal—a pH of 7.0, water temperature of 25°C, and total organic content of the water close to zero. On the other hand, if all of these variables are unfavorable—pH of 7.9, water temperature of 5°C, and high organic content—chlorine concentrations in excess of 8 mg/L with several hours of contact time may not be consistently effective. Because water conditions and water treatment plant operations (especially those related to water retention time and, therefore, to chlorine contact time) vary considerably in different parts of the United States, neither the USEPA nor the CDC has been able to identify a chlorine concentration that would be safe yet effective against *Giardia* cysts under all water conditions. For this reason, the use of chlorine as a preventive measure against waterborne giardiasis generally has been used under outbreak conditions when the amount of chlorine and contact time have been tailored to fit specific water conditions and the existing operational design of the water utility.

In an outbreak, for example, the local health department and water utility may issue an advisory to boil water, may increase the chlorine residual at the consumer's tap from 0.5 mg/L to 1 or 2 mg/L, and, if the physical layout and operation of the water treatment facility permit, increase the chlorine contact time. These are emergency procedures intended to reduce the risk of transmission until a filtration device can be installed or repaired or until an alternative source of safe water (a well, for example) can be made operational.

The long-term solution to the problem of municipal waterborne outbreaks of giardiasis involves improvements in and more widespread use of filters in the municipal water treatment process. The sand filters most commonly used in municipal water treatment today cost millions of dollars to install, which makes them unattractive for many small communities. The pore sizes in these filters are not sufficiently small to remove a *Giardia* (6 to 9 μm by 8 to 12 μm). For the sand filter to remove *Giardia* cysts from the water effectively, the water must receive some additional treatment before it reaches the filter. The flow of water through the filter bed must also be carefully regulated.

An ideal prefilter treatment for muddy water would include sedimentation (a holding pond where large suspended particles are allowed to settle out by the action of gravity) followed by flocculation or coagulation (the addition of chemicals such as alum or ammonium to cause microscopic particles to clump together). The sand filter easily removes the large particles resulting from the coagulation–flocculation process, including *Giardia* cysts bound to other microparticulates. Chlorine is then added to kill the bacteria and viruses that may escape the filtration process. If the water comes from a relatively clear source, chlorine may be added to the water before it reaches the filter.

The successful operation of a complete waterworks operation is a complex process that requires considerable training. Troubleshooting breakdowns or recognizing the potential problems in the system before they occur often requires the skills of an engineer. Unfortunately, most small water utilities with water treatment facilities that include filtration cannot afford the services of a full-time engineer. Filter operation or maintenance problems in such systems may not be detected until a *Giardia* outbreak is recognized in the community. The bottom line is that, although filtration is the best protection against waterborne giardiasis that water treatment technology has to offer for municipal water systems, it is not infallible. For municipal water filtration facilities to work properly, they must be properly constructed, operated, and maintained.

Whenever possible, persons outdoors should carry drinking water of known purity with them. When this is not practical and when water from streams, lakes, ponds, or other outdoor sources must be used, time should be taken to properly disinfect the water before drinking it.

Cryptosporidium

Ernest E. Tyzzer first described the protozoan parasite *Cryptosporidium* in 1907. Tyzzer frequently found a parasite in the gastric glands of laboratory mice. Tyzzer identified the parasite as a sporozoan but of uncertain taxonomic status, and he named it *Cryptosporidium muris*. Later, in 1910, after more detailed study, he proposed *Cryptosporidium* as a new genus and *muris* as the type of species. Amazingly, except for developmental stages, Tyzzer's original description of the life cycle (see Figure 5.10) was later confirmed by electron microscopy. In 1912, Tyzzer described another new species, *Cryptosporidium parvum* (Tyzzer, 1912).

For almost 50 years, Tyzzer's discovery of the genus *Cryptosporidium* remained (like himself) relatively obscure because it appeared to be of no medical or economic importance. Slight rumblings of the importance of the genus began to be felt in the medical community when Slavin (1955) wrote about a new species, *Cryptosporidium melagridis*, which was associated with illness and death in turkeys. Interest remained slight even when *Cryptosporidium* was found to be associated with bovine diarrhea (Panciera et al., 1971).

It was not until 1982 that worldwide interest focused on the study of organisms in the genus *Cryptosporidium*. At that time, the medical community and other interested parties were beginning a full-scale, frantic effort to find out as much as possible about acquired immune deficiency syndrome (AIDS), and the CDC reported that 21 AIDS-infected males from six large cities in the United States had severe protracted diarrhea caused by *Cryptosporidium*. It was in 1993, though, that *Cryptosporidium*—the "pernicious parasite"—made itself and Milwaukee famous (Mayo Foundation, 1996).

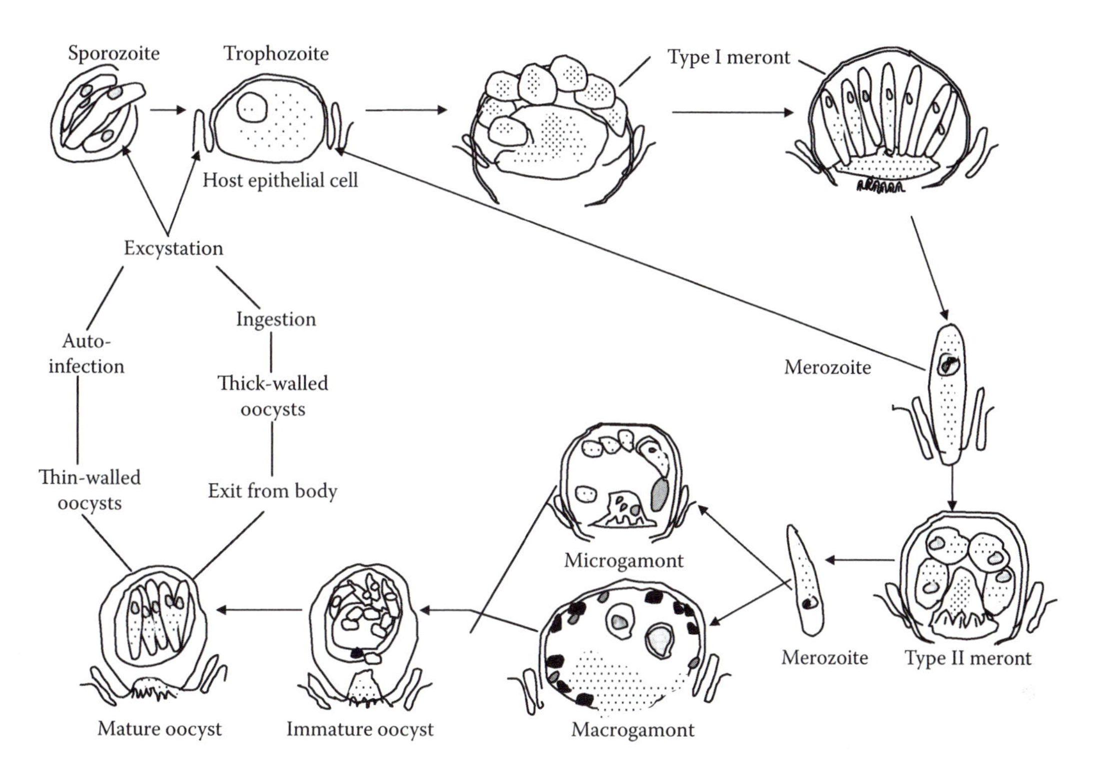

FIGURE 5.10 Life cycle of *Cryptosporidium parvum*.

> ***Note:*** The *Cryptosporidium* outbreak in Milwaukee caused the deaths of 100 people—the largest episode of waterborne disease in the United States in the 70 years since health officials began tracking such outbreaks.

The massive waterborne outbreak in Milwaukee (more than 400,000 persons developed acute and often prolonged diarrhea or other gastrointestinal symptoms) increased interest in *Cryptosporidium* to an exponential level. The Milwaukee incident spurred both public interest and the interest of public health agencies, agricultural and environmental agencies and groups, and suppliers of drinking water. This increase in interest level and concern spurred on new studies of *Cryptosporidium*, with an emphasis on developing methods for recovery, detection, prevention, and treatment (Fayer et al., 1997).

The USEPA is particularly interested in this pathogen. In its reexamination of regulations on water treatment and disinfection, the USEPA issued a maximum contaminant level goal (MCLG) and contaminant candidate list (CCL) for *Cryptosporidium*. Its similarity to *Giardia lamblia* and the need for an efficient conventional water treatment capable of eliminating viruses forced the USEPA to regulate surface water supplies in particular. The Enhanced Surface Water Treatment Rule (ESWTR) includes regulations ranging from watershed protection to specialized operation of treatment plants (certification of operators and state overview) and effective chlorination. Protection against *Cryptosporidium* includes control of waterborne pathogens such as *Giardia* and viruses (De Zuane, 1997).

Basics of *Cryptosporidium*

Cryptosporidium is one of several single-celled protozoan genera in the phylum Apircomplexa (all referred to as coccidian). *Cryptosporidium* along with other genera in the phylum Apircomplexa develop in the gastrointestinal tract of vertebrates through all of their life cycle; in short, they live

TABLE 5.4
Species of *Cryptosporidium*

Species	Host
C. baileyi	Chicken
C. felis	Domestic cat
C. meleagridis	Turkey
C. murishouse	House mouse
C. nasorium	Fish
C. parvum	House mouse
C. serpentis	Corn snake
C. wrairi	Guinea pig

Source: Fayer, R., Ed., *Cryptosporidium and Cryptosporidiosis*, CRC Press, Boca Raton, FL, 1997. With permission.

in the intestines of animals and people. This microscopic pathogen causes a disease called *cryptosporidiosis*. The dormant (inactive) form of *Cryptosporidium* is called an *oocyst* and is excreted in the feces (stool) of infected humans and animals. The tough-walled oocysts survive under a wide range of environmental conditions.

Several species of *Cryptosporidium* were incorrectly named after the host in which they were found, and subsequent studies have invalidated many species. Now, eight valid species of *Cryptosporidium* (see Table 5.4) have been named. Upton (1997) reported that *C. muris* infects the gastric glands of laboratory rodents and several other mammalian species but is not known to infect humans (even though several texts state otherwise). *C. parvum*, however, infects the small intestine of an unusually wide range of mammals, including humans, and is the zoonotic species responsible for human cryptosporidiosis. In most mammals, *C. parvum* is predominantly a parasite of neonate (newborn) animals. Upton pointed out that, even though exceptions occur, older animals generally develop poor infections, even when unexposed previously to the parasite. Humans are the one host that can be seriously infected at any time in their lives, and only previous exposure to the parasite results in either full or partial immunity to challenge infections.

Oocysts are present in most surface bodies of water across the United States, many of which supply public drinking water. Oocysts are more prevalent in surface waters when heavy rains increase runoff of wild and domestic animal wastes from the land or when sewage treatment plants are overloaded or break down. Only laboratories with specialized capabilities can detect the presence of *Cryptosporidium* oocysts in water. Unfortunately, current sampling and detection methods are unreliable. Recovering oocysts trapped on the material used to filter water samples is difficult. When a sample has been obtained, however, determining whether the oocyst is alive and if it is *C. parvum* and thus can infect humans is easily accomplished by looking at the sample under a microscope.

The number of oocysts detected in raw (untreated) water varies with location, sampling time, and laboratory methods. Water treatment plants remove most, but not always all, oocysts. Low numbers of oocysts are sufficient to cause cryptosporidiosis, but the low numbers of oocysts sometimes present in drinking water are not considered cause for alarm in the public. Protecting water supplies from *Cryptosporidium* demands multiple barriers. Why? Because *Cryptosporidium* oocysts have tough walls that can withstand many environmental stresses and are resistant to chemical disinfectants such as chlorine that are traditionally used in municipal drinking water systems.

Physical removal of particles, including oocysts, from water by filtration is an important step in the water treatment process. Typically, water pumped from rivers or lakes into a treatment plant is mixed with coagulants, which help settle out particles suspended in the water. If sand filtration is used, even more particles are removed. Finally, the clarified water is disinfected and piped to customers. Filtration is the only conventional method now in use in the United States for controlling *Cryptosporidium*.

Ozone is a strong disinfectant that kills protozoa if sufficient doses and contact times are used, but ozone leaves no residual for killing microorganisms in the distribution system, as does chlorine. The high costs of new filtration or ozone treatment plants must be weighed against the benefits of additional treatment. Even well-operated water treatment plants cannot ensure that drinking water will be completely free of *Cryptosporidium* oocysts. Water treatment methods alone cannot solve the problem; watershed protection and monitoring of water quality are critical. Land use controls such as septic system regulations and best management practices to control runoff can help keep human and animal wastes out of water.

Under the Surface Water Treatment Rule of 1989, public water systems must filter surface water sources unless water quality and disinfection requirements are met and a watershed control program is maintained. This rule, however, did not address *Cryptosporidium*. The USEPA has now set standards for turbidity (cloudiness) and coliform bacteria (which indicate that pathogens are probably present) in drinking water. Frequent monitoring must occur to provide officials with early warning of potential problems to enable them to take steps to protect public health. Unfortunately, no water quality indicators can reliably predict the occurrence of cryptosporidiosis. More accurate and rapid assays of oocysts will make it possible to notify residents promptly if their water supply is contaminated with *Cryptosporidium* and thus avert outbreaks.

Note: The collaborative efforts of water utilities, government agencies, healthcare providers, and individuals are needed to prevent outbreaks of cryptosporidiosis.

Cryptosporidiosis

> *Cryptosporidium parvum* is an important emerging pathogen in the U.S. and a cause of severe, life-threatening disease in patients with AIDS. No safe and effective form of specific treatment for cryptosporidiosis has been identified to date. The parasite is transmitted by ingestion of oocysts excreted in the feces of infected humans or animals. The infection can therefore be transmitted from person-to-person, through ingestion of contaminated water (drinking water and water used for recreational purposes) or food, from animal to person, or by contact with fecally contaminated environmental surfaces. Outbreaks associated with all of these modes of transmission have been documented. Patients with human immunodeficiency virus infection should be made more aware of the many ways that *Cryptosporidium* species are transmitted, and they should be given guidance on how to reduce their risk of exposure.
>
> **Juranek (1995)**

Since the Milwaukee outbreak, concern about the safety of drinking water in the United States has increased, and new attention has been focused on determining and reducing the risk of acquiring cryptosporidiosis from community and municipal water supplies. Cryptosporidiosis is spread by putting something in the mouth that has been contaminated with the stool of an infected person or animal. In this way, people swallow the *Cryptosporidium* parasite. As mentioned earlier, a person can become infected by drinking contaminated water or eating raw or undercooked food contaminated with *Cryptosporidium* oocysts, by direct contact with the droppings of infected animals or stools of infected humans, or by hand-to-mouth transfer of oocysts from surfaces that may have become contaminated with microscopic amounts of stool from an infected person or animal.

The symptoms may appear 2 to 10 days after infection by the parasite. Although some persons may not have symptoms, others have watery diarrhea, headache, abdominal cramps, nausea, vomiting, and low-grade fever. These symptoms may lead to weight loss and dehydration. In otherwise healthy persons, these symptoms usually last 1 to 2 weeks, at which time the immune system is able to defeat the infection. In persons with suppressed immune systems, such as persons who have AIDS or who recently have had an organ or bone marrow transplant, the infection may continue and become life threatening.

Currently, no safe and effective cure for cryptosporidiosis exists. People with normal immune systems improve without taking antibiotic or antiparasitic medications. The treatment recommended for this diarrheal illness is to drink plenty of fluids and to get extra rest. Physicians may prescribe medication to slow the diarrhea during recovery. The best way to prevent cryptosporidiosis is to

- Avoid water or food that may be contaminated.
- Wash hands after using the toilet and before handling food.
- Be sure, if you work in a daycare center, to wash your hands thoroughly with plenty of soap and warm water after every diaper change, even if you wear gloves when changing diapers.

During community-wide outbreaks caused by contaminated drinking water, drinking water practitioners should inform the public to boil drinking water for 1 minute to kill the *Cryptosporidium* parasite.

Cyclospora

Cyclospora organisms, which until recently were considered blue–green algae, were discovered at the turn of the 19th century. The first human cases of *Cyclospora* infection were reported in the 1970s. In the early 1980s, *Cyclospora* was recognized as a pathogen in patients with AIDS. We now know that *Cyclospora* is endemic in many parts of the world and appears to be an important cause of traveler's diarrhea. *Cyclospora* are two to three times larger than *Cryptosporidium* but otherwise have similar features. *Cyclospora* diarrheal illness in patients with healthy immune systems can be cured by a week of therapy with timethoprim–sulfamethoxazole (TMP–SMX).

So, what exactly is *Cyclospora?* In 1998, the CDC described *Cyclospora cayetanensis* as a unicellular parasite previously known as a cyanobacterium-like (blue–green algae-like) or coccidian-like body. The disease is known as *cyclosporasis*. *Cyclospora* infects the small intestine and causes an illness characterized by diarrhea with frequent stools. Other symptoms can include loss of appetite, bloating, gas, stomach cramps, nausea, vomiting, fatigue, muscle ache, and fever. Some individuals infected with *Cyclospora* may not show symptoms. Since the first known cases of illness caused by *Cyclospora* infection were reported in the medical journals in the 1970s, cases have been reported with increasing frequency from around the world (in part because of the availability of better techniques for detecting the parasite in stool specimens).

Huang et al. (1995) detailed what they believe is the first known outbreak of diarrheal illness associated with *Cyclospora* in the United States. The outbreak, which occurred in 1990, consisted of 21 cases of illness among physicians and others working at a Chicago hospital. Contaminated tap water from a physicians' dormitory at the hospital was the probable source of the organisms. The tap water probably picked up the organism while in a storage tank at the top of the dormitory after the failure of a water pump.

The transmission of *Cyclospora* is not a straightforward process. When infected persons excrete the oocyst state of *Cyclospora* in their feces, the oocysts are not infectious and may require from days to weeks to become so (i.e., to sporulate). Thus, transmission of *Cyclospora* directly from an infected person to someone else is unlikely; however, indirect transmission can occur if an infected person contaminates the environment and oocysts have sufficient time, under appropriate conditions, to become infectious. For example, *Cyclospora* may be transmitted by ingestion of water or food contaminated with oocysts. Outbreaks linked to contaminated water, as well as outbreaks linked to various types of fresh produce, have been reported (Herwaldt et al., 1997). The various modes of transmission and sources of infection are not yet fully understood nor is it known whether animals can be infected and serve as sources of infection for humans.

Note: *Cyclospora* organisms have not yet been grown in tissue cultures or laboratory animal models.

Persons of all ages are at risk for infection. Persons living or traveling in developing countries may be at increased risk, but infection can be acquired worldwide, including in the United States. In some countries of the world, infection appears to be seasonal. Based on currently available information, avoiding water or food that may be contaminated with stool is the best way to prevent infection. Reinfection can occur.

Note: Pathogenic parasites are not easily removed or eliminated by conventional treatment and disinfection unit processes (De Zuane, 1997). This is particularly true for *Giardia lamblia*, *Cryptosporidium*, and *Cyclospora*. Filtration facilities can be adjusted with regard to depth, prechlorination, filtration rate, and backwashing to become more effective in the removal of cysts. The pretreatment of protected watershed raw water is a major factor in the elimination of pathogenic protozoa.

THOUGHT-PROVOKING QUESTIONS

5.1 The measurement of progress toward meeting the goals of national clean water legislation has historically been dominated by measures of administrative activity and chemical water quality. These measures have increasingly come into question largely because they neither communicate about nor measure real changes and conditions in the environment. Is this an accurate statement? Explain.

5.2 Comparisons of assessments based on biological criteria and chemical criteria assessments demonstrate that relying solely on the latter carries a significant risk of making type II errors (i.e., incorrect rejection of a true null hypothesis) with regard to waterbody conditions. Do you agree or disagree with this statement? Explain.

5.3 Explain how effective and meaningful measures can be devised to determine progress toward meeting the goals of the CWA.

5.4 Water quality can easily become a confused and nebulous concept, especially when no demonstrable or tangible end product can be easily identified. Do you agree or disagree with this statement? Explain.

5.5 The health and well-being of the aquatic biota in surface waters are an important barometer of how effectively we are achieving environmental goals. Do you agree or disagree with this statement. Why?

REFERENCES AND RECOMMENDED READING

Anon. (1998). Water bugs are algae, says expert. *The Age*, September 29.

Badenock, J. (1990). *Cryptosporidium in Water Supplies*. London: Her Majesty's Stationery Office.

Bingham, A.K., Jarroll, E.L., Meyer, E.A., and Radulescu, S. (1979). Introduction to *Giardia* excystation and the effect of temperature on cyst viability compared by eosin-exclusion and *in vitro* excystation in waterborne transmission of giardiasis. In *Waterborne Transmission of Giardiasis: Proceedings of a Symposium*, September 18–20, 1978, Jakubowski, W. and Hoff, J.C., Eds., USEPA 600/9-79-001. Washington, DC: U.S. Environmental Protection Agency, pp. 217–229.

Black, R.E., Dykes, A.C., Anderson, K.E., Wells, J.G., Sinclair, S.P., Gary, G.W., Hatch, M.H., and Gnagarosa, E.J. (1981). Handwashing to prevent diarrhea in day-care centers. *American Journal of Epidemiology*, 113, 445–451.

Black-Covilli, L.L. (1992). Basic environmental chemistry of hazardous and solid wastes. In *Fundamentals of Environmental Science and Technology*, Knowles, P.C., Ed. Rockville, MD: Government Institutes, pp. 13–30.

Brodsky, R.E., Spencer, H.C., and Schultz, M.G. (1974). Giardiasis in American travelers to the Soviet Union. *Journal of Infectious Diseases*, 130, 319–323.

CDC. (1979). *Intestinal Parasite Surveillance, Annual Summary 1978*. Atlanta, GA: Centers for Disease Control and Prevention.

CDC. (1984). *Water-Related Disease Outbreaks Surveillance, Annual Summary 1983*. Atlanta, GA: Centers for Disease Control and Prevention.

CDC. (1995). *Cryptosporidiosis Fact Sheet*. Atlanta, GA: Centers for Disease Control and Prevention.

CDC. (1997a). Outbreaks of cyclosporiasis: United States and Canada. *Morbidity and Mortality Weekly Report*, 46, 521–523.

CDC. (1997b). Outbreak of cyclosporiasis: northern Virginia–Washington, DC–Baltimore, Maryland, metropolitan area. *Morbidity and Mortality Weekly Report*, 46, 689–691.

Cibas, E.S. and Ducatman, B.S. (2003). *Cytology: Diagnostic Principles and Clinical Correlates*. London: Saunders.

Craun, G.T. (1979). Waterborne giardiasis in the United States: a review. *American Journal of Public Health*, 69, 817–819.

Craun, G.F. (1984). Waterborne outbreaks of giardiasis: current status. In *Giardia and Giardiasis,* Erlandsen, S.L. and Meyer, E.A., Eds. New York: Plenum Press, pp. 243–261.

Davidson, R.A. (1984). Issues in clinical parasitology: the treatment of giardiasis. *American Journal of Gastroenterology*, 79, 256–261.

De Zuane, J. (1997). *Handbook of Drinking Water Quality*. New York: John Wiley & Sons.

DeDuve, C. (1984). *A Guided Tour of the Living Cell*. New York: W.H. Freeman.

Fayer, R., Speer, C.A., and Dubey, J.P. (1997). The general biology of *Cryptosporidium*. In *Cryptosporidium and Cryptosporidiosis,* Fayer, R. and Xiao, L., Eds. Boca Raton, FL: CRC Press, pp. 1–41.

Finean, J.B. (1984). *Membranes and Their Cellular Functions*. Oxford: Blackwell Scientific Publications.

Frank, J. et al. (1995). A model of synthesis based on cryo-electron microscopy of the *E. coli* ribosome. *Nature*, 376, 440–444.

Frost, F., Plan, B., and Liechty, B. (1984). *Giardia* prevalence in commercially trapped mammals. *Journal of Environmental Health*, 42, 245–249.

Garrett, R. et al. (2000). *The Ribosome: Structure, Function, Antibiotics, and Cellular Interactions*. Washington, DC: American Society for Microbiology.

Herwaldt, B.L. et al. (1997). An outbreak in 1996 of cyclosporiasis associated with imported raspberries. *New England Journal of Medicine*, 336, 1548–1556.

Hickman, Jr., C.P. (1997). *Biology of Animals*, 7th ed. New York: McGraw-Hill.

Hoge, C.W. et al. (1995). Placebo-controlled trail of co-trimoxazole for *Cyclospora* infections among travelers and foreign residents in Nepal. *Lancet*, 345, 691–693.

Huang, P., Weber, J.T., Sosin, D.M. et al. (1995). The first reported outbreak of diarrheal illness associated with *Cyclospora* in the United States. *Annals of Internal Medicine*, 123, 401–414.

Jarroll, Jr., E.L., Bingham, A.K., and Meyer, E.A. (1980a). *Giardia* cyst destruction: effectiveness of six small-quantity water disinfection methods. *American Journal of Tropical Medicine and Hygiene*, 29, 8–11.

Jarroll, Jr., E.L., Bingham, A.K., and Meyer, E.A. (1980b). Inability of an iodination method to destroy completely *Giardia* cysts in cold water. *Western Journal of Medicine*, 132, 567–569.

Jarroll, Jr., E.L., Bingham, A.K., and Meyer, E.A. (1981). Effect of chlorine on *Giardia lamblia* cyst viability. *Applications in Environmental Microbiology*, 41, 483–487.

Jokipii, L. and Jokippii, A.M.M. (1974). Giardiasis in travelers: a prospective study. *Journal of Infectious Diseases*, 130, 295–299.

Juranek, D.D. (1995). *Cryptosporidium parvum*. *Clinical Infectious Diseases*, 21(Suppl. 1), S57–S61.

Kemmer, F.N. (1979). *Water: The Universal Solvent*, 2nd ed. Oak Brook, IL: Nalco Chemical Co.

Keystone, J.S., Karden, S., and Warren, M.R. (1978). Person-to-person transmission of *Giardia lamblia* in day-care nurseries. *Canadian Medical Association Journal*, 119: 241–248.

Keystone, J.S., Yang, J., Grisdale, D., Harrington, M., Pillow, L., and Andrychuk, R. (1984). Intestinal parasites in metropolitan Toronto day-care centres. *Canadian Medical Association Journal*, 131, 733–735.

Kordon, C. (1993). *The Language of the Cell*. New York: McGraw-Hill.

Koren, H. (1991). *Handbook of Environmental Health and Safety: Principles and Practices*. Chelsea, MI: Lewis Publishers.

LeChevallier, M.W., Norton, W.D, and Lee, R.G. (1991). Occurrence of *Giardia* and *Cryptosporidium* spp. in surface water supplies. *Applied and Environmental Microbiology*, 57(9), 2610–2616.

Marchin, B.L., Fina, L.R., Lambert, J.L., and Fina, G.T. (1983). Effect of resin disinfectants-I3 and -I5 on *Giardia muris* and *Giardia lamblia*. *Applied and Environmental Microbiology*, 46, 965–969.

Martin, S. (1981). *Understanding Cell Structure*. New York: Cambridge University Press.

Mayo Foundation. (1996). *The "Bug" That Made Milwaukee Famous*. Rochester, MN: Mayo Foundation for Medical Education and Research.

Murray, A.W. (1993). *The Cell Cycle: An Introduction*. New York: W.H. Freeman.

NAS. (1977). *Drinking Water and Health*, Vol. 1. Washington, DC: National Research Council, National Academy of Sciences.

NAS. (1982). *Drinking Water and Health*, Vol. 4. Washington, DC: National Research Council, National Academy of Sciences.

Panciera, R.J., Thomassen, R.W., and Garner, R.M. (1971). Cryptosporidial infection in a calf. *Veterinary Pathology*, 8, 479.

Patterson, D.J. and Hedley, S. (1992). *Free-Living Freshwater Protozoa: A Color Guide*. Boca Raton, FL: CRC Press.

Pickering, L.K., Evans, D.G., Dupont, H.L., Vollet III, J.J., and Evans, Jr., D.J. (1981). Diarrhea caused by *Shigella*, *Rotavirus*, and *Giardia* in day-care centers: prospective study. *Journal of Pediatrics*, 99, 51–56.

Pickering, L.K., Woodward, W.E., Dupont, H.L., and Sullivan, P. (1984). Occurrence of *Giardia lamblia* in children in day care centers. *Journal of Pediatrics*, 104, 522–526.

Prescott, L.M., Harley, J.P., and Klein, D.A. (1993). *Microbiology*, 3rd ed. Dubuque, IA: William C. Brown.

Rendtorff, R.C. (1954). The experimental transmission of human intestinal protozoan parasites. II. *Giardia lamblia* cysts given in capsules. *American Journal of Hygiene*, 59, 209–220.

Rose, J.B., Gerb, C.P., and Jakubowski, W. (1991). Survey of potable water supplies for *Cryptosporidium* and *Giardia*. *Environmental Science & Technolology*, 25, 1393–1399.

Sealy, D.P. and Schuman, S.H. (1983). Endemic giardiasis and day care. *Pediatrics*, 72, 154–158.

Serafini, A. (1993). *The Epic History of Biology*. New York: Plenum.

Singleton, P. (1992). *Introduction to Bacteria*, 2nd ed. New York: John Wiley & Sons.

Singleton, P. and Sainsbury, D. (1994). *Dictionary of Microbiology and Molecular Biology*, 2nd ed. New York: John Wiley & Sons.

Slavin, D. (1955). *Cryptosporidium meleagridis* (s. nov.). *Journal of Comparative Pathology*, 65, 262.

Spellman, F.R. (1997). *Microbiology for Water/Wastewater Operators*. Lancaster, PA: Technomic.

Spirin, A. (1986). *Ribosomes Structure and Protein Biosynthesis*. Menlo Park, CA: Benjamin/Ammins Publishing.

Tchobanoglous, G. and Schroeder, E.D. (1987). *Water Quality*. Reading, MA: Addison-Wesley.

Thomas, L. (1974). *The Lives of a Cell*. New York: Viking Press.

Thomas, L. (1982). *Late Night Thoughts on Listening to Mahler's Ninth Symphony*. New York: Viking Press.

Thomas, L. (1995). *The Life of a Cell: Notes of a Biology Watcher*. New York: Penguin Books.

Tyzzer, E.E. (1907). A sporozoan found in the peptic glands of the common mouse. *Proceedings of the Society for Experimental Biology and Medicine*, 5, 12–13.

Tyzzer, E.E. (1912). *Cryptosporidium parvum* (sp. nov.), a coccidium found in the small intestine of the common mouse. *Archiv für Protistenkunde*, 26, 394–412.

Upton, S.J. (1997). *Basic Biology of Cryptosporidium*. Manhattan: Kansas State University.

USGS. (2014). *Bacteria in Water*. Washington, DC: U.S. Geological Survey (http://water.usgs.gov/edu/bacteria.html).

Visvesvara, G.S. et al. (1997). Uniform staining of *Cyclospora* oocysts in fecal smears by a modified safranin technique with microwave heating. *Journal of Clinical Microbiology*, 35, 730–733.

Walsh, J.A. (1981). Estimating the burden of illness in the tropics. In *Tropical and Geographic Medicine*, Warren, K.S. and Mahmoud, A.F., Eds. New York: McGraw-Hill, pp. 1073–1085.

Walsh, J.D. and Warren K.S. (1979). Selective primary health care: an interim strategy for disease control in developing countries. *New England Journal of Medicine*, 301, 976–974.

Weller, P.F. (1985). Intestinal protozoa: giardiasis. *Scientific American Medicine*, 12(4), 37–51.

Weniger, B.D., Blaser, M.J., Gedrose, H., Lippy, E.C., and Juranek, D.D. (1983). An outbreak of waterborne giardiasis associated with heavy water runoff due to warm weather and volcanic ashfall. *American Journal of Public Health*, 78, 868–872.

WHO. (1984). *Guidelines for Drinking Water Quality*. Vol. 1: *Recommendation*; Vol. 2: *Health Criteria and Other Supporting Information*. Geneva, Switzerland: World Health Organization.

Wistreich, G.A. and Lechtman, M.D. (1980). *Microbiology*, 3rd ed. New York: Macmillan.

Yoder, C.O. and Rankin, E.T. (1995). *The Role of Biological Criteria in Water Quality Monitoring, Assessment, and Regulation*. Columbus: Ohio Environmental Protection Agency.

6 Water Ecology

> The subject of man's relationship to his environment is one that has been uppermost in my own thoughts for many years. Contrary to the beliefs that seem often to guide our actions, man does not live apart from the world; he lives in the midst of a complex, dynamic interplay of physical, chemical, and biological forces, and between himself and this environment there are continuing, never-ending interactions.
>
> Unfortunately, there is so much that could be said. I am afraid it is true that, since the beginning of time, man has been a most untidy animal. But in the earlier days this perhaps mattered less. When men were relatively few, their settlements were scattered, their industries undeveloped; but now pollution has becomes one of the most vital problems of our society.
>
> **Carson (1998)**

A few years ago, my sampling partner and I were preparing to perform benthic macroinvertebrate sampling protocols in a wadeable section in one of the many reaches of the Yellowstone River. It was autumn, windy and cold. Before joining my partner knee deep in the slow-moving frigid waters, I stood for a moment at the bank and took in the surroundings. Except for a line of gold, the pallet of autumn can be austere in Yellowstone. The coniferous forests west of the Mississippi lack the bronzes, the coppers, the peach-tinted yellows, the livid scarlets that set mixed stands of the East aflame. All I could see in that line was the quaking aspen and its gold. The narrow, rounded crowns of *Populus tremuloides* display the closest thing to eastern autumn in the west. The aspen trunks stood white and antithetical against the darkness of the firs and pines. The shiny pale gold leaves, sensitive to the slightest rumor of wind, agitated and bounced the sun into my eyes. Each tree scintillated, like a heap of gold coins in free fall. The aspens tried desperately, by all of their flash and motion, to make up for the area's failure to exhibit the full spectrum of fall. But I didn't care much. I certainly wasn't disappointed. Although little is comparable to leaf-fall in autumn along the Appalachian Trail of my home, it simply didn't matter. Nature's display of gold against dark green eased the task that was before me. Bone-chilling water and all—it simply didn't matter.

Streams, the arteries of Earth, begin in capillary creeks, brooks, and rivulets. No matter the source, they move in only one direction—downhill, as the heavy hand of gravity tugs and drags the stream toward the sea. During its inexorable flow downward, now and then there is an abrupt change in geology. Boulders are mowed down by "slumping" (gravity) from their "in place" points high up on canyon walls.

As stream flow grinds, chisels, and sculpts the landscape, the effort is increased by momentum, augmented by turbulence provided by rapids, cataracts, and waterfalls. These falling waters always hypnotize us, like fire gazing or wave watching.

Before emptying into the sea, streams often pause, forming lakes. When one stares into a healthy lake, its phantom blue–green eye stares right back. Only for a moment—relatively speaking, of course, because all lakes are ephemeral, doomed. Eventually the phantom blue–green eye is close lidded by the moist verdant green of landfill.

For water that escapes the temporary bounds of a lake, most of it evaporates or moves on to the gigantic sink—the sea—where the cycles continues, forever more it is hoped.

This chapter deals primarily with the interrelationships (ecology) of biota (life forms) in placid water bodies (lakes) and running water (streams). The bias of this chapter is dictated by the author's experience and interest and by his belief that there is a need for water practitioners and students of water science to have a basic knowledge of water-related ecological processes.

Ecology is important because the environmental challenges we face today include all of the same ones we faced more than 40 years ago during the first Earth Day celebration in 1970. In spite of unflagging efforts of environmental professionals (and others), environmental problems remain. Many large metropolitan areas continue to be plagued by smog, our beaches are periodically polluted by oil spills, and many of our running and standing waters (streams and lakes) still suffer the effects of poorly treated sewage and industrial discharges. However, considerable progress has been made; for example, many of our rivers and lakes that were once unpleasant and unhealthy are now fishable and swimmable.

This is not to say that we are out of the woods yet. The problem with making progress in one area is that new problems arise that prove to be even more intractable than those we have already encountered. In restoring our running and standing waters to their original pristine state, this has been found to be the case. Those interested in the science of freshwater ecology (e.g., water practitioners and students) must understand the effects of environmental stressors, such as toxics, on the microbiological ecosystem in running and standing waters. Moreover, changes in these ecosystems must be measured and monitored.

The science of freshwater ecology is a dynamic discipline; new scientific discoveries are made daily and new regulatory requirements are almost as frequent. Today's emphasis is placed on other aspects of freshwater ecology, such as nonpoint-source pollution and total maximum daily load (TMDL). Finally, in the study of freshwater ecology it is important to remember that Mother Nature can perform wonders, but overload her and there might be hell to pay.

SETTING THE STAGE

> We poison the caddisflies in a stream and the salmon runs dwindle and die. We poison the gnats in a lake and the poison travels from link to link of the food chain and soon the birds of the lake margins become victims. We spray our elms and the following springs are silent of robin song, not because we sprayed the robins directly but because the poison traveled, step by step, through the now familiar elm leaf–earthworm–robin cycle. These are matters of record, observable, part of the visible world around us. They reflect the web of life—or death—that scientists know as ecology.
>
> **Carson (1962)**

What we do to any part of our environment has an impact on other parts. Probably the best way to state this interrelationship is to define ecology. Ecology is the science that deals with the specific interactions that exist between organisms and their living and nonliving environment. The word "ecology" is derived from the Greek word *oikos*, meaning home. Therefore, ecology is the study of the relation of an organism or a group of organisms to their environment (their "home").

Charles Darwin explained ecology in a famous passage in the *On the Origin of Species*, originally published in 1859, that helped establish the science of ecology (Darwin, 1998). According to Darwin, a "web of complex relations" binds all living things in any region. Adding or subtracting even a single species causes waves of change that race through the web, "onwards in ever-increasing circles of complexity." The simple act of adding cats to an English village would reduce the number of field mice. The reduced number of mice would benefit bumblebees, whose nests and honeycombs the mice often devour. Increasing the number of bumblebees would benefit the heartsease and red clover, which are fertilized almost exclusively by bumblebees. So, adding cats to the village could end by adding flowers. For Darwin, the whole of the Galapagos archipelago argues this fundamental lesson. The volcanoes are much more diverse in their ecology than their biology. The contrast suggests that, in the struggle for existence, species are shaped at least as much by the local flora and fauna as by the local soil and climate. "Why else would the plants and animals differ radically among islands that have the same geological nature, the same height, and climate?" (Darwin, 1998).

The environment includes everything important to the organism in its surroundings. The organism's environment can be divided into four parts:

1. Habitat and distribution (its place to live)
2. Other organisms (e.g., friendly or hostile)
3. Food
4. Weather (e.g., light, moisture, temperature, soil)

The two major subdivisions of ecology are *autecology* and *synecology*. Autecology is the study of an individual organism or a species (e.g., studying the ecology of the salmon). It emphasizes life history, adaptations, and behavior and examines communities, ecosystems, and the biosphere. Synecology, on the other hand, is the study of groups of organisms associated together as a unit. It deals with the environmental problems caused by humankind, such as the effects of discharging phosphorous-laden effluent into a stream or lake. The activities of human beings are having a major impact on many natural areas, so it is important to realize that the study of ecology must involve people.

Key Definitions

Each division of ecology has its own set of terms that are essential for communication between ecologists and those who are studying running and standing water ecological systems. Along with basic ecological terms, key terms that specifically pertain to this chapter are defined and presented below.

Abiotic factor—The nonliving part of the environment composed of sunlight, soil, mineral elements, moisture, temperature, and topography.

Aeration—A process whereby water and air or oxygen are mixed.

Bacteria—Primitive, single-celled organisms with a variety of shapes and nutritional needs; among the most common microorganisms in water.

Biochemical oxygen demand (BOD)—A widely used parameter of organic pollution applied to both wastewater and surface water; involves the measurement of the dissolved oxygen used by microorganisms in the biochemical oxidation of organic matter.

Biotic factor (community)—The living part of the environment composed of organisms that share the same area and are mutually sustaining, interdependent, and constantly fixing, utilizing, and dissipating energy.

Biotic index—Systematic survey of invertebrate aquatic organism used to correlate with river quality. The diversity of species in an ecosystem is often a good indicator of the presence of pollution: the greater the diversity, the lower the degree of pollution.

Climax community—The terminal stage of ecological succession in an area.

Community—In an ecological sense, all of the populations occupying a given area.

Competition—A critical factor for organisms in any community; animals and plants must compete successfully in the community to stay alive.

Decomposition—The breakdown of complex material into simple substances by chemical or biological processes.

Dissolved oxygen (DO)—The amount of oxygen dissolved in a stream is an indication of the degree of health of the stream and its ability to support a balanced aquatic ecosystem.

Ecosystem—The community and the nonliving environment functioning together as an ecological system.

Emigration—The departure of organisms from one place to take up residence in another area.

Eutrophication—The natural aging of a lake or land-locked body of water, which results in organic material being produced in abundance due to a ready supply of nutrients accumulated over the years.

Habitat—Refers to the place where an organism lives.

Heterotroph—Any living organism that obtains energy by consuming organic substances produced by other organisms.

Immigration—The movement of organisms into a new area of residence.

Limiting factor—A necessary material that is in short supply. Because of the lack of it, an organism cannot reach its full potential.

Niche—The role that an organism plays in its natural ecosystem, including its activities, resource use, and interaction with other organisms.

Nonpoint-source pollution—Source of pollutants found within the landscape (e.g., agricultural runoff).

Point-source pollution—Source of pollutants from an identifiable discharge point, such as a smokestack or sewage treatment plant.

Pollution—An adverse alteration to the environment by a pollutant.

Population—A group of organisms of a single species inhabiting a certain region at a particular time.

Sewage—Liquid wastes from a community; domestic sewage comes from housing, and industrial sewage normally comes from mixed industrial and residential sources.

Succession—A process that occurs subsequent to disturbance and involves the progressive replacement of biotic communities with others over time.

Surface runoff—Occurs after organic waste has been applied to a soil and some of the waste is transmitted by rainfall, snowmelt, or irrigation runoff into surface waters.

Symbiosis—A compatible association between dissimilar organisms to their mutual advantage.

Trophic level—The feeding position occupied by a given organism in a food chain measured by the number of steps removed from the producers.

LEVELS OF ORGANIZATION

Odum (1983) suggested that the best way to delimit modern ecology is to consider the concept of *levels of organization*. Levels of organization can be simplified as shown in Figure 6.1. In this relationship, organs form an organism, organisms of a particular species form a population, and populations occupying a particular area form a community. Communities, interacting with nonliving or abiotic factors, separate in a natural unit to create a stable system known as the *ecosystem* (the major ecological unit), and the part of Earth in which the ecosystem operates is known as the *biosphere*. Every community is influenced by a particular set of abiotic factors (Tomera, 1989). Inorganic substances such as oxygen and carbon dioxide, among others, and some organic substances represent the abiotic part of the ecosystem.

The physical and biological environment in which an organism lives is referred to as its *habitat*; for example, the habitat of two common aquatic insects, the backswimmer (*Notonecta*) and the water boatman (*Corixa*), is the littoral zone of ponds and lakes (shallow, vegetation-choked areas; see Figure 6.2) (Odum, 1983). Within each level of organization of a particular habitat, each organism has a special role. The role the organism plays in the environment is referred to as its *niche*. A niche might be that the organism is food for some other organism or is a predator of other organisms. Odum (1975) referred to an organism's niche as its "profession." In other words, each organism has a job or role to fulfill in its environment. Although two different species might occupy the same habitat, niche separation based on food habits differentiates two species (Odum, 1983). Comparing the niches of the water backswimmer and the water boatman reveals such niche separation. The backswimmer is an active predator, while the water boatman feeds largely on decaying vegetation (McCafferty, 1981).

Organism → Population → Communities → Ecosystem → Biosphere

FIGURE 6.1 Levels of organization.

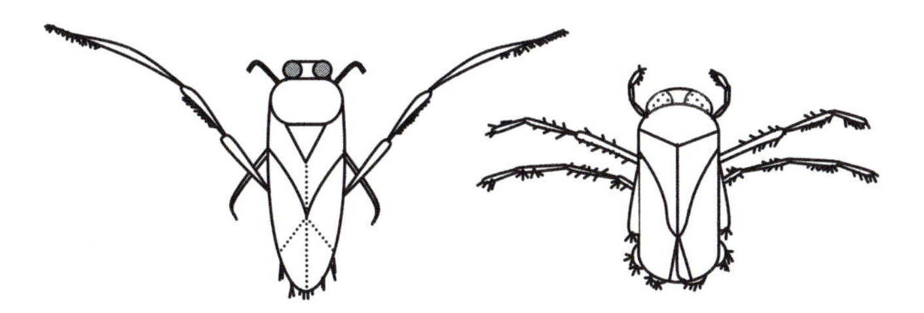

FIGURE 6.2 *Notonecta* (left) and *Corixa* (right). (Adapted from Odum, E.P., *Basic Ecology*, Saunders, Philadelphia, PA, 1983, p. 402.)

ECOSYSTEMS

An ecosystem is an area that includes all organisms therein and their physical environment. The ecosystem is the major ecological unit in nature. Living organisms and their nonliving environment are inseparably interrelated and interact upon each other to create a self-regulating and self-maintaining system. To create a self-regulating and self-maintaining system, ecosystems are homeostatic; that is, they resist any change through natural controls. These natural controls are important in ecology. This is especially the case because it is people through their complex activities who tend to disrupt natural controls.

The ecosystem encompasses both the living and nonliving factors in a particular environment. The living or biotic part of the ecosystem has two components: *autotrophic* and *heterotrophic*. The autotrophic (self-nourishing) component does not require food from its environment but can manufacture food from inorganic substances; for example, some autotrophic components (plants) manufacture needed energy through photosynthesis. Heterotrophic components, on the other hand, depend on autotrophic components for food.

The nonliving or abiotic part of the ecosystem is formed by three components: inorganic substances, organic compounds (linking biotic and abiotic parts), and climate regime. Figure 6.3 is a simplified diagram of a few of the living and nonliving components of an ecosystem found in a freshwater pond.

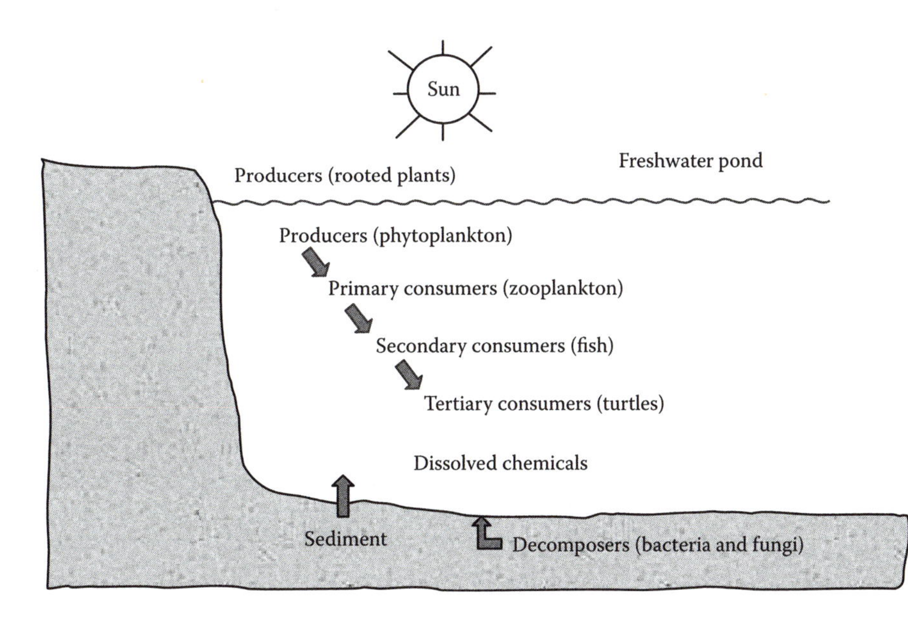

FIGURE 6.3 Major components of a freshwater pond ecosystem.

An ecosystem is a cyclic mechanism in which biotic and abiotic materials are constantly exchanged through *biogeochemical cycles*, where *bio* refers to living organisms; *geo* to water, air, rocks, or solids; and *chemical* to the chemical composition of the Earth. Biogeochemical cycles are driven by energy, directly or indirectly from the sun.

Figure 6.3 depicts an ecosystem where biotic and abiotic materials are constantly exchanged. Producers construct organic substances through photosynthesis and chemosynthesis. Consumers and decomposers use organic matter as their food and convert it into abiotic components; that is, they dissipate energy fixed by producers through food chains. The abiotic part of the pond in Figure 6.3 is formed of inorganic and organic compounds dissolved and in sediments such as carbon, oxygen, nitrogen, sulfur, calcium, hydrogen, and humic acids. Producers such as rooted plants and phytoplanktons represent the biotic part. Fish, crustaceans, and insect larvae make up the consumers. Mayfly nymphs represent detrivores, which feed on organic detritus. Decomposers make up the final biotic part. They include aquatic bacteria and fungi, which are distributed throughout the pond.

Note: An ecosystem is a cyclic mechanism. From a functional viewpoint, an ecosystem can be analyzed in terms of several factors. The factors important in this study include the biogeochemical cycles and energy and food chains.

BIOGEOCHEMICAL CYCLES

Several chemicals are essential to life and follow predictable cycles through nature. In these natural cycles, or *biogeochemical cycles*, the chemicals are converted from one form to another as they progress through the environment. The water/wastewater operator should be aware of those cycles dealing with the nutrients (e.g., carbon, nitrogen, sulfur) because they have a major impact on the performance of the plant and may require changes in operation at various times of the year to keep them functioning properly; this is especially the case in wastewater treatment. The microbiology of each cycle deals with the biotransformation and subsequent biological removal of these nutrients in wastewater treatment plants.

Note: Biogeochemical cycles can be categorized as *gaseous* or *sedimentary* (Smith, 1974). Gaseous cycles include the carbon and nitrogen cycles; the main sinks of nutrients in the gaseous cycle are the atmosphere and the ocean. Sedimentary cycles include the sulfur cycle; the main sinks for sedimentary cycles are soil and rocks of the Earth's crust.

CARBON CYCLE

Carbon, which is an essential ingredient of all living things, is the basic building block of the large organic molecules necessary for life. Carbon is cycled into food chains from the atmosphere, as shown in Figure 6.4. From Figure 6.4, it can be seen that green plants obtain carbon dioxide (CO_2) from the air and through photosynthesis, which was described by Asimov (1989) as the "most important chemical process on Earth." Photosynthesis produces the food and oxygen that all organisms require. Part of the carbon produced remains in living matter; the other part is released as CO_2 in cellular respiration. The carbon dioxide released by cellular respiration in all living organisms is returned to the atmosphere (Miller, 1988).

Some carbon is contained in buried dead and animal and plant materials. Much of these buried animal and plant materials were transformed into fossil fuels. Fossil fuels—coal, oil, and natural gas—contain large amounts of carbon. When fossil fuels are burned, stored carbon combines with oxygen in the air to form carbon dioxide, which enters the atmosphere. In the atmosphere, carbon dioxide acts as a beneficial heat screen, as it does not allow the radiation of Earth's heat out into space. This balance is important. The problem is that as more carbon dioxide from burning is released into the atmosphere the balance can and is being altered. Increases in the consumption of fossil fuels "coupled with the decrease in 'removal capacity' of the green belt is beginning to exceed

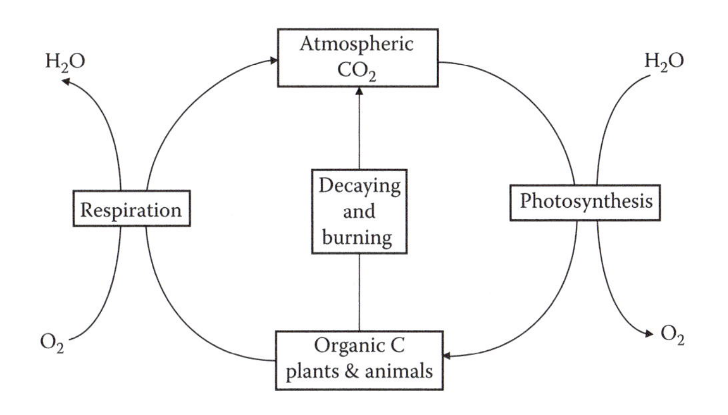

FIGURE 6.4 Carbon cycle.

the delicate balance" (Odum, 1983). Massive increases of carbon dioxide into the atmosphere tend to increase the possibility of global warming. The consequences of global warming "would be catastrophic ... and the resulting climatic change would be irreversible" (Abrahamson, 1988).

Nitrogen Cycle

Nitrogen is an essential element that all organisms need. In animals, nitrogen is a component of crucial organic molecules such as proteins and DNA and constitutes 1 to 3% dry weight of cells. Our atmosphere contains 78% by volume of nitrogen, yet it is not a common element. Although nitrogen is an essential ingredient for plant growth, it is chemically very inactive, and it must be *fixed* before the vast majority of the biomass can incorporate it. Special nitrogen-fixing bacteria found in soil and water fix nitrogen; thus, microorganisms play a major role in nitrogen cycling in the environment. These microorganisms (bacteria) have the ability to take nitrogen gas from the air and convert it to nitrate via a process known as *nitrogen fixation*. Some of these bacteria occur as free-living organisms in the soil. Others live in a *symbiotic relationship* with plants. A symbiotic relationship is a close relationship between two organisms of different species and one where both partners benefit from the association. An example of a symbiotic relationship, related to nitrogen, can be seen, for example, in the roots of peas. These roots have small swellings along their length. These contain millions of symbiotic bacteria that have the ability to take nitrogen gas from the atmosphere and convert it to nitrates that can be used by the plant. The plant is plowed back into the soil after the growing season to improve the nitrogen content. Price (1984) described the nitrogen cycle as an example "of a largely complete chemical cycle in ecosystems with little leaching out of the system." Simply, the nitrogen cycle provides various bridges between the atmospheric reservoirs and the biological communities (see Figure 6.5).

Atmospheric nitrogen is fixed by either natural or industrial means; for example, nitrogen is fixed by lightning or by soil bacteria that convert it to ammonia, then to nitrite, and finally to nitrates, which plants can use. Nitrifying bacteria make nitrogen from animal wastes. Denitrifying bacteria convert nitrates back to nitrogen and release it as nitrogen gas.

The logical question now is "What does all of this have to do with water?" The best way to answer this question is to ask another question. Have you ever dived into a slow-moving stream and had the noxious misfortune to surface right in the middle of an algal bloom? When this happens to you, the first thought that runs through your mind is, "Where is my nose plug?" Why? Because of the horrendous stench, disablement of the olfactory sense is a necessity.

If too much nitrate, for example, enters the water supply as runoff from fertilizers it produces an overabundance of algae known as an *algal bloom*. If this runoff from fertilizer gets into a body of water, algae may grow so profusely that they form a blanket over the surface. This usually happens in summer, when the light levels and warm temperatures favor rapid growth.

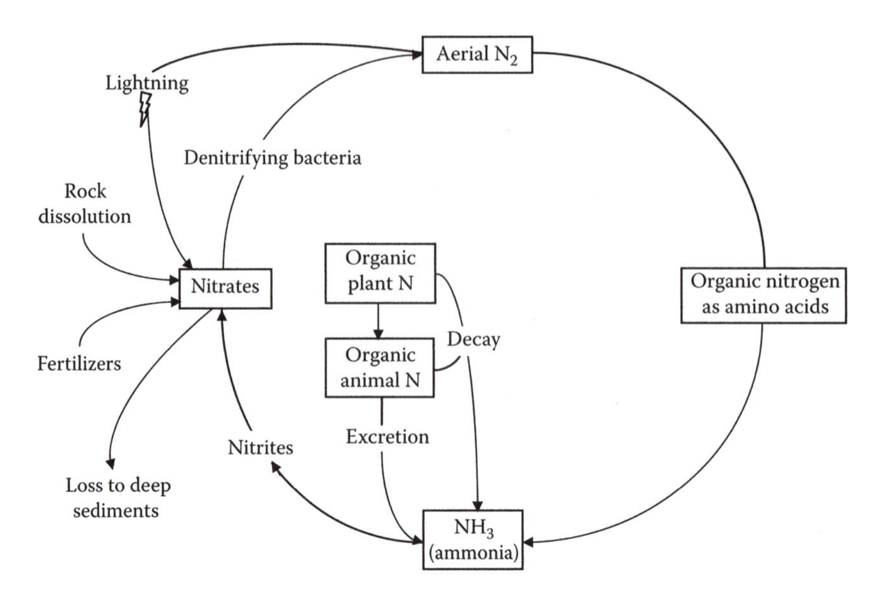

FIGURE 6.5 The nitrogen cycle.

Nitrogen is found in wastewater in the form of urea (Metcalf & Eddy, 2003). During wastewater treatment, the urea is transformed into ammonia nitrogen. Because ammonia exerts a BOD and chlorine demand, high quantities of ammonia in wastewater effluents are undesirable. The process of nitrification is utilized to convert ammonia to nitrates. *Nitrification* is a biological process that involves the addition of oxygen to the wastewater. If further treatment is necessary, another biological process called *denitrification* is used. In this process, nitrate is converted into nitrogen gas, which is lost to the atmosphere, as can be seen in Figure 6.5. From the wastewater operator's point of view, nitrogen and phosphorus are both considered limiting factors for productivity. Phosphorus discharged into streams contributes to pollution. Of the two, nitrogen is more difficult to control but is found in smaller quantities in wastewater.

Sulfur Cycle

Sulfur, like nitrogen, is characteristic of organic compounds. The sulfur cycle is both sedimentary and gaseous (see Figure 6.6). The principal forms of sulfur that are of special significance in water quality management are organic sulfur, hydrogen sulfide, elemental sulfur, and sulfate (Tchobanoglous and Schroeder, 1985). Bacteria play a major role in the conversion of sulfur from one form to another. In an anaerobic environment, bacteria break down organic matter, thereby producing hydrogen sulfide with its characteristic rotten-egg odor. The bacterium *Beggiatoa* converts hydrogen sulfide into elemental sulfur and sulfates. Other sulfates are contributed by the dissolving of rocks and some sulfur dioxide. Sulfur is incorporated by plants as proteins. Organisms then consume some of these plants. Many heterotrophic anaerobic bacteria liberate sulfur from proteins as hydrogen sulfide.

Phosphorus Cycle

Phosphorus is another chemical element that is common in the structure of living organisms (see Figure 6.7); however, the phosphorus cycle is different from the hydrologic, carbon, and nitrogen cycles because phosphorus is found in sedimentary rock. These massive deposits are gradually eroding to provide phosphorus to ecosystems. A large amount of eroded phosphorus ends up n deep sediments in the oceans and in lesser amounts in shallow sediments. Part of the phosphorus comes

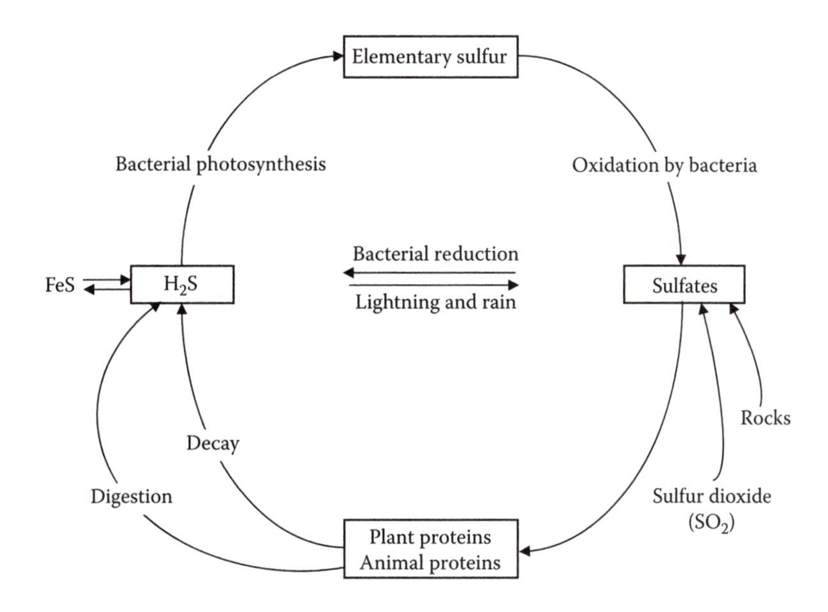

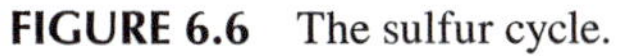
FIGURE 6.6 The sulfur cycle.

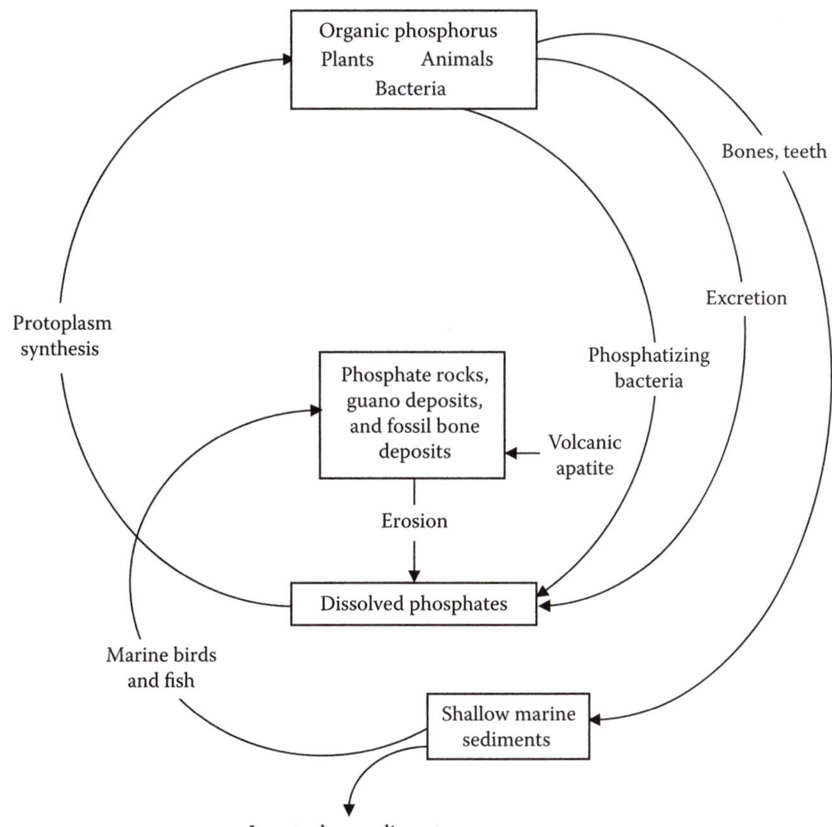

FIGURE 6.7 The phosphorus cycle.

to land when marine animals surface. Decomposing plant or animal tissue and animal droppings return organic forms of phosphorus to the water and soil. Fish-eating birds, for example, play a role in the recovery of phosphorus. Guano deposits (bird excreta) on the Peruvian coast are an example. Humans have hastened the rate of phosphorus loss through mining and the production of fertilizers, which is washed away and lost. Odum (1983) suggested, however, that there was no immediate cause for concern, as the known reserves of phosphate are quite large.

Phosphorus has become very important in water quality studies, because it is often found to be a limiting factor. Control of phosphorus compounds that enter surface waters and contribute to the growth of algal blooms is of considerable interest and has generated much study (Metcalf & Eddy, 2003). Upon entering a stream, phosphorus acts as a fertilizer, promoting the growth of undesirable algae populations or algal blooms. As the organic matter decays, dissolved oxygen levels decrease and fish and other aquatic species die.

Phosphorus discharged into streams is a contributing factor to stream pollution, but phosphorus is not the lone factor. Odum (1975) warned against what he called the *one-factor control hypothesis* (i.e., the one-problem/one-solution syndrome). He noted that environmentalists in the past have focused on one or two items, such as phosphorus contamination, and have failed to understand that the strategy for pollution control must involve reducing the input of all enriching and toxic materials.

> ***Note:*** Because of its high reactivity, phosphorus exists in combined form with other elements. Microorganisms produce acids that form soluble phosphate from insoluble phosphorus compounds. The phosphates are utilized by algae and terrestrial green plants, which in turn pass into the bodies of animal consumers. Upon the death and decay of organisms, phosphates are released for recycling.

ENERGY FLOW IN THE ECOSYSTEM

Simply defined, energy is the ability or capacity to do work. For an ecosystem to exist, it must have energy. All activities of living organisms involve work, which is the expenditure of energy. This means the degradation of a higher state of energy to a lower state. Two laws govern the flow of energy through an ecosystem: the *first and second laws of thermodynamics*. The first law, sometimes referred to as the *conservation law*, states that energy may not be created or destroyed. The second law states that no energy transformation is 100% efficient; that is, in every energy transformation, some energy is dissipated as heat. The term *entropy* is used as a measure of the non-availability of energy to a system. Entropy increases with an increase in dissipation. Because of entropy, input of energy in any system is higher than the output or work done; thus, the resultant efficiency is less than 100%.

The interaction of energy and materials in the ecosystem is important. Energy drives the biogeochemical cycles. Note that energy does not cycle as nutrients do in biogeochemical cycles; for example, when food passes from one organism to another, energy contained in the food is reduced systematically until all of the energy in the system is dissipated as heat. Price (1984) referred to this process as "a *unidirectional flow* of energy through the system, with no possibility for recycling of energy." When water or nutrients are recycled, energy is required. The energy expended in this recycling is not recyclable. The principal source of energy for any ecosystem is sunlight. Green plants, through the process of photosynthesis, transform the energy of the sun into carbohydrates, which are consumed by animals. This transfer of energy, again, is unidirectional—from producers to consumers. Often, this transfer of energy to different organisms is referred to as a *food chain*. Figure 6.8 shows a simple aquatic food chain.

FIGURE 6.8 Aquatic food chain.

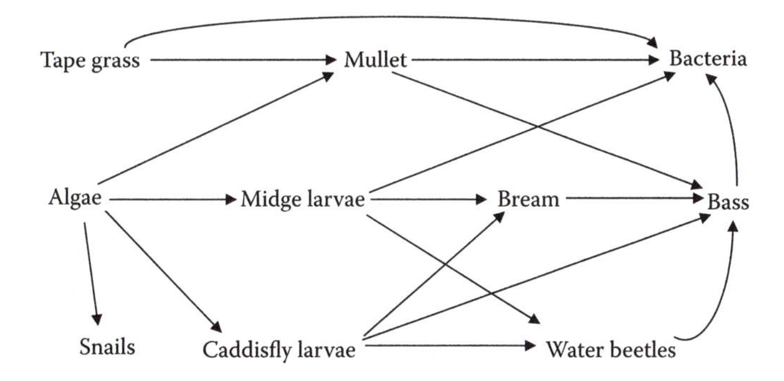

FIGURE 6.9 Simple food chain.

All organisms, alive or dead, are potential sources of food for other organisms. All organisms that share the same general type of food in a food chain are said to be at the same *trophic level* (nourishment or feeding level). Because green plants use sunlight to produce food for animals, they are the *producers*, or the first trophic level. The herbivores, which eat plants directly, are the *primary consumers*, or the second trophic level. The carnivores are flesh-eating consumers; they include several trophic levels from the third on up. At each transfer, a large amount of energy (about 80 to 90%) is lost as heat and wastes. Thus, nature normally limits food chains to four or five links. In aquatic ecosystems, however, food chains are commonly longer than those on land. The aquatic food chain is longer because several predatory fish may be feeding on the plant consumers. Even so, the built-in inefficiency of the energy transfer process prevents development of extremely long food chains.

Only a few simple food chains are found in nature. Most simple food chains are interlocked. This interlocking of food chains forms a *food web*. Most ecosystems support a complex food web. A food web involves animals that do not feed on one trophic level; for example, humans feed on both plants and animals. An organism in a food web may occupy one or more trophic levels. Trophic level is determined by an organism's role in its particular community, not by its species. Food chains and webs help to explain how energy moves through an ecosystem.

An important trophic level of the food web is comprised of the *decomposers*, which feed on dead plants or animals and play an important role in recycling nutrients in the ecosystem. Simply, there is no waste in ecosystems. All organisms, dead or alive, are potential sources of food for other organisms. An example of an aquatic food web is shown in Figure 6.9.

FOOD CHAIN EFFICIENCY

Energy from the sun is captured (via photosynthesis) by green plants and used to make food. Most of this energy is used to carry on the plant's life activities. The rest of the energy is passed on as food to the next level of the food chain. Nature limits the amount of energy that is accessible to organisms within each food chain. Not all food energy is transferred from one trophic level to the next. Only about 10% (the 10% rule) of the amount of energy is actually transferred through a food chain. If we apply the 10% rule to the diatoms–copepods–minnows–medium-fish–large-fish food chain shown in Figure 6.10, we can predict that 1000 grams of diatoms produce 100 grams of copepods, which will produce 10 grams of minnows, which will produce 1 gram of medium fish, which, in turn, will produce 0.1 gram of large fish. Thus, only about 10% of the chemical energy available at each trophic level is transferred and stored in usable form at the next level. The other 90% is lost to the environment as low-quality heat in accordance with the second law of thermodynamics.

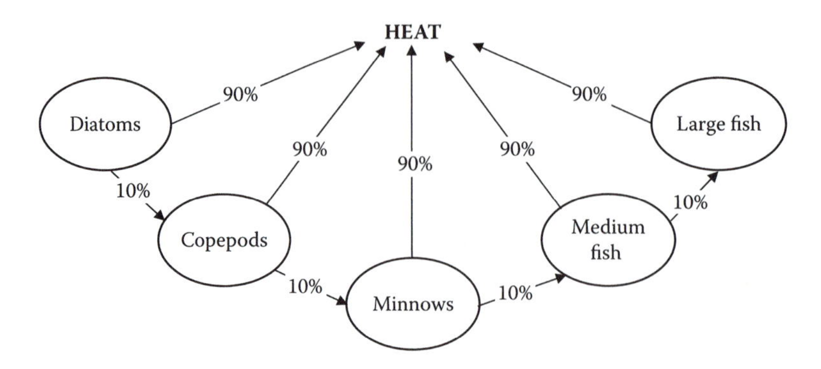

FIGURE 6.10 Sample food chain.

Ecological Pyramids

In the food chain, from the producer to the final consumer, it is clear that a particular community in nature often consists of several small organisms associated with a smaller and smaller number of larger organisms. A grassy field, for example, has a larger number of grasses and other small plants, a smaller number of herbivores such as rabbits, and an even smaller number of carnivores such as the fox. The practical significance of this is that we must have several more producers than consumers.

This pound-for-pound relationship, which requires more producers than consumers, can be demonstrated graphically by building an *ecological pyramid*. In an ecological pyramid, separate levels represent the number of organisms at various trophic levels in a food chain or bars placed one above the other with a base formed by producers and the apex formed by the final consumer. The pyramid shape is formed due to a great amount of energy loss at each trophic level. The same is true if the corresponding biomass or energy substitute numbers. Ecologists generally use three types of ecological pyramids: *number*, *biomass*, and *energy*. Obviously, differences exist among them, but here are some generalizations:

1. Energy pyramids must always be larger at the base than at the top (because of the second law of thermodynamics and the dissipation of energy as it moves from one trophic level to another).
2. Likewise, biomass pyramids (in which biomass is used as an indicator of production) are usually pyramid shaped. This is particularly true of terrestrial systems and aquatic ones dominated by large plants (marshes), in which consumption by heterotroph is low and organic matter accumulates with time. Biomass pyramids can sometimes be inverted. This is common in aquatic ecosystems where the primary producers are microscopic planktonic organisms that multiply very rapidly, have very short life spans, and are subject to heavy grazing by herbivores. At any single point in time, the amount of biomass in primary producers is less than that in larger, long-lived animals that consume primary producers.
3. Numbers pyramids can have various shapes (and not be pyramids at all), depending on the sizes of the organisms that make up the trophic levels. In forests, the primary producers are large trees and the herbivore level usually consists of insects, so the base of the pyramid is smaller than the herbivore level above it. In grasslands, the number of primary producers (grasses) is much larger than that of the herbivores above (large grazing animals).

PRODUCTIVITY

The flow of energy through an ecosystem starts with the fixation of sunlight by plants through photosynthesis. When evaluating an ecosystem, the measurement of photosynthesis is important. Ecosystems may be classified into highly productive or less productive; therefore, the study of ecosystems must involve some measure of the productivity of that ecosystem. Primary production is the rate at which the ecosystem's primary producers capture and store a given amount of energy, in a specified time interval. In simpler terms, primary productivity is a measure of the rate at which photosynthesis occurs. Four successive steps in the production process are as follows:

1. *Gross primary productivity*—The total rate of photosynthesis in an ecosystem during a specified interval.
2. *Net primary productivity*—The rate of energy storage in plant tissues in excess of the rate of aerobic respiration by primary producers.
3. *Net community productivity*—The rate of storage of organic matter not used.
4. *Secondary productivity*—The rate of energy storage at consumer levels.

When attempting to comprehend the significance of the term *productivity* as it relates to ecosystems, it is wise to consider an example. Consider the productivity of an agricultural ecosystem such as a wheat field. Often its productivity is expressed as the number of bushels produced per acre. This is an example of the harvest method for measuring productivity. For a natural ecosystem, several 1-m^2 plots are marked off, and the entire area is harvested and weighed to give an estimate of productivity as grams of biomass per square meter per given time interval. From this method, a measure of net primary production (net yield) can be measured.

Productivity, in both natural and cultured ecosystems, may vary considerably, not only between types of ecosystems but also within the same ecosystem. Several factors influence year-to-year productivity within an ecosystem. Such factors as temperature, availability of nutrients, fire, animal grazing, and human cultivation activities are directly or indirectly related to the productivity of a particular ecosystem.

Productivity can be measured in several different ways in the aquatic ecosystem. For example, the production of oxygen may be used to determine productivity. Oxygen content may be measured in several ways. One way is to measure it in the water every few hours for a period of 24 hours. During daylight, when photosynthesis is occurring, the oxygen concentration should rise. At night, the oxygen level should drop. The oxygen level can be measured by using a simple *x*–*y* graph. The oxygen level can be plotted on the *y*-axis and time plotted on the *x*-axis, as shown in Figure 6.11.

Another method of measuring oxygen production in aquatic ecosystems is to use light and dark bottles. Biochemical oxygen demand (BOD) bottles (300 mL) are filled with water to a particular height. One of the bottles is tested for the initial dissolved oxygen (DO), and then the other two

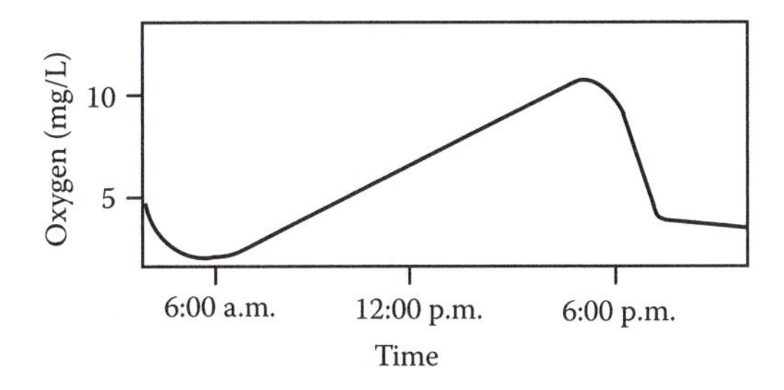

FIGURE 6.11 The diurnal oxygen curve for an aquatic ecosystem.

bottles (one clear, one dark) are suspended in the water at the depth from which they were taken. After a 12-hour period, the bottles are collected and the DO values for each bottle recorded. When the oxygen production is known, the productivity in terms of grams per meter per day can be calculated. In the aquatic ecosystem, pollution can have a profound impact upon the productivity of the system.

POPULATION ECOLOGY

Charles Darwin, in the first edition of his *On the Origin of Species*, made the theoretical potential rate of increase in population an integral part of his theory of evolution:

> The elephant is reckoned to be the slowest breeder of all known animals, and I have taken some pains to estimate its probable minimum rate of natural increase: it will be under the mark to assume that it begins breeding when thirty years old, and goes on breeding till ninety years old, bringing forth three pair of young in this interval; if this be so, at the end of the fifth century there would be alive fifteen million elephants, descended from the first pair.

Webster's Third New International Dictionary defines *population* as "the total number or amount of things especially within a given area; the organisms inhabiting a particular area or biotype; and a group of interbreeding biotypes that represents the level of organization at which speciation begins."

The term *population* is interpreted differently in various sciences. For example, in human demography a population is a set of humans in a given area. In genetics, a population is a group of interbreeding individuals of the same species which is isolated from other groups. In population ecology, a population is an interbreeding group of organisms of the same species, inhabiting the same area at a particular time.

If we want to study the organisms in a slow-moving stream or stream pond, we have two options. We can study each fish, aquatic plant, crustacean, and insect one by one, in which case we would be studying individuals. It would be relatively simple to do this if the subject were trout, but it would be difficult to separate and study each aquatic plant. The second option would be to study all of the trout, all of the insects of each specific kind, all of a certain aquatic plant type in the stream or pond at the time of the study. When ecologists study a group of the same kind of individuals in a given location at a given time, they are investigating a *population*. When attempting to determine the population of a particular species, it is important to remember that time is a factor. Time is important because populations change, whether it be at various times of the day, in different seasons of the year, or from year to year.

Population density, the number of organisms per unit of area or volume, may change dramatically. For example, if a dam is closed off in a river midway through salmon spawning season, with no provision allowed for fish movement upstream (a fish ladder), it would drastically decrease the density of spawning salmon upstream. Along with the swift and sometimes unpredictable consequences of change, it can be difficult to draw exact boundaries between various populations. Density is the characteristic of populations that has always been of greatest interest to us; for example, we might want to know if there are enough deer to hunt or if there are too many pests.

The population density or level of a species depends on *natality*, *mortality*, *immigration*, and *emigration*. Changes in population density are the result of both births and deaths. The birth rate of a population is its *natality* and the death rate is its *mortality*. In aquatic populations, two factors besides natality and mortality can affect density. In a run of returning salmon to their spawning grounds, for example, the density could vary as more salmon migrate in or as others leave the run for their own spawning grounds. The arrival of new salmon to a population from other places is immigration, and the departure of salmon from a population is emigration. Thus, natality and immigration increase population density, whereas mortality and emigration decrease it. The net increase in population is the difference between these two sets of factors.

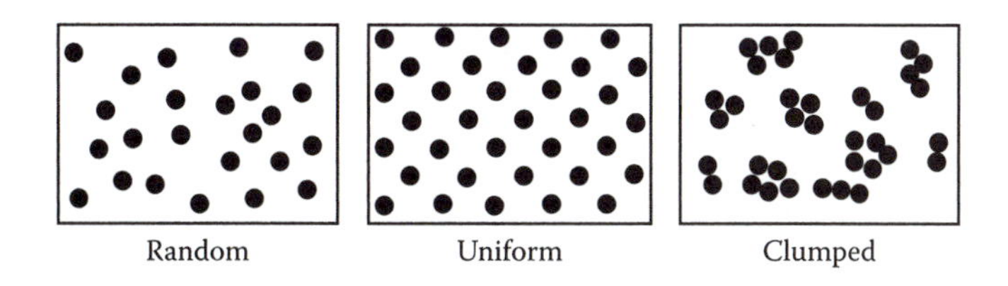

FIGURE 6.12 Basic patterns of distribution. (Adapted from Odum, E.P., *Fundamentals of Ecology*, 3rd ed., Saunders, Philadelphia, PA, 1971, p. 205).

The distribution of a population is the area in which that population can be found. Each organism occupies only those areas that can provide for its requirements, resulting in an irregular distribution. How a particular population is distributed within a given area has considerable influence on density. As shown in Figure 6.12, organisms in nature may be distributed in three ways. In a *random distribution*, there is an equal probability of an organism occupying any point in space, and each individual is independent of the others. In a *regular* or *uniform distribution*, organisms are spaced more evenly; they are not distributed by chance. Animals compete with each other and effectively defend a specific territory, excluding other individuals of the same species. In regular or uniform distribution, the competition between individuals can be quite severe and antagonistic to the point where the spacing generated is quite even. The most common distribution is the *contagious* or *clumped distribution*, where organisms are found in groups; this may reflect the heterogeneity of the habitat. Organisms that exhibit a contagious or clumped distribution may develop social hierarchies to live together more effectively. Animals within the same species have evolved many symbolic aggressive displays that carry meanings that not only are mutually understood but also prevent injury or death within the same species.

Note: Distribution is usually mapped for species of plants and animals. For example, bird guides contain distribution maps showing where each species of bird lives and reproduces.

The size of animal populations is constantly changing due to natality, mortality, emigration, and immigration. The population size will increase if the natality and immigration rates are high, and it will decrease if the mortality and emigration rates are high. Each population has an upper limit on size, often referred to as the *carrying capacity*. Carrying capacity is the optimum number of individuals of a species that can survive in a specific area over time. Stated differently, the carrying capacity is the maximum number of species that can be supported in a bioregion. A pond may be able to support only a dozen frogs depending on the food resources for the frogs in the pond. If there were 30 frogs in the same pond, at least half of them would probably die because the pond environment would not have enough food for them to live. Carrying capacity is based on the quantity of food supplies, the physical space available, the degree of predation, and several other environmental factors.

The carrying capacity can be of two types: ultimate and environmental. The *ultimate carrying capacity* is the theoretical maximum density; that is, it is the maximum number of individuals of a species in a place that can support itself without rendering the place uninhabitable. The *environmental carrying capacity* is the actual maximum population density that a species maintains in an area. Ultimate carrying capacity is always higher than environmental. Ecologists have concluded that a major factor that affects population stability or persistence is *species diversity*. Species diversity is a measure of the number of species and their relative abundance.

If the stress on an ecosystem is small, the ecosystem can usually adapt quite easily. Moreover, even when severe stress occurs, ecosystems have a way of adapting. Severe environmental change to an ecosystem can result from such natural occurrences as fires, earthquakes, and floods and from people-induced changes such as land clearing, surface mining, and pollution. One of the most important applications of species diversity is in the evaluation of pollution. Stress of any kind will reduce the species diversity of an ecosystem to a significant degree. In the case of domestic sewage pollution, for example, the stress is caused by a lack of dissolved oxygen (DO) for aquatic organisms.

Ecosystems can and do change; for example, if a fire devastates a forest, it will grow back eventually because of *ecological succession*. Ecological succession is the observed process of change (a normal occurrence in nature) in the species structure of an ecological community over time. Succession usually occurs in an orderly, predictable manner. It involves the entire system. The science of ecology has developed to such a point that ecologists are now able to predict several years in advance what will occur in a given ecosystem. Scientists know, for example, that if a burned-out forest region receives light, water, nutrients, and an influx or immigration of animals and seeds, it will eventually develop into another forest through a sequence of steps or stages. Ecologists recognize two types of ecological succession: primary and secondary. The particular type that takes place depends on the condition at a particular site at the beginning of the process.

Primary succession, sometimes referred to as *bare-rock succession*, occurs on surfaces such as hardened volcanic lava, bare rock, and sand dunes, where no soil exists, and where nothing has ever grown before (see Figure 6.13). Obviously, to grow, plants require soil; thus, soil must form on the bare rock before succession can begin. Usually this soil formation process results from weathering. Atmospheric exposure—weathering, wind, rain, and frost—produces tiny cracks and holes in rock

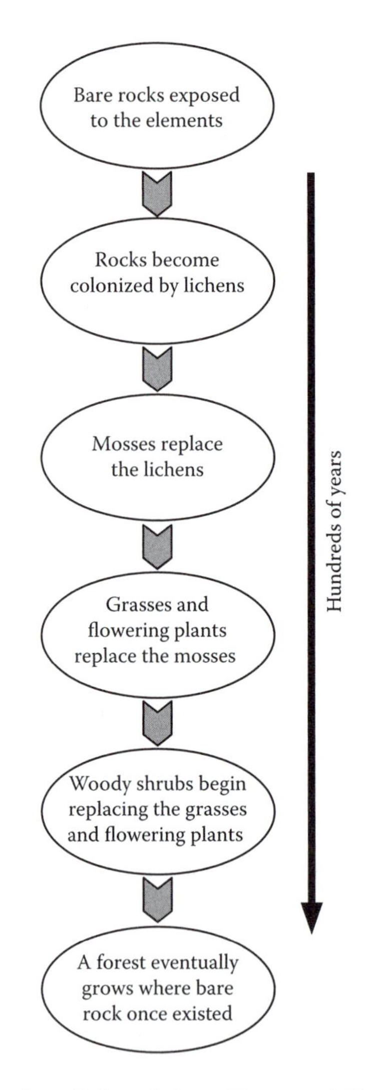

FIGURE 6.13 Bare-rock succession. (Adapted from Tomera, A.N., *Understanding Basic Ecological Concepts*, J. Weston Walch, Portland, ME, 1989, p. 67.)

surfaces. Water collects in the rock fissures and slowly dissolves the minerals out of the surface of the rock. A pioneer soil layer is formed from the dissolved minerals and supports such plants as lichens. Lichens gradually cover the rock surface and secrete carbonic acid, which dissolves additional minerals from the rock. Eventually, mosses replace the lichens. Organisms known as *decomposers* move in and feed on dead lichen and moss. A few small animals such as mites and spiders arrive next. The result is a *pioneer community*, which is defined as the first successful integration of plants, animals, and decomposers into a bare-rock community.

After several years, the pioneer community builds up enough organic matter in its soil to be able to support rooted plants such as herbs and shrubs. Eventually, the pioneer community is crowded out and is replaced by a different environment. This, in turn, works to thicken the upper soil layers. The progression continues through several other stages until a mature or climax ecosystem is developed, several decades later. In bare-rock succession, each stage in the complex succession pattern dooms the stage that existed before it. *Secondary succession* is the most common type of succession. Secondary succession occurs in an area where the natural vegetation has been removed or destroyed but the soil is not destroyed; for example, succession that occurs in abandoned farm fields, known as *old-field succession*, illustrates secondary succession. An example of secondary succession can be seen in the Piedmont region of North Carolina. Early settlers of the area cleared away the native oak–hickory forests and cultivated the land. In the ensuing years, the soil became depleted of nutrients, reducing the fertility of the soil. As a result, farming ceased in the region a few generations later, and the fields were abandoned. Some 150 to 200 years after abandonment, the climax oak–hickory forest was restored.

In a stream ecosystem, growth is enhanced by biotic and abiotic factors, including the following:

- Ability to produce offspring
- Ability to adapt to new environments
- Ability to migrate to new territories
- Ability to compete with species for food and space to live
- Ability to blend into the environment so as not to be eaten
- Ability to find food
- Ability to defend itself from enemies
- Favorable light
- Favorable temperature
- Favorable dissolved oxygen content
- Sufficient water level

The biotic and abiotic factors in an aquatic ecosystem that reduce growth include the following:

- Predators
- Disease
- Parasites
- Pollution
- Competition for space and food
- Unfavorable stream conditions (e.g., low water levels)
- Lack of food

With regard to stability of a freshwater ecosystem, the higher the species diversity the greater the inertia and resilience of the ecosystem are. When the species diversity is high within a stream ecosystem, a population within the stream can be out of control because of an imbalance between growth and reduction factors, but the ecosystem will remain stable at the same time. With regard to instability of a freshwater ecosystem, recall that imbalance occurs when growth and reduction factors are out of balance; for example, when sewage is accidentally dumped into a stream, the stream ecosystem, via the self-purification process, responds and returns to normal. This process can be described as follows:

1. Raw sewage is dumped into the stream.
2. Available oxygen decreases as the detritus food chain breaks down the sewage.
3. Some fish die at the pollution site and downstream.
4. Sewage is broken down, washes out to sea, and is finally broken down in the ocean.
5. Oxygen levels return to normal.
6. Fish populations that were deleted are restored as fish about the spill reproduce and the young occupy the real estate formerly occupied by the dead fish.
7. Populations all return to "normal."

A shift in the balance of the ecosystem of a stream (or in any ecosystem) similar to the one just described is a common occurrence. In this particular case, the stream responded (on its own) to the imbalance the sewage caused, and through the self-purification process it returned to normal. Recall that succession is the method by which an ecosystem either forms itself or heals itself; thus, we can say that a type of succession occurred in our polluted stream example, because, in the end, it healed itself. More importantly, this healing process is a good thing; otherwise, long ago there would have been few streams on Earth suitable for much more than the dumping of garbage.

In summary, through research and observation, ecologists have found that the succession patterns in different ecosystems usually display common characteristics. First, succession brings about changes in the plant and animal members present. Second, organic matter increases from stage to stage. Finally, as each stage progresses, there is a tendency toward greater stability or persistence. Remember, succession is usually predictable. This is the case unless humans interfere.

STREAM GENESIS AND STRUCTURE

Consider the following: Early in the spring, on a snow- and ice-covered high alpine meadow, the water cycle continues. The main component of the cycle—water—has been held in reserve, literally frozen over the long, dark winter months. Now, though, because of the longer, warmer spring days, the sun is higher, more direct, and of longer duration, and the frozen masses of water respond to the increased warmth. The melt begins with a single drop, then two, then more. As the snow and ice melt, the drops of water join a chorus that continues apparently unending; they fall from ice-bound lips to the bare rock and soil terrain below.

The terrain on which the snowmelt falls is not like glacial till, which is an unconsolidated, heterogeneous mixture of clay, sand, gravel, and boulders dug out, ground out, and exposed by the force of a huge, slow, inexorably moving glacier. Instead, this soil and rock ground is exposed to the falling drops of snowmelt because of a combination of wind and the tiny, enduring force exerted by drops of water as season after season they collide with the thin soil cover, exposing the intimate bones of the Earth.

Gradually, the single drops increase to a small rush—they join to form a splashing, rebounding, helter-skelter cascade, many separate rivulets that trickle, then run their way down the face of the granite mountain. At an indented ledge halfway down the mountain slope, a pool forms whose beauty, clarity, and sweet iciness provide the visitor with an incomprehensible, incomparable gift—a blessing from Earth.

The mountain pool fills slowly, tranquil under the blue sky, reflecting the pines, snow and sky around and above it, an open invitation to lie down and drink and to peer into the glass-clear, deep phantom blue–green eye, so clear that it seems possible to reach down over 50 feet and touch the very bowels of the mountain. The pool has no transition from shallow margin to depth; it is simply deep and pure. As the pool fills with more melt water, we wish to freeze time, to hold this place and this pool in its perfect state forever, it is such a rarity to us in our modern world. However, this cannot be, as Mother Nature calls, prodding, urging. For a brief instant, the water laps in the breeze against the outermost edge of the ridge, then a trickle flows over the rim. The giant hand of gravity reaches out and tips the overflowing melt onward and it continues the downward journey, following the path of least resistance to its next destination, several thousand feet below.

When the overflow, still high in altitude but with its rock-strewn bed bent downward, toward the sea, meets the angled, broken rocks below, it bounces, bursts, and mists its way against steep, V-shaped walls that form a small valley, carved out over time by water and the forces of the Earth. Within the valley confines, the melt water has grown from drops to rivulets to a small mass of flowing water. It flows through what is at first a narrow opening, gaining strength, speed, and power as the V-shaped valley widens to form a U shape. The journey continues as the water mass picks up speed and tumbles over massive boulders, and then slows again.

At a larger but shallower pool, waters from higher elevations have joined the main body—from the hillsides, crevices, springs, rills, and mountain creeks. At the influent poolsides, all appears peaceful, quiet, and restful, but not far away, at the effluent end of the pool, gravity takes control again. The overflow is flung over the jagged lip, and cascades downward several hundred feet, where the waterfall again brings its load to a violent, mist-filled meeting.

The water separates and joins repeatedly, forming a deep, furious, wild stream that calms gradually as it continues to flow over lands that are less steep. The waters widen into pools overhung by vegetation, surrounded by tall trees. The pure, crystalline waters have become progressively discolored on their downward journey, stained brown–black with humic acid, and literally filled with suspended sediments; the once-pure stream is now muddy.

The mass divides and flows in different directions over different landscapes. Small streams divert and flow into open country. Different soils work to retain or speed the waters, and in some places the waters spread out into shallow swamps, bogs, marshes, fens, or mires. Other streams pause long enough to fill deep depressions in the land and form lakes. For a time, the water remains and pauses in its journey to the sea, but this is only a short-term pause, because lakes are only a short-term resting-place in the water cycle. The water will eventually move on, by evaporation or seepage into groundwater. Other portions of the water mass stay with the main flow, and the speed of flow changes to form a river, which braids its way through the landscape, heading for the sea. As it changes speed and slows the river bottom changes from rock and stone to silt and clay. Plants begin to grow, stems thicken, and leaves broaden. The river is now full of life and the nutrients necessary to sustain life. As the river courses onward, though, it meets its destiny when the flowing rich mass slows at last and finally spills into the sea.

Freshwater systems are divided into two broad categories: running waters (*lotic systems*) and standing waters (*lentic systems*). We concentrate here on lotic systems, although many of the principles described herein apply to other freshwater surface bodies as well, which are known by common names. Some examples include seeps, springs, brooks, branches, creeks, streams, and rivers. Again, because it is the best term to use in freshwater ecology, it is the stream we are concerned with here. Although there is no standard scientific definition of a stream, it is usually distinguished subjectively as follows: A stream is of intermediate size that can be waded from one side to the other.

Physical processes involved in the formation of a stream are important to the ecology of the stream, because stream channel and flow characteristics directly influence the functioning of the ecosystem of the stream and the biota found therein. Thus, in this section, we discuss the pathways of water flow contributing to stream flow; namely, we discuss precipitation inputs as they contribute to flow. We also discuss stream flow discharge, transport of material, characteristics of stream channels, stream profile, sinuosity, the floodplain, pool–riffle sequences, and depositional features—all of which directly or indirectly impact the ecology of the stream.

Water Flow in a Stream

Most elementary students learn early in their education process that water on Earth flows downhill—from land to the sea; however, they may or may not be told that water flows downhill toward the sea by various routes. The route (or pathway) that we are primarily concerned with is the surface water route taken by surface water runoff. Surface runoff is dependent on various factors; for example, climate, vegetation, topography, geology, soil characteristics, and land use all determine how much surface runoff occurs compared with other pathways.

The primary source (input) of water to total surface runoff is, of course, precipitation. This is the case even though a substantial portion of all precipitation input returns directly to the atmosphere by *evapotranspiration*, which is a combination process, as the name suggests, whereby water in plant tissue and in the soil evaporates and transpires to water vapor in the atmosphere. A substantial portion of precipitation input returns directly to the atmosphere by evapotranspiration. It is important to point out that when precipitation occurs some rainwater is intercepted by vegetation where it evaporates, never reaching the ground or being absorbed by plants. A large portion of the rainwater that reaches the ground, lakes, and streams also evaporates directly back to the atmosphere.

Although plants display a special adaptation to minimize transpiration, plants still lose water to the atmosphere during the exchange of gases necessary for photosynthesis. Notwithstanding the large percentage of precipitation that evaporates, rain or melt water that reaches the ground surface follows several pathways to reach a stream channel or groundwater.

Soil can absorb rainfall to its *infiltration capacity* (i.e., to its maximum rate). During a rain event, this capacity decreases. Any rainfall in excess of infiltration capacity accumulates on the surface. When this surface water exceeds the depression storage capacity of the surface, it moves as an irregular sheet of overland flow. In arid areas, overland flow is likely because of the low permeability of the soil. Overland flow is also likely when the surface is frozen or when human activities have rendered the land surface less permeable. In humid areas, where infiltration capacities are high, overland flow is rare.

In rain events, where the infiltration capacity of the soil is not exceeded, rain penetrates the soil and eventually reaches the groundwater—from which it discharges to the stream slowly and over a long period. This phenomenon helps to explain why stream flow through a dry-weather region remains constant; the flow is continuously augmented by groundwater. This type of stream is known as a *perennial stream*, as opposed to an *intermittent* one, because the flow continues during periods of no rainfall.

When a stream courses through a humid region, it is fed water via the water table, which slopes toward the stream channel. Discharge from the water table into the stream accounts for flow during periods without precipitation and explains why this flow increases, even without tributary input, as one proceeds downstream. Such streams are called *gaining* or *effluent*, as opposed to *losing* or *influent streams*, which lose water into the ground (see Figure 6.14). The same stream can shift between gaining and losing conditions along its course because of changes in underlying strata and local climate.

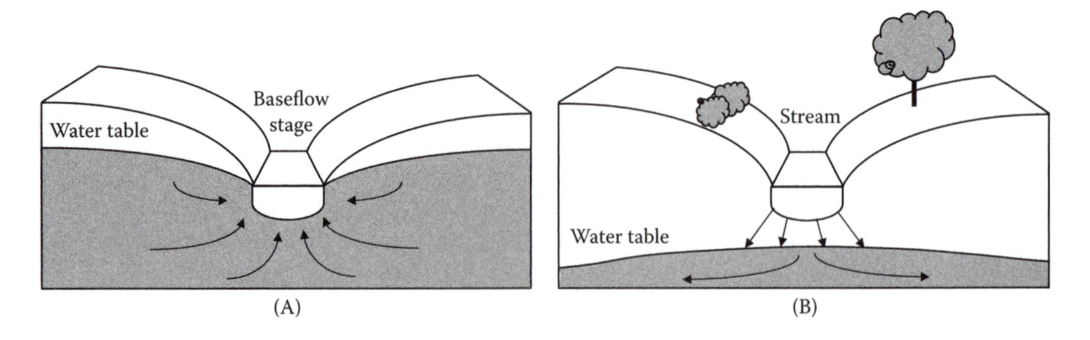

FIGURE 6.14 (A) Cross-section of a gaining stream; (B) cross-section of a losing stream.

Stream Water Discharge

The current velocity (speed) of water (driven by gravitational energy) in a channel varies considerably within the cross-section of a stream due to friction with the bottom and sides, with sediment, and with the atmosphere, as well as to sinuosity (bending or curving) and obstructions. Highest velocities, obviously, are found where friction is least, generally at or near the surface and near the center of the channel. In deeper streams, current velocity is greatest just below the surface due to the friction with the atmosphere; in shallower streams, current velocity is greatest at the surface due to friction with the bed. Velocity decreases as a function of depth, approaching zero at the substrate surface.

Transport of Material

Water flowing in a channel may exhibit *laminar flow* (parallel layers of water shear over one another vertically) or *turbulent flow* (complex mixing). In streams, laminar flow is uncommon, except at boundaries where flow is very low and in groundwater. The flow in streams is generally turbulent. Turbulence exerts a shearing force that causes particles to move along the streambed by pushing, rolling, and skipping, referred to as the *bed load*. This same shear causes turbulent eddies that entrain particles in suspension, referred to as the *suspended load* (particle size under 0.06 mm).

Entrainment is the incorporation of particles when stream velocity exceeds the *entraining velocity* for a particular particle size. The entrained particles in suspension (suspended load) also include fine sediment, primarily clays, silts, and fine sands that require only low velocities and minor turbulence to remain in suspension. These are referred to as the *wash load* (particle size under 0.002 mm). Thus, the suspended load includes the wash load and coarser materials (at lower flows). Together, the suspended load and bed load constitute the *solid load*. It is important to note that in bedrock streams the bed load will be a lower fraction than in alluvial streams where channels are composed of easily transported material.

A substantial amount of material is also transported as the *dissolved load*. Solutes are generally derived from chemical weathering of bedrock and soils, and their contribution is greatest in subsurface flows and in regions of limestone geology. The relative amount of material transported as solute rather than solid load depends on basin characteristics, lithology (i.e., the physical character of rock), and the hydrologic pathways. In areas of very high runoff, the contribution of solutes approaches or exceeds sediment load, whereas in dry regions sediments make up as much as 90% of the total load.

Deposition occurs when *stream competence*—which refers to the largest particles that a stream can move, which in turn depends on the critical erosion competent of velocity—falls below a given velocity. Simply stated: The size of the particle that can be eroded and transported is a function of current velocity.

Sand particles are the most easily eroded. The greater the mass of larger particles (e.g., coarse gravel), the higher the initial current velocities must be for movement. Smaller particles (silts and clays), however, require even greater initial velocities because of their cohesiveness and because they present smaller, streamlined surfaces to the flow. Once in transport, particles will continue in motion at somewhat slower velocities than initially required to initiate movement and will settle at still lower velocities.

Particle movement is determined by size, flow conditions, and mode of entrainment. Particles over 0.002 mm (medium–coarse sand size) tend to move by rolling or sliding along the channel bed as *traction load*. When sand particles fall out of the flow, they move by *saltation*, or repeated bouncing. Particles under 0.06 mm (silt) move as *suspended load*, and particles under 0.002 mm (clay) as *wash load*. Unless the supply of sediments becomes depleted, the concentration and amount of transported solids increase. Discharge is usually too low, throughout most of the year, to scrape or scour, shape channels, or move significant quantities of sediment in all but sand-bed streams, which can experience change more rapidly. During extreme events, the greatest scour occurs and the amount of material removed increases dramatically.

Sediment inflow into streams can be both increased and decreased because of human activities. For example, poor agricultural practices and deforestation greatly increase erosion. Fabricated structures such as dams and channel diversions, on the other hand, can greatly reduce sediment inflow.

Characteristics of Stream Channels

Flowing waters (rivers and streams) determine their own channels, and these channels exhibit relationships attesting to the operation of physical laws—laws that are not, as of yet, fully understood. The development of stream channels and entire drainage networks and the existence of various regular patterns in the shape of channels indicate that streams are in a state of dynamic equilibrium between erosion (sediment loading) and deposition (sediment deposit) and are governed by common hydraulic processes. Because channel geometry is four dimensional, with a long cross-section, depth and slope profile, and because these mutually adjust over a time scale as short as years and as long as centuries or more, cause-and-effect relationships are difficult to establish. Other variables that are presumed to interact as the stream achieves its graded state include width and depth, velocity, size of sediment load, bed roughness, and the degree of braiding (sinuosity).

Stream Profiles

Mainly because of gravity, most streams exhibit a downstream decrease in gradient along their length. Beginning at the headwaters, the steep gradient becomes less so as one proceeds downstream, resulting in a concave longitudinal profile. Although diverse geography provides for almost unlimited variation, a lengthy stream that originates in a mountainous area typically comes into existence as a series of springs and rivulets; these coalesce into a fast-flowing, turbulent mountain stream, and the addition of tributaries results in a large and smoothly flowing river that winds through the lowlands to the sea.

When studying a stream system of any length, it becomes readily apparent (almost from the start) that what we are studying is a body of flowing water that varies considerably from place to place along its length. As an example, increases in discharge cause corresponding changes in the width, depth, and velocity of the stream. In addition to physical changes that occur from location to location along the course of a stream, a legion of biological variables correlate with stream size and distance downstream. The most apparent and striking changes are in steepness of slope and in the transition from a shallow stream with large boulders and a stony substrate to a deep stream with a sandy substrate. The particle size of bed material is also variable along the course of a stream. The particle size usually shifts from an abundance of coarser material upstream to mainly finer material in downstream areas.

Sinuosity

Unless forced by humans in the form of heavily regulated and channelized streams, straight channels are uncommon. Stream flow creates distinctive landforms composed of straight (usually in appearance only), meandering, and braided channels, channel networks, and flood plains. Simply put, flowing water will follow a sinuous course. The most commonly used measure is the *sinuosity index* (SI). Sinuosity equals 1 in straight channels and more than 1 in sinuous channels. *Meandering* is the natural tendency for alluvial channels and is usually defined as an arbitrarily extreme level of sinuosity, typically a SI greater than 1.5. Many variables affect the degree of sinuosity.

Even in many natural channel sections of a stream course that appear straight, meandering occurs in the line of maximum water or channel depth (known as the *thalweg*). Keep in mind that streams have to meander; that is how they renew themselves. By meandering, they wash plants and soil from the land into their waters, and these serve as nutrients for the plants in the rivers.

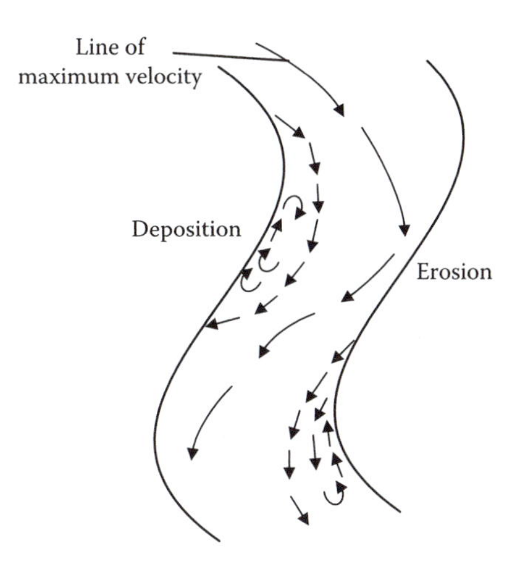

FIGURE 6.15 A meandering reach.

If rivers are not allowed to meander, if they are *channelized*, the amount of life they can support will gradually decrease. That means fewer fish, as well as fewer bald eagles, herons, and other fishing birds (Cave, 2000). Meander flow follows predictable pattern and causes regular regions of erosion and deposition (Figure 6.15). The streamlines of maximum velocity and the deepest part of the channel lie close to the outer side of each bend and cross over near the point of inflection between the banks. A huge elevation of water at the outside of a bend causes a helical flow of water toward the opposite bank. In addition, a separation of surface flow causes a back eddy. The result is zones of erosion and deposition and explains why point bars develop in a downstream direction in depositional zones.

Bars, Riffles, and Pools

Implicit in the morphology and formation of meanders are *bars*, *riffles*, and *pools*. Bars develop by deposition in slower, less competent flow on either side of the sinuous mainstream. Onward moving water, depleted of bed load, regains competence and shears a pool in the meander, reloading the stream for the next bar. Alternating bars migrate to form riffles (see Figure 6.16). As stream flow continues along its course, a pool–riffle sequence is formed. The riffle is a mound or hillock and the pool is a depression.

Flood Plain

A stream channel influences the shape of the valley floor through which it courses. The self-formed, self-adjusted flat area near the stream is the *flood plain*, which loosely describes the valley floor prone to periodic inundation during overbank discharges. What is not commonly known is that valley flooding is a regular and natural behavior of the stream. The aquatic community of a stream has several unique characteristics. Such a community operates under the same ecologic principles as terrestrial ecosystems, but the physical structure of the community is more isolated and exhibits limiting factors that are very different from the limiting factors of a terrestrial ecosystem. Certain materials and conditions are necessary for the growth and reproduction of organisms. If, for example, a farmer plants wheat in a field containing too little nitrogen, the wheat will stop growing when it has used up the available nitrogen, even if the requirements of wheat for oxygen, water, potassium, and other nutrients are met. In this particular case, nitrogen is said to be the *limiting factor*. A limiting factor is a condition or a substance (the resource in shortest supply) that limits the presence and success of an organism or a group of organisms in an area.

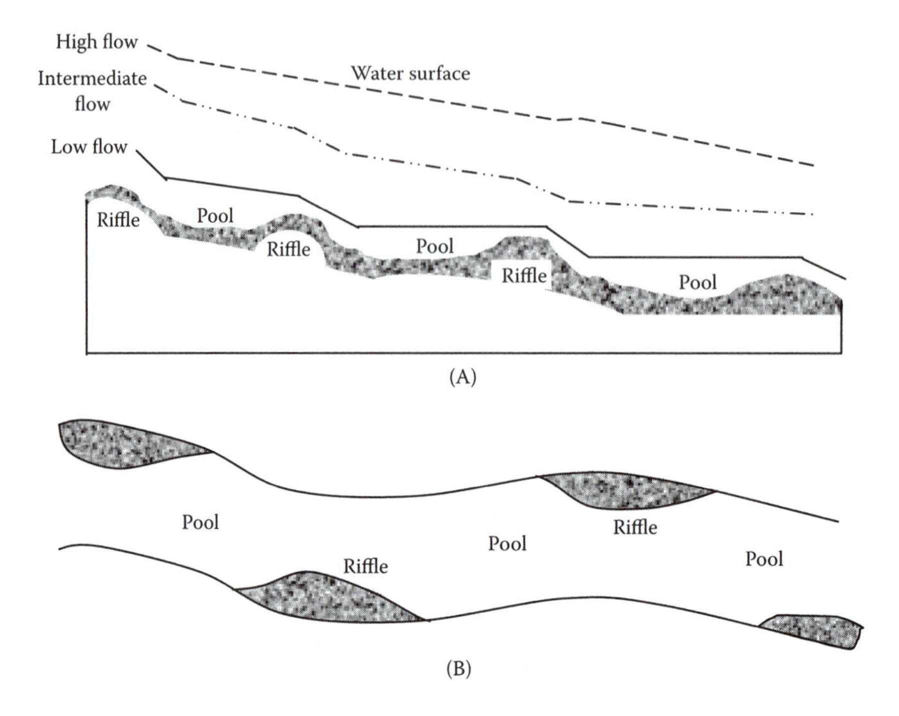

FIGURE 6.16 (A) Longitudinal profile of a riffle–pond sequence; (B) plain view of riffle–pool sequence.

Even the smallest mountain stream provides an astonishing number of different places for aquatic organisms to live, or *habitats*. If it is a rocky stream, every rock of the substrate provides several different habitats. On the side facing upriver, organisms with special adaptations, such as being able to cling to rock, do well. On the side that faces downriver, a certain degree of shelter is provided from current, but organisms can still hunt for food. The top of a rock, if it contacts air, provides a good perch for organisms that cannot breathe underwater and need to surface now and then. Underneath the rock is a popular place for organisms that hide to prevent predation. Normal stream life can be compared to that of a balanced aquarium (ASTM, 1969); that is, nature continuously strives to provide clean, healthy, normal streams. This is accomplished by maintaining the flora and fauna of the stream in a balanced state. Nature balances stream life by maintaining both the number and the type of species present in any one part of the stream. Such balance prevents an overabundance of one species compared to another. Nature structures the stream environment so plant and animal life is dependent on the existence of others within the stream.

Lotic (washed) habitats are characterized by continuously running water or current flow. These running water bodies typically have three zones: riffle, run, and pool. The *riffle zone* contains faster flowing, well-oxygenated water, with coarse sediments. In the riffle zone, the velocity of current is great enough to keep the bottom clear of silt and sludge, thus providing a firm bottom for organisms. This zone contains specialized organisms adapted to living in running water; for example, organisms adapted to living in fast streams or rapids (e.g., trout) have streamlined bodies that aid in their respiration and in obtaining food (Smith, 1996). Stream organisms that live under rocks to avoid the strong current have flat or streamlined bodies. Others have hooks or suckers to cling or attach to a firm substrate to avoid being washed away by the strong current. The *run zone* (or intermediate zone) is the slow-moving, relatively shallow part of the stream with moderately low velocities and little or no surface turbulence. The *pool zone* of the stream is usually a deeper water region where the velocity of the water is reduced, and silt and other settling solids provide a soft bottom (more homogeneous sediments), which is unfavorable for sensitive bottom dwellers. Decomposition of some of these solids leads to a reduced amount of dissolved oxygen. Some stream organisms spend part of their time in the rapids area of the stream, and at other times they can be found in the pool zone. Trout, for example, typically spend about the same amount of time in the rapid zone pursuing food as they do in the pool zone pursuing shelter.

Organisms are sometimes classified based on their mode of life:

- *Benthos* (mud dwellers)—The term originates from the Greek word for "bottom" and broadly includes aquatic organisms living on the bottom or on submerged vegetation. They live under and on rocks and in the sediments. A shallow sandy bottom has sponges, snails, earthworms, and some insects. A deep, muddy bottom will support clams, crayfish, and nymphs of damselflies, dragonflies, and mayflies. A firm, shallow, rocky bottom has nymphs of mayflies and stoneflies and larvae of water beetles.
- *Periphytons or aufwuchs*—The first term usually refers to microfloral growth on substrata (e.g., benthic-attached algae). The second term, *aufwuchs* (pronounced OWF-vooks; German for "growth upon") refers to the fuzzy, sort of furry-looking, slimy green coating that attaches or clings to stems and leaves of rooted plants or other objects projecting above the bottom without penetrating the surface. It consists of not only algae such as Chlorophyta but also diatoms, protozoa, bacteria, and fungi.
- *Planktons (drifters)*—These are small, mostly microscopic plants and animals that are suspended in the water column; movement depends on water currents. They usually float in the direction of the current. There are two types of planktons: (1) *phytoplanktons*, which are assemblages of small plants (algae) that have limited locomotion abilities and are subject to movement and distribution by water movements; and (2) *zooplanktons*, which are animals suspended in water that have limited means of locomotion (e.g., crustaceans, protozoa, rotifers).
- *Nektons or pelagic organisms* (capable of living in open waters)—Nektons are distinct from planktons in that they are capable of swimming independently of turbulence; they are swimmers that can navigate against the current. Examples of nektons include fish, snakes, diving beetles, newts, turtles, birds, and large crayfish.
- *Neustons*—These organisms float or rest on the surface of the water. Some varieties can spread out their legs so the surface tension of the water is not broken (e.g., water striders) (see Figure 6.17).
- *Madricoles*—These are organisms that live on rock faces in waterfalls or seepages.

In a stream, the rocky substrate is the home for many organisms; thus, we need to know something about the particles that make up the substrate. Namely, we need to know how to measure the particles so we can classify them by size. Substrate particles are measured with a metric ruler in centimeters (cm). Because rocks can be long and narrow, we measure them twice: first the width, then the length. By adding the width to the length and dividing by two, we obtain the average size of the rock.

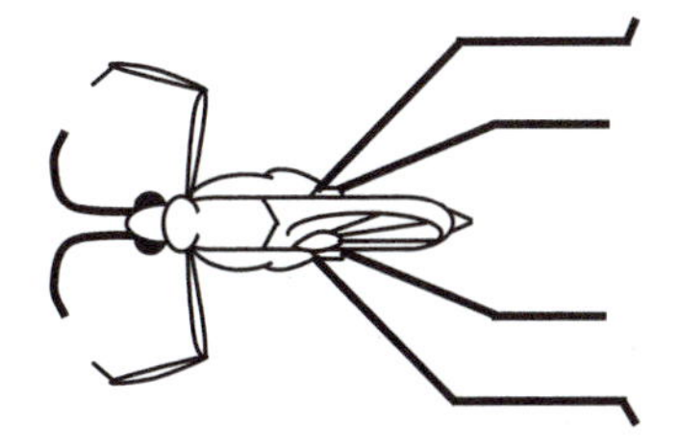

FIGURE 6.17 Water strider. (Adapted from APHA, *Standard Methods for the Examination of Water and Wastewater*, 15th ed., American Public Health Association, Washington, DC, 1981.)

It is important to randomly select the rocks that we wish to measure; otherwise, we would tend to select larger rocks, more colorful rocks, or those with unusual shapes. Instead, we should just reach down and collect those rocks in front of us and within easy reach. We then measure each of these rocks. Upon completion of measurement, each rock should be classified. Ecologists have developed a standard scale (Wentworth scale) for size categories of substrate rock and other mineral materials:

Boulder	>256 mm
Cobble	64–256 mm
Pebble	16–64 mm
Gravel	2–16 mm
Sand	0.0625–2 mm
Silt	0.0039–0.0625 mm
Clay	<0.0039 mm

Organisms that live in, on, or under rocks or in small spaces occupy what is known as a *microhabitat*. Some organisms make their own microhabitats; many caddisflies build cases about themselves to use as shelter.

Rocks are not the only physical features of streams where aquatic organisms can be found; for example, fallen logs and branches (commonly referred to as *large woody debris*, or LWD) provide an excellent place for some aquatic organisms to burrow into and for others to attach themselves, as they might to a rock. They also create areas where small detritus such as leaf litter can pile up underwater. These piles of leaf litter are excellent shelters for many organisms, including large, fiercely predaceous larvae of dobsonflies.

Another important aquatic organism habitat is found in the matter, or *drift*, that floats along downstream. Drift is important because it is the main source of food for many fish. It may include insects such as mayflies (Ephemeroptera), some true flies (Diptera), and some stoneflies (Plecoptera) and caddisflies (Trichoptera). In addition, dead or dying insects and other small organisms, terrestrial insects that fall from the trees, leaves, and other matter are common components of drift. Among the crustaceans, amphipods (small crustaceans) and isopods (small crustaceans including sow bugs and gribbles) also have been reported in the drift.

Adaptations to Stream Current

The current is the outstanding feature of streams and the major factor limiting the distribution of organisms. The current is determined by the steepness of the bottom gradient, the roughness of the streambed, and the depth and width of the streambed. The current in streams has prompted many special adaptations by stream organisms. Odum (1971) listed these adaptations as follows (Figure 6.18):

- *Attachment to a firm substrate*—Attachment is to stones, logs, leaves, and other underwater objects such as discarded tires, bottles, pipes, etc. Organisms in this group are primarily composed of the primary producer plants and animals, such as green algae, diatoms, aquatic mosses, caddisfly larvae, and freshwater sponges.
- *The use of hooks and suckers*—These organisms have the unusual ability to remain attached and withstand even the strongest rapids. Two Diptera larvae, *Simulium* and *Blepharocera*, are examples.
- *A sticky undersurface*—Snails and flatworms are examples of organisms that are able to use their sticky undersurfaces to adhere to underwater surfaces.

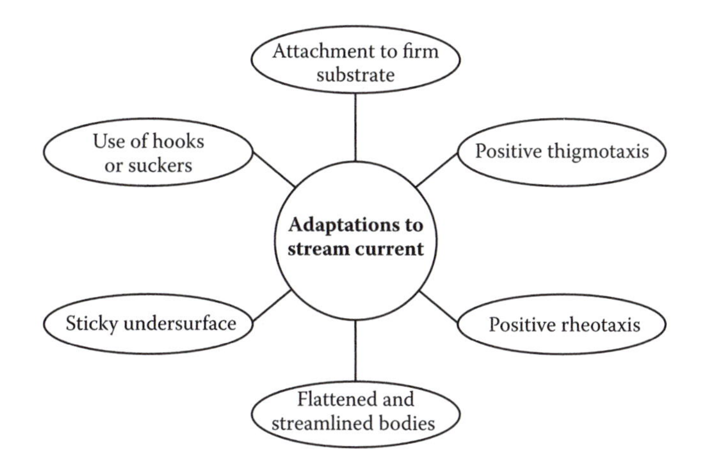

FIGURE 6.18 Adaptations to stream current.

- *Flattened and streamlined bodies*—All macroconsumers have streamlined bodies; that is, the body is broad in front and tapers posteriorly to offer minimum resistance to the current. All nektons such as fish, amphibians, and insect larvae exhibit this adaptation. Some organisms have flattened bodies, which enable them to stay under rocks and in narrow places. Examples are water penny, beetle larva, mayfly, and stonefly nymphs.
- *Positive rheotaxis* (*rheo*, current; *taxis*, arrangement)—An inherent behavioral trait of stream animals (especially those capable of swimming) is to orient themselves upstream and swim against the current.
- *Positive thigmotaxis* (*thigmo*, touch, contact)—Another inherent behavior pattern for many stream animals is to cling close to a surface or keep the body in close contact with the surface. This is the reason why stonefly nymphs (when removed from one environment and placed into another) will attempt to cling to just about anything, including each other.

It would take an entire text to describe the great number of adaptations made by aquatic organisms to their surroundings in streams. For our purposes, instead, we cover those special adaptations that are germane to this discussion. The important thing to remember is that an aquatic organism can adapt to its environment in several basic ways.

Types of Adaptive Changes

Adaptive changes are classed as genotypic, phenotypic, behavioral, or ontogenic:

- *Genotypic changes* tend to be great enough to separate closely related animals into species, such as mutations or recombination of genes. A salmonid is an example that has evolved a subterminal mouth (i.e., below the snout) to eat from the benthos.
- *Phenotypic changes* are the changes that an organism might make during its lifetime to better utilize its environment (e.g., a fish that changes sex from female to male because of an absence of males).
- *Behavioral changes* have little to do with body structure or type; for example, a fish might spend more time under an overhang to hide from predators.
- *Ontogenetic changes* take place as an organism grows and matures (e.g., coho salmon, which inhabits streams when young and migrates to the sea when older, changing its body chemistry to allow it to tolerate saltwater).

Specific Adaptations

Specific adaptations observed in aquatic organisms include their mouth, shape, color, aestivation, and schooling:

- *Mouth*—The mouth shape (morphology) of aquatic organisms such as fish varies depending on the food the fish eats. The arrangement of the jawbones and even other head bones; the length and width of gill rakers; the number, shape, and location of teeth; and the presence of barbels all change to allow fish to eat just about anything found in a stream.
- *Shape*—Changes in shape allow fish to do different things in the water. Some organisms have body shapes that push them down in the water, against the substrate, and allow them to hold their place against even strong currents (e.g., chubs, catfish, dace, sculpins). Other organisms (e.g., bass, perch, pike, trout, sunfish) have evolved an arrangement and shape of fins that allow them to lurk without moving so they can lunge suddenly to catch their prey.
- *Color*—Color may change within hours, to camouflage, or in a matter of days, or it may be genetically predetermined. Fish tend to turn dark in clear water and pale in muddy water.
- *Aestivation*—Aestivation, the ability of some fishes to burrow into the mud and wait out a dry period, helps the fish to survive in arid desert climates, where streams may dry up from time to time.
- *Schooling*—Schooling serves as protection for many fish, particularly those that are subject to predation.

AN OVERVIEW OF BENTHIC LIFE

The benthic habitat is found in the streambed, or benthos. The streambed is comprised of various physical and organic materials, and erosion and deposition are continuous factors. Erosion and deposition may occur simultaneously and alternately at different locations in the same streambed. Where channels are exceptionally deep and taper slowly to meet the relatively flattened streambed, habitats may form on the slopes of the channel. These habitats are referred to as *littoral habitats*. Shallow channels may dry up periodically in accordance with weather changes. The streambed is then exposed to open air and may take on the characteristics of a wetland.

Silt and organic materials settle and accumulate in the streambed of slowly flowing streams. These materials decay and become the primary food resource for the invertebrates inhabiting the streambed. Productivity in this habitat depends on the breakdown of these organic materials by herbivores. Bottom-dwelling organisms do not use all of the organic materials; a substantial amount becomes part of the streambed in the form of peat.

In faster moving streams, organic materials do not accumulate so easily. Primary production occurs in a different type of habitat found in the riffle regions with shoals and rocky regions for organisms to adhere to. Plants that can root themselves into the streambed dominate these regions. By plants, we are referring mostly to forms of algae, often microscopic and filamentous, that can cover rocks and debris that have settled into the streambed during summer months.

Note: If you have ever stepped into a stream, the green, slippery slime on the rocks in the streambed is representative of this type of algae.

Although the filamentous algae seem well anchored, strong currents can easily lift the algae from the streambed and carry them downstream, where they become a food resource for low-level consumers. One factor that greatly influences the productivity of a stream is the width of the channel; a direct relationship exists between stream width and richness of bottom organisms. Bottom-dwelling organisms are very important to the ecosystem, as they provide food for other, larger benthic organisms by consuming detritus.

Benthic Plants and Animals

Vegetation is not common in the streambed of slow-moving streams; however, vegetation may anchor along the banks. Algae (mainly green and blue–green) as well as common types of water moss attach themselves to rocks in fast-moving streams. Mosses and liverworts often climb up the sides of the channel onto the banks, as well. Some plants similar to the reeds of wetlands with long stems and narrow leaves are able to maintain roots and withstand the current. Aquatic insects and invertebrates dominate slow-moving streams. Most aquatic insects are in their larval and nymph forms such as the blackfly, caddisfly, and stonefly. Adult water beetles and waterbugs are also abundant. Insect larvae and nymphs provide the primary food source for many fish species, including American eel and brown bullhead catfish. Representatives of crustaceans, rotifers, and nematodes (flatworms) are sometimes present. The abundance of leeches, worms, and mollusks (especially freshwater mussels) varies with stream conditions but generally favors low-phosphate conditions. Larger animals found in slow-moving streams and rivers include newts, tadpoles, and frogs. The important characteristic of all life in streams is adaptability to withstand currents.

BENTHIC MACROINVERTEBRATES

The emphasis of aquatic insect studies, which have expanded exponentially in the last several decades, has been largely ecological. Freshwater macroinvertebrates are ubiquitous; even polluted waters contain some representative of this diverse and ecologically important group of organisms. Benthic macroinvertebrates are aquatic organisms without backbones that spend at least a part of their life cycle on the stream bottom. Examples include aquatic insects, such as stoneflies, mayflies, caddisflies, midges, and beetles, as well as crayfish, worms, clams, and snails. Most hatch from eggs and mature from larvae to adults. The majority of the insects spend their larval phase on the river bottom and, after a few weeks to several years, emerge as winged adults. The aquatic beetles, true bugs, and other groups remain in the water as adults. Macroinvertebrates typically collected from the stream substrate are either aquatic larvae or adults.

In practice, stream ecologists observe indicator organisms and their responses to determine the quality of the stream environment. A number of methods can be used to determine water quality based on biologic characteristics. A wide variety of indicator organisms (biotic groups) can be used for biomonitoring. Those used most often include algae, bacteria, fish, and macroinvertebrates. Notwithstanding their popularity, in this text we discuss benthic macroinvertebrates because they offer a number of advantages:

- They are ubiquitous, so they are affected by perturbations in many different habitats.
- They are species rich, so the large number of species produces a range of responses.
- They are sedentary, so they stay put, which allows determination of the spatial extent of a perturbation.
- They are long-lived, which allows us to follow temporal changes in abundance and age structure.
- They integrate conditions temporally so, like any biotic group, they provide evidence of conditions over long periods.

In addition, benthic macroinvertebrates are preferred as bioindicators because they are easily collected and handled by samplers; they require no special culture protocols. They are visible to the naked eye, and samplers can easily distinguish their characteristics. They have a variety of fascinating adaptations to stream life. Certain benthic macroinvertebrates have very special tolerances and thus are excellent specific indicators of water quality. Useful benthic macroinvertebrate data are easy to collect without expensive equipment. The data obtained by macroinvertebrate sampling can serve to indicate the need for additional data collection, possibly including water analysis and fish sampling.

In short, we base the focus of this discussion on benthic macroinvertebrates (with regard to water quality in streams and lakes) simply because some cannot survive in polluted water while others can survive or even thrive in polluted water. In a healthy stream, the benthic community includes a variety of pollution-sensitive macroinvertebrates. In an unhealthy stream or lake, only a few types of nonsensitive macroinvertebrates may be present; thus, the presence or absence of certain benthic macroinvertebrates is an excellent indicator of water quality.

It may be difficult to identify stream or lake pollution with water analysis, which can only provide information for the time of sampling (a snapshot of time). Even the presence of fish may not provide information about a polluted stream because fish can move away to avoid polluted water and then return when conditions improve. In contrast, most benthic macroinvertebrates cannot move to avoid pollution; thus, a macroinvertebrate sample may provide information about pollution that is not present at the time of sample collection.

Before we can use benthic macroinvertebrates to gauge water quality in a stream (or for any other reason), we must be familiar with the macroinvertebrates that are commonly used as bioindicators. Samplers must be aware of basic insect structures before they can classify the macroinvertebrates they collect. Structures that need to be stressed include head, eyes (compound and simple), antennae, mouth (no emphasis on parts), segments, thorax, legs and leg parts, gills, and abdomen. Samplers also should be familiar with insect metamorphosis—both complete and incomplete—as most of the macroinvertebrates collected are larval or nymph stages.

> ***Note:*** Information on basic insect structures is beyond the scope of this text, so the author highly recommends the standard guide to aquatic insects of North America, *An Introduction to the Aquatic Insects of North America*, 3rd ed., edited by R.W. Merritt and K.W. Cummins (Kendall/Hunt Publishing, 1996).

Identification of Benthic Macroinvertebrates

Before identifying and describing the key benthic macroinvertebrates significant to water/wastewater operators, it is important first to provide foundational information. We characterize benthic macroinvertebrates using two important descriptive classifications: *trophic groups* and *mode of existence*. In addition, we discuss their relationship in the food web—that is, what, or whom, they eat:

1. *Trophic groups*—Of the trophic groups (i.e., feeding groups) that Merritt and Cummins (1996) identified for aquatic insects, only five are likely to be found in a stream using typical collection and sorting methods:
   - *Shredders* have strong, sharp mouthparts that allow them to shred and chew coarse organic material such as leaves, algae, and rooted aquatic plants. These organisms play an important role in breaking down leaves or larger pieces of organic material to a size that can be used by other macroinvertebrates. Shredders include certain stonefly and caddisfly larvae, sowbugs, scuds, and others.
   - *Collectors* gather the very finest suspended matter in the water. To do this, they often sieve the water through rows of tiny hairs. These sieves of hairs may be displayed in fans on their heads (blackfly larvae) or on their forelegs (some mayflies). Some caddisflies and midges spin nets and catch their food in them as the water flows through.
   - *Scrapers* scrape the algae and diatoms off surfaces of rocks and debris using their mouthparts. Many of these organisms are flattened to hold onto surfaces while feeding. Scrapers include water pennies, limpets and snails, netwinged midge larvae, certain mayfly larvae, and others.
   - *Piercers* are herbivores that pierce plant tissues or cells and suck the fluids out. Some caddisflies do this.

- *Predators* eat other living creatures. Some of these are *engulfers*; that is, they eat their prey completely or in parts. This is very common in stoneflies and dragonflies, as well as caddisflies. Others are *piercers*, which are similar to the herbivorous piercers except that they eat live animal tissues.

2. *Mode of existence* (habit, locomotion, attachment, concealment)—Examples include the following:
   - *Skaters* (e.g., water striders) are adapted for skating on the surface where they feed as scavengers on organisms trapped in the surface film.
   - *Planktonic* types inhabit the open water limnetic zone of standing (lentic) waters (such as lakes, bogs, and ponds). Representatives may float and swim about in the open water but they usually exhibit a diurnal vertical migration pattern (e.g., phantom midges) or float at the surface to obtain oxygen and food, diving when alarmed (e.g., mosquitoes).
   - *Divers* (e.g., water boatmen, predaceous diving beetle) are adapted for swimming by rowing with their hind legs in lentic habitats and lotic pools. They come to the surface to obtain oxygen but dive and swim when feeding or alarmed; they may cling to or crawl on submerged objects such as vascular plants.
   - *Swimmers* (e.g., mayflies) are adapted for fishlike swimming in lotic or lentic habitats. Individuals usually cling to submerged objects, such as rocks (lotic riffles) or vascular plants (lentic), between short bursts of swimming.
   - *Clingers* (e.g., mayflies, caddisflies) have behavioral adaptations (such as fixed retreat construction) and morphological adaptations (such as long curved tarsal claws, dorsoventral flattening, ventral gills arranged as a sucker) for attachment to surfaces in stream riffles and wave-swept rocky littoral zones of lakes.
   - *Sprawlers* (e.g., mayflies, dobsonflies, damselflies) inhabit the surface of floating leaves of vascular hydrophytes or fine sediments and usually have modifications for staying on top of the substrate and maintaining the respiratory surfaces free of silt.
   - *Climbers* (e.g., dragonflies, damselflies) are adapted for living on vascular hydrophytes or detrital debris (overhanging branches, roots, and vegetation along streams and submerged brush in lakes) with modifications for moving vertically on stem-type surfaces.
   - *Burrowers* (e.g., mayflies, midges) inhabit the fine sediments of streams (pools) and lakes. Some construct discrete burrows, which may have sand grain tubes extending above the surface of the substrate or the individuals; they may ingest their way through the sediments.

Macroinvertebrates and the Food Web

In a stream or lake, the two possible sources of primary energy are (1) photosynthesis by algae, mosses, and higher aquatic plants, and (2) imported organic matter from streamside or lakeside vegetation (e.g., leaves and other parts of vegetation). Simply put, a significant portion of the food that is eaten grows right in the stream or lake—for example, algae, diatoms, nymphs and larvae, and fish. A food that originates from within the stream is *autochthonous*. Most food in a stream, however, comes from outside the stream—especially in small, heavily wooded streams, which normally have insufficient light to support substantial instream photosynthesis so energy pathways are supported largely by imported energy. Leaves provide a large portion of this imported energy. Worms drown in floods and are washed in. Leafhoppers and caterpillars fall from trees. Adult mayflies and other insects mate above the stream, lay their eggs in it, and then die in it. All of this food from outside the stream is *allochthonous*.

Units of Organization

Macroinvertebrates, like all other organisms, are classified and named. Macroinvertebrates are classified and named using a *taxonomic hierarchy*. The taxonomic hierarchy for the caddisfly (a macroinvertebrate insect commonly found in streams) is shown below:

Kingdom—Animalia (animals)
Phylum—Arthropoda ("jointed legs")
Class—Insecta (insect)
Order—Trichoptera (caddisfly)
Family—Hydropsychidae (net-spinning caddis)
Genus and species—*Hydropsyche morosa*

INSECT MACROINVERTEBRATES

The macroinvertebrates are the best-studied and most diverse animals in streams. We therefore devote our discussion here to the various macroinvertebrate groups. Non-insect macroinvertebrates, such as Oligochaeta (worms), Hirudinea (leeches), and Acari (water mites), are frequently encountered in lotic environments, but the insects are among the most conspicuous inhabitants of streams. In most cases, it is the larval stages of these insects that are aquatic, whereas the adults are terrestrial. Typically, the larval stage is much extended, while the adult lifespan is short.

The most important insect groups in streams are Ephemeroptera (mayflies), Plecoptera (stoneflies), Trichoptera (caddisflies), Diptera (true flies), Coleoptera (beetles), Hemiptera (bugs), Megaloptera (alderflies and dobsonflies), and Odonata (dragonflies and damselflies). The identification of these different orders is usually easy, and many keys and specialized references are available to help in the identification of species (e.g., Merritt and Cummins, 1996). In contrast, specialist taxonomists are often required to identify some genera and species, particularly the order Diptera. Insect macroinvertebrates are ubiquitous in streams and are often represented by many species. Although the macroinvertebrates discussed below are aquatic species, a majority of the species can be found in streams. Lotic insects are found among many different orders and brief accounts of their biology are presented in the following sections.

Mayflies (Order: Ephemeroptera)

Streams and rivers are generally inhabited by many species of mayflies, and, in fact, most mayfly species are restricted to streams. Experienced freshwater ecologist recognize mayfly nymphs by the leaf-like or feather-like gills on their abdomen, by the single tarsal claws on their legs, and generally (but not always) by the three cerci ("tails")—two cerci and between them usually a terminal filament (see Figure 6.19). The experienced ecologist knows that mayflies are hemimetabolous insects (i.e., larvae or nymphs resemble wingless adults) that go through many postembryonic molts, often

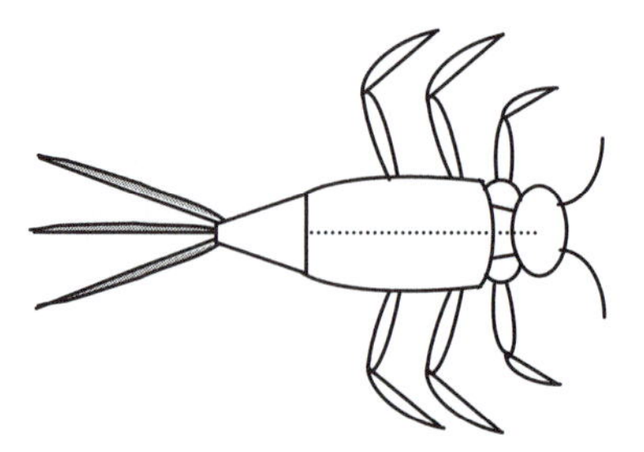

FIGURE 6.19 Mayfly (Order: Emphemeroptera).

ranging between 20 and 30. For some species, body length increases about 15% for each instar (i.e., time between each molt). During instars the nymph is very vulnerable to its principal animal, bird, fish, amphibian, and insect predators, including diving beetles, frogs, salamanders, swifts, phoebes, and dragonfly nymphs and adults.

Mayfly nymphs are mainly grazers or collector–gatherers feeding on algae and fine detritus, although a few genera are predatory. Some members filter particles from the water using hair-fringed legs or maxillary palps. Shredders are rare among mayflies. In general, mayfly nymphs tend to live primarily in unpolluted streams, where, with densities of up to 10,000 m^2, they contribute substantially to secondary producers.

Adult mayflies resemble nymphs but usually possess two pair of long, lacy wings folded upright; adults usually have only two cerci. The adult lifespan is short, ranging from a few hours to a few days, rarely up to two weeks, and the adults do not feed. Mayflies are unique among insects in having two winged stages, the subimago and the imago. The emergence of adults tends to be synchronous, thus ensuring the survival of enough adults to continue the species.

Stoneflies (Order: Plecoptera)

Although many freshwater ecologists would maintain that the stonefly is a well-studied group of insects, this is not exactly the case. Despite their importance, less than 5 to 10% of stonefly species are well known with respect to life history, trophic interactions, growth, development, spatial distribution, and nymphal behavior. Notwithstanding our lack of extensive knowledge with regard to stoneflies, enough is known to provide an accurate characterization of these aquatic insects. We know, for example, that stonefly larvae are characteristic inhabitants of cool, clean streams (most nymphs occur under stones in well-aerated streams). They are sensitive to organic pollution or, more precisely, to low oxygen concentrations accompanying organic breakdown processes, but stoneflies seem rather tolerant of acidic conditions. Lack of extensive gills at least partly explains their relative intolerance of low oxygen levels.

Stoneflies are drab-colored, small- to medium-sized (4 to 60 mm; 1/6 to 2-1/4 inches), rather flattened insects. Stoneflies have long, slender, many-segmented antennae and two long narrow antenna-like structures (cerci) on the tip of the abdomen (see Figure 6.20). The cerci may be long or short. At rest, the wings are held flat over the abdomen, giving a "square-shouldered" look compared to the roof-like position of most caddisflies and vertical position of the mayflies. Stoneflies have two pair of wings. The hind wings are slightly shorter than the forewings and much wider, having a large anal lobe that is folded fanwise when the wings are at rest. This fanlike folding of the wings gives rise to the name of the order: *pleco* ("folded or plaited") and *ptera* ("wings"). The aquatic nymphs are generally very similar to mayfly nymphs except that they have only two cerci at the tip of the abdomen. The stoneflies have chewing mouthparts. They may be found anywhere in a nonpolluted stream that food is available. Many adults, however, do not feed and have reduced or vestigial mouthparts.

Stoneflies have a specific niche in high-quality streams where they are very important as a fish food source at specific times of the year (winter to spring, especially) and of the day. They complement other important food sources, such as caddisflies, mayflies, and midges.

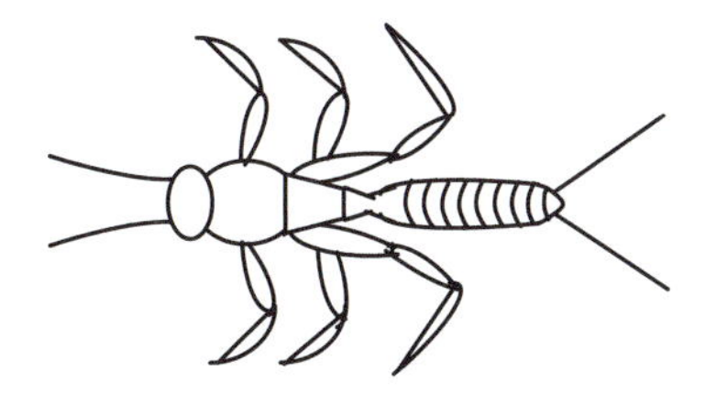

FIGURE 6.20 Stonefly (Order: Plecoptera).

Caddisflies (Order: Trichoptera)

Trichoptera (*trichos*, "hair"; *ptera*, "wings") represents one of the most diverse insect orders living in the stream environment, and caddisflies have nearly a worldwide distribution, with the exception of Antarctica. Caddisflies may be categorized broadly into free-living (roving and net spinning) and case-building species. Caddisflies are described as medium-sized insects with bristle-like and often long antennae. They have membranous hairy wings (which explains the use of *trichos* in the name), which are held tent-like over the body when at rest; most are weak fliers. They have greatly reduced mouthparts and five tarsi. The larvae are mostly caterpillar like and have a strongly sclerotized (hardened) head with very short antennae and biting mouthparts. They have well-developed legs with a single tarsi. The abdomen usually has ten segments; in case-bearing species, the first segment bears three papillae, one dorsally and the other two laterally, which helps hold the insect centrally in its case and allows a good flow of water to pass the cuticle and gills. The last or anal segment bears a pair of grappling hooks.

In addition to being aquatic insects, caddisflies are superb architects. Most caddisfly larvae (see Figure 6.21) live in self-designed, self-built houses called *cases*. They spin out silk, and either live in silk nets or use the silk to stick together bits of whatever is lying on the stream bottom. These houses are so specialized, that we can usually identify the genus of a caddisfly larva if we can see its house (case). With nearly 1400 species of caddisfly species in North America (north of Mexico), this is a good thing!

Caddisflies are closely related to butterflies and moths (Order: Lepidoptera). They live in most stream habitats, and that is why they are so diverse. Each species has particular adaptations that allow it to survive in its environment. Primarily herbivorous, most caddisflies feed on decaying plant tissue and algae. Their favorite algae are diatoms, which they scrape off rocks. Some of them, though, are predacious.

Caddisfly larvae can take a year or two to change into adults. They change into *pupae* (the inactive stage in the metamorphosis of many insects, following the larval stage and preceding the adult form) while still inside the cases built for their metamorphosis. It is interesting to note that caddisflies, unlike stoneflies and mayflies, go through a complete metamorphosis. Caddisflies remain as pupae for 2 to 3 weeks, then emerge as adults. When they split open their cases and leave, they must swim to the surface of the water to escape it. The winged adults fly evening and night, and some are known to feed on plant nectar. Most of them will live less than a month; like many other winged stream insects, their adult lives are brief compared to the time they spend in the water as larvae.

Caddisflies are sometimes grouped into five main groups according to the kinds of cases they build: (1) free-living forms that do not make cases, (2) saddle-case makers, (3) purse-case makers, (4) net-spinners and retreat makers, and (5) tube-case makers. Caddisflies demonstrate their architectural talents in the cases they design and build; for example, a caddisfly might build a perfect, four-sided box case of bits of leaves and bark or tiny bits of twigs. It may make a clumsy dome of large pebbles. Another might construct rounded tubes out of twigs or very small pebbles. The author has come to appreciate not only their architectural ability but also their flare in the selection of construction materials. It is not unusual to find caddisfly cases constructed of silk that was emitted through an opening at the tip of the labium and combined with bits of ordinary rock mixed with sparkling quartz and red garnet, green peridot, and bright fool's gold.

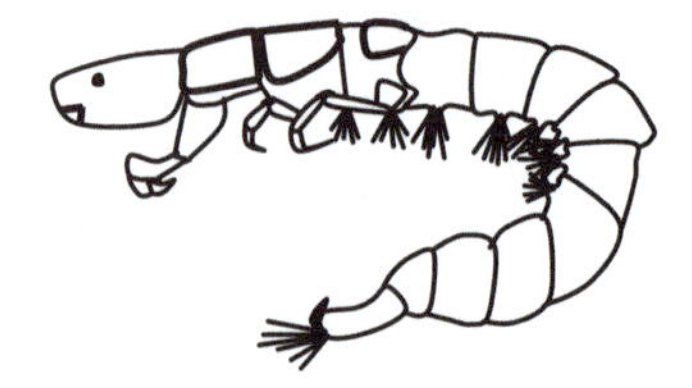

FIGURE 6.21 Caddis larvae, *Hydropsyche* sp.

In addition to the protection they provide, the cases offer another advantage in that they actually help caddisflies breathe. The caddisfly moves its body up and down, back and forth, inside its case, which produces a current that brings them fresh oxygen. The less oxygen there is in the water, the faster they have to move. It has been seen that caddisflies inside their cases get more oxygen than those that are outside of their cases—and this is why stream ecologists think that caddisflies can often be found even in still waters, where dissolved oxygen is low, in contrast to stoneflies and mayflies.

True Flies (Order: Diptera)

True or two- (*di-*) winged (*ptera*) flies include not only the flies that we are most familiar with, such as fruitflies and houseflies, but also midges (see Figure 6.22), mosquitoes, craneflies (see Figure 6.23), and others. Houseflies and fruitflies live only on land, and we do not concern ourselves with them. Some, however, spend nearly their entire lives in water; they contribute to the ecology of streams.

True flies are in the order Diptera, one of the most diverse orders of the class Insecta, with about 120,000 species worldwide. Dipteran larvae occur almost everywhere except Antarctica and deserts where there is no running water. They may live in a variety of places within a stream: buried in sediments, attached to rocks, beneath stones, in saturated wood or moss, or in silken tubes attached to the stream bottom. Some even live below the stream bottom.

True fly larvae may eat almost anything, depending on their species. Those with brushes on their heads use them to strain food out of the water that passes through. Others may eat algae, detritus, plants, and even other fly larvae.

The longest part of the true fly's life cycle, like that of mayflies, stoneflies, and caddisflies, is the larval stage. It may remain an underwater larva anywhere from a few hours to 5 years. The colder the environment, the longer it takes to mature. It pupates and emerges and then becomes a winged adult. The adult may live 4 months—or it may live for only a few days. While reproducing, it will often eat plant nectar for the energy it needs to make its eggs. Mating sometimes takes place in aerial swarms. The eggs are deposited back in the stream; some females will crawl along the stream bottom, losing their wings in the process, to search for the perfect place to put their eggs. Once they lay them, they die.

Diptera serve an important role in cleaning water and breaking down decaying material, and they are a vital food source for many of the animals living in and around streams, as they play pivotal roles in the processing of food energy. The true flies most familiar to us, however, are the midges, mosquitoes, and craneflies, because they are pests. Some midge flies and mosquitoes bite; the cranefly does not bite but looks like a giant mosquito.

Like mayflies, stoneflies, and caddisflies, true flies are mostly in larval form. Just as for caddisflies, we can find their pupae, because they are holometabolous insects; that is, they go through complete metamorphosis. Most of them are free living and travel around. Although none of the true fly larvae has the six jointed legs that we see on other insects in the stream, they sometimes have strange little prolegs to move around with. Others may move somewhat like worms do, and

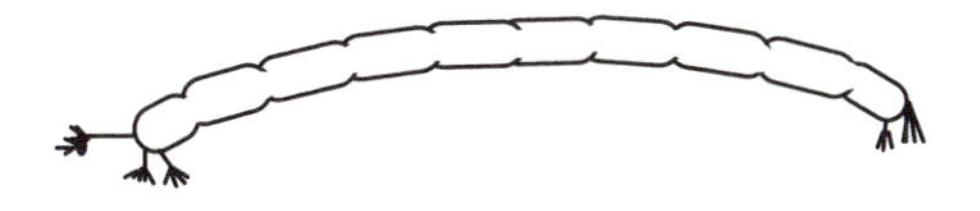

FIGURE 6.22 Midge larvae.

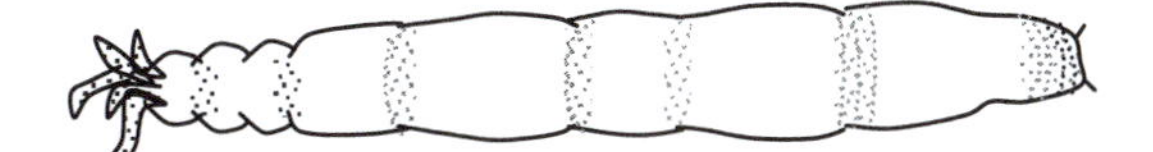

FIGURE 6.23 Cranefly larvae.

some—the ones that live in waterfalls and rapids—have a row of six suction discs that they use to move much like a caterpillar does. Many use silk pads and hooks at the ends of their abdomens to hold them fast to smooth rock surfaces.

Beetles (Order: Coleoptera)

Of the more than 1 million described species of insect, at least one third are beetles, making Coleoptera not only the largest order of insects but also the most diverse order of living organisms (Hutchinson, 1981). Even though this is the most speciose order of terrestrial insects, surprisingly their diversity is not so apparent in running waters. Coleoptera belongs to the infraclass Neoptera, division Endpterygota. Members of this order have an anterior pair of wings (the *elytra*) that are hard and leathery and not used in flight; the membranous hindwings, which are used for flight, are concealed under the elytra when the organisms are at rest. Only 10% of the 350,000 described species of beetles are aquatic.

Beetles are holometabolous. Eggs of aquatic coleopterans hatch in 1 or 2 weeks, with diapause occurring rarely. Larvae undergo from three to eight molts. The pupal phase of all coleopternas is technically terrestrial, making this life stage of beetles the only one that has not successfully invaded the aquatic habitat. A few species have diapausing prepupae, but most complete transformation to adults in 2 to 3 weeks. Terrestrial adults of aquatic beetles are typically short lived and sometimes nonfeeding, like those of the other orders of aquatic insects. The larvae of Coleoptera are morphologically and behaviorally different from the adults, and their diversity is high.

Aquatic species occur in two major suborders, the Adephaga and the Polyphaga. Both larvae and adults of six beetle families are aquatic: Dytiscidae (predaceous diving beetles), Elmidae (riffle beetles), Gyrinidae (whirligig beetles), Halipidae (crawling water beetles), Hydrophilidae (water scavenger beetles), and Noteridae (burrowing water beetles). Five families, Chrysomelidae (leaf beetles), Limnichidae (marsh-loving beetles), Psephenidae (water pennies), Ptilodactylidae (toe-winged beetles), and Scirtidae (marsh beetles), have aquatic larvae and terrestrial adults, as do most of the other orders of aquatic insects; adult limnichids, however, readily submerge when disturbed. Three families have species that are terrestrial as larvae and aquatic as adults, a highly unusual combination among insects: Curculionidae (weevils), Dryopidae (long-toed water beetles), and Hydraenidae (moss beetles). Because they provide a greater understanding of the condition of a freshwater body (i.e., they are useful indicators of water quality), we focus our discussion here on the riffle beetle, water penny, and whirligig beetle.

Riffle beetle larvae (most commonly found in running waters, hence their name) are up to 3/4 inches long (see Figure 6.24). The beetle's body is not only long but also hard, stiff, and segmented. They have six long segmented legs on the upper middle section of the body; the back end has two tiny hooks and short hairs. Larvae may take 3 years to mature before they leave the water to form a pupa; adults return to the stream. Riffle beetle adults are considered better indicators of water quality than larvae because they have been subjected to water quality conditions over a longer period. They walk very slowly under the water (on the stream bottom) and do not swim on the surface. They have small oval-shaped bodies (see Figure 6.25) and are typically about 1/4 inch in length. Both adults and larvae of most species feed on fine detritus with associated microorganisms scraped from the substrate, although others may be xylophagous—that is, wood eating (e.g., *Lara*, Elmidae). Predators do not seem to include riffle beetles in their diet, except perhaps their eggs, which are sometimes attacked by flatworms.

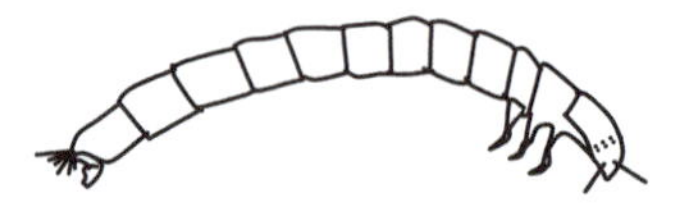

FIGURE 6.24 Riffle beetle larvae.

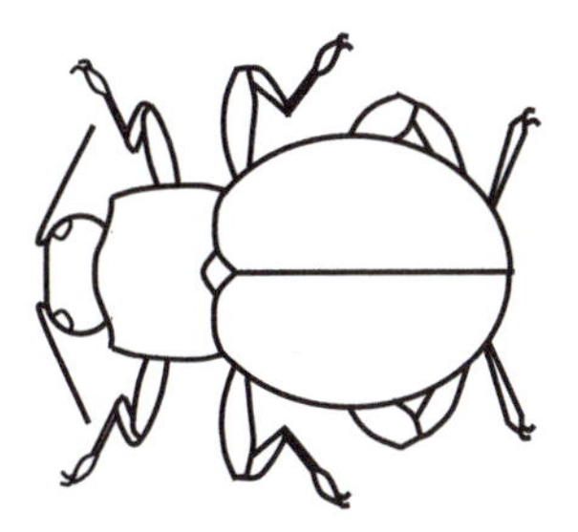

FIGURE 6.25 Riffle beetle adult.

The adult *water penny* is inconspicuous and often found clinging tightly in a sucker-like fashion to the undersides of submerged rocks, where it feeds on attached algae. The body is broad, slightly oval and flat in shape, ranging from 4 to 6 mm (about 1/4 inch) in length. The body is covered with segmented plates and looks like a tiny round leaf (see Figure 6.26). It has six tiny jointed legs (underneath). The color ranges from light brown to almost black. There are 14 water penny species in the United States. They live predominantly in clean, fast-moving streams. Aquatic larvae live a year or more; the terrestrial adults survive on land for only a few days. Water pennies scrape algae and plants from surfaces.

Whirligig beetles are common inhabitants of streams and normally are found on the surface of quiet pools. The body of a whirligig beetle has pincher-like mouthparts and six segmented legs on the middle of the body; the legs end in tiny claws. Many filaments extend from the sides of the abdomen. They have four hooks at the end of the body and no tail.

Note: When disturbed, whirligig beetles swim erratically or dive while emitting defensive secretions.

As larvae (Figure 6.27), they are benthic predators, whereas the adults live on the water surface, attacking dead and living organisms trapped in the surface film. They occur on the surface in aggregations of up to thousands of individuals. Unlike the mating swarms of mayflies, these aggregations serve primarily to confuse predators. Whirligig beetles have other interesting defensive adaptations—for example, the Johnston's organ at the base of the antennae enables them to echolocate using surface-wave signals; their compound eyes are divided into two pairs, one above and one below the water surface, enabling them to detect both aerial and aquatic predators; and they produce noxious chemicals that are highly effective at deterring predatory fish.

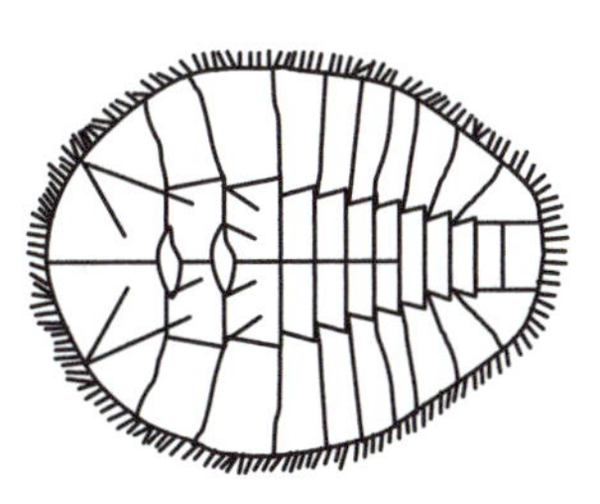

FIGURE 6.26 Water penny larva.

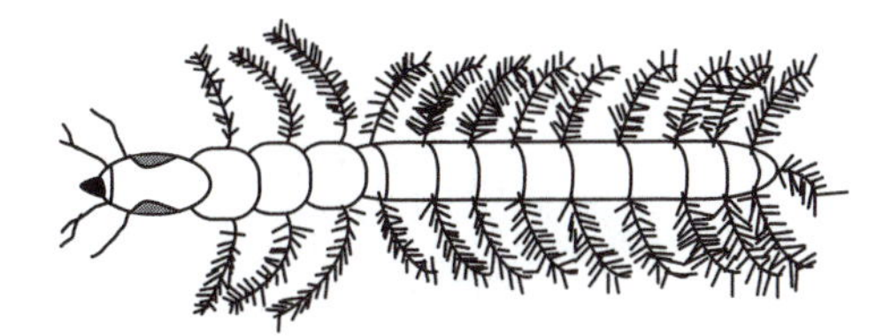

FIGURE 6.27 Whirligig beetle larva.

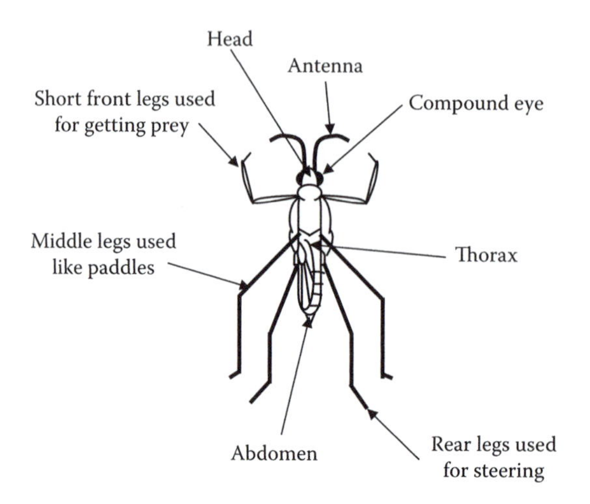

FIGURE 6.28 Water strider.

Water Strider ("Jesus Bugs") (Order: Hemiptera)

It is fascinating to sit on a log at the edge of a stream pool and watch the drama that unfolds among the small water animals. Among the star performers in small streams are the water bugs. These are aquatic members of that large group of insects known as the "true bugs," most of which live on land. Moreover, unlike many other types of water insects, they do not have gills but get their oxygen directly from the air. Most conspicuous and commonly known are the *water striders* or *water skaters*. These ride the top of the water, with only their feet making dimples in the surface film. Like all insects, the water strider has a three-part body (head, thorax, and abdomen), six jointed legs, and two antennae. It has a long, dark, narrow body (see Figure 6.28). The underside of the body is covered with water-repellent hair. Some water striders have wings, others do not. Most water striders are over 5 mm (0.2 inch) long. Water striders eat small insects that fall on the surface of the water and larvae. The water strider is very sensitive to motion and vibrations on the surface of the water. It uses this ability to locate prey. It pushes it mouth into its prey, paralyzes it, and sucks the insect dry. Predators of the water strider, such as birds, fish, water beetles, backswimmers, dragonflies, and spiders, take advantage of the fact that water striders cannot detect motion above or below the surface of the water.

Alderflies and Dobsonflies (Order: Megaloptera)

Larvae of all species of Megaloptera ("large wing") are aquatic and attain the largest size of all aquatic insects. Megaloptera is a medium-sized order with fewer than 5000 species worldwide. Most species are terrestrial; in North America, 64 aquatic species occur. In running waters, alderflies (Family: Sialidae) and dobsonflies (Family: Corydalidae; sometimes called *hellgrammites* or *toe biters*) are particularly important, as they are voracious predators, having large mandibles with sharp teeth.

Alderfly brownish-colored larvae possess a single tail filament with distinct hairs. The body is thick skinned, with six to eight filaments on each side of the abdomen; gills are located near the base of each filament. The mature body size is 0.5 to 1.25 inches (see Figure 6.29). Larvae are aggressive

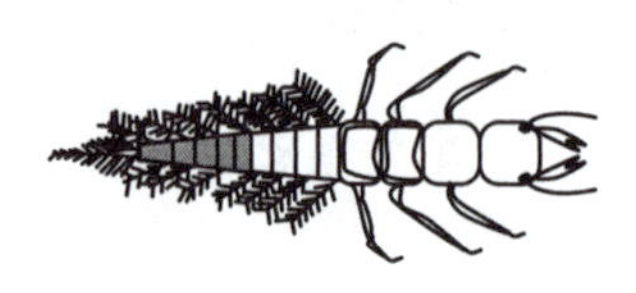

FIGURE 6.29 Alderfly larvae.

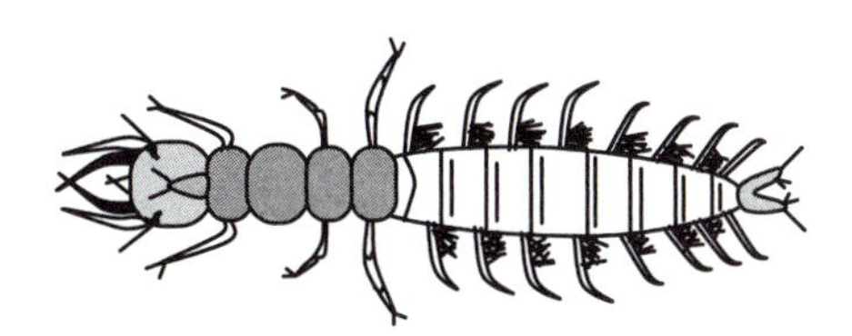

FIGURE 6.30 Dobsonfly larvae.

predators, feeding on other adult aquatic macroinvertebrates (they swallow their prey without chewing); as secondary consumers, other larger predators eat them. Female alderflies deposit eggs on vegetation that overhangs water; when the larvae hatch they fall directly into the quiet but moving water. Adult alderflies are dark with long wings folded back over the body; they live only a few days.

Dobsonfly larvae are extremely ugly (thus, they are rather easy to identify) and can be rather large, anywhere from 25 to 90 mm (1 to 3 inches) in length. The body is stout, with eight pairs of appendages on the abdomen. Brush-like gills at the base of each appendage look like hairy armpits (see Figure 6.30). The elongated body has spiracles (spines), three pairs of walking legs near the upper body, and one pair of hooked legs at the rear. The head bears four segmented antennae, small compound eyes, and strong mouth parts (large chewing pinchers). Coloration varies from yellowish, brown, gray, to black, often mottled. Dobsonfly larvae, commonly known as *hellgrammites*, are customarily found along stream banks under and between stones. As indicated by the mouthparts, they are predators and feed on all kinds of aquatic organisms. They are an important food source for larger game fish.

Dragonflies and Damselflies (Order: Odonata)

The order Odonata, which includes dragonflies (Suborder: Anisoptera) and damselflies (Suborder: Zygoptera), is a small order of conspicuous, hemimetabolous insects (lacking a pupal stage) representing about 5000 named species and 23 families worldwide. *Odonata* is a Greek word meaning "toothed one." It refers to the serrated teeth located on the insect's chewing mouthparts (mandibles). Characteristics of dragonfly and damselfly larvae include the following:

- Large eyes
- Three pairs of long segmented legs on the upper middle section (thorax) of body
- Large scoop-like lower lip that covers the bottom of the mouth
- No gills on the sides or underneath the abdomen

Note: Dragonflies and damselflies are unable to fold their four elongated wings back over the abdomen when at rest.

Dragonflies and damselflies are medium to large insects with two pairs of long equal-sized wings. The body is long and slender, with short antennae. Immature stages are aquatic, and development occurs in three stages (egg, nymph, adult).

Dragonflies are also known as *darning needles* (at one time, children were warned to keep quiet or the dragonfly's darning needles would sew the child's mouth shut). In their nymphal stage, dragonflies are grotesque creatures, robust and stoutly elongated. They do not have long tails (see Figure 6.31). They are commonly gray, greenish, or brown to black in color. They are medium to large aquatic insects, ranging in size from 15 to 45 mm; the legs are short and used for perching. They are often found on submerged vegetation and at the bottom of streams in the shallows. They are rarely found in polluted waters. Their food consists of other aquatic insects, annelids, small crustacea, and mollusks. Transformation occurs when the nymph crawls out of the water, usually onto vegetation. There it splits its skin and emerges prepared for flight. The adult dragonfly is a strong flier, capable

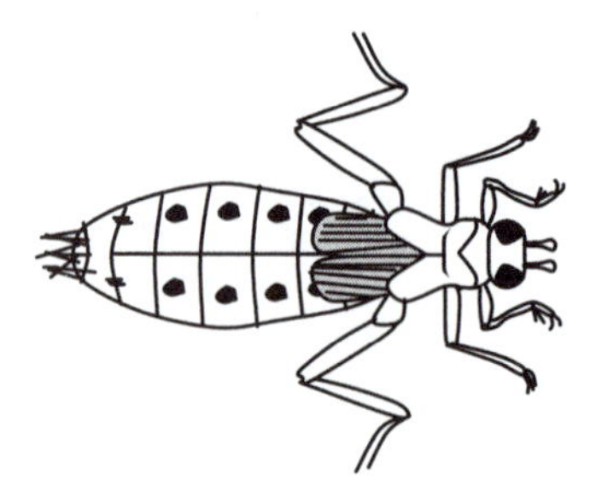

FIGURE 6.31 Dragonfly nymph.

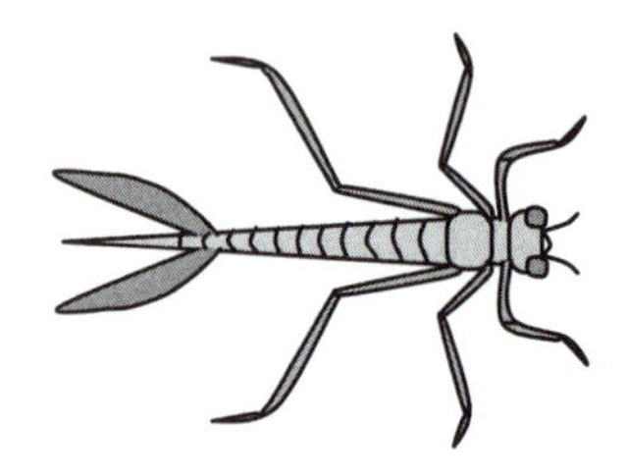

FIGURE 6.32 Damselfy.

of great speed (>60 mph) and maneuverability. (They can fly backward, stop on a dime, zip 20 feet straight up, and slip sideways in the blink of an eye!) When at rest the wings remain open and out to the sides of the body. A dragonfly's freely movable head has large, hemispherical eyes (nearly 30,000 facets each), which the insects use to locate prey with their excellent vision. Dragonflies eat small insects, mainly mosquitoes (large numbers of mosquitoes) while in flight. Depending on the species, dragonflies lay hundreds of eggs by dropping them into the water and leaving them to hatch or by inserting eggs singly into a slit in the stem of a submerged plant. The incomplete metamorphosis (egg, nymph, mature nymph, and adult) can take 2 to 3 years. Nymphs are often covered by algal growth.

> ***Note:*** Adult dragonflies are sometimes referred to as *mosquito hawks* because they eat such a large number of mosquitoes, which they catch while they are flying.

Damselflies are smaller and more slender than dragonflies. They have three long, oar-shaped feathery tails, which are actually gills, and long slender legs (see Figure 6.32). They are gray, greenish, or brown to black in color. Their habits are similar to those of dragonfly nymphs, and they emerge from the water as adults in the same manner. The adult damselflies are slow and seem uncertain in flight. Wings are commonly black or clear, and the body is often brilliantly colored. When at rest, they perch on vegetation with their wings closed upright. Damselflies mature in 1 to 4 years. Adults live for a few weeks or months. Unlike the dragonflies, adult damselflies rest with their wings held vertically over their backs. They mostly feed on live insect larvae.

> ***Note:*** Relatives of the dragonflies and damselflies are some of the most ancient of the flying insects. Fossils have been found of giant dragonflies with wingspans up to 720 mm that lived long before the dinosaurs!

NON-INSECT MACROINVERTEBRATES

Non-insect macroinvertebrates are important to our discussion of stream and freshwater ecology because many of them are used as bioindicators of stream quality. Three frequently encountered groups in running water systems are Oligochaeta (worms), Hirudinea (leeches), and Gastropoda (lung-breathing snails). They are by no means restricted to running-water conditions, and the great majority of them occupy slow-flowing marginal habitats where the sedimentation of fine organic materials takes place.

Oligochaeta (Family Tuificidae, Genus *Tubifex*)

Tubifex worms (commonly known as sludge worms) are unique in the fact that they build tubes. Sometimes we might find as many as 8000 individuals per square meter. They attach themselves within the tube and wave their posterior end in the water to circulate the water and make more oxygen available to their body surface. These worms are commonly red, because their blood contains hemoglobin. *Tubifex* worms may be very abundant in situations when other macroinvertebrates are absent; they can survive in very low oxygen levels and can live with no oxygen at all for short periods. They are commonly found in polluted streams and feed on sewage or detritus.

Hirudinea (Leeches)

Despite the many different families of leeches, they all have common characteristics. They are soft-bodied, worm-like creatures that are flattened when extended. Their bodies are dull in color, ranging from black to brown and reddish to yellow, often with a brilliant pattern of stripes or diamonds on the upper body. Their size varies within species but generally ranges from 5 mm to 45 cm when extended. Leeches are very good swimmers, but they typically move in an inchworm fashion. They are carnivorous and feed on other organisms, ranging from snails to warm-blooded animals. Leeches are found in warm protected shallows under rocks and other debris.

Gastropoda (Lung-Breathing Snail)

Lung-breathing snails (pulmonates) may be found in streams that are clean; however, their dominance may indicate that dissolved oxygen levels are low. These snails are different from *right-handed snails* because they do not breathe under water by use of gills but instead have a lung-like sac called a *pulmonary cavity*, which they fill with air at the surface of the water. When the snail takes in air from the surface, it makes a clicking sound. The air taken in can enable the snail to breathe under water for long periods, sometimes hours.

Lung-breathing snails have two characteristics that help us to identify them. First, they have no operculum or hard cover over the opening to its body cavity. Second, snails are either right-handed or left-handed; the lung-breathing snails are left-handed. We can tell the difference by holding the shell so that its tip is upward and the opening toward us. If the opening is to the left of the axis of the shell, the snail is considered to be *sinistral*—that is, it is left-handed. If the opening is to the right of the axis of the shell, the snail is *dextral*—that is, it is right-handed and it breathes with gills. Snails are animals of the substrate and are often found creeping along on all types of submerged surfaces in water from 10 cm to 2 m deep.

Before the Industrial Revolution of the 1800s, metropolitan areas were small and sparsely populated; thus, river and stream systems within or next to early communities received insignificant quantities of discarded waste. Early on, these river and stream systems were able to compensate for the small amount of wastes they received; when wounded (polluted), nature has a way of fighting back. In the case of rivers and streams, nature gives these flowing waters the ability to restore themselves through their own self-purification process. It was only when humans gathered in great numbers to form great cities that the stream systems were not always able to recover from receiving great quantities of refuse and other wastes. What exactly is it that we are doing to rivers and streams? We are upsetting the delicate balance between pollution and the purification process; that is, we are unbalancing the aquarium.

THOUGHT-PROVOKING QUESTIONS

6.1 If you arrive at a local stream to enjoy a picnic with your family and you notice that dragonflies, damselflies, and assorted other flying insects abound, what does that tell you, if anything, about the water quality of the stream? Explain.

6.2 How do stream ecosystems respond to land-use changes associated with urbanization? Explain.

6.3 How do these ecological responses vary across metropolitan areas located in different geographic settings? Explain.

REFERENCES AND RECOMMENDED READING

Abrahamson, D.E., Ed. (1988). *The Challenge of Global Warming*. Washington, DC: Island Press.

APHA. (1981). *Standard Methods for the Examination of Water and Wastewater*, 15th ed. Washington, DC: American Public Health Association.

Asimov, L. (1989). *How Did We Find Out About Photosynthesis?* New York: Walker & Company.

ASTM. (1969). *Manual on Water*. Philadelphia, PA: American Society for Testing and Materials.

Carson, R. (1962). *Silent Spring*. Boston: Houghton Mifflin.

Cave, C. (2000). *How a River Flows*, http://chamisa.freeshell.org/flow.htm.

Darwin, C. (1998). *The Origin of Species*, Suriano, G., Ed. New York: Grammercy.

Hutchinson, G.E. (1981). Thoughts on aquatic insects. *Bioscience*, 31, 495–500.

Lear, L. (1998). *Lost Woods: The Discovered Writing of Rachel Carson*. Boston: Beacon Press.

Lichatowich, J. (1999). *Salmon Without Rivers: A History of the Pacific Salmon Crisis*. Washington, DC: Island Press.

McCafferty, P.W. (1981). *Aquatic Entomology*. Boston: Jones & Bartlett.

Merrit, R.W. and Cummins, K.W. (1996). *An Introduction to the Aquatic Insects of North America*, 3rd ed. Dubuque, IA: Kendall/Hunt Publishing.

Metcalf & Eddy, Inc. (1991). *Wastewater Engineering: Treatment, Disposal, and Reuse*, 3rd ed. New York: McGraw-Hill.

Miller, G.T. (1988). *Environmental Science: An Introduction*. Belmont, CA: Wadsworth.

Odum, E.P. (1971). *Fundamentals of Ecology*, 3rd ed. Philadelphia, PA: Saunders.

Odum, E.P. (1975). *Ecology: The Link Between the Natural and the Social Sciences*. New York: Holt, Rinehart & Winston.

Odum, E.P. (1983). *Basic Ecology*, Philadelphia, PA: Saunders.

Odum, E.P. and Barrett, G.W. (2005). *Fundamentals of Ecology*, 5th ed. Belmont, CA: Thomson Brooks/Cole.

Price, P.W. (1984). *Insect Ecology*. New York: John Wiley & Sons.

Smith, R.L. (1996). *Ecology and Field Biology*. New York: HarperCollins.

Tchobanoglous, G. and Schroeder, E.D. (1985). *Water Quality*. Reading, MA: Addison-Wesley.

Tomera, A.N. (1989). *Understanding Basic Ecological Concepts*. Portland, ME: J. Weston Walch, Publisher.

Watson, L. (1988). *The Water Planet: A Celebration of the Wonders of Water*. New York: Harper & Row.

7 Water Pollution

Hang your clothes on a hickory limb but don't go near the water.

Proverb

West Virginia: As Odor Lingers, So Does Doubt About the Safety of Tap Water—Residents told smelly water isn't dangerous, despite chemical spill.

Nuckols (2014)

This country's waterways have been transformed by *omission*. Without beavers, water makes its way too quickly to the sea; without prairie dogs, water runs over the surface instead of sinking into the aquifer; without bison, there are no groundwater-recharge ponds in the grasslands and the riparian zone is trampled; without alligators, the edge between the water and land is simplified. Without forests, the water runs unfiltered to the waterways, and there is less deadwood in the channel, reducing stream productivity. Without floodplains and meanders, the water moves more swiftly, and silt carried in the water is more likely to be swept to sea.

The beaver, the prairie dog, the bison, and the alligator have been scarce for so long that we have forgotten how plentiful they once were. Beaver populations are controlled because they flood fields and forests, while wetlands acreage decrease annually. Prairie dogs are poisoned because they compete with cattle for grass, while the grasslands grow more barren year by year. Buffalo are generally seen as photogenic anachronisms, and alligators are too reptilian to be very welcome. But all of these animals once shaped the land in ways that improve water quality.

Outwater (1996)

Is water contamination really a problem—a serious problem? The answer to the first part of the question depends on where your water comes from. As to the second part of the question, refer to a book (or the film based on the book) that concerns a case of toxic contamination that you might be familiar with—*A Civil Action*, by Jonathan Harr. The book and film portray the legal repercussions connected with polluted water supplies in Woburn, Massachusetts. Two wells became polluted with industrial solvents, apparently causing 24 of the town's children who lived in neighborhoods supplied by those wells to contract leukemia and die.

Many who have read the book or have seen the movie may mistakenly get the notion that Woburn, a toxic "hot spot," is a rare occurrence. Nothing could be further from the truth. Toxic "hot spots" abound. Most striking are areas of cancer clusters, a short list of which includes not only Woburn but also Storrs, Connecticut, where wells polluted by a landfill are suspected of sickening and killing residents in nearby homes. In Bellingham, Washington, pesticide-contaminated drinking water is thought to be linked to a sixfold increase in childhood cancers. These are only a few examples of an underlying pathology that threatens many other communities. Meanwhile, cancer is now the primary cause of childhood death from disease.

Water contamination is a problem—a very serious problem. This chapter discusses a wide range of water contaminants, the sources of these contaminants, and the impact of these contaminants on drinking water supplies from both surface water and groundwater sources. In addition, the point is made that when it comes to freshwater pollution Nature is not defenseless in mitigating the situation. Nature, through its self-purification process in running water systems, is able to fight back against pollution—to a point.

SOURCES OF CONTAMINANTS

If we were to list all of the sources of contaminants and the contaminants themselves (the ones that foul our water supply systems), along with a brief description of each contaminant, we could easily fill a book. To give you some idea of the magnitude of the problem, refer to Figure 7.1, as well as the condensed list of selected sources and contaminants (our "short list") provided below.

> ***Note:*** Keep in mind that when we specify "water pollutants" we are in most cases speaking about pollutants that somehow get into the water (by whatever means) due to the interactions of the other two environmental mediums: air and soil. Probably the best example of this is the acid rain phenomenon. Pollutants originally emitted only into the atmosphere land on Earth and affect both soil and water. Consider that 69% of the anthropogenic lead and 73% of the mercury in Lake Superior have reached the lake by atmospheric deposition (Hill, 1997).

- *Subsurface percolation*—Hydrocarbons, metals, nitrates, phosphates, microorganisms, cleaning agents (e.g., trichloroethylene, or TCE)
- *Injection wells*—Hydrocarbons, metals, non-metal inorganics, organic and inorganic acids, organics, microorganisms, radionuclides
- *Land application*—Nitrogen, phosphorus, heavy metals, hydrocarbons, microorganisms, radionuclides
- *Landfills*—Organics, inorganics, microorganisms, radionuclides
- *Open dumps*—Organics, inorganics, microorganisms
- *Residential (local) disposal*—Organic chemicals, metals, non-metal inorganics, inorganic acids, microorganisms
- *Surface impoundments*—Organic chemicals, metals, non-metal inorganics, inorganic acids, microorganisms, radionuclides
- *Waste tailings*—Arsenic, sulfuric acid, copper, selenium, molybdenum, uranium, thorium, radium, lead, manganese, vanadium
- *Waste piles*—Arsenic, sulfuric acid, copper, selenium, molybdenum, uranium, thorium, radium, lead, manganese, vanadium

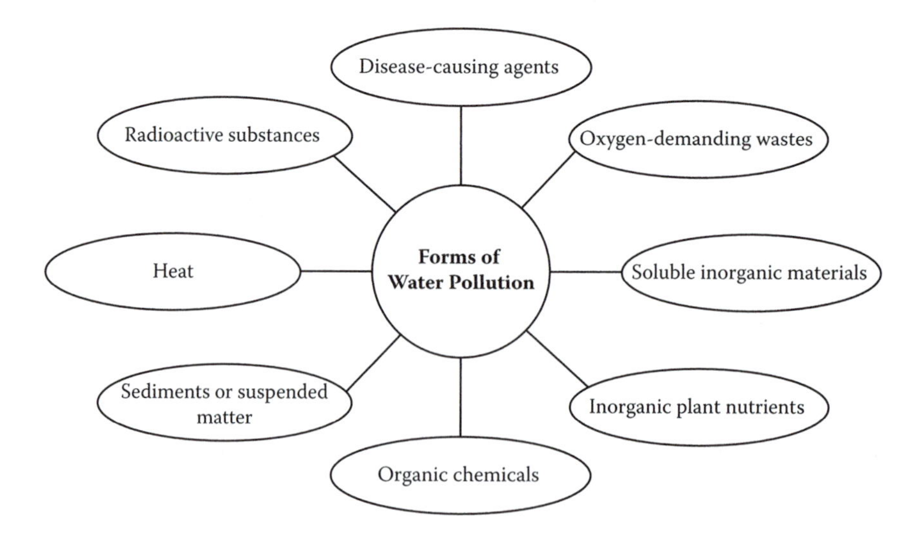

FIGURE 7.1 Biological, chemical, and physical forms of water pollution.

- *Materials stockpiles*—Aluminum, iron, calcium, manganese, sulfur, and traces of arsenic, cadmium, mercury, lead, zinc, uranium, and copper (coal piles); metals and non-metals, microorganisms (other materials piles)
- *Graveyards*—Metals, non-metals, microorganisms
- *Animal burial*—Contamination is site specific, depending on disposal practices (e.g., surface or subsurface), hydrology, proximity of the site to water sources, type and amount of disposed material, cause of death
- *Aboveground storage tanks*—Organics, metal and non-metal inorganics, inorganic acids, microorganisms, radionuclides
- *Underground storage tanks*—Organics, metals, inorganic acids, microorganisms, radionuclides
- *Containers*—Organics, metal and non-metal inorganics, inorganic acids, microorganisms, radionuclides
- *Open burning and detonating sites*—Inorganics (including heavy metals), organics (including TNT)
- *Radioactive disposal sites*—Radioactive cesium, plutonium, strontium, cobalt, radium, thorium, uranium
- *Pipelines*—Organics, metals, inorganic acids, microorganisms
- *Material transport and transfer operations*—Organics, metals, inorganic acids, microorganisms, radionuclides
- *Irrigation practices*—Fertilizers, pesticides, naturally occurring contamination, sediments
- *Pesticide applications*—Estimated at 1200 to 1400 active ingredients, including alachlor, aldicarb, atrazine, bromacil, carbofuran, cyanazine, dibromochloropropane (DBCP), dimethyl tetrachloroterephthalate (DCPA), 1,2-dichloropropane, dyfonate, ethylene dibromide (EDB), metolachlor, metribyzen, oxalyl, siazine, 1,2,3-trichloropropane
- *Concentrated animal feeding operations (CAFOs)*—Nitrogen, bacteria, viruses, phosphates
- *Deicing salts applications*—Chromate, phosphate, ferric ferocyanide, ferrocyan, chlorine
- *Urban runoff*—Suspended solids and toxic substances, especially heavy metals and hydrocarbons, bacteria, nutrients, petroleum residues
- *Percolation of atmospheric pollutants*—Sulfur and nitrogen compounds, asbestos, heavy metals
- *Mining and mine drainage*—Coal: acids, toxic inorganics (heavy metals), nutrients; phosphate: radium, uranium, fluorides; metallic ores: sulfuric acid, lead, cadmium, arsenic, sulfur, cyan
- *Production wells*—Oil wells (1.2 million abandoned production wells); farm irrigation wells; installation, operation, and plugging of all wells
- *Construction excavation*—Pesticides, diesel fuel, oil, salt, various others

Note: Before we discuss specific water pollutants, we must examine several terms important to the understanding of water pollution. One of these is *point source*. The U.S. Environmental Protection Agency defines a *point source* as "any single identifiable source of pollution from which pollutants are discharged, e.g., a pipe, ditch, ship, or factory smokestack." For example, the outlet pipes of an industrial facility or a municipal wastewater treatment plant are point sources. In contrast, *nonpoint sources* are widely dispersed sources and are a major cause of stream pollution. An example of a nonpoint source of pollution is rainwater carrying topsoil and chemical contaminants into a river or stream. Some of the major sources of nonpoint-source pollution include water runoff from farming, urban areas, forestry, and construction activities. The word *runoff* signals a nonpoint source that originated on land. Runoff may carry a variety of toxic substances and nutrients, as well as bacteria and viruses with it. Nonpoint sources now comprise the largest source of water pollution, contributing approximately 65% of the contamination in quality-impaired streams and lakes.

RADIONUCLIDES

When radioactive elements decay, they emit alpha, beta, or gamma radiations caused by transformation of the nuclei to lower energy states. In drinking water, radioactivity can be from natural or artificial radionuclides (the radioactive metals and minerals that cause contamination). These radioactive substances in water are of two types: radioactive minerals and radioactive gas. The U.S. Environmental Protection Agency (USEPA) has estimated that some 50 million Americans face increased cancer risk because of radioactive contamination of their drinking water.

Because of their occurrence in drinking water and their effects on human health, the natural radionuclides of chief concern are radium-226, radium-228, radon-222, and uranium. The source of some of these naturally occurring radioactive minerals is typically associated with certain regions of the country where mining is active or was active in the past. Mining activities expose rock strata, most of which contain some amount of radioactive ore. Uranium mining, for example, produces runoff. Radioactive contamination also occurs when underground streams flow through various rockbed and geologic formations containing radioactive materials. Other sources of radioactive minerals that may enter water supplies are smelters and coal-fired electrical generating plants. Also contributing to radioactive contamination of water are nuclear power plants, nuclear weapons facilities, radioactive materials disposal sites, and mooring sites for nuclear-powered ships. Hospitals contribute radioactive pollution when they dump low-level radioactive wastes into sewers; some of these radioactive wastes eventually find their way into water supply systems.

Although radioactive minerals such as uranium and radium in water may present a health hazard in particular areas, a far more dangerous threat exists in the form of radon. Radon is a colorless, odorless gas created by the natural decay of minerals in the soil. Normally present in all water in minute amounts, radon is especially concentrated in water that has passed through rock strata of granite, uranium, or shale. Radon enters homes from the soil beneath through cracks in the foundation or through crawl spaces and unfinished basements, as well as in tainted water. Radon is considered to be the second leading cause of lung cancer in the United States (about 20,000 cases each year), second only to cigarette smoking. Contrary to popular belief, radon is not a threat from surface water (lakes, rivers, or aboveground reservoirs), because radon dissipates rapidly when water is exposed to air. Even if the water source is groundwater, radon is still not a threat if the water is exposed to air (aerated) or if it is processed through an open tank during treatment. Studies have shown that where high concentrations of radon are detected within the air in a house most of that radon has come through the foundation and from the water; however, hot water used for showers, baths, or cooking (hot water) can release high concentrations of radon into the air. Still, radon is primarily a threat from groundwater taken directly from an underground source—either a private well or from a public water supply whose treatment of the water does not include exposure to air. Because radon in water evaporates quickly into the air, the primary danger is from inhaling it, not from drinking it.

CHEMICAL COCKTAIL

If we were to take the time to hold a full glass of water and inspect the contents, we might find that the contents appear cloudy or colored, making us think that the water is not fit to drink. Or, the contents might look fine but an odor of chlorine is prevalent. Most often, though, we simply draw water from the tap and either drink it or use it to cook dinner. The fact is that typically a glass of treated water is a chemical cocktail (Kay, 1996). Water utilities in communities seek to protect the public health by treating raw water with certain chemicals; what they are in essence doing is providing a drinking water product that is a mixture of various treatment chemicals and their byproducts. Water treatment facilities typically add chlorine to disinfect, and chlorine can produce contaminants. Another concoction is formed when ammonia is added to disinfect. Alum and polymers are added to the water to settle out various contaminants. The water distribution system and appurtenances must be protected from pipe corrosion, so the water treatment facility adds caustic soda, ferric chloride, and lime, which in turn increase the aluminum, sulfates, and salts in the water. Thus, when we hold that glass of water before us and we perceive a full glass of crystal clear, refreshing water, what we are really seeing is a concoction of many chemicals mixed with water, forming the chemical cocktail.

The most common chemical additives used in water treatment are chlorine, fluorides, and flocculants. Because we have already discussed fluorides, we focus our discussion in the following sections on the byproducts of chlorine and flocculant additives.

Byproducts of Chlorine

To lessen the potential impact of that chemical cocktail, the biggest challenge today is to make sure that the old standby, chlorine, will not produce as many new contaminants as it destroys. At the present time, arguing against the use of chlorine is difficult. Since 1908, chlorine has been used in the United States to kill off microorganisms that spread cholera, typhoid fever, and other waterborne diseases. In the 1970s, however, scientists discovered that, although chlorine does not seem to cause cancer in lab animals, when used in the water treatment process it can create a long list of byproducts that do. The byproducts of chlorine that present the biggest health concern are organic hydrocarbons, known as *trihalomethanes*, which are usually discussed as total trihalomethanes (TTHMs).

The USEPA classifies three of these trihalomethane byproducts—chloroform, bromoform, and bromodichloromethane—as probable human carcinogens. The fourth, dibromochloromethane, is classified as a possible human carcinogen. The USEPA set the first trihalomethane limits in 1979. Most water companies met these standards initially, but the standards were tightened after passage of the 1996 Safe Drinking Water Act (SDWA) Amendments. The USEPA is continuously studying the need to regulate other cancer-causing contaminants, including haloacetic acids (HAAs), also produced by chlorination.

Most people concerned with protecting public health applaud the USEPA's efforts in regulating water additives and disinfection byproducts; however, some people involved in the water treatment and supply business have expressed concern. A common concern often heard from water utilities having a tough time balancing the use of chlorine without going over the regulated limits revolves around the necessity of meeting regulatory requirements by lowering chlorine amounts to meet byproducts standards and at the same ensuring that all the pathogenic microorganisms are killed off. Many make the strong argument that, although no proven case exists that disinfection byproducts cause cancer in humans, many cases—an extensive history of cases—show that if we do not chlorinate water, then people get sick and sometimes die from waterborne diseases.

Because chlorination is now prompting regulatory pressure and compliance with new, demanding regulations, many water treatment facilities are looking for other options. Choosing an alternative disinfection chemical process is feeding a growing enterprise. One alternative that is currently being given widespread consideration in the United States is ozonation, which uses ozone gas to kill microorganisms. Ozonation is Europe's preferred method, and it does not produce trihalomethanes, but the USEPA does not yet recommend a wholesale switchover to ozone to replace chlorine or chlorination systems utilizing sodium hypochlorite or calcium hypochlorite. The USEPA points out that ozone also has problems, as it does not produce a residual disinfectant in the water distribution system, it is much more expensive, and in salty water it can produce another carcinogen, bromate.

At the present, what drinking water practitioners are doing (in the real world) is attempting to fine tune water treatment. What it all boils down to is a delicate balancing act. Drinking water professionals do not want to cut back on disinfection; if anything, they would prefer to strengthen it. So, we have to ask how we can bring into parity the microbial risks versus the chemical risks. How can both risks be reduced to an acceptable level? Unfortunately, no one is quite sure how to do this. The problem really revolves around the enigma associated with a "we don't know what we don't know" scenario.

The disinfection byproducts problem stems from the fact that most U.S. water systems produce the unwanted byproducts when the chlorine reacts to decayed organics: vegetation and other carbon-containing materials in water. Communities that take drinking water from lakes and rivers have a tougher time keeping the chlorine byproducts out of the tap than those that use clean groundwater. When a lot of debris is in the reservoir, a water utility may switch to alternative sources, such as wells. In other facilities, chlorine is combined with ammonia in a disinfection method called *chloramination*. This method is not as potent as pure chlorination, but it does prevent the production of unwanted trihalomethanes.

In communities where rains wash leaves, grasses, and trees into the local water source (such as a lake or river), hot summer days can trigger algae blooms, upping the organic matter that can produce trihalomethanes. Spring runoff in many communities exacerbates the problem. With increased runoff comes agricultural waste, pesticides, and quantities of growth falling into the water that must be dealt with. Nature's conditions in summer diminish some precursors for trihalomethanes—the bromides in salty water. Under such conditions, usually nothing unusual is visible in the drinking water; however, water that is cloudy due to silt (dissolved organics from decayed plants) could harbor trihalomethanes.

Most cities today strain out the organics from their water supplies before chlorinating to prevent the formation of trihalomethanes and haloacetic acids. In other communities, the move is already on to switch from chlorine to ozone and other disinfectant methods. The National Resources Defense Council (NRDC, 2003) believes that most U.S. systems will catch up with Europe in the next decade or so and use ozone to kill resistant microbes such as *Cryptosporidium*. When this method is employed, the finishing touch is usually accomplished by filtering the water through granulated activated carbon, which increases the cost for consumers (estimated at about $100 or more per year per hookup).

Total Trihalomethanes

Total trihalomethanes (TTHMs) are a byproduct of chlorinating water that contains natural organics (USEPA, 1998). A USEPA survey discovered that trihalomethanes are present in virtually all chlorinated water supplies. Many years ago, the USEPA required large towns and cities to reduce TTHM levels in potable water; however, national drinking water quality standards now require water treatment systems of smaller towns to reduce TTHMs. It is important to note that TTHMs do not pose a high health risk compared to waterborne diseases, but they are among the most important water quality issues to be addressed in the U.S. water supply.

A major challenge for drinking water practitioners is how to balance the risks from microbial pathogens and disinfection byproducts. Providing protection from these microbial pathogens while simultaneously ensuring decreasing health risks to the population from disinfection byproducts (DBPs) is important. The Safe Drinking Water Act (SDWA) Amendments, signed by President Clinton in 1996, required the USEPA to develop rules to achieve these goals. The Stage 1 Disinfectants and Disinfection Byproducts Rule and the Interim Enhanced Surface Water Treatment Rule are the first of a set of rules under the Amendments.

Public Health Concerns

Most Americans drink tap water that meets all existing health standards all the time. These rules were designed to further strengthen existing drinking water standards and thus increase protection for many water systems. The USEPA's Science Advisory Board concluded in 1990 that exposure to microbial contaminants such as bacteria, viruses, and protozoa (e.g., *Giardia lamblia* and *Cryptosporidium*) was likely the greatest remaining health risk management challenge for drinking water suppliers. Acute health effects from exposure to microbial pathogens is documented, and associated illness can range from mild, to moderate cases lasting only a few days, to more severe infections that can last several weeks and may result in death for those with weakened immune systems.

Although disinfectants are effective in controlling many microorganisms, they react with natural organic and inorganic matter in source water and distribution systems to form potential disinfectant byproducts. Many of these DBPs have been shown to cause cancer and reproductive and developmental effects in laboratory animals. More than 200 million people consume water that has been disinfected. Because of the large population exposed, health risks associated with DBPs, even if small, need to be taken seriously.

Existing Regulations

- *Microbial contaminants*—The 1989 Surface Water Treatment Rule applies to all public water systems using surface water sources or groundwater sources under the direct influence of surface water. It establishes maximum contaminant level goals (MCLGs) for viruses, bacteria, and *Giardia lamblia*. It also addresses treatment technique requirements for filtered and unfiltered systems specifically designed to protect against the adverse health effects of exposure to these microbial pathogens. The Total Coliform Rule, revised in 1989, applies to all public water systems and establishes a maximum contaminant level (MCL) for total coliforms.
- *Disinfection byproducts*—In 1979, the USEPA set an interim MCL for total trihalomethanes of 0.10 mg/L as an annual average. This applies to any community water system serving at least 10,000 people that adds a disinfectant to the drinking water during any part of the treatment process.

Information Collection Rule

To support the microbial and DBP rulemaking process, the Information Collection Rule establishes monitoring and data reporting requirements for large public water systems serving at least 100,000 people. This rule was intended to provide the USEPA with information on the occurrence in drinking water of microbial pathogens and DBPs. The USEPA is collecting engineering data on how public water systems currently control such contaminants as part of the Information Collection Rule.

Ground Water Rule

The USEPA developed and promulgated the final Ground Water Rule in 2006. The rule specifies the appropriate use of disinfection and, equally importantly, addresses other components of groundwater systems to ensure public health protection. More than 158,000 public or community systems serve almost 89 million people through groundwater systems. Of these, 99% of groundwater systems (157,000) serve fewer than 10,000 people; however, systems serving more than 10,000 people serve 55% (more than 60 million) of all people who get their drinking water from public groundwater systems.

Filter Backwash Recycling

The 1996 SDWA Amendments required the USEPA to set a standard on recycling filter backwash within the treatment process of public water systems. The regulation applies to all public water systems, regardless of size. In June 2001, the Filter Backwash Recycling Rule (FBRR) was published in the *Federal Register* and presented the specific regulatory requirements that must be met by affected systems. Specifically, the FBRR establishes regulatory provisions governing the way that certain recycle streams are handled within the water treatment processes of conventional filtration (series of processes including coagulation, flocculation, sedimentation, and filtration) and direct filtration (series of processes including coagulation and filtration, but excluding sedimentation) water treatment systems. The FBRR also established reporting and recordkeeping requirements for recycle practices within all states and the USEPA to better evaluate the impact of recycle practices on overall treatment plant performance. Implementation of the FBRR was designed to reduce the risk of illness from microbial pathogens in drinking water, particularly *Cryptosporidium*.

Opportunities for Public Involvement

The USEPA encourages public input into regulation development. Public meetings and opportunities for public comment on microbial and DBP rules are announced in the *Federal Register.* The USEPA's Office of Groundwater and Drinking Water also provides information regarding the rules and other programs online (www.epa.gov/safewater/standards.html).

Flocculants

In addition to chlorine and sometimes fluoride, water treatment plants often add several other chemicals, including flocculants, to improve the efficiency of the treatment process—and they all add to the cocktail mix. Flocculants are chemical substances added to water to make particles clump together, which improves the effectiveness of filtration. Some of the most common flocculants are polyelectrolytes (polymers)—chemicals with constituents that cause cancer and birth defects and are banned for use by several countries. Although the USEPA classifies them as "probable human carcinogens," it still allows their continued use. Acrylamide and epichlorohydrin are two flocculants used in the United States that are known to be associated with probable cancer risk (Lewis, 1996).

GROUNDWATER CONTAMINATION

Groundwater under the direct influence of surface water comes under the same monitoring regulations as does surface water (i.e., all water open to the atmosphere and subject to surface runoff). The legal definition of *groundwater under the direct influence of surface water* is any water beneath the surface of the ground with (1) significant occurrence of insects or microorganisms, algae, or large-diameter pathogens such as *Giardia lamblia*; or (2) significant and relatively rapid shifts in water characteristics such as turbidity, temperature, conductivity, or pH, which closely correlate to climatological or surface water conditions. Direct influence must be determined for individual sources in accordance with criteria established by the state. The state determines for individual sources in accordance with criteria established by the state, and that determination may be based on site-specific measurements of water quality and/or documentation of well construction characteristics and geology with field evaluation.

Generally, most groundwater supplies in the United States are of good quality and produce essential quantities. The full magnitude of groundwater contamination in the United States is, however, not fully documented, and federal, state, and local efforts continue to assess and address the problems (Fitts, 2012; Rail, 1985). Groundwater supplies about 25% of the freshwater used for all purposes in the United States, including irrigation, industrial uses, and drinking water (about 50% of the U.S. population relies on groundwater for drinking water). Elsewhere, groundwater aquifers beneath or close to Mexico City provide the area with more than 3.2 billion liters per day (Chilton, 1998). But, as groundwater pumping increases to meet water demand, it can exceed the aquifer rate of replenishment, and in many urban aquifers water levels are showing a long-term decline. With excessive extraction comes a variety of other undesirable effects, including the following:

- Increased pumping costs
- Changes in hydraulic pressure and underground flow directions (in coastal areas, this results in seawater intrusion)
- Saline water drawn up from deeper geological formations
- Poor-quality water from polluted shallow aquifers leaking downward

Severe depletion of groundwater resources is often compounded by a serious deterioration in its quality. Without a doubt, contamination of a groundwater supply should be a concern of drinking water practitioners responsible for supplying a community with potable water provided by groundwater.

Despite our strong reliance on groundwater, groundwater has for many years been one of the most neglected natural resources. Why? Good question. Groundwater has been ignored because it is less visible than other environmental resources—rivers or lakes, for example. What the public cannot see or observe, the public doesn't worry about, or even think about; however, recent publicity about events concerning groundwater contamination is making the public more aware of the problem, and the regulators have also taken notice.

Are natural contaminants a threat to human health—harbingers of serious groundwater pollution events? No, not really. The main problem with respect to serious groundwater pollution has been human activities. When we improperly dispose of wastes or spill hazardous substances onto the ground, we threaten our groundwater and in turn public health.

Underground Storage Tanks

If we could look at a map of the United States indicating the exact location of every underground storage tank (UST), most of us would be surprised at the large number of tanks buried underground. With so many buried tanks, it should come as no surprise that structural failures arising from a wide variety of causes have occurred over the years. Subsequent leaking has become a huge source of contamination that affects the quality of local groundwaters.

> ***Note:*** A UST is any tank, including any underground piping connected to the tank, that has at least 10% of its volume below ground (USEPA, 1987).

The fact is, leakage of petroleum and its products from USTs occurs more often than we generally realize. This widespread problem has been and continues to be a major concern and priority in the United States. In 1987, the USEPA promulgated regulations for many of the nation's USTs, and much progress has been made in mitigating this problem to date.

When a UST leak or past leak is discovered, the contaminants released to the soil and thus to groundwater, it would seem to be a rather straightforward process to identify the contaminant, which in many cases would be fuel oil, diesel, or gasoline. Other contaminants, however, also present problems; for example, in the following section, we discuss one such contaminant, a byproduct of gasoline, to help illustrate the magnitude of leaking USTs.

MtBE

In 1997, the USEPA issued a drinking water advisory titled *Consumer Acceptability Advice and Health Effects Analysis on Methyl Tertiary-Butyl Ether (MtBE)* (http://water.epa.gov/action/advisories/drinking/mtbe.cfm). The purpose of the advisory was to provide guidance to communities exposed to drinking water contaminated with MtBE.

> ***Note:*** A USEPA Advisory is usually initiated to provide information and guidance to individuals or agencies concerned with potential risk from drinking water contaminants for which no national regulations currently exist. Advisories are not mandatory standards for action and are used only for guidance. They are not legally enforceable and are subject to revision as new information becomes available. The USEPA's Health Advisory program is recognized in the Safe Drinking Water Act Amendments of 1996, which state in Section 102(b)(1)(F):
>
>> The Administrator may publish health advisories (which are not regulations) or take other appropriate actions for contaminants not subject to any national primary drinking water regulation.
>
> As its title indicates, this Advisory includes consumer acceptability advice as "appropriate" under this statutory provision, as well as a health effects analysis.

MtBE is a volatile organic compound. Since the late 1970s, MtBE has been used as an octane enhancer in gasoline. Because it promotes more complete burning of gasoline (thereby reducing carbon monoxide and ozone levels), it is commonly used as a gasoline additive in localities that do not meet the National Ambient Air Quality Standards.

In the Clean Air Act of 1990, Congress mandated the use of reformulated gasoline (RFG) in areas of the country with the worst ozone or smog problems. RFG must meet certain technical specifications set forth in the Act, including a specific oxygen content. Ethanol and MtBE are the primary oxygenates used to meet the oxygen content requirement. MtBE is used in about 84% of RFG supplies. Currently, 32 areas in a total of 18 states are participating in the RFG program, and RFG accounts for about 30% of gasoline nationwide.

Studies have identified significant air quality and public health benefits that directly result from the use of fuels oxygenated with MtBE, ethanol, or other chemicals. Refiners' fuel data submitted to the USEPA for 1995/1996 indicated that the national emissions benefits exceeded those required. The 1996 Air Quality Trends Report showed that toxic air pollutants declined significantly between 1994 and 1995, and analysis indicated that at least some of this progress could be attributed to the use of RFG. Beginning in the year 2000, required emission reductions were substantially greater, at about 27% for volatile organic compounds, 22% for toxic air pollutants, and 7% for nitrogen oxides.

Note: When gasoline that has been oxygenated with MtBE comes in contact with water, large amounts of MtBE dissolve. At 25°C, the water solubility of MtBE is about 5000 mg/L for a gasoline that is 10% MtBE by weight. In contrast, for a non-oxygenated gasoline, the total hydrocarbon solubility in water is typically about 120 mg/L. MtBE sorbs only weakly to soil and aquifer material; therefore, sorption will not significantly retard MtBE transport by groundwater. In addition, the compound generally resists degradation in groundwater (Squillace et al., 1998).

A limited number of instances of significant contamination of drinking water with MtBE have occurred because of leaks from underground and aboveground petroleum storage tank systems and pipelines. Due to its small molecular size and solubility in water, MtBE moves rapidly into groundwater, faster than do other constituents of gasoline. Public and private wells have been contaminated in this manner. Nonpoint sources (such as recreational watercraft) are most likely to be the cause of small amounts of contamination in a large number of shallow aquifers and surface waters. Air deposition through precipitation of industrial or vehicular emissions may also contribute to surface water contamination. The extent of any potential for build-up in the environment from such deposition is uncertain.

Based on the limited sampling data currently available, most concentrations at which MtBE has been found in drinking water sources are unlikely to cause adverse health effects; however, the USEPA is continuing to evaluate the available information and is doing additional research to seek more definitive estimates of potential risks to humans from drinking water. There are no data on the effects on humans of drinking MtBE-contaminated water. In laboratory tests on animals, cancer and noncancer effects have occurred at high levels of exposure. These tests are conducted by inhalation exposure or by introducing the chemical in oil directly to the stomach. The tests support a concern for potential human hazard; however, because the animals were not exposed through drinking water, significant uncertainties exist concerning the degree of risk associated with human exposure to low concentrations typical found in drinking water.

The very unpleasant taste and odor of MtBE make contaminated drinking water unacceptable to the public. Studies conducted on the concentrations of MtBE in drinking water determined the level at which individuals can detect the odor or taste of the chemical. Humans vary widely in the concentrations they are able to detect. Some who are sensitive can detect very low concentrations. Others do not taste or smell the chemical, even at much higher concentrations. The presence or absence of other natural or water treatment chemicals sometimes masks or reveals the taste or odor effects. Studies to date have not been extensive enough to completely describe the extent of this variability or to establish a population response threshold. Nevertheless, we conclude from the available studies that keeping concentrations in the range of 20 to 40 μg/L of water or below will likely avert unpleasant taste and odor effects, recognizing that some people may detect the chemical below this.

Concentrations in the range of 20 to 40 μg/L are about 20,000 to 100,000 (or more) times lower than the range of exposure levels in which cancer or noncancer effects were observed in rodent tests. This margin of exposure lies within the range of margins of exposure typically provided to protect against cancer effects by the National Primary Drinking Water Standards under the federal Safe Drinking Water Act—a margin greater than such standards typically provided to protect against noncancer effects. Protection of the water source from unpleasant taste and odor as recommended also protects consumers from potential health effects. The USEPA observed that occurrences of groundwater contamination at or above this 20-to 40-μg/L taste and odor threshold (that is, contamination at levels that may create consumer acceptability problems for water supplies) have, to date, resulted from leaks in petroleum storage tanks or pipelines, not from any other sources.

Public water systems that conduct routine monitoring for volatile organic compounds can test for MtBE at little additional cost, and some states are already moving in this direction. Public water systems detecting MtBE in their source water at problematic concentrations can remove MtBE from water using the same conventional treatment techniques that are used to clean up other contaminants originating from gasoline releases—air stripping and granular activated carbon (GAC), for example. Because MtBE is more soluble in water and more resistant to biodegradation than other chemical constituents in gasoline, air stripping and GAC treatment require additional optimization and must often be used together to effectively remove MtBE from water. The costs of removing MtBE are higher than when treating for gasoline releases that do not contain MtBE. Oxidation of MtBE using ultraviolet/peroxide/ozone treatment may also be feasible but typically has higher capital and operating costs than air stripping and GAC.

The bottom line: Because MtBE has been found in sources of drinking water many states are phasing out the sale of gasoline with MtBE. Also, MtBE is being phased out because ethanol is now favored as its replacement.

> ***Note:*** Of the 60 volatile organic compounds (VOCs) analyzed in samples of shallow ambient groundwater collected from eight urban areas from 1993 to 1994 as part of the U.S. Geological Survey's National Water Quality Assessment program, MtBe was the second most frequently detected compound, after trichloromethane (chloroform) (Squillace et al., 1998).

Industrial Wastes

Because industrial waste represents a significant source of groundwater contamination, drinking water practitioners and others expend an increasing amount of time in abating or mitigating pollution events that damage groundwater supplies. Groundwater contamination from industrial wastes usually begins with the practice of disposing of industrial chemical wastes in surface impoundments—unlined landfills or lagoons, for example. Fortunately, these practices, for the most part, are part of our past. Today, we know better; for example, we now know that what is most expedient or least expensive does not work for industrial waste disposal practices. We have found through actual experience that the long run has proven just the opposite—for society as a whole (with respect to health hazards and the costs of cleanup activities) to ensure clean or unpolluted groundwater supplies is very expensive and utterly necessary.

Septic Tanks

Seepage from septic tanks is a biodegradable waste capable of affecting the environment through water and air pollution. The potential environmental problems associated with use of septic tanks are magnified when you consider that subsurface sewage disposal systems (septic tanks) are used by almost one-third of the U.S. population. Briefly, a septic tank and leaching field system traps and stores solids while the liquid effluent flows from the tank into a leaching or absorption field, where it slowly seeps into the soil and degrades naturally. The problem with subsurface sewage disposal systems such as septic tanks is that most of the billions of gallons of sewage that enter the ground each year are not properly treated. Because of faulty construction or lack of maintenance, not all of these systems work properly. Experience has shown that septic disposal systems are frequently sources of fecal bacteria and virus contamination of water supplies taken from private wells. Many septic tank owners dispose of detergents, nitrates, chlorides, and solvents in their septic systems or use solvents to treat their sewage waste. A septic tank cleaning fluid that is commonly used contains organic solvents (trichloroethylene, or TCE) that are potential human carcinogens that pollute the groundwater in areas served by septic systems.

Landfills

Humans have been disposing of waste by burying it in the ground since time immemorial. In the past, this practice was largely uncontrolled, and the disposal sites (i.e., garbage dumps) were places where municipal solid wastes were simply dumped on and into the ground without much thought or concern. Even in this modern age, landfills have been used to dispose of trash and waste products at controlled locations that are then sealed and buried under the ground. Now such practices are increasingly seen as a less than satisfactory disposal method, because of the long-term environmental impact of waste materials in the ground and groundwater. Unfortunately, many of the older (and even some of the newer) sites have been located in low-lying areas with high groundwater tables. *Leachate* (seepage of liquid through the waste) high in biochemical oxygen demand (BOD), chloride, organics, heavy metals, nitrate, and other contaminants has little difficulty reaching the groundwater in such disposal sites. In the United States, literally thousands of inactive or abandoned dumps like this exist.

Agriculture

Fertilizers and pesticides are the two most significant groundwater contaminants that result from agricultural activities. The impact of agricultural practices wherein fertilizers and pesticides are normally used is dependent on local soil conditions. If, for example, the soil is sandy, nitrates from fertilizers are easily carried through the porous soil into the groundwater, contaminating private wells. Pesticide contamination of groundwater is a subject of national importance because groundwater is used for drinking water by about 50% of the nation's population. This especially concerns people living in the agricultural areas where pesticides are most often used, as about 95% of that population relies upon groundwater for drinking water. Before the mid-1970s, the common thought was that soil acted as a protective filter, one that stopped pesticides from reaching groundwater. Studies have now shown that this is not the case. Pesticides can reach water-bearing aquifers below ground from applications onto crop fields, seepage of contaminated surface water, accidental spills and leaks, improper disposal, and even through injection of waste material into wells.

Pesticides are mostly modern chemicals. Many hundreds of these compounds are used, and extensive tests and studies of their effect on humans have not been completed. This leads us to ask, "Just how concerned should we be about their presence in our drinking water?" Certainly, considering pesticides to be potentially dangerous and handling them with appropriate care would be wise. We can say they pose a potential danger if they are consumed in large quantities, but as any experienced scientist knows we cannot draw factual conclusions unless scientific tests have been done. Some pesticides have had a designated maximum contaminant level (MCL) in drinking water set by the USEPA, but many have not. Another serious point to consider is the potential effect of combining more than one pesticide in drinking water, which might be different than the effects of each individual pesticide alone. This is another situation where we do not have sufficient scientific data to draw reliable conclusions—in other words, we don't know what we don't know.

Saltwater Intrusion

In many coastal cities and towns as well as in island locations, the intrusion of salty seawater presents a serious water quality problem. Because freshwater is lighter than saltwater (the specific gravity of seawater is about 1.025), it will usually float above a layer of saltwater. When an aquifer in a coastal area is pumped, the original equilibrium is disturbed and saltwater replaces the freshwater (Barlow and Reichard, 2010; USGS, 2003, Viessman and Hammer, 1998). The problem is compounded by increasing population, urbanization, and industrialization, which increases the use of groundwater supplies. In such areas, while groundwater is heavily drawn upon, the quantity of natural groundwater recharge is decreased because of the construction of roads, tarmac, and parking lots, which prevent rainwater from infiltrating, decreasing the groundwater table elevation. In coastal areas, the natural interface between the fresh groundwater flowing from upland areas and the saline water from the sea is constantly under attack by human activities. Because seawater is approximately 2.5 times more dense than freshwater, a high-pressure head of seawater occurs (in relation to freshwater), which results in a significant rise in the seawater boundary. Potable water wells close to this rise in sea level may have to be abandoned because of saltwater intrusion.

Other Sources of Groundwater Contamination

To this point, we have discussed only a few of the many sources of groundwater contamination; for example, we have not discussed mining and petroleum activities that lead to contamination of groundwater or contamination caused by activities in urban areas, both of which are important. Urban activities (including spreading salt on roads to keep them free of ice during the winter) eventually contribute to contamination of groundwater supplies, as can underground injection wells used to dispose of hazardous materials. As discussed earlier, underground storage tanks are also significant contributors to groundwater pollution. Other sources of groundwater contamination include the following items on our short list:

- Waste tailings
- Residential disposal
- Urban runoff
- Hog wastes
- Biosolids
- Land-applied wastewater
- Graveyards
- Deicing salts
- Surface impoundments
- Waste piles
- Animal feeding operations
- Natural leaching
- Animal burial
- Mine drainage
- Pipelines
- Open dumps
- Open burning
- Atmospheric pollutants

Raw sewage is not listed, because for the most part raw sewage is no longer routinely dumped into our nation's wells or into our soil. Sewage treatment plants effectively treat wastewater so that it can be safely discharged to local water bodies. In fact, the amount of pollution being discharged from these plants has been cut by over one-third during the past 20 years, even as the number of people served has doubled. Yet, in some areas raw sewage spills still occur, sometimes because a underground sewer line is blocked, broken, or too small or because periods of heavy rainfall overload the capacity of the sewer line or sewage treatment plant so overflows into city streets or streams occur. Some of this sewage finds its way to groundwater supplies.

The best way to prevent groundwater pollution is to stop it from occurring in the first place. Unfortunately, a perception held by many is that natural purification of chemically contaminated ground can take place on its own—without the aid of human intervention. To a degree this is true; however, natural purification functions on its own time, not on human time. Natural purification could take decades, perhaps centuries. The alternative? Remediation. But remediation and mitigation don't come cheap. When groundwater is contaminated, the cleanup efforts are sometimes much too expensive to be practical. The USEPA established the Groundwater Guardian Program in 1994, which is a voluntary approach to improving drinking water safety. Established and managed by a nonprofit organization in the Midwest and strongly promoted by the USEPA, this program focuses on communities that rely on groundwater for their drinking water. It provides special recognition and technical assistance to help communities protect their groundwater from contamination. Groundwater Guardian programs have been established in over 150 communities in 34 states and one Canadian province (http://www.groundwater.org).

SELF-PURIFICATION OF STREAMS

Hercules, that great mythical giant and arguably the globe's first environmental engineer, discovered that the solution to stream pollution is dilution—that is, dilution is the solution. In reality, today's humans depend on various human-made water treatment processes to restore water to potable and palatable condition. However, it should be pointed out that Nature, as Hercules observed, is not defenseless in its fight against water pollution. For example, when a river or stream is contaminated, natural processes (including dilution) immediately kick in to restore the water body and its contents back to its natural state. If the level of contamination is not excessive, the stream or river can restore itself to normal conditions in a relatively short period of time. In this section, Nature's ability to purify and restore typical river systems to normal conditions is discussed. In terms of practical usefulness, the waste assimilation capacity of streams as a water resource has its basis in the complex phenomenon termed *stream self-purification*. This is a dynamic phenomenon reflecting hydrologic and biologic variations, and the interrelations are not yet fully understood in precise terms, although this does not preclude applying what is known. Sufficient knowledge is available to permit quantitative definition of resultant stream conditions under expected ranges of variation to serve as practical guides in decisions dealing with water resource use, development, and management (Spellman, 1996; Velz, 1970).

Balancing the Aquarium

An outdoor excursion to the local stream can be a relaxing and enjoyable undertaking. However, when the wayfarer arrives at the local stream, spreads a blanket on its bank, and then looks out upon its flowing mass and discovers a parade of waste and discarded rubble bobbing along cluttering the adjacent shoreline and downstream areas, he quickly loses any feeling of relaxation or enjoyment. The sickening sensation only increases as our observer closely scrutinizes the putrid flow. He recognizes the rainbow-colored shimmer of an oil slick, interrupted here and there by dead fish and floating refuse, and the slimy fungal growth that prevails. At the same time, the observer's sense of smell is also alerted to the noxious conditions. Along with the fouled water and the stench of rot-filled air, the observer notices the ultimate insult and tragedy: a sign warning, "DANGER—NO SWIMMING or FISHING." The observer soon realizes that the stream before him is not a stream at all; it is little more than an unsightly drainage ditch. He has discovered what ecologists have known and warned about for years—that is, contrary to popular belief, rivers and streams do not have an infinite capacity for taking care of pollution.

Before the early 1970s, occurrences such as the one just described were common along the rivers and streams near main metropolitan areas throughout most of the United States. Many aquatic habitats were fouled during the past because of industrialization, but our streams and rivers were not always in such deplorable condition. Before the Industrial Revolution of the 1800s, metropolitan areas were small and sparsely populated. Thus, river and stream systems within or next to early communities received insignificant quantities of discarded waste. Early on, these river and stream systems were able to compensate for the small amount of wastes they received; when wounded by pollution, nature has a way of fighting back. In the case of rivers and streams, nature provides their flowing waters with the ability to restore themselves through their own self-purification process. It was only when humans gathered in great numbers to form great cities that the stream systems were not always able to recover from having received great quantities of refuse and other wastes. What exactly is it that we are doing to rivers and streams?

Halsam (1990) observed that humans' actions are determined by their expediency. In addition, what most people do not realize is that we have the same amount of water as we did millions of years ago, and through the water cycle we continually reuse that same water—water that was used by the ancient Romans and Greeks is the same water we are using today. Increased demand has put enormous stress on our water supply. Humans are the cause of this stress; that is, humans upset the delicate balance between pollution and the purification process. We tend to unbalance the aquarium.

With the advent of industrialization, local rivers and streams became deplorable cesspools that worsened with time. During the Industrial Revolution, the removal of horse manure and garbage from city streets became a pressing concern. Moran and colleagues (1986) noted that, "None too frequently, garbage collectors cleaned the streets and dumped the refuse into the nearest river." As late as 1887, river keepers were employed full time to remove a constant flow of dead animals from a river in London (Halsam, 1990). Moreover, the prevailing attitude of that day was "I don't want it anymore—throw it into the river."

Once we came to understand the dangers of unclean waters, any threat to the quality of water destined for use for drinking and recreation has quickly angered those affected. Fortunately, since the 1970s we have moved to correct the stream pollution problem. Through scientific study and incorporation of wastewater treatment technology, we have started to restore streams to their natural condition.

Fortunately, through the phenomenon of self-purification, the stream aids us in this effort to restore a steam's natural water quality. A balance of biological organisms is normal for all streams. Clean, healthy streams have certain characteristics in common; for example, as mentioned, one property of streams is their ability to dispose of small amounts of pollution. However, if streams receive unusually large amounts of waste, the stream life will change and attempt to stabilize such pollutants; that is, the biota will attempt to balance the aquarium. If the stream biota are not capable of self-purifying, then the stream may become a lifeless body of muck.

Note: The self-purification process discussed here relates to the purification of organic matter only.

Sources of Stream Pollution

Sources of stream pollution are normally classified as point or nonpoint sources. A *point source* (PS) is a source that discharges effluent, such as wastewater from sewage treatment and industrial plants. Simply put, a point source is usually easily identified as "end of the pipe" pollution; that is, it emanates from a concentrated source or sources. In addition to organic pollution received from the effluents of sewage treatment plants, other sources of organic pollution include runoffs and dissolution of minerals throughout an area and are not from one or more concentrated sources. Non-concentrated sources are known as *nonpoint sources* (see Figure 7.2) Nonpoint-source (NPS) pollution, unlike pollution from industrial and sewage treatment plants, comes from many diffuse

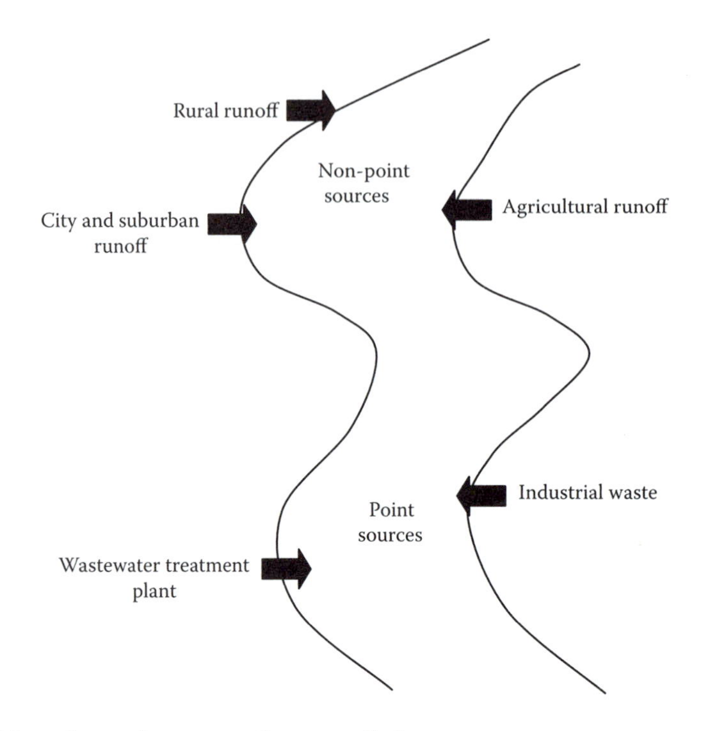

FIGURE 7.2 Point and nonpoint sources of stream pollution.

sources. Rainfall or snowmelt moving over and through the ground causes NPS pollution. As the runoff moves, it picks up and carries away natural and manmade pollutants, finally depositing them into streams, lakes, wetlands, rivers, coastal waters, and even our underground sources of drinking water. Some of these pollutants are listed below:

- Excess fertilizers, herbicides, and insecticides from agricultural lands and residential areas
- Oil, grease, and toxic chemicals from urban runoff and energy production
- Sediment from improperly managed construction sites, crop and forest lands, and eroding streambanks
- Salt from irrigation practices and acid drainage from abandoned mines
- Bacteria and nutrients from livestock, pet wastes, and faulty septic systems

Atmospheric deposition and hydromodification are also sources of nonpoint-source pollution (Spellman, 2008; USEPA, 1994).

Specific examples of nonpoint sources include runoff from agricultural fields and from cleared forest areas, construction sites, and roadways. Of particular interest to environmental practitioners in recent years have been agricultural effluents. As a case in point, take, for example, farm silage effluent, which has been estimated to be more than 200 times as potent (in terms of biochemical oxygen demand, or BOD) as treated sewage (USEPA, 1994).

Nutrients are organic and inorganic substances that provide food for microorganisms such as bacteria, fungi, and algae. Nutrients are supplemented by the discharge of sewage. The bacteria, fungi, and algae are consumed by the higher trophic levels in the community. Each stream, due to a limited amount of dissolved oxygen (DO), has a limited capacity for aerobic decomposition of organic matter without becoming anaerobic. If the organic load received is above that capacity, the stream becomes unfit for normal aquatic life, and it is not able to support organisms sensitive to oxygen depletion (Mason, 1991; Spellman, 1996).

Effluent from a sewage treatment plant is most commonly disposed of in a nearby waterway. At the point of entry of the discharge, there is a sharp decline in the concentration of DO in the stream. This phenomenon is known as *oxygen sag*. Unfortunately (for the organisms that normally occupy a clean, healthy stream), when the DO decreases, there is a concurrent massive increase in BOD because microorganisms utilize the DO as they break down the organic matter. When the organic matter is depleted, the microbial population and BOD decline, while the DO concentration increases, assisted by stream flow (in the form of turbulence) and by the photosynthesis of aquatic plants. This self-purification process is very efficient, and the stream will suffer no permanent damage as long as the quantity of waste is not too high. Obviously, an understanding of this self-purification process is important in preventing stream ecosystem overload.

As urban and industrial centers continue to grow, waste disposal problems also grow. Because wastes have increased in volume and are much more concentrated than earlier, natural waterways must have help in the purification process. Wastewater treatment plants provide this help. Wastewater treatment plants function to reduce the organic loading that raw sewage would impose on discharge into streams. Wastewater treatment plants utilize three stages of treatment: primary, secondary, and tertiary treatment. In breaking down the wastes, a secondary wastewater treatment plant uses the same type of self-purification process found in any stream ecosystem. Small bacteria and protozoa (one-celled organisms) begin breaking down the organic material. Aquatic insects and rotifers are then able to continue the purification process. Eventually, the stream will recover and show little or no effects of the sewage discharge. This phenomenon is known as natural stream purification (Spellman and Whiting, 1999).

CONCENTRATED ANIMAL FEEDING OPERATIONS*

> … don't let the folks down there know that you are looking for sites for hog facilities or they will prevaricate and try to take us to the cleaners, they will carry on with letters to various editors, every kind of meanness and so forth, as they have been brainwashed by the Sierra Club to think that hog facilities are bad, even the folks who love baby back ribs, even the ones hunting jobs. … The panhandle region is perfect for hog operations—plenty of room, low population, nice long dry seasons, good water. There is no reason why the Texas panhandle can't produce seventy-five percent of the world's pork.
>
> **Proulx (2002)**

The debate over the future of agriculture and agricultural policy is not new. It can be found in the history of such agrarian protest movements as the Grange (1870s), populism (1890s), the Farm Holiday Movement (1930s), the National Farmers Organization (1950s), and the immigration and environment groups of the last two decades. There have always been farmers and rural residents lauding progress, while others are lamenting the loss of the agrarian nature of America which allowed more to provide for themselves, and many more who are ambivalent (Schwab, 1998).

A current debate in agriculture is gaining steam over concentrated animal feeding operations (CAFOs). The debate is fueled by the serious impacts of CAFOs on the environment (streams and groundwater sources) and on the social fabric of rural living. Because livestock factories produce and store large quantities of animal waste in leak-prone lagoons, America's freshwater supply is at risk. Livestock factories also pose a threat to our air quality. Fish and wildlife also suffer from manure spills. The huge corporations that run the livestock factories are edging out family farmers, who often use more environmentally friendly techniques. Owning a home might be the American dream, but a hog operation in the backyard is a nightmare for property values and for those who live on adjoining properties.

* Adapted from Spellman, F.R. and Whiting, N.E., *Environmental Management of Concentrated Animal Feeding Operations (CAFOs)*, CRC Press, Boca Raton, FL, 2007.

Concentrated animal feeding operations are a big issue now and one that is expected to get bigger. Public awareness is rising. Hog farms in North Carolina, for example, have been in the national news because of water pollution and odor problems. Poultry farms and beef feedlots are drawing attention in other parts of the United States. The urban–rural interface side is also garnering attention. Consider, for example, the various aspects of urban–wildlife interfaces, urban sprawl, and its impact on forestry. Other issues include air pollution, antibiotics, hormones, and other chemical contaminants deposited into the environment via the manure environmental medium interface. The fact is that the popular myth behind the common image of traditional, idyllic, self-contained, and self-operated farm operations stands in stark contrast to the modern reality of factory farming, based on CAFOs.

Siting of CAFOs is a divisive issue, pitting neighbors against neighbors in rural communities amid angry debates about odors, environmental degradation, increasing concentration among producers, and a sense of injustice among those who feel unheard and disenfranchised (Barrette, 1996; Keller and Miller, 1997). Finances are of grave concern, too, for those in rural farming communities impacted by a CAFO presence.

Concentrated animal feeding operations are farming operations where large numbers (often in the thousands of animals) of livestock or poultry are housed inside buildings or in confined feedlots. How many animals? The U.S. Environmental Protection Agency defines a CAFO or industrial operation as a concentrated animal feeding operation where animals are confined for more than 45 days per year. To classify as a CAFO, such an operation must also have over 1000 animal units—a standardized number based on the amount of waste each species produces, basically 1000 lb of animal weight. Thus, dairy cattle count as 1.4 animal units each. A CAFO could house more than 750 mature dairy cattle (milking and or dry cows) or 500 horses and discharge into navigable water through a ditch or similarly man-made device. The CAFO classification sets numbers for various species per 1000 animal units; for example,

- 2500 hogs
- 700 dairy cattle
- 1000 beef cattle
- 100,000 broiler chickens
- 82,000 layer hens

Getting a grasp on the scope of the problem can be difficult. By way of comparison, let's quantify the issue: How do the amounts of CAFO-generated animal manure compare to human waste production? Let's take a look at the numbers. Here's a small-scale number: One hog per day excretes 2.5 times more waste than an adult human—nearly 3 gallons (Cantrell et al., 1996). Here's a medium-scale number: A 10,000-hog operation produces as much waste in a single day as a town of 25,000 people (Sierra Club, 2004), but the town has a treatment plant. Here's a big picture approach: The USEPA estimates that humans generate about 150 million tons (wet weight) of human sanitary waste annually in the United States, assuming a U.S. population of 285 million and an average waste generation of about 0.518 tons per person per year. The U.S. Department of Agriculture (USDA, 2001) estimated that operations that confine livestock and poultry animals generate about 500 million tons of excreted manure annually. Over 450,000 CAFOs are producing 575 billion pounds of manure annually in the United States today.

Here's the bottom line: By these estimates, all confined animals generate well over *three times* more raw waste than is generated by humans in the United States. Much of this waste undergoes no or very little waste treatment. Waste handling for any CAFO is a major business concern and expense. Unless regulation and legislation support sound environmental practices for these operations, CAFO owners have little incentive to improve their waste handling practices.

WATER SUPPLY, USE, AND WASTEWATER TREATMENT

Because wastewater treatment is so interconnected with other uses of water, it is considered a water use. Much of the water used by homes, industries, and businesses is treated prior to its release back into the environment, becoming part of the endless water cycle. The scope of treatment processes inclusive in the term "wastewater treatment" is unknown to most people, who think of it solely in terms of sewage treatment. Nature has an amazing ability to cope with small amounts of water wastes and pollution through its self-purification process. However, Nature would be overwhelmed if the billions of gallons of wastewater and sewage produced every day were not treated before being released back into the environment. Wastewater treatment plants reduce pollutants in wastewater to a level nature can handle.

We treat used water for many reasons. Principle among them is the matter of caring for our environment and our own public health. We treat used water because clean water is critical to our water supply, as well as to plants and animals that live in water. Human health, environmental health, and many commercial interests (for example, the fishing industry and sport fishing enthusiasts) depend on the biota that can only survive in clean healthy water systems—and we, of course, as today's responsible adults hold our water system in its entirety for future generations.

Our rivers and ocean waters teem with life that depends on healthy shorelines, beaches, and marshes, which provide critical habitats for hundreds of species of aquatic and semi-aquatic life. Migratory water birds use the areas for feeding and resting. Species of both flora and fauna are adapted to live in these zones that lie between or on the verge of water. These areas are extremely vulnerable to certain types of pollution.

Water is one of our most used playgrounds. The scenic and recreational values of our waters are important factors for many people when deciding where to live. Tourists are drawn to water activities: swimming, fishing, boating, hunting, and picnicking. Improper treatment of wastes impacts all of these activities.

In short, if used water is not properly cleaned, it carries waterborne disease. Because we live, work, and play so close to water, harmful pathogenic organisms must be removed or made harmless to make water safe, regardless of the aesthetic factors involved in untreated wastes in the water system. So we treat our wastewater before releasing it to the environment. The major aim of used water treatment is to remove as much of the suspended solids and other contaminants as possible before the remaining water, called *effluent*, is discharged to the environment.

Treatment involves several interrelated steps. *Primary treatment* removes about 60% of suspended solids from used water. This treatment also involves aerating (mixing up) the used water to put oxygen back in, which is essential because as solid material biodegrades it uses up the oxygen needed by the plants and animals living in the water. *Secondary treatment* removes more than 90% of suspended solids. In some cases, *tertiary treatment* takes the waste removal further or addresses the removal of specific waste elements (e.g., nutrients such as nitrogen and phosphorus) not removed by other means. After treatment, used water is returned to the water cycle as treated effluent. Whether from consumer or industrial sources, the treated water should be returned at least as clean as—if not cleaner than—the receiving body of water.

What does wastewater contain and where does it come from? Used water (wastewater) carries substances that include human waste, food scraps, oils, soaps, and chemicals. Consumer use in homes includes water from showers, sinks, toilets, bathtubs, washing machines, and dishwashers. Businesses and industries also contribute their share of used water (industrial waste) that must be cleaned and thus recycled. The treatment of both household wastewaters and industrial wastewaters is regulated and monitored. These are point-source pollutants; the sources are identifiable and limited in scope. We know what is in there, where the sources are, and, in a general way, how much will be produced; the quantity can be predicted to fall within the capacity of the system's treatment capability.

Other wastes that enter our water supply are more difficult to define, quantify, and control. Stormwater or storm runoff, for example, is also a major contributor to the endless wastewater stream. Many people might assume that rain that runs off their homes, into their yards, and then down the streets during a storm event is fairly clean. It is not. Harmful substances wash off roads, fields, lawns, parking lots, and rooftops and can harm our still waters (lakes and ponds) and running waters (rivers and streams). Stormwater also, obviously, is incident related. If it doesn't rain, stormwater cannot enter the system. Stormwater provides nonpoint-source pollutants to the wastewater stream. We know in a general way what they will carry, and we put systems in place to channel and control the stream. Complex programs and modeling are required to evaluate how to handle stormwater to avoid serious problems with treatment and control, specifically with regard to handling many different levels of force in storm incidents of differing durations.

Agricultural sources are a wastewater treatment issue that has been difficult to evaluate, identify, and control. Historically, especially before the chemical industry provided manufactured fertilizers and pesticides for crop farm production and growth hormones and antibiotics for livestock production, the techniques individual farmers used were reasonably environmentally friendly, because the size of the farm dictated the limits of production: Individual farmers have limits with regard to their physical and financial ability to do the work, in addition to effective natural limits on how many bushels of grain or animals per acre the land itself can support. In general, overloading these capacities provides negative results. As modern practices have evolved, though, our water systems have suffered. Fertilizers, pesticides, hormones, and antibiotics send nonpoint-source pollution directly into local water systems, thus creating downstream problems in the water system. A number of crop and livestock farming practices contribute to soil erosion, which, of course, affects both soil and water quality. Changes in farming practices over the last several decades, perhaps best defined as a switch from the small farmer to agribusiness or factory farms, and the increasing demand for inexpensive meat products have created a new set of problems to address. Whereas the factory farming of crops presents an entirely different set of issues, here we address problems created by the factory farming of livestock, which creates agricultural point source pollution of extreme scope.

Animal Feeding Operations and Animal Waste Treatment

Waste problems and innovative solutions to these problems are not new, even in mythical terms. For example, mythical Greek gods had animal housing. Hercules, perhaps the first environmental engineer, stumbled upon the solution to animal waste piling up in ancient Greek animal housing. The "Hercules" engineering principle states that the solution to pollution is dilution. Hercules applied this principle when he cleaned up the royal Aegean Stables that had not been cleaned for at least 30 years. He cleaned the stables by diverting the flow of two upstream rivers and directing the combined flow through the stables. Today we call this technique "flushing."

This flushing idea of Hercules worked so well in cleaning the stables that the same idea was later applied to the design of human toilets and sewer systems. As the world because industrialized, the Hercules idea was applied just as successfully in the dispersion of air pollutants through tall chimneys. These chimneys are not different from the sewer pipes that take the waste away from the source. Note that the Hercules principle that the solution to pollution is dilution has an ample scientific basis. It should also be noted that, although treating today's massive quantities of animal waste from CAFOs is certainly a Herculean task, unfortunately Hercules is no longer around to solve our current problems.

Fast-forwarding from mythical times to the present, the fact is that other types of animal waste problems and their solutions are also not new. For example, field spreading of human and animal wastes is accomplished naturally under nomadic and pasture social systems. Intentional manure conservation and reuse were practiced by early Chinese. In Iceland, at least 200 years ago slotted floors allowed waste material to drop below the floor surface. From at least the 19th century, dairy barns with wastes from scores of animals were contained in one building. Huge poultry centers with wastes concentrated in a small area have been around for decades, as have some very large swine and beef units.

Figuratively speaking and in general, animal manure deposited by animals managed by standard grazing livestock methods does not pose a serious environmental problem, especially if the farmer limits herd size to numbers the acreage can support without environmental damage, restricts livestock access to stream beds, and applies practices that include soil erosion prevention methods (e.g., greenbelts for waterways, shoreline planting). Accidentally stepping into such deposits is an occupational hazard, of course.

It is important to understand that small-farm animal manure waste is not the problem we are addressing here. The manure deposited by a large herd of animals that is not assimilated through the soil surface and is carried off by storm runoff into local streams or other water bodies represents an obvious "isolated incident" problem. Agribusiness and large-scale factory farming practices have created a different farm category in CAFOs, the livestock version of factory crop farming, which produce a massive quantity of manure. In the 1920s, no one was capable of spilling millions of gallons of manure into a local stream in a single event. Ikerd (1998) observed that such events are possible today because too much manure is being piled up in one place. Simply, the piling up is the result of greater concentration and reduced diversity in farm operations.

Agribusinesses do not use traditional pastures and feeding practices. Typically, manure is removed from the livestock buildings or feedlots and stored in stockpiles or lagoon or pond systems until it can be spread on farm fields, sold to other farmers as fertilizer, or composted. When properly designed, constructed, and managed, CAFOs produce manure that is an agronomically important and environmentally safe source of nutrients and organic matter necessary for the production of food, fiber, and good soil health. Experience has demonstrated that when applied to land at proper levels manure will not cause water quality problems. When properly stored or deposited in holding lagoons or ponds, properly conveyed to the disposal outlet, and properly applied to the appropriate end use, potential CAFO waste environmental problems can be mitigated.

Concentrated animal feeding operations must be monitored and controlled, as they are potential sources of contaminants (pollutants) to the three environmental mediums of air, water, and soil. Let's take a look at manure handling and storage practices recommended by the USEPA (2003) that should be employed to prevent water pollution from CAFOs. In addition to water pollution prevention, it should be noted that manure and wastewater handling, storage, and subsequent application and treatment practices should also consider odor and other environmental and public health problems.

- *Divert clean water*—Siting and management practices should divert clean water from contact with feed lots and holding pens, animal manure, or manure storage systems. Clean water can include rainfall falling on roofs of facilities, runoff from adjacent lands, or other sources.
- *Prevent leakage*—Construction and maintenance of buildings, collection systems, conveyance systems, and permanent and temporary storage facilities should prevent leakage of organic matter, nutrients, and pathogens to ground or surface water.
- *Provide adequate storage*—Liquid manure storage systems should safely store the quantity and contents of animal manure and wastewater produced, contaminated runoff from the facility, and rainfall. Dry manure, such as that produced in certain poultry and beef operations, should be stored n production buildings or storage facilities, or otherwise stored in such a way so as to prevent polluted runoff. Location of manure storage systems should consider proximity to water bodies, floodplains, and other environmentally sensitive areas.
- *Manure treatments*—Manure should be handled and treated to reduce the loss of nutrients to the atmosphere during storage, to make the material a more stable fertilizer when land-applied, or to reduce pathogens, vector attraction, and odors, as appropriate.

- *Management of dead animals*—Dead animals should be disposed of in a way that does not adversely affect ground or surface water or create public health concerns. Composting, rendering, and other practices are common methods used to dispose of dead animals.

Manure Treatment

"Advanced technologies are being developed for the biological, physical, and chemical treatment of manure and wastewaters. Some of these greatly reduce constituents in the treated solids and liquids that must be managed on the farm. Byproduct recovery processes are being developed that transform waste into value-added products that can be marketed off the farm" (Sutton and Humenik, 2003, p. 2).

Animal Waste Treatment Lagoons

Primarily because it is an economical means of treating highly concentrated wastes from confined livestock operations, the most widespread and common treatment technique for managing animal waste is the use of lagoons. In the late 1960s, considerable attention was paid to the impact of lagoons on surface water quality; since the 1970s, that attention has shifted to the potential impacts on groundwater quality. Unfortunately, these lagoons are prone to leaks and breakage. Groundwater has been contaminated with bacteria from them. The lagoons can also be overrun by floods that push the wastes into streams, lakes, and oceans. North Carolina, with its concentration of factory farms, has been the focus of massive water contamination due to its waste lagoons. The storage lagoons for factory farms are often stinking manure lakes the size of several football fields, containing millions of gallons of liquefied manure. A single animal factory can generate the waste equivalent of a small town.

In the past 30 years, several studies on the effectiveness of factory farm lagoons, specifically on lagoon liners, in preventing environmental damage have been conducted. Consider the following review of studies on effective lagoon construction vs. defective construction. Sewell et al. (1975) studied an anaerobic dairy lagoon and found that the lagoon bottom sealed within 2 months of start up, and few or no pollutants were found in the groundwater after this time. Ritter et al. (1984) studied a two-stage anaerobic swine lagoon for 4 years and determined that the contaminant concentration increased in wells (50 meters from the lagoon) the first year and then steadily decreased afterward. His data led him to speculate that biological sealing takes place over a period of time depending on the loading rate to a lagoon. Collins et al. (1975) studied three swine lagoons, each within a high water table area. They found no significant effect on groundwater beyond 3 meters from the lagoon edge. Miller et al. (1985) studied the performance of beef manure lagoons in sandy soil and found that the lagoons had effectively sealed to infiltration within 12 weeks of the addition of manure. Humenik et al. (1980) summarized research conducted by others on the subject on lagoon sealing, and they concluded that the studies indicated that lagoon sealing may be expected to occur within about 6 months, after which the area of seepage impact becomes restricted to approximately 10 meters.

On the other hand, Hegg et al. (1978) collected data from a dairy lagoon and from newly established swine lagoons and found that some of the monitoring wells became contaminated while others did not. This led them to conclude that seepage does not occur uniformly over the entire wetted perimeter of the lagoon but at specific unpredictable sites where sealing has not taken place. Similarly, Ritter et al. (1980) monitored an anaerobic two-stage swine lagoon for two years and found that one of the wells showed contamination that indicated localized seepage, while the other monitoring well indicated that the lagoon system produced a minimum impact on groundwater quality and that sealing had gradually taken place.

In many states, notwithstanding USEPA and USDA manure handling, storage, and treatment recommendations, lawsuits against CAFOs for unsound environmental practices demonstrate that CAFO operations are still creating problems. In short, regulations and legislation have fallen behind the increasing growth in CAFO operations, enforcement of existing regulations is spotty, and problems associated with CAFOs are still being identified—although you can be sure those who neighbor CAFOs can identify some big issues, both environmental and social.

THOUGHT-PROVOKING QUESTION

7.1 Many agricultural practices are designed to protect the soil and water. Just as agriculture is sometimes said to be a major source of water pollution, it is also a major source of practices to protect the environment. No area of activity does more to protect water than agriculture. Do you agree with this statement? Explain.

REFERENCES AND RECOMMENDED READING

Barlow, P.M. and Reichard, E.G. (2010). Saltwater intrusion in coastal regions of North America. *Hydrogeology Journal*, 18, 247–260.

Barrette, M. (1996). Hog-tied by feedlots. *Zoning News*, October, p. 4.

Cantrell, P., Perry, R., and Sturtz, P. (1996). *Hog Wars: The Corporate Grab for Control of the Hog Industry and How Citizens Are Fighting Back*. Columbia: Missouri Rural Crisis Center.

Chilton, J. (1998). Dry or drowning. *Our Planet*, 9(4).

Collins E.R., Ciravolo, T.G., Hallock, D.L., Martens, H.R., Thomas, E.T., and Kornegay, E.T. (1975). Effect of anaerobic swine lagoons on groundwater quality in high water table soils. In: *Proceedings of the 3rd International Symposium on Livestock Wastes*. St. Joseph, MI: American Society of Agricultural Engineers, pp. 366–368.

Fitts, C.R. (2012). *Groundwater Science*. Waltham, MA: Academic Press.

Halsam, S.M. (1990). *River Pollution: An Ecological Perspective*. New York: Belhaven Press.

Harr, J. (1996). *A Civil Action*. New York: Knopf Doubleday.

Hegg, R.O., King, T.G., and Wilson, T.V. (1978). *The Effect on Groundwater from Seepage of Livestock Manure Lagoons*, WRRI Technical Report #78. Clemson, SC: Clemson University.

Hill, M.K. (1997). *Understanding Environmental Pollution*. Cambridge, UK: Cambridge University Press.

Humenick, F.J., Overcash, M.R., Baker, J.C., and Westerman, P.W. (1980). Lagoons: state of the art. In: *Livestock Waste: A Renewable Resource*. St. Joseph, MI: American Society of Agricultural Engineers, pp. 211–216.

Ikerd, J. (1998). *Large Scale, Corporate Hog Operations: Why Rural Communities Are Concerned and What They Should Do*, http://archive.sare.org/sanet-mg/archives/html-home/27-html/0161.html.

Kay, J. (1996). Chemicals used to cleanse water can also cause problems. *San Francisco Examiner*, October 3.

Keller, D. and Miller, D. (1997). Neighbor against neighbor. *Progressive Farmer*, 112(11), 16–18.

Lewis, S.A. (1996). *The Sierra Club Guide to Safe Drinking Water*. San Francisco, CA: Sierra Book Club.

Mason, C.F. (1990). Biological aspects of freshwater pollution. In: *Pollution: Causes, Effects, and Control*, Harrison, R.M., Ed. Cambridge: Royal Society of Chemistry.

McBride, W.D. (1997). *Change in U.S. Livestock Production, 1969–92*, Agricultural Economic Report No. 754. Washington, DC: Economic Research Service, U.S. Department of Agriculture.

Miller, H.H., Robinson, J.B., and Gillam, W. (1985). Self-sealing of earthen liquid manure storage ponds. I. A case study. *Journal of Environmental Quality*, 14, 533–538.

Moran, J.M., Morgan, M.D., and Wiersman, J.H. (1986). *Introduction to Environmental Science*. New York: W.H. Freeman & Co.

NRDC. (2003). *What's on Tap?* Washington, DC: Natural Resources Defense Council.

Nuckols, B. (2014). As odor lingers, so does doubt about the safety of the tap water. *The Virginia-Pilot (Norfolk, VA)*, January 19.

Outwater, A. (1996). *Water a Natural History*. New York: Basic Books.

Proulx, A. (2002). *That Old Ace in the Hole*. New York: Scribner, p. 6.

Rail, C.D. (1985). Groundwater monitoring within an aquifer: a protocol. *Journal of Environmental Health*, 48(3), 128–132.

Ritter, W.R., Walpole, E.W., and Eastburn, R.P. (1980). An anaerobic lagoon for swine manure and its effect of the groundwater quality in sandy-loam soils. In: *Livestock Waste: A Renewable Resource*. St. Josephs, MI: American Society of Agricultural Engineers, pp. 244–246.

Ritter, W.R., Walpole, E.W., and Eastburn, R.P. (1984). Effect of an anaerobic swine lagoon on groundwater quality in Sussex County, Delaware. *Agricultural Wastes*, 10(4), 267–284.

Schwab, J. (1998). *Planning and Zoning for Concentrated Animal Feeding Operations*, Planning Advisory Service Report Number 482. Chicago, IL: American Planning Association.

Sewell, J.I., Mulling, J.A., and Vaigneur, H.O. (1975). Dairy lagoon system and groundwater quality. In: *Managing Livestock Wastes: Proceedings of the Third International Symposium on Livestock Wastes*, April 21–24, University of Illinois, Urbana–Champaign, pp. 286–288.

Sierra Club. (2004). *Clean Water and Factory Farms*, http://indiana.sierraclub.org/issues/edwardsport.html.

Smith, R.L. (1996). *Ecology and Field Biology*. New York: HarperCollins.

Spellman, F.R. (1996). *Stream Ecology and Self-Purification*. Boca Raton: CRC Press.

Spellman, F.R. (2008). *The Science of Air*, 2nd ed. Boca Raton: CRC Press.

Spellman, F.R. and Whiting, N.E. (1999). *Water Pollution Control Technology*. Rockville, MD: Government Institutes.

Squillace, P.J., Pando, J.F., Corte, N.E., and Zagorsk, J.S. (1998). Environmental behavior and fate of methyl tertiary-butyl ether. *Water Online*, November 4, http://www.wateronline.com/article.mvc/Environmental-Behavior-and-Fate-of-Methyl-Ter-0002.

Sutton, A. and Humenik, F., (2003). *CAFO Fact Sheet #24: Technology Options to Comply with Land Application Rules*. Ames, IA: MidWest Plan Service, Iowa State University.

USDA. (2001). *Manure and Byproduct Utilization: National Program Annual Report: FY 2001*. Washington, DC: Department of Agriculture, Agricultural Research Service.

USEPA. (1987). *Proposed Regulations for Underground Storage Tanks: What's in the Pipeline?* Washington, DC: U.S. Environmental Protection Agency.

USEPA. (1994). *What Is Nonpoint Source Pollution?* Washington, DC: U.S. Environmental Protection Agency.

USEPA. (1996). *Targeting High Priority Problems*. Washington, DC: U.S. Environmental Protection Agency.

USEPA. (1998). *Drinking Water Priority Rulemaking: Microbial and Disinfection By-products Rules*, EPA 815-F-95-0014. Washington, DC: U.S. Environmental Protection Agency.

USEPA. (2003). *National Management Measures to Control Nonpoint Source Pollution from Agriculture*, EPA 841-B-03-004. Washington, DC: U.S. Environmental Protection Agency.

USEPA. (2005). *Operating and Maintaining Underground Storage Tank Systems: Practical Help and Checklists*. Washington, DC: U.S. Environmental Protection Agency (http://www.epa.gov/oust/pubs/ommanual.htm).

USGS. (2003). *Ground Water in Freshwater–Saltwater Environments of the Atlantic Coast*. Washington, DC: U.S. Geological Survey (http://pubs.usgs.gov/circ/2003/circ1262/).

Velz, F.J. (1970). *Applied Stream Sanitation*. New York: Wiley-Interscience.

Viessman, Jr., W. and Hammer, M.J. (1998). *Water Supply and Pollution Control*, 6th ed. Menlo Park, CA: Addison-Wesley.

8 Environmental Biomonitoring, Sampling, and Testing*

During another visit to the New England Medical Center, three months after Robbie's first complaints of bone pain, doctors noted that his spleen was enlarged and that he had a decreased white-blood-cell count with a high percentage of immature cells—blasts—in the peripheral blood. A bone marrow aspiration was performed. The bone marrow confirmed what the doctors had begun to suspect: Robbie had acute lymphatic leukemia.

Harr (1995)

In an age when man has forgotten his origins and is blind even to his most essential needs for survival, water along with other resources has become the victim of his indifference.

Carson (1962)

In January, we take our nets to a no-name stream in the foothills of the Blue Ridge Mountains of Virginia to do a special kind of macroinvertebrate monitoring—looking for winter stoneflies. Winter stoneflies have an unusual life cycle. Soon after hatching in early spring, the larvae bury themselves in the streambed. They spend the summer lying dormant in the mud, thereby avoiding problems such as overheated streams, low oxygen concentrations, fluctuating flows, and heavy predation. In later November, they emerge, grow quickly for a couple of months, and then lay their eggs in January. January monitoring of winter stoneflies helps in interpreting the results of spring and fall macroinvertebrate surveys. In spring and fall, a thorough benthic survey is conducted based on Protocol II of the U.S. Environmental Protection Agency's *Rapid Bioassessment Protocols for Use in Streams and Rivers*. Some sites on various rural streams have poor diversity and sensitive families. Is the lack of macroinvertebrate diversity because of specific warm-weather conditions, high water temperature, low oxygen, or fluctuating flows, or is some toxic contamination present? In the January screening, if winter stoneflies are plentiful, seasonal conditions could probably be blamed for the earlier results; if winter stoneflies are absent, the site probably suffers from toxic contamination (based on our rural location, probably emanating from nonpoint sources) that is present all year. Though different genera of winter stoneflies are found in our region (southwestern Virginia), *Allocapnia* is sought because it is present even in the smallest streams (Spellman and Drinan, 2001).

WHAT IS BIOMONITORING?

The life in, and physical characteristics of, a stream ecosystem provide insight into the historical and current status of its quality. The assessment of a water body ecosystem based on organisms living in it is called *biomonitoring*. The assessment of the system based on its physical characteristics is called a *habitat assessment*. Biomonitoring and habitat assessments are two tools that stream ecologists use to assess the water quality of a stream.

Biological monitoring involves the use of various organisms, such as periphytons, fish, and macroinvertebrates (combinations of which are referred to as *assemblages*), to assess environmental condition. Biological observation is more representative as it reveals cumulative effects as opposed to chemical observation, which is representative only at the actual time of sampling.

* Much of the information presented in this chapter is adapted from Spellman, F.R., *Handbook of Water and Wastewater Treatment Plant Operations*, 3rd ed., CRC Press, Boca Raton, FL, 2014.

The presence of benthic macroinvertebrates (bottom-dwelling fauna) can be monitored. These are the larger-than-microscopic organisms such as aquatic insects, insect larvae, and crustaceans that are generally ubiquitous in freshwater and live in the bottom portions of a waterway for part of their life cycle. Benthic macroinvertebrates are ideal for use in biomonitoring because they are ubiquitous, relatively sedentary, and long lived. The overall community is reflective of conditions in its environment. They provide a cross-section of the situation, as some species are extremely sensitive to pollution while others are more tolerant. Like toxicity testing, biomonitoring does not tell us *why* animals are present or absent, but benthic macroinvertebrates are excellent indicators for several reasons:

1. Biological communities reflect overall ecological integrity (i.e., chemical, physical, and biological integrity); therefore, biosurvey results directly assess the status of a water body relative to the primary goal of the Clean Water Act (CWA).
2. Biological communities integrate the effects of different stressors and thus provide a broad measure of their aggregate impact.
3. Because they are ubiquitous, communities integrate the stressors over time and provide an ecological measure of fluctuating environmental conditions.
4. Routine monitoring of biological communities can be relatively inexpensive, because they are easy to collect and identify.
5. The status of biological communities is of direct interest to the public as a measure of a particular environment.
6. Where criteria for specific ambient impacts do not exist (e.g., nonpoint sources that degrade habitats), biological communities may be the only practical means of evaluation.
7. They can be used to assess non-chemical impacts on the aquatic habitat, such as thermal pollution, excessive sediment loading (siltation), or eutrophication.

Benthic macroinvertebrates act as continuous monitors of the water they live in. Unlike chemical monitoring, which provides information about water quality at the time of measurement (a snapshot), biological monitoring can provide information about past or episodic pollution (a videotape). This concept is analogous to miners who took canaries into deep mines with them to test for air quality. If the canary died, the miners knew the air was bad and they had to leave the mine. Biomonitoring a water body ecosystem uses the same theoretical approach. Aquatic macroinvertebrates are subject to pollutants in the water body; consequently, the health of the organisms reflects the quality of the water they live in. If the pollution levels reach a critical concentration, certain organisms will migrate away, fail to reproduce, or die, eventually leading to the disappearance of those species at the polluted site. Normally, these organisms will return if conditions improve in the system (Bly and Smith, 1994; USEPA, 2011).

Biomonitoring (and the related concept bioassessment) surveys are conducted before and after an anticipated impact to determine the effect of the activity on the water body habitat. Moreover, surveys are performed periodically to monitor water body habitats and watch for unanticipated impacts. Finally, biomonitoring surveys are designed to reference conditions or to set biocriteria for determining that an impact has occurred; that is, they establish monitoring thresholds that signal future impacts or necessary regulatory actions (Camann, 1996). Biological monitoring cannot replace chemical monitoring, toxicity testing, and other standard environmental measurements. Each of these tools provides the analyst with specific information available only through its respective methodology.

Note: The primary justification for bioassessment and monitoring is that degradation of water body habitats affects the biota using those habitats; therefore, the living organisms themselves provide the most direct means of assessing real environmental impacts.

BIOTIC INDEX IN STREAMS

Certain common aquatic organisms, by indicating the extent of oxygenation of a stream, may be regarded as indicators of the intensity of pollution from organic waste. The responses of aquatic organisms in water bodies to large quantities of organic wastes are well documented. They occur in a predictable cyclical manner; for example, upstream from the discharge point, a stream can support a wide variety of algae, fish, and other organisms, but in the section of the water body where oxygen levels are low (below 5 ppm), only a few types of worms survive. As stream flow courses downstream, oxygen levels recover, and those species that can tolerate low rates of oxygen (such as gar, catfish, and carp) begin to appear. In a stream, eventually, at some point further downstream, a clean water zone reestablishes itself, and a more diverse and desirable community of organisms returns. During this characteristics pattern of alternating levels of dissolved oxygen (in response to the dumping of large amounts of biodegradable organic material), a stream goes through an *oxygen sag curve* cycle. Its state can be determined using the biotic index as an indicator of oxygen content.

The biotic index is a systematic survey of macroinvertebrates organisms. Because the diversity of species in a stream is often a good indicator of the presence of pollution, the biotic index can be used to evaluate stream quality. Observation of the types of species present or missing is used as an indicator of stream pollution. The biotic index, which reflects the types, species, and numbers of biological organisms present in a stream, is commonly used as an auxiliary to biochemical oxygen demand (BOD) determination when monitoring stream pollution. The biotic index is based on two principles:

1. A large dumping of organic waste into a stream tends to restrict the variety of organisms at a certain point in the stream.
2. As the degree of pollution in a stream increases, key organisms tend to disappear in a predictable order. The disappearance of particular organisms tends to represent the water quality of the stream.

Several different forms of the biotic index are in use. In Great Britain, for example, the Trent Biotic Index (TBI), the Chandler score, the Biological Monitoring Working Party (BMWP) score, and the Lincoln Quality Index (LQI) are widely used. Most of the forms use a biotic index that ranges from 0 to 10. The most polluted stream, which therefore contains the smallest variety of organisms, is at the lowest end of the scale (0); clean streams are at the highest end (10). A stream with a biotic index of greater than 5 will support game fish; a stream with a biotic index of less than 4 will not support game fish.

Because they are easy to sample, macroinvertebrates have been prominent in biological monitoring. Macroinvertebrates are a diverse group. They demonstrate tolerances that vary between species; thus, discrete differences can be observed between tolerant and sensitive indicators. Macroinvertebrates can be easily identified using identification keys that are portable and easily used in field settings. Current knowledge of macroinvertebrate tolerances and responses to stream pollution is well documented. In the United States, for example, the U.S. Environmental Protection Agency (USEPA) required states to incorporate narrative biological criteria into their water quality standards by 1993. The National Park Service (NPS) has collected macroinvertebrate samples from American streams since 1984. Through their sampling effort, the NPS has been able to derive quantitative biological standards (Huff, 1993).

The biotic index provides a valuable measure of pollution, especially with regard to species that are very sensitive to a lack of oxygen. An example of an organism that is commonly used in biological monitoring is the stonefly. Stonefly larvae live underwater and survive best in cool, well-aerated, unpolluted waters with clean gravel bottoms. When stream water quality deteriorates due to organic

TABLE 8.1
BMWP Score System (Modified)

Family	Common Name Example	Score
Heptageniidae	Mayflies	10
Leuctridae	Stoneflies	9–10
Aeshnidae	Dragonflies	8
Polycentropidae	Caddisflies	7
Hydrometridae	Water strider	6–7
Gyrinidae	Whirligig beetle	5
Chironomidae	Mosquitoes	2
Oligochaeta	Worms	1

pollution, stonefly larvae cannot survive. The degradation of stonefly larvae has an exponential effect upon other insects and fish that feed off the larvae; when the stonefly larvae disappears, so in turn do many insects and fish (O'Toole, 1986).

Table 8.1 is a modified version of the BMWP biotic index, which takes into account the sensitivities of various macroinvertebrate species representing diverse populations that are excellent indicators of pollution. These aquatic macroinvertebrates are large enough to be seen by the unaided eye. Moreover, most aquatic macroinvertebrates species live for at least a year, and they are sensitive to stream water quality both on a short-term basis and over the long term. Mayflies, stoneflies, and caddisflies are aquatic macroinvertebrates that are considered clean-water organisms; they are generally the first to disappear from a stream if water quality declines, so they are given a high score in the index. On the other hand, tubificid worms (which are tolerant to pollution) are given a low score.

As shown in Table 8.1, a score from 1 to 10 is given for each family present. A site score is calculated by adding the individual family scores. The site score or total score is then divided by the number of families recorded to derive the average score per taxon (ASPT). High ASPT scores are obtained when such taxa as stoneflies, mayflies, and caddisflies are found in the stream. A low ASPT score is obtained for streams that are heavily polluted and dominated by tubificid worms and other pollution-tolerant organism. From Table 8.1, it can be seen that those organisms having high scores, especially mayflies and stoneflies, are the most sensitive, and others, such as dragonflies and caddisflies, are very sensitive to any pollution (deoxygenation) of their aquatic environment.

Benthic Macroinvertebrate Biotic Index

The benthic macroinvertebrate biotic index employs the use of certain benthic macroinvertebrates to determine, or gauge, the water quality (relative health) of a water body such as a stream or river. Benthic macroinvertebrates are classified into three groups based on their sensitivity to pollution. The number of taxa in each of these groups are tallied and assigned a score. The scores are then summed to yield a score that can be used as an estimate of the quality of the water body life. A sample index of macroinvertebrates with regard to their sensitivity to pollution is provided in Table 8.2.

In summary, it can be said that unpolluted streams normally support a wide variety of macroinvertebrates and other aquatic organisms with relatively few of any one kind. Any significant change in the normal population usually indicates pollution.

BIOLOGICAL SAMPLING IN STREAMS

When planning a biological sampling outing, it is important to determine the precise objectives. One important consideration is to determine whether sampling will be accomplished at a single point or at isolated points. Additionally, frequency of sampling must be determined. That is, will

TABLE 8.2
Sample Groupings of Macroinvertebrates

Group One (Sensitive)	Group Two (Somewhat Sensitive)	Group Three (Tolerant)
Stonefly larvae	Alderfly larvae	Aquatic worms
Caddisfly larvae	Damselfly larvae	Midgefly larvae
Water penny larvae	Cranefly larvae	Blackfly larvae
Riffle beetle adults	Beetle adults	Leeches
Mayfly larvae	Dragonfly larvae	Snails
Gilled snails	Sowbugs	

sampling be accomplished at hourly, daily, weekly, monthly, or even longer intervals? Whatever sampling frequency of sampling is chosen, the entire process will probably continue over a protracted period (i.e., preparing for biological sampling in the field might take several months from the initial planning stages to the time when actual sampling occurs). An experienced freshwater ecologist should be centrally involved in all aspects of planning.

In its *Monitoring Water Quality: Intensive Stream Bioassay*, the USEPA (2000a) recommended that the following issues should be considered when planning a sampling program:

- Availability of reference conditions for the chosen area
- Appropriate dates for sampling in each season
- Appropriate sampling gear
- Availability of laboratory facilities
- Sample storage
- Data management
- Appropriate taxonomic keys, metrics, or measurement for macroinvertebrate analysis
- Habitat assessment consistency
- Availability of a USGS topographical map
- Familiarity with safety procedures

When the initial objectives (issues) have been determined and the plan devised, then the sampler can move on to other important aspects of the sampling procedure. Along with the items just mentioned, it is imperative for the sampler to understand what biological sampling is all about.

Sampling is one of the most basic and important aspects of water quality management (Tchobanoglous and Schroeder, 1985). Biological sampling allows for rapid and general water quality classification. Rapid classification is possible because quick and easy cross-checking between stream biota and a standard stream biotic index is possible. Biological sampling is typically used for general water quality classification in the field because sophisticated laboratory apparatus is usually not available. Additionally, stream communities often show a great deal of variation in basic water quality parameters such as dissolved oxygen (DO), BOD, suspended solids, and coliform bacteria. This occurrence can be observed in eutrophic lakes that may vary from oxygen saturation to less than 0.5 mg/L in a single day, and the concentration of suspended solids may double immediately after a heavy rain. Moreover, the sampling method chosen must take into account the differences in the habits and habitats of the aquatic organisms.

The first step toward accurate measurement of the water quality of a stream is to make sure that the sampling targets those organisms (i.e., macroinvertebrates) that are most likely to provide the information that is being sought. Second, it is essential that representative samples be collected. Laboratory analysis is meaningless if the sample collected was not representative of the aquatic environment being analyzed. As a rule, samples should be taken at many locations, as often as possible. If, for example, we are studying the effects of sewage discharge into a stream, we should take at least six samples upstream of the discharge, six samples at the discharge, and at least six samples at several points below the discharge for two to three days (the six–six sampling rule). If these samples show wide variability, then the number of samples should be increased. On the other hand, if the initial samples exhibit little variation, then a reduction in the number of samples may be appropriate (Kittrell, 1969).

When planning the biological sampling protocol (using biotic indices as the standards), remember that when the sampling is to be conducted in a stream, findings are based on the presence or absence of certain organisms; thus, the absence of these organisms must be a function of pollution and not of some other ecological problem. The preferred aquatic group for biological monitoring in streams is the macroinvertebrates, which are usually retained by 30-mesh sieves (pond nets).

Sampling Stations

After determining the number of samples to be taken, sampling stations (locations) must be determined. Several factors determine where the sampling stations should be set up. These factors include stream habitat types, the position of the wastewater effluent outfalls, the stream characteristics, stream developments (dams, bridges, navigation locks, and other manmade structures), the self-purification characteristics of the stream, and the nature of the objectives of the study (Velz, 1970). The stream habitat types used in this discussion are macroinvertebrate assemblages in stream ecosystems. Some combination of these habitats would be sampled in a multi-habitat approach to benthic sampling (Barbour et al., 1997):

1. *Cobble (hard substrate)*—Cobble is prevalent in the riffles (and runs) that are common features throughout most mountain and piedmont streams. In many high-gradient streams, this habitat type will be dominant; however, riffles are not a common feature of most coastal or other low-gradient streams. Sample shallow areas with coarse substrates (mixed gravel, cobble, or larger) by holding the bottom of the dip net against the substrate and dislodging organisms by kicking the substrate 0.5 m upstream of the net (this is where the designated kicker, the sampling partner, comes into play).
2. *Snags*—Snags and other woody debris that have been submerged for a relatively long period (not recent deadfall) provide excellent colonization habitat. Sample submerged woody debris by jabbing medium-sized snag material (sticks and branches). The snag habitat may be kicked first to help to dislodge organisms but only after placing the net downstream of the snag. Accumulated woody material in pool areas is considered snag habitat. Large logs should be avoided because they are generally difficult to sample adequately.
3. *Vegetated banks*—When lower banks are submerged and have roots and emergent plants associated with them, they are sampled in a fashion similar to snags. Submerged areas of undercut banks are good habitats to sample. Sample banks with protruding roots and plants by jabbing into the habitat. Bank habitat can be kicked first to help dislodge organisms, but only after placing the net downstream.
4. *Submerged macrophytes*—Submerged macrophytes are seasonal in their occurrence and may not be a common feature of many streams, particularly those that are high gradient. Sample aquatic plants that are rooted on the bottom of the stream in deep water by drawing the net through the vegetation from the bottom to the surface of the water (maximum of 0.5 m each jab). In shallow water, sample by bumping or jabbing the net along the bottom in the rooted area, avoiding sediments where possible.
5. *Sand* (and other fine sediment)—Usually the least productive macroinvertebrate habitat in streams, this habitat may be the most prevalent in some streams. Sample banks of unvegetated or soft soil by bumping the net along the surface of the substrate rather than dragging the net through soft substrate; this reduces the amount of debris in the sample.

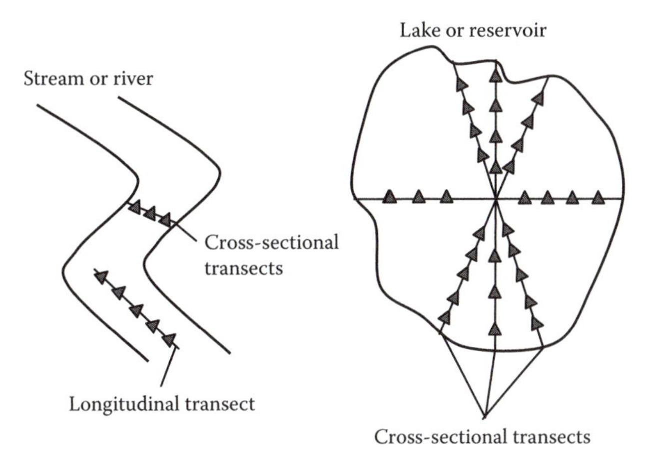

FIGURE 8.1 Transect sampling.

In a biological sampling program (based on the author's experience), the most common sampling methods are the *transect* and the *grid*. Transect sampling involves taking samples along a straight line either at uniform or at random intervals (see Figure 8.1). The transect approach samples a cross-section of a lake or stream or a longitudinal section of a river or stream. The transect sampling method allows for a more complete analysis by including variations in habitat.

In grid sampling, an imaginary grid system is placed over the study area. The grids may be numbered, and random numbers are generated to determine which grids should be sampled (see Figure 8.2). This type of sampling method allows for quantitative analysis because the grids are all of a certain size; for example, to sample a stream for benthic macroinvertebrates, grids that are 0.25 m^2 may be used. The weight or number of benthic macroinvertebrates per square meter can then be determined.

Random sampling requires that each possible sampling location have an equal chance of being selected. Numbering all sampling locations and then using a computer, calculator, or a random numbers table to collect a series of random numbers can accomplish this. An illustration of how to put the random numbers to work is provided in the following example. Given a pond that has 300 grid units, find 8 random sampling locations using the following sequence of random numbers taken from a standard random numbers table: 101, 209, 007, 018, 099, 100, 017, 069, 096, 033, 041, 011. The first eight numbers of the sequence could be selected and only those grids would be sampled to obtain a random sample.

Sample Collection

The following sample collection procedures have been suggested by USEPA (2000b). After establishing the sampling methodology and the sampling locations, the frequency of sampling must be determined. The more samples collected, the more reliable the data will be. A frequency of once a week or once a month will be adequate for most aquatic studies. Usually, the sampling period covers an entire year so yearly variations may be included. The details of sample collection will depend on the type of problem that is being solved and will vary with each study. When a sample is collected, it must be carefully identified with the following information:

1. Location (name of water body and place of study; longitude and latitude)
2. Date and time
3. Site (sampling location)

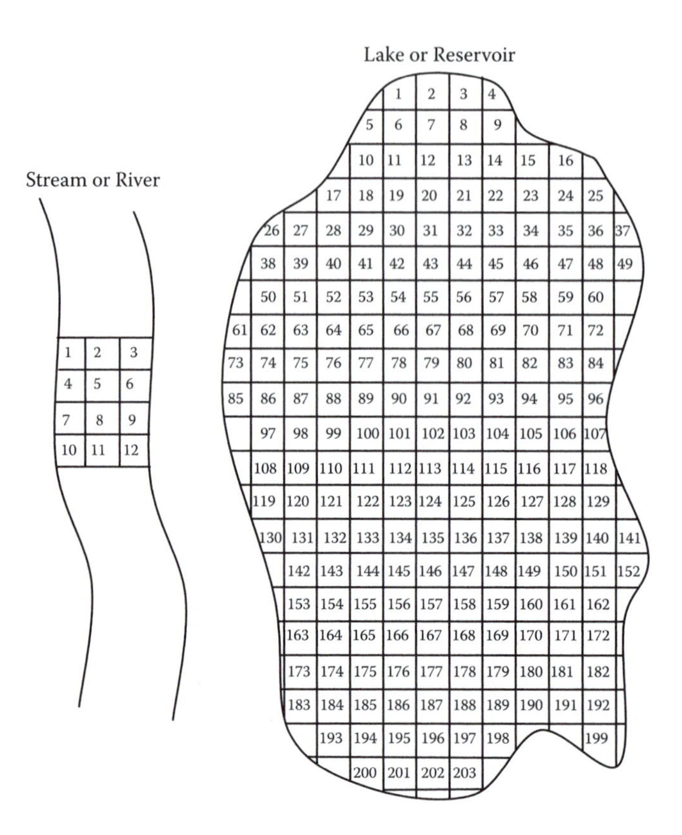

FIGURE 8.2 Grid sampling.

4. Name of collector
5. Weather (temperature, precipitation, humidity, wind, etc.)
6. Miscellaneous (any other important information, such as observations)
7. Field notebook

With regard to the last item, on each sampling day notes on field conditions should be taken; for example, miscellaneous observations and weather conditions can be entered. Additionally, notes that describe the condition of the water are also helpful (e.g., color, turbidity, odor, algae). All unusual findings and conditions should also be entered.

Macroinvertebrate Sampling Equipment

In addition to the appropriate sampling equipment, also assemble the following equipment:

1. Jars (two), at least quart size, plastic, wide-mouth with tight cap; one should be empty and the other filled about two thirds with 70% ethyl alcohol
2. Hand lens, magnifying glass, or field microscope
3. Fine-point forceps
4. Heavy-duty rubber gloves
5. Plastic sugar scoop or ice-cream scoop
6. Kink net (rocky-bottom stream) or dip net (muddy-bottom stream)
7. Buckets (two) (see Figure 8.3)

FIGURE 8.3 Sieve bucket.

8. String or twine (50 yards) and a tape measure
9. Stakes (four)
10. Orange to measure velocity (a stick, an apple, or a fish float may also be used in place of an orange)
11. Reference maps indicating general information pertinent to the sampling area, including the surrounding roadways, as well as a hand-drawn station map
12. Station ID tags
13. Spray water bottle
14. Pencils (at least two)

Most professional biological monitoring programs employ sieve buckets as a holding container for composited samples. These buckets have a mesh bottom that allows water to drain out while the organisms and debris remain. This material can then be easily transferred to the alcohol-filled jars. However, sieve buckets can be expensive. Many volunteer programs employ alternative equipment, such as the two regular buckets mentioned in this list. Regardless of the equipment, the process for compositing and transferring the sample is basically the same. The decision is one of cost and convenience.

MACROINVERTEBRATE SAMPLING IN ROCKY-BOTTOM STREAMS

Rocky-bottom streams are defined as those with bottoms made up of gravel, cobbles, and boulders in any combination. They usually have definite riffle areas. Riffle areas are fairly well oxygenated and, therefore, are prime habitats for benthic macroinvertebrates. In these streams, we use the *rocky-bottom sampling method.* This method of macroinvertebrate sampling is used in streams that have riffles and gravel/cobble substrates. Three samples are to be collected at each site, and a composite sample is obtained (i.e., one large total sample).

1. *Locate a site on a map, with its latitude and longitude indicated.*
   a. Samples will be taken in three different spots within a 100-yard stream site. These spots may be three separate riffles, one large riffle with different current velocities, or, if no riffles are present, three run areas with gravel or cobble substrate. Combinations are also possible (if, for example, the site has only one small riffle and several run areas). Mark off the 100-yard stream site. If possible, it should begin at least 50 yards upstream of any manmade modification of the channel, such as a bridge, dam, or pipeline crossing. Avoid walking in the stream, because this might dislodge macroinvertebrates and affect later sampling results.
   b. Sketch the 100-yard sampling area. Indicate the location of the three sampling spots on the sketch. Mark the most downstream site as Site 1, the middle site as Site 2, and the upstream site as Site 3.

2. *Get into place.*
   a. Always approach sampling locations from the downstream end and sample the site farthest downstream first (Site 1). This prevents biasing of the second and third collections with dislodged sediment of macroinvertebrates. Always use a clean kick seine, relatively free of mud and debris from previous uses. Fill a bucket about one third full with stream water, and fill the spray bottle.
   b. Select a 3 × 3-ft riffle area to sample at Site 1. One member of the team, the net holder, should position the net at the downstream end of this sampling area. Hold the net handles at a 45° angle to the surface of the water. Be sure the bottom of the net fits tightly against the streambed so no macroinvertebrates escape under the net. Rocks from the sampling area can be used to anchor the net against the stream bottom. Do not allow any water to flow over the net.
3. *Dislodge the macroinvertebrates.*
   a. Pick up any large rocks in the 3 × 3-ft sampling area and rub them thoroughly over the partially filled bucket so any macroinvertebrates clinging to the rocks will be dislodged into the bucket. Place each cleaned rock outside of the sampling area. After sampling is completed, rocks can be returned to the stretch of stream where they were found.
   b. The member of the team designated as the kicker should thoroughly stir up the sampling areas with his feet, starting at the upstream edge of the 3 × 3-ft sampling area and working downstream, moving toward the net. All dislodged organisms will be carried by the stream flow into the net. Be sure to disturb the first few inches of stream sediment to dislodge burrowing organisms. As a general guide, disturb the sampling area for about 3 minutes, or until the area is thoroughly worked over.
   c. Any large rocks used to anchor the net should be thoroughly rubbed into the bucket as above.
4. *Remove the net.*
   a. Remove the net without allowing any of the organisms it contains to wash away. While the net holder grabs the top of the net handles, the kicker grabs the bottom of the net handles and the bottom edge of the net. Remove the net from the stream with a forward scooping motion.
   b. Roll the kick net into a cylinder shape and place it vertically in the partially filled bucket. Pour or spray water down the net to flush its contents into the bucket. If necessary, pick debris and organisms from the net by hand. Release back into the stream any fish, amphibians, or reptiles caught in the net.
5. *Collect the second and third samples.*
   a. When all of the organisms have been removed from the net, repeat the steps above at Sites 2 and 3. Put the samples from all three sites into the same bucket. Combining the debris and organisms from all three sites into the same bucket is called *compositing.*

Note: If the bucket is nearly full of water after washing the net clean, let the debris and organisms settle to the bottom. Then, cup the net over the bucket and pour the water through the net into a second bucket. Inspect the water in the second bucket to be sure no organisms came through.

6. *Preserve the sample.*
   a. After collecting and compositing samples from all three sites, it is time to preserve the sample. All team members should leave the stream and return to a relatively flat section of the stream bank with their equipment. The next step will be to remove large pieces of debris (leaves, twigs, and rocks) from the sample. Carefully remove the debris one piece at a time. While holding the material over the bucket, use the forceps, spray bottle, and your hands to pick, rub, and rinse the leaves, twigs, and rocks to

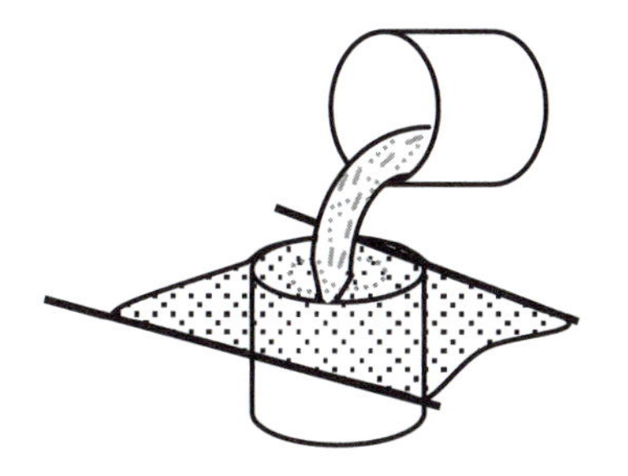

FIGURE 8.4 Pouring sample water through the net.

remove any attached organisms. Use a magnifying lens and forceps to find and remove small organisms clinging to the debris. When satisfied that the material is clean, discard it back into the stream.

b. The water will have to be drained before transferring material to the jar. This process requires two team members. Place the kick net over the second bucket, which has not yet been used and should be completely empty. One team member should push the center of the net into the second bucket, creating a small indentation or depression. Hold the sides of the net closely over the mouth of the bucket. The second person can now carefully pour the remaining contents of the first bucket onto a small area of the net to drain the water and concentrate the organisms. Use care when pouring so organisms are not lost over the side of the net (see Figure 8.4). Use the spray bottle, forceps, sugar scoop, and gloved hands to transfer all material from the first bucket onto the net. When the first bucket is empty, use your hands and the sugar scoop to transfer the material from the net into the empty jar. The second bucket captures the water and any organisms that might have fallen through the netting during pouring. As a final check, repeat the process above, but this time pour the second bucket over the net into the first bucket. Transfer any organisms on the net into the jar.
c. Now fill the jar (so all material is submerged) with the alcohol from the second jar. Put the lid tightly back onto the jar and gently turn the jar upside down two or three times to distribute the alcohol and remove air bubbles.
d. Complete the sampling station ID tag. Be sure to use a pencil, not a pen, because the ink will run in the alcohol! The tag includes the station number, stream, location (e.g., upstream from a road crossing), date, time, and names of the members of the collecting team. Place the ID tag into the sample container, writing side facing out, so the identification can be seen clearly.

Rocky-Bottom Habitat Assessment

The habitat assessment (including measuring general characteristics and local land use) for a rocky-bottom stream is conducted in a 100-yard section of stream that includes the riffles from which organisms were collected.

1. *Delineate the habitat assessment boundaries.*
   a. Begin by identifying the most downstream riffle that was sampled for macroinvertebrates. Using a tape measure or twine, mark off a 100-yard section extending 25 yards below the downstream riffle and about 75 yards upstream.
   b. Complete the identifying information of the field data sheet for the habitat assessment site. On the stream sketch, be as detailed as possible, and be sure to note which riffles were sampled.

2. *Describe the general characteristics and local land use on the field sheet.* For safety reasons as well as to protect the stream habitat, it is best to estimate the following characteristics rather than actually wading into the stream to measure them.
   a. Water appearance can be a physical indicator of water pollution:
      — Clear—colorless, transparent
      — Milky—cloudy-white or gray, not transparent; might be natural or due to pollution
      — Foamy—might be natural or due to pollution, generally detergents or nutrients (foam that is several inches high and does not brush apart easily is generally due to pollution)
      — Turbid—cloudy brown due to suspended silt or organic material
      — Dark brown—might indicate that acids are being released into the stream due to decaying plants
      — Oily sheen—multicolored reflection might indicate oil floating in the stream, although some sheens are natural
      — Orange—might indicate acid drainage
      — Green—might indicate that excess nutrients are being released into the stream
   b. Water odor can be a physical indicator of water pollution:
      — No odor, or a natural smell
      — Sewage—might indicate the release of human waste material
      — Chlorine—might indicate that a sewage treatment plant is overchlorinating its effluent
      — Fishy—might indicate the presence of excessive algal growth or dead fish
      — Rotten eggs—might indicate sewage pollution (the presence of a natural gas)
   c. Water temperature can be particularly important for determining whether the stream is suitable as habitat for some species of fish and macroinvertebrates that have distinct temperature requirements. Temperature also has a direct effect on the amount of dissolved oxygen available to aquatic organisms. Measure temperature by submerging a thermometer for at least 2 minutes in a typical stream run. Repeat once and average the results.
   d. The width of the stream channel can be determined by estimating the width of the streambed that is covered by water from bank to bank. If it varies widely along the stream, estimate an average width.
   e. Local land use refers to the part of the watershed within 1/4 mile upstream of and adjacent to the site. Note which land uses are present, as well as which ones seem to be having a negative impact on the stream. Base observations on what can be seen, what was passed on the way to the stream, and, if possible, what is noticed when leaving the stream.
3. *Conduct the habitat assessment.* The following information describes the parameters that will be evaluated for rocky-bottom habitats. Use these definitions when completing the habitat assessment field data sheet. The first two parameters should be assessed directly at the riffles or runs that were used for the macroinvertebrate sampling. The last eight parameters should be assessed in the entire 100-yard section of the stream.
   a. *Attachment sites* for macroinvertebrates are essentially the amount of living space or hard substrates (rocks, snags) available for adequate insects and snails. Many insects begin their life underwater in streams and need to attach themselves to rocks, logs, branches, or other submerged substrates. The greater the variety and number of available living spaces or attachment sites, the greater the variety of insects in the stream. Optimally, cobble should predominate, and boulders and gravel should be common. The availability of suitable living spaces for macroinvertebrates decreases as cobble becomes less abundant and boulders, gravel, or bedrock become more prevalent.

b. *Embeddedness* refers to the extent to which rocks (gravel, cobble, and boulders) are surrounded by, covered with, or sunken into the silt, sand, or mud of the stream bottom. Generally, as rocks become embedded, fewer living spaces are available to macroinvertebrates and fish for shelter, spawning, and egg incubation.

Note: To estimate the percent of embeddedness, observe the amount of silt or finer sediments overlaying and surrounding the rocks. If kicking does not dislodge the rocks or cobbles, they might be greatly embedded.

c. *Shelter* for fish includes the relative quantity and variety of natural structures in the stream, such as fallen trees, logs, and branches; cobble and large rock; and undercut banks that are available to fish for hiding, sleeping, or feeding. A wide variety of submerged structures in the stream provide fish with many living spaces; the more living spaces in a stream, the more types of fish the stream can support.
d. *Channel alteration* is a measure of large-scale changes in the shape of the stream channel. Many streams in urban and agricultural areas have been straightened, deepened (e.g., dredged), or diverted into concrete channels, often for flood control purposes. Such streams have far fewer natural habitats for fish, macroinvertebrates, and plants than do naturally meandering streams. Channel alteration is present when the stream runs through a concrete channel; when artificial embankments, riprap, and other forms of artificial bank stabilization or structures are present; when the stream is very straight for significant distances; when dams, bridges, and flow-altering structures such as combined sewer overflow (CSO) are present; when the stream is of uniform depth due to dredging; and when other such changes have occurred. Signs that indicate the occurrence of dredging include straightened, deepened, and otherwise uniform stream channels, as well as the removal of streamside vegetation to provide dredging equipment access to the stream.
e. *Sediment deposition* is a measure of the amount of sediment that has been deposited in the stream channel and the changes to the stream bottom that have occurred as a result of the deposition. High levels of sediment deposition create an unstable and continually changing environment that is unsuitable for many aquatic organisms. Sediments are naturally deposited in areas where the stream flow is reduced, such as in pools and bends, or where flow is obstructed. These deposits can lead to the formation of islands, shoals, or point bars (sediments that build up in the stream, usually at the beginning of a meander) or can result in the complete filling of pools. To determine whether these sediment deposits are new, look for vegetation growing on them; new sediments will not yet have been colonized by vegetation.
f. *Stream velocity and depth* combinations are important to the maintenance of healthy aquatic communities. Fast water increases the amount of dissolved oxygen in the water, keeps pools from being filled with sediment, and helps food items such as leaves, twigs, and algae move more quickly through the aquatic system. Slow water provides spawning areas for fish and shelters macroinvertebrates that might be washed downstream in higher stream velocities. Similarly, shallow water tends to be more easily aerated (i.e., holds more oxygen), but deeper water stays cooler longer. Thus, the best stream habitat includes all of the following velocity/depth combinations and can maintain a wide variety of organisms:
   — Slow (<1 ft/sec), shallow (<1.5 ft)
   — Slow, deep
   — Fast, deep
   — Fast, shallow

Measure stream velocity by marking off a 10-foot section of stream run and measuring the time it takes an orange, stick, or other floating biodegradable object to float the 10 feet. Repeat five times, in the same 10-foot section, and determine the average time. Divide the distance (10 feet) by the average time (seconds) to determine the velocity in feet per second. Measure the stream depth by using a stick of known length and taking readings at various points within the stream site, including riffles, runs, and pools. Compare velocity and depth at various points within the 100-yard site to see how many of the combinations are present.

g. *Channel flow status* is the percent of the existing channel that is filled with water. The flow status changes as the channel enlarges or as flow decreases because of dams and other obstructions, diversions for irrigation, or drought. When water does not cover much of the streambed, the living area for aquatic organisms is limited.

Note: For the following parameters, evaluate the conditions of the left and right stream banks separately. Define the left and right banks by standing at the downstream end of the study stretch and look upstream. Each bank is evaluated on a scale of 0 to 10.

h. *Bank vegetation protection* measures the amount of the stream bank that is covered by natural (i.e., growing wild and not obviously planted) vegetation. The root system of plants growing on stream banks helps hold soil in place, reducing erosion. Vegetation on banks provides shade for fish and macroinvertebrates and serves as a food source by dropping leaves and other organic matter into the stream. Ideally, a variety of vegetation should be present, including trees, shrubs, and grasses. Vegetation disruption can occur when the grasses and plants on the stream banks are mowed or grazed, or when the trees and shrubs are cut back or cleared.
i. *Condition of banks* measures the erosion potential and whether the stream banks are eroded. Steep banks are more likely to collapse and suffer from erosion than are gently sloping banks; therefore, they are considered to have erosion potential. Signs of erosion include crumbling and unvegetated banks, exposed tree roots, and exposed soil.
j. The *riparian vegetative zone* is defined as the width of natural vegetation from the edge of the stream bank. The riparian vegetative zone is a buffer zone to pollutants entering a stream from runoff. It also controls erosion and provides stream habitat and nutrient input into the stream.

Note: A wide, relatively undisturbed riparian vegetative zone reflects a healthy stream system; narrow, far less useful riparian zones occur when roads, parking lots, fields, lawns, and other artificially cultivated areas, bare soil, rock, or buildings are near the stream bank. The presence of old fields (i.e., previously developed agricultural fields allowed to revert to natural conditions) should rate higher than fields in continuous or periodic use. In arid areas, the riparian vegetative zone can be measured by observing the width of the area dominated by riparian or water-loving plants, such as willows, marsh grasses, and cottonwood trees.

MACROINVERTEBRATE SAMPLING IN MUDDY-BOTTOM STREAMS

In muddy-bottom streams, as in rocky-bottom streams, the goal is to sample the most productive habitat available and look for the widest variety of organisms. The most productive habitat is the one that harbors a diverse population of pollution-sensitive macroinvertebrates. Samplers should sample by using a D-frame net (see Figure 8.5) to jab at the habitat and scoop up the organisms that are dislodged. The idea is to collect a total sample that consists of 20 jabs taken from a variety of habitats. Use the following method of macroinvertebrate sampling in streams that have muddy-bottom substrates.

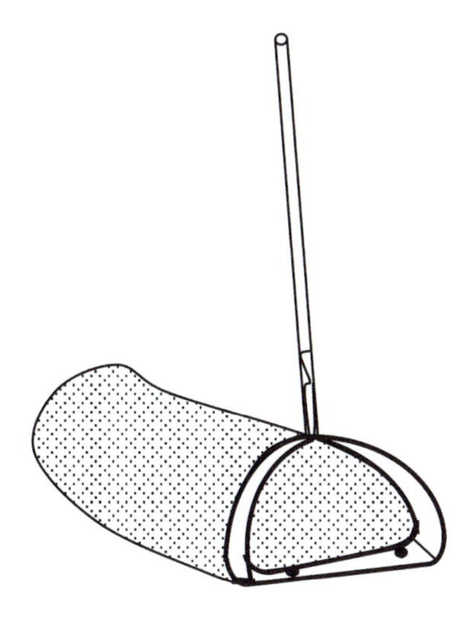

FIGURE 8.5 D-frame aquatic net.

1. *Determine which habitats are present.* Muddy-bottom streams usually have four habitats: vegetated banks margins, snags and logs, aquatic vegetation beds and decaying organic matter, and silt/sand/gravel substrate. It is generally best to concentrate sampling efforts on the most productive habitat available, yet sample other principal habitats if they are present. This ensures that you will secure as wide a variety of organisms as possible. Not all habitats are present in all streams or are present in significant amounts. If the sampling areas have not been preselected, determine which of the following habitats are present.

Note: Avoid standing in the stream while making habitat determinations.

   a. Vegetated bank margins consist of overhanging bank vegetation and submerged root mats attached to the banks. The bank margins may also contain submerged, decomposing leaf packs trapped in root wads or lining the stream banks. This is generally a highly productive habitat in a muddy stream, and it is often the most abundant type of habitat.
   b. Snags and logs consist of submerged wood, primarily dead trees, logs, branches, roots, cypress knees, and leaf packs lodged between rocks or logs. This is also a very productive muddy-bottom stream habitat.
   c. Aquatic vegetation beds and decaying organic matter consist of beds of submerged, green, leafy plants that are attached to the stream bottom. This habitat can be as productive as vegetated bank margins and snags and logs.
   d. Silt/sand/gravel substrate includes sandy, silty, or muddy stream bottoms; rocks along the stream bottom; or wetted gravel bars. This habitat may also contain algae-covered rocks (*aufwuchs*). This is the least productive of the four muddy-bottom stream habitats, and it is always present in one form or another (e.g., silt, sand, mud, or gravel might predominate).
2. *Determine how many times to jab in each habitat type.* The sampler's goal is to jab 20 times. The D-frame net (see Figure 8.5) is 1 foot wide, and a jab should be approximately 1 foot in length. Thus, 20 jabs equal 20 square feet of combined habitat.
   a. If all four habitats are present in plentiful amounts, jab the vegetated banks 10 times and divide the remaining 10 jabs among the remaining three habitats.
   b. If three habitats are present in plentiful amounts, and one is absent, jab the silt/sand/gravel substrate, the least productive habitat, five times and divide the remaining 15 jabs between the other two more productive habitats.

c. If only two habitats are present in plentiful amounts, the silt/sand/gravel substrate will most likely be one of those habitats. Jab the silt/sand/gravel substrate five times and the more productive habitat 15 times.
d. If some habitats are plentiful and others are sparse, sample the sparse habitats to the extent possible, even if it is possible to take only one or two jabs. Take the remaining jabs from the plentiful habitats. This rule also applies if a habitat cannot be reached because of unsafe stream conditions. Jab 20 times total.

Note: Because the sampler might need to make an educated guess to decide how many jabs to take in each habitat type, it is critical that the sampler note, on the field data sheet, how many jabs were taken in each habitat. This information can be used to help characterize the findings.

3. *Get into place.* Outside and downstream of the first sampling location (first habitat), rinse the dip net and check to make sure it does not contain any macroinvertebrates or debris from the last time it was used. Fill a bucket approximately one-third full with clean stream water. Also, fill the spray bottle with clean stream water. This bottle will be used to wash the net between jabs and after sampling is completed.

Note: This method of sampling requires only one person to disturb the stream habitats. While one person is sampling, a second person should stand outside the sampling area, holding the bucket and spray bottle. After every few jabs, the sampler should hand the net to the second person, who can then rinse the contents of the net into the bucket.

4. *Dislodge the macroinvertebrates.* Approach the first sample site from downstream, and sample while walking upstream. Sample in the four habitat types as follows:
   a. Sample vegetated bank margins by jabbing vigorously, with an upward motion, brushing the net against vegetation and roots along the bank. The entire jab motion should occur underwater.
   b. To sample snags and logs, hold the net with one hand under the section of submerged wood being sampled. With the other hand (which should be gloved), rub about 1 square foot of area on the snag or log. Scoop organisms, bark, twigs, or other organic matter dislodged into the net. Each combination of log rubbing and net scooping is one jab.
   c. To sample aquatic vegetation beds, jab vigorously, with an upward motion, against or through the plant bed. The entire jab motion should occur underwater.
   d. To sample a silt/sand/gravel substrate, place the net with one edge against the stream bottom and push it forward about a foot (in an upstream direction) to dislodge the first few inches of silt, sand, gravel, or rocks. To avoid gathering a net full of mud, periodically sweep the mesh bottom of the net back and forth in the water, making sure that water does not run over the top of the net. This will allow fine silt to rinse out of the net. When 20 jabs have been completed, rinse the net thoroughly in the bucket. If necessary, pick any clinging organisms from the net by hand, and put them in the bucket.
5. *Preserve the sample.*
   a. Look through the material in the bucket, and immediately return any fish, amphibians, or reptiles to the stream. Carefully remove large pieces of debris (leaves, twigs, and rocks) from the sample. While holding the material over the bucket, use the forceps, spray bottle, and your hands to pick, rub, and rinse the leaves, twigs, and rocks to remove any attached organisms. Use the magnifying lens and forceps to find and remove small organisms clinging to the debris. When satisfied that the material is clean, discard it back into the stream.
   b. Drain the water before transferring material to the jar. This process will require two people. One person should place the net into the second bucket, like a sieve (this bucket, which has not yet been used, should be completely empty), and hold it securely. The second person can now carefully pour the remaining contents of the first bucket onto

Station ID Tag

Station # ____________________

Stream ____________________

Location ____________________

Date/Time ____________________

Team Members: ____________________

____

FIGURE 8.6 Station ID tag.

the center of the net to drain the water and concentrate the organisms. Use care when pouring so organisms are not lost over the side of the net. Use the spray bottle, forceps, sugar scoop, and gloved hands to remove all of the material from the first bucket onto the net. When satisfied that the first bucket is empty, use your hands and the sugar scoop to transfer all the material from the net into the empty jar. The contents of the net can also be emptied directly into the jar by turning the net inside out into the jar. The second bucket captures the water and any organisms that might have fallen through the netting. As a final check, repeat the process above, but this time pour the second bucket over the net into the first bucket. Transfer any organisms on the net into the jar.

c. Fill the jar (so all material is submerged) with alcohol. Put the lid tightly back onto the jar and gently turn the jar upside down two or three times to distribute the alcohol and remove air bubbles.

d. Complete the sampling station ID tag (see Figure 8.6). Be sure to use a pencil, not a pen, because the ink will run in the alcohol. The tag should include the station number, the stream, location (e.g., upstream from a road crossing), date, time, and names of the members of the collecting crew. Place the ID tag into the sample container, writing side facing out, so the identification can be seen clearly.

Note: To keep samples from being mixed up, samplers should place the ID tag *inside* the sample jar.

Muddy-Bottom Stream Habitat Assessment

The muddy-bottom stream habitat assessment (which includes measuring general characteristics and local land use) is conducted in a 100-yard section of the stream that includes the habitat areas from which organisms were collected.

Note: Any references made to a field data sheet (habitat assessment field data sheet) in this chapter assume that the sampling team is using standard forms provided by the USEPA, the U.S. Geological Survey (USGS), or state water control authorities; generic forms put together by the sampling team can also be used. The source of the form and exact type of form are not important, but some type of data recording field sheet should be employed to record pertinent data.

1. *Delineate the habitat assessment boundaries.*
   a. Begin by identifying the most downstream point that was sampled for macroinvertebrates. Using a tape measure or twine, mark off a 100-yard section extending 25 yards below the downstream sampling point and about 75 yards upstream.

b. Complete the identifying information on the field data sheet for the habitat assessment site. On the stream sketch, be as detailed as possible, and be sure to note which habitats were sampled.

2. *Record general characteristics and local land use on the data field sheet.* For safety reasons as well as to protect the stream habitat, it is best to estimate these characteristics rather than to actually wade into the stream to measure them. For instructions on completing these sections of the field data sheet, see the rocky-bottom habitat assessment instructions.
3. *Conduct the habitat assessment.* The following information describes the parameters to be evaluated for muddy-bottom habitats. Use these definitions when completing the habitat assessment field data sheet.
   a. *Shelter* for fish and attachment sites for macroinvertebrates are the living space and shelter (rocks, snags, and undercut banks) available for fish, insects, and snails. Many insects attach themselves to rocks, logs, branches, or other submerged substrates. Fish can hide or feed in these areas. The greater the variety and number of available shelter sites or attachment sites, the greater the variety of fish and insects in the stream.

Note: Many of the attachment sites result from debris falling into the stream from the surrounding vegetation. When debris first falls into the water, it is termed *new fall,* and it has not yet been broken down by microbes (conditioned) for macroinvertebrate colonization. Leaf material or debris that is conditioned is called *old fall.* Leaves that have been in the stream for some time lose their color, turn brown or dull yellow, become soft and supple with age, and might be slimy to the touch. Woody debris becomes blackened or dark in color; smooth bark becomes coarse and partially disintegrated, creating holes and crevices. It might also be slimy to the touch.

b. *Pool substrate characterization* evaluates the type and condition of bottom substrates in pools. Pools with firmer sediment types (e.g., gravel, sand) and rooted aquatic plants support a wider variety of organisms than do pools with substrates dominated by mud or bedrock and no plants. In addition, a pool with one uniform substrate type will support far fewer types of organisms than will a pool with a wide variety of substrate types.

c. *Pool variability* rates the overall mixture of pool types found in the stream according to size and depth. The four basic types of pools are large-shallow, large-deep, small-shallow, and small-deep. A stream with many pool types supports a wide variety of aquatic species. Rivers with low sinuosity (few bends) and monotonous pool characteristics do not have sufficient quantities and types of habitats to support a divers aquatic community.

d. *Channel alteration*—see the rocky-bottom habitat assessment instructions.

e. *Sediment deposition*—see the rocky-bottom habitat assessment instructions.

f. *Channel sinuosity* evaluates the sinuosity or meandering of the stream. Streams that meander provide a variety of habitats (such as pools and runs) and stream velocities and reduce the energy from current surges during storm events. Straight stream segments are characterized by even stream depth and unvarying velocity, and they are prone to flooding. To evaluate this parameter, imagine how much longer the stream would be if it were straightened out.

g. *Channel flow status*—see the rocky-bottom habitat assessment instructions.

h. *Bank vegetation protection*—see the rocky-bottom habitat assessment instructions.

i. *Condition of banks*—see the rocky-bottom habitat assessment instructions.

j. *Riparian vegetative zone width*—see the rocky-bottom habitat assessment instructions.

Note: Whenever stream sampling is to be conducted, it is a good idea to have a reference collection on hand. A reference collection is a sample of locally found macroinvertebrates that have been identified, labeled, and preserved in alcohol. The program advisor, along with a professional biologist/entomologist, should assemble the reference collection, properly identify all samples, preserve them in vials, and label them. This collection may then be used as a training tool and, in the field, as an aid in macroinvertebrate identification.

POST-SAMPLING ROUTINE

After completing the stream characterization and habitat assessment, make sure that all of the field data sheets have been completed properly and that the information is legible. Be sure to include the identifying name of the site and the sampling date on each sheet. This information will function as a quality control element. Before leaving the stream location, make sure that all sampling equipment and devices have been collected and rinsed properly. Double-check to make sure that sample jars are tightly closed and correctly identified. All samples, field sheets, and equipment should be returned to the team leader at this point. Keep a copy of the field data sheets for comparison with future monitoring trips and for personal records. The next step is to prepare for macroinvertebrate laboratory work. This step includes setting up a laboratory for processing samples into subsamples and identifying macroinvertebrates to the family level. A professional biologist, entomologist, freshwater ecologist, or advisor should supervise the identification procedure. (The actual laboratory procedures that follow the sampling and collecting phase are beyond the scope of this text.)

SAMPLING DEVICES

In addition to the sampling equipment mentioned previously, it may be desirable to employ, depending on stream conditions, the use of other sampling devices. Additional sampling devices commonly used, and discussed in the following sections, include dissolved oxygen and temperature monitors, sampling nets (including the D-frame aquatic net), sediment samplers (dredges), plankton samplers, and Secchi disks. The methods described below are approved by the USEPA (1998).

Dissolved Oxygen and Temperature Monitor

The dissolved oxygen (DO) content of a stream sample can provide the investigator with vital information, as DO content reflects the ability of a stream to maintain aquatic life.

Winkler Dissolved Oxygen with Azide Modification Method

The Winkler DO with azide modification method is commonly used to measure DO content. The Winkler method is best suited for clean waters. It can be used in the field but is better suited for laboratory work where greater accuracy may be achieved. The Winkler method adds a divalent manganese solution followed by a strong alkali to a 300-mL BOD bottle of stream-water sample. Any DO rapidly oxidizes an equivalent amount of divalent manganese to basic hydroxides of higher balance states. When the solution is acidified in the presence of iodide, oxidized manganese again reverts to the divalent state, and iodine, equivalent to the original DO content of the sample, is liberated. The amount of iodine is then determined by titration with a standard, usually thiosulfate, solution.

Fortunately for the field biologist, this is the age of miniaturized electronic circuit components and devices; thus, it is not too difficult to obtain portable electronic measuring devices for DO and temperature that are of quality construction and have better than moderate accuracy. These modern electronic devices are usually suitable for laboratory and field use. The device may be subjected to severe abuse in the field; therefore, the instrument must be durable, accurate, and easy to use. Several quality DO monitors are available commercially.

When using a DO monitor, it is important to calibrate (standardize) the meter prior to use. Calibration procedures can be found in APHA (1998) or in the manufacturer's instructions for the meter. The meter can be calibrated by determining the air temperature and the DO at saturation for that temperature and then adjusting the meter so it reads the saturation value. After calibration, the monitor is ready for use. All recorded measurements, including water temperatures and DO readings, should be entered in a field notebook.

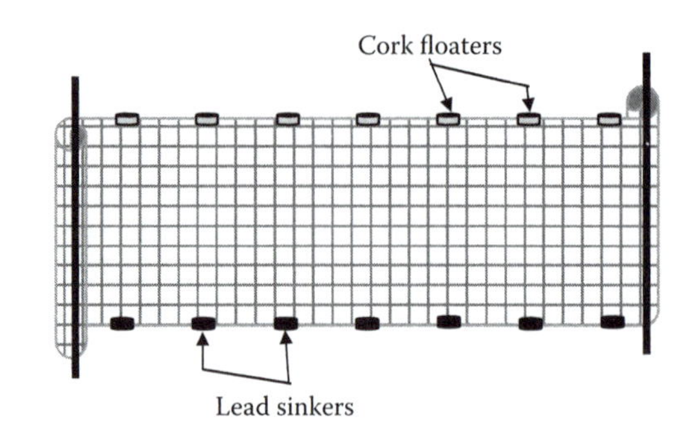

FIGURE 8.7 Two-person seine net.

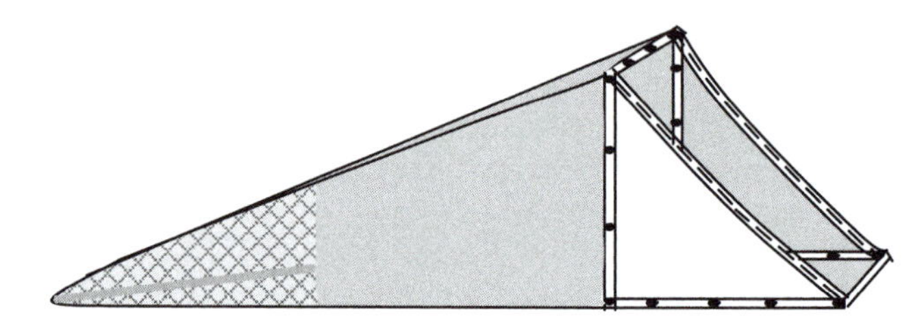

FIGURE 8.8 Surber sampler.

Sampling Nets

A variety of sampling nets is available for use in the field. The two-person seine net shown in Figure 8.7 is 20 ft long by 4 ft deep with an 1/8-inch mesh and is utilized to collect various organisms. Two people, each holding one end, walk upstream, and small organisms are gathered in the net. Dip nets are used to collect organisms in shallow streams. The Surber sampler collects macroinvertebrates stirred up from the bottom (see Figure 8.8) and can be used to obtain a quantitative sample (number of organisms per square feet). It is designed for sampling riffle areas in streams and rivers up to a depth of about 450 mm (18 in.). It consists of two folding stainless steel frames set at right angles to each other. The frame is placed on the bottom, with the net extending downstream. Using a hand or a rake, all sediment enclosed by the frame is dislodged. All organisms are caught in the net and transferred to another vessel for counting. The D-frame aquatic dip net (see Figure 8.5) is ideal for sweeping over vegetation or for use in shallow streams.

Sediment Samplers (Dredges)

A sediment sampler or dredge is designed to obtain a sample of the bottom material in a slow-moving stream and the organisms in it. The simple homemade dredge shown in Figure 8.9 works well in water too deep to sample effectively with handheld tools. The homemade dredge is fashioned from a #3 coffee can and a smaller can with a tight-fitting plastic lid (peanut cans work well). To use the homemade dredge, first invert it under water so the can fills with water and no air is trapped. Then, lower the dredge as quickly as possible with the "down" line. The idea is to bury the open end of the coffee can in the bottom. Then, quickly pull the "up" line to bring the can to the surface with a minimum loss of material. Dump the contents into a sieve or observation pan to sort. It works best in bottoms composed of sediment, mud, sand, and small gravel. The bottom sampling dredge can be used for a number of different analyses. Because the bottom sediments represent a good area in which to find macroinvertebrates and benthic algae, the communities of organisms living on or in the bottom can be easily studied quantitatively and qualitatively. A chemical analysis of the bottom sediment can be conducted to determine what chemicals are available to organisms living in the bottom habitat.

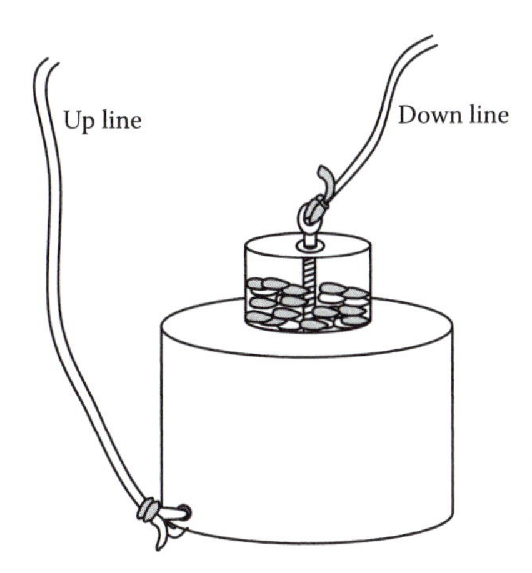

FIGURE 8.9 Homemade dredge.

Plankton Sampler

(More detailed information on plankton sampling can be found in AWRI, 2000.) *Plankton* (meaning "to drift") are distributed through the stream and, in particular, in pool areas. They are found at all depths and are comprised of plant (phytoplankton) and animal (zooplankton) forms. Plankton show a distribution pattern that can be associated with the time of day and seasons. The three fundamental sizes of plankton are *nanoplankton*, *microplankton*, and *macroplankton*. The smallest, nanoplankton, range in size from 5 to 60 μm (millionth of a meter). Because of their small size, most nanoplankton will pass through the pores of a standard sampling net. Special fine mesh nets can be used to capture the larger nanoplankton. Most planktonic organisms fall into the microplankton (or net plankton) category, where the sizes range from the largest nanoplankton to about 2 mm (thousandths of a meter). Nets of various sizes and shapes are used to collect microplankton. The nets collect the organism by filtering water through fine meshed cloth. The third group of plankton, macroplankton, are visible to the naked eye. The largest can be several meters long.

FIGURE 8.10 Plankton net.

The plankton net or sampler (see Figure 8.10) is a device that makes it possible to collect phytoplankton and zooplankton samples. For quantitative comparisons of different samples, some nets have a flow meter used to determine the amount of water passing through the collecting net. The plankton net or sampler provides a means of obtaining samples of plankton from various depths so distribution patterns can be studied. The net can be towed to sample plankton at a single depth (horizontal tow) or it can be lowered into the water to sample the water column (vertical tow). Another possibility is oblique tows where the net is lowered to a predetermined depth and raised at a constant rate as the vessel moves forward.

After towing and removal from the stream, the sides of the net are rinsed to dislodge the collected plankton. If a quantitative sample is desired, a certain quantity of water is collected. If the plankton density is low, then the sample may be concentrated using a low-speed centrifuge or some other filtering device. A definite volume of the sample is studied under the compound microscope for counting and identification of plankton.

Secchi Disk

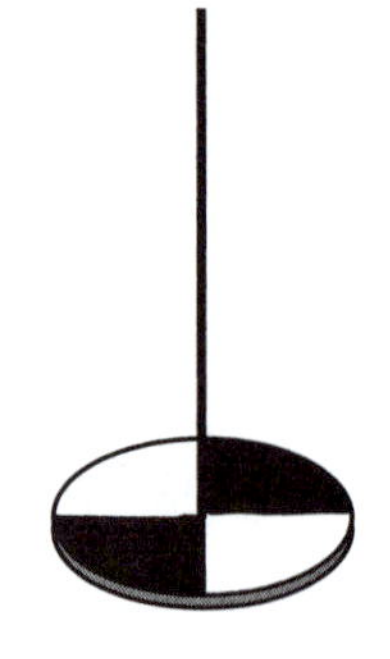

FIGURE 8.11 Secchi disk.

For determining water turbidity or degree of visibility in a stream, a Secchi disk is often used (Figure 8.11). The Secchi disk originated with Father Pietro Secchi, an astrophysicist and scientific advisor to the Pope. In 1865, when the head of the Papal Navy asked him to measure transparency in the Mediterranean Sea, Secchi used some white disks to measure the clarity of water. Various sizes of disks have been used since that time, but the most frequently used disk is an 8-inch-diameter metal disk painted in alternating black and white quadrants. The disk shown in Figure 8.11 is 20 cm in diameter; it is lowered into the stream using the calibrated line. To use the Secchi disk properly, it should be lowered into the stream water until no longer visible. At the point where it is no longer visible, a measurement of the depth is taken. This depth is called the *Secchi disk transparency light extinction coefficient.* The best results are usually obtained after early morning and before late afternoon.

Miscellaneous Sampling Equipment

Several other sampling tools and devices are available for use in sampling a stream; for example, consider the standard sand/mud sieve. Generally made of heavy-duty galvanized 1/8-inch mesh screen supported by a water-sealed 24 × 15 × 3-inch wood frame, this device is useful for collecting burrowing organisms found in soft bottom sediments. Moreover, no stream sampling kit would be complete without a collecting tray, collecting jars of assorted sizes, heavy-duty plastic bags, small pipets, large 2-ounce pipets, fine mesh straining nets, and black china marking pencils. In addition, depending on the quantity of material to be sampled, it is prudent to include several 3- and 5-gallon collection buckets in the stream sampling field kit.

Biological Sampling: The Bottom Line

This discussion has stressed the practice of biological monitoring, employing the use of biotic indices as key measuring tools. We emphasized biotic indices not only for their simplicity of use but also for the relative accuracy they provide, although their development and use can sometimes be derailed. The failure of a monitoring protocol to assess environmental condition accurately or to protect running waters usually stems from conceptual, sampling, or analytical pitfalls. Biotic indices can be combined with other tools for measuring the condition of ecological systems in ways that enhance or hinder their effectiveness. The point is that, like any other tool, these tools can be misused; however, the fact that biotic indices can be misused does not mean that the approach itself is useless. Thus, to ensure that the biotic indices approach is not useless, it is important for the practicing freshwater ecologist and water sampler to remember a few key guidelines:

1. Sampling everything is not the goal. Biological systems are complex and unstable in space and time, and samplers often feel compelled to study all components of this variation (Botkin, 1990). Complex sampling programs proliferate, but every study need not explore everything. Freshwater samplers and monitors should avoid the temptation to sample all the unique habitats and phenomena that make freshwater monitoring so interesting. Emphasis should be placed on the central components of a clearly defined research agenda (a sampling and monitoring protocol) to detect and measure the influence of human activities on the ecological system of a water body.
2. With regard to the influence of human activities on the ecological system of a water body, we must consider protecting biological condition to be a central responsibility of water resource management. One thing is certain—until biological monitoring is seen as

> essential to tracking attainment of that goal and biological criteria become enforceable standards mandated by the Clean Water Act, the diversity of life in the nation's freshwater systems will continue to decline.

Biomonitoring is only one of several tools available to the water practitioner. No matter the tool employed, all results depend on proper biomonitoring techniques. Biological monitoring must be designed to obtain accurate results, and current approaches must be strengthened. In addition, "the way it's always been done" must be reexamined, and efforts must be undertaken to do what works to keep freshwater systems alive. We can afford nothing less.

DRINKING WATER QUALITY MONITORING

When we speak of water quality monitoring, we refer to monitoring practices based on three criteria: (1) to ensure to the extent possible that the water is not a danger to public health, (2) to ensure that the water provided at the tap is as aesthetically pleasing as possible, and (3) to ensure compliance with applicable regulations. To meet these goals, all public systems must monitor water quality to some extent. The degree of monitoring employed is dependent on local needs and requirements and on the type of water system; small water systems using good-quality water from deep wells may have to conduct only occasional monitoring, but systems using surface water sources must test water quality frequently (AWWA 1995).

Drinking water must be monitored to provide *adequate control* of the entire water drawing, treatment, and conveyance system. Adequate control can be defined as monitoring employed to assess the current level of water quality, so action can be taken to maintain the required level (whatever that might be). We can define *water quality monitoring* as the sampling and analysis of water constituents and conditions. When we monitor, we collect data. As a monitoring program is developed, identifying the reasons for collecting the information is important. The reasons are defined by establishing a set of objectives that includes a description of who will collect the information.

It may come as a surprise to know that today the majority of people collecting data are not water and wastewater operators; instead, many are volunteers. These volunteers have a stake in their local stream, lake, or other water body and in many cases are proving they can successfully carry out a water quality monitoring program.

Is the Water Good or Bad?

To answer the question of whether the water is good or bad we must consider two factors. First, we return to the basic principles of water quality monitoring—sampling and analyzing water constituents and conditions. These constituents include the following:

1. Introduced pollutants, such as pesticides, metals, and oil
2. Constituents found naturally in water that can nevertheless be affected by human sources, such as dissolved oxygen, bacteria, and nutrients

The magnitude of their effects is influenced by properties such as pH and temperature; for example, temperature influences the quantity of dissolved oxygen that water is able to contain, and pH affects the toxicity of ammonia.

The second factor to be considered is that the only valid way to answer this question is to conduct tests, the results of which must then be compared to some form of water quality standards. If simply assigning a "good" and "bad" value to each test factor were possible, the meters and measuring devices in water quality test kits would be much easier to make. Instead of fine graduations, they could simply have a "good" and a "bad" zone.

TABLE 8.3
Total Residual Chlorine Effects

Total Residual Chlorine (mg/L)	Effect
0.06	Toxic to striped bass larvae
0.31	Toxic to white perch larvae
0.5–1.0	Typical drinking water residual
1.0–3.0	Recommended for swimming pools

Source: Spellman, F.R., *Spellman's Standard Handbook for Wastewater Operators*, Vol. 1, CRC Press, Boca Raton, FL, 1999.

Water quality—the difference between good and bad water—must be interpreted according to the intended use of the water; for example, the perfect balance of water chemistry that provides a sparkling clear, sanitary swimming pool would not be acceptable for drinking water and would be a deadly environment for many biota (Table 8.3). In another example, widely different levels of fecal coliform bacteria are considered acceptable, depending on the intended use of the water.

State and local water quality practitioners as well as volunteers have been monitoring water quality conditions for many years. In fact, until the past decade or so (until biological monitoring protocols were developed and began to take hold), water quality monitoring was generally considered the primary way to identify water pollution problems. Today, professional water quality practitioners and volunteer program coordinators alike are moving toward approaches that combine chemical, physical, and biological monitoring methods to achieve the best picture of water quality conditions. Water quality monitoring can be used for many purposes:

1. *To identify whether waters are meeting designated uses.* All states have established specific criteria (limits on pollutants) identifying what concentrations of chemical pollutants are allowable in their waters. When chemical pollutants exceed maximum or minimum allowable concentrations, waters may no longer be able to support the beneficial uses—such as fishing, swimming, and drinking—for which they have been designated (see Table 8.4). Designated or intended uses and the specific criteria that protect them (along with antidegredation statements prohibiting waters from deteriorating below existing or anticipated uses) together form water quality standards. State water quality professionals assess water quality by comparing the concentrations of chemical pollutants found in streams to the criteria in the state's standards and so judge whether or not streams are meeting their designated uses. Water quality monitoring, however, might be inadequate for determining whether aquatic life needs are being met in a stream. Although some

TABLE 8.4
Fecal Coliform Bacteria per 100 mL of Water

Desirable	Permissible	Type of Water Use
0	0	Potable and well water (for drinking)
<200	<1000	Primary contact water (for swimming)
<1000	<5000	Secondary contact water (boating and fishing)

Source: Spellman, F.R., *Spellman's Standard Handbook for Wastewater Operators*, Vol. 1, CRC Press, Boca Raton, FL, 1999.

constituents (such as dissolved oxygen and temperate) are important to maintaining healthy fish and aquatic insect populations, other factors (such as the physical structure of the stream and the condition of the habitat) play an equal or greater role. Biological monitoring methods are generally better suited to determining whether aquatic life is supported.
2. *To identify specific pollutants and sources of pollution.* Water quality monitoring helps link sources of pollution to water body quality problems because it identifies specific problem pollutants. Because certain activities tend to generate certain pollutants (bacteria and nutrients are more likely to come from an animal feedlot than an automotive repair shop), a tentative link to what would warrant further investigation or monitoring can be formed.
3. *To determine trends.* Chemical constituents that are properly monitored (i.e., using consistent time of day and on a regular basis using consistent methods) can be analyzed for trends over time.
4. *To screen for impairment.* Finding excessive levels of one or more chemical constituents can serve as an early warning for potential pollution problems.

State Water Quality Standards Programs

Each state has a program to set standards for the protection of each body of water within its boundaries. Standards for each body of water are developed that

1. Depend on the water's designated use
2. Are based on USEPA national water quality criteria and other scientific research into the effects of specific pollutants on different types of aquatic life and on human health
3. May include limits based on the biological diversity of the body of water (the presence of food and prey species)

State water quality standards set limits on pollutants and establish water quality levels that must be maintained for each type of water body, based on its designated use. Resources for this type of information include the following:

1. USEPA Water Quality Criteria Program
2. U.S. Fish and Wildlife Service habitat suitability index models (for specific species of local interest)

Monitoring test results can be plotted against these standards to provide a focused, relevant, required assessment of water quality.

Designing a Water Quality Monitoring Program

The first step in designing a water quality-monitoring program is to determine the purpose for the monitoring. This aids in selection of parameters to monitor. This decision should be based on such factors as

1. Types of water quality problems and pollution sources that are likely to be encountered (see Table 8.5)
2. Cost of available monitoring equipment
3. Precision and accuracy of available monitoring equipment
4. Capabilities of monitors

TABLE 8.5
Sources of Water Quality Problems and Pollution

Source	Common Associated Chemical Pollutants
Cropland	Turbidity, phosphorus, nitrates, temperature, total solids
Forestry harvest	Turbidity, temperature, total solids
Grazing land	Fecal bacteria, turbidity, phosphorus
Industrial discharge	Temperature, conductivity, total solids, toxics, pH
Mining	pH, alkalinity, total dissolved solids
Septic systems	Fecal bacteria, (e.g., *Escherichia coli*, *Enterococcus*), nitrates, dissolved oxygen/ biochemical oxygen demand, conductivity, temperature
Sewage treatment	Dissolved oxygen and BOD, turbidity, conductivity, phosphorus, nitrates, fecal bacteria, temperature, total solids, pH
Construction	Turbidity, temperature, dissolved oxygen and BOD, total solids, toxics
Urban runoff	Turbidity, phosphorus, nitrates, temperature, conductivity, dissolved oxygen, BOD

Source: Spellman, F.R., *Spellman's Standard Handbook for Wastewater Operators*, Vol. 1, CRC Press, Boca Raton, FL, 1999.

Note: In this section, we discuss in detail the parameters most commonly monitored by drinking water practitioners in streams (i.e., we assume, for illustration and discussion purposes, that our water source is a surface water stream). These parameters include dissolved oxygen, biochemical oxygen demand, temperature, pH, turbidity, total orthophosphate, nitrates, total solids, conductivity, total alkalinity, fecal bacteria, apparent color, odor, and hardness. When monitoring water supplies under the Safe Drinking Water Act (SDWA) or the National Pollutant Discharge Elimination System (NPDES), utilities must follow test procedures approved by the USEPA for these purposes. Additional testing requirements under these and other federal programs are published as amendments in the *Federal Register.*

Except when monitoring discharges for specific compliance purposes, a large number of approximate measurements can provide more useful information than one or two accurate analyses. Because water quality and chemistry continually change, making periodic, representative measurements and observations that indicate the range of water quality is necessary, rather than testing the quality at any single moment. The more complex a water system, the more time is required to observe, understand, and draw conclusions regarding the cause and effect of changes in the particular system.

General Preparation and Sampling Considerations

Sampling devices should be corrosion resistant, easily cleaned, and capable of collecting desired samples safely and in accordance with test requirements. Whenever possible, assign a sampling device to each sampling point. Sampling equipment must be cleaned on a regular schedule to avoid contamination.

Note: Some tests require special equipment to ensure that the sample is representative. Dissolved oxygen and fecal bacteria sampling requires special equipment and procedures to prevent collection of nonrepresentative samples.

Reused sample containers and glassware must be cleaned and rinsed before the first sampling run and after each run by following Method A or Method B described below. The more suitable method depends on the parameter being measured.

Method A: General Preparation of Sampling Containers

Use the following method when preparing all sample containers and glassware for monitoring conductivity, total solids, turbidity, pH, and total alkalinity. Wearing latex gloves,

1. Wash each sample bottle or piece of glassware with a brush and phosphate-free detergent.
2. Rinse three times with cold tap water.
3. Rinse three times with distilled or deionized water.

Method B: Acid Wash Procedures

Use this method when preparing all sample containers and glassware for monitoring nitrates and phosphorus. Wearing latex gloves,

1. Wash each sample bottle or piece of glassware with a brush and phosphate-free detergent.
2. Rinse three times with cold tap water.
3. Rinse with 10% hydrochloric acid.
4. Rinse three times with deionized water.

Sample Types

Two types of samples are commonly used for water quality monitoring: *grab samples* and *composite samples*. The type of sample used depends on the specific test, the reason the sample is being collected, and the applicable regulatory requirements. Grab samples are taken all at once, at a specific time and place. They are representative only of the conditions at the time of collection. Grab samples must be used to determine pH, total residual chlorine, dissolved oxygen, and fecal coliform concentrations. Grab samples may also be used for any test that does not specifically prohibit their use.

> ***Note:*** Before collecting samples for any test procedure, it is best to review the sampling requirements of the test.

Composite samples consist of a series of individual grab samples collected over a specified period in proportion to flow. The individual grab samples are mixed together in proportion to the flow rate at the time the sample was collected to form the composite sample. This type of sample is taken to determine average conditions in a large volume of water whose properties vary significantly over the course of a day.

Collecting Samples from a Stream

In general, sample away from the stream bank in the main current. Never sample stagnant water. The outside curve of the stream is often a good place to sample, because the main current tends to hug this bank. In shallow stretches, carefully wade into the center current to collect the sample. A boat is required for deep sites. Try to maneuver the boat into the center of the main current to collect the water sample. When collecting a water sample for analysis in the field or at the lab, follow the procedures provided below.

Whirl-Pak® Bags

To collect water samples using Whirl-Pak® bags, use the following procedures:

1. Label the bag with the site number, date, and time.
2. Tear off the top of the bag along the perforation above the wire tab just before sampling. Avoid touching the inside of the bag. If you accidentally touch the inside of the bag, use another one.

3. When wading, try to disturb as little bottom sediment as possible. In any case, be careful not to collect water that contains bottom sediment. Stand facing upstream. Collect the water samples in front of you. By boat, carefully reach over the side and collect the water sample on the upstream side of the boat.
4. Hold the two white pull-tabs in each hand and lower the bag into the water on your upstream side with the opening facing upstream. Open the bag midway, between the surface and the bottom by pulling the white pull-tabs. The bag should begin to fill with water. You may need to scoop water into the bag by drawing it through the water upstream and away from you. Fill the bag no more than three-quarters full!
5. Lift the bag out of the water. Pour out excess water. Pull on the wire tabs to close the bag. Continue holding the wire tabs and flip the bag over at least four to five times quickly to seal the bag. Do not try to squeeze the air out of the top of the bag. Fold the ends of the bag, being careful not to puncture the bag. Twist them together, forming a loop.
6. Fill in the bag number and site number on the appropriate field data sheet. *This is important.* It is the only way the lab specialist will know which bag goes with which site.
7. If samples are to be analyzed in a lab, place the sample in the cooler with ice or cold packs. Take all samples to the lab.

Screw-Cap Bottles

To collect water samples using screw-cap bottles, use the following procedures (see Figure 8.12):

1. Label the bottle with the site number, date, and time.
2. Remove the cap from the bottle just before sampling. Avoid touching the inside of the bottle or the cap. If you accidentally touch the inside of the bottle, use another one.
3. When wading, try to disturb as little bottom sediment as possible. In any case, be careful not to collect water that has sediment from bottom disturbance. Stand facing upstream. Collect the water sample on your upstream side, in front of you. You may also tape the bottle to an extension pole to sample from deeper water. By boat, carefully reach over the side and collect the water sample on the upstream side of the boat.
4. Hold the bottle near its base and plunge it (opening downward) below the water surface. If you are using an extension pole, remove the cap, turn the bottle upside down, and plunge it into the water, facing upstream. Collect a water sample 8 to 12 inches beneath the surface, or midway between the surface and the bottom if the stream reach is shallow.
5. Turn the bottle underwater into the current and away from you. In slow-moving stream reaches, push the bottle underneath the surface and away from you in the upstream direction.

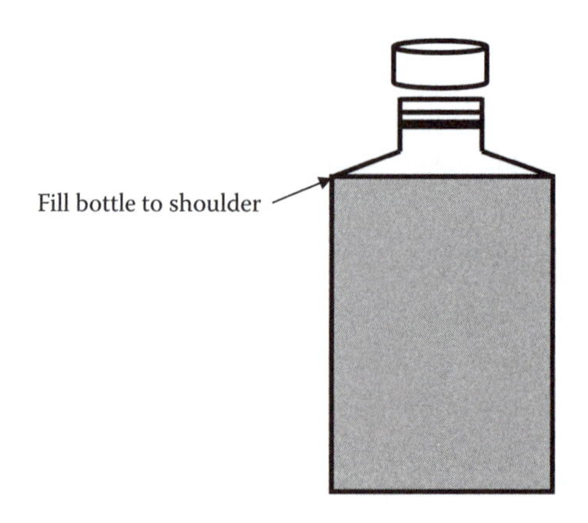

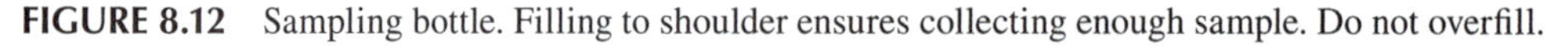

FIGURE 8.12 Sampling bottle. Filling to shoulder ensures collecting enough sample. Do not overfill.

6. Leave a 1-inch air space (except for DO and BOD samples). Do not fill the bottle completely (so the sample can be shaken just before analysis). Recap the bottle carefully, remembering not to touch the inside.
7. Fill in the bottle number and site number on the appropriate field data sheet. This is important because it tells the lab specialist which bottle goes with which site.
8. If the samples are to be analyzed in the lab, place them in the cooler for transport to the lab.

Sample Preservation and Storage

Samples can change very rapidly; however, no single preservation method will serve for all samples and constituents. If analysis must be delayed, follow the instructions for sample preservation and storage listed in *Standard Methods* (APHA, 1998) or those specified by the laboratory that will eventually process the samples (see Table 8.6). In general, handle the sample in a way that does not cause changes in biological activity, physical alterations, or chemical reactions. Cool the sample to reduce biological and chemical reactions. Store in darkness to suspend photosynthesis. Fill the sample container completely to prevent the loss of dissolved gases. Metal cations such as iron and lead and suspended particles may adsorb onto container surfaces during storage.

TABLE 8.6
Recommended Sample Storage and Preservation Techniques

Test Factor	Container Type	Preservation Recommended or Required	Maximum Storage Time (Recommended/Regulatory)
Alkalinity	Plastic Glass	Refrigerate	24 hours/14 days
BOD	Plastic Glass	Refrigerate	6 hours/48 hours
Conductivity	Plastic Glass	Refrigerate	28 days/28 days
Hardness	Plastic Glass	Lower pH to <2	6 months/6 months
Nitrate	Plastic Glass	Analyze ASAP	48 hours/48 hours
Nitrite	Plastic Glass	Analyze ASAP	None recommended/48 hours
Odor	Glass	Analyze ASAP	6 hours/none recommended
Dissolved oxygen			
Electrode	Glass	Immediately analyze	0.5 hour/none allowed
Winkler	Glass	Fix immediately	8 hours/8 hours
pH	Plastic Glass	Immediately analyze	2 hours/none allowed
Phosphate	Glass[a]	Immediately refrigerate	48 hours/none recommended
Salinity	Glass Wax seal	Immediately analyze or use wax seal	6 months/none recommended
Temperature	Plastic Glass	Immediately analyze	None allowed/none allowed
Turbidity	Plastic Glass	Analyze same day or store in dark up to 24 hours; refrigerate	24 hours/48 hours

Source: Spellman, F.R., *Spellman's Standard Handbook for Wastewater Operators*, Vol. 1, CRC Press, Boca Raton, FL, 1999.

[a] Glass rinsed with 1:1 HNO_3.

Standardization of Methods

References used for sampling and testing must correspond to those listed in the current federal regulations. For the majority of tests, to compare the results of either different water quality monitors or the same monitors over the course of time requires some form of standardization of the methods. The American Public Health Association (APHA) recognized this requirement when, in 1899, it appointed a committee to draw up standard procedures for the analysis of water. The report (published in 1905) constituted the first edition of what is now known as *Standard Methods for the Examination of Water and Wastewater* or *Standard Methods*. This book is now in its 20th edition (APHA, 1998) and serves as the primary reference for water testing methods, and as the basis for most USEPA-approved methods.

TEST METHODS

The material presented in this section is based on personal experience and adaptations from *Standard Methods* (APHA, 1998), *Federal Register*, and *The Monitor's Handbook* (LaMotte, 1992). Descriptions of general methods to help you understand how each works in specific test kits follow. Always use the specific instructions included with the equipment and individual test kits. Most water analyses are conducted by either titrimetric analyses or colorimetric analyses. Both methods are simple to use and provide accurate results.

Titrimetric Methods

Titrimetric analyses are based on adding a solution of known strength (the *titrant*, which must have an exact known concentration) to a specific volume of a treated sample in the presence of an indicator. The indicator produces a color change indicating that the reaction is complete. Titrants are generally added by a titrator (microburet) or a precise glass pipet.

Colorimetric Methods

Colorimetric standards are prepared as a series of solutions with increasing known concentrations of the constituent to be analyzed. Two basic types of colorimetric tests are commonly used:

1. The pH is a measure of the concentration of hydrogen ions (the acidity of a solution) determined by the reaction of an indicator that varies in color, depending on the hydrogen ion levels in the water.
2. Tests that determine a concentration of an element or compound are based on Beer's law. Simply, this law states that the higher the concentration of a substance, the darker the color produced in the test reaction and therefore the more light absorbed. Assuming a constant viewpath, the absorption increases exponentially with concentration.

Visual Methods

The Octet Comparator uses standards that are mounted in a plastic comparator block. It employs eight permanent translucent color standards and built-in filters to eliminate optical distortion. The sample is compared using either of two viewing windows. Two devices that can be used with the comparator are a B color reader, which neutralizes color or turbidity in water samples, and an axial mirror, which intensifies faint colors of low concentrations for easy distinction.

Electronic Methods

Although the human eye is capable of differentiating color intensity, interpretation is quite subjective. Electronic colorimeters consist of a light source that passes through a sample and is measured on a photodetector with an analog or digital readout. Besides electronic colorimeters, specific electronic instruments are manufactured for lab and field determination of many water quality factors, including pH, total dissolved solids, conductivity, dissolved oxygen, temperature, and turbidity.

Dissolved Oxygen Testing

A stream system used as a source of water produces and consumes oxygen. It gains oxygen from the atmosphere and from plants through photosynthesis. Because of the churning of running water, it dissolves more oxygen than does still water, such as in a reservoir behind a dam. Respiration by aquatic animals, decomposition, and various chemical reactions consume oxygen.

Oxygen is actually poorly soluble in water. Its solubility is related to pressure and temperature. In water supply systems, dissolved oxygen (DO) in raw water is considered the necessary element to support life of many aquatic organisms. From the drinking water practitioner's point of view, DO is an important indicator of the water treatment process and an important factor in corrosivity.

Oxygen is measured in its dissolved form as dissolved oxygen. If more oxygen is consumed than produced, DO levels decline and some sensitive animals may move away, weaken, or die. DO levels fluctuate over a 24-hour period and seasonally. They vary with water temperature and altitude. Cold water holds more oxygen than warm water (see Table 8.7), and water holds less oxygen at higher altitudes. Thermal discharges (such as water used to cool machinery in a manufacturing plant or a power plant) raise the temperature of water and lower its oxygen content. Aquatic animals are most vulnerable to lowered DO levels in the early morning on hot summer days when stream flows are low, water temperatures are high, and aquatic plants have not been producing oxygen since sunset.

Sampling and Equipment Considerations

In contrast to lakes, where DO levels are most likely to vary vertically in the water column, changes in DO in rivers and streams move horizontally along the course of the waterway. This is especially true in smaller, shallow streams. In larger, deeper rivers, some vertical stratification of dissolved oxygen might occur. The DO levels in and below riffle areas, waterfalls, or dam spillways are typically higher than those in pools and slower moving stretches. If we want to measure the effect of a dam, sampling for DO behind the dam, immediately below the spillway, and upstream of the dam would be important. Because DO levels are critical to fish, a good place to sample is in the pools that fish tend to favor or in the spawning areas they use.

An hourly time profile of DO levels at a sampling site is a valuable set of data, because it shows the change in DO levels from the low point (just before sunrise) to the high point (sometime near midday); however, this might not be practical for a volunteer monitoring program. Note the time of the DO sampling to help judge when in the daily cycle the data were collected.

Dissolved oxygen is measured either as the amount of oxygen in a liter of water (milligrams per liter, or mg/L) or as percent saturation, which is the amount of oxygen in a liter of water relative to the total amount of oxygen that the water can hold at that temperature. DO samples are collected using a special BOD bottle: a glass bottle with a "turtleneck" and a ground stopper. You can fill the bottle directly in the stream if it is possible to wade into the stream, or you can use a sampler dropped from a bridge or boat into water deep enough to submerge the sampler. Samplers can be made or purchased. Dissolved oxygen is determined using some variation of the Winkler method or by using a meter and probe.

TABLE 8.7
Maximum DO Concentrations vs. Temperature Variations

Temperature (°C)	DO (mg/L)	Temperature (°C)	DO (mg/L)
0	14.60	23	8.56
1	14.19	24	8.40
2	13.81	25	8.24
3	13.44	26	8.09
4	13.09	27	7.95
5	12.75	28	7.81
6	12.43	29	7.67
7	12.12	30	7.54
8	11.83	31	7.41
9	11.55	32	7.28
10	11.27	33	7.16
11	11.01	34	7.05
12	10.76	35	6.93
13	10.52	36	6.82
14	10.29	37	6.71
15	10.07	38	6.61
16	9.85	39	6.51
17	9.65	40	6.41
18	9.45	41	6.31
19	9.26	42	6.22
20	9.07	43	6.13
21	8.90	44	6.04
22	8.72	45	5.95

Source: Spellman, F.R., *Spellman's Standard Handbook for Wastewater Operators*, Vol. 1, CRC Press, Boca Raton, FL, 1999.

Winkler Method (Azide Modification)

The Winkler method (azide modification) involves filling a sample bottle completely with water (no air is left to bias the test). The dissolved oxygen is then fixed using a series of reagents that form a titrated acid compound. Titration involves the drop-by-drop addition of a reagent that neutralizes the acid compound, causing a change in the color of the solution. The point at which the color changes is the *endpoint* and is equivalent to the amount of oxygen dissolved in the sample. The sample is usually fixed and titrated in the field at the sample site, but preparing the sample in the field and delivering it to a lab for titration is possible. The azide modification method is best suited for relatively clean waters; otherwise, substances such as color, organics, suspended solids, sulfide, chlorine, and ferrous and ferric iron can interfere with test results. If fresh azide is used, nitrite will not interfere with the test. In testing, iodine is released in proportion to the amount of DO present in the sample. By using sodium thiosulfate with starch as the indicator, the sample can be titrated to determine the amount of DO present.

The chemicals used include the following:

1. Manganese sulfate solution
2. Alkaline azide–iodide solution
3. Sulfuric acid (concentrated)

4. Starch indicator
5. Sodium thiosulfate solution (0.025 *N*), or phenylarsine solution (0.025 *N*), or potassium biniodate solution (0.025 *N*)
6. Distilled or deionized water

The equipment used includes the following:

1. Buret, graduated to 0.1 mL
2. Buret stand
3. 300-mL BOD bottles
4. 500-mL Erlenmeyer flasks
5. 1.0-mL pipets with elongated tips
6. Pipet bulb
7. 250-mL graduated cylinder
8. Laboratory-grade water rinse bottle
9. Magnetic stirrer and stir bars (optional)

The procedure is as follows:

1. Collect sample in a 300-mL BOD bottle.
2. Add 1 mL manganous sulfate solution at the surface of the liquid.
3. Add 1 mL alkaline iodide–azide solution at the surface of the liquid.
4. Stopper bottle, and mix by inverting the bottle.
5. Allow the floc to settle halfway in the bottle, remix, and allow to settle again.
6. Add 1 mL concentrated sulfuric acid at the surface of the liquid.
7. Restopper bottle, rinse top with laboratory-grade water, and mix until precipitate is dissolved (the liquid in the bottle should appear clear and have an amber color).
8. Measure 201 mL from the BOD bottle into an Erlenmeyer flask.
9. Titrate with 0.025-*N* PAO or thiosulfate to a pale yellow color, and note the amount of titrant.
10. Add 1 mL of starch indicator solution.
11. Titrate until blue color first disappears.
12. Record the total amount of titrant.

Use Equation 8.1 to calculate the DO concentration:

$$\text{DO (mg/L)} = \frac{[\text{Buret}_{\text{Final}}\ (\text{mL}) - \text{Buret}_{\text{Start}}\ (\text{mL})] \times N \times 8000}{\text{Sample volume (mL)}} \tag{8.1}$$

Note: Using a 200-mL sample and a 0.025-*N* (*N* = normality of the solution used to titrate the sample) titrant reduces this calculation to

$$\text{DO (mg/L)} = \text{Titrant used (mL)}$$

■ Example 8.1

Problem: The operator titrates a 200-mL DO sample. The buret reading at the start of the titration was 0.0 mL. At the end of the titration, the buret read 7.1 mL. The concentration of the titrating solution was 0.025 *N*. What is the DO concentration in mg/L?

Solution:

$$DO = \frac{(7.1\ \text{mL} - 0.0\ \text{mL}) \times 0.025 \times 8000}{200\ \text{mL}} = 7.1\ \text{mg/L}$$

Dissolved oxygen field kits using the Winkler method are relatively inexpensive, especially compared to a meter and probe. Field kits run between $35 and $200, and each kit comes with enough reagents to run 50 to 100 DO tests. Replacement reagents are inexpensive and can be purchased already measured out for each test in plastic pillows. Reagents can also be purchased in larger quantities in bottles and measured out with a volumetric scoop. The advantage of the pillows is that they have a longer shelf life and are much less prone to contamination or spillage. Buying larger quantities in bottles has the advantage of considerably lower cost per test.

The major factor in the expense for the kits is the method of titration used. Eyedropper or syringe-type titration is less precise than digital titration, because a larger drop of titrant is allowed to pass through the dropper opening; on a microscale, the drop size (and thus volume of titrant) can vary from drop to drop. A digital titrator or a buret (a long glass tube with a tapered tip like a pipet) permits much more precision and uniformity with regard to the titrant it allows to pass.

If a high degree of accuracy and precision in DO results is required, a digital titrator should be used. A kit that uses an eyedropper-type or syringe-type titrator is suitable for most other purposes. The lower cost of this type of DO field kit might be attractive if several teams of samplers and testers at multiple sites at the same time are required.

Meter and Probe

A *dissolved oxygen meter* is an electronic device that converts signals from a probe placed in the water into units of DO in milligrams per liter. Most meters and probes also measure temperature. The probe is filled with a salt solution and has a selectively permeable membrane that allows DO to pass from the stream water into the salt solution. The DO that has diffused into the salt solution changes the electric potential of the salt solution, and this change is sent by electric cable to the meter, which converts the signal to milligrams per liter on a scale that the user can read.

If samples are to be collected for analysis in the laboratory, a special APHA sampler or the equivalent must be used. If the sample is exposed to or mixed with air during collection, test results can change dramatically; therefore, the sampling device must allow collection of a sample that is not mixed with atmospheric air and allows for at least 3× bottle overflow (see Figure 8.12). Again, because the DO level in a sample can change quickly, only grab samples should be used for dissolved oxygen testing. Samples must be tested immediately (within 15 minutes) after collection.

> ***Note:*** Samples collected for analysis using the modified Winkler titration method may be preserved for up to 8 hours by adding 0.7 mL of concentrated sulfuric acid or by adding all the chemicals required by the procedure. Samples collected from the aeration tank of the activated sludge process must be preserved using a solution of copper sulfate–sulfamic acid to inhibit biological activity.

The advantage of using the DO oxygen meter method is that the meter can be used to determine DO concentration directly (see Figure 8.13). In the field, a direct reading can be obtained using a probe (see Figure 8.14) or by collecting samples for testing in the laboratory using a laboratory probe (see Figure 8.15).

> ***Note:*** The field probe can be used for laboratory work by placing a stirrer in the bottom of the sample bottle, but the laboratory probe should never be used in any situation where the entire probe might be submerged.

The probe used in the determination of DO consists of two electrodes, a membrane, and a membrane filling solution. Oxygen passes through the membrane into the filling solution and causes a change in the electrical current passing between the two electrodes. The change is measured and

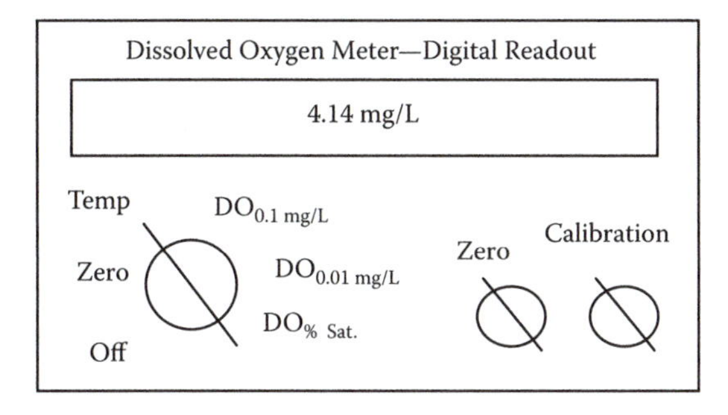

FIGURE 8.13 Dissolved oxygen meter.

displayed as the concentration of DO. For accuracy, the probe membrane must be in proper operating condition, and the meter must be calibrated before use. The only chemical used in the DO meter method during normal operation is the membrane filling solution, whereas in the Winkler DO method chemicals are required for meter calibration.

Calibration prior to use is important. Both the meter and the probe must be calibrated to ensure accurate results. The frequency of calibration is dependent on the frequency of use; for example, if the meter is used once a day, then calibration should be performed before use. Three methods are available for calibration: *saturated water*, *saturated air*, and the *Winkler method*. It is important to note that if the Winkler method is not used for routine calibration, periodic checks using this method are recommended.

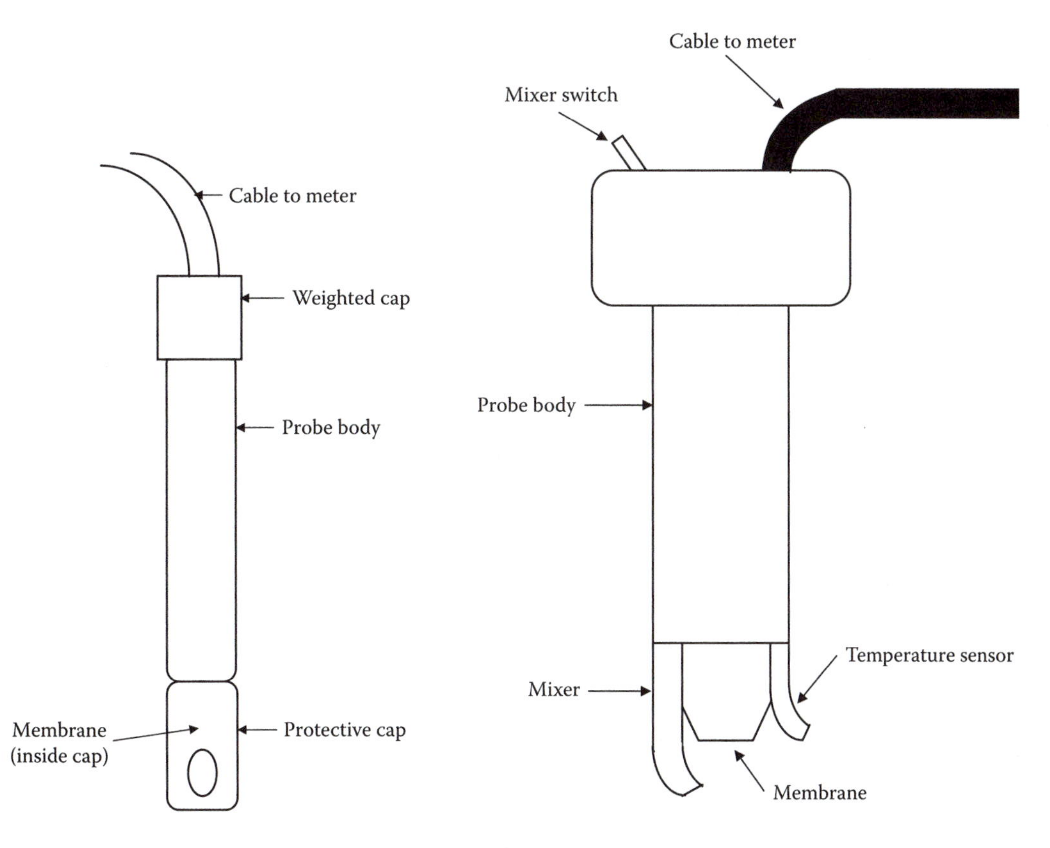

FIGURE 8.14 Dissolved oxygen-field probe.

FIGURE 8.15 Dissolved oxygen-lab probe.

It is important to keep in mind that the operating procedures of the meter and probe supplier should always be followed. Normally, the manufacturer's recommended procedure will include the following generalized steps:

1. Turn DO meter on, and allow 15 minutes for it to warm up.
2. Turn meter switch to 0, and adjust as needed.
3. Calibrate meter using the saturated air, saturated water, or Winkler azide procedure for calibration.
4. Collect sample in 300-mL bottle, or place field electrode directly in stream.
5. Place laboratory electrode in BOD bottle without trapping air against membrane, and turn on stirrer.
6. Turn meter switch to temperature mode, and measure the temperature.
7. Turn meter switch to DO mode, and allow 10 seconds for the meter reading to stabilize.
8. Read DO mg/L from meter, and record the results.

No calculation is necessary using this method because results are read directly from the meter.

Dissolved oxygen meters are expensive compared to field kits that use the titration method. Meter and probe combinations run between $500 and $1200, including a long cable to connect the probe to the meter. The advantage of a meter and probe is that DO and temperature can be quickly read at any point where the probe is inserted into the stream. DO levels can be measured at a certain point on a continuous basis. The results are read directly as milligrams per liter, unlike the titration methods, in which the final titration result might have to be converted to milligrams per liter. DO meters, however, are more fragile than field kits, and repairs to a damaged meter can be costly. The meter and probe must be carefully maintained and must be calibrated before each sample run and, if many tests are done, between sampling. Because of the expense, a small water/wastewater facility might have only one meter and probe, which means that only one team of samplers can sample DO, and they must test all the sites. With field kits, on the other hand, several teams can sample simultaneously.

Biochemical Oxygen Demand Testing

Biochemical oxygen demand measures the amount of oxygen consumed by microorganisms in decomposing organic matter in stream water. BOD also measures the chemical oxidation of inorganic matter (the extraction of oxygen from water via chemical reaction). A test is used to measure the amount of oxygen consumed by these organisms during a specified period of time (usually 5 days at 20°C). The rate of oxygen consumption in a stream is affected by a number of variables: temperature, pH, the presence of certain kinds of microorganisms, and the type of organic and inorganic material in the water. BOD directly affects the amount of dissolved oxygen in water bodies. The greater the BOD, the more rapidly oxygen is depleted in the water body, leaving less oxygen available to higher forms of aquatic life. The consequences of high BOD are the same as those for low dissolved oxygen: aquatic organisms become stressed, suffocate, and die. Most river waters used as water supplies have a BOD less than 7 mg/L; therefore, dilution is not necessary.

Sources of BOD include leaves and wood debris; dead plants and animals; animal manure; effluents from pulp and paper mills, wastewater treatment plants, feedlots, and food-processing plants; failing septic systems; and urban stormwater runoff.

Note: To evaluate the potential for raw water to be used as a drinking water supply, it is usually sampled, analyzed, and tested for biochemical oxygen demand when turbid, polluted water is the only source available.

Sampling Considerations

Biochemical oxygen demand is affected by the same factors that affect dissolved oxygen. Aeration of stream water—by rapids and waterfalls, for example—will accelerate the decomposition of organic and inorganic material. BOD levels at a sampling site with slower, deeper waters might be higher for a given column of organic and inorganic material than the levels for a similar site in highly aerated waters. Chlorine can also affect BOD measurement by inhibiting or killing the microorganisms that decompose the organic and inorganic matter in a sample. When sampling in chlorinated waters (such as those below the effluent from a sewage treatment plant), it is necessary to neutralize the chlorine with sodium thiosulfate (APHA, 1998).

Biochemical oxygen demand measurement requires taking two samples at each site. One is tested immediately for dissolved oxygen, and the second is incubated in the dark at 20°C for 5 days and then tested for dissolved oxygen remaining. The difference in oxygen levels between the first test and the second test (in milligrams per liter) is the amount of BOD. This represents the amount of oxygen consumed by microorganisms and used to break down the organic matter present in the sample bottle during the incubation period. Because of the 5-day incubation, the tests are conducted in a laboratory.

Sometimes by the end of the 5-day incubation period, the dissolved oxygen level is 0. This is especially true for rivers and streams with a lot of organic pollution. It is not possible to know when the 0 point was reached, so determining the BOD level is also impossible. In this case, diluting the original sample by a factor that results in a final dissolved oxygen level of at least 2 mg/L is necessary. Special dilution water should be used for the dilutions (APHA, 1998).

Some experimentation is necessary to determine the appropriate dilution factor for a particular sampling site. The result is the difference in dissolved oxygen between the first measurement and the second, after multiplying the second result by the dilution factor. *Standard Methods* (APHA, 1998) prescribes all phases of procedures and calculations for BOD determination. A BOD test is not required for monitoring water supplies.

BOD Sampling, Analysis, and Testing

The approved biochemical oxygen demand sampling and analysis procedure measures the DO depletion (biological oxidation of organic matter in the sample) over a 5-day period under controlled conditions (20°C in the dark). The test is performed using a specified incubation time and temperature. Test results are used to determine plant loadings, plant efficiency, and compliance with NPDES effluent limitations. The duration of the test (5 days) makes it difficult to use the data effectively for process control.

The standard BOD test does not differentiate between oxygen used to oxidize organic matter and that used to oxidize organic and ammonia nitrogen to more stable forms. Because many biological treatment plants now control treatment processes to achieve oxidation of the nitrogen compounds, it is possible that BOD testing for plant effluent and some process samples may produce BOD test results based on both carbon and nitrogen oxidation. To avoid this situation, a nitrification inhibitor can be added. When this is done, the tests measure *carbonaceous BOD* (CBOD). A second uninhibited BOD should also be run whenever CBOD is determined.

When taking a BOD sample, no special sampling container is required. Either a grab or composite sample can be used. BOD_5 samples can be preserved by refrigeration at or below 4°C (not frozen); composite samples must be refrigerated during collection. Maximum holding time for preserved samples is 48 hours.

Using the incubation of dissolved approved test method, a sample is mixed with dilution water in several different concentrations (dilutions). The dilution water contains nutrients and materials to provide optimum environment. Chemicals used include dissolved oxygen, ferric chloride, magnesium sulfate, calcium chloride, phosphate buffer, and ammonium chloride.

TABLE 8.8
BOD_5 Test Procedure

1.	Fill two bottles with BOD dilution water; insert stoppers.
2.	Place sample in two BOD bottles; fill with dilution water; insert stoppers.
3.	Test for dissolved oxygen (DO).
4.	Incubate for 5 days.
5.	Test for DO.
6.	Add 1 mL $MnSO_4$ below surface.
7.	Add 1 mL alkaline KI below surface.
8.	Add 1 mL H_2SO_4.
9.	Transfer 203 mL to flask.
10.	Titrate with PAO or thiosulfate.

Note: Remember that all chemicals can be dangerous if not used properly and in accordance with the recommended procedures. Review appropriate sections of the individual chemical Material Safety Data Sheet (MSDS) to determine proper methods for handling and for safety precautions that should be taken.

Sometimes it is necessary to add healthy organisms to the sample—that is, to seed the sample. The dissolved oxygen levels of the dilution and the dilution water are determined. If seed material is used, a series of dilutions of seed material must also be prepared. The dilutions and dilution blanks are incubated in the dark for 5 days at 20 ± 1°C. At the end of 5 days, the DO levels of each dilution and the dilution blanks are determined. For the test results to be valid, certain criteria must be met:

1. Dilution water blank DO change must be ≤0.2 mg/L.
2. Initial DO must be >7.0 mg/L but ≤9.0 mg/L (or saturation at 20°C and test elevation).
3. Sample dilution DO depletion must be ≥2.0 mg/L.
4. Sample dilution residual DO must be ≥1.0 mg/L.
5. Sample dilution initial DO must be ≥7.0 mg/L.
6. Seed correction should be ≥0.6 but ≤1.0 mg/L.

The BOD_5 test procedure consists of 10 steps (for unchlorinated water) as shown in Table 8.8.

Note: BOD_5 is calculated individually for all sample dilutions that meet the criteria. Reported result is the average of the BOD_5 of each valid sample dilution.

BOD_5 Calculation (Unseeded)

Unlike the direct reading instrument used in the DO analysis, BOD results require calculation. There are several criteria used in selecting which BOD_5 dilutions should be used for calculating test results. Consult a laboratory testing reference manual, such as *Standard Methods* (APHA, 1998), for this information.

Currently, two basic calculations are used for BOD_5. The first is used for samples that have not been seeded. The second must be used whenever BOD_5 samples are seeded. In this section, we illustrate the calculation procedure for unseeded samples:

$$\text{BOD}_5 \text{ (unseeded)} = \frac{[\text{DO}_{\text{Start}} \text{ (mg/L)} - \text{DO}_{\text{Final}} \text{ (mg/L)}] \times 300 \text{ mL}}{\text{Sample volume (mL)}} \tag{8.2}$$

■ **EXAMPLE 8.2**

Problem: The BOD_5 test is completed. Bottle 1 of the test had a DO of 7.1 mg/L at the start of the test. After 5 days, bottle 1 had a DO of 2.9 mg/L. Bottle 1 contained 120 mg/L of sample.

Solution:

$$\text{BOD}_5 \text{ (unseeded)} = \frac{(7.1 \text{ mg/L} - 2.9 \text{ mg/L}) \times 300 \text{ mL}}{120 \text{ mL}} = 10.5 \text{ mg/L}$$

BOD_5 Calculation (Seeded)

If the BOD_5 sample has been exposed to conditions that could reduce the number of healthy, active organisms, the sample must be seeded with organisms. Seeding requires use of a correction factor to remove the BOD_5 contribution of the seed material:

$$\text{Seed correction} = \frac{\text{Seed material BOD}_5 \times \text{Seed in dilution (mL)}}{300 \text{ mL}} \tag{8.3}$$

$$\text{BOD}_5 \text{ (seeded)} = \frac{\left[(\text{DO}_{\text{Start}} \text{ (mg/L)} - \text{DO}_{\text{Final}} \text{ (mg/L)}) - \text{Seed correction}\right] \times 300 \text{ mL}}{\text{Sample volume (mL)}} \tag{8.4}$$

■ **EXAMPLE 8.3**

Problem: Using the data provided below, determine the BOD_5:

BOD_5 of seed material = 90 mg/L
Dilution #1:
Seed material = 3 mL
Sample = 100 mL
Start DO = 7.6 mg/L
Final DO = 2.7 mg/L

Solution:

$$\text{Seed correction} = \frac{90 \text{ mg/L} \times 3 \text{ mL}}{300 \text{ mL}} = 0.90 \text{ mg/L}$$

$$\text{BOD}_5 = \frac{[(7.6 \text{ mg/L} - 2.7 \text{ mg/L}) - 0.90] \times 300 \text{ mL}}{100 \text{ mL}} = 12 \text{ mg/L}$$

TEMPERATURE MEASUREMENT

An ideal water supply should have, at all times, an almost constant temperature or one with minimum variation. Knowing the temperature of the water supply is important because the rates of biological and chemical processes depend on it. Temperature affects the oxygen content of the water (oxygen levels become lower as temperature increases), the rate of photosynthesis by aquatic plants, the metabolic rates of aquatic organisms, and the sensitivity of organisms to toxic wastes, parasites, and diseases. Causes of temperature change include weather, removal of shading stream-bank vegetation, impoundments (a body of water confined by a barrier, such as a dam), and discharge of cooling water, urban stormwater, and groundwater inflows to the stream.

Sampling and Equipment Considerations

Temperature—for example, in a stream—varies with width and depth, and the temperature of well-sunned portions of a stream can be significantly higher than the shaded portion of the water on a sunny day. In a small stream, the temperature will be relatively constant as long as the stream is uniformly in sun or shade. In a large stream, temperature can vary considerably with width and depth, regardless of shade. If it is safe to do so, temperature measurements should be collected at varying depths and across the surface of the stream to obtain vertical and horizontal temperature profiles. This can be done at each site at least once to determine the necessity of collecting a profile during each sampling visit. Temperature should be measured at the same place every time.

Temperature is measured in the stream with a thermometer or a meter. Alcohol-filled thermometers are preferred over mercury-filled because they are less hazardous if broken. Armored thermometers for field use can withstand more abuse than unprotected glass thermometers and are worth the additional expense. Meters for other tests, such as pH (acidity) or dissolved oxygen, also measure temperature and can be used instead of a thermometer.

Hardness Measurement

Hardness refers primarily to the amount of calcium and magnesium in the water. Calcium and magnesium enter water mainly by leaching of rocks. Calcium is an important component of aquatic plant cell walls and the shells and bones of many aquatic organisms. Magnesium is an essential nutrient for plants and is a component of the chlorophyll molecule. Hardness test kits express test results in ppm of $CaCO_3$, but these results can be converted directly to calcium or magnesium concentrations:

$$\text{Calcium hardness as ppm } CaCO_3 \times 0.40 = \text{ppm Ca} \tag{8.5}$$

$$\text{Magnesium hardness as ppm } CaCO_3 \times 0.24 = \text{ppm Mg} \tag{8.6}$$

Note: Because of less contact with soil minerals and more contact with rain, surface raw water is usually softer than groundwater.

As a rule, when hardness is greater than 150 mg/L, softening treatment may be required for public water systems. Hardness determination via testing is required to ensure efficiency of treatment.

In the hardness test, the sample must be carefully measured, and then a buffer is added to the sample to correct pH for the test and an indicator to signal the titration endpoint. The indicator reagent is normally blue in a sample of pure water, but if calcium or magnesium ions are present in the sample the indicator combines with them to form a red complex. The titrant in this test is ethylenediaminetetraacetic acid (EDTA), used with its salts in the titration method; it is a *chelant* that pulls the calcium and magnesium ions away from the red-colored complex. The EDTA is added dropwise to the sample until all the calcium and magnesium ions have been chelated away from the complex and the indicator returns to its normal blue color. The amount of EDTA required to cause the color change is a direct indication of the amount of calcium and magnesium ions in the sample.

Some hardness kits include an additional indicator that is specific for calcium. This type of kit will provide three readings: total hardness, calcium hardness, and magnesium hardness. For more information, consult the latest edition of *Standard Methods* (APHA, 1998).

pH Measurement

pH is defined as the negative log of the hydrogen ion concentration of the solution. This is a measure of the ionized hydrogen in solution. Simply, it is the relative acidity or basicity of the solution. The chemical and physical properties and the reactivity of almost every component in water are dependent on pH. It relates to corrosivity, contaminant solubility, and conductance of the water and has a secondary maximum contaminant level (MCL) range set at 6.5 to 8.5.

Analytical and Equipment Considerations

The pH can be analyzed in the field or in the lab. In the lab, it must be measured within 2 hours of the sample collection, because the pH will change due to carbon dioxide in the air dissolving in the water, moving the pH toward 7. If your program requires a high degree of accuracy and precision in pH results, the pH should be measured with a laboratory-quality pH meter and electrode. Meters of this quality range in cost from around $250 to $1000. Color comparators and pH "pocket pals" are suitable for most other purposes. The cost of either of these is in the $50 range. The lower cost of the alternatives might be attractive if multiple samplers are used to sample several sites at the same time.

pH Meters

A pH meter measures the electric potential (millivolts) across an electrode when immersed in water. This electric potential is a function of the hydrogen ion activity in the sample; therefore, pH meters can display results in either millivolts (mV) or pH units. A pH meter consists of a *potentiometer*, which measures electric potential where it meets the water sample; a reference electrode, which provides a constant electric potential; and a *temperature-compensating device,* which adjusts the readings according to the temperature of the sample (because pH varies with temperature). The reference and glass electrodes are frequently combined into a single probe called a *combination electrode.* A wide variety of meters is available, but the most important part of the pH meter is the electrode; thus, purchasing a good, reliable electrode and following the manufacturer's instructions for proper maintenance are important. Infrequently used or improperly maintained electrodes are subject to corrosion, which makes them highly inaccurate.

pH Pocket Pals and Color Comparators

pH pocket pals are electronic handheld "pens" that are dipped in the water to obtain a digital readout of the pH. They can be calibrated to only one pH buffer. (Lab meters, on the other hand, can be calibrated to two or more buffer solutions and thus are more accurate over a wide range of pH measurements.) Color comparators involve adding a reagent to the sample that colors the sample water. The intensity of the color is proportional to the pH of the sample and is matched against a standard color chart. The color chart equates particular colors to associated pH values, which can be determined by matching the colors from the chart to the color of the sample. For instructions on how to collect and analyze samples, refer to *Standard Methods* (APHA, 1998).

Turbidity Measurement

Turbidity is a measure of water clarity—how much the material suspended in water decreases the passage of light through the water. Turbidity consists of suspended particles in the water and may be caused by a number of materials, organic and inorganic. These particles are typically in the size range of 0.004 mm (clay) to 1.0 mm (sand). The occurrence of turbid source waters may be permanent or temporary. It can affect the color of the water. Higher turbidity increases water temperatures, because suspended particles absorb more heat. This in turn reduces the concentration of dissolved oxygen because warm water holds less DO than cold. Higher turbidity also reduces the amount of light penetrating the water, which reduces photosynthesis and the production of DO. Suspended materials can clog fish gills, reducing resistance to disease in fish, lowering growth rates, and affecting egg and larval development. As the particles settle, they can blanket the stream bottom (especially in slower waters) and smother fish eggs and benthic macroinvertebrates.

Turbidity also affects treatment plant operations; for example, turbidity hinders disinfection by shielding microbes, some of them pathogens, from the disinfectant. Obviously, this is the most important significance of turbidity monitoring; testing for turbidity provides an indication of the effectiveness of filtration of water supplies. It is important to note that turbidity removal is the principal reason for chemical addition, settling, coagulation, settling, and filtration in potable water treatment. Sources of turbidity include the following:

1. Soil erosion
2. Waste discharge
3. Urban runoff
4. Eroding stream banks
5. Large numbers of bottom feeders (such as carp), which stir up bottom sediments
6. Excessive algal growth

Sampling and Equipment Considerations

Turbidity can be useful as an indicator of the effects of runoff from construction, agricultural practices, logging activity, discharges, and other sources. Turbidity often increases sharply during rainfall, especially in developed watersheds, which typically have relatively high proportions of impervious surfaces. The flow of stormwater runoff from impervious surfaces rapidly increases stream velocity, which increases the erosion rates of stream banks and channels. Turbidity can also rise sharply during dry weather if earth-disturbing activities occur in or near a stream without erosion control practices in place.

Regular monitoring of turbidity can help detect trends that might indicate increasing erosion in developing watersheds; however, turbidity is closely related to stream flow and velocity and should be correlated with these factors. Comparisons of the change in turbidity over time, therefore, should be made at the same point at the same flow. Keep in mind that turbidity is not a measurement of the amount of suspended solids present or the rate of sedimentation of a stream because it measures only the amount of light that is scattered by suspended particles. Measurement of total solids is a more direct measurement of the amount of material suspended and dissolved in water.

Turbidity is generally measured by using a turbidity meter or *turbidimeter.* The turbidimeter is a modern *nephelometer.* Early nephelometers were comprised of a box containing a light bulb that directed light at a sample. The amount of light scattered at right angles by the turbidity particles was measured and registered as nephelometric turbidity units (NTUs). Today's turbidimeters use a photoelectric cell to register the scattered light on an analog or digital scale, and the instrument is calibrated with permanent turbidity standards composed of the colloidal substance formazin. Meters can measure turbidity over a wide range—from 0 to 1000 NTUs. A clear mountain stream might have a turbidity of around 1 NTU, whereas a large river such as the Mississippi might have a dry-weather turbidity of 10 NTUs. Because these values can jump into hundreds of NTUs during runoff events, the turbidity meter should be reliable over the range in which you will be working. Meters of this quality cost about $800. Many meters in this price range are designed for field or lab use.

An operator may also take samples to a lab for analysis. Another approach, discussed previously, is to measure transparency (an integrated measure of light scattering and absorption) instead of turbidity. Water clarity and transparency can be measured using a Secchi disk or transparency tube. The Secchi disk can only be used in deep, slow-moving rivers; the transparency tube is gaining acceptance but is not yet in wide use.

Using a Secchi Disk

A Secchi disk is a black and white disk that is lowered by hand into the water to the depth at which it vanishes from sight (see Figure 8.11). The distance to vanishing is then recorded—the clearer the water, the greater the distance. Secchi disks are simple to use and inexpensive. For river monitoring, they have limited use, because in most cases the river bottom will be visible and the disk will not reach a vanishing point. Deeper, slower moving rivers are the most appropriate places for Secchi disk measurement, although the current might require that the disk be weighted so it does not sway and make measurement difficult. Secchi disks cost about $50 but can be homemade.

The line attached to the Secchi disk must be marked in waterproof ink according to units designated by the sampling program. Many programs require samplers to measure to the nearest 1/10 meter. Meter intervals can be tagged (e.g., with duct tape) for ease of use. To measure water clarity with a Secchi disk:

1. Check to make sure that the Secchi disk is securely attached to the measured line.
2. Lean over the side of the boat and lower the Secchi disk into the water, keeping your back to the sun to block glare.
3. Lower the disk until it disappears from view. Lower it one third of a meter and then slowly raise the disk until it just reappears. Move the disk up and down until you find the exact vanishing point.
4. Attach a clothespin to the line at the point where the line enters the water. Record the measurement on your data sheet. Repeating the measurement provides you with a quality control check.

The key to consistent results is to train samplers to follow standard sampling procedures and, if possible, to have the same individual take the reading at the same site throughout the season.

Transparency Tube

Pioneered by Australia's Department of Conservation, the transparency tube is a clear, narrow plastic tube marked in units with a dark pattern painted on the bottom. Water is poured into the tube until the pattern painted on the bottom disappears. Some U.S. volunteer monitoring programs, such as the Tennessee Valley Authority (TVA) Clean Water Initiative and the Minnesota Pollution Control Agency (MPCA), are testing the transparency tube in streams and rivers. MPCA uses tubes marked in centimeters and has found that the tube readings relate fairly well to lab measurements of turbidity and total suspended solids, although it does not recommend the transparency tube for applications where precise and accurate measurement is required or in highly colored waters. The TVA and MPCA suggest the following sampling techniques:

1. Collect the sample in a bottle or bucket in midstream and at mid-depth if possible. Avoid stagnant water and sample as far from the shoreline as is safe. Avoid collecting sediment from the bottom of the stream.
2. Face upstream as you fill the bottle or bucket.
3. Take readings in open but shaded conditions. Avoid direct sunlight by turning your back to the sun.
4. Carefully stir or swish the water in the bucket or bottle until it is homogeneous, taking care not to produce air bubbles (these scatter light and affect the measurement). Pour the water slowly in the tube while looking down the tube. Measure the depth of the water column in the tube at the point where the symbol just disappears.

Orthophosphate Measurement

Earlier we discussed the nutrients phosphorus and nitrogen. Both phosphorus and nitrogen are essential nutrients for the plants and animals that make up the aquatic food web. Because phosphorus is the nutrient in short supply in most freshwater systems, even a modest increase in phosphorus can (under the right conditions) set off a whole chain of undesirable events in a stream, including accelerated plant growth, algal blooms, low dissolved oxygen, and the death of certain fish, invertebrates, and other aquatic animals. Phosphorus comes from many sources, both natural and human. These include soil and rocks, wastewater treatment plants, runoff from fertilized lawns and cropland, failing septic systems, runoff from animal manure storage areas, disturbed land areas, drained wetlands, water treatment, and commercial cleaning preparations.

Forms of Phosphorus

Phosphorus has a complicated story. Pure, elemental phosphorus (P) is rare. In nature, phosphorus usually exists as part as part of a phosphate molecule (PO_4). Phosphorus in aquatic systems occurs as organic phosphate and inorganic phosphate. Organic phosphate consists of a phosphate molecule associated with a carbon-based molecule, as in plant or animal tissue. Phosphate that is not associated with organic material is inorganic, the form required by plants. Animals can use either organic or inorganic phosphate. Both organic and inorganic phosphate can be dissolved in the water or suspended (attached to particles in the water column).

Phosphorus Cycle

Phosphorus cycles through the environment (refer to Figure 6.7), changing form as it does so. Aquatic plants take in dissolved inorganic phosphorus, as it becomes part of their tissues. Animals get the organic phosphorus they need by eating aquatic plants, other animals, or decomposing plant and animal material. In water bodies, as plants and animals excrete wastes or die, the organic phosphorus they contain sinks to the bottom, where bacterial decomposition converts it back to inorganic phosphorus, both dissolved and attached to particles. This inorganic phosphorus gets back into the water column when animals, human activity, interactions, or water currents stir up the bottom. Plants then take it up and the cycle begins again. In a stream system, the phosphorus cycle tends to move phosphorus downstream as the current carries decomposing plant and animal tissue and dissolved phosphorus. It becomes stationary only when it is taken up by plants or is bound to particles that settles to the bottom of ponds.

In the field of water quality chemistry, phosphorus is described by several terms. Some of these terms are chemistry based (referring to chemically based compounds), and others are methods based (they describe what is measured by a particular method). The term *orthophosphate* is a chemistry-based term that refers to the phosphate molecule all by itself. More specifically, orthophosphate is simple phosphate, or reactive phosphate—that is, Na_3PO_4 (sodium phosphate, tribasic) and NaH_2PO_4 (sodium phosphate, monobasic). Orthophosphate is the only form of phosphate that can be directly tested for in the laboratory and is the form that bacteria use directly for metabolic processes. *Reactive phosphorus* is a corresponding method-based term that describes what is actually being measured when the test for orthophosphate is being performed. Because the lab procedure is not quite perfect, mostly orthophosphate is obtained along with a small fraction of some other forms. More complex inorganic phosphate compounds are referred to as *condensed phosphates* or *polyphosphates*. The method-based term for these forms in *acid hydrolyzable*.

Testing Phosphorus

Testing phosphorus is challenging because it involves measuring very low concentrations, down to 0.01 mg/L or even lower. Even such very low concentrations of phosphorus can have a dramatic impact on streams. Less sensitive methods should be used only to identify serious problem areas. Although many tests for phosphorus exist, only four are likely to be performed by most samplers. The *total orthophosphate test* is largely a measure of orthophosphate. Because the sample is not filtered, the procedure measures both dissolved and suspended orthophosphate. The USEPA-approved method for measuring phosphorus is known as the *ascorbic acid method*. Briefly, a reagent (either liquid or powder) containing ascorbic acid and ammonium molybdate reacts with orthophosphate in the sample to form a blue compound. The intensity of the blue color is directly proportional to the amount of orthophosphate in the water.

The *total phosphate test* measures all the forms of phosphorus in the sample (orthophosphate, condensed phosphate, and organic phosphate) by first digesting (heating and acidifying) the sample to convert all the other forms to orthophosphate. The orthophosphate is then measured by the ascorbic acid method. Because the sample is not filtered, the procedure measures both dissolved and suspended orthophosphate. The *dissolved phosphorus test* measures that fraction of the total phosphorus that is in solution in the water (as opposed to being attached to suspended particles). It is determined by first filtering the sample, then analyzing the filtered sample for total phosphorus. Insoluble phosphorus is calculated by subtracting the dissolved phosphorus result from the total phosphorus result.

All of these tests have one thing in common—they all depend on measuring orthophosphate. The total orthophosphate test measures the orthophosphate that is already present in the sample. The others measure that which is already present and that which is formed when the other forms of phosphorus are converted to orthophosphate by digestion. Monitoring phosphorus involves two basic steps:

1. Collect a water sample.
2. Analyze it in the field or lab for one of the types of phosphorus described above.

Sampling and Equipment Considerations

Sample containers made of either some form of plastic or Pyrex® glass are acceptable to the USEPA. Because phosphorus molecules have a tendency to adsorb (attach) to the inside surface of sample containers, containers that are to be reused must be acid-washed to remove adsorbed phosphorus. The container must be able to withstand repeated contact with hydrochloric acid. Plastic containers, either high-density polyethylene or polypropylene, might be preferable to glass from a practical standpoint because they are better able to withstand breakage. Some programs use disposable, sterile, plastic Whirl-Pak® bags. The size of the container depends on the sample amount required for the phosphorus analysis method chosen and the amount required for other analyses to be performed.

All containers that will hold water samples or come into contact with reagents used in the orthophosphate test must be dedicated. They should not be used for other tests, to eliminate the possibility that reagents containing phosphorus will contaminate the labware. All labware should be acid-washed.

The only form of phosphorus this text recommends for field analysis is total orthophosphate, which uses the ascorbic acid method on an untreated sample. Analysis of any of the other forms requires adding potentially hazardous reagents, heating the sample to boiling, and using too much time and too much equipment to be practical. In addition, analysis for other forms of phosphorus is prone to errors and inaccuracies in field situations. Pretreatment and analysis for these other forms should be handled in a laboratory.

Ascorbic Acid Method for Determining Orthophosphate

In the ascorbic acid method, a combined liquid or prepackaged powder reagent consisting of sulfuric acid, potassium antimonyl tartrate, ammonium molybdate, and ascorbic acid (or comparable compounds) is added to either 50 or 25 mL of the water sample. This colors the sample blue in direct proportion to the amount of orthophosphate in the sample. Absorbence or transmittance is then measured after 10 minutes but before 30 minutes, using a color comparator with a scale in milligrams per liter that increases with the increase in color hue, or an electronic meter that measures the amount of light absorbed or transmitted at a wavelength of 700 to 880 nanometers (again, depending on manufacturer's directions).

A color comparator may be useful for identifying heavily polluted sites with high concentrations (greater than 0.1 mg/L); however, matching the color of a treated sample to a comparator can be very subjective, especially at low concentrations, and can lead to variable results. A field spectrophotometer or colorimeter with a 2.5-cm light path and an infrared photocell (set for a wavelength of 700 to 880 nm) is recommended for accurate determination of low concentrations (between 0.2 and 0.02 mg/L). Use of a meter requires that a prepared known standard concentration be analyzed ahead of time to convert the absorbence readings of a stream sample to milligrams per liter or that the meter reads directly in milligrams per liter.

For information on how to prepare standard concentrations and on how to collect and analyze samples, refer to *Standard Methods* (APHA, 1998) and USEPA's *Methods for Chemical Analysis of Water and Wastes* (USEPA, 1991, Method 365.2).

Nitrates Measurement

Nitrates are a form of nitrogen found in several different forms in terrestrial and aquatic ecosystems. These forms of nitrogen include ammonia (NH_3), nitrates (NO_3), and nitrites (NO_2). Nitrates are essential plant nutrients, but excess amounts can cause significant water quality problems. Together with phosphorus, nitrates in excess amounts can accelerate eutrophication, causing dramatic increases in aquatic plant growth and changes in the types of plants and animals that live in the stream. This, in turn, affects dissolved oxygen, temperature, and other indicators. Excess nitrates can cause hypoxia (low levels of dissolved oxygen) and can become toxic to warm-blooded animals at higher concentrations (10 mg/L or higher) under certain conditions. The natural level of ammonia or nitrate in surface water is typically low (less than 1 mg/L); in the effluent of wastewater treatment plants, it can range up to 30 mg/L. Conventional potable water treatment plants cannot remove nitrate. High concentrations must be prevented by controlling the input at the source. Sources of nitrates include wastewater treatment plants, runoff from fertilized lawns and cropland, failing onsite septic systems, runoff from animal manure storage areas, and industrial discharges that contain corrosion inhibitors.

Sampling and Equipment Considerations

Nitrates from land sources end up in rivers and streams more quickly than other nutrients such as phosphorus; they dissolve in water more readily than phosphorus, which has an attraction for soil particles. As a result, nitrates serve as a better indicator of the possibility of a source of sewage or manure pollution during dry weather. Water that is polluted with nitrogen-rich organic matter might show low nitrates. Decomposition of the organic matter lowers the dissolved oxygen level, which in turn slows the rate at which ammonia is oxidized to nitrite (NO_2) and then to nitrate (NO_3). Under such circumstances, monitoring for nitrites or ammonia (considerably more toxic to aquatic life than nitrate) might be also necessary. For appropriate nitrite methods, see *Standard Methods*, Sections 4500-NH3 and 4500-NH2 (APHA, 1998). Water samples to be tested for nitrate should be collected in glass or polyethylene containers that have been prepared by using Method B (described previously). Two methods are typically used for nitrate testing: the cadmium reduction method and the nitrate electrode. The more commonly used cadmium reduction method produces a color reaction measured by comparison to a color wheel or by use of a spectrophotometer. A few programs also use a nitrate electrode, which can measure in the range of 0 to 100 mg/L nitrate. A newer colorimetric immunoassay technique for nitrate screening is also now available.

Cadmium Reduction Method

In the cadmium reduction method, nitrate is reduced to nitrite by passing the sample through a column packed with activated cadmium. The sample is then measured quantitatively for nitrite. More specifically, the cadmium reduction method is a colorimetric method that involves contact of the nitrate in the sample with cadmium particles, which cause nitrates to be converted to nitrites. The nitrites then react with another reagent to form a red color, in proportional intensity to the original amount of nitrate. The color is measured either by comparison to a color wheel with a scale in milligrams per liter that increases with the increase in color hue or by use of an electronic spectrophotometer that measures the amount of light absorbed by the treated sample at a 543-nanometer wavelength. The absorbence value converts to the equivalent concentration of nitrate against a standard curve. Methods for making standard solutions and standard curves are presented in *Standard Methods* (APHA, 1998).

Before each sampling run, the sampling/monitoring supervisor should create this curve. The curve is developed by making a set of standard concentrations of nitrate, reacting them, and developing the corresponding color, then plotting the absorbence value for each concentration against concentration. A standard curve could also be generated for the color wheel. Use of the color wheel is appropriate only if nitrate concentrations are greater than 1 mg/L. For concentrations below 1 mg/L, use a spectrophotometer. Matching the color of a treated sample at low concentrations to a color wheel (or cubes) can be very subjective and can lead to variable results. Color comparators, however, can be effectively used to identify sites with high nitrates.

This method requires that the samples being treated are clear. If a sample is turbid, filter it through a 0.45-μm filter. Be sure to test to make sure the filter is nitrate free. If copper, iron, or others metals are present in concentrations above several milligrams per liter, the reaction with the cadmium will slow down and the reaction time must be increased.

The reagents used for this method are often prepackaged for different ranges, depending on the expected concentration of nitrate in the stream. Hack, for example, provides reagents for the following ranges: low (0 to 0.40 mg/L), medium (0 to 15 mg/L), and high (0 to 30 mg/L). Determining the appropriate range for the stream being monitored is important.

Nitrate Electrode Method

A nitrate electrode (used with a meter) is similar in function to a dissolved oxygen meter. It consists of a probe with a sensor that measures nitrate activity in the water; this activity affects the electric potential of a solution in the probe. This change is then transmitted to the meter, which converts the electric signal to a scale that is read in millivolts; the millivolts are then converted to mg/L of nitrate by plotting them against a standard curve. The accuracy of the electrode can be affected by high concentrations of chloride or bicarbonate ions in the sample water. Fluctuating pH levels can also affect the meter reading.

Nitrate electrodes and meters are expensive compared to field kits that employ the cadmium reduction method. (The expense is comparable, however, if a spectrophotometer is used rather than a color wheel.) Meter and probe combinations run between $700 and $1200, including a long cable to connect the probe to the meter. A pH meter that displays readings in millivolts can be used with a nitrate probe, and no separate nitrate meter is needed. Results are read directly as milligrams per liter.

Although nitrate electrodes and spectrophotometers can be used in the field, they have certain disadvantages. These devices are more fragile than the color comparators and are therefore more at risk of breaking in the field. They must be carefully maintained and must be calibrated before each sample run and, if many tests are being run, between samplings. This means that samples are best tested in the lab. Note that samples to be tested with a nitrate electrode should be at room temperature, whereas color comparators can be used in the field with samples at any temperature.

Solids Measurement

Solids in water are defined as any matter that remains as residue upon evaporation and drying at 103°C. They are separated into two classes: *suspended solids* and *dissolved solids*:

$$\text{Total solids} = \begin{matrix}\text{Suspended solids} \\ \text{(nonfilterable residue)}\end{matrix} + \begin{matrix}\text{Dissolved solids} \\ \text{(filterable residue)}\end{matrix}$$

As shown above, *total solids* are dissolved solids plus suspended and settleable solids in water. In natural freshwater bodies, dissolved solids consist of calcium, chlorides, nitrate, phosphorus, iron, sulfur, and other ions—particles that will pass through a filter with pores of around 2 μm (0.002 cm) in size. Suspended solids include silt and clay particles, plankton, algae, fine organic debris, and other particulate matter. These are particles that will not pass through a 2-μm filter.

The concentration of total dissolved solids affects the water balance in the cells of aquatic organisms. An organism placed in water with a very low level of solids (distilled water, for example) swells because water tends to move into its cells, which have a higher concentration of solids. An organism placed in water with a high concentration of solids shrinks somewhat, because the water in its cells tends to move out. This in turn affects the organism's ability to maintain the proper cell density, making keeping its position in the water column difficult. It might float up or sink down to a depth to which it is not adapted, and it might not survive.

Higher concentrations of suspended solids can serve as carriers of toxics, which readily cling to suspended particles. This is particularly a concern where pesticides are being used on irrigated crops. Where solids are high, pesticide concentrations may increase well beyond those of the original application as the irrigation water travels down irrigation ditches. Higher levels of solids can also clog irrigation devices and might become so high that irrigated plant roots will lose water rather than gain it.

A high concentration of total solids will make drinking water unpalatable and might have an adverse effect on people who are not used to drinking such water. Levels of total solids that are too high or too low can also reduce the efficiency of wastewater treatment plants, as well as the operation of industrial processes that use raw water.

Total solids affect water clarity. Higher solids decrease the passage of light through water, thereby slowing photosynthesis by aquatic plants. Water heats up more rapidly and holds more heat; this, in turn, might adversely affect aquatic life adapted to a lower temperature regime.

Sources of total solids include industrial discharges, sewage, fertilizers, road runoff, and soil erosion. Total solids are measured in milligrams per liter (mg/L).

Solids Sampling and Equipment Considerations

When conducting solids testing, many things can affect the accuracy of the test or result in wide variations in results for a single sample, including the following:

1. Drying temperature
2. Length of drying time
3. Condition of desiccator and desiccant
4. lack of consistency in test procedures for nonrepresentative samples
5. Failure to achieve constant weight prior to calculating results

Several precautions can be taken to improve the reliability of test results:

1. Use extreme care when measuring samples, weighing materials, and drying or cooling samples.
2. Check and regulate oven and furnace temperatures frequently to maintain the desired range.
3. Use an indicator drying agent in the desiccator that changes color when it is no longer good; change or regenerate the desiccant when necessary.
4. Keep desiccator cover greased with the appropriate type of grease to seal the desiccator and prevent moisture from entering the desiccator as the test glassware cools.
5. Check ceramic glassware for cracks and glass-fiber filter for possible holes; a hole in a glass filter will cause solids to pass through and give inaccurate results.
6. Follow the manufacturer's recommendation for care and operation of analytical balances.

Total solids are important to measure in areas where discharges from sewage treatment plants, industrial plants, or extensive crop irrigation may occur. In particular, streams and rivers in arid regions where water is scarce and evaporation is high tend to have higher concentrations of solids and are more readily affected by human introduction of solids from land-use activities.

Total solids measurements can be useful as an indicator of the effects of runoff from construction, agricultural practices, logging activities, sewage treatment plant discharges, and other sources. As with turbidity, concentrations often increase sharply during rainfall, especially in developed watersheds. They can also rise sharply during dry weather if earth-disturbing activities occur in or near the stream without erosion control practices in place. Regular monitoring of total solids can help detect trends that might indicate increasing erosion in developing watersheds. Total solids are closely related to stream flow and velocity and should be correlated with these factors. Any change in total solids over time should be measured at the same site at the same flow.

Total solids are measured by weighing the amount of solids present in a known volume of sample; this is accomplished by weighing a beaker, filling it with a known volume, evaporating the water in an oven to completely dry the residue, and weighing the beaker with the residue. The total solids concentration is equal to the difference between the weight of the beaker with the residue and the weight of the beaker without it. Because the residue is so light in weight, the lab should have a balance that is sensitive to weights as light as 0.0001 gram. Balances of this type are called *analytical* or *Mettler* balances, and they are expensive (around $3000). The beakers must be kept in a desiccator, a sealed glass container that contains material that absorbs moisture and ensures that the weighing is not biased by water condensing on the beaker. Some desiccants change color to indicate moisture content. Measurement of total solids cannot be done in the field. Samples must be collected using clean glass or plastic bottles or Whirl-Pak® bags and taken to a laboratory where the test can be run.

Total Suspended Solids Testing

Total Suspended Solids Samples

The term *solids* means any material suspended or dissolved in water and wastewater. Although normal domestic wastewater contains a very small amount of solids (usually less than 0.1%), most treatment processes are designed specifically to remove or convert solids to a form that can be removed or discharged without causing environmental harm. When sampling for total suspended solids (TSS), samples may be either grab or composite and can be collected in either glass or plastic containers. TSS samples can be preserved by refrigeration at or below 4°C (not frozen); however, composite samples must be refrigerated during collection. The maximum holding time for preserved samples is 7 days.

Total Suspended Solids Test Procedure

To conduct a TSS test procedure, a well-mixed measured sample is poured into a filtration apparatus and, with the aid of a vacuum pump or aspirator, is drawn through a preweighed glass-fiber filter. After filtration, the glass-fiber filter is dried at 103 to 105°C, cooled, and reweighed. The increase in weight of the filter and solids compared to the filter alone represents the total suspended solids. An example of the specific test procedure used for total suspended solids is given below:

1. Select a sample volume that will yield between 10 and 200 mg of residue with a filtration time of 10 minutes or less.

   Note: If filtration time exceeds 10 minutes, increase filter area or decrease volume to reduce filtration time.

   Note: For nonhomogeneous samples or samples with very high solids concentrations (e.g., raw wastewater, mixed liquor), use a larger filter to ensure that a representative sample volume can be filtered.

2. Place a preweighed glass-fiber filter on the filtration assembly in a filter flask.
3. Mix the sample well, and measure the selected volume of sample.
4. Apply suction to the filter flask, and wet the filter with a small amount of laboratory-grade water to seal it.
5. Pour the selected sample volume into the filtration apparatus.
6. Draw the sample through the filter.
7. Rinse the measuring device into the filtration apparatus with three successive 10-mL portions of laboratory-grade water. Allow complete drainage between rinsings.

8. Continue suction for 3 minutes after filtration of the final rinse is completed.
9. Remove the glass filter from the filtration assembly (membrane filter funnel or clean Gooch crucible). If using the large disks and membrane filter assembly, transfer the glass filter to a support (aluminum pan or evaporating dish) for drying.
10. Place the glass filter with solids and support (pan, dish, or crucible) in a drying oven.
11. Dry the filter and solids to a constant weight at 103 to 105°C (at least 1 hour).
12. Cool to room temperature in a desiccator.
13. Weigh the filter, and support and record the constant weight in the test record.

Total Suspended Solids Calculations

To determine the total suspended solids concentration in mg/L, we use the following equations:

1. To determine weight of dry solids in grams:

$$\text{Dry solids (g)} = \text{Weight of dry solids and filter (g)} - \text{weight of dry filter (g)} \tag{8.7}$$

2. To determine weight of dry solids in milligrams (mg):

$$\text{Dry solids (mg)} = \text{Weight of dry solids and filter (mg)} - \text{weight of dry filter (mg)} \tag{8.8}$$

3. To determine the TSS concentration in mg/L:

$$\text{TSS (mg/L)} = \frac{\text{Dry solids (mg)} \times 1000 \text{ mL}}{\text{Sample (mL)}} \tag{8.9}$$

■ **Example 8.4**

Problem: Using the data provided below, calculate total suspended solids (TSS):

Sample volume = 250 mL
Weight of dry solids and filter = 2.305 g
Weight of dry filter = 2.297 g

Solution:

Dry solids (g) = 2.305 g – 2.297 g = 0.008 g
Dry solids (mg) = 0.008 g × 1000 mg/g = 8 mg

$$\text{TSS} = \frac{8.0 \text{ mg} \times 1000 \text{ mL}}{250 \text{ mL}} = 32.0 \text{ mg/L}$$

Volatile Suspended Solids Testing

When the total suspended solids are ignited at 550 ± 50°C, the volatile (organic) suspended solids of the sample are converted to water vapor and carbon dioxide and are released to the atmosphere. The solids that remain after the ignition (ash) are the inorganic or fixed solids. In addition to the equipment and supplies required for the total suspended solids test, we need the following:

1. Muffle furnace (550 ± 50°C)
2. Ceramic dishes
3. Furnace tongs
4. Insulated gloves

Volatile Suspended Solids Test Procedure

An example of the test procedure used for volatile suspended solids is given below:

1. Place the weighed filter with solids and support from the total suspended solids test in the muffle furnace.
2. Ignite the filter, solids, and support at 550 ± 50°C for 15 to 20 minutes.
3. Remove the ignited solids, filter, and support from the furnace, and partially air cool.
4. Cool to room temperature in a desiccator.
5. Weigh the ignited solids, filter, and support on an analytical balance.
6. Record the weight of the ignited solids, filter, and support.

Total Volatile Suspended Solids Calculations

To calculate total volatile suspended solids (TVSS) requires the following information:

1. Weight of dry solids, filter, and support (g)
2. Weight of ignited solids, filter, and support (g)

$$\text{TVSS (mg/L)} = \frac{(A - C) \times 1000 \text{ mg/g} \times 1000 \text{ mL/L}}{\text{Sample volume (mL)}} \tag{8.10}$$

where

A = Weight of dried solids, filter, and support.
C = Weight of ignited solids, filter, and support.

■ Example 8.5

Problem: Using the data provided below, calculate the total volatile suspended solids:

Weight of dried solids, filter, and support = 1.6530 g
Weight of ignited solids, filter, and support = 1.6330 g
Sample volume = 100 mL

Solution:

$$\text{TVSS} = \frac{(1.6530 \text{ g} - 1.6330 \text{ g}) \times 1000 \text{ mg/g} \times 1000 \text{ mL/L}}{100 \text{ mL}} = \frac{0.02 \times 1{,}000{,}000 \text{ mg/L}}{100} = 200 \text{ mg/L}$$

Note: The concentration of *total fixed suspended solids* (TFSS) is the difference between the total volatile suspended solids (TVSS) and the total suspended solids (TSS) concentrations:

$$\text{TFSS (mg/L)} = \text{TTS} - \text{TVSS} \tag{8.11}$$

■ Example: 8.6

Problem: Using the data provided below, calculate the total fixed suspended solids:

Total fixed suspended solids = 202 mg/L
Total volatile suspended solids = 200 mg/L

Solution:

$$\text{TFSS (mg/L)} = 202 \text{ mg/L} - 200 \text{ mg/L} = 2 \text{ mg/L}$$

Conductivity Testing

Conductivity is a measure of the capacity of water to pass an electrical current. Conductivity in water is affected by the presence of inorganic dissolved solids such as chloride, nitrate, sulfate, and phosphate anions (ions that carry a negative charge), or sodium, magnesium, calcium, iron, and aluminum cations (ions that carry a positive charge). Organic compounds such as oil, phenol, alcohol, and sugar do not conduct electrical current very well and therefore have a low conductivity when in water. Conductivity is also affected by temperature: the warmer the water, the higher the conductivity.

Conductivity in streams and rivers is affected primarily by the geology of the area through which the water flows. Streams that run through areas with granite bedrock tend to have lower conductivity because granite is composed of more inert materials that do not ionize (dissolve into ionic components) when washed into the water. On the other hand, streams that run through areas with clay soils tend to have higher conductivity because of the presence of materials that ionize when washed into the water. Groundwater inflows can have the same effects, depending on the bedrock through which they flow. Discharges to streams can change the conductivity depending on their make-up. A failing sewage system would raise the conductivity because of the presence of chloride, phosphate, and nitrate; an oil spill would lower conductivity.

The basic unit of measurement of conductivity is the mho or siemens. Conductivity is measured in micromho per centimeter (μmho/cm) or microsiemens per centimeter (μS/cm). Distilled water has conductivity in the range of 0.5 to 3 μmho/cm. The conductivity of rivers in the United States generally ranges from 50 to 1500 μmho/cm. Studies of inland freshwaters indicated that streams supporting good mixed fisheries have a range between 150 and 500 μmho/cm. Conductivity outside this range could indicate that the water is not suitable for certain species of fish or macroinvertebrates. Industrial waters can range as high as 10,000 μmho/cm.

Sampling, Testing, and Equipment Considerations

Conductivity is useful as a general measure of source water quality. Each stream tends to have a relatively constant range of conductivity that, once established, can be used as a baseline for comparison with regular conductivity measurements. Significant changes in conductivity could indicate that a discharge or some other source of pollution has entered a stream. The conductivity test is not routine in potable water treatment, but when performed on source water it is a good indicator of contamination. Conductivity readings can also be used to indicate wastewater contamination or saltwater intrusion.

Note: Distilled water used for potable water analyses at public water supply facilities must have a conductivity of no more than 1 μmho/cm.

Conductivity is measured with a probe and a meter. Voltage is applied between two electrodes in a probe immersed in the sample water. The drop of voltage caused by the resistance of the water is used to calculate the conductivity per centimeter. The meter converts the probe measurement to micromhos per centimeter (μmhos/cm) and displays the result for the user.

Note: Some conductivity meters can be used to test for total dissolved solids and salinity. The total dissolved solids concentration in milligrams per liter (mg/L) can also be calculated by multiplying the conductivity result by a factor between 0.55 and 0.9, which is empirically determined; see *Standard Methods* (APHA, 1998, Method 2510).

Suitable conductivity meters cost about $350. Meters in this price range should also measure temperature and automatically compensate for temperature in the conductivity reading. Conductivity can be measured in the field or the lab. In most cases, collecting samples in the field and taking them to a lab for testing probably represent a better approach. In this way, several teams can collect samples simultaneously. If testing in the field is important, meters designed for field use can be obtained for around the same cost mentioned above. If samples will be collected in the field for later measurement, the sample bottle should be a glass or polyethylene bottle that has been washed in phosphate-free detergent and rinsed thoroughly with both tap and distilled water. Factory-prepared Whirl-Pak® bags may be used.

Total Alkalinity

Alkalinity is defined as the ability of water to resist a change in pH when acid is added; it relates to the pH buffering capacity of the water. Almost all natural waters have some alkalinity. Alkaline compounds in the water, such as bicarbonates (baking soda is one type), carbonates, and hydroxides, remove H^+ ions and lower the acidity of the water (which means increased pH). They usually do this by combining with the H^+ ions to make new compounds. Without this acid-neutralizing capacity, any acid added to a stream would cause an immediate change in the pH. Measuring alkalinity is important in determining a stream's ability to neutralize acidic pollution from rainfall or wastewater—one of the best measures of the sensitivity of the stream to acid inputs. Alkalinity in streams is influenced by rocks and soils, salts, certain plant activities, and certain industrial wastewater discharges.

Total alkalinity is determined by measuring the amount of acid (e.g., sulfuric acid) required to bring the sample to a pH of 4.2. At this pH all of the alkaline compounds in the sample are used up. The result is reported as milligrams per liter of calcium carbonate (mg/L $CaCO_3$).

Testing for alkalinity in potable water treatment is most important with regard to its relation to coagulant addition; that is, we must know that enough natural alkalinity exists in the water to buffer chemical acid addition so floc formation will be optimum and turbidity removal can proceed. In water softening, proper chemical dosage will depend on the type and amount of alkalinity in the water. For corrosion control, the presence of adequate alkalinity in a water supply neutralizes any acid tendencies and prevents it from becoming corrosive. For total alkalinity, a double endpoint titration using a pH meter (or pH pocket pal) and a digital titrator or buret is recommended. This can be done in the field or in the lab. If alkalinity must be analyzed in the field, a digital titrator should be used instead of a buret, because burets are fragile and more difficult to set up. The alkalinity method described below was developed by the Acid Rain Monitoring Project of the University of Massachusetts Water Resources Research Center (River Watch Network, 1992).

Burettes, Titrators, and Digital Titrators for Measuring Alkalinity

The total alkalinity analysis involves titration. In this test, titration is the addition of small, precise quantities of sulfuric acid (the reagent) to the sample, until the sample reaches a certain pH (known as an *endpoint*). The amount of acid used corresponds to the total alkalinity of the sample. Alkalinity can be measured using a buret, titrator, or digital titrator (described below):

1. A buret is a long, graduated glass tube with a tapered tip like a pipet and a valve that opens to allow the reagent to drop out of the tube. The amount of reagent used is calculated by subtracting the original volume in the buret from the column left after the endpoint has been reached. Alkalinity is calculated based on the amount used.
2. Titrators forcefully expel the reagent by using a manual or mechanical plunger. The amount of reagent used is calculated by subtracting the original volume in the titrator from the volume left after the endpoint has been reached. Alkalinity is then calculated based on the amount used or is read directly from the titrator.
3. Digital titrators have counters that display numbers. A plunger is forced into a cartridge containing the reagent by turning a knob on the titrator. As the knob turns, the counter changes in proportion to the amount of reagent used. Alkalinity is then calculated based on the amount used. Digital titrators cost approximately $100.

Digital titrators and burets allow for much more precision and uniformity in the amount of titrant that is used.

Fecal Coliform Bacteria Testing

Much of the information in this section is from USEPA (1985, 1986). Fecal coliform bacteria are non-disease-causing organisms found in the intestinal tract of all warm-blooded animals. Each discharge of body wastes contains large amounts of these organisms. The presence of fecal coliform bacteria in a stream or lake indicates the presence of human or animal wastes. The number of fecal coliform bacteria present is a good indicator of the amount of pollution present in the water. The USEPA's 2001 Total Coliform Rule 816-F-01-035

1. Is intended to improve public health protection by reducing fecal pathogens to minimal levels through control of total coliform bacteria, including fecal coliforms and *Escherichia coli* (*E. coli*).
2. Establishes a maximum contaminant level (MCL) based on the presence or absence of total coliforms, modifies monitoring requirements including testing for fecal coliforms or *E. coli*, requires use of a sample siting plan, and requires sanitary surveys for systems collecting fewer than five samples per month.
3. Applies to all public water systems.
4. Has resulted in a reduction in the risk of illness from disease-causing organisms associated with sewage or animal wastes; disease symptoms may include diarrhea, cramps, nausea, and possibly jaundice, and associated headaches and fatigue.

Fecal coliforms are used as indicators of possible sewage contamination because they are commonly found in human and animal feces. Although they are not generally harmful themselves, they indicate the possible presence of pathogenic (disease-causing) bacteria and protozoa that also live in human and animal digestive systems. Their presence in streams suggests that pathogenic microorganisms might also be present and that swimming in or eating shellfish from the waters might present a health risk. Because testing directly for the presence of a large variety of pathogens is difficult, time consuming, and expensive, water is usually tested for coliforms and fecal streptococci instead. Sources of fecal contamination to surface waters include wastewater treatment plants, onsite septic systems, domestic and wild animal manure, and storm runoff. In addition to the possible health risks associated with the presence of elevated levels of fecal bacteria, they can also cause cloudy water, unpleasant odors, and an increased oxygen demand.

Note: In addition to the most commonly tested fecal bacteria indicators, total coliforms, fecal coliforms, and *E. coli*, fecal streptococci and enterococci are also commonly used as bacteria indicators. The focus of this presentation is on total coliforms and fecal coliforms.

Fecal coliforms are widespread in nature. All members of the total coliform group can occur in human feces, but some can also be present in animal manure, soil, and submerged wood, and in other places outside the human body. The usefulness of total coliforms as an indicator of fecal contamination depends on the extent to which the bacteria species found are fecal and human in origin. For recreational waters, total coliforms are no longer recommended as an indicator. For drinking water, total coliforms are still the standard test, because their presence indicates contamination of a water supply by an outside source.

Fecal coliforms, a subset of total coliform bacteria, are more fecal specific in origin; however, even this group contains a genus, *Klebsiella*, with species that are not necessarily fecal in origin. *Klebsiella* are commonly associated with textile and pulp- and papermill wastes. If these sources discharge to a local stream, consideration should be given to monitoring more fecal and human-specific bacteria. For recreational waters, this group was the primary bacteria indicator until relatively recently, when the USEPA began recommending *E. coli* and enterococci as better indicators of health risk from water contact. Fecal coliforms are still being used in many states as indicator bacteria.

Sampling and Equipment Considerations

Under the USEPA's Total Coliform Rule, sampling requirements are specified as follows.

Routine Sampling Requirements

1. Total coliform samples must be collected at sites that are representative of water quality throughout the distribution system according to a written sample siting plan subject to state review and revision.
2. Samples must be collected at regular time intervals throughout the month, except that groundwater systems serving 4900 persons or fewer may collect them on the same day.
3. Monthly sampling requirements are based on population served (see Table 8.9 for the minimum sampling frequency).
4. A reduced monitoring frequency may be available for systems serving 1000 persons or fewer and using only groundwater if a sanitary survey within the past 5 years shows the system is free of sanitary defects (the frequency may be no less than one sample per quarter for community and one sample per year for non-community systems).
5. Each total coliform-positive routine sample must be tested for the presence of fecal coliforms or *Escherichia coli.*

Repeat Sampling Requirements

1. Within 24 hours of learning of a total coliform-positive *routine* sample result, at least three *repeat* samples must be collected and analyzed for total coliforms.
2. One *repeat* sample must be collected from the same tap as the original sample.
3. One *repeat* sample must be collected within five service connections upstream.
4. One *repeat* sample must be collected within five service connections downstream.
5. Systems that collect 1 *routine* sample per month or fewer must collect a fourth *repeat* sample.
6. If any *repeat* sample is total coliform positive:
   a. The system must analyze that total coliform-positive culture for fecal coliforms or *E. coli.*
   b. The system must collect another set of *repeat* samples, as before, unless the MCL has been violated and the system has notified the state.

Additional Routine Sample Requirements

A positive *routine* or *repeat* total coliform result requires a minimum of five *routine* samples to be collected the following month the system provides water to the public unless waived by the state. Public system routine monitoring frequency is shown in Table 8.9.

Other Total Coliform Rule Provisions

1. Systems collecting fewer than five *routine* samples per month must have a sanitary survey every 5 years (or every 10 years if it is a non-community water system using protected and disinfected groundwater).
2. Systems using surface water or groundwater under the direct influence (GWUDI) of surface water and meeting filtration avoidance criteria must collect and have analyzed one coliform sample each day the turbidity of the source water exceeds 1 NTU. This sample must be collected from a tap near the first service connection.

Compliance

Compliance is based on the presence or absence of total coliforms. Moreover, compliance is determined each calendar month the system serves water to the public (or each calendar month that sampling occurs for systems on reduced monitoring). The results of *routine* and *repeat* samples are

TABLE 8.9
Public Water System Routine Monitoring Frequencies

Population	Minimum Samples/Month	Population	Minimum Samples/Month
25–1000[a]	1	59,001–70,000	70
1001–2500	2	70,000–83,000	80
2501–3300	3	83,001–96,000	90
3301–4100	4	96,001–130,000	100
4101–4900	5	130,000–220,000	120
4901–5800	6	220,001–320,000	150
5801–6700	7	320,001–450,000	180
6701–7600	8	450,001–600,000	210
7601–8500	9	600,001–780,000	240
8501–12,900	10	780,001–970,000	270
12,901–17,200	15	970,001–1,230,000	330
17,201–21,500	20	1,520,001–1,850,000	360
21,501–25,000	25	1,850,001–2,270,000	390
25,001–33,000	30	2,270,001–3,020,000	420
33,001–41,000	40	3,020,001–3,960,000	450
41001–50,000	50	≥3,960,001	480
50,001–59,000	60		

[a] Includes PWSs that have at least 15 service connections but serve <25 people.

used to calculate compliance. With regard to violations, a monthly MCL violation is triggered if a system collecting fewer than 40 samples per month has more than one *routine/repeat* sample per month that is total coliform positive. In addition, a system collecting at least 40 samples per month for which more than 5% of the *routine/repeat* samples are total coliform positive is technically in violation of the Total Coliform Rule. An acute MCL violation is triggered if any public water system has any fecal coliform- or *E. coli*-positive *repeat* sample or has a fecal coliform- or *E. coli*-positive *routine* sample followed by a total coliform-positive *repeat* sample.

The Total Coliform Rule also has requirements for public notification and reporting; for example, for a monthly MCL violation, the violation must be reported to the state no later than the end of the next business day after the system learns of the violation. The public must be notified within 14 days. For an acute MCL violation, the violation must be reported to the state no later than the end of the next business day after the system learns of the violation. The public must be notified within 72 hours. Systems with *routine* or *repeat* samples that are fecal coliform or *E. coli* positive must notify the state by the end of the day they are notified of the result or by the end of the next business day if the state office is already closed.

Sampling and Equipment Considerations

Bacteria can be difficult to sample and analyze, for many reasons. Natural bacteria levels in streams can vary significantly; bacteria conditions are strongly correlated with rainfall, making the comparison of wet and dry weather bacteria data a problem. Many analytical methods have a low level of precision, yet can be quite complex to accomplish, and absolutely sterile conditions are essential to maintain while collecting and handling samples. The primary equipment decision to make when sampling for bacteria is what type and size of sample container to use. Once that decision has been made, the same straightforward collection procedure is used, regardless of the type of bacteria being monitored.

When monitoring bacteria, it is critical that all containers and surfaces with which the sample will come into contact are sterile. Containers made of either some form of plastic or Pyrex® glass are acceptable to the USEPA; however, if the containers are to be reused, they must be sturdy enough to survive sterilization using heat and pressure. The containers can be sterilized by using an autoclave, a machine that sterilizes with pressurized steam. If an autoclave is used, the container material must be able to withstand high temperatures and pressure. Plastic containers, either high-density polyethylene or polypropylene, might be preferable to glass from a practical standpoint because they will better withstand breakage. In any case, be sure to check the manufacturer's specifications to see whether the container can withstand 15 minutes in an autoclave at a temperature of 121°C without melting. (Extreme caution is advised when working with an autoclave.) Disposable, sterile, plastic Whirl-Pak® bags are used by a number of programs. The size of the container depends on the sample amount required for the bacteria analysis method chosen and the amount required for other analyses. The two basic methods for analyzing water samples for bacteria in common use are the *membrane filtration* and *multiple-tube fermentation* methods (described later).

Given the complexity of the analysis procedures and the equipment required, field analysis of bacteria is not recommended. Bacteria can be analyzed at a well-equipped lab or sent to a state-certified lab for analysis. If a bacteria sample is sent to a private lab, be sure the lab is certified by the state for bacteria analysis. Consider state water quality labs, university and college labs, private labs, wastewater treatment plant labs, and hospitals. On the other hand, if the treatment plant has a modern lab with the proper equipment and properly trained technicians, the fecal coliform testing procedures described in the following section will be helpful. *A note of caution:* If you decide to analyze samples in your own lab, be sure to carry out a quality assurance/quality control program.

Testing for Fecal Coliform

Federal regulations cite two approved methods for the determination of fecal coliform in water: (1) multiple-tube fermentation or most probable number (MPN) procedure, and (2) membrane filter (MF) procedure.

Note: Because the MF procedure can yield low or highly variable results for chlorinated wastewater, USEPA requires verification of results using the MPN procedure to resolve any controversies; however, do not attempt to perform the fecal coliform test using the summary information provided in this handbook. Instead, refer to the appropriate reference cited in the federal regulations for a complete discussion of these procedures.

Equipment and Techniques

Whenever microbiological testing of water samples is performed, certain general considerations and techniques will be required. Because these are basically the same for each test procedure, they are reviewed here prior to discussion of the two methods.

1. *Reagents and media*—All reagents and media utilized in performing microbiological tests on water samples must meet the standards specified in the reference cited in federal regulations.
2. *Reagent-grade water*—Deionized water that is tested annually and found to be free of dissolved metals and bactericidal or inhibitory compounds is preferred for use in preparing culture media and test reagents, although distilled water may be used.
3. *Chemicals*—All chemicals used in fecal coliform monitoring must be ACS reagent grade or equivalent.
4. *Media*—To ensure uniformity in the test procedures, the use of dehydrated media is recommended. Sterilized, prepared media in sealed test tubes, ampoules, or dehydrated media pads are also acceptable for use in this test.
5. *Glassware and disposable supplies*—All glassware, equipment, and supplies used in microbiological testing must meet standards specified in the references cited in federal regulations.

Preparation of Equipment and Chemicals

All glassware used for bacteriological testing must be thoroughly cleaned using a suitable detergent and hot water. The glassware should be rinsed with hot water to remove all traces of residual from the detergent and, finally, should be rinsed with distilled water. Laboratories should use a detergent certified to meet bacteriological standards or, at a minimum, rinse all glassware after washing with two tap water rinses followed by five distilled water rinses. For sterilization of equipment, a hot-air sterilizer or autoclave can be used. When a hot-air sterilizer is used, all equipment should be wrapped in high-quality (Kraft) paper or placed in containers prior to hot-air sterilization. All glassware, except glassware in metal containers, should be sterilized for a minimum of 60 minutes at 170°C. Sterilization of glassware in metal containers should require a minimum of 2 hours. Hot-air sterilization cannot be used for liquids. An autoclave can be used to sterilize sample bottles, dilution water, culture media, and glassware at 121°C for 15 minutes.

Sterile Dilution Water Preparation

The dilution water used for making sample serial dilutions is prepared by adding 1.25 mL of stock buffer solution and 5.0 mL of magnesium chloride solution to 1000 mL of distilled or deionized water. The stock solutions of each chemical should be prepared as outlined in the reference cited by the federal regulations. The dilution water is then dispensed in sufficient quantities to produce 9 or 99 mL in each dilution bottle following sterilization. If the membrane filter procedure is used, additional 60- to 100-mL portions of dilution water should be prepared and sterilized to provide the rinse water required by the procedure.

Serial Dilution Procedure

At times, the density of the organisms in a sample makes it difficult to accurately determine the actual number of organisms in the sample. When this occurs, the sample size may need to be reduced to as small as one millionth of a milliliter. To obtain such small volumes, a technique known as *serial dilutions* has been developed.

Bacteriological Sampling

To obtain valid test results that can be utilized in the evaluation of process efficiency of water quality, proper technique, equipment, and sample preservation are critical. These factors are especially critical in bacteriological sampling:

1. *Sample dechlorination*—When samples of chlorinated effluents are to be collected and tested, the sample must be dechlorinated. Prior to sterilization, place enough sodium thiosulfate solution (10%) in a clean sample container to produce a concentration of 100 mg/L in the sample (for a 120-mL sample bottle, 0.1 mL is usually sufficient). Sterilize the sample container as previously described.
2. *Sample procedure:*
   a. Keep the sample bottle unopened after sterilization until the sample is to be collected.
   b. Remove the bottle stopper and hood or cap as one unit. Do not touch or contaminate the cap or the neck of the bottle.
   c. Submerge the sample bottle in the water to be sampled.
   d. Fill the sample bottle approximately three quarters full, but not less than 100 mL.
   e. Aseptically replace the stopper or cap on the bottle.
   f. Record the date, time, and location of sampling, as well as the sampler's name and any other descriptive information pertaining to the sample.

3. *Sample preservation and storage*—Examination of bacteriological water samples should be performed immediately after collection. If testing cannot be imitated within 1 hour of sampling, the sample should be iced or refrigerated at 4°C or less. The maximum recommended holding time for fecal coliform samples from wastewater is 6 hours. The storage temperature and holding time should be recorded as part of the test data.

Multiple-Tube Fermentation Technique

The multiple-tube fermentation technique for fecal coliform testing is useful in determining the fecal coliform density in most water, solid, or semisolid samples. Wastewater testing normally requires use of the presumptive and confirming test procedures. It is recognized as the method of choice for any samples that may be controversial (enforcement related). The technique is based on the most probable number (MPN) of bacteria present in a sample that produces gas in a series of fermentation tubes with various volumes of diluted sample. The MPN is obtained from charts based on statistical studies of known concentrations of bacteria.

The technique utilizes a two-step incubation procedure. The sample dilutions are first incubated in lauryl (sulfonate) tryptose broth for 24 to 48 hours (presumptive test). Positive samples are then transferred to EC broth and incubated for an additional 24 hours (confirming test). Positive samples from this second incubation are used to statistically determine the MPN from the appropriate reference chart. A single-medium, 24-hour procedure is also acceptable. In this procedure, sample dilutions are inoculated in A-1 medium, incubated for 3 hr at 35°C, and then incubated the remaining 20 hr at 44.5°C. Positive samples from these inoculations are then used to statistically determine the MPN value from the appropriate chart.

Fecal Coliform MPN Presumptive Test Procedure

1. Prepare dilutions and inoculate five fermentation tubes for each dilution.
2. Cap all tubes, and transfer to incubator.
3. Incubate 24 ± 2 hr at 35 ± 0.5°C.
4. Examine tubes for gas:
   a. Gas present (positive test)—transfer.
   b. No gas—continue incubation.
5. Incubate for a total time of 48 ± 3 hr at 35 ± 0.5°C
6. Examine tubes for gas.
   a. Gas present (positive test)—transfer.
   b. No gas (negative test).

Note: Keep in mind that the fecal coliform MPN-confirming procedure or fecal coliform procedure using the A-1 broth test is used to determine the MPN/100 mL. The MPN procedure for fecal coliform determinations requires a minimum of three dilutions with five tubes per dilution.

Calculation of Most Probable Number per 100 mL

Calculation of the most probable number (MPN) test results requires selection of a valid series of three consecutive dilutions. The number of positive tubes in each of the three selected dilution inoculations is used to determine the MPN/100 mL. When selecting the dilution inoculations to be used in the calculation, each dilution is expressed as a ratio of positive tubes per tubes inoculated in the dilution (e.g., three positive/five inoculated, or 3/5). Several rules should be followed when determining the most valid series of dilutions. In the following examples, four dilutions were used for the test.

1. Using the confirming test data, select the highest dilution showing all positive results (no lower dilution showing less than all positive) and the next two higher dilutions.
2. If a series shows all negative values with the exception of one dilution, select the series that places the only positive dilution in the middle of the selected series.
3. If a series shows a positive result in a dilution higher than the selected series (using rule 1), it should be incorporated into the highest dilution of the selected series. After selecting the valid series, the MPN/100 mL is determined by locating the selected series on the MPN reference chart. If the selected dilution series matches the dilution series of the reference chart, the MPN value from the chart is the reported value for the test. If the dilution series used for the test does not match the dilution series of the chart, the test result must be calculated.

$$\text{MPN/100 mL} = \text{MPN from chart} \times \frac{\text{Sample volume of first column of chart}}{\text{Sample volume of first dilution of selected series}} \tag{8.12}$$

Membrane Filtration Technique

The membrane filtration technique can be useful for determining the fecal coliform density in wastewater effluents, except for primary treated wastewater that has not been chlorinated or wastewater containing toxic metals or phenols. Chlorinated secondary or tertiary effluents may be tested using this method, but results are subject to verification by MPN technique. The membrane filter technique utilizes a specially designed filter pad with uniformly sized pores (openings) that are small enough to prevent bacteria from entering the filter. Another unique characteristic of the filter allows liquids, such as the media, placed under the filter to pass upward through the filter to provide nourishment required for bacterial growth.

Note: In the membrane filter method, the number of colonies grown estimates the number of coliforms.

The procedure for the membrane filter method is described below.

1. Sample filtration:
   a. Select a filter, and aseptically separate it from the sterile package.
   b. Place the filter on the support plate with the grid side up.
   c. Place the funnel assembly on the support; secure as needed.
   d. Pour 100 mL of sample or serial dilution onto the filter; apply vacuum.

Note: The sample size and necessary serial dilution should produce a growth of 20 to 60 fecal coliform colonies on at least one filter. The selected dilutions must also be capable of showing permit excursions.

   e. Allow all of the liquid to pass through the filter.
   f. Rinse the funnel and filter with three portions (20 to 30 mL) of sterile, buffered dilution water. (Allow each portion to pass through the filter before the next addition.)

Note: Filtration units should be sterile at the start of each filtration series and should be sterilized again if the series is interrupted for 30 minutes or more. A rapid interim sterilization can be accomplished by a 2-minute exposure to ultraviolet (UV) light, flowing steam, or boiling water.

2. Incubation:
   a. Place absorbent pad into culture dish using sterile forceps.
   b. Add 1.8 to 2.0 mL M-FC media to the absorbent pad.
   c. Discard any media not absorbed by the pad.

d. Filter sample through sterile filter.
e. Remove filter from assembly, and place on absorbent pad (grid up).
f. Cover culture dish.
g. Seal culture dishes in a weighted plastic bag.
h. Incubate filters in a water bath for 24 hours at 44.5 ± 0.2°C.

Colony Counting

Upon completion of the incubation period, the surface of the filter will have growths of both fecal coliform and nonfecal coliform bacterial colonies. The fecal coliform will appear blue in color, while nonfecal coliform colonies will appear gray or cream colored. When counting the colonies, the entire surface of the filter should be scanned using a 10× to 15× binocular, wide-field dissecting microscope. The desired range of colonies, for the most valid fecal coliform determination is 20 to 60 colonies per filter. If multiple sample dilutions are used for the test, counts for each filter should be recorded on the laboratory data sheet.

1. *Too many colonies*—Filters that show a growth over the entire surface of the filter with no individually identifiable colonies should be recorded as confluent growth. Filters that show a very high number of colonies (greater than 200) should be recorded as TNTC (too numerous to count).
2. *Not enough colonies*—If no single filter meets the desired minimum colony count (20 colonies), the sum of the individual filter counts and the respective sample volumes can be used in the formula to calculate the colonies/100 mL.

Note: In each of these cases, adjustments in sample dilution volumes should be made to ensure that future tests meet the criteria for obtaining a valid test result.

Determining Coliform Density

The fecal coliform density can be calculated using the following formula:

$$\text{Colonies/100 mL} = \frac{\text{Colonies counted}}{\text{Sample volume (mL)}} \times 100 \text{ mL} \tag{8.13}$$

■ Example 8.8

Problem: Using the data shown below, calculate the colonies per 100 mL for the influent and effluent samples noted.

	Influent Sample Dilutions			Effluent Sample Dilutions		
Sample (mL)	1.0	0.1	0.01	10	1.0	0.1
Colonies counted	97	48	16	10	5	3

Solution:

1. *Influent sample.* Select the influent sample filter that has a colony count in the desired range (20 to 60). Because one filter meets this criterion, the remaining influent filters that did not meet the criterion are discarded.

$$\text{Colonies/100 mL} = \frac{\text{48 colonies}}{\text{0.1 mL}} \times 100 \text{ mL} = 48{,}000 \text{ colonies/100 mL}$$

2. *Effluent sample.* Because none of the filters for the effluent sample meets the minimum test requirement, the colonies per 100 mL must be determined by totaling the colonies on each filter and the sample volumes used for each filter.

$$\text{Total colonies} = 10 + 5 + 3 = 18 \text{ colonies}$$

$$\text{Total sample} = 10.0 \text{ mL} + 1.0 \text{ mL} + 0.1 \text{ mL} = 11.1 \text{ mL}$$

$$\text{Colonies/100 mL} = \frac{18 \text{ colonies}}{11.1 \text{ mL}} \times 100 = 162 \text{ colonies/100 mL}$$

Note: The USEPA criterion for fecal coliform bacteria in bathing waters is a logarithmic mean of 200 per 100 mL, based on the minimum of 5 samples taken over a 30-day period, with not more than 10% of the total samples exceeding 400 per 100 mL. Because shellfish may be eaten without being cooked, the strictest coliform criterion applies to shellfish cultivation and harvesting. The USEPA criterion states that the mean fecal coliform concentration should not exceed 14 per 100 mL, with not more than 10% of the samples exceeding 43 per 100 mL.

Interferences

Large amounts of turbidity, algae, or suspended solids may interfere with this technique by blocking filtration of the sample through the membrane filter. Dilution of these samples to prevent this problem may make the test inappropriate for samples with low fecal coliform densities because the sample volumes after dilution may be too small to give representative results. The presence of large amounts of non-coliform group bacteria in the samples may also prohibit the use of this method.

Note: Many NPDES discharge permits require fecal coliform testing. Results for fecal coliform testing must be reported as a geometric mean (average) of all the test results obtained during a reporting period. A geometric mean, unlike an arithmetic mean or average, dampens the effect of very high or low values that might otherwise cause a non-representative result.

Apparent Color Testing and Analysis

Color in water often originates from organic sources: decomposition of leaves, and other forest debris such as bark, pine needles, etc. Tannins and lignins, organic compounds, dissolve in water. Some organics bond to iron to produce soluble color compounds. Biodegrading algae from a recent bloom may cause significant color. Though less likely a source of color in water, possible inorganic sources of color are salts of iron, copper, and potassium permanganate added in excess at the treatment plant.

Note: Noticeable color is an objectionable characteristic that makes the water psychologically unacceptable to the consumer.

Recall that *true color* is dissolved. It is measured colorimetrically and compared against a USEPA color standard. *Apparent color* may be caused by suspended material (turbidity) in the water. It is important to point out that even though it may also be objectionable in the water supply, it is not meant to be measured in the color analysis or test. Probably the most common cause of apparent color is particulate oxidized iron.

By using established color standards, people in different areas can compare test results. Over the years, several attempts have been made to standardize the method of describing the apparent color of water using comparisons to color standards. *Standard Methods* (APHA, 1998) recognizes visual comparison methods as reliable methods for analyzing water from the distribution system.

The Forel–Ule color scale consists of a dozen shades ranging from deep blue to khaki green that are typical of offshore and coastal bay waters. Another visual comparison method is the Borger color system, which provides an inexpensive, portable color reference for shades typically found in natural waters; it can also be used for its original purpose—describing the colors of insects and larvae found in streams of lakes. The Borger Color System also allows recording of the color of algae and bacteria on streambeds. To ensure reliable and accurate descriptions of apparent color, use a system of color comparisons that is reproducible and comparable to the systems used by other groups.

Note: Do not leave color standard charts and comparators in direct sunlight.

Measured levels of color in water can serve as indicators for a number of conditions; for example, transparent water with a low accumulation of dissolved minerals and particulate matter usually appears blue and indicates low productivity. Yellow to brown color normally indicates that the water contains dissolved organic materials, humic substances from soil, peat, or decaying plant material. Deeper yellow to reddish colors indicate the presence of some algae and dinoflagellates. A variety of yellows, reds, browns, and grays is indicative of soil runoff.

Note: Color by itself has no health significance in drinking waters; however, a secondary MCL is set at 15 color units, and it is recommended that community supplies provide water that has less color.

When treating for color in water, alum and ferric coagulation is often effective. This process removes apparent color and often much of the true color. Oxidation of color-causing compounds to a noncolored version is sometimes effective. Activated carbon treatment may adsorb some of the organics causing color. For apparent color problems, filtration is usually effective in trapping the colored particles.

Odor Analysis of Water

Odor is expected in wastewater—the fact is, any water containing waste, especially human waste, has a detectable (expected) odor associated with it. Odor in a raw water source (for potable water) is caused by a number of constituents; for example, chemicals that may come from municipal and industrial waste discharges or natural sources such as decomposing vegetable matter or microbial activity may cause odor problems. Odor affects the acceptability of drinking water, the aesthetics of recreation water, and the taste of aquatic foodstuffs.

The human nose can accurately detect a wide variety of smells and serves as the best odor-detection and testing device currently available. To measure odor, collect a sample in a large-mouthed jar. After waving off the air above the water sample with your hand, smell the sample. Use the list of odors provided in Table 8.10, which is a system of qualitative descriptions that can help monitors describe and record odors they detect. Record all observations (APHA, 1998).

When treating for odor in water, removal depends on the source of the odor. Some organic substances that cause odor can be removed with powdered activated carbon. If the odor is of gaseous origin, scrubbing (aeration) may remove it. Some odor-causing chemicals can be oxidized to odorless chemicals with chlorine, potassium permanganate, or other oxidizers. Settling may remove some material which, when later dissolved in the water, may have potential odor-causing capacity. Unfortunately, the test for odor in water is subjective and is not very accurate, but no scientific means of measurement exists.

To test odor in water intended for potable water use, a sample is generally heated to 60°C. Odor is observed and recorded. A threshold odor number (TON) is assigned. TON is found by using the following equation:

$$\text{TON} = \frac{\text{Total volume of water sample}}{\text{Lowest sample volume with odor}} \tag{8.14}$$

TABLE 8.10
Descriptions of Odors

Nature of Odor	Description	Examples
Aromatic	Spicy	Camphor, cloves, lavender
Balsamic	Flowery	Geranium, violet, vanilla
Chemical	Industrial wastes or chlorinous treatments	Chlorine
Hydrocarbon	Oil refinery wastes	
Medicinal	Phenol and iodine	
Sulfur	Hydrogen sulfide	
Disagreeable	Fishy	Dead algae
Pigpen	Algae	
Septic	Stale sewage	
Earthy	Damp earth	
Peaty	Peat	
Grassy	Crushed grass	
Musty	Decomposing straw; moldy	
Vegetable	Root vegetables	Damp cellar

Source: Adapted from APHA, *Standard Methods for the Examination of Water and Wastewater*, 20th ed., American Public Health Association, Washington, DC, 1998.

Chlorine Residual Testing and Analysis

Chlorination is the most widely used means of disinfecting water in the United States. When chlorine gas is dissolved into pure water, it forms hypochlorous acid, hypochlorite ion, and hydrogen chloride (hydrochloric acid). The total concentration of HOCl and OCl ion is known as *free chlorine residual.* Current federal regulations cite the following approved methods for determination of total residual chlorine (TRC):

1. DPD–spectrophotometric
2. Titrimetric–amperometric direct
3. Titrimetric–iodometric direct
4. Titrimetric–iodometric back
   a. Starch iodine endpoint, iodine titrant
   b. Starch iodine endpoint, iodate titrant
5. Amperometric endpoint
6. DPD–FAS titration
7. Chlorine electrode

All of these test procedures are approved methods and, unless prohibited by the plant's NPDES discharge permit, can be used for effluent testing. Based on current most popular method usage in the United States, discussion is limited to

1. DPD–spectrophotometric
2. DPD–FAS titration
3. Titrimetric–amperometric direct

Note: Treatment facilities required to meet nondetectable total residual chlorine limitations must use one of the test methods specified in the plant's NPDES discharge permit. For information on any of the other approved methods, refer to the appropriate reference cited in the federal regulations.

Fluorides

It has long been accepted that a moderate amount of fluoride ions (F^-) in drinking water contributes to good dental health; it has been added to many community water supplies throughout the United States to prevent dental caries in children's teeth. Fluoride is seldom found in appreciable quantities of surface waters and appears in groundwater in only a few geographical regions. Fluorides are used to make ceramics and glass. Fluoride is toxic to humans in large quantities and to some animals. The chemicals added to potable water in treatment plants include the following:

- NaF (sodium fluoride, solid)
- Na_2SiF_6 (sodium silicofluoride, solid)
- H_2SiF_6 (hyrofluosilicic acid; most widely used)

Analysis of the fluoride content of water can be performed using the colorimetric method. In this test, fluoride ion reacts with zirconium ion and produces zirconium fluoride, which bleaches an organic red dye in direct proportion to its concentration. This can be compared to standards and read colorimetrically.

THOUGHT-PROVOKING QUESTIONS

8.1 Do you think biomonitoring is an old or new concept? Explain.

8.2 Are environmental monitoring and biomonitoring linked? What is the difference between them? Explain.

8.3 Biomonitoring is a finite process. Do you agree with this statement? Explain.

REFERENCES AND RECOMMENDED READING

APHA. (1971). *Standard Methods for the Examination of Water and Wastewater*, 17th ed. Washington, DC: American Public Health Association.

APHA. (1998). *Standard Methods for the Examination of Water and Wastewater*, 20th ed. Washington, DC: American Public Health Association, pp. 4–129.

AWWA. (1995). *Water Treatment*, 2nd ed. Denver, CO: American Water Works Association.

Barbour, M.T., Gerritsen, J., Snyder, B.D., and Stibling, J.B. (1997). *Revision to Rapid Bioassessment Protocols for Use in Streams and Rivers: Periphytons, Benthic Macroinvertebrates, and Fish*. Washington, DC: U.S. Environmental Protection Agency.

Bly, T.D. and Smith, G.F. (1994). *Biomonitoring Our Streams: What's It All About*? Nashville, TN: U.S. Geological Survey.

Botkin, D.B. (1990). *Discordant Harmonies*. New York: Oxford University Press.

Camann, M. (1996). *Freshwater Aquatic Invertebrates: Biomonitoring*. Arcata, CA: Humboldt State University (http://www.humboldt.edu).

Carson, R.L. (1962). *Silent Spring*. Boston: Houghton Mifflin.

Harr, J. (1995). *A Civil Action*. New York: Vantage Books.

Hillborn, R. and Mangel, M. (1997). *The Ecological Detective: Confronting Models with Data*. Princeton, NJ: Princeton University Press.

Huff, W.R. (1993). Biological indices define water quality standard. *Water Environment & Technology*, 5, 21–22.

Huston, M.A. (1994). *Biological Diversity: The Coexistence of Species on Changing Landscapes*. New York: Cambridge University Press.

Kittrell, F.W. (1969). *A Practical Guide to Water Quality Studies of Streams*. Washington, DC: U.S. Department of Interior.

O'Toole, C., Ed. (1986). *The Encyclopedia of Insects*. New York: Facts on File.

Pimm, S. (1991). *The Balance of Nature: Ecological Issues in the Conservation of Species and Communities*. Chicago, IL: University of Chicago.

Spellman, F.R. (1999). *Spellman's Standard Handbook for Wastewater Operators*, Vol. 1. Boca Raton, FL: CRC Press.

Spellman, F.R. (2003). *Handbook of Water and Wastewater Treatment Plant Operations*. Boca Raton, FL: Lewis Publishers.

Spellman, F.R. (2014). *Handbook of Water and Wastewater Treatment Plant Operations*, 3rd ed. Boca Raton, FL: CRC Press.

Spellman, F.R. and Drinan, J.E. (2001). *Stream Ecology and Self-Purification: An Introduction*. Boca Raton, FL: CRC Press.

Tchobanoglous, G. and Schroeder, E.D. (1985). *Water Quality*. Reading, MA: Addison-Wesley.

USEPA. (2000a). *Monitoring Water Quality: Intensive Stream Bioassay*. Washington, DC: U.S. Environmental Protection Agency.

USEPA. (2000b). *Volunteer Stream Monitoring: A Methods Manual*. Washington, DC: U.S. Environmental Protection Agency, pp. 1–35.

USEPA. (2011). *A Primer on Using Biological Assessments to Support Water Quality Management*, EPA 810-R-11-01. Washington, DC: U.S. Environmental Protection Agency.

Velz, C.J. (1970). *Applied Stream Sanitation*. New York: Wiley Interscience.

9 Water Economics

> The market price of a good does not necessarily reflect its true value.
>
> **USEPA (2012)**

Question: Have you ever thought about how expensive it can be to transport 1000 or more gallons of water (at 8.34 lb per gallon) from point A to point B when the two points are separated by 350 miles? Does the transportation of water from point A to point B make it a valuable commodity? Do these transportation costs make the water valuable?

Answer: Maybe.

INTRODUCTION

A fundamental understanding of water economics is foundational to any study related to the science of water. This is because water is vital to a productive and growing economy in the United States, directly and indirectly affecting the production of goods and services in many sectors (USEPA, 2012). Griffin (2006, p. xiii) made the point that, "Due to other pressing demands, few water managers or planners have invested in economics. Because they come mainly from the engineering and science disciplines, most water professionals have limited exposure to economic fundamentals." You can't maximize the management of a water source, a water treatment plant, and/or a water distribution system without some knowledge of economics.

This chapter examines the value of water to the U.S. economy from a number of perspectives. Among the most important of these are the following:

- *Microeconomic efficiency*—The value of water is related to its relative scarcity, its alternative uses, and the opportunity costs of those uses. To maximize social welfare, scarce resources must be used in ways that provide the greatest value. Markets can produce this economically efficient use of resources, provided that they are competitive and well informed. When they are not—for example, when trade is restricted, when prices are distorted by taxes or subsidies, or when the cost to the consumer fails to incorporate externalities such as environmental impacts—an inefficient use of resources is likely to occur.
- *Sustainability*—The value of water must be considered within the context of dynamic and integrated environmental, economic, and social systems. Within an integrated system, the value of a resource is a function of both the direct and indirect impacts associated with its use. The impacts of interest include not only those that are reflected in markets but also those that affect the production or consumption of non-market goods and services. This includes both immediate effects and those that may not be realized until well into the future.

In addition, the concepts that follow provide important background for the water use and value information discussed in subsequent sections of this chapter.

From an economist's viewpoint, the value of water can be analyzed through both a microeconomic and a macroeconomic lens. *Microeconomics* provides a framework for examining the value of water to an individual household, firm, or industry. In contrast, *macroeconomics* provides a multi-sector framework for understanding how water resources contribute to economic activity at a regional or national level, as measured by such indicators as employment and gross domestic product. As discussed below, macroeconomic analysis also offers a means of understanding the complex interrelationships between the use of water in one sector of the economy and economic activity in others.

MICROECONOMIC CONCEPTS

Before beginning a detailed discussion of the microeconomic concepts of water, it is first important for the reader to be familiar with and to understand some key terms that he or she may not be familiar with. Familiarity with (and later understanding of) these terms provides for a smoother introduction to the important aspects of water economics. The following terms are provided to impart recognition and understanding of the material to be presented throughout this chapter. The reader with little or no training in economics will find an understanding of these important terms helpful. These terms are presented in the order in which they appear in the discussion that follows.

- *Value in use*—What we call *utility* (and its *value in exchange*); in other words, its *price*
- *Benefit*—Derived from its use
- *Willingness to pay*—Maximum amount a person would be willing to pay, sacrifice, or exchange in order to receive a good to avoid something undesired, such as environmental damage
- *Total economic value*—In cost–benefit analysis, refers to the value derived by people from a natural resource or a man-made heritage resource or infrastructure system, compared to not having it
- *Marginal value*—A value that holds true given particular constraints
- *Consumer surplus*—The difference between the total amount consumers would be willing to pay to consume the quantity of goods transacted on the market and the amount they actually have to pay for those goods
- *Opportunity cost*—Cost of an alternative that must be foregone in order to pursue a certain action
- *Scarcity value*—An economic factor that increases an item's relative price based more on its relatively low supply
- *Externalities*—A cost of a benefit that results from an activity or transaction and that affects an otherwise uninvolved party who did not choose to incur that cost or benefit

DID YOU KNOW?

We have stated that if someone possesses a good, he will use it to satisfy some need or want. Which one? As you would expect, the one that takes highest priority. Eugen von Bohm-Bawerk (1888) illustrated this with the example of a farmer having five sacks of grain. With the first, he will make bread to survive. With the second, he will make more bread in order to be strong enough to work. With the next, he will feed his farm animals. The next is used to make whiskey, and the last one he feeds to the pigeons. If one of those bags is stolen, he will not reduce each of those activities by one-fifth; instead, he will stop feeding the pigeons. The point is the value of the fifth bag of grain is equal to the satisfaction he gets from feeding the pigeons. If he loses that bag and neglects the pigeons, his least productive use of the remaining grains is to make whiskey; accordingly, the value of a fourth bag of grain is the value of the whiskey. Only if he loses four bags of grain will he start eating less; that is the most productive use of his grain. The bottom line: The last bag is worth his life.

- *Marginal product value of water*—An analytical method developed in which the impact of small economic changes is evaluated
- *Marginal rate of technical substitution*—The amount by which the quantity of one input has to be reduced when one extra unit of another input is used, so the output remains constant
- *Marginal utility*—An item's most important use to a person; if someone possesses a good, he will use it to satisfy some need or want

The question of water's value has been the subject of inquiry since the first formulation of modern economic theory. Indeed, much has been written about the value of water. Debates about the value of water, however, are nothing new, and in several ancient sources we can find opinions on this issue:

> All you who are thirsty,
> come to the water!
> You who have no money,
> come, buy grain and eat!
> Come, buy grain without money,
> wine and milk without cost!
>
> **—Isaiah 55:1**

> For only what is rare is valuable; and "water," which, as Pindar says, is the "best of all things," is also the cheapest.
>
> **—Plato (*Euthydemus*, 304 BCE)**

Adam Smith (1776) described the *paradox of value* (also known as the *diamond–water paradox*):

> Nothing is more useful than water: but it will purchase scarce anything; scarce anything can be had in exchange for it. A diamond, on the contrary, has scare any use-value in use; but a very great quantity of other goods may frequently be had in exchange for it.

There is criticism of Smith's statement about the paradox; some say it is flawed because it consisted of a comparison between heterogeneous goods, and such comparison would have required using the concept of marginal utility of income. And because this concept was not known in Smith's time, then the *value in use* and *value in exchange* judgment may be meaningless. George Stigler (1950) pointed out:

> The paradox—that value in exchange may exceed or fall short of value in use—was, strictly speaking, a meaningless statement, for Smith had no basis (i.e., no concept of marginal utility of income or marginal price of utility) on which he could compare such heterogeneous quantities. On any reasonable interpretation, moreover, Smith's statement that value in use could be less than value in exchange was clearly a moral judgment, not shared by the possessors of diamonds. To avoid the incomparability of money and utility, one may interpret Smith to mean that the ratio of values of two commodities is not equal to the ratio of their total utilities. Or, alternatively, that the ratio of the prices of two commodities is not equal to the ratio of their total utilities; but this also required an illegitimate selection of units: The price of what quantity of diamonds is to be compared with the price of one gallon of water?

Note: This text takes the view that the genius Adam Smith was way ahead of his time and spot on with his predictions and views; therefore, they are projected throughout this chapter.

Smith attempted to resolve this seeming paradox by differentiating between a good's *value in use*—again, what we might call its *utility*—and its *value in exchange*—in other words, its price. Although later economists abandoned this distinction in favor of a more elegant resolution, Smith was simply articulating a common thought: that the market price of a good does not necessarily reflect its true value. This is particularly obvious in the case of a good like water, which is essential to human life. Indeed, nothing is more useful than water, yet it is bought and sold every day, often at a very low price. What does this tell us about the value of water, and how are we to use this and other economic data in making choices about the use and management of such an essential resource? The discussion that follows attempts to address this question (USEPA, 2012).

Water's Price, Cost, Value, and Essential Nature

Why is the price of diamonds high and the price of water low? The key insight, arrived at by later economists, is that a commodity's price is related not to its total value in use but rather to the usefulness of the last unit consumed. More specifically, price is determined by the simultaneous interaction of two market forces, supply and demand, which reflect, respectively, the *cost* of producing the commodity and the *benefit* derived from its use. When water is abundant, as it was in Smith's native Britain, both the cost of supplying another gallon of water and the benefit derived from consuming that gallon are low; thus, the price of water is low. Conversely, the scarcity of diamonds relative to consumer demand gives them a high price. If the situation were reversed—if, for example, consumers awoke to find themselves in a sparkling desert in which water was scarce but diamonds were plentiful—the relative value of the two commodities would quickly be reversed (Nicholson, 1978).

As mentioned, the price of a commodity is determined by the interaction of supply and demand. In contrast, the concept of *value* reflects the net difference between the gross benefit received from the use of that commodity and the cost of that use. In this sense, the value of a commodity is determined largely by the nature of demand for it, as measured by the amount of money that prospective purchasers of the commodity would be willing to pay to acquire a specific amount. In turn, *willingness to pay* is determined by the marginal benefit that these purchasers would derive from each increment of consumption. In the case of water, these prospective purchasers include the following two groups:

- Households and similar users for whom water serves as a final good
- Farmers, utilities, manufacturers, and others for whom water serves as an input to production

From the perspective of either group, the *total economic value* of water may be quite high, if not infinite, while its *marginal value* at current levels of supply may be quite low. Hanemann (2005, pp. 81–82) made this point clearly when discussing the essential nature of water:

> Water is essential for all life—human, animal, or plant. In economics, there is a concept … that formalizes this notion. The concept can be applied either to something that is an input to production or to something that is directly enjoyed by people as a consumption commodity. In the case of an input, if an item has the property that no production is possible when this input is lacking, the item is said to be an essential input. In the case of a final good, if it has the property that *no* amount of any *other* final good can compensate for having a zero level of consumption of this commodity, then it is said to be an essential commodity. Water obviously fits the definition of an essential final good: human life is not possible without access to 5 to 10 liters of water per person per day. Water fits the definition of an essential input in agriculture, and also in several manufacturing industries. … However, essentialness conveys to information about the productivity or value of water. It implies nothing about the marginal value associated with, say, applying 76 versus 89 cm of water to irrigate cotton in the Central Valley of California. It says nothing about the marginal value of residential water use at the levels currently experienced in Western Europe or the United States—the latter averages … more than two orders of magnitude larger than the minimum quantity that is needed for human survival.

Therefore, in discussing the value of water, it is important to be clear about terms and to focus on the appropriate dimensions of value. As Hanemann noted, most people have access to some water, and most policy interventions involve changing the quantity and/or quality of access, rather than transforming the situation from no access to some access. In most public policy applications within the United States, the relevant consideration is the impact of a decision on the marginal use of water and the change in net benefits associated with that change in use. These are the values upon which this book focuses.

Average, Marginal, and Total Economic Values

The simplified linear supply and demand curves presented in Figure 9.1 help to illustrate the relationship between marginal and total economic values in the supply of water. Consider, in this case, a group of farmers served by an irrigation district. The downward sloping demand curve (*D*) indicates the farmers' willingness to pay for each incremental unit of water, which declines as the amount of water supplied increases; this reflects the declining marginal benefit of using additional quantities of water for irrigation purposes. The two horizontal supply curves indicate the unit cost to the irrigation district of supplying water under two scenarios: normal weather conditions, which allow the district to draw water from a nearby river (S_1), and drought conditions, which force it to draw and transport water from a more distant and expensive source (S_2). As Figure 9.1 indicates, farmers will consume more water (Q_1) under normal conditions, when the irrigation district can supply water at a lower price (P_1), and the total economic value realized is represented by the total area under the demand curve, from the origin to Q_2. In contrast, farmers will consume less water (Q_2) if drought conditions force the irrigation district to raise its price (P_2) to cover the added cost of drawing water from its secondary source. In this case, the marginal value realized from the farmers' use of irrigation water (P_2) will be higher, because the farmers will restrict their use of water to applications in which the marginal return is equal to or greater than P_2; however, the total economic value derived from the use of water will be reduced to the area beneath the demand curve from the origin to Q_2, reflecting the elimination of uses with a marginal return between P_1 and P_2.

The example presented above assumes that farmers will consume water up to the point at which the marginal benefit (or marginal value) of consumption equals its price. It is important to point out, however, that the total economic value derived from the use of water exceeds the group's total expenditures on it. Under the first supply scenario, the farmers' total expenditures on water are represented by the rectangle bounded by Q_1, S_1, and the *x*- and *y*-axes. The total economic value realized, however, also includes both the dark gray and light gray areas that together form

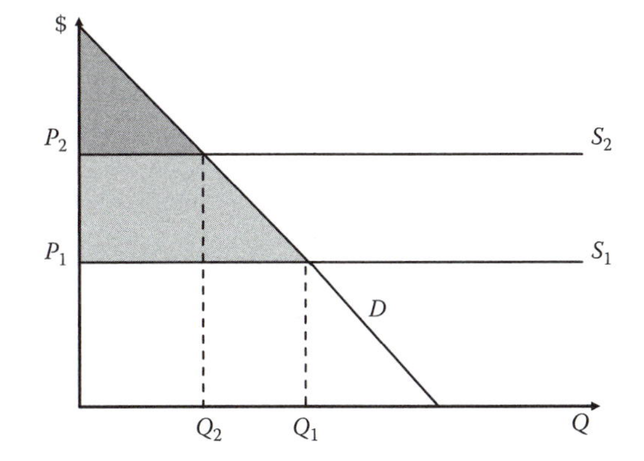

FIGURE 9.1 Illustration of water's economic value under two supply scenarios.

a triangle above P_1. Likewise, under the second supply scenario, the farmers' total expenditures on water are represented by the rectangle bounded by the two axes, Q_2, and S_2; the total economic value includes this rectangle, as well as the dark gray triangle above P_2. In both cases, the total economic value exceeds the farmers' expenditures because, as the demand curve indicates, they would be willing to pay more than they actually pay for all but the last unit of water they consume. This difference, which economists call *consumer surplus*, represents the net benefit associated with the consumption of water.

The example presented above also helps to illustrate the relationship between the marginal value of water and its average value—that is, the total economic value derived from the use of water divided by the quantity employed. Under drought conditions, for example, the average value of water can be determined by dividing the total value derived from its use by the total volume used (Q_2). Consistent with the downward-sloping demand function, the average value of water in this application will exceed its marginal value (P_2), reflecting the marginal return realized from its use.

The relationship between the marginal value of water and its average value may seem moot, but it can be quite important in a decision-making context. In many cases, available estimates of the value of water reflect average values, in large part because they are relatively simple to estimate. In contrast, estimates of marginal values are more difficult to derive, particularly in the absence of price data; as Hanemann noted, estimation of marginal values in such cases requires a formal or informal model of how water generates value in the production of a particular good. Reliance on average values, however, can significantly overestimate the marginal value of water as an input to production and can lead to decisions that appropriate consideration of marginal values would not support (USEPA, 2012).

Why Water Prices May Fail to Encourage Efficient Use

A most important principle of economics is that competitive and well-informed markets will produce an optimal use of resources, reaching an equilibrium at which the market price of a good is equal to both the marginal cost of supplying it and its marginal value to consumers. In most cases, however, the price at which water is sold in the United States is not a product of market forces that will yield this optimal use. In part, this is because the prices charged for water do not, in most instances, reflect the full *opportunity cost* of its use—that is, the cost of forgoing the use of the water for its best alternative purpose. Stated differently, it is what you have to forgo when you choose to do A rather than B. As Hanemann (2005, pp. 81–82) noted:

> It is important to emphasize that the prices which most users pay for water reflect, at best, its physical supply cost and *not its scarcity value*. Users pay for the capital and operating costs of the water supply infrastructure but … there is no charge for the water *per se*. Water is owned by the state, and the right to use it is given away for free. Water is thus treated differently than … minerals for which the … government requires payment of a royalty to extract the resource.

The opportunity cost of water may be negligible when its supply is abundant but significant when water is scarce. Beyond this limitation, the prices charged for water may not even reflect full supply costs. In some cases, this is the result of explicit government subsidies, as is the case with some federal irrigation projects. In others, this may be the result of the common practice of establishing prices to recover the historic cost of public water supply systems, rather than their long-run future replacement costs. Again, Hanemann (2005, pp. 81–82) has provided a helpful explanation:

> There is typically a large gap between these two costs because of the extreme … longevity of the surface water supply infrastructure. The capital intensity of the infrastructure exacerbates the problem because, after a major surface water project is completed, … supply capacity so far exceeds current demand there is a strong economic incentive to set price to cover just the short-run marginal cost (essentially, the

operating cost), which is typically minuscule. As demand eventually grows and the capacity becomes more fully utilized, it is economically optimal to switch to pricing based on long-run (i.e., replacement) marginal cost, but by then water agencies are often politically locked into a regime of low water prices focused narrowly on the recovery of the historical cost of construction.

Thus, we cannot assume in all cases that the prices currently charged for water reflect its true (long-run) marginal cost or that the resulting use of water resources is economically efficient. To the extent that water is underpriced, it will be used in quantities that exceed the economically efficient amount and in applications in which its marginal value is less than its true opportunity cost.

In addition to the issues discussed above, it is important to note other factors that may lead markets to fail to account for the full cost of water's use. Chief among these are *externalities*, costs imposed on third parties that are not reflected in market prices. One clear example of an externality is the impact of water use on the quality of water available for other purposes (e.g., the impact of pollutants contained in irrigation return flows on the quality of water available downstream). These pollutants may affect the costs that downstream municipalities incur to treat and supply drinking water to their residents; the costs of other market uses of water may also be affected. Such externalities may also affect the provision of non-market services, such as recreational fishing opportunities. To the extent that this occurs, the failure of the market to reflect the true costs will lead to inefficiencies in the use of water resources.

Marginal Product Value of Water

An additional challenge in assessing the value of water is the fact that, for many economic purposes—in manufacturing, in agriculture, in mining, or in generating electricity—water is not purchased from an external provider but is instead self-supplied. As a result, no market information on the user's willingness to pay for water is available. Even in the absence of market data, however, it is possible to draw inferences about a producer's willingness to pay for water, based on its value as an input to production. Specifically, the marginal value of water to the producer is equal to the associated gain in the value of the producer's output: the *marginal product value of water.* All else being equal, a profit-maximizing producer would be expected to use water up to the point at which the value of water's marginal product is equal to its marginal cost. Thus, when combined with information on product prices, information on the impact of water on a producer's output (e.g., the impact of various levels of irrigation on crop yields) can provide a meaningful indicator of the marginal value of water in particular applications.

It is important to note that the value of the marginal product of water in a given application is likely to depend on multiple factors, including the overall mix of inputs used in the production process. In agriculture, for example, the value of the marginal product of water may depend in part on fertilizer or pesticide application rates. The value of water may also be affected by the *marginal rate*

DID YOU KNOW?

Two commodities are said to be a *perfect substitutes* in consumption if the consumer is willing to trade off one for the other at a fixed rate of exchange, regardless of how much or how little is consumed; in the consumer's eyes, each commodity can always be used in exactly the same way, with exactly the same outcome. When two commodities are perfect substitutes, they have essentially the same value. The polar opposite of a perfect substitutes is a *perfect complement.* Two commodities are perfect complements if they are valued in fixed proportions to one another; consequently, they will always be purchased together in fixed proportions. In this case, no value is placed on increasing one of the items unless there is a corresponding increase in the other; an old-fashioned example (in England) was tea and milk.

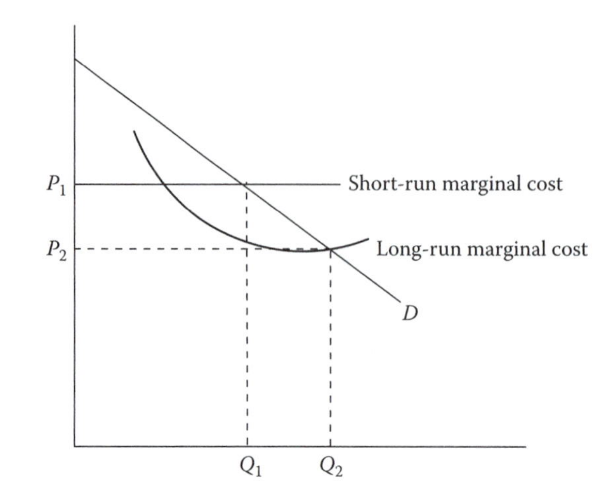

FIGURE 9.2 Short-run vs. long-run marginal costs.

of technical substitution between inputs—in other words, the extent to which the use of one input, such as water, may be substituted for another, such as labor, while maintaining the same level of production. Within a given industry, different proceeds may employ different suits of inputs; thus, the value of the marginal product of water may vary from case to case. This is particularly true in the case of agriculture, where regional differences in climate, soils, or other factors may dictate significant variation in the mix of inputs employed and contribute to marked disparities in the value of the marginal product of water.

Long-Run vs. Short-Run Values

In addition to the variation described above, the value of water as an input to production depends on the temporal perspective employed (i.e., the specific point of view or attitude one holds about time, such as long-run vs. short-run). In the short run (when certain inputs, such as equipment and other capital stock, are fixed), the total economic value derived from the use of water may be constrained. In the long-run, however, when such constraints are eliminated, the total economic value derived from the use of water may increase. This might be the case for a manufacturer who, in the short-run, must purchase water from a municipal utility at a set price but, in the long-run, can reduce the marginal cost of the water it requires by drilling a well. As Figure 9.2 illustrates, the reduction in the marginal cost of water results in an increase in the manufacturer's demand for water and an increase in the total economic value derived from use of the water.

Dimensions of Water That Influence Its Value

Water is not a one-dimensional commodity. A user's willingness to pay for water from a particular source may depend upon the following:

- *Quantity*—The total volume of water the source can supply
- *Time*—When the water will be supplied
- *Space*—The location at which the water will be supplied
- *Reliability*—The likelihood that the supply will not be interrupted
- *Quality*—The extent to which the water is free of contaminants and otherwise suitable for the intended use

MACROECONOMIC CONCEPTS

To fully appreciate water's role in the U.S. economy, we must consider not only microeconomic principles but also the use of water in a macroeconomic context. Our dependence on a clean, reliable supply of water becomes evident when we consider the sectors that use water directly and their relationship to the economy at large. The discussion that follows presents a wide-angle view of the structure of the U.S. economy, noting not just where water use is concentrated but also how the flow of goods and services between sectors links the use of water in one sector to economic productivity in others.

Sector View of the Economy

Economists have long sought to characterize the diverse and complex elements of economic systems by grouping activities into major economic sectors. Some of the earliest distinctions were established in ancient Greece, which recognized agriculture, household management, and trade. Later systems classified manufacturing activity by industry and expanded to recognize other forms of activity, such as education and public administration. In the 20th century, economists such as Zoltan Kenessey arrived at the following designations, which are still commonly used today:

- *Primary sectors*, including agriculture, forestry, fishing, and mining
- *Secondary sectors*, including utilities, manufacturing, and construction
- *Tertiary sectors*, including transportation, wholesale trade, and retail trade
- *Quaternary sectors*, including finance, insurance, real estate, and public administration

Kenessey (1987, p. 367) noted that these sector groupings, or "mega-sectors," are "anchored to the four major elements of the work process: *extraction*, *processing*, *delivery*, and *information*." As described below, this logical outline of the work process provides a useful framework for illustrating how the direct use of water—which occurs predominantly in the primary and secondary sectors of the economy—ultimately affects the production of goods and services in sectors in which water is not a direct input. It also provides a basis for evaluating how shocks in the availability or quality of water may affect the structure and performance (e.g., efficiency) of the economy as a whole

Water Use in Major Economic Sectors

The U.S. Department of Commerce (DOC) instituted formal use of a sector-based industrial classification system in the 1930s. The DOC's Standard Industrial Classification (SIC) system categorized commercial establishments by predominant economic activity. Over several decades, government economists revised the SIC system to accommodate the changing composition of the U.S. economy. The last update to the SIC system occurred in 1987.

In 1997, the DOC adopted a new scheme: the North American Industry Classification System (NAICS). The NAICS corrected methodological weaknesses in the SIC system and established a consistent classification structure for the United States, Canada, and Mexico, the partners under the North American Free Trade Agreement (U.S. Census Bureau, 2014). Table 9.1 summarizes the major NAICS sectors (at the two-digit level). The correspondence of the NAICS codes to Kenessey's view of the economy is evident. Primary activities, such as agriculture and mining, are assigned the lowest numbered codes. Secondary activities, such as manufacturing, construction, and provision of water and electricity, constitute the next tier. The highest numbered sectors represent tertiary and quaternary activities, such as transportation, trade, the provision of business or personal services, and publish administration.

TABLE 9.1
Major Sectors in North American Industry Classification System

Mega-Sector	NAICS Code	Sector	USGS Water Use Categories
Primary	11	Agriculture, forestry, fishing, and hunting	Irrigation, livestock, and aquaculture
	21	Mining	Mining
Secondary	22	Utilities	Public water supply, thermoelectric power
	23	Construction	NA
	31, 32, 33	Manufacturing	Industrial (e.g., paper, petroleum, metals, chemicals)
Tertiary	42	Wholesale trade	NA
	44, 45	Retail trade	NA
	48, 49	Transportation and warehousing	NA
Quaternary	51	Information	NA
	52, 53	Finance, insurance, real estate, rental, and leasing	NA
	54, 55, 56	Professional and business services	NA
	60	Educational service, health care, and social assistance	NA
	70	Arts, entertainment, recreation, accommodation, and food services	NA
	81	Other services, except government	NA
	92	Government	NA

Source: Bureau of Economic Analysis, *Input–Output Accounts Data*, U.S. Department of Commerce, Washington, DC, 2014 (http://www.bea.gov/industry/io_annual.htm).

Central to this report is the understanding that direct use of water is heavily concentrated in the lower numbered primary and secondary tiers of the economy. The final column of Table 9.1 identifies the water-use sectors corresponding to the NAICS industries, as defined by the U.S. Geological Survey (USGS, 2009). Subsequent chapters provide a more detailed analysis of water use by sector and characterize specific water applications in key economic activities. Here, we simply note that agriculture (including the use of water for livestock and aquaculture) and mining (including the use of water for oil and gas extraction) together account for approximately 35% of all water withdrawals in the United States; these are primary sectors at the top of the NAICS listings. Manufacturing and electric utilities, generally considered secondary sectors, account for another 53% of total withdrawals. In 2010, these four sectors together generated approximately 16% of the nation's gross domestic product (Bureau of Economic Analysis, 2013).

Water Use and Sector Interactions

Economic theorists such as Piero Sraffa and Wassily Leontief studied the interaction of major economic sectors. Leontief earned a Nobel Prize for his work in the field of input–output analysis, modeling the process by which one industry supplied inputs to others. Much of Sraffa's work stressed the role of "basic commodities" in the overall economy and the ways in which the loss of basic commodities could undermine the functioning of a closed economic system (Kenessey, 1987). The fundamental concepts in Sraffa and Leontief's work are reflected in current input–output data for the U.S. economy. The U.S. Bureau of Economic Analysis (BEA) publishes Industry Economic Accounts (http://www.bea.gov/Industry/Index.htm) that trace the flow of goods and services between economic sectors.

TABLE 9.2
Flow of Intermediate Inputs between Mega-Sectors ($ Millions)

	Mega-Sectors Purchasing Intermediate Commodities			
Source of Commodities	**Primary ($ millions)**	**Secondary ($ millions)**	**Tertiary ($ millions)**	**Quaternary ($ millions)**
Primary	$99,107	$719,764	$6671	$16,306
Secondary	$157,254	$1,883,746	$339,373	$1,117,804
Tertiary	$45,090	$427,768	$241,127	$314,851
Quaternary	$95,241	$647,749	$644,544	$4,373,758

Source: Bureau of Economic Analysis, *Input–Output Accounts Data*, U.S. Department of Commerce, Washington, DC, 2014 (http://www.bea.gov/industry/io_annual.htm).

One simple observation is that industries in the secondary sector, such as manufacturing, rely heavily on inputs from industries in the primary sector. Table 9.2 helps to clarify this point, consolidating the flow of intermediate products to the mega-sector level. The output of the primary sector flows predominantly to the secondary sector; in turn, the output of the secondary sector supports both higher level manufacturing and activity in the tertiary and quaternary sectors of the economy.

A second observation is that the sectors of the economy that make the greatest direct use of water (i.e., agriculture, mining, utilities, and manufacturing) are located at its base, in its primary and secondary tiers. The goods and services that these sectors produce are used extensively by the intermediate sectors, which in turn sell their output to the rest of the economy. As such, the economy as a whole is indirectly dependent on the output of industries for which water is a critical input.

The role of water-intensive sectors in supporting the economy becomes evident when we consider how a major water supply shortage could affect the broader economy. For example, a major shortage affecting U.S. agricultural output would result in a shortage of inputs for a variety of industries, with the greatest impact on certain categories of manufacturing. Consider food and beverage manufacturers, which purchase about $194 billion in inputs from U.S. farms. To the extent that food and beverage manufacturers curtailed production, an array of other costs would be affected: the food and beverage makers would purchase less packaging from the paper and plastics industries; transporters of food and beverage products (primarily rail and truck) would haul less freight; wholesalers would sell fewer food products; and so on. The economic repercussions of water shortages are not hypothetical; they can be readily observed in current events. From 2009 to 2011, large parts of west Texas and neighboring states experienced their worst drought on record. The drought depleted storage reservoirs and severely limited water availability for cotton, wheat, and peanut cultivation; beef cattle operations; and other agricultural activity. Immediate effects in the regional agricultural sector included failed crops and a sell-off of cattle (Galbraith, 2011). In the longer run, economists anticipate additional impacts, both domestically and internationally. Domestically, winter wheat shortages are expected to produce price spikes and affect producers and consumers of bread and other wheat-based products. Likewise, once the short-run glut of beef cattle passes, reduced activity at domestic slaughterhouses and increases in domestic beef prices are likely. In addition, much of the cotton crop from the region is exported to textile mills in China, Mexico, and Southeast Asia. These buyers are likely to seek other sources of cotton, endangering the long-term viability of U.S. cotton operations (Hylton, 2011).

Water Use in an Open Economy

Although attempts to describe the structure of a nation's economy may, for simplicity, depict a closed system, it is unquestionably clear today that globalization has increased interdependencies between the U.S. economy and the economics of other nations. As such, domestic water shortages can have far-reaching international repercussions (as illustrated in the case of the Texas drought described above). Likewise, water supply shocks in other nations could affect the availability and prices of imported goods purchased by U.S. consumers. Because water is an essential input into the economic system, major water shortages have the potential to affect not only the balance of trade, but also the structure and composition of different nations' economies, including the U.S. economy. Although disruptions in water supply have yet to play a key role in the mix of goods and services produced in the United States, such outcomes are conceivable in certain scenarios. If the United States were to become an importer of water or water-intensive products, U.S economic security could be affected. This highlights the importance of efficient and sustainable management of domestic water resources, as well as the importance of global cooperation in water resource management.

In summary, Figure 9.3 integrates the key concepts surrounding water use and macroeconomic interactions. Water and other natural resources (this characterization of a natural resource use is not unique to water resources) serve as essential inputs to activity in the primary (i.e., extractive) and

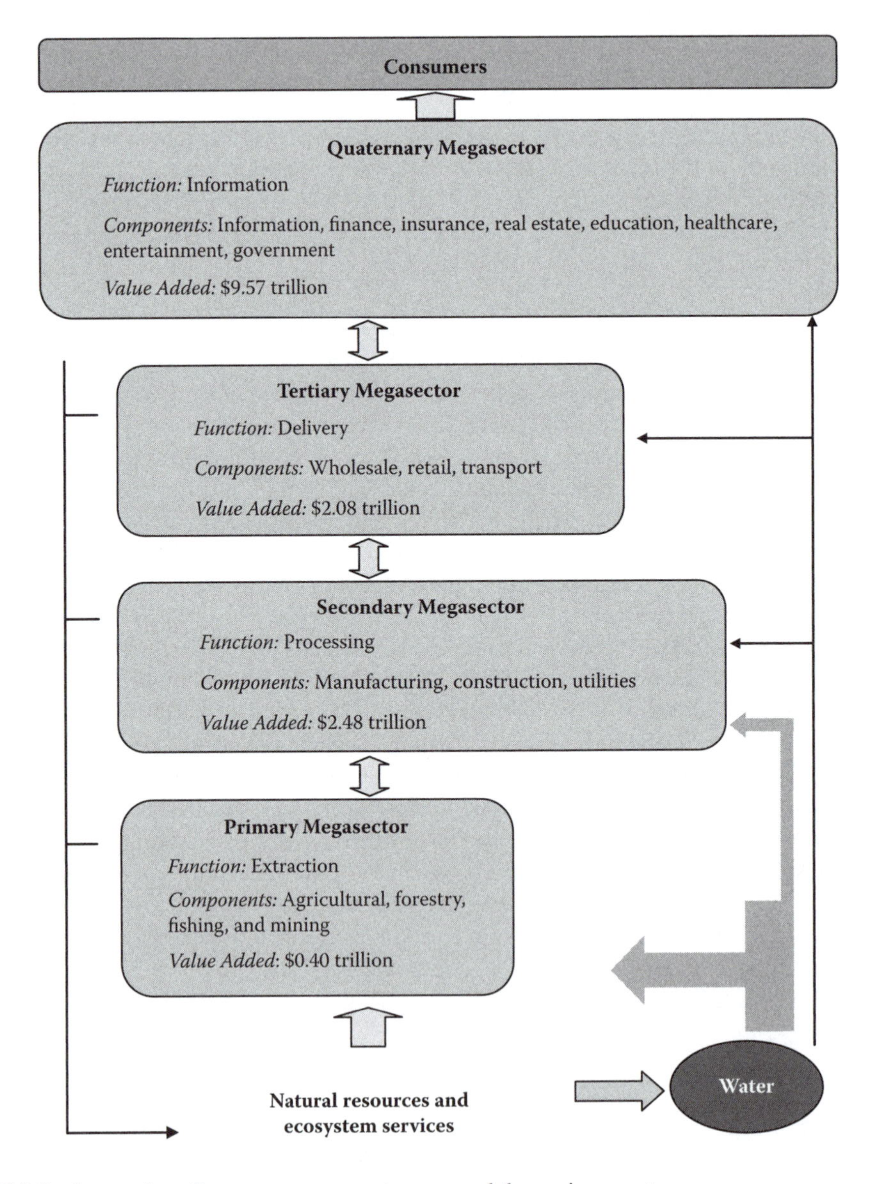

FIGURE 9.3 Interaction of macroeconomy, water use, and the environment.

secondary (i.e., processing) mega-sectors of the economy. All four levels of the economy interact, exchanging goods and services and delivering final goods to consumers at a fundamental level; however, the extraction and use of natural resources lie at the base of much economic activity. Moreover, economic activity in all four sectors has an iterative or repetitive effect on the abundance and quality of natural resources. This is most readily observed in the environmental impacts of resource extraction and industrial pollution. A less obvious recursive impact concerns the positive effect the information sector can have on the environmental, supporting decisions that will ensure the sustainability of natural resources and ecosystems.

THOUGHT-PROVOKING QUESTIONS

9.1 In your opinion, what is the value of water? Explain.

9.2 Because water is constantly resupplied by the water cycle, should water be free to everyone? Explain.

REFERENCES AND RECOMMENDED READING

Bohm-Bawerk, E. von (1888). *The Positive Theory of Capital*. Book III, Chapter IV. *The Marginal Utility*. London: Macmillan & Co. (http://www.econlib.org/library/BohmBawerk/bbPTC.html).

Brown, T.C. (2004). *The Marginal Economic Value of Streamflow from National Forests*, Discussion Paper DP-04-1, RMRS-4851. Fort Collins, CO: Rocky Mountain Research Station, U.S. Forest Service.

Bureau of Economic Analysis. (2013). *Gross Domestic Product (GDP) by Industry Data*. Washington, DC: U.S. Department of Commerce, http://www.bea.gov/industry/gdpbyind_data.htm.

Galbraith, K. (2011). Catastrophic drought in Texas causes global economic ripples. *The New York Times*, October 30.

Griffin, R.C. (2006). *Water Resource Economics*. Cambridge, MA: The MIT Press.

Hanemann, W.H. (2005). The economic conception of water. In: *Water Crisis: Myth or Reality?*, Rogers, P.P., Llamas, M.R., and Martine-Cortina, L., Eds. London: Taylor & Francis, Ch. 4.

Hylton, H. (2011). Forget Irene: the drought in Texas is the catastrophe that could really hurt. *Time*, August 31.

Kenessey, Z. (1987). The primary, secondary, tertiary, and quaternary sectors of the economy. *Review of Income and Wealth*, 33(4), 359–385.

Nicholson, W. (1978). *Microeconomic Theory: Basic Principles and Extensions*. Hinsdale, IL: Dryden Press.

Raucher, R.S., Chapman, D., Henderson, J., Hagenstad, M.L., Rice, J. et al. (2006). *The Value of Water: Concepts, Estimates, and Applications for Water Managers*, Project #2855. Denver, CO: AWWA Research Foundation.

Smith, A. (1776). *An Inquiry into the Nature and Causes of the Wealth of Nations*. London: Methuen & Co.

Stigler, G. (1950). The development of utility theory. I. *Journal of Political Economy*, 58(4), 307–327.

U.S. Census Bureau. (2014). *Classifying Businesses*. Washington, DC: U.S. Census Bureau, http://www.census.gov/history/www/programs/economic/classifying_businesses.html.

USEPA. (2012). *The Importance of Water to the U.S. Economy*. Part 1. *Background Report*. Washington, DC: Office of Water, U.S. Environmental Protection Agency.

USGS. (2009). *Estimated Water Use in the United States in 2005*. Washington, DC: U.S. Geological Survey.

10 Water Use and Availability

For more than 20 years industry has been moving south looking for cheaper labor. I'm hoping that now they'll start coming back looking for cheaper water.

—Rickard Meeusen, *Wave* founder

Water use is generally not publicized much outside of droughts. Water sort of has a technical side that often doesn't get communicated well to the public.

—Drew Beckwith, Water Policy Manager

Water is vital to American households, farms, and businesses. In the United States, rainfall averages approximately 4250×10^9 gallons a day. About two-thirds of this returns to the atmosphere through evaporation directly from the surface of rivers, streams, and lakes and transpiration from plant foliage. This leaves approximately 1250×10^9 gallons a day to flow across or through the Earth to the sea.

In the United States, it has been estimated that about 408 billion gallons per day (1000 million gallons per day, abbreviated BGD) were withdrawn for all uses during 2000. This total has varied less than 3% since 1985 as withdrawals have stabilized for the two largest uses—thermoelectric power and irrigation. Fresh groundwater withdrawals (83.3 BGD) during 2000 were 14% more than during 1985. Fresh surface water withdrawals for 2000 were 262 BGD, varying less than 2% since 1985 (USGS, 2004).

Major *off-stream* uses of water (i.e., uses for which water is withdrawn or diverted from its source) range from domestic consumption to the use of water in agriculture, mining, energy resource extraction, manufacturing, and the production of thermoelectric power. Major *in-stream* uses are similarly diverse, ranging from the generation of hydroelectric power to commercial fishing, commercial navigation, swimming, boating, and other forms of recreation. These uses draw on the nation's natural endowment in water resources and on private and public investment in the infrastructure required to employ and manage those resources. The discussion that follows presents an overview of water use and availability in the United States. The discussion draws primarily on data published by the U.S. Environmental Protection Agency and other federal agencies.

OFF-STREAM WATER USE

In 2005, off-stream water use in the United States totaled an estimated 410 billion gallons per day (BGD). Approximately 80% of this water (327.5 BGD) was drawn from lakes, rivers, oceans, and other surface water sources. The remaining 20% (82.6 BGD) was groundwater. More than 85% of the water used in 2005 was fresh; 15% was saline. The U.S. Geological Survey (USGS) separates water withdrawals into eight water use categories, or sectors: public supply, domestic self-supply, irrigation, livestock, aquaculture, industrial, mining, and thermoelectric power. Note that all sectors draw from both surface water and groundwater sources. Most sectors, however, rely exclusively on freshwater. Only the industrial, mining, and thermoelectric power sectors make use of saline water.

DID YOU KNOW?

Public water supply systems withdraw about 44.2 billion gallons of water per day, accounting for approximately 13% of all freshwater withdrawals. Most of this water—about 58%—is delivered to residential users. Approximately 86% of the U.S. population is served by a public system; the rest of the population supplies its own water, relying primarily upon private wells.

Public Supply and Domestic Self-Supply

In the United States, access to clean and safe drinking water offers substantial economic and public health benefits. Most residential users receive their drinking water from public supplies, while a smaller fraction extract their own water from private sources, primarily wells. This section examines residential water use, focusing on the following topics:

- Economic importance of the public water supply sector
- Water use in the public supply and domestic self-supply sectors
- Investment in the public water supply and treatment infrastructure that is likely to be needed to meet future demands
- Available estimates of the value of water in domestic use, including the value of improvements in residential water quality and reliability

More than 86% of the U.S. population receives its household water from public water supply systems—establishments involved in water extraction, water treatment, and/or water distribution (USGS, 2009). As mentioned, these systems are part of the secondary mega-sector of the economy. Some are owned and operated by private sector utilities, but most—especially the largest systems—are operated by municipalities or state and regional authorities (USEPA, 2009a). The U.S. Census Bureau compiles data on output and employment for water utilities in the private sector, while the U.S. Census of Governments provides economic data on government-operated systems.

According to the 2007 census, privately held water supply and irrigation systems generated annual revenues of nearly $8 billion dollars. Approximately 92% of private utility revenues reflect the sale of water to customers. The remainder is comprised of various user charges and fees. The industry includes establishments that operate water treatment plants, pumping stations, aqueducts, and distribution mains. Table 10.1 presents a basic profile of the sector. Note that these figures represent total revenue and employment for privately held water supply and irrigation systems, including revenue derived from the sale of water to non-residential customers. More detailed data on the share of revenues or employment attributable solely to the sale of water for domestic use are not available.

The data available from the U.S. Census of Governments on government-operated water supply systems include revenues from the sale of water to residential, industrial, and commercial customers, as well as connection, meter inspection, and late payment fees. Table 10.2 provides an overview of this sector. Notably, government-operated water suppliers serve substantially more people and generate significantly more revenues than do water utilities in the private sector. In 2009, water sales by government-operated systems amounted to $45.5 billion. Taken together, privately operated and

DID YOU KNOW?

Economies of scale are the cost advantages that enterprises obtain due to size, with cost per unit of output generally decreasing with increasing scale as fixed costs are spread out over more units of output.

TABLE 10.1
Economic Profile of Private Sector Water Supply and Irrigations Systems

Sector Characteristic	Estimate
Total number of systems	24,271
Population served	24 million
Employment	33,871
Annual payroll	$1.6 billion
Total revenues	$7.6 billion
Revenues from water sales	$7.0 billion

Sources: Data from U.S. Census Bureau (2007a) and USEPA (2009a).

TABLE 10.2
Economic Profile of Government-Operated Water Supply Systems

Sector Characteristics	Estimate
Total number of systems	24,861
Population served	256 million
Employment (full-time equivalent)	166,854
Annual payroll	$8.1 billion
Total annual expenditures	$57.4 billion
Total annual revenues	$45.5 billion

Sources: Data from U.S. Census Bureau (2007a) and USEPA (2009a).

government-operated public water supply systems employ approximately 200,725 people and generate annual revenues of more than $53 billion. The U.S. public water supply sector has grown steadily for several decades, reflecting the combined effects of population growth, economic development, increased urbanization, and economies of scale. These forces have spurred demand for water from centralized sources. As a result, the percentage of the population served by public water supply systems has increased from 62% in 1950 to 86% in 2005 (USGS, 2009).

SIDEBAR 10.1 Publicly Owned Treatment Works: Cash Cows or Cash Dogs?*

Water and wastewater treatment facilities are usually owned, operated, and managed by the community (municipality) where they are located. Although many of these facilities are privately owned, the majority of water treatment plants (WTPs) and wastewater treatment plants (WWTPs) are publicly owned treatment works (POTW) operated by local government agencies. These publicly owned facilities are managed by professionals in the field. Onsite management, however, is usually controlled by a board of elected, appointed, or hired directors or commissioners, who set policy, determine budgets, plan for expansion or upgrades, hold decision-making power for large purchases, set rates for ratepayers, and in general control the overall direction of the operation.

When final decisions on matters that affect plant performance are in the hands of, for example, a board of directors comprised of elected and appointed city officials, their knowledge of the science, the engineering, and the hands-on problems that those who are onsite must solve can range from comprehensive to none.

* Adapted from Spellman, F.R., *Handbook of Water and Wastewater Treatment Plant Operations*, 3rd ed., CRC Press, Boca Raton, FL, 2013.

DID YOU KNOW?

Efficient supply of safe drinking water depends heavily upon high-quality treatment, storage, and delivery infrastructure. Analysts anticipate that an annual investment of $10 billion to $20 billion over the next 20 years is needed to maintain and upgrade public water supply systems in the United States.

Matters that are of critical importance to those in onsite management may mean little to those on the board. The board of directors may also be responsible for other city services and have an agenda that encompasses more than just the water or wastewater facility. Thus, decisions that affect onsite management can be affected by political and financial concerns that have little to do with the successful operation of a WTP or POTW.

Finances and funding are always of concern, no matter how small or large, well-supported or underfunded the municipality. Publicly owned treatments works are generally funded from a combination of sources. These include local taxes, state and federal monies (including grants and matching funds for upgrades), and usage fees for water and wastewater customers. In smaller communities, in fact, their water/wastewater (W/WW) plants may be the only city services that actually generate income. This is especially true in water treatment and delivery, which is commonly looked upon as the cash cows of city services. As a cash cow, the water treatment works generates cash in excess of the amount of cash necessary to maintain the treatment works. These treatment works are "milked" continuously with as little investment as possible, and funds generated by the facility do not always stay with the facility. Funds can be reassigned to support other city services, so when facility upgrade time rolls around funding for renovations can be problematic. On the other end of the spectrum, spent water (wastewater) treated in a POTW is often looked upon as one of the cash dogs of city services. Typically, these units make only enough money to sustain operations. This is the case, of course, because managers and oversight boards or commissions are fearful, for political reasons, of charging ratepayers too much for treatment services. Some progress has been made, however, in marketing and selling treated wastewater for reuse in industrial cooling applications and some irrigation projects. Moreover, wastewater solids have been reused as soil amendments; also, ash from incinerated biosolids has been used as a major ingredient in forming cement revetment blocks used in areas susceptible to heavy erosion from river and sea inlets and outlets.

Water Use and the Public Supply

A total of 49.6 million acre-feet of water were withdrawn for public supply in 2005, representing 13% of all freshwater withdrawals (USGS, 2009). Two-thirds of these withdrawals were from surface water sources, while the remainder was from groundwater. As noted above, public water supply systems serve residential, industrial, and commercial users. Deliveries to residential users account for approximately 58% of the water that public suppliers withdraw, or 28.7 million acre-feet. The remaining 20.9 million acre-feet are used in commercial, industrial, and public services (e.g., firefighting, municipal buildings) or are unaccounted for due to system losses (e.g., leaks, flushing). The discussion that follows focuses specifically on water in the residential sector.

Residential water use pertains to any indoor and outdoor water use at households. Typical indoor uses include drinking, food preparation, washing clothes and dishes, and flushing toilets; common outdoor uses include watering lawns, maintaining swimming pools, and washing cars. In the United

DID YOU KNOW?

Water that is lost or wasted (i.e., unaccounted for) is attributable to leaks in the distribution system, inaccurate meter readings, and unauthorized connections. Loss and waste of water can be expensive. In order to reduce loss and waste, a regular program that includes maintenance of the system and replacement and/or recalibration of meters is required.

States, the typical four-person household consumes 400 gallons of water per day; approximately 70% of this amount is used for indoor purposes, and the rest is devoted to outdoor uses (USEPA, 2008b). In general, toilets account for the largest share of indoor use, while landscaping accounts of the largest share of outdoor use (USEPA, 2008a,b).

In 2005, withdrawals for domestic self-supply and residential deliveries from public supply amounted to 29.4 billion gallons per day (USGS, 2009). This implies that residential water use accounted for roughly 7% of total withdrawals and that the average person used 98 gallons of water per day. Although aggregate residential water use has increased over time, water use per capita in recent years has remained relatively constant. In 1985, the average person consumed 100 gallons per day; similarly, in 1995, the average person consumed 101 gallons per day (USGS, 2009).

Both aggregate water use and per capita water use are subject to regional variation. Not surprisingly, the states with the largest populations—California, Texas, New York, Florida, and Illinois—use the most water. Per capita use ranges from 54 gallons per person per day in Maine to 190 gallons per person per day in Nevada. Climate is one important determinant of the per-capita use, as drier regions require more water for landscaping. Other factors, such as population density and household income, also play a role.

As water supplies tighten, domestic water users may consider substitution and efficiency measures to limit their consumption. A variety of options exist for limiting household water use, including installation of low-flow plumbing fixtures, use of water-efficient appliances, installation of less water-intensive landscaping, and behavioral changes (e.g., shorter showers) (USEPA, 2000). Many households have instituted such practices, although data showing relatively constant water use per household suggest that further efficiency gains are possible. These observations have implications for the elasticity of demand for water in residential use. As mentioned below, short-run demand is relatively inelastic, whereas long-run demand is more elastic, reflecting the ability in the long run to make capital investments to improve the efficiency of household water use (USEPA, 2012a).

Future Supply

The reliability of future supplies of water for residential use is dependent on producers' ability to meet quality standards in the face of increasing demand. Satisfying future demand will largely hinge on improvements to the public water supply infrastructure, most of which was constructed in the early 20th century (USEPA, 2009c). Each year, more than 240,000 water mains break, and leaking pipes lose an estimated 7 billion gallons of clean drinking water every day (ASCE, 2013; Olmstead, 2010). From an economic perspective, repairing this failing infrastructure would reduce costs and enhance the efficiency of the public water supply system.

Overall Investment Needs

In 2009, the nation's drinking water infrastructure received a score of D– from the American Society of Civil Engineers, reflecting the lowest grade awarded on the organization's Infrastructure Report Card. The poor grade is indicative of the need to replace aging facilities and to comply with existing and future federal water regulations. Although specific estimates of the investment required to maintain the nation's public water supply vary, the overarching message is the same—the system is currently underfunded and will require significant investment over the coming decades to prevent supply shortages and drinking water contamination.

Investment in water supply infrastructure is necessary both to enhance or repair existing infrastructure and to comply with specific Safe Drinking Water Act (SDWA) regulations. Although all infrastructure projects facilitate continued provision of safe drinking water, certain projects are directly attributable to SDWA regulations. According to USEPA estimates, 16% of the investment required is directly attributable to SDWA mandates; the remaining 84% is associated with projects that will enable water utilities to meet expected demand (USEPA, 2009a).

SIDEBAR 10.2 Frequently Asked Questions on Water Infrastructure and Sustainability*

- *What does the USEPA mean by "sustainable water infrastructure"?* In 1987, the World Commission on Environment and Development released *Our Common Future*, also known as the Bruntland Report, which defined sustainable development as that which "meets the needs of the present generation without compromising the ability of future generations to meet their needs." The current U.S. population benefits from the investments that were made over the past several decades to build our nation's water infrastructure. The USEPA seeks to promote practices that encourage water sector utilities and their customers to address existing needs so future generations will not be left to address the approaching wave of needs resulting from aging water infrastructures. To be on a sustainable path, our investments need to result in efficient infrastructure and infrastructure systems and be at a pace and level that allow our water sector to provide the desired levels of service over the long term.
- *What can be said generally about the age of the nation's water infrastructure?* Some of our water infrastructure is over 100 years old, but much of it was built in the period following World War II. Looking at pipes only, the USEPA's 2000 survey on community water systems found that, in systems that serve more than 100,000 people, about 40% of drinking water pipes are greater than 40 years old. However, it is important to note that age, in and of itself, does not necessarily point to problems. If a system is well maintained, it can operate over a long time period.
- *What is the useful life of infrastructure*? Treatment plants typically have a useful life of 20 to 50 years before they require retrofitting, rehabilitation, or expansion. Pipes have life cycles that can range from 15 to over 100 years, depending on the type of material and where they are laid. With pipes, the material used and proper installation of the pipe can be a greater indicator of failure than age.
- *Why is it important to properly maintain infrastructure assets?* Water infrastructure is expensive, and the monetary and social costs incurred with infrastructure failures can be large. Therefore, it only makes sense to ensure that our infrastructure assets are properly managed. If a system is well maintained, it can operate safely over a long time period. Utilities need to carry out an ongoing process of oversight, evaluation, maintenance, and replacement of their assets to maximize the useful life of infrastructure. The USEPA encourages utilities to actively plan for and manage their infrastructure by understanding its condition and making risk-based decisions on maintaining and improving infrastructure.
- *What is the role of infrastructure planning?* Planning is essential for water and wastewater infrastructure sustainability. The infrastructure we build today will be with us for a long time, so it must be efficient to operate, offer the best solution in meeting the needs of ac community, and be coordinated with infrastructure investments in other sectors such as transportation and housing. It is both important and challenging to ensure that a plan is in place to renew and replace it at the right time, which may be years away. Replacing an infrastructure asset too soon means not benefiting from the remaining useful life of that asset. Replacing an asset too late can lead to expensive, emergency repairs that are significantly more expensive than those which are planned.
- *How can aging drinking water and wastewater pipes affect water quality and the cost of water service?* Long-term corrosion reduces a pipe's carrying capacity, requiring increasing investments in power and pumping. When water or sewer pipes are to the point of failure, the result can be contamination of drinking water, the release of sewage into our surface waters or basements, and high costs both to replace the pipes and repair any resulting damage.
- *What does the "water infrastructure gap" mean?* The term references a 2002 USEPA report that compared current spending trends at the nation's drinking water and wastewater utilities to the expenses they can expect to incur for both capital and operations and maintenance costs. The "gap" is the difference between projected and needed spending and was found to be over $500 billion over a 20-year period.

Investment Needs by Category

In general, infrastructure projects can be grouped into four major categories: transmission and distribution projects, treatment projects, storage projects, and source projects. The first of these, transmission and distribution, accounts for 60% of projected infrastructure requirements over the

* Adapted from USEPA, *Frequently Asked Questions: Water Infrastructure & Sustainability*, U.S. Environmental Protection Agency, Washington, DC, 2012 (http://water.epa.gov/infrastructure/sustain/si_faqs.cfm).

DID YOU KNOW?

Most of the assets of drinking water treatment systems are comprised of pipe. The useful life of pipe varies considerably based on a number of factors. Some of these factors include the material of which the pipe is made, the conditions of the soil in which it is buried, and the character of the water flowing through it. In addition, pipes do not deteriorate at a constant rate. During the initial period following installation, the deterioration rate is likely to be slow, and repair and upkeep expenses are low. For pipe, this initial period may last several decades. Later in the life cycle, pipe will deteriorate more rapidly.

next 20 years. This is not surprising. Depending on soil conditions, climate, capacity requirements, and the material of which they are constructed, the pipes employed in water transmission and distribution systems are expected to remain in service for 60 to 95 years. Given that most U.S. cities' water distribution systems were constructed in the early 20th century, many pipes have already reached or are nearing the end of their useful lives. A significant investment is needed to replace or rehabilitate these pipes in order to prevent delivery system failures and contamination.

Treatment projects account for 22% of the projected infrastructure investment needed over the next 20 years. This category of need includes most of the projects directly attributable to SDWA mandates. Treatment projects include the construction, expansion, and rehabilitation of treatment facilities that provide filtration, disinfection, and corrosion control in order to reduce drinking water contamination.

Storage needs account for 11% of projected infrastructure requirements over the next 20 years. This category of need includes projects to construct, rehabilitate, or cover finished water storage tanks. These projects are necessary to ensure that systems have sufficient storage to provide the public with treated water, especially during periods of peak demand. Again, adequate storage enables systems to maintain minimum pressure that prevents intrusion of contaminants into the distribution network.

Source needs account for the remaining 6% of projected infrastructure requirements over the next 20 years. This category of need reflects investments in the construction or rehabilitation of surface water intake structures, drilled wells, and spring collectors. The objective of many of these projects is to obtain higher quality raw water that results in lower treatment costs to comply with SDWA standards. In addition, expanded capacity at intake structures enable systems to maintain minimum pressure that prevents intrusion of contaminants into the distribution network.

Future Demand and Availability

Although the United States enjoys abundant water resources compared to many nations, residential water use in some parts of the country is confronting the reality of limited water supplies. Population growth in some areas has brought municipalities into direct competition with other water users such as agriculture and hydropower. For example, the 2007–2008 drought in the southeast United States created legal conflicts between states, dam operators, and irrigators, as well as environmentalists seeking in-stream flows for wildlife. Atlanta's primary water source became dangerously depleted, and the condition was ultimately linked to rapid population growth and associated municipal water demands (Seager et al., 2009). Other U.S. municipalities have encountered similar issues.

Climate change is expected to greatly increase the risk that water supplies will not be able to keep pace with withdrawals. A study of water supply vulnerability found that in the coming decades approximately 35% of all U.S. counties face greater risk of water shortages as a result of climate change (Roy et al., 2012). The anticipated impact of climate change on future water supplies is driven by reduced precipitation in some regions and an increase in potential evapotranspiration in most of

DID YOU KNOW?

Forecasts suggest that climate change, population growth, and other factors could make current levels of domestic water use unsustainable in some parts of the United States. In several of these areas, municipal water supply needs already compete with the needs of farmers, hydropower facilities, and other users.

the country. Declining precipitation is expected to be most pronounced in eastern Texas, the Lower Mississippi Basin, California's Central Valley, and the southwestern United States. Furthermore, a rise in sea level may allow saltwater intrusion into shallow coastal aquifers that serve as important municipal water supplies (USEPA, 2010).

VALUE OF WATER USE

Compared to most categories of water use, there is an extensive literature on the economics of domestic water use. The available studies can be grouped into several broad categories according to the methodology used and the issues the studies are designed to assess. The first group of studies simply examines the rates that public water supply systems charge residential users; the second estimates the price elasticity of demand for domestic water; the third considers willingness to pay figures from market transfers between municipalities and other water use sectors; and the fourth uses non-market methods to determine consumer surplus associated with key features of domestic water use (quality and reliability).

Why Domestic Water Rates May Fail to Encourage Efficient Use

The price at which water is sold in the United States is generally not the product of market forces that will promote an economically efficient use of the resource (i.e., the point at which the long-run marginal cost of supplying water equals its marginal value to consumers). Water prices are not determined in competitive markets and as a result do not reflect water scarcity (Olmstead, 2010). Instead, water rates are typically set by elected councils and public utility commissions. Even in times of scarcity, management officials are reluctant to raise prices. As a result, the rate paid by consumers often falls below the long-run marginal cost of water supply, its efficient price (Olmstead, 2007).

According to a cost-recovery analysis by Global Water Intelligence (GWI, 2004), water systems in industrialized nations are able to cover their operation and maintenance costs (O&M) by charging $0.40 to $1.00 for every cubic meter of water delivered. If water systems charge more than $1.00 per cubic meter, they are able to cover both O&M and capital costs. This amounts to approximately $45.42 per month for a four-person household that consumes 400 gallons per day. As illustrated in the discussion below, water prices fall well below these rates in many U.S. cities.

Most U.S. households served by a public water supply system face one of three water rate structures: uniform rates, increasing block rates, or decreasing block rates. With uniform pricing, consumers are charged the same price per gallon regardless of the amount of water they consume. Increasing (or decreasing) block rate pricing means that prices per gallon increase (or decrease) with the amount of water consumed. Over the past few decades, increasing block rate price structures have become more common, with nearly a third of the population facing this type of fee schedule (Olmstead et al., 2007). In effect, this pricing structure means that marginal water uses (i.e., lawn watering and care washing) are charged a price that more closely equates to the efficient price; in most instances, however, first-priority uses (i.e., drinking and bathing) reflect subsidized prices (Olmstead et al., 2007). Across the country there is wide variation in use and prices for water consumption in major urban areas, with residential rates being lowest in the Great Lakes region (Walton, 2010).

The practice of subsidizing public water supply systems enables residents of many U.S. cities to purchase water at prices well below the public water supply system's long-run marginal cost. Moreover, the variation in water rates between cities can be substantial—ranging from a low of $19.64 per month for a four-person household that consumes 400 gallons per month in San Antonio to a high of $121.42 per month for a similar household in Santa Fe. This is due in part to underlying variation in a number of factors that affect unit costs to suppliers, including the energy required to pump and transport water, treatment costs, and infrastructures costs. Nonetheless, non-market factors such as government subsidies are also important determinants of prices. For example, many cities in the arid southwest enjoy inexpensive water because expensive infrastructure projects have been financed by government funding as opposed to the utility companies themselves (Walton, 2010). As a result, consumers in areas with high rainfall and lower rates of consumption may actually face higher water rates than those in areas with little rainfall and higher rates of consumption.

Demand Elasticity for Domestic Water

The price elasticity of demand for residential water captures the relationship between the quantity of water demanded and its price, providing insight into how residential users value water. Specifically, it measures the percentage change in quantity demanded for a given percentage change in prices. Both economic theory and empirical evidence suggest that domestic water is inelastic at current prices, meaning that when its price increases the quantity demanded falls but at a correspondingly lower rate than the price increase (Olmstead, 2010). This is not surprising given that there are no readily available and comparably priced substitutes for water in most residential applications. However, as the data show, long-run demand elasticity is generally higher than short-run; that is, if prices remain high, residential users may invest in water efficiency measures, thereby reducing their total water demand.

Importantly, price elasticities provide an indicator of how the average consumer will respond to marginal price changes (e.g., 1%) given current local prices. Thus, an estimated price elasticity that relies on data for the East Coast cannot be extrapolated to a tenfold increase in prices on the West Coast. Additionally, elasticities reflect price responsiveness to actual prices, not efficient prices. Because consumers are paying prices that fall well below their reservation price, even large rate increases may have little or no impact on the quantity consumed.

A household's demand for water is a function of the price of water, household income, and household preferences. Non-household characteristics such as season and weather also affect residential demand (Olmstead, 2010). Research on the demand for residential water began in the 1960s. Since then, several economists have estimated empirical price elasticities using data on domestic water use and rate schedules. A review of 124 estimates obtained from studies completed between 1963 and 1993 revealed that the average price elasticity of demand for residential water in the United States was –0.51, and that short-run estimates are more inelastic than long-run estimates (Espey, 1997). Similarly, a meta-analysis of more than 300 estimates suggested that the average price elasticity of demand for residential water in the United States is –0.41 (Dalhuisen, 2003). This means that when the price of residential water increases by 1%, the quantity demanded falls by 0.41% (USEPA, 2012a).

Notably, the estimates of the price elasticity of residential water demand obtained from various research studies vary widely from study to study. Differences in the specification of the demand model, characteristics of the underlying data, and econometric techniques make it challenging to compare results across regions and studies. Perhaps most importantly, the specification of demand models in these studies varies according to assumptions made about the water prices faced by the consumer. Whereas the earliest studies on price elasticity relied on simple demand models that employed uniform average prices, more recent research suggests that incorporating information about the rate structure faced by consumers is important. In particular, the current literature focuses on estimating demand under increasing block rate prices. Analyses that allow for discontinuous prices tend to result in higher price elasticities, suggesting that consumers are more price sensitive to water prices than indicated by the earlier literature (Dalhuisen, 2003; Olmstead, 2010). One intuitive explanation for this result is that increasing block rates make prices more salient to consumers because the price changes with the amount consumed, thereby increasing their price responsiveness (Olmstead et al., 2007).

Values from Water Transfer Programs

The research discussed above is based on consumer behavior in response to water rates, which are generally considered a poor indicator of the true economic value of domestic water. Only recently have U.S. economists been able to observe transactions with the potential to capture the full value of water in specific applications. In areas of water competition (e.g., the arid southwest), water transfer programs allow one party to sell or lease water rights to another part. These water transfers harness the power of market pricing signals to achieve more efficient use of water.

Water transfer programs often entail the transfer of water rights from agricultural to municipal interests. Brewer et al. (2007) compiled detailed information from western U.S. water transactions implemented between 1987 and 2005. The price that municipalities paid in temporary lease arrangements averaged $199 per acre-foot, but the permanent purchase of water rights entailed prices averaging over $4500 per acre-foot.

The relatively high price paid by municipal water supply entities for additional water provides an indicator of the willingness to pay for water for municipal use and suggests the potential for economic welfare gains through mechanisms that provide for the transfer of water when municipal supplies are inadequate to meet demand. However, the obstacles to water transfers limit the availability of willingness-to-pay data. Transfer markets tend to be active only where water law, institutional flexibility, and physical infrastructure are sufficient for implementing trades.

Methods for Valuing Water Supply Reliability

Economists have also developed studies to value the reliability of domestic water supplies. These studies use either stated preference or revealed preference methods to assess consumers' willingness to pay (WTP)—the maximum amount a person would be willing to pay in order to receive a good or to avoid something undesired, such as pollution—for greater water supply reliability. Stated preference methods use survey responses to hypothetical choices about water supply reliability, whereas revealed preference studies infer value from market data on expenditures to increase the reliability of supply. Taken together, these studies suggest that consumers are willing to pay to avoid supply shortages (USEPA, 2012a).

One analysis of the value of water supply reliability, a stated preference study of California residents, estimated that they would be willing to pay an additional $12 to $17 on monthly household water bills to avoid water shortages of varying degrees; statewide, this amounted to more than $1 billion in 1994 (California Urban Water Agencies, 1994). A similar study of Texas residents incorporated information on household characteristics to determine the WTP for avoided supply shortages by demographic. Not surprisingly, the study suggested that more affluent households would be willing to pay more to avoid shortages than low-income households. In particular, a low-income household would pay $17.19 to avoid a three-week shortfall with a 20% chance of occurring, whereas a high-income household would pay $44 (Griffin and Mjelde, 2000). Similarly, a study of Colorado residents estimated the effect of the length of a shortage and its probability of occurrence on WTP. The authors found that residents would pay a base of $18.41 to avoid any shortage; this payment would increase with the anticipated length of the shortage and its chance of occurrence (Howe and Smith, 1994).

These WTP studies are subject to a number of criticisms related to their validity and reliability (Venkatachalam, 2004). With respect to residential water supply, two criticisms are particularly relevant. First, the results of WTP studies may be biased if consumers have imperfect information

DID YOU KNOW?

Estimates of the value of water in residential use vary with the approach employed and components of value measured. Public water supply systems in the United States often charge rates that fall below the long-term marginal cost of supply. As a result, these rates, which range anywhere from $20 to more than $120 per month for a typical household of four, are poor indicators of the efficient price for water in domestic use. Other information on residential values can be inferred from studies of demand elasticity and willingness to pay for reliable supplies, as measured by stated preference methods. Perhaps the most reliable information comes from studies of market transfers between public water supply systems and other water users; these transactions suggest that, on average, municipalities will pay more than $4500 per acre-foot for water rights.

about the good in question. Second, they may be biased if respondents do not have experience valuing the good in question. Imperfect information is problematic because most consumers are not fully cognizant of their monthly water usage and bills. Water bills do not garner much attention because they represent a small share of monthly household spending and are often bundled with charges for other utilities, such as electricity and natural gas. If consumers do not understand how usage affects expenditures, it is difficult to imagine that they will accurately gauge how much they would spend to prevent a shortage. Additionally, most consumers have never actually had to pay to avoid a water shortage. With no prior experience, consumers' responses may not reflect their true WTP.

Methods for Valuing Domestic Water Quality

The quality of the water supplied for domestic use can have a significant effect on its value. The USEPA regulates drinking water quality according to standards developed under SDWA (CBO, 2002). As a result of these regulations and additional state and local standards, the quality of water provided by public water supply systems in the United States is generally very good. Isolated instances of contamination, however, illustrate the potentially catastrophic implications of deterioration in the quality of domestic water supplies. The largest recorded waterborne disease outbreak in the United States took place in Milwaukee, Wisconsin, in 1993. *Cryptosporidium* oocysts—transported by runoff from cattle pastures—passed through the filtration systems at one of the city's treatment plants, resulting in more than 403,000 cases of illness (25% of the population) and 104 deaths in just 2 weeks (Corso et al., 2003). According to an analysis by the Centers for Disease Control and Prevention, the total cost associated with the outbreak was $96.2 million, including $31.7 million in medical costs and $64.6 million in productivity losses (Corso et al., 2003). Note that these estimates likely reflect a lower bound on the true economic cost of the outbreak, as they do not consider willingness to pay to avoid the deaths and illnesses the outbreak caused.

As the example above indicates, protecting the quality of public water supplies provides substantial economic benefits, including reduced morbidity and mortality, avoided worker and school absences, and lower medical expenditures. Although the literature on these benefits is sparse for the United States, many economists have considered the impacts of improved water supplies in developing countries. In particular, expanded access to high-quality water supplies is strongly correlated with improved health outcomes that subsequently reduce costs associated with death, malnutrition, stunting, and productivity losses (Esrey et al., 1991; Galiani et al., 2005; Listorti, 1996). High-quality water also confers educational benefits by reducing barriers to school attendance (Komives et al., 2005).

Premiums paid for bottled water also provide an indication of individuals' willingness to pay for perceived higher quality water. According to research by the Natural Resources Defense Council, more than half of the U.S. population consumes bottled water and more than one-third of Americans drink it regularly (Olson, 1999). In 2007, U.S. bottled water sales reached 8.8 billion gallons, worth $11.7 billion (Giesi, 2008). Per gallon, bottled water costs from 240 to over 10,000 times as much as tap water (Olson, 1999).

OFF-STREAM WATER USE BY AGRICULTURE

A reliable supply of clean water is vital to the success of the U.S. agricultural sector. Irrigation allows cultivation of otherwise non-arable land and increases the productivity of farms, especially in the Great Plains and West. Likewise, water is an essential element in livestock and aquaculture operations. This section characterizes the role of water in agriculture, focusing on the following topics:

- The economic importance of U.S. agriculture and its place in the global economy
- The use of water in the U.S. agricultural sector
- Supply issues affecting agriculture water use and ways in which competition for water is influencing its use
- Water quality requirements associated with agricultural water use
- Available estimates of the value of water used in agriculture

Agriculture is a major component of the U.S. economy. In 2010, agriculture accounted for approximately 1.1% of U.S. gross domestic production (CIA, 2011). Although the relative significance of the agricultural sector has decreased over the last century, U.S. output of crops and livestock has grown steadily, the result of major productivity advances associated with new machinery, soil science, and other agricultural technology (USEPA, 2012a).

U.S. Agricultural Sector

The National Agricultural Statistics Service (NASS) within the U.S. Department of Agriculture conducts a periodic census of U.S. agriculture. Table 10.3 draws on the most recent census (2007) to provide a basic economic profile of the U.S. agricultural sector. As shown, the total market value of all agricultural products in 2007 was approximately $297 billion, with crops and livestock making up roughly equal shares of that output. This production occurred on about 2.2 million farms. Most of these farms are small, family operations that account for a relatively small share of total production. In contrast, about 64% of all market value is generated by a small number of large farming operations having annual sales of $500,000 or more (USDA, 2009).

TABLE 10.3
Overview of the U.S. Agricultural Sector (2007)

Sector Characteristic	Estimate
Total number of farms	2,204,792
Employment	2,903,797
Land used in farming (acres)	922,095,840
Irrigated land (acres)	56,599,305
Market value of all agricultural products	$297.2 billion
Market value of crops	$143.6 billion
Market value of livestock	$153.6 billion

Sources: Data from USDA (2009) and BLS (2010).

Neither the Bureau of Labor Statistics (BLS) nor the USDA maintains comprehensive data on employment in the agriculture sector. A rough estimate can be obtained by adding the number of farm proprietors (2.15 million based on the 2007 census of agriculture) and the number of paid employees in crop and animal production (0.754 million based on 2010 BLS data). These figures suggest that the agriculture sector employs approximately 2.9 million individuals.

The market value of U.S. farm products has grown over time and saw particularly marked growth from 2002 to 2007. Much of this growth has come as a result of productivity increases and strong markets—both domestic and export—for key products. A specific driver has been growth in U.S. corn production and increases in corn prices. Demand for corn has grown as a result of its application in new food products (e.g., sweeteners), its use as livestock feed, and its subsidized use for biofuels such as ethanol (Leibtag, 2008).

SIDEBAR 10.3 Corn: Ethanol Production or Food Supply?*

In 2007, record U.S. trade driven by economic growth in developing countries and favorable exchange rates, combined with tight global grain supplies, resulted in record or near-record prices for corn, soybeans, and other food and feed grains. For corn, these factors, along with increased demand for ethanol, helped push prices from under $2 per bushel in 2005 to $3.40 per bushel in 2007. By the end of the 2006–2007 crop year, over 2 billion bushels of corn (19% of the harvested crop) were used to produce ethanol, a 30% increase from the previous year. Higher corn prices motivated farmers to increase corn acreage at the expense of other crops, such as soybeans and cotton, raising their prices as well.

The pressing and pertinent question becomes: What effect do these higher commodity costs have on retail food prices? In general, retail food prices are much less volatile than farm-level prices and tend to rise by a fraction of the change in farm prices. The magnitude of response depends on both the retailing costs beyond the raw food ingredients and the nature of competition in retail food markets. The impact of ethanol on retail food prices depends on how long the increased demand for corn drives up farm corn prices and the extent to which higher corn prices are passed through to retail.

Retail food prices adjust as the cost of inputs into retail food production change and the competitive environment in a given market evolves. Strong competition among three to five retail store chains in most U.S. markets has had a moderating effect on food price inflation. Overall, retail food prices have been relatively stable over the past 20 years, with prices increasing an average of 3.0% per year from 1987 through 2007, just below the overall rate of inflation. Since then, food price inflation has averaged just 2.5% per year.

Field corn is the predominant corn type grown in the United States, and it is primarily used for animal feed. Currently, less than 10% of the U.S. field corn crop is used for direct domestic human consumption in corn-based foods such as corn meal, corn starch, and corn flakes; the remainder is used for animal feed, exports, ethanol production, seed, and industrial uses. Sweet corn, both white and yellow, is usually consumed as immature whole-kernel corn by humans and also as an ingredient in other corn-based foods, but it makes up only about 1% of the total U.S. corn production.

Because U.S. ethanol production uses field corn, the most direct impact of increased ethanol production should be on field corn prices and on the price of food products based on field corn. However, even for those products heavily based on field corn, the effect of rising corn prices is dampened by other market factors. For example, an 18-ounce box of corn flakes contains about 12.9 ounces of milled field corn. When field corn is priced at $2.28 per bushel (the 20-year average), the actual value of corn represented in the box of corn flakes is about 3.3 cents (1 bushel = 56 pounds). (The remainder of the price represents packaging, processing, advertising, transportation, and other costs.) At $3.40 per bushel, the average price in 2007, the value is about 4.9 cents. The 49% increase in corn prices would be expected to raise the price of a box of corn flakes by about 1.6 cents, or 0.5%, assuming no other cost increases.

In 1985, Coca-Cola made the shift from sugar to corn syrup in most of its U.S.-produced soda, and many other beverage makers followed suit. Currently, about 4.1% of U.S.-produced corn is made into high-fructose corn syrup. A 2-liter bottle of soda contains about 15 ounces of corn in the form of high-fructose corn syrup. At $3.40 per bushel, the actual value of corn represented is 5.7 cents, compared with 3.8 cents when corn is priced at $2.28 per bushel. Assuming no other cost increases, the higher corn price in 2007 would be expected to raise soda prices by 1.9 cents per 2-liter bottle, or 1%. These are notable changes in terms of price measure and inflation, but relatively minor changes in the average household budget.

* Adapted from Leibtag, E., Corn prices near record high, but what about food costs? *Amber Waves*, February, 2008.

In addition to the impact on corn flakes and soda beverage prices, livestock prices are also impacted. This stands to reason when you consider that livestock rations traditionally contain a large amount of corn; thus, a bigger impact would be expected in meat and poultry prices due to higher feed costs than in other food products. Currently, 55% of corn produced in the United States is used as animal feed for livestock and poultry. However, estimating the actual corn used as feed to produce retail meat is a complicated calculation. Livestock producers have many options when deciding how much corn to include in a feed ration. For example, at one extreme, grass-fed cattle consume no corn, while other cattle may have a diet consisting primarily of corn. For hog and poultry producers, ration variations may be less extreme but can still vary quite a bit. To estimate the impact of higher corn prices on retail meat prices, it is necessary to make a series of assumptions about feeding practices and grain conversion rates from animal to final retail meat products. To avoid downplaying potential impacts, this analysis uses upper-bound conversion estimates of 7 pounds of corn to produce 1 pound of beef, 6.5 pounds of corn to produce 1 pound of pork, and 2.6 pounds of corn to produce 1 pound of chicken.

Using these ratios and data from the Bureau of Labor Statistics, a simple pass-through model provides estimates of the expected increase in meat prices given the higher corn prices. The logic of this model is illustrated by an example using chicken prices. Over the past 20 years, the average price of a bushel of corn in the United States has been $2.28, implying that a pound of chicken at the retail level uses 8 cents worth of corn, or about 4% of the $2.05 average retail price for chicken breasts. Using the average price of corn for 2007 ($3.40 per bushel) and assuming producers do not change their animal-feeding practices, retail chicken prices would rise 5.2 cents, or 2.5%. Using the same corn data, retail beef prices would go up 14 cents per pound, or 8.7%, while pork prices would rise 13 cents per pound, or 4.1%.

These estimates for meat, poultry, and corn-related foods, however, assume that the magnitude of the corn price change does not affect the rate at which cost increases are passed through to retail prices. It could be the case that corn price fluctuations have little impact on retail food prices until corn prices rise high enough for a long enough time to elicit a large price adjustment by food producers and notably higher retail food prices.

On the other hand, these estimates may be overstating the effect of corn price increases on retail food prices because they do not account for potential substitution by producers from more expensive to less costly inputs. Such substitution would dampen the effect of higher corn prices on retail meat prices. Even assuming the upper-bound effects outlined above, the impact of rising corn commodity prices on overall food prices is limited. Given that less than a third of retail food contains corn as a major ingredient, these rising prices for corn-related products would raise overall U.S. retail food prices less than 1 percentage point per year above the normal rate of inflation.

Continuing elevated prices for corn will depend on the extent to which corn remains the most efficient feedstock for ethanol production and ethanol remains a viable source of alternative energy. Both of these conditions may change over time as other crops and biomass are used to produce ethanol and other alternative energy sources develop.

Even if these conditions do not change in the near term, market adjustments may dampen long-run impacts. In 1996, when field corn prices reached an all-time high of $3.55 per bushel due to drought-related tighter supplies in the United States and strong demand for corn from China and other parts of Asia, the effect on food prices was short lived. At that time, retail prices rose for some foods, including pork and poultry, but these effects did not extend beyond the middle of 1997. For the most part, food markets adjusted to the higher corn prices and corn producers increased supply, bringing down price.

Food producers, manufacturers, and retailers may also adjust to the changing market conditions by adopting more efficient production methods and improved technologies to counter higher costs. For example, soft drink manufacturers may consider substituting sugar for corn syrup as a sweetener if corn prices remain high, while livestock and poultry producers may develop alternative feed rations that minimize the corn needed for animal feed. Adjustments by producers, manufacturers, and retailers, along with continued strong retail competition, imply that U.S. retail food prices will remain relatively stable.

A diverse climate and other factors enable the U.S. agricultural sector to produce a broad array of goods. Major crops such as corn, soybeans, and wheat remain significant, but numerous other products are grown on U.S. farms. Livestock operations are dominated by cattle and dairy, but pigs and poultry play a large role as well.

DID YOU KNOW?

While higher commodity costs may have a relatively modest impact on U.S. retail food prices, there may be a greater effect on retail food prices in lower-income developing countries. As a relatively low-priced food, grains have historically accounted for a large share of the diet in less developed countries. Even with incomes rising, consumers in such countries consume a less processed diet than is typical in the United States and other industrialized countries, so food prices are more closely tied to swings in both domestic and global commodity prices.

The importance of the U.S. agricultural sector should be considered in light of its linkages to other sectors of the U.S. economy. Key farming inputs from other economic sectors include energy, fertilizer, pesticides, and machinery. The USDA's Economic Research Service has estimated that each dollar of U.S. output added to agricultural exports stimulates $1.31 in activity in related economic sectors (USDA, 2009). Furthermore, as noted earlier, agriculture is a key component of the primary mega-sector of the U.S. economy. Agricultural output supports activity in other mega-sectors, especially the secondary mega-sector, which includes manufacturers that process raw agricultural inputs into final consumer goods.

U.S. Agriculture in a Global Context

The U.S. agricultural sector is part of a complex global system of agriculture, food processing, and food marketing. Both exports and imports of agricultural products have grown over time, but growth in the export sector has been more rapid, yielding a consistently positive trade surplus in agriculture. Exports have undergone an important transformation in recent years. Although bulk grains (e.g., corn, wheat) historically accounted for growth in U.S. exports, high-value products now play a larger role. These products include meats, poultry, fruits, and vegetables. This change is the direct result of growth in population and incomes worldwide; demand from this growth has driven U.S. exports. In addition, trade agreements have played a central role in expanding U.S. export markets. In particular, since full implementation of the North American Free Trade Agreement (NAFTA), Canada and Mexico have become the largest export destinations for U.S. agricultural products, eclipsing Japan and other trading partners. Overall, exports now represent roughly 30% of the total market value of U.S. agricultural production.

Numerous factors influence the competitiveness of the U.S. agricultural sector in the global economy. These factors include the prices of key inputs such as labor, seed, fertilizer, and machinery; the status of trade agreements with various nations; changes in subsidies and other agricultural policies; technological improvements; and access to capital. It is difficult to summarize competitiveness because the U.S. position varies greatly depending on the trading partners and the individual agricultural products under consideration; however, productivity serves as a good, stand-alone indicator of the resiliency and competitiveness of U.S. agriculture. The USDA maintains an index measuring agricultural productivity, defined as the difference between the growth in farm output

DID YOU KNOW?

Agricultural operations—crop irrigation, livestock watering, and aquaculture—withdraw approximately 140 billion gallons of water per day. Irrigation accounts for over 90% of this use and is the largest single consumptive water use in the United States.

DID YOU KNOW?

Agricultural use is a major consideration in the competition for water, especially in western states. Increasingly, agricultural interests that hold water rights are participating in water transfer initiatives through which they sell water to municipalities and other users.

and the growth of all agricultural inputs; this measure of agricultural productivity has increased steadily over time, rising by 152% in the last 60 years. Through productivity gains and efficient use of inputs, the U.S. agricultural sector continues to improve its competitive position and maintain a positive balance of trade.

In the context of this text, a key question is whether a reliable supply of water includes the productivity and competitiveness of the nation's agricultural sector. Arguably, U.S. producers enjoy a comparative advantage as a result of several water-related factors, including the following:

- A relatively temperate climate that allows for non-irrigated production in many parts of the country
- A relatively efficient and technologically advanced system of irrigation in areas in which it is required
- A reliable supply of adequate-quality groundwater and surface water for irrigation and livestock watering

A detailed comparison of the United States to other nations with respect to these factors in beyond the scope of this book, but the literature suggests that few nations enjoy the advantages enjoyed by U.S. producers. First, most nations are more reliant upon irrigation for food production than is the United States. Approximately 11% of all U.S. cropland is irrigated. In contrast, 19% of cropland worldwide is irrigated. Some major nations such as China and Pakistan irrigate over half of their cropland (USEPA, 2012a). Another obvious indicator of U.S. competitive advantage in agriculture is our food security relative to other nations. Studies suggest that over 850 million people worldwide do not have a reliable source of food; most of these people live in the arid nations of southern Asia and Sub-Saharan Africa, where water access is poor. Finally, recommendations for increasing food security emphasize the need for expanding and improving the efficiency of irrigation systems worldwide (Molden, 2007).

Studies suggest that the competitive advantage of U.S. agriculture could be reinforced by emerging climate trends. In general, climate change is anticipated to yield increased temperatures and altered precipitation patterns, intensifying competition for limited water resources in many areas. An analysis by Lobell et al. (2011) estimated that, between 1980 and 2008, global maize production declined by 3.8% and global wheat production declined by 5.5%, both as a result of climate change. The warming pattern that caused the production decrease, however, was absent in U.S. farming areas. Strzepek and Boehlert (2010) analyzed how future climate change patterns could increase

DID YOU KNOW?

The absence of formal markets complicates efforts to value water applied in agriculture, and studies yield vastly different estimates depending upon the methodology used. Estimates based on delivery cost and factor input methods yield relatively low estimates (generally less than $100 per acre-foot). Data from water transfers suggest much higher values. Transfers between farms show values averaging about $1800 per acre-foot. Transfers from farms to municipalities show even higher average values of over $4000 per acre-foot.

competition for irrigation water. The study modeled likely changes in demand for municipal/industrial water, environmental flows (i.e., in-stream flows), and agricultural water under various climate change scenarios. The authors found that threats to agricultural water supplies are likely to intensify in Africa, Latin America, and the Caribbean. In contrast, the threat to agricultural water supplies in North America is expected to decline, primarily due to wetter climate conditions and decreased need for water to maintain in-stream flows (USEPA, 2012a).

Water Use by Agriculture

Agricultural applications account for approximately 34% of all water withdrawals. Irrigation uses represent over 90% of these withdrawals, while watering of livestock and supplies to aquaculture make up the balance. The discussion below provides a more detailed review of how water is used in these three agricultural applications.

Irrigation

With regard to quantity and sources of water used, estimates of the total quantity used for irrigation on U.S. farms vary, depending on year and data source. Table 10.4 shows that, although the USGS has estimated that 144 million acre-feet were used in irrigation in 2005, the USDA estimated that only 91.2 million acre-feet were used for irrigation in 2008. The USDA survey is more recent and provides more detail on irrigation methods and patterns; therefore, the discussion below focuses on the USDA figures when possible.

Just over half of all irrigation water is pumped from on-farm groundwater sources (USDA, 2010). Surface water sources are also common, with most coming from off-farm suppliers (e.g., irrigation districts). As discussed below, the source of irrigation water can play an important role in creating incentives to improve the efficiency of agricultural water use.

In all forms of irrigation, the preferred outcome is to deliver water directly to the crop. In practice, however, the ultimate disposition of irrigation water is more complex. In addition to being taken up by crops, water may leak to the ground in non-cultivated areas, evaporate during delivery or after initial application, or return overland or through drainage systems to surface water and thus be available for reuse. Studies by the USGS and USDA typically consider the "consumptive" component of irrigation water use to include the fraction that is "evaporated transported, incorporated into products or crops … or otherwise removed from the immediate water environment" (USGS, 2009). Based on data from 1995, the USDA estimated that 61% of irrigation water is consumed, making irrigation the most dominant consumptive water use (Wiebe and Gollehon, 2006).

Farmers in the United States generally distribute irrigation water using one of three methods:

- *Gravity systems*—Gravity-fed irrigation systems include traditional methods that deliver water to fields via ditches, furrows, or pipes. Although this is a relatively lower cost approach, it is generally considered less efficient than other distribution method.
- *Sprinkler systems*—In a sprinkler irrigation system, water is delivered via a series of perforated pipes. Variations of sprinkler system include rolling systems, which can be moved

TABLE 10.4
Irrigation Water Use: U.S. Geological Survey vs. U.S. Department of Agriculture

Source	Year	Acre-Feet Applied	Acres Irrigated	Average Acre-Feet per Acre
USGS	2005	144,000,000	66,100,000	2.35
USDA Farm and Ranch Irrigation Survey	2008	91,235,036	54,929,915	1.70

Sources: Data from USGS (2009) and USDA (2010).

DID YOU KNOW?

Water use efficiency has improved through increased use of new irrigation technologies, but room for improvement still exists. Historically, efficiency incentives were often reduced by abundant water supplies made available through large public infrastructure projects. These projects created interdependencies between agriculture and other water users, such as municipality and power generators.

across fields, and rotating or pivot systems that direct water over a wide area by moving the water stream. Sprinkler systems were applied on over half of the irrigated acreage in the United States in 2008.

- *Drip/trickle or micro sprinkler systems*—Drip/trickle systems use small-diameter tubes placed on or below the soil surface, applying water frequently and slowly. Water is applied directly to the root zone of the plants. Micro sprinkler systems use low-volume sprinkler heads positioned just above the soil surface. These low-flow methods are relatively high-cost but are considered highly efficient and are generally applied to high-value, perennial corps (Wiebe and Gollehon, 2006). Low-flow drip irrigation systems are used selectively and account for only about 7% of all acreage.

Various factors have combined to encourage improvements in the efficiency of irrigation practices. A major driver has been the increasing scarcity of water in key agricultural areas and increases in the explicit or implicit price paid for water. In addition, technological shifts and advances have allowed for greater irrigation efficiency. One technologic change is simply the method of water delivery. Data show an increase in the use of sprinkler systems and a reduction in the use of flood or gravity systems (which lose water in conveyance). Between 1985 and 2005, the amount of land irrigated by sprinkler systems increased from 22 million to more than 30 million acres. In that same period, the amount of land irrigated by flood systems fell from 35.0 million to 26.6 million acres (USGS, 2009).

Irrigation water application rates further demonstrate trends in water use efficiency. The USGS estimated that in 1950 the average application rate was 3.55 acre-feet per acre; by 2005, this figure had fallen to 2.35 acre-feet per acre. The 2008 USDA irrigation survey estimated per-acre application rates to be as low as 1.7 acre-feet per acre. It is important to note that increases in irrigation efficiency do not necessarily translate into decreases in water consumption. Because gravity systems deliver more water to fields than is required by crops, a significant portion of that water runs off and may be available to downstream users through irrigation return flows. By contrast, drip irrigation systems are much more precise in the amount of water delivered to crops, meaning that a much higher percentage of water used for irrigation is consumed by plants. To the extent that efficiency gains reduce losses to evaporation, they can decrease total consumptive use of water in this sector.

Despite recent increases in irrigation efficiency, significant room for improvement remains. As discussed below, distortions in the prices that some irrigators pay for water can limit the economic incentive to improve efficiency. Comparative studies suggest that the United States is not among the most efficient users of irrigation water. Many operations that could employ precision technologies are still using inefficient gravity systems (Gleick, 2006). Likewise, few farmers are taking advantage of scheduling and measurement technologies in irrigation. Most irrigators still rely heavily on the feel of the soil or the condition of the crop. The 2008 USDA irrigation survey suggested that only about 9% use soil moisture sensors; likewise, only 2% use plant moisture sensors (USDA, 2010).

TABLE 10.5
Value of Irrigated Crops

	Farms with Any Irrigated Land		Farms with Entire Crop Irrigated	
Market Value of All Crops	Crop Market Value ($ Thousands)	Percent of All Crops	Crop Market Value ($ Thousands)	Percent of All Crops
$143,657,928	$78,297,158	54.5%	$46,872,638	32.6%

Source: USDA, *2007 Census of Agriculture*, U.S. Department of Agriculture, Washington, DC, 2009.

Notwithstanding the continuing tendency of many farmers to use personal judgment and old habits such as personal feel of the soil or condition of the crop in determining proper irrigation treatments, one thing is certain: There is little doubt that irrigation has and does contribute significantly to the productivity of U.S. agriculture. First, as shown in Table 10.5, over 50% of the market value of all corps is generated on farms where at least some irrigation is used. Farms that irrigate their entire crop account for about one-third of total crop value.

The importance of irrigation is also reflected in its effect on crop yields. Table 10.6 lists major irrigated crops in the United States and the effect that irrigation has on yields of these crops. As shown, ranked by total acreage at farms where the entire crop is irrigated, major irrigated crops include corn, alfalfa, soybeans, cotton, and wheat. (Note that several irrigated products for which yield information is not available, including forage, orchard crops, rice, and vegetables, are excluded from this discussion.) In all cases, irrigation improves the recorded yield per acre. This improvement is most striking for alfalfa and wheat, where yields double when the entire crop is irrigated.

The value of irrigated crops reflects two key trends: (1) the importance of irrigation for farm productivity, and (2) increased efficiency in the use of irrigation water. The market value of irrigated crops has steadily increased despite relatively unchanged water inputs. Moreover, it is important to note that irrigation water use is concentrated in a limited number of states. This is the result of some obvious factors, such as arid climate, as well as complex factors that include demographic patterns and institutional decisions that have enhanced the availability of water for irrigation. Seven states account for about two thirds of all irrigation water use: California, Arkansas, Texas, Nebraska, Idaho, Arizona, and Colorado. California is by far the largest user, accounting for about one quarter of all irrigation water.

TABLE 10.6
Major Irrigated Products and Yield Improvements on Irrigated Land

	Entire Crop Irrigated		Part of Crop Irrigated		None of Crop Irrigated	
Product	**Acres**	**Average Yield**	**Acres**	**Average Yield**	**Acres**	**Average Yield**
Corn for grain (bushels)	6,103,769	180	7,053,000	150	66,656,287	144.3
Alfalfa hay (tons, dry)	5,746,037	4.9	810,615	3.3	12,846,779	2.5
Soybeans for beans (bushels)	2,175,069	45.3	3,062,006	40.8	55,282,030	40.2
Cotton (bales)	2,046,094	2.5	1,989,516	1.8	4,214,480	1.5
Wheat for grain (bushels)	1,806,902	80.3	1,557,177	42.7	43,865,291	37

Source: USDA, *2007 Census of Agriculture*, U.S. Department of Agriculture, Washington, DC, 2009.

TABLE 10.7
USGS Livestock Water Use Coefficients

Animal Type	Water-Use Coefficient (gallons/day)
Dairy cows	35
Beef and other cattle	12
Hogs and pigs	3.5
Laying hens	0.06
Broilers and other chickens	0.06
Turkeys	0.1
Sheep and lambs	2
Goats	2
Horses	12

Source: Lovelace, J.K., *Method for Estimating Water Withdrawals for Livestock in the United States, 2005*, U.S. Geological Survey, Washington, DC, 2009.

Livestock

The USGS has estimated that U.S. farmers and ranchers withdraw approximately 2.1 billion gallons of water per day to maintain livestock. These withdrawals accounted for less than 1% of total U.S. water withdrawals in 2005. Approximately 60% of this supply is drawn from groundwater sources. The vast majority of the water is consumed rather than returned to surface water or aquifers. USGS studies of water consumption have estimated consumptive-use coefficients ranging from 84% to 100%, depending on the predominant animal type, geographic region, and other factors (Shaffer and Runkle, 2007).

Few states require that livestock operations report water usage. As a result, the USGS method relies on water-use coefficients to estimate livestock water withdrawals. These coefficients reflect total daily water usage for each major category of livestock; the coefficients are combined with USDA data on animal inventories to estimate water use. Where available, the USGS applies water-use coefficients specific to a given state. When state-specific coefficients are not available, the USGS applies the median values shown in Table 10.7.

As shown, water requirements vary greatly by animal type. This is partly due to the water consumption requirements of larger vs. smaller animals; it also reflects differences in the use of water for purposes other than consumption. These purposes include waste disposal, sanitation, cooling, and other needs. Water use in waste disposal can vary greatly, depending on the manure management method applied on a given farm. Dairy operations are especially water intensive; they require water for cleaning cow udders prior to milking, sanitation of equipment, and cooling of storage tanks (Lovelace, 2009a). This intensive water use is reflected in the high water-use coefficient for dairy cows.

Livestock operations such as dairy procedures require water for such activities as cleaning cow udders and sanitation of equipment. Obviously, in cleaning cow udders and in sanitizing diary operation equipment, the quality of the water used is important (or it certainly should be!). With water quality in mind, let's focus this discussion, for the purpose of brief illustration, on other livestock such as range cattle.

Water is the most important nutrient for range cattle. It is required for all life processes. A loss of 20% of a body's water will be fatal. Total body water of cattle is 56 to 81% of body weight. Physiologic state and body composition affect the body's water content. Lactating cows possess the greatest amount of body water (from 62 to 69%). Fat cows contain less water than thin lactating cows, and younger animals have higher water content than older animals (USDA, 2013).

Loss of body water occurs through milk production, fecal excretion, urine excretion, sweat, and vapor loss. Water losses through milk production are equal to approximately 33% of the cow's daily consumption. Fecal losses are usually comparable to milk losses, and urinary losses are approximately 50% of fecal losses. Fecal water losses are affected by intake, dry matter content of the diet, and digestibility. Urinary losses range from 1 to 9 gallons per day. Urinary water excretion is related to water availability, water absorbed, nitrogen and potassium content of urine, and dry matter content of the diet.

Five criteria might be combined to encompass the characteristics of livestock water quality (a more in-depth discussion of water quality in general is provided later in the text):

1. Odor and taste
2. Chemical properties (pH, dissolved solids, total dissolved oxygen, and hardness)
3. Toxic compounds (e.g., heavy metals, toxic minerals, organophosphates, hydrocarbons)
4. Presence of excess minerals (e.g., nitrates, sodium sulfate, iron)
5. Presence of living organism (e.g., bacteria)

Cattle can meet their water requirement from three sources:

- Drinking free water
- Ingestion of water contained in feed
- Water produced by the body's metabolism

It has been estimated that

- Cows require 2.6 to 3.0 pounds of water for every pound of milk produced.
- In moist pastures, cattle may consume only 40% by drinking.
- Salt availability influences consumption. Cattle need to consume an average of 11 to 15 grams of salt every day to maintain good nutrition. An adequate intake of salt also contributes to better productivity and overall herd health.
- High salt or protein may stimulate water consumption. For every gram of sodium consumed (28 grams = 1 ounce) water intake generally will increase 0.1 pound.

Water is especially important during periods of heat stress:

- Properties for cooling include heat conductivity and vaporization.
- As effective temperatures rises from 65 to 86°F, consumption increases by 30%.
- Losses from urine, sweating, and respiration increase 15, 59, and 50%, respectively, with temperature increases.
- No shade during the summer will increase water consumption by 18%.

An important water quality parameter is salinity, which is expressed as total dissolved solids (TDS) expressed as milligrams/liter (mg/L) or parts per million (ppm). Such solids include sodium chloride, carbonates, nitrates, sulfates, calcium, magnesium, and potassium.

The primary symptom of high-salinity water is diarrhea. If the TDS level is high, cattle will be reluctant to drink but will then drink a large amount at once. This can cause the animal to become very sick and potentially die. Guidelines include the following:

- Less than 1000 ppm is generally safe.
- Greater than 1000 ppm may cause some diarrhea, may reduce the availability of other minerals, and may reduce performance.

- Levels of 3000 to 5000 ppm are marginally safe. The cattle may be reluctant to drink, and such levels may reduce performance and affect health. The water should also be tested for sulfates.
- Levels of 5000 to 7000 ppm indicate poor water. Performance and health slump, especially when temperatures are high. The water should be tested for sulfates and should be used only with low-value stock.
- Levels of 7000 ppm and over are unsuitable. Performance and health effects can be expected. Limit the use of such water with lactating and pregnant stock. Also, sulfates are mostly likely high.

Another important livestock water quality parameter is sulfates, which are commonly high in groundwater in South Dakota and Montana. Adult cattle seem to be more resistant to the effects of sulfates than calves. Sulfates can cause secondary deficiencies of copper, zinc, iron, and manganese. The form of sulfates can vary. The most common forms are sodium sulfate and magnesium sulfate. All forms are a laxative. Cattle will drink less and have diarrhea at a lower sulfate concentration with sodium. At over 2000 ppm, diarrhea may start but cattle will adapt. The presence of iron sulfate may cause the most severe rejection of drinking water. Guidelines for interpretation of sulfate results are as follows:

- Less than 500 ppm—*Safe.*
- 500 to 1500 ppm—*Generally safe.* Trace mineral availability may be reduced; decreased performance of confined cattle may be observed.
- 1500 to 3000 ppm—*Marginal.* Poor for confined cattle during hot weather; sporadic cases of polio may be seen in confined cattle; performance maybe reduced.
- 3000 to 4000 ppm—*Poor.* Sporadic cases of polio are probable, especially in confined cattle; performance of grazing cattle may be affected.
- 4000-plus ppm—*Dangerous.* Health problems and reduction in performance can be expected.

Following are the results of two South Dakota studies evaluating three groups of yearling cattle performance in pens fed a roughage diet and different qualities of water:

1. Good—400 ppm sulfate per 1000 ppm TDS
2. Poor—3100 ppm sulfates per 4800 ppm TDS
3. Poorer—3900 ppm per 6200 ppm TDS

Steer average daily gain declined from 1.4 lb to 1 lb per day as the amount of sulfates increased. The steers drinking the 3100- and 3900-ppm sulfate water exhibited reduced water intake and reduced dry matter intake feed efficiency, and polio was seen in 15% of the cattle on the high-sulfate water. Daily sulfur contents of the diet were 0.27, 0.74, and 0.93%, respectively. The requirement is near 0.2%.

In another set of studies, steers had access to four sources of water:

1. 400 ppm sulfates per 1000ppm TDS
2. 1700 ppm sulfates per 3000ppm TDS
3. 2900 ppm sulfates per 5000ppm TDS
4. 4600 ppm sulfates per 7000 TDS

Steer gains were 1.8, 1.6, 1.5, and 0.6 lb/day, respectively, for each sulfate treatment; 47% of steers on the two highest sulfates levels had symptoms of polio and 33% died (USDA, 2013).

Although some technological changes have improved water quality and livestock water use efficiencies, the changes in the latter are minor. The drinking component of water use dominates, and animals' water requirements are obviously constant. As a result, livestock water use has varied little over the last 60 years. The growth in water use largely reflects growth in the overall size of the U.S. livestock inventory. The geographic distribution of withdrawals corresponds to where livestock are raised. In particular, water use is high in states with major dairy and beef cattle operations; these states include Texas, California, and Oklahoma, the top three users of livestock water in the United States.

Aquaculture

Aquaculture operations account for roughly 2% of all U.S. water withdrawals (Lovelace, 2009b). The USGS considers all the withdrawals to be self-supplied, with about 78% taken from surface water sources and the remainder from groundwater. Available data are limited but suggest that aquaculture water use is growing rapidly, as indicated by an estimated 52% increase in withdrawals for this purpose from 2000 to 2005. Aquaculture water use is largely non-consumptive but depends on the aquaculture method applied. Major methods include the following:

- *Raceways*—Some production (e.g., trout, salmon) occurs in flow-through raceways. In raceways, water is temporarily diverted from a spring or stream to maintain a flowing environment for the fish. This growing method is a major water user but is non-consumptive, as virtually all of the water is returned to its source. About 10% of all aquaculture operations use raceways (Lovelace, 2009b).
- *Ponds*—Ponds, another common aquaculture method, are used to grow species such as catfish. The amount of water added to ponds varies greatly, depending on precipitation, evaporation, leakage, and the species grown. Therefore, ponds can represent a somewhat more consumptive water use in comparison to raceways, but generally they do not require large water withdrawals. About 54% of all aquaculture operations use ponds (Lovelace, 2009b).
- *Tanks*—Tanks are used to grow a diverse array of fish species, including trout, salmon, bass, tilapia, perch, and others. Water is removed during waste management and through evaporation; however, the method generally consumes little water because many operations use recirculating filtration systems. About 17% of all aquaculture operations use tanks (Lovelace, 2009b).
- *Other methods*—Other types of aquaculture operations include pens and cages (used for both fish and shellfish) as well as egg incubators.

Five states account for nearly two thirds of aquaculture water use: Idaho, North Carolina, Alaska, Oregon, and California. Idaho, by far the largest user, is the nation's leading producer of rainbow trout; hence, the use of raceways there is common.

Aquaculture is growing rapidly in the United States, although comprehensive data are not readily available to characterize this growth. The USDA only began reporting complete aquaculture information in its 2007 census. Table 10.8 summarizes the market value of major products and the associated number of aquaculture operations. Products from the aquaculture industry had a market value of approximately $1.4 billion in 2007. Catfish, trout, and mollusks were the most significant products, both with respect to market value and in terms of the number of operations. It is important to note that the species listed differ greatly in terms of aquaculture method and water use. For example, whereas catfish are predominantly growing in man-made ponds, trout are frequently raised in in-stream raceways that effectively consume little or no water.

Supply and Pricing Issues

In a textbook microeconomic system, scarcity and pricing combine to produce an efficient use of finite resources. The use of water in agriculture is at odds with this textbook system in a variety of ways. This section examines the economic setting for agricultural water use by first summarizing

TABLE 10.8
Major Aquaculture Products (2007)

Product	Number of Operations	Market Value ($ Thousands)
Catfish	1725	$455,378
Trout	1124	$210,568
Other food fish	643	$187,711
Baitfish	358	$40,341
Crustaceans	788	$50,855
Mollusks	1097	$243,007
Ornamental fish	684	$61,049
Sport or game fish	815	$80,568
Other aquaculture products	341	$85,793
Total		$1,415,272

Source: USDA, *2007 Census of Agriculture*, U.S. Department of Agriculture, Washington, DC, 2009.

the current system of pricing and distribution. The discussion then reviews how heightened competition of scarce water supplies is encouraging stakeholders to pursue more efficient approaches to the use of water resources (USEPA, 2012a).

Pricing and Distribution of Agricultural Water

The ways in which farmers acquire water greatly affect the cost of water and the economic signals to which they respond. Table 10.9 summarizes key information on the three major sources of irrigation water: self-supplied groundwater, self-supplied surface water, and surface water delivered from off-farm sources. Self-supplied surface water is generally the least costly water supply method, although this option is available on limited acreage, mostly in eastern areas where surface water is plentiful. Groundwater pumping costs vary greatly, depending on local conditions. Water purchases from off-farm retailers also vary in cost, depending on the prices charged by the supplier; this water acquisition approach is common in the western states. As discussed below, these costs are very low when compared to both true delivery costs and the full social cost of irrigation water use.

TABLE 10.9
Irrigation Water Sources and Costs

Water Source	Share of Irrigated Acres (2008)	Average Cost ($ per acre, 2003)	Cost Range ($ per acre, 2003)	Cost Considerations
Groundwater, self-supplied	53%	$39.50	$7–$176	Pumping cost varies with energy prices, depth to water, and pump efficiency.
Surface water, self-supplied	15%	$26.39	$10–$82	Cost reflects expense for lifting or pressurizing surface water; when pumping is not required, effective cost is zero.
Surface water, off-farm supply	32%	$41.73	$5–$86	Costs reflect costs charged by water supply intermediaries; most common in western United States

Sources: Data from Wiebe and Gollehon (2006), Heimlich (2003), and USDA (2010).

These modes of water delivery and their associated costs are a product of numerous institutional factors that have evolved over time. First, laws governing water rights are complex. Riparian water rights (i.e., the right to divert surface water on one's own property) have long applied in areas with abundant surface water supplies. In arid regions, however, the "appropriation doctrine" has generally prevailed. Introduced during the period of western expansion, this system allows water users to stake claims to water sources; this has allowed a complex system of seniority rights to develop (see Sidebar 10.4).

Beginning with the Reclamation Act of 1902, the federal government took a key role in supplying irrigation water. The Bureau of Reclamation (BOR) acts as a water wholesaler by developing large water supply infrastructure projects. Through 1994, the BOR had built 133 projects that supply irrigation water at a total cost of $21.8 billion. State governments have funded and developed additional water supply projects. BOR and other water developers typically make the water available to intermediate water retailers such as irrigation districts. These intermediaries are generally non-profit entities that seek to supply water at the lowest possible cost (Wiebe and Gollehon, 2006).

Over the past century, state governments have become increasingly involved in the management of water rights and resources. Although state policies vary, they generally grant water-use rights to individuals, charging nothing other than minor administrative fees for these rights (Wiebe and Gollehon, 2006). These approaches also extend to the management of groundwater resources. Some states actively manage groundwater use through permit systems, but others place no limits on pumping.

The sum effect of all these institutional factors is that, in many cases, the cost that farmers face for the use of irrigation water has been established entirely outside of conventional economic markets. Water costs for farmers may partially reflect access and delivery costs, but they have little or no relationship to (1) total supply costs, (2) the on-farm value of irrigation water (e.g., increased yield), and (3) the potential social value of water in its highest and best use. For example, water from federal supply projects is often heavily subsidized by other project beneficiaries (e.g., hydropower products), allowing farmers to obtain water at prices far below the actual cost of delivery.

These subsidies can result in a variety of economic distortions. First, studies show that increasing water scarcity has driven up the value of water in industrial and municipal applications (discussed later); as such, the lack of pricing mechanisms results in an economically inefficient use of water in some regions of the United States. In addition, subsidized water may reduce incentives for water-use efficiency in irrigation and may encourage the cultivation of water-intensive crops poorly suited to natural and marker conditions.

SIDEBAR 10.4 Reclamation Act and Appropriation Doctrine*

Less than 1% of water on Planet Earth is suitable for people to use, and that small percentage must be shared with nature. The other 99+% is either too salty or brackish, or it is unavailable in icecaps, glaciers, the atmosphere, and groundwater. Because the American West is generally arid, water has always been a major concern of Native Americans and settlers who relied upon the relatively meager supply for agriculture, settlement, and industry.

As settlers moved into the West, they had to watch the gush of spring and early summer runoff flow away from their towns and crops. They knew they had lost water that would not be available in the dry days of late summer when water shortages are common in the West. Settlers responded by developing relatively simple and inexpensive water projects and creating complicated water law systems. Varied in details, western water law systems generally permanently allocated the right to use water based on the concept of prior appropriation (the *appropriation doctrine*—first in time, first in right) for beneficial use.

At first, water development projects were simple. Settlers diverted water from a stream or river and used it nearby, but in many areas the demand for water outstripped supply. As demands for water increased, settlers wanted to store runoff, which they considered "wasted," for later use. Storage projects would help

* Adapted from Bureau of Reclamation, *Brief History of the Bureau of Reclamation*, Bureau of Reclamation, U.S. Department of the Interior, Washington, DC, 2012.

maximize water use and make more water available for use when needed. Unfortunately, private and state-sponsored irrigation ventures often failed because of lack of money or lack of engineering skill. This resulted in mounting pressure for the federal government to develop water resources.

In the jargon of the day, advocates called irrigation projects *reclamation* projects. The concept was that irrigation would "reclaim" or "subjugate" western arid lands for human use. Many pressures contributed to the discussions that influenced American public opinion, Congress, and the executive branch to support of "reclamation." Among them were

- John Wesley Powell's scientific explorations in the West, including in 1869 the first exploration of the dangerous and almost uncharted canyons of the Green and Colorado Rivers and his published articles and reports
- Private pressures through publications, irrigation organizations, and irrigation "congresses"
- Nonpartisan political pressures in the West
- Federal government studies conducted by the U.S. Army Corps of Engineers and the U.S. Geological Survey (USGS)

Before 1900, the U.S. Congress had already invested heavily in American's infrastructure. Roads, river navigation, harbors, canals, and railroads had all received major subsidies. A tradition of government subsidization of settlement of the West was longstanding when the Congress in 1866 passed "An Act Granting the Right-of-Way to Ditch and Canal Owners over the Public Lands, and for Other Purposes." A sampling of subsequent congressional actions promoting irrigation includes passage of the Desert Land Act in 1877 and the Carey Act in 1894 (also known as the Federal Desert Land Act)—both intended to encourage irrigation projects in the West. In addition, beginning in 1888, Congress appropriated money to the USGS to study irrigation potential in the West. Then, in 1890 and 1891, while that irrigation study continued, Congress passed legislation reserving rights-of-way for reservoirs, canals, and ditches on lands then in the public domain. However, westerners wanted more; they wanted the federal government to invest directly in irrigation projects. The reclamation movement demonstrated its strength when pro-irrigation planks found their way into both Democratic and Republican platforms in 1900. In 1901, the idea of reclamation gained a powerful and aggressive supporter in Theodore Roosevelt when he became President after the assassination of William McKinley.

President Roosevelt supported the reclamation movement because of his personal experience in the West and because of his "conservation" ethic. At that time, "conservation" meant a movement for sustained exploitation of natural resources through careful management for the good of the many. Roosevelt also believed that reclamation would permit homemaking and support the agrarian Jeffersonian ideal. Reclamation supporters believed the program would make homes for Americans on family farms. Passed in both houses of Congress by wide margins, President Roosevelt signed the Reclamation Act on June 17, 1902.

In July of 1902, Secretary of the Interior Ethan Allen Hitchcock established the U.S. Reclamation Service (USRS) within the USGS. There, Charles D. Walcott, director of the USGS and the first director of the USRS, placed the new activity in the USRS's Division of Hydrography. Frederick Newell became the first chief engineer while continuing his responsibilities as chief of the Division of Hydrography.

The Reclamation Act required that:

> Nothing in this act shall be construed as affecting or intended to affect or in any way interfere with the laws of any state or territory relating to the control, appropriation, use, or distribution of water … or any vested right acquired there under, and the Secretary of the Interior … shall proceed in conformity with such laws.

That meant that implementation of the act required that Reclamation comply with numerous and often widely varying state and territorial legal codes. Development and ratification over the years of numerous interstate compacts governing the sharing of stream flows between states, as well as several international treaties governing the sharing of streams by the United States with Mexico or Canada, made the Reclamation Service's efforts to comply with U.S., state, and territorial water law even more complex.

In its early years, the Reclamation Service relied heavily on the USGS Division of Hydrography's previous studies of potential projects in each western state or territory. As a result, between 1903 and 1906, about 25 projects were authorized throughout the West. These projects were funded by a Reclamation Fund into which Congress ordered the deposit of revenues from public land sales in the West. Later, other revenue sources were also directed into the Reclamation Fund. Because Texas had no federal lands, it was not one of the original reclamation states. It became a reclamation state only in 1906.

During its early years, the Reclamation Service, at the direction of the various administrations and the Congress, developed several basic principles for the reclamation program. The details have changed over the years, but the general principles remain:

- Federal monies on reclamation water development projects would be repaid by the water users who benefited.
- Projects remain federal property even when the water users repay their share of federal construction costs.
- Reclamation generally contracts with the private sector for construction work.
- Reclamation employees administer contracts and inspect construction to ensure that contractors' work meets government specifications.
- In the absence of acceptable bids on a contract, the Reclamation Service, especially in its early years, would complete a project by "force account" (that is, would use reclamation employees to do the construction work).
- Hydroelectric power revenues could be used to repay project construction charges.

In 1907, the USRS separated from the USGS to become an independent bureau within the Department of the Interior. The Congress and the executive branch, including USRS, were then just beginning a learning period during which the economic and technical needs of reclamation projects became clearer. Initially overly optimistic about the ability of water users to repay construction costs, Congress established a 10-year repayment period. Subsequently, Congress increased the repayment period to 20 years, then to 40 years, and ultimately to an indefinite period based on "ability to pay." Other issues that arose included soil science problems related both to construction and to the ability of soils to grow crops; economic viability of projects (repayment potential), including climatic limitations on the value of crops; waterlogging of irrigated lands on projects, resulting in the need for expensive drainage projects; and the need for practical farming experience for people to successfully take up project farms. Many projects were far behind their repayment schedules and settlers were vocally discontented.

The learning period for the Reclamation Service and Congress resulted in substantial changes when the USRS was renamed the Bureau of Reclamation in 1923. Then, in 1924, the Fact Finder's Act began making major adjustments to the basic reclamation program. Those adjustments were suggested by the Fact Finder's Report, which resulted from an in-depth study of the economic problems and settler unrest regarding the Bureau of Reclamation's 20-plus projects. Elwood Mead, one of the members of the Fact Finder's Commission, was appointed Commissioner of Reclamation in 1924 as the reshaping of reclamation continued. A signal of the changes came in 1928, such as when Congress authorized the Boulder Canyon Project (Hoover Dam), and, for the first time, large appropriates began to flow to the Bureau of Reclamation from the general funds of the United States instead of from the Reclamation Fund or loans to the Fund.

In 1928, the Boulder Canyon Act ratified the Colorado River Compact and authorized construction of Hoover Dam and the All-American Canal—key elements in implementation of the Compact. Subsequently, during the Great Depression, Congress authorized almost 40 projects for the dual purposes of promoting infrastructure development and providing public works jobs. Among these projects were the beginnings of the Central Valley Project in California, the Colorado–Big Thompson Project in Colorado, and the Columbia Basin Project in Washington.

Ultimately, of the Bureau of Reclamation's more than 180 projects, about 70 were authorized before World War II. The remainder were authorized during and after World War II in small and major authorizations, such as the Pick–Sloan Missouri Basin Program (1944), the Colorado River Storage Project (1956), and the Third Powerplant at Grand Coulee Dam (1966). The Bureau of Reclamation's last really big project construction authorization occurred in 1968 when Congress approved the Colorado River Basin Project Act, which included, among others, the Central Arizona Project, the Dolores Project, the Animas–La Plata Project, and parts of the Central Utah Project.

During construction, one problem confronted by the Bureau of Reclamation was laboratory testing of special problems. The Bureau of Reclamation conducted model testing at various locations such as Montrose and Estes Park, Colorado; Colorado State University; and various locations in downtown Denver. In 1946, the Bureau of Reclamation located its primary laboratory at the Denver Federal Center. A sampling of the work of these research laboratories included model and design studies for hydraulic structures, concrete technology, electrical problems, construction design innovations, groundwater, weed control in canals and reservoirs, various environmental issues, water quality, ecology, drainage, control of evaporation and other water losses, and other technical subjects.

DID YOU KNOW?

About 195 billion gallons per day, or 8% of all freshwater and saline water withdrawals for 2000, were used for thermoelectric power that year. Most of this water was derived from surface water and used for once-through cooling at power plants. About 52% of fresh surface water withdrawals and about 96% of saline water withdrawals were for thermoelectric power use. Withdrawals for thermoelectric power have been relatively stable since 1985.

Passage and implementation of the various reclamation projects and the associated laboratory testing described above all have contributed significantly to or are a result of and in line with the Bureau of Reclamation's mission statement: "The mission of the Bureau of Reclamation is to manage, develop, and protect water and related resources in an environmentally and economically sound manner in the interest of the American public." Under this mission statement, Reclamation's number one priority is always to deliver water, but that priority is often affected by the available water supply and the constraints imposed by various laws, regulations, and court rulings. During an average water year, more that 180 reclamation projects deliver agricultural water that irrigates about 10,000,000 acres of land in the arid West—about one–third of the irrigated acreage in the West. Reclamation also delivers water used by about one third of the people in the West.

The second major product delivered to the American public by Reclamation is hydroelectricity. Although the earliest hydroelectric plants on reclamation projects went into operation in 1908 and 1909, it was only during the 1930s that generation of hydroelectric power became a principle benefit of reclamation projects. The Bureau of Reclamation built the large hydroelectric powerplants at Hoover, Grand Coulee, and Shasta dams only after a hard public debate about whether the federal government should become involved in public power production or whether private power production should be the rule. It was the Hoover Dam precedent that ultimately allowed the Bureau of Reclamation to become a major hydroelectric producer. Once the issues received public airing during discussions on approval of Hoover Dam, hydroelectric projects became a feature of many additional Bureau of Reclamation dams. Hydroelectric revenues have subsequently proven an important source for funding repayment of Bureau of Reclamation project costs. The water available in a system determines how much power it is possible to generate, but in 1993, for example, the Bureau of Reclamation had 56 powerplants online and generated 34.7 billion kilowatt hours of electricity. In 1999, revenues from Grand Coulee hydroelectric generation equaled about two-thirds of the Bureau of Reclamation's entire appropriated budget.

Commissioners representing Arizona, California, Colorado, Nevada, New Mexico, Utah, and Wyoming met in Santa Fe in 1922, with Secretary of Commerce Herbert Hoover moderating, to address allocation of the water of the Colorado River. The commissioners developed and signed the Colorado River Compact to divide and allocate waters of the Colorado River. The Compact then had to be ratified by each state legislature before going to Congress for ratification. For the Bureau of Reclamation, this is the most complex and difficult of the interstate compacts, and it was ratified by the Congress in 1928 without the concurrence of Arizona. California and Arizona argued for years over how to calculate Arizona's share of the waters of the lower Colorado River. The Arizona legislature ratified the Compact only in 1944 and then later sued California over how to interpret the Compact. The lawsuit lasted until 1964. Concern over the Compact has only heighted over the years as it became increasingly apparent that there is not consistently as much water in the Colorado River system as was presumed by the signers and ratifiers. In addition, the Compact required the Upper and Lower Basins to share the burden of delivering 1.5 million acre-feet of water promised to Mexico in a later 1944 treaty. The Bureau of Reclamation is deeply involved in these complicated Colorado River issues because the Bureau reservoirs largely store and regulate the flow of the Colorado River. Reclamation dams in the Upper Colorado River Basin deliver water to Glen Canyon Dam, which then stores the water in Lake Powell. From Lake Powell, the water is delivered in accordance with the terms of the Colorado River Compact and operating agreements among the Colorado River Basin states, to the Lower Colorado River Basin states. Once delivered to the Lower Colorado River Basin, the water is held until needed in Lake Mead behind Hoover Dam.

It is important to point out that all over the West the Bureau of Reclamation is affected and guided by other compacts under the terms of which states share the waters of interstate streams. It is also important to note that, even though the Bureau of Reclamation's traditional area of operation is the 17 arid states of the American West, it has at times been assigned work outside that traditional operational area. For instance, during the later 1920s, the Reclamation studied "planned group settlement" in the South in cut over areas and swamps. The project was supposed to create new farms, but it ultimately died as impacts of the Depression on the farm economy were recognized. Other projects in the eastern United States were also undertaken, and the Bureau of Reclamation's photograph collection includes hundreds of photographs from areas outside the arid West. Reclamation was also involved in the technical issues and design of many of the Tennessee Valley Authority's dams.

Beginning in the 1930s, the Bureau of Reclamation studied possible projects in Hawaii, and in 1954 the Congress authorized investigations on Oahu, Hawaii, and Molokai among the Hawaiian Islands. In the 1940s and 1950s, the Bureau of Reclamation studied water development projects in Alaska and ultimately built the Eklutna Project (hydroelectric project) outside Anchorage. The Department of the Interior transferred the Eklutna Project out of the Bureau of Reclamation in 1967.

Since 2000, the Bureau of Reclamation has supervised additional studies in Hawaii, and Commissioner Bob Johnson was tasked to work with the states of Georgia, Alabama, and Florida with regard to Atlanta's water supply, the flows of the Chattahoochee River, and affected endangered species.

Another important concern of the Bureau of Reclamation, especially in its early years, was its active involvement with the Indian Service to develop irrigation projects for Indian tribes, including the San Carlos, Blackfeet, and Yuma; however, the majority of reclamation project water went to non-Indians. In the early years, the Bureau of Reclamation's mission to develop water supplies appeared to carry the potential for injuring the rights of tribes. If non-Indians began using the Bureau of Reclamation-provided water, it was feared they would establish a senior right under the appropriation doctrine, leaving little or no water for the tribes when they were ready to develop their reservation lands. In a landmark 1908 decision, *Winters v. United States*, the Supreme Court attempted to reconcile this potential conflict and laid out the principles that came to be known as the Winters Doctrine. This case concerned the Milk River in Montana and delayed development of the Bureau of Reclamation's Milk River Project for several years. The Winters Doctrine established the principle of reserved rights. Indian tribes with reservations have reserved water rights in sufficient quantities to fulfill the purposes for which the reservation was established, and the date of the reserved right is the date of the treaty or executive order setting aside the reservation. The dates of reserved rights generally are very early in relation to non-Indian settlement and, thus, establish very high priority for Indian water rights. Further, unlike appropriated water rights, a reserved water right does not have to have been used to remain in effect, regardless of how many years have passed. A congressionally authorized and funded Bureau of Reclamation project could not take precedence over senior or reserved water rights. However, for various reasons, some tribes have encountered difficulties in attempting to develop their senior reserved water rights.

In recent years, the Bureau of Reclamation has designed projects that include provisions for delivering water to tribes. Among notable examples are the Central Arizona Project, the Dolores Project, and the Animas–La Plata Project, as well as rural water distribution systems such as the Mni Wiconi and Mid-Dakota in the Dakotas which provide a rural culinary water supply in a large area that includes several reservations. Bureau of Reclamation staff members often serve on negotiating teams or provide technical expertise to negotiating teams working for the Secretary of the Interior to develop water solutions for Native Americans.

The Bureau of Reclamation has also been involved with conservation projects that are environmentally sensitive and important; however, the nature of conservation and environmental issues and how they have affected reclamation efforts have changed considerably. For example, between 1908 and 1912, there was a public outcry about conservation of Lake Tahoe's natural lake level and scenic beauty. The Bureau of Reclamation had proposed to build a dam both to increase storage capacity and to occasionally lower the existing lake level to benefit the Newlands Project. Due to public opposition that project was not built. In a distinctly different direction, the Bello Fourche Project in South Dakota was specifically designed to avoid mixing hazardous industrial mining wastes in Whitewood Creek with canal irrigation water.

Subsequently, proposals for reclamation projects raised public consciousness about major dams and their impacts on various resources. By the mid-1930s, the Bureau of Reclamation was looking at fishery issues as it addressed construction of Grand Coulee and other dams. On another front, in the mid- to late-1930s, Coloradoans and their congressional representatives pushed the Bureau of Reclamation to build the Colorado–Big Thompson Project, which would require construction on the fringe of and under Rocky Mountain National Park. The project was ultimately built because Rocky Mountain National Park was created with a provision in the enabling law specifically authorizing a water development project infringing on the national park. In the 1950s, the controversy over construction of Echo Park Dam in Dinosaur National Monument heightened public awareness of issues surrounding construction of a dam in a National Park

Service-managed area. Before public awareness the importance of preserving scenic beauty did not make sense to many state residents, who saw the monument as a barren wasteland, or to Mormons, who believed that creating an oasis in the desert was their mission and God's will.

At the same time, as many westerners demanded equal water rights, members of the growing nation "wilderness movement" saw the Echo Park Dam development as an opportunity to prevent a loss equivalent to that of Yosemite's Hetch-Hetchy Valley. The Wilderness Society, National Parks Association, Sierra Club, National Wildlife Federation, Izaak Walton League, American Planning and Civic Association, Wildlife Management Institute, Audubon Society, and National Council of State Garden Clubs were among the approximately 30 conservation groups involved in the Echo Park issue. Ultimately, these groups and public opinion forced cancellation of plans for Echo Park Dam and resulted in construction of an alternative, Glen Canyon Dam. By the 1960s, Marble Canyon and Bride Canyon dams were proposed, but Secretary of the Interior Stewart Udall canceled those dams because of public pressure to preserve parts of the Grand Canyon. Ironically, opposition was based at least partly on the public and environmentalists' belief that nuclear power generation was a viable alternative to meet growing electric power needs in the West.

During the 1960s, the Bureau of Reclamation's work began to change substantially as public awareness changed. Americans became increasingly concerned and proprietary about the use and protection of natural resources. This change resulted, in part, from improved communication and, in part, from improved science resulting in clearer understanding of the complex interactions of the communities of nature with western water issues. Americans were beginning to better understand issues about the West and to consider the West "mine" or "ours"—even though they lived elsewhere.

Rachel Carson's *Silent Spring*, published in 1962, helped build public support for more environmentally sensitive project development. Even popular music expressed growing environmental concerns, and increased public consciousness and support manifested itself in political action when Congress passed the Wilderness Act in 1964, amendments to the Fish and Wildlife Coordination Act in 1965, the National Historic Preservation Act in 1966, the Wild and Scenic Rivers Act of 1968, the National Environmental Policy Act (NEPA) of 1969, and the Endangered Species Act in 1973, among many other laws. Accompanying and buttressing these federal laws were presidential executive orders, federal regulations, and state and local laws, orders, and regulations.

The specific effects of reclamation projects were also better understood in this period. Dam construction affected fish populations and often altered the flow characteristics and ecology of rivers and streams. Land reclamation and construction projects affected plant, animal, fish, and bird populations through displacement or destruction of habitat. In addition, land development for agriculture or subdivisions often destroyed historic or archeological resources. Destruction of non-arable wetlands was a special environmental problem. Hydroelectric production, often considered to be pollution free, was recognized as causing environmental effects by altering water temperatures, flow regimes, and natural fluctuation patterns, thus affecting native fish populations' environment, migration, and spawning. Environmental issues that conflicted with the traditional Bureau mission were not unique to reclamation. Americans identified and looked toward resolution of many environmental effects caused by construction and natural resources exploitation programs in both the government and private sectors.

Because of the new laws and regulations and increasing public/political pressure, the Bureau of Reclamation hired new staff to deal with environmental and historic preservation issues. The Bureau now invests a great deal of time and money in issues such as endangered species, in-stream flows, preservation and enhancement of quality freshwater fisheries below dams, preservation of wetlands, conservation and enhancement of fish and wildlife habitat, dealing with Endangered Species Act issues, controlling water salinity and sources of pollution, groundwater contamination, and the recovery of salmon populations, particularly on the Trinity/Klamath, Columbia/Snake, and San Joaquin/Sacramento River systems. The Bureau of Reclamation implemented "reoperation" (a revision of the way hydroelectric power generation is scheduled and carried out) of hydroelectric facilities at Glen Canyon Dam on the Colorado River to better achieve environmental objectives. Costly modifications have been made to dams such as Shasta and Flaming Gorge to achieve environmental goals. There is also a major effort underway among federal and state agencies and other interest groups to improve environmental and water quality in the delta at the mouth of the Central Valley of California where the San Joaquin and Sacramento rivers flow into San Francisco Bay.

Ironically, the Bureau of Reclamation's attempts, in partnership with the Fish and Wildlife Service, to use drainage water to support environmental objectives at the Kesterson National Wildlife Refuge in the Central Valley of California resulted in unexpected and difficult environmental problems. The drainage water mobilized and concentrated selenium in the water of the refuge, which caused death and deformity among affected animal populations. The selenium issue was a problem, now resolved, that neither the Bureau nor the Fish and Wildlife Service foresaw.

In addition to public awareness of the potential negative impact of reclamation projects on the surrounding environment, the public also recognized that reclamation reservoirs provide a positive impact all over the West (particularly with regard to flatwater recreation opportunities). While westerners quickly identified and began to enjoy recreation opportunities on and in the water captured behind Bureau of Reclamation dams, recreation was not recognized legally as a project use until 1937. The Bureau transferred Lake Mead, behind Hoover Dam, to the National Park Service for recreation management in 1936 and initiated the still-existing pattern of seeking other agencies to manage recreation at reclamation facilities. That pattern means that today the Bureau of Reclamation directly manages only about one-sixth of the recreation areas on its projects. From the 1930s to the early 1960s, recreation for specific projects was authorized; however, in the mid-1960s, Congress began to give the Bureau more generalized authorities for funding recreation on all projects. Fishing, hunting, boating, picnicking, swimming, and other recreational opportunities have developed over the years. In addition to flatwater recreation opportunities, many stretches of "blue ribbon" fishing developed all over the West in favorable conditions below reclamation dams.

In 2010, the Bureau of Reclamation had 289 recreation areas located on about 6,5 million acres of land, most of which are open for public outdoor recreation. In recent years, the Bureau has "reoperated" some facilities seeking to improve fishing and whitewater recreational opportunities. Three recreation areas managed by the National Park Service—Lake Roosevelt behind Grand Coulee Dam, Lake Mead behind Hoover Dam, and Lake Powell behind Glen Canyon Dam—as well as the U.S. Forest Service's Lake Shasta behind Shasta Dam are among the most prominent national recreation areas on reclamation projects. Other managing partners for recreation areas include other federal agencies, state agencies, counties, and cities. These partnerships result in millions of recreation days of use on reclamation projects annually and raise numerous issues in terms of interagency coordination, water quality, public safety, public access, cost-sharing, law environment, etc. As water is converted from rural to urban uses in the West, and urban population increases, recreation visits to reclamation projects are expected to continue to increase.

Other benefits gained from reclamation projects are flood control and drought relief. The Bureau of Reclamation operates its facilities to prevent millions of dollars of flood damage. Between 1950 and 1992, reclamation projects prevented in excess of $8.3 billion in flood damage. During periods of drought, the Bureau of Reclamation becomes involved in drought management activities. Reclamation projects have carryover storage which often can provide water during a few consecutive years of drought. In some areas, however, growing demand stresses the water supply even in normal water years. Water shortages, often drought-influenced, will probably increase in the West, thus forcing more effective and efficient use of water supplies. Bureau of Reclamation drought activities are quite varied and include, for example, assisting water users with planning for use and allocation of limited water supplies, participating in cooperative contingency planning for future droughts, water conservation, loans, cooperation in water banking, deepening wells, and water purchases.

The Bureau of Reclamation currently has more than 180 projects in the 17 western states that are managed out of over 20 area offices. The area offices are within five regions which are organized around western watersheds. Many projects are actually operated and maintained by the water users. The Bureau's projects provide agricultural, municipal, and industrial water to about one third of the population of the west. Farmers on Bureau of Reclamation projects produce a significant percentage of the value of all crops in the United States, including about 60% of vegetables and 25% of the fruit and nut crops. Because of continuing initiatives begun under the presidency of Bill Clinton, the Bureau's staffing level has trended downward in recent years. In 2010, the Bureau's staff was about 29% smaller than in 1993.

Responses to Increased Competition for Water

Agricultural interests face increased competition for water from a number of sectors, particularly in the West. The sources of this competition include the following (CBO, 1997):

- *Municipal water use*—Growth in many arid and semi-arid cities in the western United States has required increasing water deliveries for municipal water systems.
- *Energy sector*—Hydropower projects may store water and make it available for agricultural use, creating interdependencies between farms and power generators. If water

shortages arise, however, power generation may compete with agriculture for access to adequate supplies of water.
- *Conservation and recreation*—Increasingly, environmental regulations require the maintenance of in-stream flows for the benefit of key wildlife species, as well as for recreational and other uses.

Data and modeling suggest that competition among these interests has increased the potential for irrigation water shortages. USDA survey data indicate that, in 2008, 33,000 farms reported diminished crop yields due to irrigation problems. Of these, over 17,000 highlighted a shortage of surface water as the specific problem encountered, the single largest cause reported. Another 3400 farms reported a shortage of ground water (USDA, 2010).

In recent years, stakeholders have responded to the increased competition for water in a variety of ways. Some of these efforts are essentially regulatory in nature. For instance, Congress passed the Central Valley Project Improvement Act (CVPIA) in 1992. Established in 1935, the Central Valley Project (CVP) in California is the largest of the Bureau of Reclamation's water resource projects. It delivers water to the Sacramento Valley in northern California and the San Joaquin Valley in Central California. About 85% of the CVP's water supply is used for irrigation, while 15% is delivered to municipal and industrial users. The CVPIA sought to address problems associated with dewatering of wetlands and rivers in the region. It mandates the release of more water to supplement rivers and wetlands, habitat restoration, water temperature control, water conservation, and other steps. These actions have increased water prices for Central Valley producers and encourage increased irrigation efficiency.

Other initiatives seek to harness the power of market pricing signals to achieve more efficient water use. Many of these efforts fall under the general umbrella of "water transfers" or "voluntary water marketing." In general, these arrangements introduce flexibility into traditional water rights systems, bringing regional water users together in a collaborative trading setting. Specifically, water transfers involve transactions where one party sells or leases water rights to another party. These transactions frequently entail transfer of water rights from agricultural to municipal interests. They may involve retirement of irrigated cropland, or they may focus on improved irrigation efficiency and transfer of surplus water generated by these efficiency gains.

Research has clearly demonstrated the potential for economic welfare gains from water transfers. One comprehensive study of water transfers between 1987 and 2005 showed consistently high prices when irrigators sold to municipalities. The sample of over 1000 agriculture-to-municipal sales showed a median sales price of over $2600 per acre-foot (Brewer et al., 2007). Economists have concluded that because water can command higher prices for municipal and other non-irrigation uses there is significant economic gain to be found in agriculture-to-urban transfers (Eden et al., 2008).

DID YOU KNOW?

Within 20 years, the population of the South Platte River basin outside Denver, Colorado, is projected to grow by 1.9 million. This growth is expected to strain available water supplies, significant shares of which are currently used to irrigate alfalfa and other forage crops. Through the Lower South Platte River irrigation and Research Demonstration Project, agriculture experts from Colorado State University are collaborating with the Parker Water and Sanitation District to address this issue. Initiated in 2007, the project is identifying and implementing irrigation efficiency techniques designed to generate surplus water that can be traded to the municipal water authority. The project also involves proceedings in Colorado's Water Court to demonstrate that agriculture-to-municipal trades will have no net impact on in-stream flows in the South Platte River (Lytle et al., 2008).

But, there are many obstacles to water transfer and marketing. Markets tend to be active only where water law, institution flexibility, and physical infrastructure combine to allow trading. Likewise, trading may generate unintended externalities for third parties (e.g., reduced groundwater supplies for farms neighboring the irrigation involved in the trading arrangement). Studies suggest that current water markets involve only 1 to 2% of all irrigation withdrawals (Wiebe and Gollehon, 2006).

Long-Term Challenges

Long-term global changes are likely to intensify water competition and the need for innovative response worldwide. Foremost among these challenges is climate change. Climate change is likely to lead to rising temperatures, shifting patterns of precipitation, and more extreme weather events. Most studies predict that agriculture in lower latitude countries will suffer the greatest harm (Molden, 2007). These studies show that areas such as Sub-Saharan Africa and parts of India and China have high vulnerability (in terms of climate and hydrological conditions) and low adaptability (e.g., in terms of water supply and storage options) (United Nations, 2009). As noted earlier, modeling of future climate change scenarios also supports the idea that threats to agricultural water could intensify in parts of Africa, Latin America, and the Caribbean (Strzepek and Boehlert, 2010).

Climate change poses a threat to agricultural water supplies in the United States but to a lesser degree relative to some other nations. Increased temperature and drought risk in portions of the southwest and Great Plains may render current irrigation withdrawals unsustainable. Studies have forecast that increased drought risk in southwestern and Rocky Mountain regions may necessitate development of storage capacity to better manage variable water supplies (Strzepek et al., 2010).

In addition to climate change, other global trends may increase water competition and affect agricultural water supplies. Population growth worldwide has the combined impact of increasing urban water demand while adding to the demand for food. As economic globalization proceeds, increased incomes in developing countries will continue to enhance the demand for meat and other foods that are resource intensive. A legitimate comparison of the water resource demands of meat vs. crop production is the subject of considerable debate; however, the available literature consistently demonstrates that much more water is required to produce a given quantity of meat than a calorie-equivalent quantity of virtually any crop (Peden et al., 2007). Competition for water by agriculture in the United States could increase as producers strive to meet food demand in export markets. Finally, the surge in production of certain crops for biofuels may place additional stress on agricultural water supplies. Worldwide, the production of bioethanol from sugarcane, corn, sugar beets, wheat, and sorghum tripled between 2000 and 2007. Along with Brazil, the United States is the major producer meeting this demand, primarily through corn cultivation (United Nations, 2009).

In the United States, in order to meet increasing biofuel demands, agriculture will require greater land and water resources. This will likely require (1) conversion of existing crop land to grow biofuel corps, (2) changes in other land uses (such as forest and pastureland) to grow biofuel crops, and (3) increasing the use of fertilizer and agrochemicals (Uhlenbrook, 2007). Ultimately, all these actions will heighten potential agricultural impacts on natural resources. If local agriculture shifts to biofuel/bioenergy crops that require more than current agricultural water supplies, there is a likelihood of deleterious impacts on limited water resources. To be sustainable, bioenergy production must conserve and protect natural resources, including freshwater. The bottom line: Future trends in biofuels markets may have important consequences for U.S. agricultural water supplies.

Modeling efforts have attempted to integrate all these long-term economic and demographic factors. A study of worldwide trends in water demand determined that competition among municipal/industrial users, environmental flow requirements, and the agricultural sector would result in an 18% reduction in the availability of irrigation water by 2050 (Strzepek and Boehlert, 2010).

Water Quality Issues

The adequacy of water resources for agricultural purposes depends not only on the quantity of water supplied but also on its quality. Data suggest that water quality issues do not currently pose a major problem for U.S. producers. Of the 33,000 farms that reported interruptions in irrigation operations in 2008, only about 2% attributed the problem to poor water quality (USDA, 2010). As competition for water grows, however, producers may seek out new water supplies of less quality, increasing the potential for quality-related problems. The following discussion briefly reviews water quality requirements for irrigation, livestock, and aquaculture applications.

Irrigation

The United Nations Food and Agriculture Organization (FAO) has compiled a comprehensive reference document on water quality issues affecting agriculture (Ayers and Westcot, 1994). The FAO study highlights four major categories of water quality problems affecting irrigation:

- *Salinity*—Salts contained in irrigation water can accumulate in the root zone of crops. Yield reduction may result when salt levels grow so great that plants are unable to extract sufficient water from the soil. The salts that contribute to salinity problems are readily transported in irrigation water. Salinity problems may be exacerbated in the presence of a shallow water table, where salts can accumulate and remain in contact with the crop root zone.
- *Water infiltration rate*—The rate at which irrigation water infiltrates soil can be slowed under certain conditions, causing water to remain on the soil surface and possibly evaporate. Although soil factors affect infiltration rates, water quality plays a role as well. High sodium relative to calcium and magnesium content in the water will decrease infiltration (as will high salinity).
- *Toxicity*—Plants can absorb certain toxic constituents in irrigation water, and these contaminants can concentrate to levels that damage the plants or reduce yields. Permanent, perennial crops such as trees are most sensitive to these effects. The constituents of greatest concern include chloride, sodium, and boron.
- *Other*—The FAO study cites additional water quality problems that may affect irrigation, but which are less common. High nitrogen concentrations can cause excessive vegetative growth and delayed crop maturity. Likewise, sprinkler water containing excessive concentrations of bicarbonate, gypsum, or iron can create unsightly deposits on fruit or leaves. Finally, excessive pH can lead to various plant abnormalities.

Livestock

Guidance documents for livestock operations emphasize the importance of safe, healthy water supplies. Ensuring the quality of the water supply can be especially challenging given the impacts that livestock operations themselves can have on nearby source waters. Categories of water quality concerns include the following:

- *Salinity and associated minerals*—Salinity is the problem most commonly encountered in livestock water. Salinity is correlated with the presence of many specific compounds, including sodium, chloride, calcium, magnesium, and bicarbonate. Available guidance suggests that impacts on animals' digestive systems may begin at total salt levels of 3000 mg/L, and some animals may refuse water with this level of salt. Concentrations above 10,000 mg/L are considered highly saline and unacceptable for use (Faries et al., 1998).
- *Nitrogen*—Nitrogen in the form of nitrate is digested and converted to nitrite in animals. The nitrite reduces the blood's ability to carry oxygen and can therefore poison animals. Guidance recommends against using livestock water with greater than 300 ppm nitrate (Pfost and Fulhage, 2001).

- *Blue–green algae*—High nutrient loadings in farm runoff can result in excessive growth of blue–green algae in livestock water. When ingested by animals, the algae can be toxic and may result in muscle tremors, liver damage, and possibly death (Pfost and Fulhage, 2001). Guidance recommends chlorination of livestock water to reduce algae growth (Faries et al., 1998).
- *Suspended solids*—Water with high levels of suspended solids may discourage animals from drinking adequately. The desired range for suspended solids is below 500 mg/L, whereas concentrations above 3000 mg/L are considered to be problematic (Prost and Fulhage, 2001).

Aquaculture

Because water constitutes the environment in which fish live, water quality is of utmost importance in aquaculture. Aquaculture operations are often carefully calibrated to create the optimum setting for fish and shellfish to thrive. Although water quality requirements are highly specific to the species being cultured, key considerations include the following:

- *Temperature*—Every species has a water temperature in which it grows most readily; for example, catfish thrive in temperatures between 70 and 85°F. As a result, pond-based catfish farming has been highly successful in southern states (Swann, 1992).
- *Dissolved oxygen*—To prevent stress or death, dissolved oxygen (DO) in aquaculture systems must be maintained at levels conducive to the species in question. In general, warmwater species (e.g., tilapia, carp, catfish) require DO levels above 3.0 ppm, and coldwater species (e.g., trout) require DO levels above 5.0 ppm (Buttner, et al., 1993).
- *Ammonia, nitrates, and nitrites*—Most fish and shellfish excrete ammonia as their primary nitrogenous waste. Fish exposed to total ammonia nitrogen levels above 0.02 ppm may grow slower and become more susceptible to disease (Buttner et al., 1993).
- *pH*—Fish survive and grow best in water with a pH between 6 and 9.

Table 10.10 summarizes the primary water quality concerns associated with irrigation, livestock, and aquaculture.

TABLE 10.10
Summary of Agriculture Water Quality Issues

Agriculture Sector	Primary Water Quality Concerns
Irrigation	Salinity
	Toxicity
Livestock	Salinity and minerals
	Nitrogen
	Algae
	Suspended solids
Aquaculture	Temperature
	Dissolved oxygen
	Ammonia, nitrates, nitrites
	pH

Source: USEPA, *The Importance of Water to the U.S. Economy.* Part 1. *Background Report*, U.S. Environmental Protection Agency, Washington, DC, 2012.

Value of Water Use by Agriculture

A variety of factors influence the availability of water for agriculture and the economic incentives faced by farmers. These factors include irrigation technology, water rights and water use law, subsidizing of public water supply projects, competition from other users, climate change, commodity prices, and the structure and nature of global food demand. These and other factors create a system that strays far from a conventional microeconomic pricing framework for water. The result is that traditional information for assessing commodity value (e.g., consumer and producer surplus estimates) is frequently lacking.

In response to the lack of conventional microeconomic data, economists have implemented a variety of approaches that seek to establish the value of a unit of water used in the agricultural sector. One method for valuing irrigation water simply considers the cost that farmers incur in acquiring the water. Some of the lowest rates are paid by farms acquiring water from large surface water supply projects in the western United States. Farmers in California's Imperial Irrigation District, for example, paid only about $15.50 per acre-foot for water in 2003 (Brewer et al., 2007). Historically, these projects are subsidized and can offer irrigation water at low rates.

Average acquisition costs nationwide are somewhat higher. By multiplying published information on the per-acre cost of irrigation by the average quantity of water applied per acre, we can estimate a rough average for the cost of acquiring water through different sources. Specifically, in this text, the acquisition costs using the per-acre irrigation costs are determined using the technique reported by Wiebe and Gollehon (2006). To obtain a lower-bound cost per acre-foot, we multiply the per-acre cost by the average acre-feet of water applied to irrigated land (1.7 acre-feet) as reported by the USDA (Leibtag, 2008). The upper-bound estimate is based on the USGS water application estimate (2.35 acre-feet per acre). The cost of self-supplied groundwater is roughly $80 to $110 per acre-foot; this is also true for the nationwide average cost of surface water delivery from non-farm sources. Self-supplied surface water costs about $53 to $73 per acre-foot on average.

Acquisition cost is an imperfect reflection of the true value of irrigation water. At best, acquisition cost represents a lower-bound estimate of the water's value; that is, farmers pay the implicit or explicit price to acquire the water, so it must be worth at least that amount. Other studies have attempted to establish more reliable measures of the actual value of irrigation water. One method for doing so is the factor input method. This method incorporates the relationship between crop yield and water input. As discussed earlier, irrigation has a demonstrable positive impact on the yield of many crops, depending on the growing conditions. Under the factor input method, yield increases can be valued by commodity prices to provide an estimate of the value of water as an input to production.

Frederick et al. (1996) assembled data from over 170 individual studies of irrigation water value. The average value masks wide variation, depending on the crop in question, the region of the country, and other factors. In the lower bound, some studies found irrigation water values of zero, while upper-bound values approached $1000 per acre-foot for some crops.

The value of water in irrigation can be further characterized by examining data from water transfers. As discussed above, water transfers involve one party selling or leasing water rights to another party. When properly structured, such trades provide the only direct, market-based evidence of the value of irrigation water. Brewer et al. (2007) assembled data on numerous water transfer arrangements in the western United States. Water transfers yield a wide range of values, depending on the specific conditions of the transfer. When an agricultural entity leases temporary water rights to another agricultural entity, the average price per acre-foot is approximately $30. The price rises to an average of over $1800 per acre-foot for the permanent sale of water rights. This differential highlights the value of securing a reliable or certain source of water, an important facet of the overall value of water.

Water transfers also occur between agricultural entities and municipal water suppliers. Lease arrangements averaged $119 per acre-foot, but permanent sale of water rights entailed prices averaging over $4500 per acre-foot. These transactions reveal a gap between the value of water in irrigation and its value in domestic use. In economic terms, irrigators' willingness to accept compensation in exchange for their water rights offers an indication (at the high end) of the value of water in an agricultural setting.

Finally, economists have implemented other methods for valuing irrigation water, although the literature supporting these methods is limited. Most notably Faux and Perry (1999) tested a hedonic price approach for valuing irrigation water in Malheur County, Oregon. This study involved estimating the implicit price paid for water by studying the sale price of properties with varying access to irrigation water. The study found that property sales imply a value per acre-foot of between $12 and $56. A similar hedonic analysis conducted by Petrie and Taylor (2007) estimated a comparable value of $39 per acre-foot for irrigation water in Georgia's Flint River basin. Although these studies involve permanent purchase of water rights, the estimated price per acre-foot is much lower than that observed in water transfers involving sales of permanent water rights. Numerous factors could account for this difference, although the relatively abundant supply of water in the hedonic study areas is like a key contributor.

OFF-STREAM WATER USE BY MANUFACTURING

Water use in manufacturing varies greatly across industries. In industries such as chemical, paper, petroleum, primary metal, and food product manufacturing, water is vital to the production process. In some instances, these industries use water to fabricate, process, wash, dilute, cool, or transport a product; in others, they incorporate water directly into the product. Water is also used for sanitation needs within manufacturing facilities (USGS, 2009). This section characterizes the role of water in manufacturing, focusing on the following topics:

- The economic importance of U.S. manufacturing and its place in the global economy
- The use of water in the U.S. manufacturing sector
- Water quality requirements associated with the use of water in manufacturing
- Available estimates of the value of water used in manufacturing

U.S. Manufacturing Sector

Manufacturing is a major component of the secondary mega-sector of the U.S. economy. In 2007, manufacturing accounted for approximately 17% of U.S. gross domestic production. The U.S. Census Bureau's Economic Census provides a detailed portrait of the United States' economy once every 5 years. Table 10.11 draws on the most recent census (2007) to provide a basic economic profile of the U.S. manufacturing sector. As shown, the total value added by manufacturing industries in 2007 was in excess of $2.38 trillion. Approximately 288,000 firms engaged in manufacturing in 2007, employing approximately 13.4 million workers.

TABLE 10.11
Overview of the U.S. Manufacturing Sector (2007)

Sector Characteristic	Estimate
Number of companies	287,654
Number of establishments	332,536
Number of employees	13,395,670
Value added	$2382 billion
Total value of shipments	$5319 billion

Source: U.S. Census Bureau, *2007 Economic Census*, U.S. Census Bureau, Washington, DC, 2007.

DID YOU KNOW?

Manufacturers withdraw approximately 18.2 billion gallons of water per day, which represents 4% of total water withdrawals in the United States. In addition to these direct withdrawals, the most recent data available suggest that approximately 12% of publicly supplied water withdrawals (4.74 billion gallons of water per day in 1995) were used for manufacturing.

The value of U.S. manufacturing has grown in recent years, but the importance of manufacturing as a share of total GDP has diminished. In addition, employment within the sector has declined steadily, from 17.8 million workers in 1982 to 13.3 million workers in 2007 (U.S. Census Bureau, 2007a). The declines in employment can be explained by gains in productivity and increased competition from foreign producers (Brauer, 2008). The gains in productivity between 1995 and 2007 are particularly notable; during this period, productivity growth in manufacturing averaged 4.1% annually, up from an average of 2.7% from 1973 to 1995 (Brauer, 2008).

The United States produces a wide range of durable and non-durable goods. Chemical manufacturing contributes the largest share of value added to the U.S. economy followed by transportation equipment and food products. Overall, these three industries account for approximately 38% of the value added by manufacturing.

U.S. Manufacturing in a Global Context

In 2007, the United States led the world in manufacturing value added, followed by China and Japan. U.S. manufacturing has long held this position. Between 1970 and 2000, the U.S. share of global manufacturing fluctuated little, varying from 21 to 29%. Since 2000, however, the United States has seen a steady decline in its relative contribution to manufacturing worldwide, while China has seen a sharp increase. Between 2000 and 2007, the U.S. share of global manufacturing declined from 26% to 19%, while China's share increased from 8% to 16% (United Nations, 2010).

In recent years, the U.S. manufacturing sector has been greatly affected by foreign competition, especially competition from emerging economics. Between 1999 and 2007, the nominal trade deficit in manufactured goods doubled. Over this period exports of manufactured goods from the U.S. rose by $334 billion (58%), but imports grew by $692 billion (78%) (Brauer, 2008). The deficit narrowed in both 2008 and 2009, reflecting reductions in both the import and export of manufactured goods, but began to increase again in 2010, as both imports and exports began to rise. Increased exposure to competition from low-wage countries, such as China, has had a negative effect on U.S. manufacturing plant survival and growth. In response to competition from abroad, the United States has shifted manufacturing activity toward capital- and skill-intensive products (Bernard et al., 2006). Recent analysis has shown that even these high-tech industries are not immune to competition from abroad. In some high-tech industries, increased international trade has led to decreasing demand for skilled labor (Silva, 2008). Even in high-tech industries where demand for skilled labor has increased, wages have not necessarily risen in response (Silva, 2008). Competition from abroad is likely to continue to affect wages in the U.S. manufacturing sector for an extended period of time, reflecting the comparative advantage currently enjoyed by manufacturers based in lower-wage countries.

DID YOU KNOW?

Since 1985, direct withdrawals of water by the manufacturing sector have declined by 30%. This change is due in large part to increasing efficiency in water use, including recycling and/or reuse of water.

DID YOU KNOW?

The pollution control requirements introduced under the Clean Water Act are one factor that has contributed to the decline in withdrawals of water by the manufacturing sector. The cost of complying with these requirements provides a strong economic incentive to reduce effluent discharges which in turn encourages greater efficiency in water use.

Water Use by Manufacturing

Well-supplied manufacturing water withdrawals account for approximately 4% of total water withdrawals in the United States (18,200 MGD). In addition, some industries use publicly supplied water. The most recent published estimate of the use of publicly supplied water in manufacturing dates from 1995. At that time, 12% of publicly supplied water withdrawals (4750 MGD) were used for manufacturing (USGS, 1998). This amount represented 18% of water use in manufacturing in 1995.

The withdrawal of water for manufacturing purposes is heavily concentrated in a limited number of states; 11 states account for approximately 70% of self-supplied manufacturing water use. Louisiana, Indiana, and Texas are the top three users; combined they account for 38% of self-supplied manufacturing withdrawals. The withdrawal of water for manufacturing purposes is greatest east of the Mississippi and in the Gulf Coast states.

The withdrawal of water for manufacturing purposes does not necessarily correspond to state-level manufacturing output. California has the highest level of manufacturing output, but manufacturers in this state withdraw relatively little water. On the other hand, states such as Texas and Pennsylvania host a large manufacturing sector and report significant withdrawals of water for manufacturing purposes.

The inconsistently in manufacturing output and water withdrawals across states can be explained in part by the varying levels of water used across industries. Table 10.12 provides data on water use by industry group in 1983, as reported in the 1987 Statistical Abstract of the United States (the last time that the U.S. Census Bureau published manufacturing water use data at this level of disaggregation). At that time, chemical, paper, petroleum and coal, primary metal, and food manufacturing accounted for approximately 90% of water used for manufacturing. Other major uses included transportation equipment, which accounted for approximately 3% of water used for manufacturing; the remaining 12 industries each accounted for less than 1%. Thus, the variation in the intensity with which water is used in the manufacturing sectors of different states likely reflects underlying differences in each state's industrial base. Keep in mind that the composition of the nation's manufacturing sector has changed since 1983. This change in composition may have contributed, at least in part, to the approximately 30% reduction in direct withdrawals of water by the manufacturing sector between 1985 and 2005 (USGS, 1988, 2009). Given the available data, however, it is reasonable to assume that disparities in water use across industries persist and that the industries that were major users of water in 1983 remain major users today. More recent data have been collected on water pollution abatement operating costs, which give a sense of the levels of water use across industries. The data on pollution abatement costs are similar to the 1983 water use data, with the five industries representing the highest water pollution abatement costs being chemical, food, paper, petroleum and coal, and primary metal manufacturing, in that order.

Overall withdrawals of water by the manufacturing sector declined approximately 30% between 1985 and 2005 (USGS, 1988, 2009). This change is due in large part to increasing efficiency in water use, including recycling and reuse of water. The efficiency gains have been driven by a variety of factors, including the diminishing availability of sources of raw water of sufficient quality, increasing water purchase costs, and strict environmental effluent standards (Ellis et al., 2003). Many uses of water in manufacturing (e.g., cooling and processing) do not require water of particularly high

TABLE 10.12
Water Use by Industry (1983)

North American Industry Classification System (NAICS) Code	Industry	Total Gross Water Used[a] (Billion Gallons)	Water Intake[b] (Billion Gallons)	Water Recycled[c] (Billion Gallons)
311	Food and kindred products	1406	648	759
313 and 314	Textile mill products	333	133	200
316	Leather and leather products	7	6	1
321	Lumber and wood products	218	86	132
321	Tobacco products	34	5	29
322	Paper and allied products	7436	1899	5537
324	Petroleum and coal products	6177	818	5359
325	Chemicals and allied products	9630	1401	6229
326	Rubber, misc. plastic products	328	76	252
327	Stone, clay, and glass products	337	155	182
331	Primary metal product	5885	2363	3523
332	Fabricated metal products	258	65	193
333	Machinery, excluding electrical	307	120	186
335	Electric and electronic equipment	335	74	261
336	Transportation equipment	1011	153	859
337	Furniture and fixtures	7	3	3
339	Miscellaneous manufacturing	15	4	11
N/A	Instruments and related products	112	30	82

Source: U.S. Census Bureau, *Statistical Abstract of the United States: 1987*, 107th ed., U.S. Census Bureau, Washington, DC, 1988.

[a] Based on establishments reporting water intake of 20 million gallons. This represents 96% of the total water use estimated for manufacturing industries.

[b] Refers to water used/consumed in the production and processing operations and for sanitary services.

[c] Refers to water recirculated and water reused.

quality, allowing manufacturers to reuse water multiple times before treating and discharging it as wastewater. In addition, substitutes, such as using air for cooling instead of water, allow manufacturers to decrease their water use. Sector-specific gains in water use efficiency are described in greater detail below.

Although most industries in the manufacturing sector are primarily concerned with access to adequate supplies of water, some are also concerned with the water's quality. These industries include the food and beverage sector and the electronics sector. Water used in the processing of food and beverages must meet health and safety standards. For example, we know that water is essential for living. If the quality of water is not satisfactory it may result in ill health. A large number of diarrheal illnesses are caused by consumption of contaminated water. Diarrhea illnesses constitute one of the leading infectious causes of death and disability worldwide, particularly in areas with poor sanitation; thus, access to clean water is an important consideration when locating bottling and food processing facilities. Ironically—that is, as compared to water quality or purity requirements for food—in the electronics sector, the water used in the manufacture of high-tech products such as semiconductors and microchips must be ultrapure water (UPW). In these sectors, water quality is so important that manufacturers are likely to rely on sophisticated systems to purify their source water, regardless of its initial quality (Strzepek et al., 2010). Nonetheless, high-quality source water can, at least to some extent, reduce the costs associated with water purification and treatment.

TABLE 10.13
Chemical Manufacturing Use by Product (1983)

Product	Total Gross Water Used (million gallons per day)	Water Intake (million gallons per day)	Water Recycled (million gallons per day)
Industrial organic chemicals	11,300	4150	7150
Industrial inorganic chemicals	5930	2420	3510
Plastics and synthetics	3930	1170	2760
Agricultural chemicals	3780	836	2944
Drugs	658	249	409
Soaps, cleaners, and toilet goods	285	178	107
Paints and allied products	11	5	6
Miscellaneous chemical products	490	301	189
Total	26,400	9310	17,090

Source: Adapted from David, E.L., in *USGS National Water Summary 1987—Hydrologic Events and Water Supply and Use*, Carr, J.E. et al., Eds., Water-Supply Paper 2350, U.S. Geological Survey, Denver, CO, 1990.

Chemical Manufacturing

Chemical and allied products manufacturing likely represents the leading use of water in the U.S. manufacturing sector. It accounted for 29% of industrial water use in 1983 (U.S. Census Bureau, 2008). The primary uses of water in chemical manufacturing are for non-contact cooling, steam applications, and product processing (USEPA, 2008c). Water use varies by product—silicone-based chemicals require larger quantities of water to produce, while many of the top manufactured chemicals by volume (including nitrogen, ethylene, ammonia, phosphoric acid, propylene, and polyethylene) require less water during production (USEPA, 2008c). Table 10.13 provides data on water use by product within the chemical manufacturing industry for 1983. The table suggests that much of the water used for chemical manufacturing can be recycled. In fact, water use per unit of production decreased steadily from 1954 to 1973 in part due to water recycling. In addition, improvements in overall efficiency and the substitution of air for water during certain cooling processes have contributed to the reduction in water use per unit of production (David, 1990).

The importance of water quality varies across use within the chemical manufacturing industry. Cooling accounts for approximately 88% of the gross water used in chemical manufacturing; 67% of this water is recycled (Ellis et al., 2003). An excellent example of the use of recycled water for cooling in manufacturing is the trend toward using treated wastewater effluent for machinery cooling (Spellman, 2013). In general, the quality of water used for cooling is not of great importance. On the other hand, certain processes, such as rinsing parts, may require high-quality water (MCC, 2008a).

Paper Manufacturing

Paper and allied products manufacturing accounted for 22% of industrial water use in 1983 (U.S. Census Bureau, 2008). Paper is produced from raw materials containing cellulose fibers, such as wood, recycled paper, and agricultural residues. The main steps in paper manufacturing are raw material preparation (e.g., wood debarking, chip making), pulp manufacturing, paper manufacturing, and fiber recycling (World Bank Group, 1999). Water is used at various points in this process. In the initial step of raw material preparation, much water is needed to clean the wood, transport wood from one place to another in the facility, cool machinery used for conveyors, debark, and chip. To produce pulp more water is used to steam-cook wood chips, which are then washed and screened. In the paper manufacturing step, the pulp is further diluted before being drained, heat-dried, and pressed. The water that drains off can be reused in the paper manufacturing step (David, 1990). Water use in paper manufacturing decreased from 26,700 gal/ton-product in 1975 to 16,000 gal/ton-product in 1996 (Ellis et al., 2003). Decreases in water use are due to recycling of water at various points in the production process and improved technology, such as high-pressure, low-volume showers (MCC, 2008b). Although positive from the perspective of total resource use, water use reductions can increase the concentration of contaminants in process water, leading to high rates of scale deposition and other unwanted effects. This build-up of contaminants may increase production costs and decrease product quality (Ellis et al., 2003).

Petroleum and Coal Products Manufacturing

Petroleum and coal products manufacturing accounted for approximately 18% of industrial water use in 1983 (U.S. Census Bureau, 2008). Within this industry, petroleum refining accounts for between 1000 and 2000 million gallons of water used daily, compared to 400 million gallons per day for natural gas processing and pipeline operations. Refineries use about 1 to 2.5 gallons of water for processing and cooling per gallon of product (USDOE, 2006). This figure has decreased more than 95% since 1975 (Ellis et al., 2003). Cooling is the primary use of water in petroleum refining; a typical refinery may use 10 times as much cooling water as process water (David, 1990). However, petroleum refineries have the highest rate of water recycling of any major industry (Ellis et al., 2003). Water is used approximately 7.5 times before being discharged (David, 1990). As with chemical manufacturing, water quality is not a critical consideration in the use of water for cooling. The use of water for other purposes, however, may be more sensitive to water quality considerations.

Primary Metal Manufacturing

Primary metal products manufacturing accounted for approximately 17% of industrial water use in 1983 (U.S. Census Bureau, 2008). Information on water use in primary metal manufacturing focuses largely on iron and steel manufacturing. Water is used in the steel industry for three purposes: material conditioning, air pollution control (e.g., in wet scrubbers), and cooling (CH2M Hill, 2003). Table 10.14 provides data on water use for each of these three purposes. In addition, this table provides information on the percentage of water recycled or reused for each process. Overall, cooling represents the primary use of water within the steel industry (approximately 75%). In addition, 12% of water is used for material conditioning and 13% is used for air pollution control. Cooling represents over 70% of the water used in most processes. The exceptions are sinter plants and pickling, where air pollution control represents the largest water use. The intensity of water use within the steel industry has declined in recent years, principally due to recycling and reuse of water in production facilities. In addition, process changes in steel production, such as the replacement of basic oxygen furnaces with electric arc furnaces, have decreased water demand. Water use within the industry is expected to continue to decline, and facilities may reduce their use of surface and groundwater by moving toward reuse of treated municipal effluent. In addition, internal treatment and recycling of water are expected to increase (CH2M Hill, 2003).

Food and Beverage Manufacturing

Food and kindred products manufacturing accounted for approximately 4% of industrial water use in 1983 (U.S. Census Bureau, 2008). Although this industry accounts for a relatively small portion of overall industrial water use, water quality is particularly important in food and beverage manufacturing. Within the food and beverage industry, water is used as an ingredient in products, as a mixing and seeping medium in food processing, and as a medium for cleaning and sanitizing operations (USEPA, 2008c). Water use varies by food product. Table 10.15 provides estimates of water used for the processing of various water-intensive foods. In addition to the products listed in Table 10.15, sugar refining is a large user of water within the food processing industry. Unlike most food products, which require water primarily for processing, sugar refineries use about half of their intake water for cooling, and beverage manufacturers also use large quantities of water for cooling (David, 1990). Unlike the other industries discussed above, water use in the food products industry has not declined dramatically over the last several decades (David, 1990). Food processing

TABLE 10.14
Water Use For Various Unit Operations in Steel Manufacturing

		Water Use Purpose			
Process Area	**Water Use Unit**	**Material Conditioning**	**Air Pollution Control**	**Cooling**	**Percent Recycled/Reused**
Coke-making	Gallons/ton coke	200	250–300	8000–8500	0% (newer plants may recycle cooling water)
Boilers for converting coke oven gas, tars, and light oils	Gallons/ton coke	—	—	40,000–120,000	Varies depending on age of boiler
Sinter plant	Gallons/ton sinter	20–30	900–1000	200	80%
Blast furnace	Gallons/ton molten iron	100–200	800–1000	2500–3000	90%
Boilers for converting blast furnace gas	Gallons/ton molten iron	—	—	20,000–60,000	Varies depending on age of boiler
Basic oxygen furnace	Gallons/ton liquid steel	100–200	800–1000	2500–3000	50%
Direct reduced iron processes	Gallons/ton iron	70–80	Negligible	200–250	80%
Electric arc furnace	Gallons/ton liquid steel	Negligible	Negligible	2000–2500	80%
Continuous caster	Gallons/ton cast product	Negligible	Negligible	3000–3500	70%
Plate mill	Gallons/ton plate	1000–2000	Negligible	7000–8000	30%
Hot strip mill	Gallons/ton hot-rolled strip	400–600	Negligible	7000–8000	60%
Pickling	Gallons/ton steel pickled	30–40	80–100	20–30	70%
Cold rolling	Gallons/ton cold-rolled strip	50–100	Negligible	2500–3000	90%
Coating	Gallons/ton coated steel	60–70	1–10	1200–1800	80%

Source: CH2M Hill, in *Industrial Water Management: A Systems Approach*, 2nd ed., Byers, W. et al., Eds., American Institute of Chemical Engineers, New York, 2003.

techniques have changed little, and only minimal water recycling and reuse occur within the industry (Ellis et al., 2003). Water recycling and reuse are limited by safety concerns. The water used in food processing must meet human health and safety standards.

Value of Water Use by Manufacturing

It is difficult to develop estimates of the value of water in the manufacturing sector largely because most water used within the sector is self-supplied; where industries have made purchases on water markets, estimates can be derived from price data (AWWA, 2005). Where information on the price paid for water does not exist, but analysts have access to information on the quantity of water withdrawn, methods have been developed to infer the value of water used (Renzetti and Dupont, 2002). These methods consider the relationship between the value of industry output and the quantity of water used. Where data on neither price nor quantity of water used exist, analysts have examined the marginal cost of in-plant water recycling as a proxy for the marginal value of intake water (Renzetti and Dupont, 2002).

TABLE 10.15
Water Use in Processing of Food Products

Product	Water Use (gallons/ton-product)
Beer	2400–3840
Milk products	2400–4800
Meat packing	3600–4800
Bread	440–960
Whisky	14,400–19,200
Green beans (canned)	12,000–17,000
Peaches and pears (canned)	3600–4800
Other fruits and vegetables (canned)	960–8400
Industry-wide average	8.6[a]

Source: Ellis, M. et al., *Industrial Water Use and Its Energy Implications*, U.S. Department of Energy, Washington, DC, 2003.

[a] In units of gallons/unit output (e.g., 1 gallon of milk).

Available Estimates

Although it is difficult to estimate the value of water to the manufacturing sector, studies have indicated that manufacturing may be among the highest value uses (Frederick et al., 1996). One method of valuing manufacturing water simply considers the cost that firms incur in acquiring the water. Because the majority of water used in manufacturing is self-supplied, the data on water purchases are limited. A 1991 survey of manufacturers in California indicated that the price paid to utilities for publicly supplied water ranged from $219 to $1113 per acre-foot, with an average price of $736 per acre-foot. The cost of self-supplied groundwater ranged from $107 per acre-foot for food manufacturers to $280 for petroleum refiners and averaged approximately $206 per acre-foot (Wade et al., 1991). Regional variations in the cost of both publicly supplied and self-supplied water are great. It is important to note that the prices presented here were collected from manufacturers in California.

As with other sectors, acquisition cost is an imperfect reflection of the true value of water in manufacturing. At best, acquisition cost represents a lower-bound estimate of the water's value; that is, manufacturers pay the implicit or explicit price to acquire the water, so its value to them must be at least that great. Other studies have attempted to establish more reliable measure of the marginal value of water to manufacturers. One method for doing so uses information on manufacturing inputs, including the quantity of water used, and information on the value of industry output to estimate the value of water. Following this general method, studies have derived the value of manufacturing water from the production function, the cost function, and the input distance function.

Methods that rely on the quantity of water used and output value provide a range of average values per acre-foot of water from $74 to $1527. This wide range in values may depend in part on the country of study. Renzetti and Dupont (2002), who examined the value of manufacturing water use in Canada, noted that their estimate was much lower than that found in previous American studies; they attributed the difference in part to differences between the Canadian and American regulatory environments. In addition, they noted that most manufacturing water intake in Canada is self-supplied and is available at almost zero external cost; therefore, it follows that the marginal value derived from the use of water would also be very low (Renzetti and Dupont, 2002). It is unclear whether it would be appropriate to transfer marginal values derived from manufacturing experience in foreign countries to the United States. Economic conditions in the United States differ

TABLE 10.16
Values for Manufacturing Water by Industry

	Value Per Acre-Foot of Water (2010$)		
Industry	**Wang and Lall (1999) (China)**	**Renzetti and Dupont (2002) (Canada)**	**Kumar (2006) (India)**
Chemicals	$297	$115	$141
Paper and allied products	$254	$49	$1358
Petroleum and coal	$1643	$460	—
Primary metal	$1156	$171	—
Food	$778[a]	$27	—
Beverage	—	$61	—

Source: USEPA, *The Importance of Water to the U.S. Economy*. Part 1. *Background Report*, U.S. Environmental Protection Agency, Washington, DC, 2012.

[a] Value provided is for the food and beverage industry.

from conditions in other countries, especially developing nations such as China, Korea, and India. In addition, water supply and regulations governing water use may differ from country to country. All of these factors may affect the value of water.

The third method explored here requires data on neither price nor quantity of water used. When these data do not exist, analysts have examined the marginal cost of in-plant water recycling as a proxy for the marginal value of intake water. Following this method, Gibbons (1986) reported that the marginal cost of recirculation for cooling water was lower ($14 to $23 per acre-foot) than that used for processing ($37 to $174 per acre-foot). This finding is intuitive, as processing generally requires higher quality water and additional treatment before reuse. It should be noted that these estimates come from studies performed before recent technological advances and therefore may no longer be applicable.

Finally, Frederick et al. (1996) reviewed seven studies that provide a value for industrial water use and concluded that industrial processing is one of the highest value uses. Their review of these studies found an average value of $282 per acre-foot and a median value of $132 per acre-foot. The authors noted that, “At the national level the median values may provide a better indication of the relative values of water in various uses under relatively normal hydrologic conditions.”

Water use across industries within the manufacturing sector differs greatly; clearly, the value of water also differs by industry. Table 10.16 provides estimates of the value of water for the five industries discussed above. The table shows a wide range of estimates within and across industries. Both Wang and Lall (1999) and Renzetti and Dupont (2002) reported the highest value for water used in petroleum and coal products manufacturing, followed by the value of water in primary metal manufacturing. All of these values, however, are based on studies conducted outside the United States. Their relevance to the value of water in U.S. manufacturing is unclear.

Although the body of literature on the value of water in manufacturing is small, the literature that investigates the effect of water quality on the value of water in manufacturing is even smaller. A study by Renzetti and Dupont (2002) investigated this relationship. By incorporating water treatment expenditures into their model, the authors were able to examine the relationship between changing water treatment costs (most likely caused by changes in water quality) and firms’ valuation of intake water. It is expected that decreases in water quality will increase water treatment costs, thereby decreasing the value of raw intake water. In other words, firms are willing to pay less for low-quality intake water as they will have to spend additional funds on internal treatment. The authors’ study confirmed this hypothesis, as Canadian manufacturing firms’ valuation of intake water is positively related to the quality of water.

DID YOU KNOW?

The distinction between mining and energy resource extraction is a simple one. Mining refers to the extraction of any solid mineral other than coal and uranium, while energy resource extraction includes coal and uranium, as well as raw petroleum and natural gas.

Demand for Water in Manufacturing

The price elasticity of demand for water in manufacturing captures the relationship between the quantity of water demanded and its price—specifically, it measures the percentage change in quantity demanded for a given percentage change in prices. Both economic theory and the empirical evidence suggest that the demand for water in manufacturing is inelastic at current prices, meaning that when its price increases the quantity demanded falls but at a correspondingly lower rate. The available evidence suggests that industrial firms' water demands may be relatively more price sensitive than agricultural or domestic water demands (Renzetti, 2005).

OFF-STREAM WATER USE BY MINING AND ENERGY RESOURCE EXTRACTION

Compared to other sectors, the mining and energy resource extraction sector uses a small amount of water: approximately 4.0 billion gallons per day, only 1% of the nation's total use (USGS, 2009). This figure has remained relatively constant since 1985. Nonetheless, water plays a crucial role in many production processes, including hydraulic fracturing (fracking), secondary oil recovery, and the extraction and processing of oil shale. This sector examines the economic importance of water in mining and energy resource extraction. It includes the following:

- Background information on the mining and energy resource extraction industry
- A discussion of the role of water in the production of minerals, crude oil, and natural gas
- Available estimates of the value of water to this sector

The mining and energy resource extraction sector of the U.S. economy is part of the primary megasector described earlier. Most of its output flows to the manufacturing or utility sectors, with relatively little output flowing directly to other sectors of the economy.

Mining

Mining includes the extraction of two general categories of mineral: metals (iron, gold, copper, etc.) and industrial materials (clays, feldspar, salt, sand, gravel, etc.). In 2009, the United States removed approximately 5.0 billion metric tons of material in mining operations, including approximately 3.6 billion metric tons (73%) of crude ore and 1.4 billion metric tons (27%) of waste material. Of this total, 51% was associated with the production of metals and 49% was associated with the production of industrial materials (USGS, 2011). In 2009, the average revenue per metric ton of material extracted during mining operations was $15.42, with an average of $21.51 for metals and an average of $12.90 for industrial materials. Excluding sand, gravel, and stone, the average revenue per metric ton of material extracted was $47.66. Dimension stone (i.e., stone cut to specific size and shape specifications) had the highest yield, earning $202.61 per metric ton of material extracted. This high yield reflects the minimal amount of waste material extracted during the stone-cutting process. Among metals, iron had the highest revenue per metric ton of material extracted, averaging $92.76. In comparison, the revenue derived from gold mines averaged $26.68 per metric ton of material extracted, reflecting the high ratio of material extracted (both waste and crude ore) to final product in the gold mining process (USGS, 2011).

Energy Resource Extraction

The United States is one of the leading producers of energy resources in the world; as of 2010, the United States ranked third worldwide in crude oil production, second in natural gas production, and second in coal production (world statistics for uranium production were unavailable). The United States is the only nation to rank in the top three in each of these categories.

Crude Oil

The U.S. Department of Energy (USDOE) reported that the United States produced approximately 2.0 billion barrels of crude oil in 2010. The DOE tracks this production by region, identifying each area as one of five Petroleum Administration for Defense Districts (PADDs). The Gulf region produced the most crude oil in 2010, accounting for 58% (1.2 billion barrels) of total U.S. production. The West (22%), Midwest (13%), and Rocky Mountain (7%) regions also contributed significantly to crude oil production. In comparison, the production of crude oil in the East (7.7 million barrels) was minor (EIA, 2011a).

In 2010, onshore wells accounted for approximately 68% of total production. Offshore production (32% of the nation's total) was concentrated in the Gulf region. There, approximately 49% (573.7 million barrels) of the region's production was accounted for by offshore drilling. This represents 91% of U.S. offshore production.

Total U.S. crude oil production was 5% higher in 2010 than in 2005, increasing from approximately 1.9 billion barrels to 2.0 billion barrels per year. The Midwest region experienced the greatest growth in that period, with production increasing 56%, from 161.6 million barrels to 251.9 million barrels per year. In contrast, the West experienced a 23% decrease in production, with output falling from 572.8 million barrels in 2005 to 442.7 million barrels in 2010. This decline can be attributed to the dramatic dip in the region's offshore production, which fell from 148.1 million barrels in 2005 to only 57.0 million barrels in 2010. In particular, production in Alaskan state waters (0 to 3 miles offshore) saw the greatest drop-off (a fall of 79%) and was the driving force behind the West's overall decline in production (EIA, 2011a).

Natural Gas

In 2009, the United States extracted approximately 26.0 trillion cubic feet of natural gas. Approximately 22.9 trillion cubic feet (88%) of this total was produced onshore, while 2.5 trillion cubic feet (10%) came from federal offshore sources and 587.0 billion cubic feet (2%) came from state offshore sources. Texas was the nation's leading producer of natural gas, accounting for 7.7 trillion cubic feet (29%) of the national total. Texas's production helped make the Gulf region the leading regional producer of natural gas. The region's total production in 2009 was 55% of the U.S. total, approximately 14.4 trillion cubic feet (EIA, 2011b).

The U.S. production of natural gas rose 11% from 2005 to 2009, from 23.5 trillion cubic feet to 26.0 trillion cubic feet. During this period, the only region that did not see overall growth was the West, which experienced a 9% decrease in natural gas production. Much of the growth in natural gas production can be attributed to the increased extraction of shale gas and gas from coal beds. Prior to 2007, no production of gas from coal beds was reported, and prior to 2008 no production of shale gas was reported. By 2009, however, shale gas and coal bed extractions added a combined 5.4 trillion cubic feet to U.S. production. Conversely, the extraction of gas from conventional wells fell 15%, from 17.5 trillion cubic feet to 14.8 trillion cubic feet (EIA, 2011b).

Coal

In 2010, the United States produced approximately 1.1 billion tons of coal; surface mines accounted for approximately 68% (745.4 million tons) of this total. Ninety-three percent (1175) of the nation's coal mines are located east of the Mississippi River; however, eastern mines produced only 41% (444.3 million tons) of the nation's coal. In contrast, Wyoming, with only 19 mines, alone accounted

DID YOU KNOW?

The use of water in mining and energy resource extraction is relatively insensitive to source water quality. For example, much of the water used in oil and gas extraction is water that is withdrawn during the drilling process. Reuse of this water, which is often unsuitable for other purposes, helps to offset the demands of water-intensive processes such as secondary oil recovery.

for 41% (442.5 million tons) of the nation's coal production. Next to Wyoming, West Virginia produced the most coal, generating approximately 12% (135.2 million tons) of the national total (EIA, 2011c). Recently, U.S. coal production has declined slightly. Production totaled 1131.5 million tons in 2005; by 2010, it had fallen to 1084.3 million tons, a 4% reduction.

Sector Employment

Aggregated sector employment in 2010 was approximately 652,000, with 24% employed in oil and gas extraction, 31% employed in mining (which includes coal mining), and 44% employed in mining support activities (which includes the drilling of oil and gas wells). The state with the greatest aggregate employment was Texas, which accounted for approximately 31% of the national total. Other states accounting for a substantial share of employment in the sector include Louisiana (8%) and Oklahoma (7%).

Value Added

Value added for the industry as a whole totaled approximately $417.8 billion. Oil and gas extraction accounted for approximately 66% of this total. Mining accounted for an additional 17%, as did mining support activities. Of the $72.8 billion in value added generated from mining, $27.6 billion was attributable to coal production, $26.6 billion was attributable to the mining and quarrying of non-metallic minerals, and $18.6 billion was attributable to the mining of metal ores. The state with the highest value added for the industry was Texas, which, at $111.6 billion, represented 27% of the national total.

Sector in a Global Context

The United States is one of the leading producers of energy resources in the world, ranking third in 2010 in world production of petroleum, second in world production of natural gas, and second in world production of coal. The United States, however, is also the world's leading consumer of energy resources. In 2010, the United States ranked first in natural gas consumption, utilizing approximately 24.1 trillion cubic feet; second in coal consumption, utilizing 1.0 billion tons; and first in petroleum consumption, utilizing 19.1 million barrels per day (no other country consumed more than 10.0 million barrels per day). U.S. petroleum imports (3.4 billion barrels) far outstripped exports (15.2 million barrels). The source of U.S. petroleum imports was almost evenly split between OPEC and non-OPEC countries. The United States was also a net importer of natural gas in 2010, importing approximately 3.7 trillion cubic feet (primarily from Canada) and exporting approximately 1.1 trillion cubic feet (primarily to Canada). In contrast, the United States has a favorable balance of trade in coal, as it imported only 9.9 million tons in 2010 and exported approximately 39.8 million tons. South America was the leading source of U.S. coal imports (7.5 million tons), while Europe was the top destination for U.S. coal exports (19.4 million tons).

Water Use by Mining and Energy Resource Extraction

Water is used in mining and energy resource extraction for a variety of purposes. In mineral extraction, it is used for such processes as milling, wet-screening, and hydraulic mining, while in petroleum and natural gas extraction it is used for procedures such as hydraulic fracturing, secondary oil

DID YOU KNOW?

Scarcity of water in the West may constrain exploitation of the region's oil shale deposits, a potentially significant source of petroleum. This issue has encouraged research into less water-intensive processes.

recovery, and the extraction and processing of oil shale. In 2005, withdrawals of water by the mining and energy resource extraction sector were estimated at approximately 4.0 billion gallons per day (BGD), representing 1% of total U.S. withdrawals. The 2005 figure is consistent with estimates since 1985, when USGS first reported withdrawals for the mining sector as a separate category. Since then, estimated withdrawals have ranged from 3.4 to 4.9 BGD, with a mean of 4.1 BGD (USGS, 2009). Oil and gas operations in Texas, California, Oklahoma, Wyoming, and Louisiana were responsible for the large volume of saline groundwater withdrawals in these states; these withdrawals are a byproduct of the resource extraction process. In contrast, sand and gravel operations in Indiana and iron ore mining in Michigan and Minnesota accounted for the largest volume of withdrawals from fresh surface water sources. Mineral salt extraction from the Great Salt Lake in Utah accounted for the largest volume of saline surface water withdrawals for mining purposes, while withdrawals of fresh groundwater for mining purposes were highest in Florida, Ohio, Nevada, Arizona, and Pennsylvania (USGS, 2009).

SIDEBAR 10.5 Fracking and the Water Supply*

The drilling and hydraulic fracturing of a horizontal shale gas well requires water—2 to 4 million gallons of water (Satterfield et al., 2008), with about 3 million gallons being most common. Note that the volume of water required may vary substantially between wells. In addition, the volume of water required per foot of wellbore appears to be decreasing as technology and methods improve over time. Table S10.5.1 presents data regarding estimated per-well water needs for four shale gas plays currently being developed.

The water required for drilling and hydraulic fracturing of these wells frequently comes from surface water bodies such as rivers and lakes, but it can also come from groundwater, private water sources, municipal water, and reused produced water. Most of the producing gas shale basins contain large amounts of local water sources.

Even though water volumes needed to drill and stimulate shale gas wells are large, they occur in areas with moderate to high levels of annual precipitation. However, even in areas of high precipitation, due to growing populations, other industrial water demands, and seasonal variation in precipitation, it can be difficult to meet the needs of shale gas development and still satisfy regional needs for water.

Although the water volumes required to drill and stimulate shale gas wells are large, they generally represent a small percentage of total water resource use in the shale gas basins. Calculations indicate that water use will range from less than 0.1 to 0.8% by basin (Satterfield, 2008). This volume is small in terms of the overall surface water budget for an area; however, operators need this water when drilling activity is occurring (on demand), requiring that the water be procured over a relatively short period of time. Water withdrawals during periods of low stream flow could affect fish and other aquatic life, fishing and other recreational activities, municipal water supplies, and other industries such as power plants. To put shale gas water use in perspective, the consumptive use of freshwater for electrical generation in the Susquehanna River Basin alone is nearly 150 million gallons per day, whereas the projected total demand for peak Marcellus Shale activity in the same area is 8.4 million gallons per day (Gaudlip et al., 2008).

One alternative that states and operators are pursuing is to make use of seasonal changes in river flow to capture water when surface water flows are greatest. Utilizing seasonal flow differences allows planning of withdrawals to avoid potential impacts to municipal drinking water supplies or to aquatic or riparian communities. In the Fayetteville Shale play of Arkansas, one operator is constructing a 500-ac-ft impoundment to store water withdrawals from the Little Red River obtained during periods of high flow (storm events or hydroelectric power generation releases from Greer's Ferry dam upstream of the intake) when excess

* Adapted from Spellman, F.R., *Environmental Impacts of Hydraulic Fracturing*, CRC Press, Boca Raton, FL, 2012.

TABLE S10.5.1
Estimated Water Needs for Drilling and Fracturing Wells in Select Shale Gas Plays

Shale Gas Play (gal)	Volume of Drilling Water per Well (gal)	Volume of Fracturing Water per Well (gal)	Total Volume of Water per Well (gal)
Barnett Shale	400,000	2,300,000	2,700,000
Fayetteville Shale	60,000[a]	2,900,000	3,060,000
Haynesville Shale	1,000,000	2,700,000	3,700,000
Marcellus Shale	80,000[a]	3,800,000	3,880,000

Source: Adapted from Satterfield, J.M. et al., Chesapeake Energy Corp. Managing Water Resource's Challenges in Select Natural Gas Shale Plays, presentation at the GWPC Annual Meeting by ALL Consulting, Lisle, IL, 2008.

[a] Drilling performed with an air "mist" and/or water-based or oil-based muds for deep horizontal well completions.

Note: These volumes are approximate and may vary substantially between wells.

water is available (Chesapeake Energy, 2008a). The project is limited to 1550 acre-ft of water annually. As additional mitigation, the company has constructed extra pipelines and hydrants to provide portions of this rural area with water for fire protection. Also included is monitoring of in-stream water quality as well as game and non-game fish species in the reach of rivers surrounding the intake. This design provides a water recovery system similar in concept to what some municipal water facilities use. It will minimize the impact on local water supplies because surface water withdrawals still be limited to times of excess flow in the Little Red River. This project was developed with input from a local chapter of Trout Unlimited, an active conservation organization in the area, and represents an innovative environmental solution that serves both the community and the gas developer.

These water needs may challenge supplies and infrastructure in new areas of shale gas development—that is, in areas where the impact of shale gas operations is new and the potential impact is unknown to local inhabitants and governing officials. As operators look to develop new shale gas plays, a failure to communicate with local officials is not a viable option. Communication with local water planning agencies can help operators and communities to coexist and effectively manage local water resources. Understanding local water needs can help operators develop a water storage or management plan that will meet with acceptance in neighboring communities. Although the water needed for drilling an individual well may represent a small volume over a large area, the withdrawals may have a cumulative impact to watersheds over the short term. This potential impact can be avoided by working with local water resource managers to develop a plan outlining when and where withdrawals will occur (i.e., avoiding headwaters, tributaries, small surface water bodies, or other sensitive sources).

DID YOU KNOW?

For very deep wells, up to 4.5 million gal of water may be needed to drill and fracture a shale gas well; this is equivalent to the amount of water consumed by (Chesapeake Energy, 2014):

- New York City in approximately 7 minutes
- A 1000 megawatt coal-fired power plant in 12 hours
- A golf course in 25 days
- 7.5 acres of corn in a season

DID YOU KNOW?

One acre-foot of water is equivalent to the volume of water required to cover one acre with one foot of water.

Before a gas shale play hydraulic fracking operation is developed, not only is it a good idea to communicate with state and local government officials but it is also prudent to obtain information and data related to the urban water cycle (i.e., if the gas shale operation is near or has an impact on the surrounding area). Moreover, a study of the effect of gas shale hydraulic fracking operations on the indirect water reuse process is called for. In the planning stage, close scrutiny must be paid to water sources, outfalls, annual precipitation levels, drought histories, indirect water reuse, and (if located in or near an urban area) the urban water cycle.

Coal and Other Mineral Extraction

Extracting minerals such as coal, hard rock, sand and gravel, and metal ores from the earth can involve a number of procedures that are water intensive, particularly those that involve reducing the size of the extracted material. Wet screening, for example, in which the mined material is filtered by water through a series of screens, can use anywhere from 30 to 250 gallons of water per ton of mined material. Milling (or grinding), in which the mined material is broken down into smaller particles, also uses a large amount of water—anywhere from 125 to 300 gallons per ton of mined material. Water is also used in drilling for minerals; however, usage is highly variable, depending on the diameter of the hole, depth, orientation, etc. In general, water use ranges from 2 to 5 gallons per meter drilled (Mavis, 2003).

The use of water in mineral extraction is greatest when processing softer minerals, such as kaolin (a type of clay) and silica sand. Kaolin clay, a material used primarily in papermaking, goes through several processing procedures, such as suspension and dispersion, screening, grit removal, brightening, and flocculation, all of which are water intensive. Although the exact amount varies from facility to facility, roughly 2000 gallons of water are used to process a ton of kaolin. Kaolin is also shipped via slurry pipelines in a mixture of 70% kaolin and 30% water (Mavis, 2003).

In coal mining, water is used for several purposes, mainly for the cooling of machinery, dust suppression, and safety (i.e., dousing fires). The majority of water used in mining is for dust control, which utilizes approximately 5.2 gallons per ton of coal produced. Water use statistics for other coal mining processes are not readily available. Water quality is of little concern in most mining operations; much of the water that is utilized is later reused for the same process. Some procedures, particularly in increasing mineral concentration, are sensitive to water quality, yet these are so rare as to be inconsequential (Mavis, 2003).

Crude Oil

In petroleum extraction, water is used in many of the same ways that it is for mineral extraction: dust suppression, cooling of machinery, etc. Much of the water used in oil extraction (and in gas extraction) is produced water, or water that is generated during the drilling process. This water is generally saline but can range from fresh to hypersaline. In 1995, the American Petroleum Institute estimated that oil and gas operations produced roughly 49 million gallons of water per day, much of which was reused in oil and gas production. Only some of this water was sent offsite for treatment (USDOE, 2006).

Water is a primary input to the process known as secondary oil recovery, which is used to maintain production at wells that would otherwise be abandoned. Water (or steam) is injected into the wells in order to extract additional oil. The amount of water used in secondary oil recovery varies greatly, as anywhere from 2 to 350 gallons of water can be used per gallon of oil extracted. This process can be extremely water intensive, but much of the water used is produced water and otherwise unusable (USDOE, 2006).

DID YOU KNOW?

With regard to retorting oil shale, Shell Oil is currently developing an *in situ* conversion process (ICP). The process involves heating underground oil shale by using electric heaters placed in deep vertical holes drilled through a section of oil shale. The volume of oil shale is heated over a period of 2 to 3 years, until it reaches 650 to 700°F, at which point oil is released from the shale. The released product is gathered in collection wells positioned within the heated zone (USDOI, 2013).

In addition to conventional crude oil reserves, the United States holds one of the largest deposits of oil shale in the world. These reserves are not heavily utilized at present but could have a significant impact on the nation's future oil output. By 2035, shale oil could account for 2% of total U.S. oil production (EIA, 2011d). A potential impediment to that production, however, is the resource-intensive process for turning mined oil shale into useable crude oil. The process of retorting (heating process that separates the oil fractions of oil shale from the mineral fractions) requires 2 to 5 gallons of water per gallon of refinery-ready oil. Moreover, the majority of oil shale deposits are located in areas of the West in which water is scarce. The scarcity of water increases the costs of retorting and may, in some cases, render it economically infeasible. To address this issue, the industry is attempting to develop *in situ* retorting processes that are less water intensive. These processes could dramatically reduce reliance on water for oil shale production (USDOE, 2006).

Natural Gas

As with mineral extraction and oil extraction, natural gas extraction utilizes water in a variety of ways, such as the cooling of machinery and dust suppression. Water is also used extensively in the mining of unconventional natural gas sources, such as tight gas (gas that is trapped beneath sandstone formations), coal bed gas (gas that is generated by coal, then stored within its seams), and shale gas (gas stored within low-porosity shale, which also acts as a source). The extraction of gas from these unconventional sources often (almost always in the case of shale gas) requires a water-intensive process known as hydraulic fracturing (Ground Water Protection Council, 2009).

Hydraulic fracturing involves high-pressure injection of a solution—typically made up of 98% water and sand and 2% chemical additives—into the shale where the gas is trapped. This creates cracks or fractures in the shale, allowing for easier extraction of the gas. This process allows for access to gas that is otherwise unreachable, greatly increasing natural gas production (Ground Water Protection Council, 2009). Although shale gas accounted for only 16% of U.S. natural gas production in 2009, by 2035 its share of total production may increase to roughly 43% (EIA, 2011d).

Much of the projected increase in shale gas production is contingent on the use of hydraulic fracturing, a process that is under increasing scrutiny due to environmental concerns, mainly the potential contamination of groundwater resources. The greatest concern is that fracturing may allow gas and other contaminants, such as those in the fracturing fluid, to seep into underground sources of drinking water. The USEPA is now investigating these issues and conducting a study that has not yet been completed; however, the agency recently released a draft report detailing tests done at hydraulic fracturing sites in Wyoming's Pavillion gas field. In this report, the USEPA noted that domestic groundwater sources located near hydraulic fracturing sites had a high number of organic and inorganic contaminants associated with hydraulic fracturing (DiGulio et al., 2011). These findings are not definitive and the USEPA has reached no conclusions about the safety of hydraulic fracturing, but the findings suggest that environmental externalities associated with the production of shale gas may, at least in some cases, affect the use of groundwater for other purposes.

TABLE 10.17
Western Water Market Transactions (1990–2003): Purchases for Mining Purposes (2010$, N = 28)

Parameter	$/Acre-Foot/Year
Mean price	$482.24
Median price	$201.62
Minimum price	$40.09
Maximum price	$2662.35

Source: Brown, T.C., *The Marginal Economic Value of Streamflow from National Forests*, Rocky Mountain Research Station, U.S. Forest Service, Denver, CO, 2004.

Value of Water Use by Mining and Energy Resource Extraction

Much of the water used in mining and energy resource extraction is produced water—that is, water that is generated during the mining process. In these cases, water is not purchased by an external provider, and no market information on the user's willingness to pay for water is available. In some cases, however, produced water is unavailable, and mining operators must obtain it from other sources, including the purchase of water in water markets. A study by the U.S. Forest Service (USFS) of more than 2000 water market transactions that occurred between 1990 and 2003 identified 28 transactions that involved the lease of water for the purpose of mining; 25 of the 28 leases were entered into by mining interests in Texas's Rio Grande basin. The amount of water purchased through these arrangements totaled 3241 acre-feet per year.

Table 10.17 summarizes the data on purchases of water by mining operations, which provide some insight to the value of water in the mining sector. All values are reported in 2010 dollars. As the table indicates, prices ranged from $40 per acre-foot per year to nearly $2700 per acre-foot per year, with a median value of approximately $202 and a mean value of $482. The median price paid for water by mining concerns was more than double the median price paid by any other group considered in the USFS study, including municipalities, environmental interests, farmers, water districts, public agencies, power plants, developers, and others. The small number of transactions involved makes it difficult to draw definitive conclusions from these findings, particularly given the concentration of purchases in a relatively small geographical area. Nonetheless, this price information provides an indication that at least in some circumstances the marginal value of water to mining interests is relatively high. Presumably, this reflects

- The relatively high value of the marginal product of water
- The limited ability to substitute other inputs (such as labor) for water
- A relatively inelastic demand for water, at least in the quantities acquired

The literature reviewed provides no empirical information on any of these presumptions. It is reasonable to infer, however, that under circumstances similar to those surrounding this particular suite of transactions, the marginal value of water for mining purposes by be extremely high.

OFF-STREAM AND IN-STREAM WATER USE FOR ELECTRIC POWER GENERATION

Water plays a vital role in the U.S. electric power sector, both as a coolant for thermoelectric power generation and as an energy source for the generation of hydroelectricity. The direct use of water as a power source at hydroelectric facilities is generally considered a non-consumptive, in-stream use.

DID YOU KNOW?

Thermoelectric power plants withdraw over 200 billion gallons of water per day, more than that withdrawn by any other sector. Thermoelectric plants use most of this water for cooling. Those with once-through cooling systems return most of what they use to its source, while those with recirculating systems retain the water for future use. The power sector also makes significant use of in-stream water by employing it to generate electricity at hydroelectric plants.

In contrast, the use of water as a coolant in thermoelectric power generation is considered an off-stream use, one that accounted for 49% of total water withdrawals in 2005 (USGS, 2009). The vast majority of water used by the electric power sector is not consumed; instead, it is either returned to its source or retained by the power generator for future use.

This section discusses various aspects of the use of water in electric power generation, including the following topics:

- An overview of the electric power generation sector
- A summary of the sector's use of water
- A description of the potential effects of water resource constraints on the sector's current and future operations
- Available estimates of the value of water used in electric power generation

Electricity is vital to the economy of the United States, which produces (and consumes) more electricity than any other country. The discussion that follows presents an overview of the electric power generation sector, which is part of the U.S. economy's secondary mega-sector. It focuses in particular on thermoelectric power and hydropower generation. The discussion draws primarily on data published by the Department of Energy's Energy Administration (EIA), including the *Annual Energy Outlook*.

Electricity is generated at facilities that convert energy from a variety of sources into electrical energy that can be distributed and used in the residential, commercial, and industrial sectors. The sources of power used to generate electricity include fossil fuels (primarily coal and natural gas) and nuclear fission, as well as renewable energy sources such as hydropower, wind, solar radiation, biomass (e.g., forest residues and municipal solid waste), and geothermal energy (heat from within the Earth).

Generating Capacity and Generation

In 2010, the generating capacity of the electric power sector in the United States totaled 1.1 million megawatts (MW), and net generation of electricity totaled 4125 million megawatt hours (MWh). Table 10.18 presents more detailed information on generating capacity and net generation, including the distribution of net generation by power source. Within the electric power generation sector, the two primary water users, thermoelectric power generation and hydropower generation, together accounted for the vast majority of net generation in 2010. Thermoelectric power generation, which encompasses all power generated from combustion of fossil fuels (e.g., coal, natural gas, petroleum), nuclear power, and some renewable power sources such as biomass and geothermal, produced more than 90% of net power generated in 2010; hydropower accounted for an additional 6%.

TABLE 10.18
Generating Capacity and Net Generation by Power Source (2010)

Power Source	Generating Capacity (MW)	Net Generation (Million MWh)
Coal	342,296	1847
Petroleum	62,504	17
Natural gas	467,214	988
Other gases	3130	11
Nuclear	106,731	807
Hydroelectric conventional	78,204	206
Wind	39,516	95
Solar thermal and photovoltaic	912	1
Wood and wood-derived fuels	7949	37
Geothermal	3498	15
Other biomass	5043	19
Pumped storage[a]	20,538	–6
Other[b]	1027	13
Total	1,138,563	4125

Source: EIA, *Annual Energy Outlook 2011 with Projections to 2035*, U.S. Energy Information Administration, Washington, DC, 2010.

[a] Pumped storage refers to hydroelectric facilities that use electricity during periods of low electricity demand to store energy by pumping water into reservoirs for later use during high-demand periods.

[b] Other includes non-biogenic municipal solid waste, batteries, chemicals, hydrogen, and other sources.

Distribution of Generation by State

Electric power is generated throughout the United States, but the quantity of power produced and the methods of power production vary widely by region. Factors such as proximity to fuel sources, construction costs, and federal and state regulations have played a role in how the electric power generation sector has developed in each state. Texas, Florida, Pennsylvania, Illinois, and California were the largest generators of thermoelectric power in 2010; Washington, California, Oregon, New York, and Montana were the largest producers of hydropower.

Economic Importance of the Electric Power Generation Sector

Electrical power in the United States is generated both at private sector and government-operated facilities. This complicates the compilation of data on economic activity related to power generation. The 2007 Economic Census provides economic data for electric power generation, but only for private sector facilities. The U.S. Census of Governments reports economic data for government-operated electric utilities, but this includes electric power transmission and distribution, in addition to generation. Table 10.19 summarizes Economic Census data on the number of privately own establishments in each subsector within the electric power generation sector, the total employment at such establishments, total payroll, and annual revenue. The table also summarizes economic data for government-operated electric utilities, including total employment, payroll, and revenues.

TABLE 10.19
Economic Profile of the U.S. Electric Power Generation Sector (2007)

Subsector	Number of Establishments	Employment	Annual Payroll ($ Billion)	Annual Revenue ($ Billion)
Private hydroelectric	295	4086	$0.3	$2.2
Private fossil fuel	1248	74,860	$6.4	$85.4
Private nuclear	79	37,972	$4.1	$29.0
Private other[a]	312	5875	$0.5	$4.4
Total private	1934	122,793	$11.3	$121.0
Total public[b]	Not available	79,697	$5.1	$76.0
Total	>1934	202,490	$16.4	$197.0

Source: Data for private facilities are from the U.S. Census Bureau Economic Census (2007a); the data were summarized for NAICS codes 21111, 21113, and 21119. Data for public facilities are from the U.S. Census of Government Employment (2007b) and the U.S. Census of Governments Survey of State and Local Government Finances (2009).

[a] Includes solar, wind, tidal, and geothermal power generation.

[b] Includes employment, payroll, and revenue associated with electric power transmission and distribution, as well as generation.

DID YOU KNOW?

Government-operated facilities range from small municipal power stations to complex federal installations, such as the 31 hydroelectric dams in the Columbia River basin that generate power for much of the Pacific Northwest. Power from these dams is marketed by the U.S. Department of Energy's Bonneville Power Administration.

Thermoelectric Power

Thermoelectric power plants produce more than 90% of the electricity generated in the United States. Large thermoelectric power plants—particularly coal-fired and nuclear plants—require long startup times and operate at highest efficiency at relatively constant levels of output. Accordingly, most thermoelectric power generation is used to meet "base load" demand, or the minimal amount of electricity that must be available at all times. By contrast, "peak load" demand, or the electricity required to meet the highest daily, weekly, and yearly demand, depends on power plants that can come online rapidly, in response to sudden increases in demand. Smaller thermoelectric plants, such as gas-turbine plants, are often used to meet peak demand.

DID YOU KNOW?

The withdrawal of water for thermoelectric power generation peaked in 1980 and has remained relatively constant since that time, despite a significant increase in power production. The heightened efficiency in the use of water is largely due to a shift away from once-through cooling systems to recirculating cooling systems, a change that was triggered by Clean Water Act restrictions on cooling water discharge. Recirculating systems withdraw less water than do once-through systems; however, their use of evaporative cooling increases overall water consumption.

DID YOU KNOW?

As the largest off-stream user of water, thermoelectric power generation competes with other major water users, particularly agriculture and public supply. In contrast, the development of a hydropower project can act as a complement to other uses of water, as reservoirs created for large storage dams improve the availability of water for other purposes, including agriculture, public supply, and recreation. In some cases, however, concerns about the environmental impact of hydropower, particularly the impact of dams on fish and other wildlife, have restricted project development or constrained operations.

Hydropower

Although hydropower accounts for only 6% of national electric power generation, it has a much larger share of total generation in some states, particularly in the western United States. In Washington, Oregon, Idaho, and South Dakota, more than 50% of total generation comes from conventional hydropower. In addition, hydropower generation plays an important role in ensuring the reliability of electricity supply, both by meeting peak load demand and by storing excess electricity during low-demand periods. Because hydroelectric generators can be activated or deactivated very rapidly, hydropower is well suited for meeting peak demand (Gillian and Brown, 1997). The three primary types of hydropower facilities are storage, run-of-the-river, and pumped storage. A storage facility uses a dam to create a reservoir in a water body, creating a *head*, or difference in elevation between the reservoir and the water body beneath the dam (Bureau of Reclamation, 2005). Run-of-the-river plants do not rely on reservoirs and do not substantially interfere with the flow of the rivers in which they are located. Pumped-storage facilities use electricity to pump water into a storage reservoir when electricity demand is low and release the stored water to generate power when demand is high, essentially serving as batteries for the electric power grid. Storage and pumped-storage facilities can control the timing of electricity production and are therefore used to meet peak demand. Because run-of-the-river facilities are subject to seasonal variation in river flows, they are primarily used to meet base demand.

International Trade

The United States is the world's largest electricity consumer, but also the largest electricity producer (Enerdata, 2013). International trade in electricity is generally limited to countries with shared borders, and the United States is no exception, trading electricity only with Canada and—to a lesser extent—Mexico. The United States in recent years has been a net importer of electricity, with net imports ranging between 34 million megawatt-hours (MWh) in 2000 and 2009 and 6 million MWh in 2003. The trade deficit is small, however, relative to total domestic production. In 2010, U.S. net imports of electricity from Mexico and Canada were about 26 million MWh, less than 1% of the 4125 million MWh produced in the United States that year.

Projected Future Generation

Each year, the EIA's Annual Energy Outlook (AEO) projects energy production and consumption for the next 25 years, producing forecasts linked to alternative economic growth scenarios. In the 2011 AEO (EIA, 2011d), the EIA projected that electric power generation will increase steadily

DID YOU KNOW?

Hydropower is a relatively inexpensive source of electricity, primarily because hydroelectric plants incur no costs for fuel.

TABLE 10.20
Thermoelectric Water Withdrawals and Consumption (1995)

	Total Withdrawals (MGD)		
Water Use	**Saline**	**Fresh**	**Consumptive Freshwater (MGD)**
Thermoelectric power	57,900	132,000	3210
Total	60,800	341,000	100,000

Source: Solley, W.B. et al., *Estimated Use of Water in the United States in 1995*, U.S. Geological Survey, Washington, DC, 1998.

between 2011 and 2015, ranging from a 10.7% increase in its "low economic growth" scenario to a 27.2% increase in its "high economic growth" scenario. To meet increased demand for electricity in the future, the EIA expects that increased generation will come primarily from increased utilization of existing capacity at coal-fired plants, as well as increased reliance on plants powered by natural gas or renewable sources (EIA, 2011e).

Water Withdrawals

Thermoelectric power generation is the largest off-stream water user, accounting for approximately 49% of all water withdrawals, though a much smaller share of total water consumption. In addition, hydropower is a significant user of in-stream water, particularly in the western United States. The discussion below provides a more detailed review of how water is used in the generation of electricity.

Water Use in Cooling for Thermoelectric Power Generation

In all thermoelectric power plants, heat sources are used to generate steam, which turns a turbine to generate electricity. Cooling is then required to condense the steam back into boiler feed water before it can be used again; most plants have wet cooling systems, which use water as a cooling agent, although a small number of plants use dry cooling systems, which do not. Although thermoelectric cooling is the largest user of water, the vast majority of water used in this way is not consumed but is instead returned to its source or retained for further use. Although the USGS study of water use in 2005 did not track water consumption, the 1995 edition of the study found that only 2.5% of fresh water withdrawn for thermoelectric cooling was consumed (Solley et al., 1998). Total water consumption by thermoelectric cooling was just 3.3% of total water consumption in 1995 (see Table 10.20), representing a much smaller share than the water consumed in the domestic, industrial, and agriculture sectors. More recently, the DOE's National Energy Technology Laboratory (NETL) estimated water withdrawal and consumption factors for several types of thermoelectric power plants and found that consumptive water use in this sector totaled 3600 MGD in 2005 (NETL, 2009a), which again was about 2.5% of total water withdrawals for thermoelectric cooling in that year (USGS, 2009).

Geographic Distribution of Water Use

Just as thermoelectric power generation varies by state, so does the use of water for thermoelectric cooling (see Table 10.21). Unsurprisingly, the states generating the largest amount of electricity from thermoelectric power plants—California, Florida, Illinois, New York, and Texas—are the largest users of water for thermoelectric cooling. However, several coastal states, including California,

TABLE 10.21
Distribution of Cooling Systems by Generation Type (2005)

	Cooling System			
Generation Type	**Once-Through**	**Wet Recirculating (Towers)**	**Wet Recirculating (Ponds)**	**Dry**
Coal	39.1%	48.0%	12.7%	0.2%
Fossil non-coal	59.2%	23.8%	17.1%	0.0%
Combined cycle	8.6%	30.8%	1.7%	59.0%
Nuclear	38.1%	43.6%	18.3%	0.0%
Total	42.7%	41.9%	14.5%	0.9%

Source: NETL, *Estimating Freshwater Needs to Meet Future Thermoelectric Generation Requirements (2009 Update)*, National Energy Technology Laboratory, U.S. Department of Energy, Washington, DC, 2009.

Note: Data for combined cycle plants is limited to only 7% of the total plants in operation. Because of the small sample size, the percentage of combined cycle plants using dry cooling system may be overestimated.

Florida, Maryland, and New Jersey, use saline water for the majority of their thermoelectric cooling needs. Looking just at freshwater withdrawals, Illinois and Texas remain among the largest users, joined by Michigan, Tennessee, and Ohio.

Types of Cooling Systems

Thermoelectric power plants use two types of wet cooling systems: once-through cooling and recirculating, or closed-cycle, cooling. In once-through cooling, water is withdrawn from a water body, passed through heat exchangers (also called condensers) to cool the boiler steam used to power the generator, and then returned to the water body, usually at a temperature about 10 to 20 degrees higher than the receiving water. In recirculating cooling, water is withdrawn from a water body, passed through heat exchangers, cooled using ponds or towers, and then recirculated within the system. A small number of thermoelectric plants in the United States have dry cooling systems, in which an air-cooled condenser uses ambient air to dissipate steam heat without the use of water. Although once-through cooling systems originally predominated, most power plants constructed since the passage of the Clean Water Act in 1972 have been built using recirculating or dry cooling systems. The two cooling systems have different implications for water withdrawals and consumption. Once-through systems require high water withdrawals, though only a very small fraction of total withdrawals are consumed. Because water is recycled after cooling, total water withdrawals for recirculating cooling systems are much lower than for once-through systems. However, the process of using ponds or towers to cool the water involves high rates of evaporative losses.

Water Use per Unit of Electricity Produced

The cooling system used by a thermoelectric power plant affects the rate of water use per unit of electricity generated. A 2007 study by the American Water Resources Association found that once-through cooling systems withdrew about 570 gal/kWh but consumed less than 1 gal/kWh, whereas recirculating cooling systems withdrew less than 20 gal/kWh and consumed about 7 to 10 gal/kWh generated (Yang and Dziegielewski, 2007). The authors of the study also found that rates of both water withdrawal and water consumption varied widely among plants that use similar cooling systems, suggesting that water use is also affected by factors such as fuel type, water source, operation

TABLE 10.22
Unit Consumption of Water for Electricity Generation by Power Source

Power Source	Evaporative Cooling Water at Power Plant (gal/MWh)	Other Water Used for Power Plant Operations Used (gal/MWh)	Total Water Used (gal/MWh)
Coal	243–449	53–68	296–517
Natural gas	192	0	192
Nuclear	720	30	750
Biomass/waste	300–480	30	330–510
Geothermal	0	175–585	175–585
Concentrating solar power	750–920	80–90	840–920

Source: Carter, N.T., *Energy's Water Demands: Trends, Vulnerabilities and Management*, Congressional Research Service, Washington, DC, 2010.

conditions, and cooling system efficiency. A report prepared by the Congressional Research Service on the water demands of domestic energy production collected estimates of the amount of water consumed per unit of electricity produced at different types of thermoelectric power plants. The estimates include both water consumed for cooling and water consumed in other plant processes, such as equipment washing, emission treatment, and human use. These estimates, which all assume the use of recirculating cooling systems, are presented in Table 10.22. Of the three main thermoelectric power sources, natural gas-fired plants appear to be much more efficient in their use of water than coal-fired and nuclear plants, although some of that difference may be due to the fact that natural gas-fired plants are, on average, newer than coal-fired plants and may use more efficient wet cooling systems.

Future Use

Although electric power generation is projected to increase steadily over the next few decades, it is less clear whether water withdrawals and consumption for thermoelectric cooling will experience similar growth. As noted earlier, water withdrawals for thermoelectric power generation peaked in 1980 and declined by 11% afterwards, though they have gradually increased since 1985. As new thermoelectric plants increasingly rely on recirculating cooling systems and older plants are retired, future water withdrawals for thermoelectric cooling may actually decrease, although water consumption in the sector would increase. A 2009 NETL study of thermoelectric water use trends assumes that all new thermoelectric generating capacity will use recirculating cooling. Based on this assumption, the analysis projects that water withdrawals for thermoelectric power will decrease by about 4.4% between 2005 and 2030 (from 146,300 MGD to 139,900 MGD), while water consumption in the sector will increase by about 22.2% (from 3600 MGD to 4400 MGD) (NETL, 2009a).

Hydropower

It is difficult to quantify the amount of water used for hydroelectric power generation, as this process generally does not require water to be withdrawn from its source. Nonetheless, the production of hydropower often requires disruption of river flow regimes, which can affect the availability of water for other uses. This section briefly discusses available estimates of the use of water for hydropower generation including expected changes in future use.

Overview of Hydropower Water Use

The USGS study of water use in 2005 did not estimate the use of water for hydroelectric power. The USGS last provided this figure in its report on water use in 1995; at that time, it estimated that a total of 3,160,000 MGD was used in the generation of hydroelectric power. This number, which exceeds the average annual runoff in the United States by a factor of 2.6, is misleading because it over-counts water that is used several times as it passes through multiple hydroelectric dams on a single river (Solley et al., 1998). The USGS study of water use in 1995 also reported that 90,000 MGD were used for off-stream hydroelectric power generation (i.e., hydropower relying on diversions of water away from primary river channels), which would represent more than 25% of total water withdrawals from all other sources in that year. It is not clear whether that number also over-counts the total amount of water withdrawn for hydroelectric use. Although water is not consumed in the generation of hydroelectric power, the reservoirs created for storage and pumped-storage facilities can lead to water loss in the affected water bodies through increased evaporation rates. A study by the USDOE's National Renewable Energy Laboratory examined the 120 largest hydroelectric facilities in the United States and concluded that the reservoirs created for those dams evaporated 9063 MGD more than would be evaporated from free-running rivers. Evaporation rates varied significantly across facilities. Overall, however, evaporative losses averaged 18,000 gal/kWh. This rate of water consumption is several orders of magnitude greater than the rates for thermoelectric power plants reported in Table 10.22 (Torcellini et al., 2003). It would be inappropriate, however, to ascribe the full amount of water lost to evaporation at reservoirs to hydropower generation, as these reservoirs frequently serve multiple purposes, including recreation, flood control, and providing a reliable water supply for agricultural and domestic uses.

Future Use

The 2011 Annual Energy Outlook anticipates that hydropower generation will increase at the same rate as total electric power generation, implying the need to develop between 1600 and 3000 MW of additional hydropower capacity by 2035 (EIA, 2011d). The construction of new large dams would create new reservoirs, which could reduce the amount of freshwater available for downstream use. It may be possible, however, to increase the generation of hydroelectricity through improved efficiency or expansion of power plants at existing dams. The Bureau of Reclamation, which currently generates about 40 million MWh at its hydropower facilities, reviewed 530 sites currently under its jurisdiction to evaluate their potential for additional hydropower development. The study found that 191 sites could be developed with a total potential capacity of 268.3 MW, although not all sites were economically viable to develop (Bureau of Reclamation, 2011). In addition, a 2007 study by the Electric Power Research Institute (EPRI) estimated that 10,000 MW of additional hydropower capacity could be developed by 2025 without construction of any new dams (EPRI, 2007a). Nonetheless, development of new hydropower capacity has slowed in recent years, due to rising awareness of the harmful impacts of large dams and reservoirs on fish and wildlife, Native American communities, and competing use of in-stream water. It is possible, therefore, that the predicted growth in hydropower capacity will not take place.

Water Resource Constraints

Because once-through cooling systems require large quantities of water, thermoelectric power plants using such systems are particularly vulnerable to drought conditions and other water shortages. In recent years, water shortages have curtailed power generation at a number of facilities; for example,

- In 1999, drought in the Susquehanna River basin in New York and Pennsylvania prevented power plants in the region from obtaining sufficient water supplies to meet operations needs (GAO, 2003).

- In 2006, drought along the Mississippi River caused power plants in Illinois and Minnesota to restrict operations (NETL, 2009b).
- In 2007, drought in the southeastern United States caused several nuclear power plants to reduce output by up to 50%, due to low river levels (NETL, 2009b).

The move away from once-through cooling systems has somewhat mitigated this vulnerability, but a substantial portion of the country's electricity generating capacity still relies on regular access to large quantities of water.

In recognition of this challenge, EPRI launched a 10-year research plan in 2007 aimed at helping the U.S. electricity industry adapt to current and future water supply constraints (EPRI, 2007b). The proposed areas of research include improving dry cooling technology, reducing water loss from cooling towers, using impaired water, and developing decision support tools to anticipate and respond to water shortages and climate change. Dry cooling systems, or hybrid dry–wet cooling systems, could drastically reduce water withdrawals and consumption, but they currently have much higher costs than wet cooling systems and can negatively affect plant operating efficiency. Reduced operating efficiency in turn leads to higher fuel consumption per unit of electricity produced, with associated environmental consequences of fossil fuel extraction and combustion. Use of impaired water for thermoelectric cooling—effluent from wastewater treatment plants or low-quality groundwater, for example—could also reduce the sector's use of freshwater, but such water might require pretreatment in order to prevent damage to cooling equipment (Carter, 2010). Future research into both dry cooling and use of impaired water could help reduce the dependence of the electric power generation sector on reliable access to large quantities of water.

Water Quality Constraints

Thermoelectric cooling generally does not have high water quality requirements, as demonstrated by the fact that about 30% of total water withdrawals for thermoelectric power generation in 2005 involved saline water. Water discharged from thermoelectric plants with once-through cooling systems, however, can have a detrimental impact on the quality of the receiving water. In addition to the temperature difference between discharged cooling water and receiving water (which can disrupt aquatic habitats), chemicals used to protect cooling equipment can also affect downstream water quality and use (Carter, 2010). Section 316 of the Clean Water Act (CWA) gives the USEPA the authority to regulate the use of water for industrial cooling, with Section 316(a) regulating the temperature of discharged water and Section 316(b) regulating cooling water intake structures. These regulations have played a large role in driving the shift from once-through cooling systems to recirculating cooling systems. New plants are already required to install recirculating cooling systems, and the USEPA is currently developing regulations to update requirements for existing plants with once-through systems. As noted in the previous section, the use of impaired water for cooling could ease water quantity constraints but impose new water quality constraints. Without adequate pretreatment, impaired water could lead to scaling, corrosion, and fouling of cooling equipment (Carter, 2010).

Interaction with Other Uses of Water

As a major off-stream user of water, the thermoelectric power generation sector competes for water with several other sectors, particularly agriculture and domestic supply. In western states, prior appropriation water rights laws give precedence to those that first obtained legal rights to use the water, which typically include agricultural and municipal users. Under drought conditions, users that obtained legal rights to use water at a later date—typically including thermoelectric power generators—are the first to suffer restrictions in water supply (NETL, 2009b). In contrast, the development of hydroelectric power has often served as a complement to other water use sectors, as large reservoirs created by dams are often used to provide water for domestic and agricultural users, in addition to serving as a setting for recreational activities such as boating, swimming, and fishing. As noted previously, however, these facilities can also have negative impacts on wildlife, such as salmon populations in the Pacific Northwest. Regulatory constraints have prevented development of hydropower in some areas, and allowing for fish passage through large dams can significantly increase operating costs (Bureau of Reclamation, 2011).

Future Considerations

The factors that in the future are likely to have the greatest effect on the use of water in the electric power generation sector are projected limitations on the availability of water, the potential impacts of climate change on both power demand and water supply, and changing water demands due to increased reliance on renewable energy sources. A 2011 EPRI study attempted to identify the regions of the United States most likely to face future constraints on thermoelectric power generation as a result of constraints on water supplies (EPRI, 2011). The authors first projected water use through 2030 and developed a water sustainability risk index that evaluated water supply constraints according to several dimensions, including the extent of development of available renewable water and groundwater, susceptibility to drought, expected growth in water demand, and the likelihood of increased need for storage (to ensure water availability during seasonal dry periods). The study found that about 250,000 MW of thermoelectric generation capacity, or 22% of total generating capacity in 2010, is located in counties with either high or extreme levels of water supply sustainability risk.

Climate change, and responses to climate change, could affect the relationship between water and electric power generation in several ways. First, the increased frequency of floods and droughts predicted by many climate models could significantly compromise the reliability of water access for both thermoelectric cooling and hydropower. Increased temperatures could also increase demand for electricity (e.g., for air conditioning during summer months). Second, regulations to reduce greenhouse gas emissions through carbon capture and storage could potentially increase demand for water in electricity generation. Carbon capture and storage increases water demands at fossil fuel-burning power plants, because operating carbon capture equipment requires both energy (thereby reducing a plant's generating efficiency) and additional cooling. A 2009 NETL study estimated that installing carbon capture systems at fossil fuel-burning plants could, by 2030, increase water withdrawals by between 1300 and 3700 MGD, and water consumption by between 900 and 2300 MGD (NETL, 2009a). Finally, increased development of electric power generation from renewable sources could affect future water demand in this sector. Although sources such as wind and photovoltaic solar have no cooling requirements, other sources such as biomass, geothermal, and concentrating solar power (CSP) all involve thermoelectric generation and therefore require cooling. In particular, CSP facilities, which are often located in dry, arid regions to maximize exposure to solar energy, may face significant water constraints (Carter, 2010).

Federal and state regulation could help mitigate the effects of future water constraints on electric power generation. A study by the Government Accountability Office (GAO) found that states' regulatory authority over water use by thermoelectric plants varied widely (GAO, 2009). California and Arizona, for example, have actively worked to minimize the use of freshwater in thermoelectric power generation, while other states have no official policies or permitting requirements for power plant water use.

Challenges to Estimating the Value of Water

In addition to the difficulties in estimating the value of water discussed earlier, valuing the water used in electric power generation faces several additional challenges:

1. Electricity prices are subject to government regulation and may not in all cases fully reflect the long-run margin cost of supply. Attempts to derive the value of the marginal product of water from the price of electricity will reflect any distortions introduced by government policy.
2. Because of the constantly changing nature of electricity demand (i.e., the difference between peak load demand and base load demand), the value of electricity can change depending on the season or time of day. As a result, the marginal value of water in this sector also varies, depending on whether it is used in the production of electricity to meet peak or base load demand.
3. Much of the water used in electric power generation is "non-rivalrous" (i.e., the use of the resource does not diminish its availability to others), because water used in hydropower generation and thermoelectric generation with once-through cooling can be withdrawn again by downstream users. In this regard, the use of water for electric power generation has some characteristics of public goods, which are also difficult to value using market mechanisms (Young, 2005).
4. Where multiple hydroelectric dams are located on the same river, the value of water varies widely according to its location, as the electricity generation potential of a given unit of water depends on its *developed head*, the height of a retained body of water. For example, the cumulative developed head of water at the mouth of the Snake River in the Pacific Northwest is more than 36 times the developed head of water at the last dam along the Columbia River (Frederick et al., 1996).

Despite these difficulties, it is possible to estimate the value of water to a given electricity-generating facility by using the "shadow price" of electricity or the cost of obtaining the same amount of electric power from a different facility. Once the value of the electricity produced by a facility is estimated, the marginal value of water used to generate that electricity can be derived by comparing the total cost per kWh at that facility to the cost per kWh generated from the next cheapest source of electricity (that does not use water). All else equal, the difference between the cost of electricity generation with water and electricity generation without water can be interpreted as the marginal value of water used in electric power generation. In practice, however, ensuring that "all else is equal" is nearly impossible.

Estimates of the Value of Water in the Electric Power Sector

Despite the challenges discussed above, several attempts have been made to estimate the marginal value of water used in hydropower generation and thermoelectric cooling:

- A 1996 study by Kenneth Frederick and others at Resources for the Future collected 57 water valuation estimates for the production of hydropower and 6 estimates for the production of thermoelectric power. The hydropower value estimates came from four water resource regions—Tennessee, the Upper Colorado, the Lower Colorado, and the Pacific Northwest—and reflected the average values of the cumulative upstream generating capability at each dam along a particular river.
- A 2005 report by the American Water Works Association (AWWA) also discussed estimates of the value of water used for hydropower on the Colorado River.
- A 2004 study by Thomas Brown at the U.S. Forest Service estimated the value of water used for hydropower generation on two stretches of the Colorado River by comparing hydropower costs to the costs of peaking power from thermoelectric plants.
- A 2011 analysis by Stacy Tellinghuisen at Western Resource Advocates estimated the value of water used in thermoelectric cooling by assuming that the only alternative to using water for this purpose would be the use of a more expensive dry cooling system. On an economy-wide scale, this assumption is not valid, because the electricity that would

be generated at a wet-cooling thermoelectric plant could always be replaced by increased electricity generation from a different source. From the perspective of a private developer, however, it may be valid to assume that the only alternative to using water for thermoelectric cooling is to install and operate a dry cooling system. This estimate can therefore be interpreted to represent the value of water for thermoelectric cooling to the developer or owner/operator of a particular plant.

IN-STREAM WATER USE BY COMMERCIAL FISHING

Commercial fishing is the last major component of the global food system that involves the capture and harvest of animals from their natural environment. As such, commercial fisheries are uniquely dependent on water resources. Although many economic sectors use water as an input, the very existence of commercial fisheries depends on a complex web of ecological interactions in the aquatic environment. Maintenance of this environment through management of water quality and other variables is fundamental to the sustainability of wild-capture fisheries. This section describes the relationship between water and commercial fisheries and addresses the following topics:

- The economic importance of the commercial fishing sector, including landings, revenue, employment, and links to other parts of the economy
- The way in which management of fishing effort and management of water quality combine to ensure that long-term sustainability of key commercial species

The commercial fishing industry is part of the primary (extractive) mega-sector described earlier. Some of its output is sold directly to consumers. Most, however, is sold to seafood processors in the secondary mega-sector or to wholesale and retail establishments in the tertiary mega-sector. According to the Food and Agriculture Organization, annual landings by the U.S. commercial fishing fleet rank third worldwide, behind only China and Peru. There is also a growing international trade in seafood and other fish products. The United States is currently the world's second-leading importer of such products and its fourth-leading exporter (FAO, 2009).

Landings and Ex-Vessel Revenues

The U.S. commercial fishing industry in 2010 reported total domestic landings of approximately 8.2 billion pounds, an ex-vessel value (i.e., price received by fishermen for fish, shellfish, and other aquatic plants animals landed at the dock) of $4.5 billion. Eighty-five percent of total landings by weight were accounted for by finfish, with Alaskan pollock and menhaden being the leading contributors. In contrast, shellfish accounted for only 15% of total landings by weight but 52% of ex-vessel revenues. Crabs ($572.8 million) were the leading sources of revenue, representing approximately 13% of the total. The Alaskan region reported the greatest landings in 2010, accounting for 53% of total landings by weight (4.3 billion pounds) and 35% of total landings by value ($1.6 billion). The New England region ranked second in revenue, accounting for approximately 21% ($954.0 million) of the total (FUS, 2011).

Landings in 2010 were up slightly from 2009, when the industry reported a 20-year low in total catch. Landings remained relatively low in 2010 due to a significant decline in the catch of Alaska pollock. In contrast, ex-vessel revenues in 2010 (in nominal dollars) were the highest reported in over 20 years. The increase in ex-vessel revenues can be attributed to higher prices for a number of key species. For example, in 2007 the United States landed roughly 885.0 million pounds of salmon at an ex-vessel value of $381.3 million, an average ex-vessel price of $0.43 per pound. In 2010, salmon landings totaled only 787.7 million pounds but were valued at $554.8 million, an average price of $0.70 per pound. The 63% increase in prices netted salmon fishermen a 43% increase in ex-vessel revenues (FUS, 2011).

The National Marine Fisheries Service (NMFS) data reflect harvests in marine waters (including estuarine waters) and the Great Lakes. The Great Lakes harvest, however, is relatively minor; landings in this region totaled 19.2 million pounds in 2010, with an ex-vessel value of approximately $18.0 million (less than 0.5% of all ex-vessel revenues). Whitefish and perch account for over 80% of ex-vessel revenues in the Great Lakes region (FUS, 2011).

Additional commercial harvesting of freshwater species occurs throughout the United States, but these landings are poorly traced in most states and represent a minor increment to the landings characterized by NMFS. For example, in many states, individuals harvest minnows and other species for sale to recreational anglers as baitfish. Some states are also home to small but regionally important specialty freshwater fisheries. For instance, freshwater commercial fisheries in Louisiana reported $16.2 million in sales in 2009; sales of crawfish accounted for the vast majority of this total (LSU, 2011).

Employment

Jobs in the commercial fishing industry are often transitory and poorly documented, making it difficult to track employment. As a result, the U.S. Economic Census does not report employment in the commercial fishing industry. The Bureau of Labor Statistics (BLS) does track employment in the industry; however, its data exclude jobs that are exempt from or not covered by unemployment insurance. To provide a more comprehensive estimate of employment, this report relies on both the BLS data and the U.S. Census Bureau's non-employer data regarding the number of commercial fishing firms that have no paid employees or are exempt from unemployment insurance. According to BLS data, as of 2009 approximately 6321 people were employed in the finfishing and shellfishing industries. In addition, the U.S. Census Bureau estimated that in 2009 there were 64,531 non-employer commercial fishing firms in the United States. Of these, 3546 were listed as corporations, 725 as partnerships, and 60,260 as individual proprietorships. Assuming that each of these firms represents at least one commercial fisherman, employment in the commercial fishing sector in 2009 likely totaled approximately 71,000. At the state level, Alaska boasts the highest employment with 8305 jobs, the vast majority of which are accounted for by non-employer firms. Regionally, the Gulf states (excluding the west coast of Florida) account for the greatest percentage of jobs in the industry, approximately 17% of the national total. If the west coast of Florida were included in that figure it would likely increase significantly, as at the state level Florida is second only to Alaska in the estimated number of jobs in the commercial fishing sector.

Links to Other Economic Sectors

The nation's commercial fisheries support a number of industries dedicated to the processing or sale of fish and fish products. According to the BLS, in 2010 approximately 36,469 people were employed at 846 establishments engaged in seafood product preparation and packaging. Also linked are seafood wholesalers, which in 2010 employed approximately 22,495 people in 2344 establishments. Not included in that number are wholesalers of canned or packaged frozen fish, who are counted under a different NAICS code, grouped with other wholesalers of packaged frozen and canned foods. The commercial fishing harvest is processed into both edible and non-edible products. Edible fish and shellfish are sold fresh, frozen, canned, or cured. Non-edible products are used as bait or animal food or in an industrial capacity (e.g., manufactured into fish oils, fish meals, fertilizers). In 2010, approximately 79% of all domestic landings, by weight, were put toward human consumption; 93% of this total was sold fresh or frozen, 6% was canned, and 1% was cured. The National Marine Fisheries Service estimated that revenues from the sale of fishery products by U.S. processors totaled $9.0 billion in 2010. The sale of edible domestic and imported fish products accounted for $8.5 billion of this total. Non-edible domestic and imported fish products generated estimated sales of $508.8 million, with 46% of that total accounted for by bait and animal food, 43% by fish meals and oils, and 11% by other products, such as fertilizers, agar–agar, oyster-shell products, kelp products, and animal feeds (FUS, 2011).

U.S. Commercial Fishing and the Global Economy

As of 2006, traditional capture fisheries accounted for 64% (92.0 million metric tons) of global fish production (aquaculture accounted for the remaining 36%). The United States played a significant role in that production, ranking third globally behind only China and Peru. Most of the U.S. catch, however, went to domestic use (FAO, 2009). In 2010, the United States exported only 1.2 billion pounds of edible fish products, an export value of $4.4 billion. The value of U.S. exports is much higher when industrial products (such as fertilizers) are included; the addition of this category raises the total value of fish product exports in 2010 to $22.4 billion. In comparison, U.S. imports of fish products total $27.4 billion, including $14.8 billion of edible products (FUS, 2011). As of 2006, the United States was the second leading importer of fish products in the world, and the fourth leading exporter (FAO, 2009). Asia was the source of 52% ($14.2 billion) of U.S. imports, and the destination of 39% ($8.8 billion) of U.S. exports. At a national level, China was the leading source of foreign fish products, accounting for $4.5 billion in imports, while Canada was the top destination for U.S. fish products, accounting for $4.0 billion in exports (FUS, 2011).

Commercial Fishing and the Environment

In comparison to other economic sectors examined in this text, water plays a different role in commercial fish harvesting. Rather than being an input into a production process, water is one element in a complex biological system. Likewise, the harvested species are themselves elements in this same system. It is the maintenance of this system that supports commercial activity such as sustainable fisheries. As described below, fisheries management agencies directly regulate commercial fishing activity to help ensure the long-term sustainability of the industry. The future productivity of the nation's commercial fisheries also depends on responsible management of the nation's coastal waters and on the long-term impact of climate change on habitat and fish stocks.

Fisheries Regulation

Under the authority of the Magnuson–Stevens Fishery Conservation and Management Act and other federal statutes, the National Oceanic and Atmospheric Administration (NOAA) manages the nation's marine fisheries through regulations governing ocean resources, fishing gear, and fishing effort. NOAA employs two distinct terms in assessing the health of fish stocks: *overfishing* and *overfished*. Specifically,

- A stock is subject to *overfishing* when the harvest rate is above the level that allows for maximum sustainable yield (i.e., the rate of removal is too high).
- A stock is *overfished* when its population has a biomass level below a biological threshold specified in its fishery management plant (i.e., the population is too low).

DID YOU KNOW?

The productivity and long-run sustainability of the commercial fishing industry depend in part on appropriate management of aquatic ecosystems, as well as management of fisheries to maintain fish and shellfish stocks.

In 2010, NOAA reviewed 528 stocks to determine their status. For 275 of these stocks, overfishing thresholds are unknown or cannot be determined, but sufficient information was available to evaluate the remaining 253. Of these, NOAA classified 40 (16%) as subject to overfishing. Some key stocks considered subject to overfishing in 2010 were Atlantic cod in the New England region and bluefin tuna in the Pacific region, although the latter was not fished exclusively by U.S. fishermen.

With respect to overall population, NOAA was able to assess the status of only 207 stocks. Of these, it classified 48 (23%) as overfished and identified five others that are approaching that status. Key stocks that were classified as overfished were Atlantic cod and Chinook and coho salmon in the Pacific region and blue king crab in Alaska. The commercial importance of these species is clear. For example, Chinook and coho salmon collectively accounted for 6% (46.6 million pounds) of all 2010 salmon landings by weight, and 13% ($73.9 million) of all salmon landings by value. Similarly, blue king crab makes up a significant portion of the Alaska king crab catch, which in 2010 had landings valued at $122.4 million.

The National Marine Fisheries Service's Fish Stock Sustainability Index (FSSI) measures the sustainability of 230 key stocks. The FSSI assesses each stock's sustainability on a four-point scale, where

- Half a point is awarded if the stock's overfishing status is known.
- Half a point is awarded if the stock's overfished status is known.
- One point is awarded if overfishing is not occurring.
- One point is awarded if the stock's biomass is above the level prescribed for it.
- One point is awarded if the stock is at or above 80% of the biomass required for maximum sustainable yield.

When totaled, the maximum FSSI value for all 230 stocks is 920. As of 2010, the value of the index stood at 583, 63% of the maximum. This is a significant increase since 2000, when the index stood at 357.5. This rise in the index, however, has been driven mainly by an increase in the number of stocks whose overfishing or overfished status is known, not by reductions in overfishing or increases in fish stocks.

In response to the fish stock assessments, NOAA administers a broad range of regulations and programs designed to restore stocks of overfished species or sustain the stocks of healthy species. NOAA's Office of Sustainable Fisheries (OSF) implements these measures, including the following:

- Catch limits on key commercial species
- Catch shares that limit access to key fisheries
- International cooperation programs

NOAA is assisted by Regional Fishery Management councils (which develop fishery management plans) and state agencies (which typically focus on permitting and other support tasks) (NOAA, 2011).

Habitat Quality

Commercial Fisheries Habitat Protection

Although commercial fishing occurs in both inshore and offshore areas, coastal waters play an especially vital role in maintaining fish stocks. Bays, estuaries, and coastal wetlands are essential to the life cycle of many commercial fish species. These areas serve as spawning grounds, nurseries for juvenile fish, and feeding areas for both juvenile and adult fish. Coastal areas also represent the interface between the marine environment and the built, human environment. As such, most efforts to manage the habitat of commercial fish species focus on the coastal zone. Water quality management is one key aspect of habitat protection. As it relates to commercial fish species, water quality is especially important in estuarine areas where rivers meet ocean waters. Environmental agencies are central to water quality management, administering an array of programs under the authority of the Clean Water Act and other state and federal environmental statutes. These efforts include the regulation of effluent discharged by conventional point sources such as manufacturing facilities or municipal sewage treatment plants. Additional programs address stormwater management, management of agricultural runoff, and other pollution sources. Many of these programs are based on collaborative relationships between and among state and federal agencies, local governments, conservation organizations, and the private sector. Pollution that can affect commercial fish habitat may originate in coastal areas or in areas remote from ocean waters. Hypoxia in the Gulf of Mexico provides an excellent illustration of the linkages between inland water quality management and commercial fishing impacts in marine waters. The northern Gulf of Mexico receives large loadings of nutrients from agricultural operations and other runoff sources that drain to the Mississippi River, depleting oxygen levels in coastal areas and disrupting food webs. First documented in 1972, the resulting "dead zone" has been growing in size over the last several decades (USEPA, 2011). Recent studies have demonstrated a direct statistical correlation between the size of the hypoxic area and landings of brown shrimp on the Texas and Louisiana coasts (O'Connor and Whithall, 2007). The action plan for addressing hypoxia in the Gulf calls for collaborative stakeholder efforts to reduce nitrogen and phosphorus loadings from farms and other sources (USEPA, 2008d).

Habitat quality considerations extend well beyond basic water quality concerns. Numerous other aspects of habitat structure and function are influenced by human activities, which themselves are the focus of an array of management efforts. For example, a variety of regulatory and conservation programs are designed to manage coastal development and prevent wetland loss. Under one such initiative, the National Coastal Zone Management Program, states partner with federal agencies to protect coastal resources and manage shoreline development. Likewise, unimpeded access to coastal rivers is vital to salmon and other anadromous commercial species that migrate upstream to spawn. Increasing attention has been paid to the impact of hydropower projects, and permitting of such facilities now routinely incorporates requirements for improved fish passage. Dam removal has also become more common as resource management officials consider competing uses of river flow. Finally, invasive species—including plants, fish, and shellfish—can proliferate in aquatic environments and undermine native species. Control of invasive species that threaten commercial and recreational fisheries is an increasing concern for natural resource managers.

Assessment of U.S. Coastal Habitat

Given the range of habitat quality considerations discussed above, reliable characterization of commercial fish habitat requires an integrated analysis of coastal resources. In 2008, the USEPA published the National Coastal Condition Report III (NCCR III), an assessment of the condition of the United States' estuaries and coastal waters (all waters from 0 to 3 miles offshore). The report measures coastal quality based on five factors, each of which is scored on a five-point scale from

DID YOU KNOW?

Harmful algal blooms (HABs) are events involving the proliferation of toxic or otherwise harmful phytoplankton. The events may occur naturally or may be the result of human activity (e.g., nutrient runoff). In the United States, HABs frequently cause shellfish bed closures due to concerns over health risks associated with consumption of contaminated shellfish. A study conducted by researchers at the Woods Hole Oceanographic Institute found that HABs result in average annual commercial fishing losses of $18 million. Apart from this long-term average, HABs can result in acute losses to local fisheries; for example, blooms of a particular brown tide organism eliminated the $2 million bay scallop industry off Long Island, New York (Anderson et al., 2000).

poor to good, where less than 2.0 is poor, 2.0 to 2.3 is fair to poor, 2.3 to 3.7 is fair, 3.7 to 4.0 is fair to good, and greater than 4.0 is good. To determine the overall score for a particular region, the scores of the five factors are averaged:

- *Water quality*—Water quality is measured by assessing the levels of five indicators: dissolved inorganic nitrogen (DIN), dissolved inorganic phosphorus (DIP), chlorophyll *a*, water clarity, and dissolved oxygen. A poor score indicates that more than 20% of the coastal area is in poor condition; fair indicates that 10 to 20% of the coastal area is in poor condition or more than 50% combined is in fair or poor condition; and good indicates that less than 10% of the coastal area is rated poor or more than 50% rated good.
- *Sediment quality*—Sediment quality is determined by looking at three factors: sediment toxicity, sediment contaminants, and sediment total organic carbon (TOC). Sediment showing high levels of any of these could contaminate or be inhospitable to benthic organisms. A poor score indicates that more than 15% of the coastal area is in poor condition; fair indicates that 5 to 15% is in poor condition or more than 50% combined is in poor or fair condition; and good indicates that less than 5% of the coastal area is in poor condition or more than 50% is in good condition.
- *Benthic quality*—Benthic quality assesses the health of a coastal area's benthic population (i.e., bottom-dwelling organisms). Quality is graded based on species diversity; high species diversity, as well as a large proportion of pollution-sensitive species, leads to a high benthic score, whereas low species diversity and a high proportion of pollution-tolerant species leads to lower scores. The scoring criteria differ depending on the region.
- *Coastal habitat quality*. The coastal habitat index assesses the status of the nation's marine wetlands and estuaries, many of which are adversely affected by human activities (flood control, real estate development, agriculture, etc.). The index is scored by combining two rates of wetland loss for the region being considered: the historical, an average of decadal loss from 1780 to 1990; and the modern, from 1990 to 2000 (data past 2000 were unavailable). The two scores are averaged and then multiplied by 100. The resulting figure is then used to grade the region. A poor score is given if the loss indicator is greater than 1.25, a fair score if the indicator is between 1.0 and 1.25, and a good score if it is less than 1.0.
- *Fish tissue contamination*—Fish tissue contamination is assessed by measuring the levels of certain contaminants (such as arsenic, mercury, or DDT) in samples of fish taken off the coasts of the subject regions. A poor score indicates that more than 20% of the samples are in poor conditions; a fair score indicates that 10 to 20% of the samples are in poor condition or more than 50% combined are in fair condition; and a good score indicates that less than 10% of samples are in poor condition or more than 50% are in good condition.

Regional scores are determined by averaging the scores for the five indicators. The national scores for each indicator, however, are not determined by simply averaging the regional indicator scores. Instead, each region is weighted based on the percentage of the coastline it represents. The overall national score is determined by averaging the five national indicator scores. In 2008, the overall coastal condition of the United States scored 2.8, or fair, on the coastal condition index. The United States scored a 3.9 on the water quality index, a 2.8 on the sediment quality index, a 1.7 on the coastal habitat index, a 2.1 on the benthic index, and a 3.4 on the fish tissue contaminants index. Table 10.23 summarizes the scores and illustrates the distribution of scores by region.

As Table 10.23 indicates, there are some particularly low results at the regional level. In the Northeastern region, for example, both benthic quality and fish tissue quality earned a poor rating. In this region, 31% of all fish sampled rated poor on the fish tissue contamination index, and 28% rated fair. This rating was due primarily to the presence of two contaminants: polychlorinated biphenyls (PCBs) and DDT. These are the most common contaminants in the Pacific region, which also earned a poor score on the fish tissue contamination index. There, 26% of all fish sampled

TABLE 10.23
Coastal Condition Index Scores by Region

Index	Northeast Coast	Southeast Coast	Gulf Coast	West Coast	Great Lakes	South Central Alaska	Hawaii	Puerto Rico	U.S. Total
Water quality index	3	3	3	3	3	5	5	3	3.9
Sediment quality index	2	3	1	2	1	5	4	1	2.8
Coastal habitat index	4	3	1	1	2	—	—	—	1.7
Benthic index	1	5	1	5	2	—	—	1	2.1
Fish tissue contamination index	1	4	5	1	3	5	—	—	3.4
Overall condition	2.2	3.6	2.2	2.4	2.2	5	4.5	1.7	2.8

Source: USEPA, *National Coastal Condition Report III*, U.S. Environmental Protection Agency, Washington, DC, 2008.

rated poor, and 11% rated fair. Overall, the findings of the NCCR III study highlight the potential vulnerability of commercial fish stocks to the degradation of coastal habitat, particularly along the northeast, western, and Gulf coasts.

Climate Change

Note to the reader: In fairness to the reader and user of this text it is important to point out that the author of this text takes the view that global climate change is an ongoing, cyclical event that has occurred countless times in the history of Earth (and one would hope that the cycle is perpetual). Is mankind causing global climate change? I simply do not know. Does anyone know, for certain? One thing seems certain to me: Humankind is exacerbating the global climate situation by polluting Mother Earth; thus, it can be said that humans are contributors to local climatic conditions that cannot be viewed in any positive light, even by me. Period.

Climate and atmospheric conditions are an important influence on the aquatic ecosystem supporting commercial fishing. Although not an immediate threat to the viability of the commercial fishing industry, climate change could have significant long-term effects. Several physical and ecology changes have been observed or are anticipated for the marine environment (FAO, 2008):

- Water temperatures are warming, particularly surface temperatures; however, effects will vary across geographic areas, with deeper warming possible in the Atlantic.
- Changes in salinity are already being observed, especially in low-latitude areas with rapid evaporation rates.
- Acidity is increasing, undermining the viability of coral reefs.
- Many models predict shifts to smaller species of phytoplankton, altering food webs.
- In the longer term, most models predict declines in the stocks of cold-water fish species and poleward migration of warm-water species.

Globally, these environmental changes will likely redistribute commercial fisheries. A major study by Cheung et al. (2009) predicted a 30 to 70% catch increase in high-latitude areas and a 40% decrease in the tropics. This study predicts little net change in global fishery productivity overall. Other studies, however, forecast net economic losses. A World Bank study projected that by 2050 climate change could cause anywhere from a 10 to 40% reduction in global catch relative to 2010 levels and a global revenue loss of $10 to $33 billion per year (World Bank Group, 2010).

Like the global predictions, anticipated climate change effects on U.S. commercial fisheries vary by region. The Cheung et al. (2009) study concluded that by 2100 the contiguous United States will see an approximately 13% decrease in potential catch. Alaska, on the other hand, may see a roughly

25% increase in potential catch over that same time frame. It is unlikely, however, that the potential increase in Alaska's catch would reflect the diversity of species currently landed in the U.S. mainland (Pew Environment Group, 2009).

Climate change may also influence commercial fisheries through freshwater habitat impacts. As previously noted, salmon and other anadromous species account for a significant share of U.S. commercial fishing revenues. To the extent that climate change exacerbates competition for water in the western United States, these species could be negatively affected. For example, studies have highlighted the potential for warming trends to reduce snowpack in the Pacific Northwest, reducing streamflow in the Columbia River basin. These studies acknowledge that increased water temperatures and reduced in-stream flow are a threat to the survival of Columbia River salmonids (NRC, 2004).

IN-STREAM WATER USE BY COMMERCIAL NAVIGATION

United States ports and waterways are an important element of the nation's commercial transportation infrastructure. As a non-consumptive activity, the use of water for commercial navigation generally does not affect its availability for other uses. Nonetheless, commercial navigation can raise issues that require the consideration of water resource managers, including the need to dredge or maintain sufficient in-stream flows to ensure adequate channel depths. In addition, the development of canals and seaways and the maintenance of channels to facilitate shipping can have negative environmental consequences, such as creating pathways for the introduction of non-native species. Thus, the economic importance of commercial navigation, the use of water by this sector, and the impacts of commercial navigation on other uses are key concerns in management of the nation's water resources. The discussion that follows addresses these issues:

- The role of the commercial navigation sector in transportation and shipping nationwide
- The economic importance of commercial navigation
- The use of water by this sector, including infrastructure requirements to maintain channel depths
- Available estimates of the value of water used for navigation

Commercial navigation encompasses the movement of cargo and passengers by water. It is part of the tertiary (delivery) mega-sector described earlier and is particularly vital to industries that rely on the bulk shipment of goods. The following discussion provides an overview of this sector, drawing on data from the U.S. Army Corps of Engineers and from the Department of Transportation's Maritime Administration and Bureau of Transportation Statistics.

CARGO SHIPPING

Cargo is shipped to, from, and throughout the United States by ship, rail, truck, pipeline, and airplane. According to Bureau of Transportation statistics, domestic shipments of cargo in the United States in 2007 totaled more than 4.6 trillion ton-miles (USDOT, 2011a). Table 10.24 shows the distribution of this shipping by mode, along with similar data for the three previous years. The table

DID YOU KNOW?

In 2007, commercial navigation accounted for 78% of international trade by weight and 45% of trade by value. Within the United States, commercial navigation handles approximately 10 to 15% of cargo shipments by ton-mile, including large volumes of crude oil and petroleum products, coal, chemicals, sand, grave, stone, food and farm prices, iron ore and scrap, manufactured goods, and other commodities.

TABLE 10.24
Distribution of Freight Shipping in the United States by Mode (2004–2007)

Transportation Mode	Ton-Miles of Freight (Millions)			
	2004	2005	2006	2007
Air	16,451	15,745	15,361	15,142
Truck	1,281,367	1,291,308	1,291,244	1,317,061
Railroad	1,684,407	1,733,329	1,855,902	1,819,633
Domestic water	621,170	591,276	561,629	533,143
Pipeline	937,442	938,659	906,656	904,079
Total	4,540,837	4,570,316	4,630,792	4,609,079

Source: USDOT, *Bureau of Transportation Statistics*, U.S. Department of Transportation, Washington, DC, 2011.

shows that shipping on domestic waterways accounted for between 10 and 15% of total freight ton-miles during this period, less than that reported for shipping by rail, truck, and pipeline, but more than that reported for shipping by air. As Table 10.24 shows, overall shipments of freight remained relatively constant during this 4-year period. Shipping by water, however, showed an 11% decline, from 621 billion ton-miles in 2004 to 553 billion in 2007.

Geographic Distribution

In 2009, U.S. waterborne shipments of freight totaled 2210.8 million tons, with 1353.7 million tons (61%) representing the shipment of imports and exports by sea. Domestic shipments accounted for the remaining 857.1 million tons (39%) (USACE, 2010). Internal riverways supported most of the domestic traffic, accounting for nearly 61% of domestic tonnage. Coastal shipping (20%), shipping on the Great Lakes (7%), and intra-port shipping (12%) accounted for smaller shares of the domestic total.* Water transportation is limited in its points of origin and destination by the availability of ports equipped to handle the loading and unloading of cargo. U.S. coastal ports are essential to facilitating both overseas and domestic trade. Port calls along the Gulf of Mexico accounted for 36% of the U.S. total, followed by port calls to the South Atlantic (20%), North Atlantic (16%), Pacific Southwest (15%), and Pacific Northwest (11%), as well as to Puerto Rico (2%).

Types of Cargo

Table 10.25 shows the distribution of domestic cargo shipments in 2009 by commodity and waterway. Energy-related commodities, such as coal and petroleum-related products, make up the largest share of shipments by water, over 62% by weight. In addition, a significant amount of iron ore, a key input in the manufacture of steel, is shipped on the Great Lakes. According to the 2007 U.S. Economic Census, 182 of the nation's 352 iron and steel mills are located in the Great Lakes region; thus, the lakes are an important waterway for this industry (U.S. Census Bureau, 2007a). As the exhibit shows, commercial shipping also plays a vital role in the transport of many other commodities, including chemicals, manufactured goods, stone, and agricultural products.

* Note that, for the purposes of this book, internal riverways include all flowing bodies of freshwater that feed into a lake or ocean. Coastal shipping refers to shipping between ports along the Atlantic or Pacific coasts or on the Gulf of Mexico. Shipping on the Great Lakes includes traffic on the lakes themselves and vital connection canals, channels, and locks. Finally, intra-port transport refers to the shipment of cargo by vessel from one point within a port to another, usually for the purpose of storage.

TABLE 10.25
U.S. Domestic Waterborne Traffic by Major Commodity (2009)

	Million of Tons Shipped				
Commodity	**Coastal**	**Lakes**	**Internal Riverways**	**Other**[a]	**Total**
Coal	9.2	18.8	158.5	20.3	206.8
Coal coke	0	0.4	3.3	0.2	3.9
Crude petroleum	35.2	0	28.0	1.0	64.2
Petroleum products	88.6	0.6	117.8	48.5	255.5
Chemical and related products	9.4	0	42.5	10.3	62.2
Forest products, wood, and chips	1.1	0	3.5	0.4	5.0
Sand, gravel, and stone	6.9	16.2	49.3	17.0	89.4
Iron ore and scrap	0.2	22.4	6.0	1.7	30.3
Non-ferrous ores and scrap	0.6	0	4.9	0	5.5
Sulfur, clay, and salt	0	1.0	8.7	0.3	10.0
Primary manufactured goods	5.2	3.0	15.0	1.0	24.2
Food and farm products	4.7	0.3	75.0	1.2	81.2
All manufactured equipment	6.5	0	6.8	0.8	14.1
Waste and scrap	0	0	1.0	0.8	1.8
Total	167.6	62.7	520.3	103.5	854.1

Source: USACE, *The U.S. Waterway System*, U.S. Army Corps of Engineers, Alexandria, VA, 2010.

[a] Other includes intra-port and intra-territory traffic.

Passenger Transportation

In addition to the movement of cargo, the nation's waterways are also used to transport passengers. The two main categories of passenger transportation relying on commercial navigation are cruises and ferries.

Cruises

Cruises, by design, launch from and return to the same port; they are meant as a form of recreation, not a mode of transportation. A cruise ship may make stops at various ports before returning to its point of origin, or it may not make any. The Department of Transportation's Maritime Administration reports that 4208 cruises carrying 10.6 million passengers made at least one stop in the United States in 2010 (USDOT, 2011b).

Ferries

Ferries help to connect island communities to the mainland but also, in some cases, provide faster, more direct routes than roads. In many communities west of Seattle, for example, it is faster to ferry across Puget Sound than to access the nearest bridge by road. The Washington State Ferry System is the largest ferry system in the country with respect to number of both passengers and vehicles transported (WSDOT, 2011). This system transports over 22 million riders and 10 million vehicles annually, linking growing residential communities with nearby urban economic centers. For comparison, the nation's second largest ferry system, which primarily connects mainland North Carolina to the Outer Banks, services only 2.5 million passengers and 1.3 million vehicles each year (NCDOT, 2014). Other regions that rely to a significant extent on ferry transportation include the eastern end of Long Island, where ferry services provide access to southern New England, and the islands of Martha's Vineyard and Nantucket off the coast of Massachusetts. Even though ferry services exist only in regions with specific needs, their use has increased over the last two decades. Although current ferry use is less than 0.01% of total passenger miles by all modes of transportation, it provides a service that, in most cases, would be impractical for other modes to provide.

Economic Importance

A large number of industries rely on commercial navigation directly or indirectly, making it a vital sector of the economy. In addition to the economic activity directly related to commercial navigation, the sector also drives economic activity in supporting industries, such as shipbuilding and repair. This section discusses the economic importance of commercial navigation; it presents the total value of goods shipped by water, compares waterborne shipping to shipping by other modes, and discusses employment and wages in commercial navigation and related industries.

Value of Goods Shipped

The total value of all U.S. cargo freight in 2007, regardless of mode, was $11.7 trillion, including international trade (USDOT, 2011a). When compared to the U.S. gross domestic product (GDP), which totaled $14 trillion in 2007, the importance of cargo shipping is immediately apparent (World Bank Group, 2012). Water transport accounted for 77.7% of international shipments by weight in 2007 but only 44.9% of shipments by value, illustrating the competitive advantage that water transport offers in moving large quantities of lower value goods over long distances. Conversely, air transport accounts for 0.4% of shipments by weight but 25.1% of shipments by value. This demonstrates that even within the shipping sector not all modes of transport compete for the same business. In general, different modes of shipping are not perfect substitutes for one another. They will compete, however, when circumstances and available infrastructure allow them to move cargo between two points at comparable costs.

Comparison of Waterborne Shipping to Alternatives

Most cargo uses multiple modes of transport to arrive at its final destination, with each mode providing different services. For example, cargo that can be brought to a U.S. port by ship safely and inexpensively might then require rail or truck transport, or both, to reach its final inland destination. Different modes of transportation can either be complements or substitutes, depending on the particular requirements and destination of the cargo. A large volume of waterborne cargo consists of bulk commodities that are shipped long distance where speed is not a high priority. Shipping freight by water offers a number of advantages over alternative methods of transport. As Table 10.26 shows, shipping by inland water is the most fuel-efficient mode, as measured by gallons of fuel used per ton-mile; consequently, it releases the smallest amount of greenhouse gases per ton-mile. Waterborne shipping is also the safest mode with respect to the number of injuries per ton-mile and

TABLE 10.26
Comparison of Shipping Methods

Shipping Method	Gallons of Fuel Used per Million Ton-Miles	Tons of Greenhouse Gases per Million Ton-Miles	Gallons of Oil Spilled per Million Ton-Miles	Injuries per Billion Ton-Miles
Truck freight	6452	71.6	6.06	99.0
Railroads	2421	26.9	3.86	5.8
Inland marine	1736	19.3	3.60	0.045

Source: Center for Ports & Waterways, *A Modal Comparison of Domestic Freight Transportation Effects on the General Public*, Texas Transportation Institute, Houston, TX, 2007.

the number of gallons of oil spilled per ton-mile (Center for Ports & Waterways, 2007). Waterborne shipping, however, also suffers a number of disadvantages, the most obvious of which is its inability to deliver cargo where navigation is not feasible. Additionally, water transport tends to be much slower than transport by truck, rail, or air, making it undesirable for perishable or time-sensitive shipments. In addition, waterborne shipping requires the development and maintenance of port facilities to dock and load/unload shipments (Young, 2005).

When comparing shipping by inland waterway to shipping by truck or rail it is important to note that the natural course of rivers can force ships to take a more circuitous route to their destination than would shipments by other means. This can lengthen a trip and reduce the competitiveness of shipping by water. In some instances, however, the situation is reversed. This is the case with the Great Lakes, where a ship may be able to travel a direct route between ports, while a train or truck may have to travel a greater distance to circumvent a large body of water. Comparisons of miles traveled should, therefore, be considered carefully.

Shipbuilding and Repair

Though not strictly part of the commercial navigation sector, shipbuilding and repair are a closely linked industry. Demand for water transportation services increase demand for ship construction and maintenance. According to the DOT Maritime Administration, capital investments in the industry totaled $270 million in 2006. The passage of the Oil Pollution Act of 1990 has been a major driver for growth in the shipbuilding industry. The act requires the phase-in of double-hull vessels through 2015 to reduce the risk of an oil spill in the event of a collision or some other accident. By the time the phase-in is complete, almost $5 billion will have been spent on construction to comply with this requirement. In addition to serving commercial navigation, the shipbuilding and repair industry is a major contractor for the U.S. Navy. In 1998, 70% of industry revenues came from the military (USDOC, 2001). These revenues are obviously critical to the industry's long-term sustainability.

Employment in Commercial Navigation and Related Industries

The Bureau of Labor Statistics classifies commercial navigation under six different North American Industry Classification System (NAICS) codes based on the type of shipping (freight or passenger) and the type of waterway (deep sea, coastal and Great Lakes, and inland). Table 10.27 summarizes total employment and wages in these sectors in 2010, as well as the number of establishments, both private and government, operating in each industry. The table includes similar data for ship and boat building and water transportation support activities. As the table shows, employment in the ship and boat building industry or in support activities of waterborne transportation is significantly greater than direct employment in commercial navigation. Within the commercial navigation sector itself, the transport of freight accounts for a greater share of employment and wages than does passenger transportation.

Water Use by Commercial Navigation

Commercial navigation is an in-stream, non-consumptive use of water. The importance of water to this sector is primarily related to its depth at important junctures—namely, ports, rivers, locks, and channels. The U.S. Army Corps of Engineers (USACE) regularly dredges these areas to maintain a minimum navigable depth. The USACE is also responsible for the construction and maintenance of locks, which allow ships to travel on waterways that might not otherwise be navigable.

Infrastructure Requirements of Commercial Navigation

The USACE was originally charged to clear, deepen, and otherwise improve and maintain selected waterways by the General Survey Act of 1824. Since that time, its mission has expanded to include the creation of canals to expand transportation routes and link previously unconnected bodies of

TABLE 10.27
Employment in Commercial Navigation and Related Industries (2010)

Category	Subcategory	Establishments	Total Employment	Total Wages ($ Millions)
Freight	Deep sea	500	11,616	$1160
	Coastal and Great Lakes	312	10,355	$864
	Inland water	548	20,998	$1466
Total freight		1360	42,969	$3491
Passenger	Deep sea	123	8375	$510
	Coastal and Great Lakes	187	7064	$357
	Inland water	238	4115	$202
Total passenger		548	19,554	$1069
Other	Support activities for water transportation[a]	2828	93,557	$6004
	Ship and boat building	1826	151,837	$14,946
Total other		4654	245,394	$14,950
Total		6562	307,917	$19,510

Source: BLS, *Quarterly Census of Employment and Wages*, Bureau of Labor Statistics, Washington, DC, 2014.

[a] Support activities for water transportation include port and harbor operations, marine cargo handling, navigation services to shipping and other support activities for water transportation.

water. Today, the USACE oversees and provides maintenance nationally for 12,000 miles of inland and intracoastal waterways, as well as 13,000 miles of coastal waters and navigable channels greater than 14 feet deep, including nearly 200 locks and dams. Its jurisdiction in this area reaches 40 states.

Dredging

Sediment, such as silt, sand, or gravel, is picked up and carried by currents of the faster flowing segments of a river and deposited where the current slows. Over time, these deposits build up and can be a hazard to passing ships. Dredging, the removal of these buildups, is essential to maintaining access to water bodies and ports. The depth of water at the shallowest point determines how much cargo a vessel can safely carry without grounding. According to the Lake Carriers' Association, the Great Lakes fleet gives up 200,000 tons of cargo for each foot of draft lost (*draft* refers to the vertical distance from the bottom of a ship's hull to the waterline) (USACE, 2009). According to the Navigation Data Center at the USACE, 263.6 million cubic yards of total material were dredged nationally at a cost of $1.3 billion in the 2009 fiscal year (USACE, 2010). One area where this service is needed is the Port of New York and New Jersey, where navigation and commerce generate about $20 billion annually in direct and indirect benefits (USACE, 2011b). Each year, USACE

DID YOU KNOW?

The commercial navigation sector relies on maintenance of adequate water depths at ports, locks, and channels. At the federal level, the Army Corps of Engineers has primary responsibility for maintaining commercial shipping channels. It spent $1.3 billion on dredging in fiscal 2009, but at current appropriation levels it is unable to dredge all waterways and ports in need of maintenance.

maintenance dredging removes between 1 and 2 million cubic yards of sediment from New York Harbor, which is comprised of about 24 separate channels. Additional dredging will be required in the future to deepen some channels, allowing larger vessels access to the harbor.

Locks and Dams

A lock is an area on a waterway, or connecting two waterways, that has the ability to raise or lower boats to allow passage between bodies of water at different levels. Dams allow for a degree of control over river flows so that depth can be increased during periods that would otherwise experience low flows. Both are vitally important to navigation in rivers and canals that connect major shipping routes and link the Great Lakes to each other and to rivers that travel further inland. Critical lock sites are among the areas with failing infrastructure. Locks are essential to shipping on the Great Lakes because they allow vessels to transit otherwise impassable stretches separating the lakes from each other. The Soo Locks are a set of parallel locks along the Saint Mary's River that connect Lake Superior to the lower Great Lakes by allowing ships to safely avoid rapids and a 21-foot drop. Only one of these locks, the Poe Lock, is large enough to accommodate all vessels operating on the Great Lakes (USACE, 2009). Closure of the Poe Lock would create a barrier to 70% of the commercial cargo capacity that currently utilizes the waterway. Estimates put the cost to industry of an unplanned 30-day shutdown of the Soo Locks at $160 million. Another example illustrating the importance of lock maintenance is the Upper Mississippi River Basin, comprised of the Upper Mississippi River, Illinois Waterway, and Missouri River system. The Waterways Council has estimated that waterborne shipping in this basin saves consumers approximately $1 billion in annual transportation costs (Waterways Council, 2007). Most of the system's 38 locks, however, are 600 feet long, half the length of an average barge tow (an average barge tow is typically 15 barges pushed by a towboat). Using these locks requires splitting a tow into segments and bringing each segment through the locks separately, which causes delays and backups. Additionally, many of the locks were constructed over 70 years ago and, while still operable, have begun to experience elevated failure rates. These malfunctions translate into longer delivery times and increased shipping costs throughout the region.

Potential Effects of Climate Change on Navigation in the Great Lakes

The effects of climate change are relevant to any discussion about the future availability of water resources. With regard to navigation, its impact will be most apparent in inland bodies of water, such as the Great Lakes region. The discussion below addresses the two greatest potential impacts on navigation in this region: decreased water levels and reduced ice coverage.

Decreased Water Levels

Unlike the oceans, the Great Lakes are expected to experience decreased water levels as a result of climate change (Quinn, 2002). While most of the area of the lakes will continue to have more than sufficient depth for navigation, there are critical points where reduced depths will have a significant impact on vessels—namely, locks, harbors, and channels. Depending on vessel size, the loss of an inch of draft on the Great Lakes can translate to lost cargo capacity of from 100 to 270 tons per trip. Table 10.28 shows the predicted reduction in the mean base level of each lake at various points in the future. Even as early as 2030, these estimates suggest potential problems in maintaining sufficient depths without adversely affecting shipping. As vessels are forced to carry less cargo per trip, traffic will increase to accommodate demand, increasing the likelihood of backups at locks and ports.

Reduced Ice Coverage

Another attribute of the Great Lakes that will be affected by climate change is seasonal ice coverage. Depending on annual temperature variability, the Great Lakes become unnavigable for 11 to 16 weeks each winter. Industries dependent on a year-round supply of certain commodities, such as coal-fueled power plants, must stockpile goods that cannot be delivered during this period. To

TABLE 10.28
Potential Impact of Climate Change on Water Levels in the Great Lakes

Lake	Δ 2030 (ft)[a]	Δ 2050 (ft)	Δ 2090 (ft)
Superior	–0.72	–1.02	–1.38
Michigan–Huron	–2.36	–3.31	–4.52
Erie	–1.97	–2.72	–3.71
Ontario	–1.15	–1.74	–3.25

Source: USEPA, *The Importance of Water to the U.S. Economy: EPA's Background Report*, U.S. Environmental Protection Agency, Washington, DC, 2012.

[a] Changes in water levels are calculated relative to recent means.

extend the shipping season as much as possible, the U.S. and Canadian Coast Guard jointly provide ice-breaking services. Reduced winter ice from climate change could extend the shipping season by 1 to 3 months (Quinn, 2002). This can have a twofold economic benefit. First, it will reduce the cost of warehousing commodities while shipping is unavailable, creating a steadier flow of cargo. Second, it will reduce the need for the Coast Guard to provide ice-breaking services to maintain navigable channels. This will help offset some of the increased costs associated with lower lake levels.

Influence on Other Uses of Water

Because commercial navigation relies on water simply as the medium by which ships travel, it is generally unaffected by the potential impacts of other uses of water on water quality; however, commercial navigation can affect other water uses in several ways:

- *Impacts on benthic habitat and water quality*—As noted above, the maintenance of shipping channels demands regular dredging, which can have an adverse impact on benthic habitat and increase turbidity in the water column. Similarly, the disposal of dredged material requires careful management to avoid or reduce environmental impacts. These issues become particularly critical if the dredged materials contain heavy metals, PCBs, or other potential contaminants.
- *Impacts on water supply and fish habitat*—The use of locks and dams to maintain instream flows can compete with other demands for water, such as the use of water for irrigation. Dams can also have a variety of effects on fish habitat, such as reducing the amount of dissolved oxygen in the water, increasing its temperature, and creating obstacles to the migration of anadromous species.
- *Introduction of invasive species*—The development of canals and seaways to facilitate shipping can also create pathways for the introduction of non-native species that can have a variety of ecological and economic impacts. The development of the Welland Canal, for example, led to the introduction of the sea lamprey in the Great Lakes above Niagara Falls, contributing to the decline of native species important to both commercial and recreational fishing. More recently, the discharge of ballast water from a trans-Atlantic freighter on the Great Lakes provided a vector for the introduction of the zebra mussel, an invasive species that has altered the ecology of the lakes and forced water users in both the public and private sectors to retool their systems to prevent the mussels from clogging water intake pipes (USGS, 2013).

Value of Water Use by Commercial Navigation

Because commercial navigation is an in-stream, non-consumptive use of water, it is difficult to estimate the value of water used for this purpose. Companies that operate barges on waterways in the United States do not pay any fees for their use of water for navigation, as there is no functioning market that could allow one to infer that value of water. Even if there were markets for water used in navigation, the fact that this use is not strictly "rivalrous" (i.e., water used for navigation can be used again downstream for other purposes) would suggest that its true value would be underestimated by markets. Several other factors complicated any effort to estimate the value of water for navigation:

1. Commercial navigation generally requires that water levels remain within a certain range. Too little water means that channel width and/or depth are inadequate for vessel traffic, and too much water can interfere with loading and unloading of cargo. As a result, the marginal value of water for navigation is generally zero, unless the increment in question is the specific amount that determines whether or not a waterway is navigable for vessels of a particular kind.
2. Seasonal variation in river flows affects the baseline navigability of waterways, so the value of additional water for navigation may be negative during springtime high-flow periods and positive during summertime low-flow periods.
3. Comparing the value of cargo shipping by barge to cargo shipping by alternative means is made difficult by the fact that alternative modes of shipping are not directly comparable. Shipping by truck or by air, although relatively expensive, is faster and therefore more appropriate for time-sensitive cargo. And, although rail shipping is more closely comparable to barge shipping, railway pricing differs by route, so railway companies may charge less for routes that compete directly with barge shipping, or employ seasonal price discrimination by charging more for routes during seasons when competing waterways are not navigable (Young, 2005).

Because of these difficulties, a few studies have attempted to estimate the value of water in support of commercial navigation. A 1986 study by Resources for the Future estimated the average value of water used in commercial navigation by comparing the costs of barge transportation to the costs of rail transportation. Using Army Corp of Engineers data on six river systems, the study estimated the cost savings of barge transport vs. rail transport, subtracted the operation and maintenance costs for each waterway, and divided the remaining savings by the amount of water required for each river to support barge traffic, yielding estimates of the average annual values per acre-foot of water used for commercial navigation (Gibbons, 1986). Table 10.29 presents these values, inflated to 2010 dollars. This valuation method assumes that the difference between the cost of rail transport and the cost of barge transport on these rivers is entirely attributable to the value provided by the water that allows the rivers to remain accessible to commercial shipping. As Table 10.29 shows, the resulting estimates of the value per acre-foot of water, which vary from less than $1 per acre-foot on the Missouri River to over $670 per acre-foot on the Ohio River, depend in large part on the flow required to maintain navigation in each river, which is a function of the river's physical characteristics. For example, although commercial navigation on the Mississippi is estimated to provide the greatest annual savings relative to rail transportation ($1.8 billion per year), it has the third lowest estimated value per acre-foot because of the large volume of water needed to maintain navigation (over 131 million acre-feet per year).

IN-STREAM WATER USE BY RECREATION AND TOURISM

Water is a vital resource for the recreation and tourism sector. Water-based activities such as fishing, boating, and swimming rely upon water resources to create recreational opportunities, and recreational pursuits such as hiking, hunting, and wildlife viewing can be enhanced by proximity

TABLE 10.29
Estimates of the Value of Water Used for Commercial Navigation

Waterway	Annual Cost Savings Attributable to Commercial Navigation (Thousand 2010$)	Water Requirement (Thousand Acre-Feet per Year)	Value of Water per Acre-Foot (2010$)
Ohio	$406,000	605	$671.0
Tennessee	$52,000	412	$126.0
Illinois	$70,000	120	$583.0
Mississippi	$1,819,000	131,040	$14.0
Missouri	$8000	23,968	$0.3
Columbia/Snake	$50,000	7168	$7.0

Source: Gibbons, D.C, *The Economic Value of Water*, Resources for the Future, Washington, DC, 1986.

to water. This section analyzes the role of water in recreation and tourism, focusing in particular on how the characteristics of a water resource affect people's willingness to pay for recreational activities, which in turn affects consumption of market goods. The section discusses the following topics:

- The relationship between participation in water-based recreation and market expenditures in the recreation and tourism sector
- Economic data related to water-based recreation, including participation and expenditure data
- Issues that currently affect, or in the future may affect, the ability of the nation's waters to support recreational activity
- Available estimates of the economic value of participating in water-based recreational activities, as well as the potential impacts of changes in in-stream conditions on these values

The recreation and tourism sector is unique in that a wide range of recreational activities are not typically priced in conventional markets. Access to activities such as swimming and wildlife viewing is often free in public recreation areas, and other activities such as fishing and hunting can frequently be pursued for nominal license fees. Thus, although some demand for recreational activities may be explicitly reflected in market transactions, the information provided by the direct purchase of recreational opportunities is incomplete. Demand for recreational activity, however, can be indirectly reflected in market transactions for complementary goods and services (e.g., expenditures on transportation, food, lodging, and recreational equipment). These expenditures, along with the GDP and employment impacts associated with them, are at least in part attributable to demand to participate in recreational activities.

DID YOU KNOW?

Historically, water regimes gave greater priority to off-stream water uses, such as irrigation or municipal supply, than to the preservation of in-stream flow or water levels for recreational purpose. In recent years, however, states have begun to enact legislations designed to preserve flows or levels that support recreational activities, as well as other in-stream uses.

Travel and Tourism Industry

Although the full extent of demand for water-based recreational activities such as swimming and fishing is not explicitly reflected in market transactions, the costs that people incur to pursue these recreation activities (e.g., hotel accommodations, transportation costs, equipment expenditures) are reflected in national income accounts. In this manner, recreational demand contributes to market activity in the travel and tourism industry, elements of which span the tertiary and quaternary mega-sectors of the economy. Neither the U.S. Economic Census nor the Bureau of Labor Statistics provides data exclusively for the tourism industry. Thus, to develop a basic economic profile, it is necessary to rely on the Bureau of Economic Analysis' (BEA) Travel and Tourism Satellite Accounts. The BEA Satellite Account data reveal that the travel and tourism sector accounted for $379 billion in value added to the economy in 2009, which translated to approximately 2.68% of U.S. gross domestic production. The real direct output of the travel and tourism industry, as measured by goods and services sold directly to visitors, increased 3.1% in 2010 to a total of $650.9 billion (2005 dollars). This represented a reversal in recent trends in the travel and tourism industry, which had declined by 9.3% in 2009 and 4.4% in 2008 (Zemanek, 2011). Tables 10.30 and 10.31 provide highlights of the real output from the travel and tourism sector over the last 5 years; the goods and services highlighted in the exhibit are not intended to be comprehensive but are shown as examples

TABLE 10.30
Annual Real Output of Travel and Tourism Industry (Millions of 2005$)

Commodity	2006	2007	2008	2009	2010
All tourism goods and services	712,684	728,563	696, 417	631,366	650,898
Traveler accommodations	128,211	134,915	136,922	122,717	130,084
Food and beverage services	116,309	118,200	110,637	96,272	95,563
Passenger air transportation	109,834	112,377	108,535	101,092	110,830
Passenger water transportation	11,272	12,044	12,717	12,317	12,283
Highway tolls	608	580	532	562	516
Gasoline	59,420	59,746	53,017	48,942	48,498
All other recreation and entertainment	17,361	17,550	16,842	14,733	15,106

Source: Zemanek, S., U.S. travel and tourism satellite accounts for 2007–2010, *Survey of Current Business*, June, 2011.

TABLE 10.31
Annual Growth in Real Output of Travel and Tourism Industry

Commodity	2006	2007	2008	2009	2010
All tourism goods and services	2.9%	2.2%	(4.4%)	(9.3%)	3.1%
Traveler accommodations	3.5%	5.2%	1.5%	(10.4%)	6.0%
Food and beverage services	3.0%	1.6%	(6.4%)	(13.0%)	0.3%
Passenger air transportation	1.7%	2.3%	(3.4%)	(6.3%)	9.0%
Passenger water transportation	8.2%	6.9%	5.6%	(3.1%)	(0.3%)
Highway tolls	(11.2%)	(4.6%)	(8.3%)	5.6%	(8.1%)
Gasoline	2.8%	0.5%	(11.3%)	(7.7%)	(0.9%)
All other recreation and entertainment	(1.3%)	1.1%	(4.0%)	(12.5%)	2.5%

Source: Zemanek, S., U.S. travel and tourism satellite accounts for 2007–2010, *Survey of Current Business*, June, 2011.

of areas in which demand for recreational activities such as beach visits or fishing trips could drive industry output. The BEA Satellite Accounts show that direct employment in the tourism industry decreased 0.45% in 2010, to 5,382,000 jobs. This rate of loss was much lower than experienced in 2009 (–8.14%) or 2008 (–3.45%). Nonetheless, direct employment in the tourism industry remained well below the 2007 peak of 6,096,000 jobs (Zemanek, 2011).

When considering the role of the travel and tourism industry in the U.S. economy, it is important to consider how tourism expenditures affect other economic sectors. The BEA estimated that each dollar of U.S. tourism output stimulated $0.69 in nominal output in related economic sectors; thus, the $746.2 billion in direct nominal output for tourism in 2010 stimulated $514.9 billion in additional economic activity, Further, for every 100 direct tourism jobs generated, 41 jobs are indirectly generated in related sectors (Zemanek, 2011). The overall data for the travel and tourism industry are not solely reflective of demand for water-based recreation and tourism; however, costs that recreational participants incur in order to realize demand for water-based recreational activities contribute to overall output for travel and tourism.

Water-Based Recreation: Participation and Expenditures

Beach Recreation

The National Ocean Economics Program estimated that tourism and recreation accounted for 1,737,156 jobs and contributed $69.65 billion in GDP to the economy of coastal regions of the United States in 2004. The majority of this economic output came from the food and accommodations sectors, which combined to account for 92% of sector employment and 85% of sector GDP (Kildow et al., 2009). This economic output is driven in part by demand for ocean-based recreation in beach settings. The 2000 National Survey on Recreation and the Environment (NSRE) provided data on beach visitation by state. The NSRE data on participation rates indicated the percentage of the U.S. population over the age of 16 that participated in recreational activities or visited recreational settings over the course of the year. Table 10.32 summarizes this information for the 10 states that reported the highest rates of beach visitation.

TABLE 10.32
Beach Visitation by State (2000)

State	Percent of U.S. Adults Visiting a Beach in This Location	Number of Participants (Millions)	Number of Days (Millions)
Florida	7.39	15.246	177.153
California	6.11	12.598	151.429
South Carolina	2.15	4.434	33.302
New Jersey	1.92	3.965	40.881
Texas	1.87	3.851	35.239
Hawaii	1.75	3.598	101.149
North Carolina	1.55	3.185	27.936
New York	1.44	2.964	29.225
Massachusetts	1.35	2.779	28.681
Maryland	1.23	2.530	18.696
U.S. total	30.03	61.922	853.288

Source: Leeworthy, V.R. and Wiley, P., *Current Participation Patterns in Marine Recreation*, National Oceanic and Atmospheric Administration, Silver Spring, MD, 2001.

TABLE 10.33
Coastal Recreational Participation by Activity (2000)

Activity	Participation Rate (Percent of U.S. Adults)	Number of Participants
Visit beaches	30.03	61,922,234
Swimming	25.53	52,637,390
Snorkeling	5.07	10,459,568
Scuba diving	1.35	2,786,215
Surfing	1.59	3,285,611
Wind surfing	0.39	800,016
Any coastal activity	43.30	89,270,965

Source: Leeworthy, V.R., *Preliminary Estimates from Versions 1–6: Coastal Recreation Participation*, National Oceanic and Atmospheric Administration, Silver Spring, MD, 2001.

As the leading travel destination for tourists, beaches are a key contributor to the economic output of the U.S. travel and tourism industry (Houston, 2008). According to the 2000 NSRE, beach visits were the number one recreational pursuit of participants in coastal recreation. The survey reported that 61.9 million Americans, or 30% of Americans ages 16 or older, visited a beach in 1999 (Leeworthy, 2001). Popular recreation activities pursued in conjunction with beach visits include swimming, snorkeling, scuba diving, surfing, and wind surfing. Table 10.33 provides an overview of participation in these activities. Note that boating and fishing are also popular pursuits that may be associated with a visit to a beach. These activities will be explored in more detail in later sections.

The pursuit of these and other coastal recreation activities drives economic output in the market economy, particularly in the travel and tourism sector. This is highlighted by the fact that in 2006 coastal states accounted for approximately 85% of U.S. tourism revenues (Houston, 2008). Although there is no national database for economic output related to beach recreation, there have been several case studies that have analyzed expenditures (e.g., parking, lodging, rental equipment) associated with beach visits. For example, studies analyzing beaches in Southern California have estimated that beach trip expenditures ranged from $20.33 per person-day for day trips (Wiley et al., 2006) to $170 per person-day for overnight trips (Department of Boating and Waterways and State Coastal Conservancy, 2002). Such expenditures in turn contribute to economic output and employment in the tourism industry.

Fishing

Recreational fishing is one of the most popular outdoor recreation activities in the United States. In 2006, according to the U.S. Fish and Wildlife Service's National Survey of Fishing, Hunting, and Wildlife-Associated Recreation (USFWS, 2006a), 30.0 million Americans ages 16 and older participated in recreational fishing in the United States. Freshwater fishing accounted for the majority of this fishing effort, with 25.4 million participants. In the same period, saltwater fishing attracted 7.7 million anglers. (Note that some individuals participate in both freshwater and saltwater fishing, creating an overlap in participant estimates). The U.S. Fish and Wildlife Service found that, collectively, these 30.0 million anglers accounted for 516.8 million angler days and 403.5 million fishing trips over the course of 2006, which translated to $42.0 billion in recreational fishing-related expenditures. Table 10.34 displays a breakdown of these expenditures, including trip-related expenses, equipment purchases, and other miscellaneous expenditures.

Recreational fishing is especially important in that it is considered a "gateway" recreation activity. A 2008 joint report by the Recreational Boating and Fishing Foundation (RBFF) and the Outdoor Foundation (OF), based on a national survey of recreation participants, found that over 77% of anglers participate in additional outdoor recreational activities (RB&FF and OF, 2009). Fishing is

TABLE 10.34
Recreational Fishing Expenditures (2006)

Expenditure Category	Amount ($ Billion)
Total trip-related	17.9
Food and lodging	6.3
Transportation	5.0
Other trip costs	6.6
Total equipment expenditures	18.8
Fishing equipment	5.3
Auxiliary equipment	0.8
Special equipment	12.6
Total other fishing expenditures	5.4
Magazines, books	0.1
Membership due and contributions	0.2
Land leasing and ownership	4.6
Licenses, stamps, tags, and permits	0.5
Total fishing expenditures	42.0

Source: USFWS, *2006 National Survey of Fishing, Hunting, and Wildlife-Associated Recreation*, U.S. Fish and Wildlife Service, Washington, DC, 2006.

particularly significant in driving demand for boating, as the survey found that 33% of anglers own a boat and 67% of anglers went boating in 2008. Fishing is therefore important not only for fishing-related economic impacts but also for its contribution to participation in other recreational activities.

Boating

Recreational boating encompasses a broad range of activities, including float-based recreation (e.g., kayak and canoe trips), non-motorized boating (e.g., sailing), and motorized boating (e.g., power boats). The U.S. Forest Service estimated in 2009 that approximately 89.1 million Americans, or 35.6% of the population, participate in some form of recreational boating (Cordell et al., 2009). Table 10.35 displays a breakdown of recreational boating activity based on data from the 2000 NSRE. As mentioned earlier, fishing activity in the United States is a primary driver for participation in boating: 25.8 million anglers, or 67% of all anglers in the RB&FF and OF (2009) survey, participated in 427 million boating days in 2008. This correlation between fishing and boating activity implies that any restrictions to

TABLE 10.35
Recreational Boating Participation (2000)

Activity	Participation Rate (Percent of U.S. Adults)	Number of Participants
Motor boating	24.79	51,113,437
Sailing	5.07	10,445,548
Personal watercraft use	9.42	19,423,722
Canoeing	9.71	20,027,169
Kayaking	3.26	6,723,240
Rowing	4.48	9,234,883
Water-skiing	8.05	16,604,129

Source: Leeworthy, V.R., *Preliminary Estimates from Versions 1–6: Coastal Recreation Participation*, National Oceanic and Atmospheric Administration, Silver Spring, MD, 2001.

TABLE 10.36
Wildlife and Nature Viewing by Setting (2000)

	Water-Based Settings		All Natural Settings	
Activity	**Participation Rate (Percent of U.S. Adults)**	**Number of Participants**	**Participation Rate (Percent of U.S. Adults)**	**Number of Participants**
Bird watching	30.2	62,200,000	31.8	67,800,000
Viewing other wildlife	22.4	46,200,000	44.1	93,900,000
Viewing or photographing scenery	37.0	76,300,000	59.5	126,800,000

Sources: Leeworthy, V.R., *Preliminary Estimates from Versions 1–6: Coastal Recreation Participation*, National Oceanic and Atmospheric Administration, Silver Spring, MD, 2001; NSRE, *American's Participation in Outdoor Recreation: Results from NSRE (with Weighted Data) (Versions 1 to 13)*, U.S. Department of Agriculture, Washington, DC, 2001.

fishing activity, whether due to poor water quality or other concerns, could negatively affect boating as well. According to data collected by the National Marine Manufacturers Association (NMMA), the recreational boating industry reported $30.8 billion in sales of goods and services in 2008, including over $21 billion in trip expenditures. In 2007, recreational boating expenditures helped support 18,940 boating businesses that employed over 154,300 people (quoted in Haas, 2010).

Wildlife and Nature Viewing

According to the National Survey of Fishing, Hunting, and Wildlife-Associated Recreation (USFWS, 2006a), 71.1 million Americans, or 31% of the U.S. population ages 16 and older, participated in wildlife and nature viewing in 2006. Of these 71.1 million participants, 23.0 million engaged in trips away from home for wildlife viewing purposes. To the extent that wildlife and nature viewing occurs in environments near water resources, water attributes that can influence both the abundance of wildlife and the aesthetic quality of the environment can affect such activity. The 2000 NSRE analyzed recreational viewing activity in all natural settings and in water-based environments. Table 10.36 summarizes this information, providing participation data for wildlife and nature viewing across the United States. As Table 10.36 indicates, a significant share of those who participate in wildlife and nature viewing do so in water-based settings. Data on the market impacts of wildlife and nature viewing in water-based surroundings are not available; however, the 2006 USFWS survey provides expenditure data for all wildlife and nature viewing activity in the United States. The survey results indicate that wildlife and nature viewing expenditures for 2005 totaled $45.7 billion, including $12.9 billion for trip-related expenditures and $23.2 billion for equipment expenditures. Table 10.37 provides details on the distribution of expenditures across different expense categories.

Hunting

The National Survey of Fishing, Hunting, and Wildlife-Associated Recreation (USFWS, 2006a) found that 12.5 million people ages 16 and older pursued hunting in 2006. These hunting participants took 185 million trips that accounted for 220 million hunting days. Similar to wildlife-viewing, hunting is a wildlife-dependent recreational activity. Therefore, to the extent that water attributes such as quality and availability influence natural habitats and wildlife populations, these attributes can affect participation. Hunting for waterfowl, such as geese and duck, may be particularly sensitive to the quality of the aquatic environment. According to an addendum to the USFWS survey, waterfowl hunting accounted for 1.3 million unique hunters and more than 13 million hunting days in 2006. These waterfowl hunters incurred over $900 million in trip-related and equipment expenditures (USFWS, 2006b). Table 10.38 provides a more detailed look at waterfowl hunting participation and related expenditures.

TABLE 10.37
Wildlife and Nature Viewing Expenditures (2006)

Expenditure Category	Amount ($ Billion)
Total trip-related	12.9
Food and lodging	7.5
Transportation	4.5
Other trip costs	0.9
Total equipment expenditures	23.2
Wildlife-watching equipment	9.9
Auxiliary equipment	1.0
Special equipment	12.3
Total other expenditures	9.6
Land leasing and owning	6.6
Plantings	1.6
Membership dues and contributions	1.1
Magazines, books	0.4
Total wildlife-watching expenditures	45.7

Sources: USFWS, *2006 National Survey of Fishing, Hunting, and Wildlife-Associated Recreation*, U.S. Fish and Wildlife Service, Washington, DC, 2006.

TABLE 10.38
Waterfowl Hunters, Days, and Expenditures (2006)

Category	Amount
Hunters	
Duck	1,147,000
Geese	700,000
Total hunters	1,306,000
Hunting days	
Duck	12,173,000
Geese	6,008,000
Total hunting days	13,071,000
Waterfowl hunting expenditures	
Trip expenditures	
Food and lodging	$177,125,000
Transportation	$184,329,000
Other trip costs	$132,533,000
Total	$493,987,000
Equipment expenditures	$406,298,000
Total waterfowl hunting expenditures	$900,285,000

Source: USFWS, *Economic Impact of Waterfowl Hunting in the United States: Addendum to the 2006 National Survey of Fishing, Hunting, and Wildlife-Associated Recreation*, U.S. Fish and Wildlife Service, Washington, DC, 2006.

DID YOU KNOW?

Hunting expenditures across all species and environments totaled $22.9 billion in 2006, including $56.7 billion for trip-related expenditures, $10.7 billion for equipment expenditures, and $5.5 billion for other expenditures (e.g., licenses, membership dues) (USFWS, 2006a).

Overview of Water-Based Recreation

In contrast to off-stream water uses such as irrigation which involve the withdrawal and consumption of water resources, water use for recreational activities is considered a non-consumptive, in-stream use. Recreational fishermen, boaters, and others rely on surface water to engage in recreational activities, but none of these pursuits requires the diversion or withdrawal of water from a water resource. The surface water resources used to support recreation and tourism can be divided into two main categories: freshwater and saltwater. Freshwater recreation consists of recreational activity occurring in or on freshwater resources such as river, streams, and lakes. Saltwater recreation involves the use of saltwater resources such as oceans, bays, and tidal portions of rivers. Table 10.39 draws on data from the 2000 NRSE to illustrate the distribution of water-based recreational activity across freshwater and saltwater resources.

Competition in Recreational Water Use

As population growth and other demographic trends intensify demand for water resources, the competition between recreation and other uses of water, as well as potential conflicts between or among various forms of water-based recreation, is likely to increase (CBO, 1997).

Recreation vs. Other Water Uses

Historically, water law has given greater priority to off-stream water uses (e.g., irrigation) than to in-stream water uses such as recreation. The traditional water rights regime was reinforced in part because the economic values of in-stream flows, whether for recreational purposes, ecosystem services, or natural habitat protection, were not well understood. In-stream water uses were thus marginalized in favor of consumptive water uses such as irrigation, which provides market benefits by supporting crop production, and municipal water uses, which provide essential water supplies to industrial, commercial, and residential users. This traditional system, particularly in western states with scarce water resources, commonly resulted in significant reductions in water levels and in-stream flows, which in turn negatively affected water resources' ability to support ecosystem services, natural habitats, and recreational activities (Zellmer, 2006).

As the economic value of ecosystem services, recreational activity, and habitat protection have become better understood in recent decades, state governments have begun to modify their approaches to water resource management. In particular, states have begun to enact protective in-stream flow legislation designed to preserve water flows and support ecological habitats and recreational activities (Zellmer, 2006). This in-stream flow protection represents progress in protecting in-stream flows for recreational uses, but the effort to adopt this legislation has not yet been comprehensive. As of 2009, "over 90% of stream miles in most states do not have full in-stream flow protection," and "in more than half of all states and provinces, over 75% of all streams have no legally recognized in-stream flow protection" (Annear et al., 2009). Thus, although the spread of this legislation has begun to help restore and protect water flows for recreational uses, pressure from competing water uses is likely to persist. With a large portion of the economic value of recreational activity consisting of non-market impacts, recreational water use of in-stream flows is likely to remain at risk of being marginalized in favor of in-stream or off-stream water uses that support crop production, manufacturing, or other market-based activities.

TABLE 10.39
Recreation Activity in Fresh- and Saltwater Resources (2000)

	Freshwater		Saltwater	
Activity	Participation Rate (Percent of U.S. Adults)	Number of Participants	Participation Rate (Percent of U.S. Adults)	Number of Participants
Visit beaches	17.12	35,294,236	30.03	61,922,234
Visit waterside beside beaches	24.71	50,943,698	4.50	9,269,685
Swimming	28.51	58,771,631	25.53	52,637,390
Snorkeling	1.90	3,922,436	5.07	10,459,568
Scuba diving	0.66	1,350,584	1.35	2,786,215
Surfing	0.00	0	1.59	3,285,611
Wind surfing	0.46	939,651	0.39	800,016
Fishing	29.63	61,091,330	10.32	21,283,808
Motor boating	20.52	42,306,567	7.11	14,660,277
Sailing	2.70	5,563,676	2.98	6,136,163
Personal watercraft use	7.60	15,665,261	2.57	5,304,476
Canoeing	9.07	18,708,611	1.05	2,171,666
Kayaking	2.23	4,593,991	1.33	2,746,502
Rowing	4.08	8,411,523	0.53	1,098,999
Water-skiing	7.22	14,894,922	1.15	2,375,709
Bird watching in water-based surroundings	16.84	34,718,973	7.17	14,784,752
Viewing other wildlife in water-based surroundings	20.20	41,641,844	6.45	13,303,288
Viewing or photographing scenery in water-based surroundings	24.76	51,046,395	9.19	18,943,684
Hunting waterfowl	2.21	4,558,051	0.33	680,380

Sources: Leeworthy, V.R., *Preliminary Estimates from Versions 1–6: Coastal Recreation Participation*, National Oceanic and Atmospheric Administration, Silver Spring, MD, 2001.

Competition among Recreational Users

Competition between recreational and alternative uses of water is not the only factor that affects demand for water-based recreation; inter-activity and intra-activity competition also affects participation in recreation activities (Kakoyannis and Stankey, 2003). Inter-activity conflict consists of competition among recreational activities for scarce water resources; an example of this would be recreational boaters and swimmers competing for access to river or lake resources. Intra-activity conflict consists of competition among recreational participants engaging in the same activity; crowding, which can be defined as a "negative evaluation of a certain density or number of encounters," is the most common example of intra-activity conflict (Shelby et al, 1989). The potential for inter-activity and intra-activity conflicts represents an additional challenge for water resource managers when determining how to provide for recreational uses in a water management framework.

Long-Term Challenges

Two of the greatest long-term challenges to water resource management worldwide are climate change and population growth. Although the projected impacts of climate change on U.S. water supplies are not as significant as those for low-latitude and low-precipitation countries, climate change is expected to affect both water temperatures and streamflow or water levels (Morris et al., 2009). Water temperature changes as a result of climate change could negatively affect habitat conditions in cold-water fisheries, such as valuable trout fisheries in New England (Kimball, 1997). With anglers, ranking fish abundance as a key attribute in determining their demand for recreational fishing, any negative effects of water temperature increases upon fish populations in cold-water fisheries could result in decreased recreational fish activity and corresponding economic losses in the local or regional economy (Freeman, 1995). In other regions, water temperature increases could have mixed impacts on fishing populations as temperature increases affect different species in different ways. A study in North Carolina found that, although increased water temperatures could reduce rainbow trout populations, brook trout populations could grow as their range of suitable habitat increases (Morris and Walls, 2009). Climate change also has the potential to affect flow rates and water levels, as higher temperatures can result in reduced snowpack and therefore reduced snowmelt. Studies have shown that recreational boating is "sensitive to lake, reservoir, and stream levels," thus reductions in water flows or levels due to climate change could alter recreational boating demand (Morris and Walls, 2009). Reductions in streamflow could also affect demand for water-enhanced recreational activities such as hiking, camping, and hunting, where participants have shown that proximity to water resources positively influences recreational demand. The projected impacts vary by region, with western states being most vulnerable because of their reliance upon snowmelt to supply streamflow (Morris and Walls, 2009). Population growth represents another long-term challenge for water management regimes. As population growth drives increasing demand for food and water, water demand from the agricultural and municipal use sectors is projected to increase, resulting in even more competition for scarce water resources. This could present an additional strain on in-stream flows that support recreational activity and natural habitat preservation.

Water Quality Issues Affecting Recreational Water Use

The Clean Water Act mandates that each state develop and implement water quality standards designed to support the national goal of "fishable/swimmable" waters. The supply of water for recreation is dependent on the application of these standards to determine whether a water resource can support recreational uses. If these standards cannot be met, the resource may be deemed unsuitable for recreational use, and public health authorities may restrict recreational access. In the context of recreational uses of water, these water quality standards focus on physical, chemical, and biological attributes of water quality that impair the aquatic environment and/or pose health risks to people engaging in water-based recreation activities. The discussion below briefly summarizes the nature of potential impairments to both fishing and swimming.

Water Quality Issues Affecting Fishing

Water quality can have significant impacts on the supply of recreational fishing. With fish populations requiring water of sufficient quality to survive and thrive, and recreational fishermen rating fish abundance among the most important factors affecting fishing demand, it is important to note the types of issues that can impair a water resource's ability to support recreational fishing (Freeman, 1995):

- *Bioaccumulative substances*—Toxic substances such as metals, PCBs, polycyclic aromatic hydrocarbons (PAHs), chlorophenols, and organochlorine pesticides (OCs) which are found in only trace amounts in water, can accumulate to elevated levels in sediments and bioaccumulate in fish tissues; as larger fish or animals consume contaminated fish, the contamination is passed through the food web in a process known as biomagnification. Contamination of fisheries from bioaccumulative substances represents a threat to people and to wildlife who consume fish (USEPA, 2002).
- *Eutrophication*—Nutrient-rich pollution from urban and rural sources such as sewage, stormwater, and agricultural fertilizers fuels biomass production in aquatic ecosystems. This biomass production depletes the dissolved oxygen concentrations of the nutrient-enriched

water resources, which in turn decreases the ability of these aquatic habitats to support fish populations (Selman and Greenhalgh, 2009). Further, biomass production in the form of algal blooms can decrease water clarity and give rise to unpleasant odors in the water resource (Dodds et al., 2008).
- *Pathogens*—Pathogenic microorganisms from inadequately treated sewer and other wastewater discharges can cause disease from ingestion of contaminated water. These risks can be severe in the context of primary contact recreation (i.e., activities that involve submersion in water, such as swimming). In addition, the recreational harvest of shellfish from waters containing bacterial or viral contaminants poses a health risk to those who consume them (NY-NJ HEP, 1996).

These contamination issues can negatively impact fish populations and frequently result in advisories that restrict or ban the consumption of fish in affected waters. Based on state and federal data, 4598 fish advisories were in place in 2010 covering 17.7 million lake acres and 1.3 million river miles in the United States. This means that 42% of national lake acreage and 36% of national river miles were affected by sufficient contamination problems to require advisories that ban or restrict fish consumption (USEPA, 2010b).

Water Quality Issues Affecting Swimming

Similar to fishing, water quality requirements determine the supply of water resources than can support recreational swimming. For the purposes of water quality standards, swimming falls into the category of "primary contact recreation," which encompasses activities that involve submersion in water. The two main contamination issues affecting recreational swimming are pathogens and eutrophication:

- *Pathogens*—Pathogenic contamination results from discharges that introduce microorganisms such as bacteria, viruses, and protozoans to water bodies. The presence of pathogenic contamination significantly elevates the human health risks associated with primary contact recreation in a water body, as diseases stemming from pathogenic bacteria and viruses include typhoid fever, cholera, hepatitis A, and dysentery. To determine if water quality is sufficient for primary contact recreation, state environmental agencies monitor fecal and total coliform bacteria in water resources. Fecal and total coliform bacteria are considered "indicator microorganisms" that signal the existence of fecal contamination, which in turn indicates the potential presence of pathogenic microorganisms (Anderson et al., 2000).
- *Eutrophication*—As discussed earlier, eutrophication results from pollution from sources such as sewage, stormwater, and agricultural fertilizers. Runoff or discharges from these sources can create nutrient-rich water environments that spur biomass growth such as algae. In the context of primary contact recreational activity such as swimming, the important pollution implications from eutrophication involve unattractive odors and diminished clarity (Dodds et al., 2008). Although aesthetic impacts such as these do not necessarily represent significant human health risks, they do have important implications for recreational demand at affected water resources. Studies have shown that the general public makes judgments about water quality based "primarily on vision … and secondarily on smell and touch" (Kakoyannis and Stankey, 2003). Unpleasant odors and reductions in water clarity can thus diminish public perceptions about water quality and negatively impact demand for recreation.

Sufficient contamination of water resources results in the implementation of swimming advisories that ban or restrict swimming in order to preserve public health and safety. The majority of these advisories involve beach closures or restrictions resulting from bacteria-related contamination, but freshwater resources such as rivers, streams, and lakes are also affected by swimming bans and advisories due to other concerns. State water quality monitoring data through 2011 indicate that, for 97,220 miles of assessed rivers and streams in the United States, 39.4% are impaired with respect to primary contact recreation. For lakes and reservoirs, the data indicated that 13.9% of the 3,077,549 acres assessed are impaired with respect to primary contact recreation (USEPA, 2014c). As for beaches, 2010 witnessed 24,091 "closing and advisory days" at beaches in the United States (Dorfman and Rosselot, 2011). Although the Gulf of Mexico spill contributed in part to a 51% increase in the number of precautionary beach closures or advisories (7223 days in 2010), the leading cause of beach closures or advisories in 2010 was violation of water quality standards for bacteria and other pathogens (16,828 days) (Dorfman and Rosselot, 2011).

Market Value of Water Use

Water attributes such as quality and flow are important factors in supporting water-based recreational activities that drive output in the tourism sector. Understanding the values placed on these attributes helps explain how changes in water resources influence demand for recreational activity, which in turn affects consumption of market goods. In the context of recreation and tourism, however, attempts to derive a value for these attributes are complicated by the fact that a great deal of recreational activity occurs outside conventional markets. Because access to many water-based recreational activities and settings is not priced in competitive markets, and because water-based recreation represents a non-consumptive in-stream water use, it is difficult to use market data to estimate a monetary value for water attributes that serve as inputs to demand for recreation and tourism (Raucher et al., 2005). In response to these analytic challenges, economists have developed alternative methods to evaluate and determine the non-market value (or benefit) of the attributes of a water resource that affect demand for water-based recreation. These methods rely on revealed and stated preference techniques that analyze willingness to pay for water attributes that support recreation (Hanemann, 2005).

Non-Market Value Estimates for Water-Based Recreational Activities

Recreational pursuits such as fishing, boating, and swimming provide benefits above and beyond the costs of participating in these activities. To the recreational participant, these benefits represent non-market values known as *consumer surplus.* To derive monetary estimates of these benefits, researchers use stated and revealed preference techniques to empirically analyze the consumer surplus that the public enjoys while engaged in recreation. Comparability between individual empirical analyses on this subject is limited because values can fundamentally differ depending on factors such as geographic region, socioeconomic conditions, and model choice; however, researchers can use meta-analyses to provide a broader perspective. Meta-analyses, which involve the collection and analysis of existing studies, allow researchers to "statistically measure systematic relationships between reported valuation estimates," thereby "capturing heterogeneity within and across studies" (Bergstrom and Taylor, 2006). Researchers can thus use meta-analyses to gain a more comprehensive understanding of the economic value of changes in the attributes of natural resources that support recreational activities, as well as a better understanding of the economic welfare benefits attributable to participation in activities themselves.

Several meta-analyses have analyzed the value of outdoor recreational activities, including Loomis (1999, 2005) and Rosenberger and Loomis (2001). Loomis (2005), the most recent meta-analysis, covers 1239 observations across more than 30 years of economic research. Table 10.40 displays the results of this meta-analysis, presenting average consumer surplus values per person-day of activity. The activities reported in the following text are limited to those commonly accepted as water-based or water-enhanced recreational activities. (It is important to point out that, while participation in winter sports such a skiing and snowboarding also relies on water, these activities are not ordinarily included in discussions of water-based or water-enhanced recreation.)

TABLE 10.40
Average Consumer Surplus Values per Person-Day of Activity (2004$)

Activity	No. of Studies	Estimates	Mean	Range	
Bird watching	4	8	$29.60	$5.80	$78.46
Fishing	129	177	$47.16	$2.08	$556.82
Float boating, rafting, canoeing	20	81	$100.91	$2.70	$390.82
Going to the beach	5	33	$39.43	$3.78	$117.82
Hiking	21	68	$30.84	$0.40	$262.04
Hunting	192	277	$46.92	$2.60	$250.90
Motor boating	15	32	$46.27	$3.78	$203.62
Swimming	11	26	$42.68	$2.20	$134.34
Waterskiing	1	4	$49.02	$15.13	$70.07
Wildlife viewing	69	240	$42.36	$2.40	$347.88
Windsurfing	1	1	$395.47	$395.47	$395.47

Source: Loomis, J., *Average Consumer Surplus Values per Person-Day of Activity, 2004*, Colorado State University, Ft. Collins, 2005.

Impact of Water Supply on Non-Market Recreational Use Values

The amount or supply of water available to support recreational activity (e.g., streamflow or lake levels) can have a significant impact on people's willingness to pay for recreational activities. Supply factors influence recreational demand "by altering the safety of recreational activities and recreationists' perceptions of crowding, scenic beauty, and recreational satisfaction or quality" (Kakoyannis and Stankey, 2003, p. 36). Studies analyzing the influence of water flows or levels on recreation have generally shown that recreationists' preferences follow an inverted U-shaped curve, with recreational users most valuing intermediate amounts and finding low or high amounts to be less preferable (Brown, 2004a; Brown et al., 1991; Kakoyannis and Stankey, 2003; Shelby and Whittaker, 1995). These preferences, however, vary by location and activity (e.g., boating vs. fishing) and even within recreational activities (e.g., flow levels affect elements of a rafting trip such as safety and challenge-level differently); thus, it is not possible to derive a single estimate for an optimal flow rate or water level across all activities and settings (Brown, 2004a; Kakoyannis and Stankey, 2003).

Fishing

Studies analyzing water flow impacts on fishing demand have found that increased streamflow and water levels provide benefits (i.e., increases in consumer surplus) to anglers up to a certain flow level. Water flow levels help to shape recreational fishing opportunities by affecting the habitat conditions of fish populations and influencing recreational access and safety. Eiswerth et al. (2000) evaluated the recreational benefits of increasing water levels at Nevada's Walker Lake State Park. The lake is a rare perennial lake of the Great Basin but is in danger of drying up and is one of only three lakes in Nevada that support recreational fishing. Results indicated that lake users valued a 1-foot increase in lake level in the range of $12 to $18 per user per year. Non-users of the lake maintained an option value in the range of $0.60 to $0.90 per person per year for each additional foot of water (2000$). Similarly, Loomis et al. (1986) found that potential streamflow reductions as a result of hydropower development could substantially reduce both recreational benefits and angling trips on an Idaho river that is popular with anglers. In a national study of streamflow benefits, Hansen and Hallam (1991) found that for recreational fishing the benefits of a marginal increase in stream flow can sometimes exceed the marginal value of agricultural water use. Although these studies indicate that increases in water flow have the potential to increase benefits to anglers, streamflow beyond a certain level can negatively affect recreational opportunities by reducing the suitability of fish habitats and decreasing fish abundance. This maximum-benefit flow level varies depending on the water source and fish type, but all else being equal and "given a certain fish population, fishing quality tends to increase with flows up to a point and then decrease with further flow increases, exhibiting the familiar inverted-U relation" (Brown, 2004a).

Boating

Access to recreational boating opportunities is dependent upon streamflow and water level conditions. Water supply determines what boating activities (e.g., power boating, sailing, kayaking, canoeing) can take place by influencing factors such as recreational access, safety, and *floatability*, which is defined as the "capacity of the river to support boating without excessive hits, stops, drags and portages" (Brown, 2004a). Studies analyzing boaters' preferences for flow levels have also generally found that preferences follow the shape of an inverted U-curve, with intermediate flows being preferred above either low or high flow levels (Brown et al., 1991; Shelby and Whittaker, 1995; Shelby et al., 1989). As with fishing, the exact flow level that provides maximum benefits depends on the water body and on the type of boating activity. For certain boating or paddling activities, such as canoeing, this benefit-maximizing flow level may be lower than other activities, such as whitewater rafting, where users can value higher challenge-levels as part of the recreational experience (Shelby and Whittaker, 1995). Overall, the economic benefits of different stream flows to participants in recreational boating are similar to those for other water-based recreational opportunities. Up to a certain point, marginal increases in flows increase benefits; beyond a certain level, however, marginal increases diminish recreational benefits as concerns such as safety come to outweigh increased access opportunities.

Swimming

In-stream flows and water levels in streams and lakes influence the benefits provided to recreational swimmers by influencing variables such as "water depth, velocity and temperatures" (Brown, 2004a). Although preferences vary depending on user-specific factors such as skill level, studies have generally found that preferences for flow follow the familiar inverted U-curve. A case study on the Clavey River in California found that swimmers considered flows ranging from 10 to 250 ft^3/sec to be acceptable, but they rated the range from 20 to 50 ft^3/sec as optimal. Flows over 350 ft^3/sec were deemed unsafe, and flows below 20 ft^3/sec were found to create water quality issues, particularly if the low flow levels persisted for an extended period of time (Brown, 2004a). In general, high flows create safety hazards and can decrease water temperature to uncomfortable levels, while low flows can create water quality issues. Thus, intermediate flows are generally most preferred.

Wildlife and Nature Viewing

While wildlife and nature viewing is not a water-dependent recreational activity, proximity to water resources has the potential to enhance the quality of a user's recreational experience. In an analysis of streamflow impacts on aesthetic appeal, results indicated that moderated flow levels maximize aesthetic quality. Intermediate flows were most preferred because, among other factors, flow levels that are too high can wash away sand bars, create excess turbidity, and "create an unwelcome sense that events are out of control," while flow levels that are too low can limit the aesthetic appeal of waterfalls and rapids (Brown, 2004a). Another study focusing on recreation in the San Joaquin Valley in California analyzed how increases in flow up to an ecologically "optimal level" (as determined by biologists) affected recreational benefits for hunters, anglers, and wildfire viewers. This study found that increases in flows, particularly in dry areas, could provide recreational benefits in the range of $303 to $348 per acre-foot of water (1992$) (Creel and Loomis, 1992). The value estimates of these benefits were found to be competitive with other uses of water such as irrigation.

IMPACT OF WATER QUALITY ON NON-MARKET RECREATIONAL USE VALUES

The quality of water resources is also a key factor in determining supply and demand for water-based recreational activities. As noted earlier, contamination problems can force public health authorities to restrict or ban recreational use of a water resource; in these cases, the economic benefits provided by water-based recreation can be lost to the local economy as recreational participants travel to other sites or make the decision not to recreate at all. On the demand side, the literature indicates that water quality can have significant effects on how recreational users perceive the quality of their recreational experience. In this manner, water quality will directly influence the non-market benefits that users experience from participating in various recreational activities. These non-market benefits influence demand for recreation, which in turn affects consumption of complementary goods and services in the market economy; thus, water quality can impact economic output related to water-based recreation.

Although benefits associated with water quality improvements may vary depending on factors such as initial water quality, the recreational activity of interest, and location, the literature generally shows that improvements in water quality increase the quality of recreation experiences and the economic benefits associated with these experiences. For example, Ribaudo and Epp (1984) analyzed the recreation benefits of restoring water quality in Lake Champlain's St. Albans Bay. The bay had historically provided water-based recreational opportunities for swimming, fishing, boating, and more, before eutrophication problems caused a significant decline in recreational demand. Results indicated that restoration of water quality would provide a mean level of annual benefits of $123 to current users and $97 to former users (1984$).

Fishing

The quality of water resources directly affects supply and demand for recreational fishing. On the supply side, elevated levels of bioaccumulative contaminants (e.g., PCBs, metals) can require that fish consumption be restricted or banned. Studies have shown that fish consumption advisories implemented due to contamination concerns can negatively affect angler welfare (Jakus et al., 2002). As previously noted, contamination problems have the potential to negatively affect recreational fishing demand by diminishing the ability of water resources to support fish populations (Freeman, 1995). In a 2003 study of the effect of water quality improvements on recreational use benefits in six northeastern states, Parsons et al. (2003) found that average benefits for recreational fishing ranged from approximately $3 to $8 per person (1994$), depending on the level of water quality achieved (see Table 10.41).

TABLE 10.41
Average Annual per Capita Benefits from Water Quality Improvements

Activity	All Sites Attain Medium Water Quality (1994$)	All Sites Attain High Water Quality (1994$)
Fishing	$3.14	$8.26
Boating	$0.04	$8.25
Swimming	$5.44	$70.47
Viewing	$0.00	$31.45

Source: Parsons, G. et al., *Measuring the Economic Benefits of Water Quality Improvements to Recreational Users in Six Northeastern States: An Application of the Random Utility Maximization Mode*, U.S. Environmental Protection Agency, Washington, DC, 2003.

Boating

The water quality standards that determine if water resources can support non-contact recreation such as boating are not as stringent as those for fishing and swimming. There are cases where debris or excessive biomass growth can inhibit boating, and there are some secondary contact recreation guidelines for bacteria levels, but the presence of contaminants in water generally does not require the restriction of boating activity. However, to the extent that recreational boaters participate in boating in conjunction with other water-based recreational activities, such as fishing or swimming, water quality issues can affect demand for boating. In a case study focusing on the value of improved water quality in Chesapeake Bay, participants in recreational boating estimated water quality on a scale of one to five and were asked to give a willingness-to-pay value for a one-step improvement in water quality. Results indicated that boaters' median willingness to pay was $17.50 per year (mean of $63 per year in 2003 dollars) for one-step improvements in water quality (Lipton, 2003). The Lipton (2003) study quoted an earlier study by Bockstael et al. (1989) that found that fishing drove a significant amount of demand for recreational boating in Chesapeake Bay; 72% of boaters who stored their boats on trailers and 38% of boaters who kept their boats in-water stated that they used their boats "always or usually for fishing." This would indicate that at least a portion of boaters' willingness to pay for water quality improvements could be related to how improved water quality would affect the quality of recreational fishing trips, which in turn affects boating pressure. Returning to the Parsons et al. (2003) study cited above, the analysis found that moderate improvements in water quality had relatively little effect on boater benefits, but that significant improvements in water quality provided recreational benefits that were similar to those found for recreational fishing.

Swimming

Because water quality standards directly determine the ability of a water resource to support full-contact recreation, water quality has the potential to affect both supply and demand for recreational swimming. In a national study focusing on the benefits of water quality improvements, Carson and Mitchell (1993) found the national benefits of achieving the Clean Water Act's swimmable water quality goal to be between $24 billion and $40 billion per year (1990$). On a regional scale, water quality improvements have the potential to significantly enhance swimmers' welfare. In fact, when comparing these benefits to other recreational activities included in the Parsons et al. (2003) analysis, the results indicate that swimming is the activity that would benefit most from improvements in water quality.

Wildlife and Nature Viewing

Water resources have the potential to enhance the recreational experience of wildlife and nature viewing. The aesthetic quality of the environment is a key input in determining demand for viewing activity, and water resources have been found to enhance the aesthetic quality of environmental settings. Studies have found that participants in water-based recreation judge water quality in large part based on visual indicators and smell despite the fact that many potential contaminants, such as PCBs, metals, and fecal coliform, are not detectable by sight or odor. This would indicate that water quality issues such as eutrophication, which can diminish water clarity and produce unpleasant odors, are very influential in recreational users' perceptions of water quality (Kakoyannis and Stankey, 2003). Returning once more to the results of Parsons et al. (2003), which analyzed water quality impacts on recreational activities in six northeastern states, moderate water quality improvements were found to have no impact on welfare associated with recreational viewing, but greater water quality improvements could have a significant impact on user welfare.

TABLE 10.42
Welfare and Visitation Impacts Due to Improved Water Quality at Long Beach, California

Measurement	Day Trips	Multi-Day Trips	All Beach Use
Annual person-days	5633	1353	6986
Annual economic value (2007$)	$602,781	$321,305	$924,086

Source: Leeworthy, V.R. and Wiley, P. (2007). *Southern California Beach Valuation Project: Economic Value and Impact of Water Quality Change for Long Beach in Southern California*, National Oceanic and Atmospheric Administration, Silver Spring, MD, 2007.

Beach Use

Beaches, as the leading travel destinations for tourists, are a significant source of demand for recreation and tourism (Houston, 2008). With beaches offering a variety of recreational opportunities such as swimming, boating, and fishing, water quality can influence both supply and demand for recreational beach use. Hanemann (2005) analyzed the impacts of five scenarios of water quality change at Southern California beaches and found that in scenarios where water quality improved visitation and consumer surplus were both projected to increase; in contrast, decreases in water quality were projected to result in declines in visitation and recreational user welfare. In another study focusing on Long Beach in Southern California, Leeworthy and Wiley (2007) used the Southern California Beach Valuation Model (SCBVM) to estimate the effects of improvements in water quality on annual visitation and economic welfare. The water quality improvement scenario used in the study called for water quality at Long Beach to improve from its rating of 2.8545 to the 3.9150 rating (on a scale of 0 to 4) of nearby Huntington City Beach, Table 10.42 shows how this improvement in water quality is projected to affect visitation and welfare for day trips and multi-day trips across users in four Southern California counties.

As these studies show, water quality has the potential to affect recreational demand for beach use with respect to both visitation and economic welfare. With coastal economies relying a great deal upon beach-oriented recreation and tourism, water quality can be critically important to determine the success of these economies at the local and regional scale. This is illustrated by the results of a study by Parsons et al. (2007), which analyzed the economic impacts resulting from a closure of the Padre Island National Seashore due to a contamination event. The results of this study suggest that beach closures can cause significant losses in the output of the market economy, with reductions in economic output ranging from $172,000 per weekend day in July to $26,000 per week day in September (in 2007$).

THOUGHT-PROVOKING QUESTIONS

10.1 What is the true value of water? Explain.
10.2 What would you be willing to pay, as fair market value, for a clean, safe glass of ice-cold water? Explain.
10.3 Why do many judge the state of water cleanliness and safety by sight and smell only?

REFERENCES AND RECOMMENDED READING

Adelman, B., Herberlein, T., and Bronnicksen, T. (1982). Social psychological explanations for the persistence of conflict between paddling canoeists and motorcraft users in the boundary waters canoe area. *Leisure Sciences*, 5, 45–61.

Anderson, D.M. et al. (2000). *Estimated Annual Impacts from Harmful Algae Blooms in the United States*, Woods Hole Oceanography Institute Technical Report WHOI-2000-11. Woods Hole, MA: Woods Hole Oceanography Institute.

Anderson, K. and Davidson, M. (1997). *Drinking Water & Recreational Water Quality: Microbiological Criteria*. Moscow: University of Idaho.

Angele, F.J., Sr. (1974). *Cross Connections and Backflow Protection*, 2nd ed. Denver, CO: American Water Works Association.

Annear, T., Lobb, D., Coomer, C., Woythal, M., Hendry, C., Estes, C., and Williams, K. (2009). *International In-Stream Flow Protection Initiative: A Status Report of State and Provincial Fish and Wildlife Agency Instream Flow Activities and Strategies for the Future*. Cheyenne, WY: Instream Flow Council (http://www.instreamflowcouncil.org/docs/IIFPI-final-report-with-covers.pdf).

Anon. (2014). *Water Encyclopedia: Science and Issues*, http://www.waterencyclopedia.com/.

ASCE. (2013). *Report Card for America's Infrastructure: Drinking Water*. Reston, VA: American Society of Civil Engineers (http://www.infrastructurereportcard.org/).

August, Jr., J.L. (1999). *Vision in the Desert: Carl Hayden and Hydropolitics in the American Southwest*. Fort Worth: Texas Christian University Press.

AWWA. (2001). *Financial and Revenue 1999 Survey* [CD]. Denver, CO: American Water Works Association.

AWWA. (2005). *The Value of Water: Concepts, Estimates, and Applications for Water Managers*. Denver, CO: American Water Works Association.

Ayers, R.S. and Westcot, D.W. (1994). *Water Quality for Agriculture*. New York: United Nations Food and Agriculture Organization.

Baxter, J.O. (1997). *Dividing New Mexico's Waters, 1700-1912*. Albuquerque: University of New Mexico Press.

Bergstrom, J. and Taylor, L. (2006). Using meta-analysis for benefits transfer: theory and practice. *Ecological Economics*, 60, 351–360.

Bernard, A.B., Jensen, J.B., and Schott, P.K. (2006). Survival of the best fit: exposure to low-wage countries and the (uneven) growth of U.S. manufacturing plants. *Journal of International Economics*, 68, 219–37.

Billington, D.P., Jackson, D.C., and Melosi, M.V. (2005). *The History of Large Federal Dams: Planning, Design, and Construction.* Denver, CO: Bureau of Reclamation.

BLS. (2010). *Quarterly Census of Employment and Wages*. Washington, DC: Bureau of Labor Statistics.

BLS. (2014). *Quarterly Census of Employment and Wages*. Washington, DC: Bureau of Labor Statistics (http://www.bls.gov/cew/#databases).

Bockstael, N.E., McConnell, K.E., and Stand, I.E. (1989). Measuring the benefits of improvements in water quality: the Chesapeake Bay. *Marine Resource Economics*, 6(1), 1–18.

Brauer, D. (2008). *Factors Underlying the Decline in Manufacturing Employment Since 2000*, Congressional Budget Office Economic and Budget Issue Brief. Washington, DC: U.S. Government Printing Office.

Brewer, J., Glennon, R., Ker, A., and Libecap, G.D. (2007). *Water Markets in the West: Prices, Trading, and Contractual Forms*, NBER Working Paper No. 13002. Cambridge, MA: National Bureau of Economic Research.

Brown, T.C. (2004a). Water availability and recreational opportunity. In: *Riparian Areas of the Southwestern Untied States: Hydrology, Ecology, and Management*, Baker, M.B. et al., Eds. Boca Raton, FL: CRC Press, Chap. 14.

Brown, T.C. (2004b). *The Marginal Economic Value of Streamflow from National Forests*. Denver, CO: Rocky Mountain Research Station, U.S. Forest Service.

Brown, T.C., Taylor, J., and Shelby, B. (1991). Assessing the direct effects of streamflow on recreation: a literature review. *Water Resources Bulletin*, 27(6), 979–988.

Bureau of Reclamation. (2005). *Hydroelectric Power*. Washington, DC: Bureau of Reclamation, U.S. Department of the Interior.

Bureau of Reclamation. (2011). *Hydropower Resources Assessment at Existing Reclamation Facilities*. Washington, DC: Bureau of Reclamation, U.S. Department of the Interior.

Bureau of Reclamation. (2012). *Brief History of Bureau of Reclamation*. Washington, DC: Bureau of Reclamation, U.S. Department of the Interior.

Buttner, J.K. et al. (1993). *An Introduction to Water Chemistry in Freshwater Aquaculture*, NRAC Fact Sheet No. 170-1993. College Park, MD: Northeast Regional Aquaculture Center.

California Urban Water Agencies. (1994). *The Value of Water Supply Reliability: Results of a Contingent Valuation Survey of Residential Customers*. Oakland, CA: Prepared by Barakat & Chamberlain, Inc.

Carson, R.T. and Mitchell, R.C. (1993). The value of clean water: the public's willingness to pay for boatable, fishable and swimmable quality water. *Water Resources Research*, 29(7), 2445–2454.

Carter, N.T. (2010). *Energy's Water Demands: Trends, Vulnerabilities and Management*. Washington, DC: Congressional Research Service.

CBO. (1997). *Water Use Conflicts in the West: Implications of Reforming the Bureau of Reclamation's Water Supply Policies*. Washington, DC: Congressional Budget Office.

CBO. (2002). *Future Investment in Drinking Water and Wastewater Infrastructure*. Washington, DC: Congressional Budget Office.

Center for Ports & Waterways. (2007). *A Modal Comparison of Domestic Freight Transportation Effects on the General Public*. Houston, TX: Texas Transportation Institute.

CH2M Hill. (2003). Water use in industries of the future. In: *Industrial Water Management: A Systems Approach*, 2nd ed., Byers, W. et al., Eds. New York: American Institute of Chemical Engineers.

Chesapeake Energy. (2008a). *Little Red River Project*, presentation at Trout Unlimited.

Chesapeake Energy. (2008b). *Components of Hydraulic Fracturing*, presentation to New York Department of Environmental Conservation.

Chesapeake Energy. (2014). *Hydraulic Fracturing*. Oklahoma City, OK: Chesapeake Energy (http://www.chk.com/Operations/Process/Hydraulic-Fracturing/Pages/Information.aspx).

Cheung, W.W.L. et al. (2009). Large-scale redistribution of maximum fisheries catch potential in the global ocean under climate change. *Global Change Biology*, 16, 24–35.

CIA. (2012). *The World Factbook*. Washington, DC: Central Intelligence Agency (https://www.cia.gov/library/publications/the-world-factbook/).

Combs, S. (2012). *The Impact of the 2011 Drought and Beyond*. Austin: Texas Comptroller of Public Accounts.

Cordell, H.K., Green, G., and Betz, C. (2009). *Long-Term National Trends in Outdoor Recreation Activity Participation—1980 to Now*. Washington, DC: Forest Service Southern Research Station and Forestry Sciences Laboratory, U.S. Department of Agriculture.

Corso, P. et al. (2003). Cost of illness in the 1993 waterborne *Cryptosporidium* outbreak, Milwaukee, Wisconsin. *Emerging Infectious Disease*, 9(4), 426–431.

Creel, M. and Loomis, J. (1992). Recreation value of water to wetlands in the San Joaquin Valley: linked multinomial logit and count data trip frequency models. *Water Resources Research*, 28(10), 2597–2606.

CRS. (2011). *Modern Shale Gas Development in the United States: A Primer*. Washington, DC: U.S. Department of Energy.

Dalhuisen, J. et al. (2003). Price and income elasticities of residential water demand: a meta analysis. *Land Economics*, 79, 292–308.

David, E.L. (1990). Trends and associated factors in off-stream water use: manufacturing and mining water use in the United States, 1954–83. In: *USGS National Water Summary 1987—Hydrologic Events and Water Supply and Use*, Carr, J.E. et al., Eds., Water-Supply Paper 2350. Denver, CO: U.S. Geological Survey.

Department of Boating and Waterways and State Coastal Conservancy. (2002). *California Beach Restoration Study*. Sacramento: State of California.

DiGulio, D.C. et al. (2011). *Investigation of Groundwater Contamination Near Pavillion, Wyoming*. Washington, DC: U.S. Environmental Protection Agency.

Dodds, W., Bouska, W., Eitzmann, J., Pilger, T., Pitts, K., Riley, A., Schloesser, J., and Thornbrugh, D. (2008). Eutrophication of U.S. freshwaters: analysis of potential economic damages. *Environmental Science and Technology*, 43(1), 12–19.

Dorfman, M. and Rosselot, K.S. (2011). *Testing the Waters—A Guide to Water Quality at Vacation Beaches*, 21st Annual Report. Chicago, IL: Natural Resources Defense Council (http://www.nrdc.org/water/oceans/ttw/ttw2011.pdf).

Eden, S. et al. (2008). Agriculture water use to municipal use: the legal and institutional context for voluntary transactions in Arizona. *The Water Report*, 58, 9–20.

EIA. (2010). *Annual Energy Outlook 2011 with Projections to 2035*. Washington, DC: U.S. Energy Information Administration (http://www.electricdrive.org/index.php?ht=a/GetDocumentAction/id/27843).

EIA. (2011a). *Crude Oil Production*. Washington, DC: U.S. Energy Information Administration (http://www.eia.gov/dnav/pet/pet_crd_crpdn_adc_mbbl_m.htm).

EIA. (2011b). *Natural Gas Gross Withdrawals and Production*. Washington, DC: U.S. Energy Information Administration (http://www.eia.gov/dnav/ng/ng_prod_sum_dcu_nus_m.htm).

EIA. (2011c). *Coal*. Washington, DC: U.S. Energy Information Administration (http://www.eia.gov/coal/).

EIA. (2011d). *Annual Energy Outlook 2011*. Washington, DC: U.S. Energy Information Administration (http://www.eia.gov/todayinenergy/detail.cfm?id=1110).

EIA. (2011e). *Electric Power Annual 2010*. Washington, DC: U.S. Energy Information Administration (http://large.stanford.edu/courses/2012/ph240/nam2/docs/epa.pdf).

Eiswerth, M., Englin, J., Fadali, E., and Shaw. W.D. (2000). The value of water levels in water-based recreation: a pooled revealed preference/contingent behavior model. *Water Resources Research*, 36(4), 1079–1086.

Ellis, M. et al. (2003). *Industrial Water Use and Its Energy Implications*. Washington, DC: U.S. Department of Energy.

Enerdata. (2013). *Global Energy Statistical Yearbook*. London: Enerdata (http://yearbook.enerdata.net/).

EPRI. (2007a). *Assessment of Waterpower Potential and Development Needs*. Palo Alto, CA: Electric Power Research Institute.

EPRI. (2007b). *Program on Technology Innovation: Power Generation and Waster Sustainability*. Palo Alto, CA: Electric Power Research Institute.

EPRI. (2011). *Water Use for Electricity Generation and Other Sectors: Recent Changes (1985–2005) and Future Projections (2005–2030)*. Palo Alto, CA: Electric Power Research Institute.

Espey, M. et al. (1997). Price elasticity of residential demand for water: a meta analysis. *Water Resources Research*, 33, 1369–1374.

Esrey, S. et al. (1991). Effects of improved water supply and sanitation on ascariasis, diarrhea, dracunculiasis, hookworm infection, schistomsomiasis, and trachoma, *Bulletin of the World Health Organization*, 69(5), 609–621.

FAO. (2008). *Report of the FAO Expert Workshop on Climate Change Implications for Fisheries and Aquaculture*. Geneva, Switzerland: Food and Agricultural Organization of the United Nations.

FAO. (2009). *The State of World Fisheries and Aquaculture*. Geneva, Switzerland: Food and Agricultural Organization of the United Nations.

Faries, Jr., F.C., Sweeten, J.M., and Reagor, J.C. (1998). *Water Quality: Its Relationship to Livestock*. Odessa: Texas Agricultural Extension Service.

Faux, J. and Perry, G.M. (1999). Estimating irrigation water value using hedonic price analysis: a case study in Malheur County, Oregon. *Land Economics*, 75(3), 440–452.

Frederick, K.D., VandenBerg, T., and Hanson, J. (1996). *Economic Values of Freshwater in the United States*, Discussion Paper 97-03. Washington, DC: Resources for the Future.

Freeman, A. (1995). The benefits of water quality improvements for marine recreation: a review of the empirical evidence. *Marine Resource Economics*, 10, 385–406.

FUS. (2011). *Current Fishing Statistics No. 2011*. Silver Spring, MD: Fisheries of the United States.

Galiani, S. et al. (2005). Water for life: the impact of the privatization of water service on child mortality. *Journal of Political Economy*, 113: 83–120.

GAO. (2003). *Freshwater Supply: States' Views of How Federal Agencies Could Help Them Meet the Challenges of Expected Shortages*, Report to Congressional Requestors. Washington, DC: General Accounting Office.

GAO. (2009). *Energy–Water Nexus: Improvements to Federal Water Use Data Would Increase Understanding of Trends in Power Plant Water Use*, Report to the Chairman, Committee on Science and Technology, House of Representatives. Washington, DC: General Accounting Office.

Gaudlip, A., Paugh, L., and Hayes T. (2008). Marcellus Shale Water Management Challenges in Pennsylvania, paper presented at 2008 SPE Shale Gas Production Conference, November 16–18, Irving, Texas.

Gibbons, D.C. (1986). *The Economic Value of Water*. Washington, DC: Resources for the Future.

Giesi, E. (2008). Rising sales of bottled water trigger strong reaction for U.S. conservationists, *The New York Times*, March 19 (http://www.nytimes.com/2008/03/19/business/worldbusiness/19iht-rbogbottle.html).

Gillian, D.M. and Brown, T.C. (1997). *In-Stream Flow Protection: Seeking a Balance in Western Water Use*. New York: Island Press.

Gleick, P.H. (2006). *Critical Issues on Water and Agriculture in the United States: New Approaches for the 21st Century*, prepared for the NAREEE Advisory Board. Washington, DC: U.S. Department of Agriculture.

Griffin, R. and Mjelde, J. (2000). Valuing water supply reliability. *American Journal of Agricultural Economics*, 82(2), 414.

Ground Water Protection Council. (2009). *Modern Shale Gas Development in the United States: A Primer*. Washington, DC: Office of Fossil Energy, U.S. Department of Energy.

GWI. (2004). *Tariffs: Half Way There*. Oxford: Global Water Intelligence.

Haas, G. (2010). *A Snapshot of Recreational Boating in America*. Chicago, IL: National Marine Manufacturers Association (http://www.nmma.org/assets/cabinets/Cabinet214/USCG%20Recreational%20Boating%20in%20America.doc).

Halich, G. and Stephenson, K. (2009). Effectiveness of residential water-use restrictions under varying levels of municipal effort. *Land Economics*, 85(4): 614–626.

Hanemann, W.H. (2005). The economic conception of water. In: *Water Crisis: Myth or Reality?*, Rogers, P.O., Llamas, M.R., and Martinez-Cortina, L., Eds. London: Taylor & Francis, Chap. 4.

Hansen, L. and Hallam, A. (1991). National estimates of the recreational value of streamflow. *Water Resources Research*, 27(2), 167–175.

Heimlich, R. (2003). *Agricultural Resources and Environmental Indicators, 2003*, Agriculture Handbook No. AH722. Washington, DC: U.S. Department of Agriculture (http://www.ers.usda.gov/publications/ah-agricultural-handbook/ah722.aspx#.U2EO18fQIfI).

Hellin, D., Wiggin, J., Uiterwyk, K., Starbuck, K., Napoli, N., Terkla, D., Watson, C., Roman, A., Roach, L., and Welch, T. (2011). *2010 Massachusetts Recreational Boater Survey*, Technical Report #03.uhi.11. Boston: Massachusetts Ocean Partnership.

Houston, J. (2008). The economic value of beaches—a 2008 update. *Shore & Beach*, 76(3), 22–26.

Howe, C. and Smith, M. (1994). The value of water supply reliability in urban water systems. *Journal of Environmental Economics and Management*, 26, 19–30.

Hundley, Jr., N. (1975). *Water and the West: The Colorado River Compact and the Politics of Water in the American West.* Berkeley: University of California Press.

Jakus, P., McGuinness, M., and Krupnick, A. (2002). *The Benefits and Costs of Fish Consumption: Advisories for Mercury*. Washington, DC: Resources for the Future.

Jones, F.E. (1992). *Evaporation of Water*. Chelsea, MI: Lewis Publishers.

Kakoyannis, C. and Stankey, G. (2003). *Assessing and Evaluating Recreational Uses of Water Resources: Implications for an Integrated Management Framework*, General Technical Report PNW-GTR-536. Washington, DC: U.S. Department of Agriculture.

Kildow, J. et al. (2009). *State of the U.S. Ocean and Coastal Economies*. Monterey, CA: National Ocean Economics Program.

Kimball, K. (1997). New England regional climate change impacts on recreation and tourism. In: *Workshop Summary Report, New England Regional Climate Change Impacts Workshop*. Durham, NH: Institute for the Study of Earth, Oceans, and Space, University of New Hampshire.

Kollgaard, E.B. and Chadwick, W.L. (1988). *Development of Dam Engineering in the United States*. New York: Pergamon Press.

Komives, K. et al. (2005). *Water, Electricity, and the Poor: Who Benefits from Utility Subsidies?* Washington, DC: The World Bank.

Kumar, S. (2006). Analyzing industrial water demand in India: an input distance function approach. *Water Policy*, 8, 15–29.

Leeworthy, V.R. (2001). *Preliminary Estimates from Versions 1–6: Coastal Recreation Participation*. Silver Spring, MD: National Oceanic and Atmospheric Administration.

Leeworthy, V.R. and Wiley, P. (2001). *Current Participation Patterns in Marine Recreation*. Silver Spring, MD: National Oceanic and Atmospheric Administration.

Leeworthy, V.R. and Wiley, P. (2007). *Southern California Beach Valuation Project: Economic Value and Impact of Water Quality Change for Long Beach in Southern California*. Silver Spring, MD: National Oceanic and Atmospheric Administration.

Leibtag, E. (2008). Corn prices near record high, but what about food costs? *Amber Waves*, February.

Lewis, S.A. (1996). *The Sierra Club Guide to Safe Drinking Water*. San Francisco, CA: Sierra Club Books.

Linenberger, T.R. (2002). *Dams, Dynamos, and Development: The Bureau of Reclamation's Power Program and Electrification of the West*. Washington, DC: Bureau of Reclamation.

Lipton, D. (2003). *The Value of Improved Water Quality to Chesapeake Bay Boaters*. College Park, MD: The University of Maryland.

Listorti, J.A. (1996). *Bridging Environmental Health Gaps: Lessons for Sub-Saharan Africa Infrastructure Projects*, AFTES Working Paper No. 20. Washington, DC: World Bank.

Lobell, D.B., Schlenker, W., and Costa-Roberts, J. (2011). Climate trends and global crop production since 1980. *Science*, 333(6042), 616–620.

Loomis, J. (1999). *Meta-Analysis of Non-Market Recreation Studies*. Ft. Collins: Colorado State University.

Loomis, J. (2005). *Average Consumer Surplus Values per Person-Day of Activity, 2004*. Ft. Collins: Colorado State University.

Loomis, J., Sorg, C., and Donnelly, D. (1986). Economic losses to recreation fisheries due to small-head hydropower development: a case study of Henry Fork in Idaho. *Journal of Environmental Management*, 22, 85–94.

Lovelace, J.K. (2009a). *Method for Estimating Water Withdrawals for Livestock in the United States, 2005.* Washington, DC: U.S. Geological Survey.
Lovelace, J.K. (2009b). *Methods for Estimating Water Withdrawals for Aquaculture in the United States, 2005.* Washington, DC: U.S. Geological Survey.
LSU. (2011). *Louisiana Summary: Agriculture and Natural Resources: Agriculture Is Backbone of Louisiana's Economy*. Baton Rouge: LSU AgCenter (http://www.lsuagcenter.com/agsummary/).
Lytle, B.A. et al. (2008). A win-win scenario for urban/rural water supplies. *The Water Report*, February.
Mavis, J. (2003). Water use in industries of the future: mining industry. In: *Industrial Water Management: A Systems Approach*, 2nd ed., prepared by CH2M Hill for the Center for Waste Reduction Technologies American Institute of Chemical Engineers.
Mayer, P. and DeOreo, W. (1999). *Residential End Uses of Water*. Denver, CO: American Water Works Association.
MCC. (2008a). *Water Conservation Plan: A Member of the Chemical Manufacturing Sector*. Lansing: Michigan Chamber of Commerce.
MCC. (2008b). *Water Conservation Plan: A Member of the Pulp and Paper Sector*. Lansing: Michigan Chamber of Commerce.
McGhee, T.J. (1991). *Water Supply and Sewerage*, 6th ed. New York: McGraw-Hill.
Meyer, W.B. (1996). *Human Impact on Earth*. New York: Cambridge University Press.
Molden, D., Ed. (2007). *Water for Food, Water for Life: A Comprehensive Assessment of Water Management*. London: Earthscan, and Colombo: International Water Management Institute.
Morris, D. and Walls, M. (2009). *Climate Change and Outdoor Recreation Resources*. Washington, DC: Resources for the Future.
NCDOT. (2014). *Fast Facts*. Raleigh: Ferry Division, North Carolina Department of Transportation (http://www.ncdot.gov/download/newsroom/FastFacts.pdf).
NETL. (2009a). *Estimating Freshwater Needs to Meet Future Thermoelectric Generation Requirements (2009 Update)*. Washington, DC: National Energy Technology Laboratory, U.S. Department of Energy.
NETL. (2009b). *Impact of Drought on U.S. Steam Electric Power Plant Cooling Water Intakes and Related Water Resource Management Issues*. Washington, DC: National Energy Technology Laboratory, U.S. Department of Energy.
NOAA. (2011). *NOAA Fisheries Office of Sustainable Fisheries*. Washington, DC: National Oceanic and Atmospheric Administration (http://www.nmfs.noaa.gov/sfa/).
NRC. (2004). *Managing the Columbia River: In-Stream Flows, Water Withdrawals, and Salmon Survival*. Washington, DC: National Research Council.
NSRE. (2001). *American's Participation in Outdoor Recreation: Results from NSRE (with Weighted Data) (Versions 1 to 13)*. Washington, DC: U.S. Department of Agriculture (http://www.srs.fs.usda.gov/trends/Nsre/Rnd1t13weightrpt.pdf).
NY-NJ HEP. (1996). Management of pathogenic contamination. *Pathogenic Contamination*, March, pp. 161–180 (http://www.harborestuary.org/pdf/hep_pathomod.pdf).
O'Connor, T. and Whithall, D. (2007). Linking hypoxia to shrimp catch in the northern Gulf of Mexico. *Marine Pollution Bulletin*, 54(4), 460–463.
Olmstead, S. (2010). The economics of managing scarce water resources. *Review of Environmental Economics and Policy*, 4, 179–198.
Olmstead, S. et al. (2007). Water demand under alternative price structures. *Journal of Environmental Economics and Management*, 54, 181–198.
Olmstead, S. and Stavins, R. (2009). Comparing prices and non-price approaches to urban water conservation. *Water Resources Research*, 45, 1–10.
Olson, E. (1999). *Bottled Water: Pure Drink or Pure Hype?* New York: Natural Resources Defense Council.
Parsons, G.R., Leggett, C., Boyle, K., and Kang, A. (2007). *Valuing Beach Closures on the Padre Island National Seashore*, Working Paper No. 2008-10. Newark, DE: Department of Economics, Alfred Lerner College of Business & Economics, University of Delaware.
Parsons, G.R., Helm, E., and Bondelid, T. (2003). *Measuring the Economic Benefits of Water Quality Improvements to Recreational Users in Six Northeastern States: An Application of the Random Utility Maximization Mode*. Washington, DC: U.S. Environmental Protection Agency.
Peavy, H.S. et al. (1985). *Environmental Engineering*. New York: McGraw-Hill.
Peden, D. et al. (2007). Water and livestock for human development. In: *Water for Food, Water for Life: A Comprehensive Assessment of Water Management*, Molden, D., Ed. London: Earthscan, and Colombo: International Water Management Institute.
Pereira, A.M. (2000). Is all public capital created equal? *Review of Economics and Statistics*, 82(3), 513–518.

Petrie, R.A. and Taylor, L.O. (2007). Estimating the value of water use permits: a hedonic approach applied to farmland in the southeastern United States. *Land Economics*, 83(3), 302–318.

Pew Environment Group. (2009). *Redistribution of Fish Catch by Climate Change*, Ocean Science Series. Washington, DC: Pew Environment Group.

Pfost, D.L. and Fulhage, C.D. (2001). *Water Quality for Livestock Drinking*. Columbia, MO: University of Missouri–Colombia (http://extension.missouri.edu/p/EQ381).

Pielou, E.C. (1998). *Fresh Water*. Chicago, IL: University of Chicago Press.

Pisani, D.J. (1984). *From the Family Farm to Agribusiness: The Irrigation Crusade in California and the West, 1850–1931*. Berkeley: University of California Press.

Pitzer, P.C. (1994). *Grand Coulee: Harnessing a Dream*. Pullman: Washington State University Press.

Powell, J.W. (1904). *Twenty-Second Annual Report of the Bureau of American Ethnology to the Secretary of the Smithsonian Institution, 1900–1901*. Washington, DC: U.S. Government Printing Office.

Quinn, F.H. (2002). The potential impacts of climate change on Great Lakes transportation. In: *The Potential Impacts of Climate Change on Transportation: Summary and Discussion Papers*. Washington, DC: U.S. Department of Transportation Center for Climate Change and Environmental Forecasting.

Raucher, R., Chapman, D., Henderson, J., Hagenstad, M. Rice, J. Goldstein, J., Huber-Lee, A., DeOreo, W., Mayer, Pl, Hurd, B., Linsky, R., Means, E., and Renwick, M. (2005). *The Value of Water: Concepts, Estimates, and Applications for Water Managers*. Denver, CO: American Water Works Association.

RB&FF and OF. (2009). *Special Report on Boating and Fishing*. Washington, DC: Recreational Boating & Fishing Foundation and Outdoor Foundation.

Reilly, J. et al. (2003). U.S. agriculture and climate change: new results. *Climate Change*, 57, 43–69.

Reisner, M.P. (1986). *Cadillac Desert: The American West and Its Disappearing Water*. New York: Viking.

Renzetti, S. (2005). Economic instruments and Canadian industrial water use. *Canadian Water Resources Journal*, 30(1), 21–30.

Renzetti, S. and Dupont, D.P. (2002). *The Value of Water in Manufacturing*, CSERGE Working Paper ECM 03-03. Swindon, UK: Economic and Social Research Council Centre for Social and Economic Research on the Global Environment.

Ribaudo, M. and Epp, D. (1984). The importance of sample discrimination in using the travel cost method to estimate the benefits of improved water quality. *Land Economics*, 60(4), 397–403.

Rockaway, T. et al. (2011). Residential water use trends in North America. *Journal of the American Water Works Association*, 103(2), 76–89.

Rosenberger, R. and Loomis, J. (2001). *Benefit Transfer of Outdoor Recreation Use Values: A Technical Document Supporting the Forest Service Strategic Plan*, General Technical Report RMRS-GTR-72. Ft. Collins, CO: U.S. Department of Agriculture, Forest Service, Rocky Mountain Research Station.

Rowley, W.D. (2006). *The Bureau of Reclamation: Origins and Growth to 1945*, Vol. 1. Denver, CO: Bureau of Reclamation.

Roy, S.B. et al. (2012). Projecting water withdrawal and supply for future decades in the U.S. under climate change scenarios. *Environmental Science and Technology*, 46(5), 2545–2556.

Satterfield, J.M. et al. (2008). Chesapeake Energy Corp. Managing Water Resource's Challenges in Select Natural Gas Shale Plays, paper presented at the GWPC Annual Meeting by ALL Consulting, Lisle, IL.

Seager, R., Tzanova, A., and Nakamura, J. (2009). Drought in the southeastern United States: causes, variability over the last millennium, and the potential for future hydroclimate change. *Journal of Climate*, 22, 5021–5045.

Selman, M. and Greenhalgh, S. (2009). Eutrophication: sources and drivers of nutrient pollution. *WRI Policy Note: Water Quality: Eutrophication and Hypoxia*, June, pp. 1–8 (http://pdf.wri.org/eutrophication_sources_and_drivers.pdf).

Shaffer, K.H. and Runkle, D.L. (2007). *Consumptive Water Use Coefficient for the Great Lakes Basin and Climatically Similar Areas*. Washington, DC: U.S. Geological Survey.

Shelby, B. and Whittaker, D. (1995). Flows and recreation quality on the Dolores River: integrating overall and specific evaluations. *Rivers*, 5(2), 121–131.

Shelby, B., Vaske, J.J., and Heberlein, T.A. (1989). Comparative analysis of crowding in multiple locations: results from fifteen years of research. *Leisure Sciences*, 11, 269–291.

Silva, J.A. (2008). International trade and the changing demand for skilled workers in high-tech manufacturing. *Growth and Change*, 39(2), 225–251.

Smith, K.L. (1986). *The Magnificent Experiment: Building the Salt River Reclamation Project, 1890–1917*. Tucson: University of Arizona Press.

Solley, W.B., Pierce, R.R., and Perlman, H.A. (1998). *Estimated Use of Water in the United States in 1995*. Washington, DC: U.S. Geological Survey.

Spellman, F.R. (2003). *Handbook of Water and Wastewater Treatment Plant Operations*. Boca Raton, FL: Lewis Publishers.
Spellman, F.R. (2013). *Handbook of Water and Wastewater Treatment Plant Operations*, 3rd ed. Boca Raton, FL: CRC Press.
Spellman, F.R. and Bieber, R. (2010). *The Science of Renewable Energy*. Boca Raton, FL: CRC Press.
Stevens, J.E. (1988). *Hoover Dam: An American Adventure*. Norman: University of Oklahoma Press.
Strzepek, K. and Boehlert, B. (2010). Competition for water for the food system. *Philosophical Transactions of the Royal Society*, 365, 2927–2940.
Strzepek, K., Yohe, G., Neumann, J., and Boehlert, B. (2010). Characterizing changes in drought risk for the United States from climate change. *Environmental Research Letters*, 5(4).
Swann, L. (1992). *A Basic Overview of Aquaculture*. Washington, DC: U.S. Department of Agriculture.
Tellinghuisen, S. (2011). *Every Drop Counts: Valuing the Water Used to Generate Electricity*. Boulder, CO: Western Resource Advocates.
Terrell, J.U. (1965). *War for the Colorado River: The California–Arizona Controversy*, Vol. 1. Glendale, CA: Arthur H. Clark.
Torcellini, P., Long, N., and Judkoff, R. (2003). *Consumptive Water Use for U.S. Power Production*. Washington, DC: National Renewable Energy Laboratory, U.S. Department of Energy.
TTI. (2007). *A Modal Comparison of Domestic Freight Transportation Effects on the General Public*. Houston: Texas A&M Transportation Institute Center for Ports & Waterways.
Turk, J. and Turk, A. (1988). *Environmental Science*, 4th ed. Philadelphia, PA: Saunders College Publishing.
Tyler, D. (1992). *The Last Water Hole in the West: The Colorado–Big Thompson Project and the Northern Colorado Water Conservancy District*. Niwot: University Press of Colorado.
Tyler, D. (2003). *Silver Fox of the Rockies: Delphus E. Carpenter and Western Water Compacts*. Norman: University of Oklahoma Press.
Uhlenbrook, S. (2007). Biofuel and water cycle dynamics: what are the related challenges for hydrological processes research? *Hydrological Processes*, 21, 3647–3650.
United Nations. (2009). *Water Assessment Programme: Water in a Changing World*. New York: United Nations.
United Nations. (2010). *National Accounts Main Aggregates Database*. New York: Statistics Division, United Nations (https://unstats.un.org/unsd/snaama/Introduction.asp).
U.S. Census Bureau. (1988). *Statistical Abstract of the United States: 1987*, 107th ed. Washington, DC: U.S. Census Bureau.
U.S. Census Bureau. (2007a). *2007 Economic Census*. Washington, DC: U.S. Census Bureau.
U.S. Census Bureau. (2007b). *U.S. Census of Government Employment, 2007*. Washington, DC: U.S. Census Bureau.
U.S. Census Bureau. (2008). *Pollution Abatement Costs and Expenditures, 2005*. Washington, DC: U.S. Census Bureau.
U.S. Census Bureau. (2009). *U.S. Census of Governments Survey of State and Local Government Finances*. Washington, DC: U.S. Census Bureau.
U.S. Census Bureau. (2011). *2009 Non-Employer Statistics*. Washington, DC: U.S. Census Bureau (http://www.census.gov/eocn/nonemployr/index.html).
USACE. (2009). *Great Lakes Navigation System: Economic Strength to the Nation*. Alexandria, VA: U.S. Army Corps of Engineers.
USACE. (2010). *The U.S. Waterway System*. Alexandria, VA: U.S. Army Corps of Engineers.
USACE. (2011a). *Dredged Material Management Plan for the Port of New York and New Jersey*. Alexandria, VA: U.S. Army Corps of Engineers.
USACE. (2011b). *Great Lakes Navigation* System. Alexandria, VA: U.S. Army Corps of Engineers.
USDA. (2009). *2007 Census of Agriculture*. Washington, DC: U.S. Department of Agriculture.
USDA. (2010). *Farm and Ranch Irrigation Survey, 2008*. Washington, DC: U.S. Department of Agriculture.
USDA. (2013). *Livestock Water Quality*. Fort Keogh, MT: U.S. Department of Agriculture (http://www.ars.usda.gov/SP2UserFiles/Place/54340000/Research/WATERQUALITYMKP6-09.pdf).
USDOC. (2001). *National Security Assessment of the U.S. Shipbuilding and Repair Industry*. Washington, DC: U.S. Department of Commerce.
USDOC. (2011). *Fisheries of the United States, 2010*. Washington, DC: U.S. Department of Commerce.
USDOE. (2006). *Energy Demands on Water Resources: Report to Congress on the Interdependency of Energy and Water*. Washington, DC: U.S. Department of Energy.
USDOI. (2013). *Oil Shale & Tar Sands Programmatic EIS*. Washington, DC: Bureau of Land Management, U.S. Department of Interior (http://ostseis.anl.gov/).

USDOT. (2010). *Freight Transportation: Global Highlights*. Washington, DC: U.S. Department of Transportation (http://www.rita.dot.gov/bts/sites/rita.dot.gov.bts/files/publications/freight_transportation/index.html).
USDOT. (2011a). *Bureau of Transportation Statistics*. Washington, DC: U.S. Department of Transportation (http://www.bts.gov/publications/national_transportation_statisitics/).
USDOT. (2011b). *Maritime Administration*. Washington, DC: U.S. Department of Transportation (http://www.marad.dot.gov/).
USDOT. (2011c). *U.S. Water Transportation Statistical Snapshot*. Washington, DC: U.S. Department of Transportation (http://www.marad.dot.gov/documents/US_Water_Transportation_Statistical_snapshot.pdf).
USDOT. (2013). *Vessel Calls Snapshot 2011*. Washington, DC: U.S. Department of Transportation (http://www.marad.dot.gov/documents/Vessel_Calls_at_US_Ports_Snapshot.pdf).
USEPA. (2000). *Using Water Efficiently: Ideas for Residences*. Washington, DC: U.S. Environmental Protection Agency (http://www.epa.gov/watersense/docs/residence_508.pdf).
USEPA. (2002). *A Guidance Manual to Support the Assessment of Contaminated Sediments in Freshwater Ecosystems*. Vol. 1. *An Ecosystem-Based Framework for Assessing and Managing Contaminated Sediments*, EPA-905-B02-001-A. Washington, DC: U.S. Environmental Protection Agency (http://www.cerc.usgs.gov/pubs/sedtox/volumeI.pdf).
USEPA. (2006). *What Is a Watershed?* Washington, DC: U.S. Environmental Protection Agency (http://water.epa.gov/type/watersheds/whatis.cfm).
USEPA. (2008a). *Outdoor Water Use in the United States*, EPA-832-F-06-005. Washington, DC: U.S. Environmental Protection Agency (http://www.epa.gov/WaterSense/docs/ws_outdoor508.pdf).
USEPA. (2008b). *Indoor Water Use in the United States*, EPA-832-F-06-004. Washington, DC: U.S. Environmental Protection Agency (http://www.epa.gov/WaterSense/docs/ws_indoor508.pdf).
USEPA. (2008c). *2008 Sector Performance Report*, EPA-100-R-08-002 Washington, DC: U.S. Environmental Protection Agency (http://www.epa.gov/sectors/pdf/performance-rpt-2008.pdf).
USEPA. (2008d). *Gulf Hypoxia Action Plan 2008*. Washington, DC: U.S. Environmental Protection Agency (http://water.epa.gov/type/watersheds/named/msbasin/upload/2008_8_28_msbasin_ghap2008_update082608.pdf).
USEPA. (2008e). *National Coastal Condition Report III*. Washington, DC: U.S. Environmental Protection Agency.
USEPA. (2009a). *2006 Community Water System Survey*. Vol. I. *Overview*, EPA-815-R-09-001. Washington, DC: U.S. Environmental Protection Agency (http://water.epa.gov/infrastructure/drinkingwater/pws/upload/cwssreportvolumeI2006.pdf).
USEPA. (2009b). *2006 Community Water System Survey*. Vol. II. *Detailed Tables and Survey Methodology*, EPA-815-R-09-002. Washington, DC: U.S. Environmental Protection Agency (http://water.epa.gov/infrastructure/drinkingwater/pws/upload/cwssreportvolumeII2006.pdf).
USEPA. (2009c). *2007 Drinking Water Infrastructure Needs Survey and Assessment*. Washington, DC: U.S. Environmental Protection Agency (http://water.epa.gov/infrastructure/drinkingwater/dwns/).
USEPA. (2010). *2010 Advisory Listing*. Washington, DC: U.S. Environmental Protection Agency (http://water.epa.gov/scitech/swguidance/fishshellfish/fishadvisories/upload/nlfa_slides_2011.pdf).
USEPA. (2011). *2010 Biennial Listing of Fish Advisories*, EPA-820-F-11-014. Washington, DC: U.S. Environmental Protection Agency (http://water.epa.gov/scitech/swguidance/fishshellfish/fishadvisories/upload/technical_factsheet_2010.pdf).
USEPA. (2012a). *The Importance of Water to the U.S. Economy*. Part I. *Background Report*. Washington, DC: U.S. Environmental Protection Agency (http://water.epa.gov/action/importanceofwater/upload/2-EPA-s-Background-Report-The-Importance-of-Water-to-the-US-Economy.pdf).
USEPA. (2012b). *Frequently Asked Questions: Water Infrastructure & Sustainability*. Washington, DC: U.S. Environmental Protection Agency (http://water.epa.gov/infrastructure/sustain/si_faqs.cfm).
USEPA. (2014a). *Mississippi River Gulf of Mexico Watershed Nutrient Task Force*. Washington, DC: U.S. Environmental Protection Agency (http://water.epa.gov/type/watersheds/named/msbasin/index.cfm).
USEPA. (2014b). *National Water Program 2012 Strategy: Response to Climate Change*. Washington, DC: U.S. Environmental Protection Agency (http://water.epa.gov/scitech/climatechange/2012-National-Water-Program-Strategy.cfm).
USEPA. (2014c). *Watershed Assessment, Tracking & Environmental Results: National Summary of State Information*. Washington, DC: U.S. Environmental Protection Agency (http://iaspub.epa.gov/waters10/attains_nation_cy.control).

USFWS. (2006a). *2006 National Survey of Fishing, Hunting, and Wildlife-Associated Recreation*. Washington, DC: U.S. Fish and Wildlife Service.

USFWS. (2006b). *Economic Impact of Waterfowl Hunting in the United States: Addendum to the 2006 National Survey of Fishing, Hunting, and Wildlife-Associated Recreation*. Washington, DC: U.S. Fish and Wildlife Service.

USGS. (1988). *Estimated Use of Water in the United States in 1985*. Washington, DC: U.S. Geological Survey.

USGS. (1998). *Estimated Use of Water in the United States in 1995*. Washington, DC: U.S. Geological Survey.

USGS. (2004). *Estimated Use of Water in the United States in 2000*. Washington, DC: U.S. Geological Survey.

USGS. (2006). *Water Science in Schools*. Washington, DC: U.S. Geological Survey.

USGS. (2009). *Estimated Water Use in the United States in 2005*. Washington, DC: U.S. Geological Survey.

USGS. (2011). *2009 Minerals Yearbook*. Washington, DC: U.S. Geological Survey.

USGS. (2013). *Invasive Species*. Washington, DC: Great Lakes Science Center, U.S. Geological Survey (http://www.glsc.usgs.gov/invasive-species).

Venkatachalam, L. (2004). The contingent valuation method: a review. *Environmental Impact Assessment Review*, 24, 89–124.

Wade, W.W., Hewitt, J.A., and Nussbaum, M.T. (1991). *Cost of Industrial Water Shortages*, prepared by Spectrum Economics, Inc., for California Urban Water Agencies.

Walton, B. (2010). *The Price of Water: A Comparison of Water Rates, Usage in 30 U.S. Cities*. Traverse City, MI: Circle of Blue (http://www.circleofblue.org/waternews/2010/world/the-price-of-water-a-comparison-of-water-rates-usage-in-30-u-s-cities/).

Wang, H. and Lall, S. (1999). Valuing water for Chinese industries: a marginal productivity analysis. *Sinosphere*, 2(3), 27–50.

Waterways Council. (2007). *Economic Competitiveness Demands Waterways Modernization*. Houston, TX: Waterways Council, Inc.

Wiebe, K. and Gollehon, N., Eds. (2006). *Agricultural Resources and Environmental Indicators, 2006 Edition*. Washington, DC: Economic Research Service, U.S. Department of Agriculture (http://www.ers.usda.gov/publications/arei/eib16/).

Wiley, P., Leeworthy, V., and Stone, E. (2006). *Southern California Beach Valuation Project: Economic Impact of Beach Closures and Changes in Water Quality for Beaches in Southern California*. Washington, DC: National Oceanic and Atmospheric Administration.

Wilkinson, C.F. (1992). *Crossing the Next Meridian: Land, Water, and the Future of the West*. Washington, DC: Island Press.

World Bank Group. (1999). Pulp and paper mills. In: *Pollution Prevention and Abatement Handbook 1998: Toward Cleaner Production*. Washington, DC: World Bank Group.

World Bank Group. (2010). *The Cost to Developing Countries of Adapting to Climate Change: New Methods and Estimates*. Washington, DC: World Bank Group.

World Bank Group. (2012). *World Development Indicators*. Washington, DC: World Bank Group (http://data.worldbank.org/indicator/NY.GDP.MKTP.CD).

Worster, D. (1985). *Rivers of Empire: Water, Aridity, and the Growth of the American West*. New York: Pantheon.

WSDOT. (2014). *Nation's Largest Ferry System*. Seattle: Washington State Department of Transportation (http://www.wsdot.wa.gov/NR/rdonlyres/6C78A08B-19A1-4919-B6E6-E9EF83E6376D/96747/WSFFactSheet_Feb2014_Web.pdf).

Yang, X. and Dziegielewski, B. (2007). Water use by thermoelectric power plants in the United States. *Journal of the American Water Resources Association*, 43, 160–169.

Young, R.A. (2005). *Determining the Economic Value of Water: Concepts and Methods*. Washington, DC: Resources for the Future.

Zellmer, S. (2006). *In-Stream Flow Legislation*. Lincoln: University of Nebraska.

Zemanek, S. (2011). U.S. travel and tourism satellite accounts for 2007–2010. *Survey of Current Business*, June (https://www.bea.gov/scb/pdf/2011/06%20June/0611_travel.pdf).

11 Water Treatment

> If you visit American city,
> You will find it very pretty.
> Just two things of which you must beware:
> Don't drink the water and don't breathe the air!
>
> **—Tom Lehrer (*Pollution*, 1965)**

INTRODUCTION

He wandered the foggy, filthy, garbage-strewn, corpse-ridden streets of 1854 London searching, making notes, always looking … seeking a murderous villain. No, not Jack the Ripper, but a killer just as insidious and unfeeling. He finally found the miscreant and took action; that is, he removed the handle from a water pump. And, fortunately for untold thousands of lives, his was a lifesaving action. He was a detective—of sorts. Not the real Sherlock Holmes, but absolutely as clever, as skillful, as knowledgeable, as intuitive—and definitely as driven. His name was Dr. John Snow. His middle name? Common Sense. The master criminal Snow sought? A mindless, conscienceless, brutal killer: cholera. Let's take a closer look at this medical super sleuth and his quarry. More to the point, let's look at Dr. Snow's impact on water treatment (disinfection) of raw water used for potable and other purposes.

Dr. John Snow

Dr. Snow (1813–1858) was an unassuming but creative London obstetrician who achieved prominence in the mid-19th century for proving his theory (published in his *On the Mode of Communication of Cholera*) that cholera is a contagious disease caused by a "poison" that reproduces in the human body and is found in the vomitus and stools of cholera patients. He theorized that the main (though not the only) means of transmission was water contaminated with this poison. Many opinions about cholera's cause had been offered, but Dr. Snow's theory was not held in high regard at first because a commonly held and popular theory was that diseases are transmitted by the inhalation of vapors. In the beginning, Snow's argument did not cause a great stir; it was only one of the many hopeful theories proposed during a time when cholera was causing great distress. Eventually, Snow was able to prove his theory. We describe how Snow accomplished this later, but for now let's take a look again at Snow's target: cholera.

Cholera

According to the U.S. Centers for Disease Control and Prevention (CDC), cholera is an acute, diarrheal illness caused by infection of the intestine with the bacterium *Vibrio cholerae*. The infection is often mild or without symptoms, but it can sometimes be quite severe. Approximately 1 in 20 infected persons experience severe disease symptoms such as profuse watery diarrhea, vomiting, and leg cramps. In these persons, rapid loss of body fluids leads to dehydration and shock. Without treatment, death can occur within hours.

> ***Note:*** You don't need to be a rocket scientist to figure out just how deadly cholera was during the London cholera outbreak of 1854. Comparing the state of medicine at that time to ours is like comparing the speed potential of a horse and buggy to a state-of-the-art NASCAR race car today.

> Simply stated, cholera was the classic epidemic disease of the 19th century, as the plague had been for the 14th century. Its defeat was a reflection of both common sense and of progress in medical knowledge—and of the enduring changes in European and American social thought.

How does a person contract cholera? Again, we refer to the CDC for our answer. A person may contract cholera (even today) by drinking water or eating food contaminated with the cholera bacterium. In an epidemic, the source of the contamination is usually feces of an infected person. The disease can spread rapidly in areas with inadequate treatment of sewage and drinking water. Disaster areas often pose special risks; for example, the aftermath of Hurricane Katrina in New Orleans raised concerns about a potential cholera outbreak.

The cholera bacterium also lives in brackish river and coastal waters. Raw shellfish have been a source of cholera, and a few people in the United States have contracted it from eating shellfish from the Gulf of Mexico. The disease is not likely to spread directly from one person to another; therefore, casual contact with an infected person does not pose a risk for transmission of the disease.

Flashback to 1854 London

Basically, for our purposes, the CDC confirms the fact that cholera is a waterborne disease. Today, we know quite a lot about cholera and its transmission, how to prevent infection, and how to treat it. But what did they know about cholera in the 1850s? Not much, but one thing is certain: They knew cholera was a deadly killer. That was just about all they knew, until Dr. John Snow proved his theory. Recall that Snow theorized that cholera is a contagious disease caused by a poison that reproduces in the human body and is found in the vomitus and stools of cholera victims. He also believed that the main means of transmission was water contaminated with this poison.

Dr. Snow's theory was correct, of course, as we know today. The question is, how did he prove his theory correct? The answer to that question provides us with an account of one of the all-time legendary quests for answers in epidemiological research—and an interesting story.

Dr. Snow did his research during yet another severe cholera epidemic in London. Though ignorant of the concept of bacteria carried in water (germ theory), he traced an outbreak of cholera to a water pump located at the intersection of Cambridge and Broad Street. How did he do that? He began his investigation by determining where in London persons with cholera lived and worked. He then used this information to map the distribution of cases on what epidemiologists call a "spot map." His map indicated that the majority of the deaths occurred within 250 yards of that communal water pump. The water pump was used regularly by most of the area residents. Those who did not use the pump remained healthy. Suspecting that the Broad Street pump was the source of the plague, Snow had the water pump handle removed, which ended the cholera epidemic.

Sounds like a rather simple solution, doesn't it? Remember, though, that in that era aspirin had not yet been formulated, to say nothing of the development of other medical miracles we now take for granted (such as antibiotics). Dr. Snow, through the methodical process of elimination and linkage (Sherlock Holmes would have been impressed—and was), proved his point and his theory. Specifically, he painstakingly documented the cholera cases and correlated the comparative incidence of cholera among subscribers to the city's two water companies. He learned that one company drew water from the lower Thames River, and the other company obtained water from the upper Thames. Dr. Snow discovered that cholera was much more prevalent among customers of the water company that drew its water from the lower Thames, where the river had become contaminated with London sewage. Dr. Snow tracked the source of the Broad Street pump, and it was the contaminated lower Thames, of course.

Dr. Snow, an obstetrician, became the first effective practitioner of scientific epidemiology. His creative use of logic, common sense (removing the handle from the pump), and scientific observation enabled him to solve a major medical mystery and earned him the title of "father of field epidemiology." Today, Dr. John Snow is known as the father of modern epidemiology.

From Pump Handle Removal to Water Treatment (Disinfection)

Dr. John Snow's major contribution to the medical profession, to society, and to humanity in general can be summarized rather succinctly: He determined and proved that the deadly disease cholera is a waterborne disease. (Incidentally, Dr. Snow's second medical accomplishment was that he was the first doctor to administer anesthesia during childbirth.)

What does all of this have to do with water treatment (disinfection)? Actually, Dr. Snow's discovery has quite a lot to do with water treatment. Combating any disease is difficult without a determination of how the disease is transmitted—how it travels from vector or carrier to receiver. Dr. Snow established this connection, and from his work, and the work of others, progress was made in understanding and combating many different waterborne diseases.

Today, sanitation problems in developed countries (those with the luxury of adequate financial and technical resources) deal more with the consequences that arise from inadequate commercial food preparation, and the results of bacteria becoming resistant to disinfection techniques and antibiotics. We simply flush our toilets to rid ourselves of unwanted wastes, and we turn on our taps to access a high-quality drinking water supply, from which we've all but eliminated cholera and epidemic diarrheal diseases. This is generally the case in most developed countries today—but it certainly wasn't true in Dr. Snow's time.

The progress in water treatment from that notable day in 1854 when Snow made the connection between deadly cholera and its means of transmission to the present reads like a chronology of discoveries. This makes sense, of course, because with the passage of time, pivotal events and discoveries occur that have a profound effect on how we live today. Let's take a look at a few elements of the important chronological progression that evolved from the simple removal of a pump handle to the advanced water treatment (disinfection) methods we employ today to treat our water supplies.

After Dr. Snow's discovery, subsequent events began to drive the water/wastewater treatment process. In 1859, the British Parliament was suspended during the summer because the stench coming from the Thames was unbearable. According to one account, the river began to "seethe and ferment under a burning sun." As was the case in many cities at this time, storm sewers carried a combination of storm water, sewage, street debris, and other wastes to the nearest body of water. In the 1890s, Hamburg, Germany, suffered a cholera epidemic. Detailed studies by Dr. Robert Koch tied the outbreak to a contaminated water supply. In response to the epidemic, Hamburg was among the first cities to use chlorine as part of a wastewater treatment regimen. About the same time, the town of Brewster, New York, became the first U.S. city to disinfect its treated wastewater. Chlorination of drinking water was used on a temporary basis in 1896, and its first known continuous use for water supply disinfection occurred in Lincoln, England, and in Chicago in 1905. Jersey City became one of the first routine users of chlorine in 1908.

Time marched on and with it came increased realization of the need to treat and disinfect both water supplies and wastewater. Between 1910 and 1915, technological improvements in gaseous and then solution feed of elemental chlorine (Cl_2) made the process more practical and efficient. Disinfection of water supplies and chlorination of treated wastewater for odor control increased over the next several decades. In the United States, disinfection, in one form or another, is now being used by more than 15,000 out of approximately 16,000 publicly owned treatment works (POTW). The significance of this number becomes apparent when you consider that fewer than 25 of the 600-plus POTW in the United States in 1910 were using disinfectants.

CONVENTIONAL WATER TREATMENT

This section focuses on water treatment operations and the various unit processes currently used to treat raw source water before it is distributed to the user. In addition, the reasons for water treatment and the basic theories associated with individual treatment unit processes are discussed. Water treatment systems are installed to remove those materials that cause disease or create nuisances.

At its simplest level, the basic goal of water treatment operations is to protect public health, with a broader goal to provide potable and palatable water. The bottom line is that the water treatment process functions to provide water that is safe to drink and is pleasant in appearance, taste, and odor.

In this chapter, *water treatment* is defined as any unit process that changes or alters the chemical, physical, or bacteriological quality of water with the purpose of making it safe for human consumption or appealing to the customer. Treatment also is used to protect the water distribution system components from corrosion. Many water treatment unit processes are commonly used today. The treatment processes used depend on the evaluation of the nature and quality of the particular water to be treated and the desired quality of the finished water. For water treatment unit processes employed to treat raw water, one thing is certain: As new U.S. Environmental Protection Agency (USEPA) regulations take effect, many more processes will come into use in an attempt to produce water that complies with all current regulations, despite source water conditions.

Small water systems tend to use a smaller number of the wide array of unit treatment processes available, in part because they usually rely on groundwater as the source and because their small size makes many sophisticated processes impractical (e.g., too expensive to install, too expensive to operate, too sophisticated for limited operating staff). This chapter concentrates on those individual treatment unit processes usually found in conventional water treatment systems, corrosion control methods, and fluoridation.

Waterworks Operators

Operation of a water treatment system, no matter the size or complexity, requires operators. To perform their functions at the highest knowledge and experience level possible, operators must understand the basic principles and theories behind many complex water treatment concepts and treatment systems. Under new regulations, waterworks operators must be certified or licensed. Actual water treatment protocols and procedures are important; however, without proper implementation, they are nothing more than hollow words occupying space on reams of paper. This is where the waterworks operator comes in. To successfully treat water requires skill, dedication, and vigilance. The waterworks operator must be not only highly trained and skilled but also conscientious—the ultimate user demands nothing less.

Purpose of Water Treatment

The purpose of water treatment is to condition, modify, or remove undesirable impurities and to provide water that is safe, palatable, and acceptable to users. This may seem an obvious, expected purpose of treating water, but various regulations also require water treatment. Some regulations state that if the contaminants listed under the various regulations are found in excess of maximum contaminant levels (MCLs) then the water must be treated to reduce the levels. If a well or spring source is surface influenced, treatment is required, regardless of the actual presence of contamination. Some impurities affect the aesthetic qualities (taste, odor, color, and hardness) of the water; if they exceed secondary MCLs established by the USEPA and the state, the water may have to be treated.

If we assume that the water source used to feed a typical water supply system is groundwater (usually the case in the United States), a number of common groundwater problems may require water treatment. Keep in mind that water that must be treated for any one of these problems may also exhibit several other problems:

- Bacteriological contamination
- Hydrogen sulfide odors
- Hard water
- Corrosive water
- Iron and manganese

STAGES OF WATER TREATMENT

Figure 11.1 presents the conventional model discussed throughout this text. The figure clearly illustrates that water treatment is made up of various stages or unit processes combined to form one treatment system. Note that a given waterworks may contain all the unit processes discussed in the following or any combination of them. One or more of these stages may be used to treat any one or more of the source water problems listed earlier. Also note that the model shown in Figure 11.1 does not necessarily apply to very small water systems. In some small systems, water treatment may consist of nothing more than removal of water via pumping from a groundwater source to storage to distribution. In some small water supply operations, disinfection may be added because it is required. Although it is likely that the basic model shown in Figure 11.1 does not reflect the type of treatment process used in most small systems, we use it in this handbook for illustrative and instructive purposes because higher level licensure requires operators, at a minimum, to learn these processes.

PRETREATMENT

Simply stated, water pretreatment (also called *preliminary treatment*) is any physical, chemical, or mechanical process used before main water treatment processes. It can include screening, presedimentation, and chemical addition. Pretreatment in water treatment operations usually consists of oxidation or other treatment for the removal of tastes and odors, iron and manganese, trihalomethane precursors, or entrapped gases (such as hydrogen sulfide). Unit processes may include chlorine, potassium permanganate or ozone oxidation, activated carbon addition, aeration, and presedimentation. Pretreatment of surface water supplies accomplishes the removal of certain constituents and materials that interfere with or place an unnecessary burden on conventional water treatment facilities.

Based on experience and according to the Texas Water Utilities Association (TWUA, 1988), typical pretreatment processes include the following:

- Removal of debris from water from rivers and reservoirs that would clog pumping equipment
- Destratification of reservoirs to prevent anaerobic decomposition that could result in reducing iron and manganese in the soil to a state that would be soluble in water which can cause subsequent removal problems in the treatment plant; the production of hydrogen sulfide and other taste- and odor-producing compounds also results from stratification
- Chemical treatment of reservoirs to control the growth of algae and other aquatic growths that could result in taste and odor problems
- Presedimentation to remove excessively heavy silt loads prior to the treatment processes
- Aeration to remove dissolved odor-causing gases such as hydrogen sulfide and other dissolved gases or volatile constituents and to aid in the oxidation of iron and manganese, although manganese or high concentrations of iron are not removed during detention provided in conventional aeration units

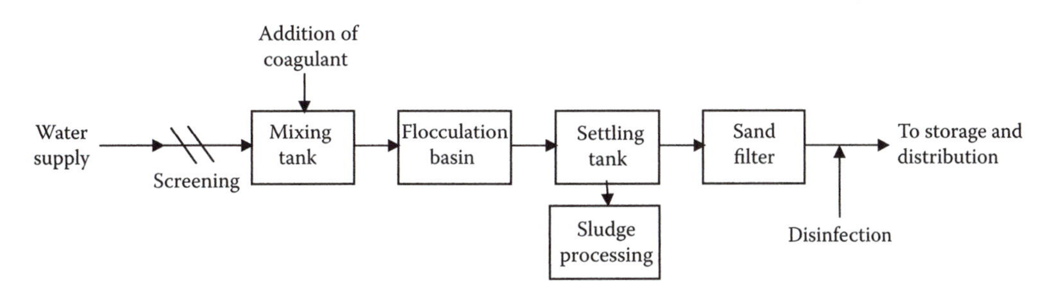

FIGURE 11.1 The conventional water treatment model.

- Chemical oxidation of iron and manganese, sulfides, taste- and odor-producing compounds, and organic precursors that may produce trihalomethanes upon the addition of chlorine
- Adsorption for removal of tastes and odors

Note: An important point to keep in mind is that, in small systems using groundwater as a source, pretreatment may be the only treatment process used. Pretreatment may be incorporated as part of the total treatment process or may be located adjacent to the source before the water is sent to the treatment facility.

Aeration

Aeration is commonly used to treat water that contains trapped gases (such as hydrogen sulfide) that can impart an unpleasant taste and odor to the water. Just allowing the water to rest in a vented tank will (sometimes) drive off much of the gas, but usually some form of forced aeration is needed. Aeration works well (about 85% of the sulfides may be removed) whenever the pH of the water is less than 6.5. Aeration may also be useful in oxidizing iron and manganese, oxidizing humic substances that might form trihalomethanes when chlorinated, eliminating other sources of taste and odor, or imparting oxygen to oxygen-deficient water.

Note: Iron is a naturally occurring mineral found in many water supplies. When the concentration of iron exceeds 0.3 mg/L, red stains will occur on fixtures and clothing. This increases customer costs for cleaning and replacement of damaged fixtures and clothing.

Manganese, like iron, is a naturally occurring mineral found in many water supplies. When the concentration of manganese exceeds 0.05 mg/L, black stains occur on fixtures and clothing. As with iron, this increases customer costs for cleaning and replacement of damaged fixtures and clothing. Iron and manganese are commonly found together in the same water supply. Iron and manganese are discussed in more detail later.

Screening

Screening is usually the first major step in the water pretreatment process (see Figure 11.1). It is defined as the process whereby relatively large and suspended debris is removed from the water before it enters the plant. River water, for example, typically contains suspended and floating debris varying in size from small rocks to logs. Removing these solids is important, not only because these items have no place in potable water but also because this river trash may cause damage to downstream equipment (e.g., clogging and damaging pumps), increase chemical requirements, impede hydraulic flow in open channels or pipes, or hinder the treatment process. The most important criteria used in the selection of a particular screening system for water treatment technology are the screen opening size and flow rate. Other important criteria include costs related to operation and equipment, plant hydraulics, debris handling requirements, and operator qualifications and availability. Large surface water treatment plants may employ a variety of screening devices including rash screens (or trash rakes), traveling water screens, drum screens, bar screens, or passive screens.

Chemical Addition

Two of the major chemical pretreatment processes used in treating water for potable use are iron and manganese and hardness removal. Another chemical treatment process that is not necessarily part of the pretreatment process but is also discussed in this section is corrosion control. Corrosion prevention is accomplished through chemical treatment—not only in the treatment process but also in the distribution process. Before discussing each of these treatment methods in detail, however, it is important to describe chemical addition, chemical feeders, and chemical feeder calibration.

When chemicals are used in the pretreatment process, they must be the proper ones, fed in the correct concentration and introduced to the water at the proper locations. Determining the proper amount of chemical to use is accomplished by testing. The operator must test the raw water periodically to determine if the chemical dosage should be adjusted. For surface supplies, checking must be done more frequently than for groundwater (remember, surface water supplies are subject to change on short notice, while groundwaters generally remain stable). The operator must be aware of the potential for interactions between various chemicals and how to determine the optimum dosage (e.g., adding both chlorine and activated carbon at the same point will minimize the effectiveness of both processes, as the adsorptive power of the carbon will be used to remove the chlorine from the water).

Note: Sometimes using too many chemicals can be worse than not using enough.

Prechlorination (distinguished from chlorination used in disinfection at the end of treatment) is often used as an oxidant to help with the removal of iron and manganese; however, a concern for systems that prechlorinate is the potential for the formation of total trihalomethanes (TTHMs), which form as a byproduct of the reaction between chlorine and naturally occurring compounds in raw water. The USEPA TTHM standard does not apply to water systems that serve fewer than 10,000 people, but operators should be aware of the impact and causes of TTHMs. Chlorine dosage or application point may be changed to reduce problems with TTHMs.

Note: TTHMs such as chloroform are known or suspected to be carcinogenic and are limited by water and state regulations.

All chemicals intended for use in drinking water must meet certain standards. Thus, when ordering water treatment chemicals, the operator must be confident that they meet all appropriate standards for drinking water use.

Note: To be effective, pretreatment chemicals must be thoroughly mixed with the water. Short-circuiting or plug flows of chemicals that do not come in contact with most of the water will not result in proper treatment.

Chemicals are normally fed with dry chemical feeders or solution (metering) pumps. Operators must be familiar with all of the adjustments required to control the rate at which the chemical is fed to the water (wastewater). Some feeders are manually controlled and must be adjusted by the operator when the raw water quality or the flow rate changes; other feeders are paced by a flow meter to adjust the chemical feed so it matches the water flow rate. Operators must also be familiar with chemical solution and feeder calibration.

As mentioned, a significant part of the waterworks operator's daily functions is measuring quantities of chemicals and applying them to water at preset rates. Normally accomplished semiautomatically by use of electromechanical–chemical feed devices, waterworks operators must still know what chemicals to add, how much to add to the water (wastewater), and the purpose of the chemical addition.

Chemical Solutions

A *water solution* is a homogeneous liquid made of the *solvent* (the substance that dissolves another substance) and the *solute* (the substance that dissolves in the solvent). Water is the solvent (see Figure 11.2). The solute (whatever it may be) may dissolve up to a certain limit. This level is its *solubility*—that is, the solubility of the solute in the particular solvent (water) at a particular temperature and pressure.

Remember, in chemical solutions, the substance being dissolved is the solute, and the liquid present in the greatest amount in a solution (that does the dissolving) is the solvent. The operator should also be familiar with another term—*concentration*, which is the amount of solute dissolved in a given amount of solvent. Concentration is measured as

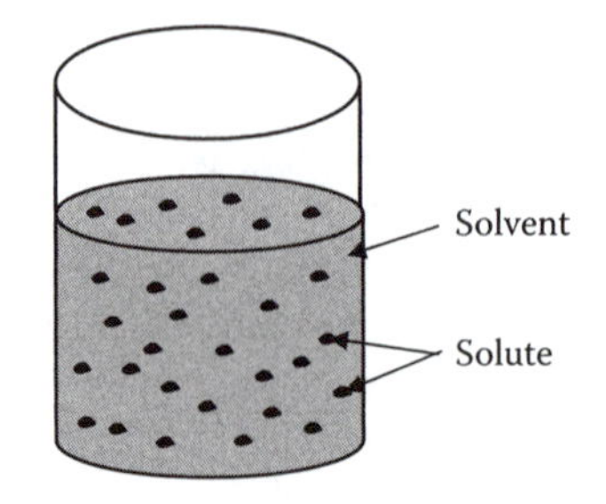

FIGURE 11.2 Solution with two components: solvent and solute.

$$\% \text{ Strength} = \frac{\text{Wt. of solute}}{\text{Wt. of solution}} \times 100 = \frac{\text{Wt. of solute}}{\text{Wt. of solute} + \text{solvent}} \times 100 \tag{11.1}$$

■ Example 11.1

Problem: If 30 lb of chemical are added to 400 lb of water, what is the percent strength (by weight) of the solution?

Solution:

$$\% \text{ Strength} = \frac{30 \text{ lb solute}}{400 \text{ lb water}} \times 100 = \frac{30 \text{ lb solute}}{30 \text{ lb solute} + 400 \text{ lb water}} \times 100 = 7.0 \text{ (rounded)}$$

Important to the process of making accurate computations of chemical strength is a complete understanding of the dimensional units involved; for example, operators should understand exactly what *milligrams per liter* (mg/L) signify:

$$\text{Milligrams per liter (mg/L)} = \frac{\text{Milligrams of solute}}{\text{Liters of solution}} \tag{11.2}$$

Another important dimensional unit commonly used when dealing with chemical solutions is *parts per million* (ppm):

$$\text{Parts per million (ppm)} = \frac{\text{Parts of solute}}{\text{Million parts of solution}} \tag{11.3}$$

Note: "Parts" is usually a weight measurement.

An example is

$$9 \text{ ppm} = \frac{9 \text{ lb solids}}{1{,}000{,}000 \text{ lb solution}} \qquad 9 \text{ ppm} = \frac{9 \text{ mg solids}}{1{,}000{,}000 \text{ mg solution}}$$

This leads us to two important parameters that operators should commit to memory:

- 1 mg/L = 1 ppm
- 1% = 10,000 mg/L

When working with chemical solutions, it is also necessary to be familiar with two important chemical properties: *density* and *specific gravity*. Density is defined as the weight of a substance per a unit of its volume—for example, pounds per cubic foot or pounds per gallon. Specific gravity is defined as the ratio of the density of a substance to a standard density.

$$\text{Density} = \frac{\text{Mass of substance}}{\text{Volume of substance}} \tag{11.4}$$

Here are a few key facts about density:

- Density is measured in units of lb/ft^3, lb/gal, or mg/L.
- Density of water = 62.5 lb/ft^3 = 8.34 lb/gal.
- Density of concrete = 130 lb/ft^3.
- Density of alum (liquid at 60°F) = 1.33.
- Density of hydrogen peroxide (35%) = 1.132.

$$\text{Specific gravity} = \frac{\text{Density of substance}}{\text{Density of water}} \tag{11.5}$$

Here are a few facts about specific gravity:

- Specific gravity has no units.
- Specific gravity of water = 1.0.
- Specific gravity of concrete = 2.08.
- Specific gravity of alum (liquid at 60°F) = 1.33.
- Specific gravity of hydrogen peroxide (35%) = 1.132.

Chemical Feeders

Simply put, a chemical feeder is a mechanical device for measuring a quantity of chemical and applying it to water at a preset rate.

Types of Chemical Feeders

Two types of chemical feeders are commonly used: solution (or liquid) feeders and dry feeders. Liquid feeders apply chemicals in solutions or suspensions, and dry feeders apply chemicals in granular or powdered forms. In a solution feeder, the chemical enters and leaves the feeder in a liquid state; in a dry feeder, the chemical enters and leaves the feeder in a dry state.

Solution Feeders

Solution feeders are small, positive-displacement metering pumps of three types: (1) reciprocating (piston-plunger or diaphragm type), (2) vacuum type (e.g., gas chlorinator), or (3) gravity feed rotameter (e.g., drip feeder). Positive-displacement pumps are used in high-pressure, low-flow applications; they deliver a specific volume of liquid for each stroke of a piston or rotation of an impeller.

Dry Feeders

Two types of dry feeders are *volumetric* and *gravimetric*, depending on whether the chemical is measured by volume (volumetric) or weight (gravimetric). Simpler and less expensive than gravimetric pumps, volumetric dry feeders are also less accurate. Gravimetric dry feeders are extremely accurate, deliver high feed rates, and are more expensive than volumetric feeders.

Chemical Feeder Calibration

Chemical feeder calibration ensures effective control of the treatment process. Obviously, chemical feed without some type of metering and accounting of chemical used adversely affects the water treatment process. Chemical feeder calibration also optimizes economy of operation; that is, it ensures the optimum use of expensive chemicals. Finally, operators must have accurate knowledge of the capabilities of each individual feeder at specific settings. When a certain dose must be administered, the operator must rely on the feeder to feed the correct amount of chemical. Proper calibration ensures chemical dosages can be set with confidence. At a minimum, chemical feeders must be calibrated on an annual basis. During operation, when the operator changes chemical strength or chemical purity or makes any adjustment to the feeder, or when the treated water flow changes, the chemical feeder should be calibrated. Ideally, any time maintenance is performed on chemical feed equipment, calibration should be performed.

What factors affect chemical feeder calibration (i.e., feed rate)? For solution feeders, calibration is affected any time solution strength changes, any time a mechanical change is introduced in the pump (change in stroke length or stroke frequency), or whenever flow rate changes. In the dry chemical feeder, calibration is affected any time chemical purity changes or mechanical damage occurs (e.g., belt change), or whenever flow rate changes. In the calibration process, calibration charts are usually used or made up to fit the calibration equipment. The calibration chart is also affected by certain factors, including change in chemical, changes in the flow rate of the water being treated, or a mechanical change in the feeder.

■ Example 11.2

To demonstrate that performing a chemical feed procedure is not necessarily as simple as opening a bag of chemicals and dumping the contents into the feed system, we provide a real-world example below.

Problem: Consider the chlorination dosage rates below.

	Setting	Dosage
100%	111/121	0.93 mg/L
70%	78/121	0.66 mg/L
50%	54/121	0.45 mg/L
20%	20/121	0.16 mg/L

Solution: This is not a good dosage setup for a chlorination system. Maintenance of a chlorine residual at the ends of the distribution system should be within 0.5 to 1.0 ppm. At 0.9 ppm, dosage will probably result in this range—depending on the chlorine demand of the raw water and detention time in the system. However, the pump is set at its highest setting. We have room to decrease the dosage but no ability to increase the dosage without changing the solution strength in the solution tank. In this example, doubling the solution strength to 1% provides the ideal solution, resulting in the following chart changes.

	Setting	Dosage
100%	222/121	1.86 mg/L
70%	154/121	1.32 mg/L
50%	108/121	0.90 mg/L
20%	40/121	0.32 mg/L

This is ideal, because the dosage we want to feed is at the 50% setting for our chlorinator. We can now easily increase or decrease the dosage, whereas the previous setup only allowed the dosage to be decreased.

Iron and Manganese Removal

Iron and manganese are frequently found in groundwater and in some surface waters. They do not cause health-related problems but are objectionable because they may cause aesthetic problems. Severe aesthetic problems may cause consumers to avoid an otherwise safe water supply in favor of one of unknown or questionable quality, or they may cause customers to incur unnecessary expenses for bottled water. Aesthetic problems associated with iron and manganese include the discoloration of water (iron, reddish water; manganese, brown or black water), staining of plumbing fixtures, a bitter taste, and the growth of microorganisms.

Although no health concerns are directly associated with iron and manganese, the growth of iron bacteria slimes may cause indirect health problems. Economic problems include damage to textiles, dye, paper, and food. Iron residue (or tuberculation) in pipes increases pumping head, decreases carrying capacity, may clog pipes, and may corrode through pipes.

> ***Note:*** Iron and manganese are secondary contaminants. Their secondary maximum contaminant levels (SMCLs) are 0.3 mg/L for iron and 0.05 mg/L for manganese.

Iron and manganese are most likely found in groundwater supplies, industrial waste, and acid mine drainage and are byproducts of pipeline corrosion. They may accumulate in lake and reservoir sediments, causing possible problems during lake/reservoir turnover. They are not usually found in running waters (e.g., streams, rivers).

Iron and Manganese Removal Techniques

Chemical precipitation treatments for iron and manganese removal are called *deferrization* and *demanganization*, respectively. The usual process is *aeration*, where dissolved oxygen in the chemical causes precipitation; chlorine or potassium permanganate may also be required.

Precipitation

Precipitation (or pH adjustment) of iron or manganese from water in their solid forms can be performed in treatment plants by adjusting the pH of the water through the addition of lime or other chemicals. Some of the precipitate will settle out with time, while the rest is easily removed by sand filters. This process requires the pH of the water to be in the range of 10 to 11.

> ***Note:*** Although the precipitation or pH adjustment technique for treating water containing iron and manganese is effective, note that the pH level must be adjusted higher (10 to 11) to cause the precipitation, which means that the pH level must also then be lowered (to 8.5 or a bit lower) to use the water for consumption.

Oxidation

One of the most common methods for removing iron and manganese is the process of oxidation (another chemical process), usually followed by settling and filtration. Air, chlorine, or potassium permanganate can oxidize these minerals. Each oxidant has advantages and disadvantages, as each operates slightly differently:

- *Air*—To be effective as an oxidant, the air must come in contact with as much of the water as possible. Aeration is often accomplished by bubbling diffused air through the water by spraying the water up into the air or by trickling the water over rocks, boards, or plastic packing materials in an aeration tower. The more finely divided the drops of water, the more oxygen comes in contact with the water and the dissolved iron and manganese.
- *Chlorine*—This is one of the most popular oxidants for iron and manganese control because it is also widely used as a disinfectant; controlling iron and manganese by prechlorination can be as simple as adding a new chlorine feed point in a facility already feeding chlorine. It also provides a predisinfecting step that can help control bacterial growth throughout

the rest of the treatment system. The downside to using chorine is that when chlorine reacts with the organic materials found in surface water and some groundwaters it forms TTHMs. This process also requires that the pH of the water be in the range of 6.5 to 7; because many groundwaters are more acidic than this, pH adjustment with lime, soda ash, or caustic soda may be necessary when oxidizing with chlorine.

- *Potassium permanganate*—This is the best oxidizing chemical to use for manganese control removal. An extremely strong oxidant, it has the additional benefit of producing manganese dioxide during the oxidation reaction. Manganese dioxide acts as an adsorbent for soluble manganese ions. This attraction for soluble manganese provides removal to extremely low levels.

The oxidized compounds form precipitates that are removed by a filter. Note that sufficient time should be allowed from the addition of the oxidant to the filtration step; otherwise, the oxidation process will be completed after filtration, creating insoluble iron and manganese precipitates in the distribution system.

Ion Exchange

The ion exchange process is used primarily to soften hard waters, but it will also remove soluble iron and manganese. The water passes through a bed of resin that adsorbs undesirable ions from the water, replacing them with less troublesome ions. When the resin has given up all of its donor ions, it is regenerated with strong salt brine (sodium chloride); the sodium ions from the brine replace the adsorbed ions and restore the ion exchange capabilities.

Sequestering

Sequestering or stabilization may be used when the water contains mainly low concentrations of iron and the volumes required are relatively small. This process does not actually remove the iron or manganese from the water but complexes it (binds it chemically) with other ions in a soluble form that is not likely to come out of solution (i.e., not likely oxidized).

Aeration

The primary physical process uses air to oxidize the iron and manganese. The water is either pumped up into the air or allowed to fall over an aeration device. The air oxidizes the iron and manganese, which are then removed by use of a filter. To raise the pH, lime is often added to the process. Although this is referred to as a physical process, removal is accomplished by chemical oxidation.

Potassium Permanganate Oxidation and Manganese Greensand

The continuous regeneration potassium greensand filter process is another commonly used filtration technique for iron and manganese control. Manganese greensand is a mineral (gluconite) that has been treated with alternating solutions of manganous chloride and potassium permanganate. The result is a sand-like (zeolite) material coated with a layer of manganese dioxide—an adsorbent for soluble iron and manganese. Manganese greensand has the ability to capture (adsorb) soluble iron and manganese that may have escaped oxidation, as well as the capability of physically filtering out the particles of oxidized iron and manganese. Manganese greensand filters are generally set up as pressure filters, totally enclosed tanks containing the greensand. The process of adsorbing soluble iron and manganese uses up the greensand by converting the manganese dioxide coating to manganic oxide, which does not have the adsorption property. The greensand can be regenerated in much the same way as ion exchange resins by washing the sand with potassium permanganate.

TABLE 11.1
Classification of Hardness

Classification	mg/L as $CaCO_3$
Soft	0–75
Moderately hard	75–150
Hard	150–300
Very hard	Over 300

Source: Spellman, F.R., *Spellman's Standard Handbook for Wastewater Operators*, Vol. 1, CRC Press, Boca Raton, FL, 1999.

Hardness Treatment

Hardness in water is caused by the presence of certain positively charged metallic irons in solution in the water. The most common of these hardness-causing ions are calcium and magnesium; others include iron, strontium, and barium. As a general rule, groundwaters are harder than surface waters, so hardness is frequently of concern to the small water system operator. This hardness is derived from contact with soil and rock formations such as limestone. Although rainwater itself will not dissolve many solids, the natural carbon dioxide in the soil enters the water and forms carbonic acid (H_2CO_3), which is capable of dissolving minerals. Where soil is thick (contributing more carbon dioxide to the water) and limestone is present, hardness is likely to be a problem. The total amount of hardness in water is expressed as the sum of its calcium carbonate ($CaCO_3$) and its magnesium hardness; however, for practical purposes, hardness is expressed as calcium carbonate. This means that, regardless of the amount of the various components that make up hardness, they can be related to a specific amount of calcium carbonate (e.g., hardness is expressed as "mg/L as $CaCO_3$," or milligrams per liter as calcium carbonate).

> ***Note:*** The two types of water hardness are *temporary hardness* and *permanent hardness*. Temporary hardness is also known as *carbonate hardness* (hardness that can be removed by boiling), and permanent hardness is also known as *noncarbonate hardness* (hardness that cannot be removed by boiling).

Hardness is of concern in domestic water consumption because hard water increases soap consumption, leaves a soapy scum in the sink or tub, can cause water heater electrodes to burn out quickly, can cause discoloration of plumbing fixtures and utensils, and is perceived as being less desirable water. In industrial water use, hardness is a concern because it can cause boiler scale and damage to industrial equipment.

The objection of customers to hardness is often dependent on the amount of hardness they are used to. People familiar with water with a hardness of 20 mg/L might think that a hardness of 100 mg/L is too much. On the other hand, a person who has been using water with a hardness of 200 mg/L might think that 100 mg/L is very soft. Table 11.1 lists the classifications of hardness.

Hardness Calculation

Recall that hardness is expressed as mg/L as $CaCO_3$. The mg/L of calcium and manganese must be converted to mg/L as $CaCO_3$ before they can be added. The hardness (in mg/L as $CaCO_3$) for any given metallic ion is calculated using the following formula:

$$\text{Hardness (mg/L CaCO}_3\text{)} = M \text{ (mg/L)} \times \frac{50}{\text{Eq. wt. of } M} \tag{11.6}$$

where

M = Metal ion concentration (mg/L).

Eq. wt. = Equivalent weight (gram molecular weight ÷ valence).

Treatment Methods

Two common methods are used to reduce hardness:

- *Ion exchange*—The ion exchange process is the process most frequently used for softening water. As a result of charging a resin with sodium ions, the resin exchanges the sodium ions for calcium or magnesium ions. Naturally occurring and synthetic cation exchange resins are available. Natural exchange resins include such substances as aluminum silicate, zeolite clays (zeolites are hydrous silicates found naturally in the cavities of lavas [greensand], glauconite zeolites, or synthetic, porous zeolites), humus, and certain types of sediments. These resins are placed in a pressure vessel. Salt brine is flushed through the resins. The sodium ions in the salt brine attach to the resin. The resin is now said to be charged. Once charged, water is passed through the resin, and the resin exchanges the sodium ions attached to the resin for calcium and magnesium ions, thus removing them from the water. The zeolite clays are most common because they are quite durable, can tolerate extreme ranges in pH, and are chemically stable. They have relatively limited exchange capacities, however, so they should be used only for water with a moderate total hardness. One of the results is that the water may be more corrosive than before. Another concern is that addition of sodium ions to the water may increase the health risk of those with high blood pressure.
- *Cation exchange*—The cation exchange process takes place with little or no intervention from the treatment plant operator. Water containing hardness-causing cations (Ca^{2+}, Mg^{2+}, Fe^{3+}) is passed through a bed of cation exchange resin. The water coming through the bed contains hardness near zero, although it will have elevated sodium content. (The sodium content is not likely to be high enough to be noticeable, but it could be high enough to pose problems to people on highly restricted salt-free diets.) The total lack of hardness in the finished water is likely to make it very corrosive, so normal practice bypasses a portion of the water around the softening process. The treated and untreated waters are blended to produce an effluent with a total hardness around 50 to 75 mg/L as $CaCO_3$.

Corrosion Control

Water operators add chemicals (e.g., lime, sodium hydroxide) to water at the source or at the waterworks to control corrosion. Using chemicals to achieve a slightly alkaline chemical balance prevents the water from corroding distribution pipes and consumers' plumbing and keeps substances such as lead from leaching out of plumbing and into the drinking water. For our purposes, we define *corrosion* as the conversion of a metal to a salt or oxide with a loss of desirable properties such as mechanical strength. Corrosion may occur over an entire exposed surface or may be localized at micro- or macroscopic discontinuities in metal. In all types of corrosion, a gradual decomposition of the material occurs, often due to an electrochemical reaction. Corrosion may be caused by (1) stray current electrolysis, (2) dissimilar metals (i.e., galvanic corrosion), or (3) differential concentration cells. Corrosion begins at the surface of a material and moves inward.

The adverse effects of corrosion can be categorized according to health, aesthetics, and economic effects, among others. The corrosion of toxic metal pipe made from lead creates a serious *health hazard*. Lead tends to accumulate in the bones of humans and animals. Signs of lead intoxication include gastrointestinal disturbances, fatigue, anemia, and muscular paralysis. Lead is not a natural contaminant in either surface waters or groundwaters, and the MCL of 0.005 mg/L in source waters is rarely exceeded. It is a corrosion byproduct from high lead solder joints in copper and lead piping. Small dosages of lead can lead to developmental problems in children. The USEPA's Lead and Copper Rule addresses the matter of lead in drinking water exceeding specified action levels.

Note: The USEPA's Lead and Copper Rule requires that a treatment facility achieve optimum corrosion control.

Cadmium is the only other toxic metal found in samples from plumbing systems. Cadmium is a contaminant found in zinc. Its adverse health effects are best known for being associated with severe bone and kidney syndrome in Japan. The proposed maximum contaminant level (PMCL) for cadmium is 0.01 mg/L.

Aesthetic effects that are a result of corrosion of iron include pitting and are a consequence of the deposition of ferric hydroxide and other products and the solution of iron—*tuberculation*. Tuberculation reduces the hydraulic capacity of the pipe. Corrosion of iron can cause customer complaints of reddish or reddish-brown staining of plumbing fixtures and laundry. Corrosion of copper lines can cause customer complaints of bluish or blue–green stains on plumbing fixtures. Sulfide corrosion of copper and iron lines can cause a blackish color in the water. The byproducts of microbial activity (especially iron bacteria) can cause foul tastes or odors in the water.

The *economic effects* of corrosion may include water main replacement, especially when tuberculation reduces the flow capacity of the main. Tuberculation increases pipe roughness, causing an increase in pumping costs and reducing distribution system pressure. Tuberculation and corrosion can cause leaks in distribution mains and household plumbing. Corrosion of household plumping may require extensive treatment, public education, and other actions under the Lead and Copper Rule.

Other effects of corrosion include short service life of household plumbing caused by pitting. A build-up of mineral deposits in a hot-water system may eventually restrict hot-water flow. Also, the structural integrity of steel water storage tanks may deteriorate, causing structural failures. Steel ladders in clearwells or water storage tanks may corrode, introducing iron into the finished water. Steel parts in flocculation tanks, sedimentation basins, clarifiers, and filters may also corrode.

Types of Corrosion

Three types of corrosion occur in water mains:

- *Galvanic* occurs when two dissimilar metals come into contact and are exposed to a conductive environment; a potential exists between them, and current flows. This type of corrosion is the result of an electrochemical reaction when the flow of electric current itself is an essential part of the reaction.
- *Tuberculation* refers to the formation of localized corrosion products scattered over the surface in the form of knob-like mounds. These mounds increase the roughness of the inside of the pipe, increasing resistance to water flow and decreasing the *C* factor of the pipe.
- *Pitting* is localized corrosion that is classified as pitting when the diameter of the cavity at the metal surface is the same or less than the depth.

Factors Affecting Corrosion

The primary factors affecting corrosion are pH, alkalinity, hardness (calcium), dissolved oxygen, and total dissolved solids. Secondary factors include temperature, velocity of water in pipes, and carbon dioxide (CO_2).

Determination of Corrosion Problems

To determine if corrosion is taking place in water mains, materials removed from the distribution system should be examined for signs of corrosion damage. A primary indicator of corrosion damage is pitting. (Measure the depth of pits to gauge the extent of damage.) Another common method used to determine if corrosion or scaling is taking place in distribution lines is to insert special steel specimens of known weight (called *coupons*) in the pipe and examine them for corrosion after a period of time. Detecting evidence of leaks, conducting flow tests and chemical tests for dissolved oxygen and toxic metals, and receiving customer complaints (e.g., red or black water, laundry and fixture stains) can also reveal corrosion problems.

Formulas can also be used to determine corrosion (to an extent). The *Langlier Saturation Index* (LSI) and the *Aggressive Index* (AI) are two of the most commonly used indices. The LSI determines whether water is corrosive. The AI is used for waters that have low natural pH, are high in dissolved oxygen, are low in total dissolved solids, and have low alkalinity and low hardness. These waters are very aggressive and can be corrosive. Both of these indices are typically used as starting points in determining the adjustments required to produce a film:

- LSI approximately 0.5
- AI value of 12 or higher

Note: The LSI and AI are based on the dissolving of and precipitation of calcium carbonate; therefore, the respective indices may not actually reflect the corrosive nature of the particular water for a specific pipe material. They can be useful tools, however, in selecting materials or treatment options for corrosion control.

Corrosion Control

As mentioned, one method used to reduce the corrosive nature of water is *chemical addition*. Selection of the chemicals depends on the characteristics of the water, where the chemicals can be applied, how they can be applied and mixed with water, and the cost of the chemicals. Another corrosion control method is *aeration*. Aeration works to remove carbon dioxide (CO_2), which can be reduced to about 5 mg/L. *Cathodic protection*, often employed to control corrosion, involves applying an outside electric current to the metal to reverse the electromechanical corrosion process. The application of DC current prevents normal electron flow. Cathodic protection uses a sacrificial metal electrode (a magnesium anode) that corrodes instead of the pipe or tank. *Linings*, *coatings*, and *paints* can also be used in corrosion control. Slip-line with a plastic liner, cement mortar, zinc or magnesium, polyethylene, epoxy, and coal tar enamels are some of the materials that can be used.

Note: Before using any protective coatings, consult the district engineer first!

Several *corrosive-resistant pipe materials* are used to prevent corrosion:

- PVC plastic pipe
- Aluminum
- Nickel
- Silicon
- Brass
- Bronze
- Stainless steel
- Reinforced concrete

In addition to internal corrosion problems, waterworks operators must also be concerned with external corrosion problems. The primary culprit involved with external corrosion of distribution system pipe is soil. The measure of corrosivity of the soil is the *soil resistivity*. If the soil resistivity is greater than 5000 ohm/cm, serious corrosion is unlikely. Steel pipe may be used under these conditions. If soil resistivity is less than 500 ohm/cm, plastic PVC pipe should be used. For intermediate ranges of soil resistivity (500 to 5000 ohm/cm), ductile iron pipe, linings, and coatings should be used.

- $CaCO_3$ not depositing a film is usually a result of poor pH control (out of the normal range of 6.5 to 8.5). This may also cause excessive film deposition.
- Persistence of red water problems is most probably a result of poor flow patterns, insufficient velocity, tuberculation of pipe surface, and the presence of iron bacteria:
  1. *Velocity*—Chemicals must make contact with the pipe surface. Dead ends and low-flow areas should have flushing program; dead ends should be looped.
  2. *Tuberculation*—The best approach is to clean with *pigs*. In extreme cases, clean pipe with metal scrapers and install cement-mortar lining.
  3. *Iron bacteria*—Slime prevents film contact with the pipe surface. Slime will grow and the coating will be lost. Pipe cleaning and disinfection programs are necessary.

Coagulation

The primary purpose in surface water treatment is chemical clarification by coagulation and mixing, flocculation, sedimentation, and filtration. These unit processes, along with disinfection, work to remove particles; natural organic matter (NOM), such as bacteria, algae, zooplankton, and organic compounds; and microbes from water to produce water that is noncorrosive. Specifically, coagulation–flocculation work to destabilize particles and agglomerate dissolved and particulate matter. Sedimentation removes solids and provides 1/2-log *Giardia* and 1-log virus removal. Filtration removes solids and provides 2-log *Giardia* and 1-log virus removal. Finally, disinfection provides microbial inactivation and 1/2-log *Giardia* and 2-log virus removal.

From Figure 11.3, it can be seen that following screening and the other pretreatment processes the next unit process in a conventional water treatment system is a mixer where chemicals are added in what is known as *coagulation*. The exception to this unit process configuration occurs in small systems using groundwater, when chlorine or other taste and odor control measures are introduced at the intake and are the extent of treatment.

Materials present in raw water may vary in size, concentration, and type. Dispersed substances in the water may be classified as *suspended*, *colloidal*, or *solution*. Suspended particles may vary in mass and size and are dependent on the flow of water. High flows and velocities can carry larger material. As velocities decrease, the suspended particles settle according to size and mass.

Other material may be in solution; for example, salt dissolves in water. Matter in the colloidal state does not dissolve, but the particles are so small they will not settle out of the water. Color (as in tea-colored swamp water) is mainly due to colloids or extremely fine particles of matter in suspension. Colloidal and solute particles in water are electrically charged. Because most of the charges are alike (negative) and repel each other, the particles stay dispersed and remain in the colloidal or soluble state.

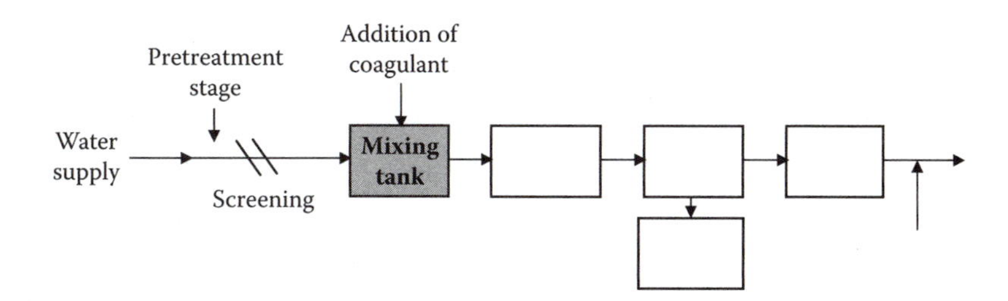

FIGURE 11.3 Coagulation.

Suspended matter will settle without treatment, if the water is still enough to allow it to settle. The rate of settling of particles can be determined, as this settling follows certain laws of physics; however, much of the suspended matter may be so slow in settling that the normal settling processes become impractical, and if colloidal particles are present settling will not occur. Moreover, water drawn from a raw water source often contains many small unstable (unsticky) particles; therefore, sedimentation alone is usually an impractical way to obtain clear water in most locations, and another method of increasing the settling rate must be used: coagulation, which is designed to convert stable (unsticky) particles to unstable (sticky) particles.

Coagulation is a series of chemical and mechanical operations by which coagulants are applied and made effective. These operations are comprised of two distinct phases: (1) rapid mixing to disperse coagulant chemicals by violent agitation into the water being treated, and (2) flocculation to agglomerate small particles into well-defined floc by gentle agitation for a much longer time.

Coagulation results from adding salts of iron or aluminum to the water. The coagulant must be added to the raw water and perfectly distributed into the liquid; such uniformity of chemical treatment is reached through rapid agitation or mixing. Common coagulants (salts) include the following:

- Alum (aluminum sulfate)
- Sodium aluminate
- Ferric sulfate
- Ferrous sulfate
- Ferric chloride
- Polymers

Coagulation is the reaction between one of these salts and water. The simplest coagulation process occurs between alum and water. Alum, or aluminum sulfate, is produced by a chemical reaction between bauxite ore and sulfuric acid. The normal strength of liquid alum is adjusted to 8.3%, while the strength of dry alum is 17%. When alum is placed in water, a chemical reaction occurs that produces positively charged aluminum ions. The overall result is the reduction of electrical charges and the formation of a sticky substance—the formation of *floc*, which, when properly formed, will settle. These two destabilizing factors are the major contributions of coagulation toward the removal of turbidity, color, and microorganisms.

Liquid alum is preferred in water treatment because it has several advantages over other coagulants:

- Ease of handling
- Lower costs
- Less labor required to unload, store, and convey
- Elimination of dissolving operations
- Less storage space required
- Greater accuracy in measurement and control
- Elimination of the nuisance and unpleasantness of handling dry alum
- Easier maintenance

The formation of floc is the first step of coagulation; for greatest efficiency, rapid, intimate mixing of the raw water and the coagulant must occur. After mixing, the water should be slowly stirred so the very small, newly formed particles can attract and enmesh colloidal particles, holding them together to form larger floc. This slow mixing is the second stage of the process (flocculation), covered later in this chapter.

A number of factors influence the coagulation process—pH, turbidity, temperature, alkalinity, and the use of polymers. The degree to which these factors influence coagulation depends on the coagulant use. The raw water conditions, optimum pH for coagulation, and other factors must be considered before deciding which chemical is to be fed and at what levels.

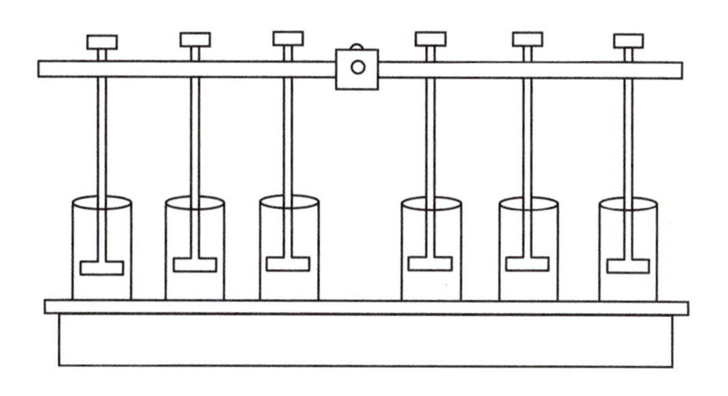

FIGURE 11.4 Variable-speed paddle mixer used in jar testing procedure.

To determine the correct chemical dosage, a *jar test* or *coagulation test* is performed. Jar tests (widely used for many years by the water treatment industry) simulate full-scale coagulation and flocculation processes to determine optimum chemical dosages. It is important to note that jar testing is only an attempt to achieve a ballpark approximation of correct chemical dosage for the treatment process. The test conditions are intended to reflect the normal operation of a chemical treatment facility. The test can be used to

- Select the most effective chemical.
- Select the optimum dosage.
- Determine the value of a flocculant aid and the proper dose.

The testing procedure requires a series of samples to be placed in testing jars (see Figure 11.4) and mixed at 100 ppm. Varying amounts of the process chemical or specified amounts of several flocculants are added (one volume/sample container). The mix is continued for 1 minute. Next, the mixing is slowed to 30 rpm to provide gentle agitation, and then the floc is allowed to settle. The flocculation period and settling process are observed carefully to determine the floc strength, settleability, and clarity of the *supernatant liquor* (the water that remains above the settled floc). Additionally, the supernatant can be tested to determine the efficiency of the chemical addition for removal of total suspended solids (TSS), biochemical oxygen demand (BOD), and phosphorus.

The equipment required for the jar test includes a six-position, variable-speed paddle mixer (see Figure 11.4); six 2-quart wide-mouthed jars; an interval timer; and assorted glassware, pipets, graduates, and so forth. The jar test procedure follows:

1. Place an appropriate volume of water sample in each of the jars (250- to 1000-mL samples may be used, depending on the size of the equipment being use). Start mixers and set for 100 rpm.
2. Add previously selected amounts of the chemical being evaluated. (Initial tests may use wide variations in chemical volumes to determine the approximate range; this is then narrowed in subsequent tests.)
3. Continue mixing for 1 minute.
4. Reduce the mixer speed to a gentle agitation (30 rpm), and continue mixing for 20 minutes. Again, time and mixer speed may be varied to reflect the facility.

Note: During this time, observe the floc formation—that is, how well the floc holds together during the agitation (floc strength).

5. Turn off the mixer and allow solids to settle for 20 to 30 minutes. Observe the settling characteristics, the clarity of the supernatant, the settleability of the solids, the flocculation of the solids, and the compactability of the solids.

6. Perform phosphate tests to determine removals.
7. Select the dose that provided the best treatment based on observations made during the analysis.

After initial ranges and chemical selections are determined, repeat the test using a smaller range of dosages to optimize performance.

Flocculation

As we see in Figure 11.5, flocculation follows coagulation in the conventional water treatment process. Flocculation is the physical process of slowly mixing the coagulated water to increase the probability of particle collision; unstable particles collide and stick together to form fewer larger flocs. Through experience, we have found that effective mixing reduces the required amount of chemicals and greatly improves the sedimentation process, resulting in longer filter runs and higher quality finished water. The goal of flocculation is to form a uniform, feather-like material similar to snowflakes—a dense, tenacious floc that entraps the fine, suspended, and colloidal particles and then carries them down rapidly in the settling basin. Proper flocculation requires from 15 to 45 minutes. The time required is based on water chemistry, water temperature, and mixing intensity. Temperature is the key component in determining the amount of time necessary for floc formation. To increase the speed of floc formation and the strength and weight of the floc, polymers are often added.

Sedimentation

After raw water and chemicals have been mixed and the floc formed, the water containing the floc (because it has a higher specific gravity than water) flows to the sedimentation or settling basin (see Figure 11.6). Sedimentation is also called *clarification*. Sedimentation removes settleable solids by gravity. Water moves slowly though the sedimentation tank/basin with a minimum of turbulence at entry and exit points with minimum short-circuiting. Sludge accumulates at the bottom of the tank or basin. Typical tanks or basins used in sedimentation include conventional rectangular basins, conventional center-feed basins, peripheral-feed basins, and spiral-flow basins.

In conventional treatment plants, the amount of detention time required for settling can vary from 2 to 6 hours. Detention time should be based on the total filter capacity when the filters are passing 2 gpm per square foot of superficial sand area. For plants with higher filter rates, the detention time is based on a filter rate of 3 to 4 gpm per square foot of sand area. The time requirement is dependent on the weight of the floc, the temperature of the water, and how quiescent the basin is.

A number of conditions affect sedimentation: (1) uniformity of flow of water through the basin; (2) stratification of water due to difference in temperature between water entering and water already in the basin; (3) release of gases that may collect in small bubbles on suspended solids, causing them to rise and float as scum rather than settle as sludge; (4) disintegration of previously formed floc; and (5) size and density of the floc.

Filtration

In the conventional water treatment process, filtration usually follows coagulation, flocculation, and sedimentation (see Figure 11.7). At present, filtration is not always used in small water systems; however, regulatory requirements under the USEPA Interim Enhanced Surface Water Treatment rules may make water filtering necessary at most water supply systems. Water filtration is a physical process of separating suspended and colloidal particles from water by passing water through a granular material. The process of filtration involves straining, settling, and adsorption. As floc passes into the filter, the spaces between the filter grains become clogged, reducing this opening

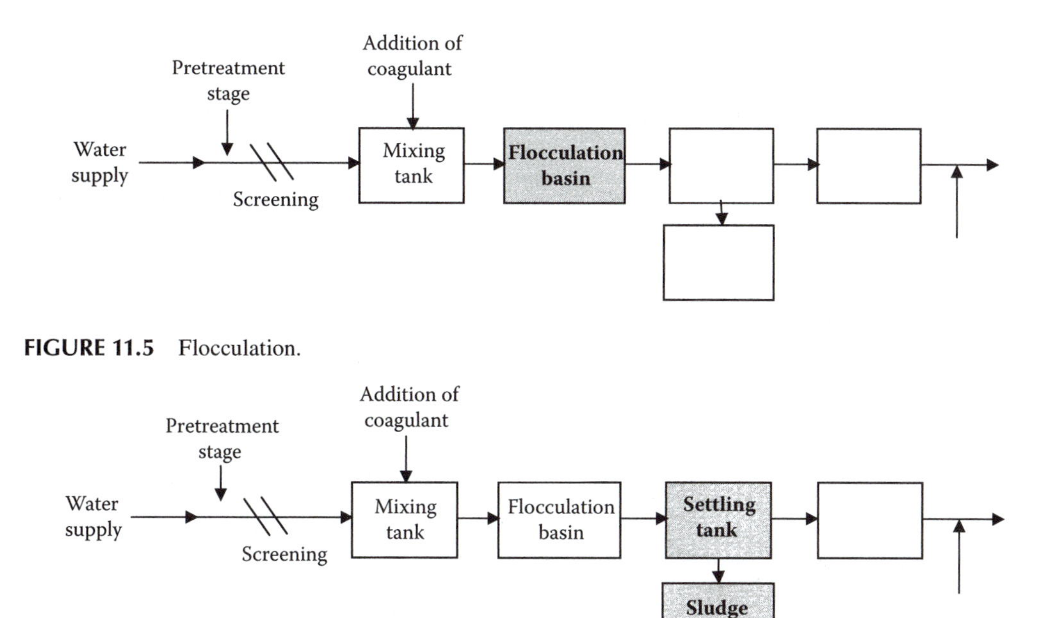

FIGURE 11.5 Flocculation.

FIGURE 11.6 Sedimentation.

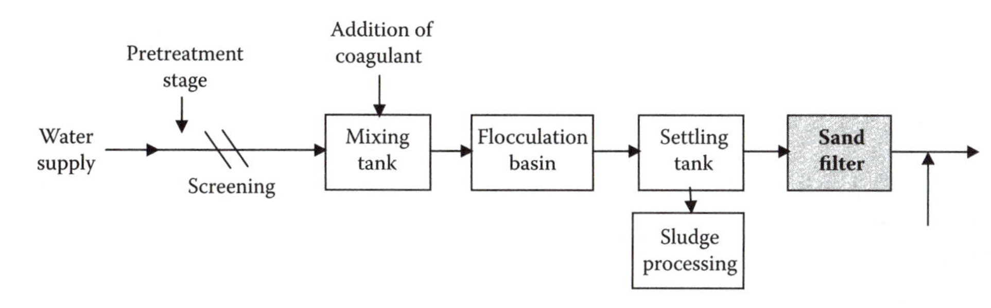

FIGURE 11.7 Filtration.

and increasing removal. Some material is removed merely because it settles on a media grain. One of the most important processes is adsorption of the floc onto the surface of individual filter grains. This helps collect the floc and reduces the size of the openings between the filter media grains. In addition to removing silt and sediment, floc, algae, insect larvae, and any other large elements, filtration also contributes to the removal of bacteria and protozoa such as *Giardia lamblia* and *Cryptosporidium*. Some filtration processes are also used for iron and manganese removal.

Types of Filter Technologies

The Surface Water Treatment Rule (SWTR) specifies four filtration technologies, although it also allows the use of alternative filtration technologies (e.g., cartridge filters). The specified technologies are (1) slow sand filtration/rapid sand filtration, (2) pressure filtration, (3) diatomaceous earth filtration, and (4) direct filtration. Of these, all but rapid sand filtration are commonly employed in small water systems that use filtration. Each type of filtration system has advantages and disadvantages. Regardless of the type of filter, however, filtration involves the processes of *straining* (where particles are captured in the small spaces between filter media grains), *sedimentation* (where the particles land on top of the grains and stay there), and *adsorption* (where a chemical attraction occurs between the particles and the surface of the media grains).

Slow Sand Filters

The first slow sand filter was installed in London in 1829, and the technique was used widely throughout Europe, although not in the United States. By 1900, rapid sand filtration began taking over as the dominant filtration technology, although a few slow sand filters are still in operation today. With the advent of the Safe Drinking Water Act (SDWA) and its regulations (especially the Surface Water Treatment Rule) and recognition of the problems associated with *Giardia lamblia* and *Cryptosporidium* in surface water, the water industry is reexamining the use of slow sand filters. The low technology requirements may prevent many state water systems from using this type of equipment.

On the plus side, slow sand filtration is well suited for small water systems. It is a proven, effective filtration process with relatively low construction costs and low operating costs (it does not require constant operator attention). It is quite effective for water systems as large as 5000 people; beyond that, the surface area requirements and manual labor required to recondition the filters make rapid sand filters the more effective choice. The filtration rate is generally in the range of 45 to 150 gallons per day per square foot. Components of a slow sand filter include the following:

- A covered structure to hold the filter media
- An underdrain system
- Graded rock that is placed around and just above the underdrain
- The filter media, consisting of 30 to 55 inches of sand with a grain size of 0.25 to 0.35 mm
- Inlet and outlet piping to convey the water to and from the filter and the means to drain filtered water to waste

The area above the top of the sand layer is flooded with water to a depth of 3 to 5 feet, and the water is allowed to trickle down through the sand. An overflow device prevents excessive water depth. The filter must have provisions for filling it from the bottom up, and it must be equipped with a loss-of-head gauge, a rate-of-flow control device (such as an orifice or butterfly valve), a weir or effluent pipe that ensures that the water level cannot drop below the sand surface, and filtered waste sample taps.

When the filter is first placed in service, the head loss through the media caused by the resistance of the sand is about 0.2 feet (i.e., a layer of water 0.2 feet deep on top of the filter will provide enough pressure to push the water downward through the filter). As the filter operates, the media become clogged with the material being filtered out of the water, and the head loss increases. When it reaches about 4 to 5 feet, the filter must be cleaned.

For efficient operation of a slow sand filter, the water being filtered should have a turbidity averaging less than 5 TU, with a maximum of 30 TU. Slow sand filters are not backwashed the way conventional filtration units are. One to 2 inches of material must be removed on a periodic basis to keep the filter operating.

Rapid Sand Filters

The rapid sand filter, which is similar in some ways to the slow sand filter, is one of the most widely used filtration units. The major difference is in the principle of operation—that is, in the speed or rate at which water passes through the media. In operation, water passes downward through a sand bed that removes the suspended particles. The suspended particles consist of the coagulated matter remaining in the water after sedimentation, as well as a small amount of uncoagulated suspended matter.

Some significant differences exist in construction, control, and operation between slow sand filters and rapid sand filters. Because of the design and construction of the rapid sand filtration, the land area required to filter the same quantity of water is reduced. Components of a rapid sand filter include the following:

- Structure to house media
- Filter media
- Gravel media support layer
- Underdrain system
- Valves and piping system
- Filter backwash system
- Waste disposal system

Usually 2 to 3 feet deep, the filter media are supported by approximately 1 foot of gravel. The media may be fine sand or a combination of sand, anthracite coal, and coal (dual- or multimedia filter). Water is applied to a rapid sand filter at a rate of 1.5 gallons per minute per square foot (gpm/ft^2) of filter media surface. When the rate is between 4 and 6 gpm/ft^2, the filter is referred to as a *high-rate filter*; at a rate over 6 gpm/ft^2, the filter is referred to as a *ultra-high-rate filter.* These rates compare to the slow sand filtration rate of 45 to 150 gallons per day per square foot. High-rate and ultra-high-rate filters must meet additional conditions to assure proper operation.

Generally, raw water turbidity is not that high; however, even if raw water turbidity values exceed 1000 TU, properly operated rapid sand filters can produce filtered water with a turbidity or well under 0.5 TU. The time the filter is in operation between cleanings (filter runs) usually ranges from 12 to 72 hours, depending on the quality of the raw water; the end of the run is indicated by the head loss approaching 6 to 8 feet. Filter *breakthrough* (when filtered material is pulled through the filter into the effluent) can occur if the head loss becomes too great. Operation with head loss too high can also cause *air binding* (which blocks part of the filter with air bubbles), increasing the flow rate through the remaining filter area.

Rapid sand filters have the advantage of a lower land requirement, and they have other advantages, as well; for example, rapid sand filters cost less, are less labor intensive to clean, and offer higher efficiency with highly turbid waters. On the downside, the operation and maintenance costs of rapid sand filters are much higher in comparison because of the increased complexity of the filter controls and backwashing system.

When *backwashing* a rapid sand filter, the filter is cleaned by passing treated water backward (upward) through the filter media and agitating the top of the media. The need for backwashing is determined by a combination of filter run time (i.e., the length of time since the last backwashing), effluent turbidity, and head loss through the filter. Depending on the raw water quality, the run time varies from one filtration plant to another (and may even vary from one filter to another in the same plant).

Note: Backwashing usually requires 3 to 7% of the water produced by the plant.

Pressure Filter Systems

When raw water is pumped or piped from the source to a gravity filter, the head (pressure) is lost as the water enters the floc basin. When this occurs, pumping the water from the plant clearwell to the reservoir is usually necessary. One way to reduce pumping is to place the plant components into pressure vessels, thus maintaining the head. This type of arrangement is known as a pressure filter system. Pressure filters are also quite popular for iron and manganese removal and for filtration of water from wells. They may be placed directly in the pipeline from the well or pump with little head loss. Most pressure filters operate at a rate of about 3 gpm/ft^2.

Although pressure filtration is operationally the same as rapid sand filtration and consists of components similar to those of a rapid sand filter, the main difference between a rapid sand filtration system and a pressure filtration system is that the entire pressure filter is contained within a pressure vessel. These units are often highly automated and are usually purchased as self-contained units with all necessary piping, controls, and equipment contained in a single unit. They are backwashed in much the same manner as the rapid sand filter.

The major advantage of the pressure filter is its low initial cost. They are usually prefabricated, with standardized designs. A major disadvantage is that the operator is unable to observe the filter in the pressure filter and so is unable to determine the condition of the media. Unless the unit has an automatic shutdown feature for high effluent turbidity, driving filtered material through the filter is possible.

Diatomaceous Earth Filters

Diatomaceous earth is a white material made from the skeletal remains of diatoms. The skeletons are microscopic and in most cases porous. Diatomaceous earth is available in various grades, and the grade is selected based on filtration requirements. These diatoms are mixed in water slurry and fed onto a fine screen called a *septum*, usually made of stainless steel, nylon, or plastic. The slurry is fed at a rate of 0.2 lb/ft^2 of filter area. The diatoms collect in a precoat over the septum, forming an extremely fine screen. Diatoms are fed continuously with the raw water, causing the buildup of a filter cake approximately 1/8 to 1/5 inch thick. The openings are so small that the fine particles that cause turbidity are trapped on the screen. Coating the septum with diatoms gives it the ability to filter out very small microscopic material. The fine screen and the buildup of filtered particles cause a high head loss through the filter. When the head loss reaches a maximum level (30 psi on a pressure-type filter or 15 inches of mercury on a vacuum-type filter), the filter cake must be removed by backwashing.

The slurry of diatoms is fed with raw water during filtration in a process called *body feed*. The body feed prevents premature clogging of the septum cake. These diatoms are caught on the septum, increasing the head loss and preventing the cake from clogging too rapidly by the particles being filtered. Although the body feed increases head loss, head loss increases are more gradual than if body feed were not used.

Diatomaceous earth filters are relatively low in cost to construct, but they have high operating costs and can cause frequent operating problems if not properly operated and maintained. They can be used to filter raw surface waters or surface-influenced groundwaters with low turbidity (<5 NTU) and low coliform concentrations (no more than 50 coliforms per 100 mL) and may also be used for iron and manganese removal following oxidation. Filtration rates are between 1.0 and 1.5 gpm/ft^2.

Direct Filtration

Direct filtration is a treatment scheme that omits the flocculation and sedimentation steps prior to filtration. Coagulant chemicals are added, and the water is passed directly onto the filter. All solids removal takes place on the filter, which can lead to much shorter filter runs, more frequent backwashing, and a greater percentage of finished water used for backwashing. The lack of a flocculation process and sedimentation basin reduces construction costs but increases the requirement for skilled operators and high-quality instrumentation. Direct filtration must be used only where the water flow rate and raw water quality are fairly consistent and where the incoming turbidity is low.

Alternative Filters

A *cartridge filter system* can be employed as an alternative filtering system to reduce turbidity and remove *Giardia*. A cartridge filter is made of a synthetic media contained in a plastic or metal housing. These systems are normally installed in a series of three or four filters. Each filter contains media successively smaller than the previous filter. The media sizes typically range from 50 to 5 μm or less. The filter arrangement is dependent on the quality of the water, the capability of the filter, and the quantity of water needed. The USEPA and state agencies have established criteria for the selection and use of cartridge filters. Generally, cartridge filter systems are regulated in the same manner as other filtration systems.

Because of new regulatory requirements and the need to provide more efficient removal of pathogenic protozoa (e.g., *Giardia* and *Cryptosporidium*) from water supplies, *membrane filtration systems* are finding increased application in water treatment systems. A *membrane* is a thin film separating two different phases of a material acting as a selective barrier to the transport of matter operated by some driving force. Simply, a membrane can be regarded as a sieve with very small pores. Membrane filtration processes are typically pressure, electrically, vacuum, or thermally driven. The types of drinking water membrane filtration systems include microfiltration, ultrafiltration, nanofiltration, and reverse osmosis. A typical membrane filtration process has one input and two outputs. Membrane performance is largely a function of the properties of the materials to be separated and can vary throughout operation.

Common Filter Problems

Two common types of filter problems occur: those caused by filter runs that are too long (infrequent backwash) and those caused by inefficient backwash (cleaning). A filter run that is too long can cause *breakthrough* (the pushing of debris removed from the water through the media and into the effluent) and *air binding* (the trapping of air and other dissolved gases in the filter media). Air binding occurs when the rate at which water exits the bottom of the filter exceeds the rate at which the water penetrates the top of the filter. When this happens, a void and partial vacuum occur inside the filter media. The vacuum causes gases to escape from the water and fill the void. When the filter is backwashed, the release of these gases may cause a violent upheaval in the media and destroy the layering of the media bed, gravel, or underdrain. Two solutions to the problems are to (1) check the filtration rates to be sure they are within the design specifications, and (2) remove the top 1 inch of media and replace with new media. This keeps the top of the media from collecting the floc and sealing the entrance into the filter media.

Another common filtration problem is associated with poor backwashing practices: the formation of *mudballs* that get trapped in the filter media. In severe cases, mudballs can completely clog a filter. Poor agitation of the surface of the filter can form a crust on top of the filter; the crust later cracks under the water pressure, causing uneven distribution of water through the filter media. Filter cracking can be corrected by removing the top 1 inch of the filter media, increasing the backwash rate, or checking the effectiveness of the surface wash (if installed). Backwashing at too high a rate can cause the filter media to wash out of the filter over the effluent troughs and may damage the filter underdrain system. Two possible solutions are to (1) check the backwash rate to be sure that it meets the design criteria, and (2) check the surface wash (if installed) for proper operation.

Disinfection*

The process used to control waterborne pathogenic organisms and prevent waterborne disease is disinfection. The goal in proper disinfection in a water system is to destroy all disease-causing organisms. Disinfection should not be confused with *sterilization*, which is the complete killing of all living organisms. Waterworks operators disinfect by destroying organisms that might be dangerous; they do not attempt to sterilize water. In water treatment, disinfection is almost always accomplished by adding chlorine or chlorine compounds after all other treatment steps (see Figure 11.8), although in the United States ultraviolet (UV) light and potassium permanganate and ozone processes may be encountered.

The effectiveness of disinfection in a drinking water system is measured by testing for the presence or absence of coliform bacteria. *Coliform bacteria* found in water are generally not pathogenic, although they are good indicators of contamination. Their presence indicates the possibility of contamination, and their absence indicates the possibility that the water is potable—if the source is adequate, the waterworks history is good, and acceptable chlorine residual is present.

* Disinfection is a unit process used in both water and wastewater treatment. Many of the terms, practices, and applications discussed in this section apply to both water and wastewater treatment. There are also some differences, mainly in the types of disinfectants used and the applications, between the disinfection method used in water treatment and that used in wastewater treatment. Thus, in this section we discuss disinfection as it applies to water treatment.

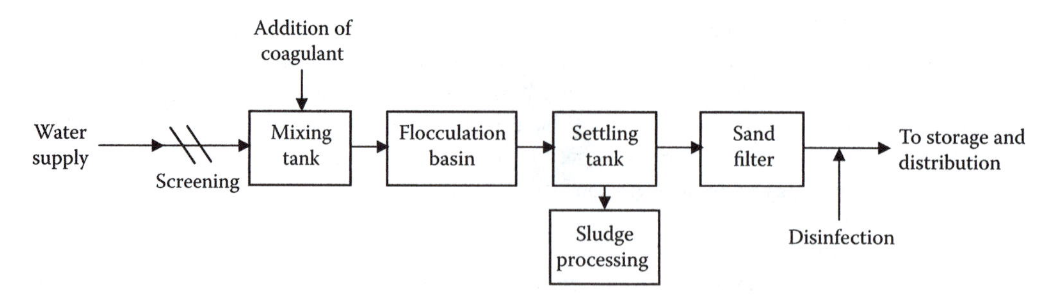

FIGURE 11.8 Disinfection.

Desired characteristics of a disinfectant include the following:

- It must be able to deactivate or destroy any type or number of disease-causing microorganisms that may be in a water supply, in reasonable time, within expected temperature ranges, and despite changes in the character of the water (pH, for example).
- It must be nontoxic.
- It must not add unpleasant taste or odor to the water.
- It must be readily available at a reasonable cost and be safe and easy to handle, transport, store, and apply.
- It must be quick and easy to determine the concentration of the disinfectant in the treated water.
- It should persist within the disinfected water at a high enough concentration to provide residual protection through the distribution.

Note: Disinfection is effective in reducing waterborne diseases because most pathogenic organisms are more sensitive to disinfection than are nonpathogens; however, disinfection is only as effective as the care used in controlling the process and ensuring that all of the water supply is continually treated with the amount of disinfectant required producing safe water.

Methods of disinfection include the following:

- *Heat*—Possibly the first method of disinfection, which is accomplished by boiling water for 5 to 10 minutes; good, obviously, only for household quantities of water when bacteriological quality is questionable
- *Ultraviolet (UV) light*—A practical method of treating large quantities, but adsorption of UV light is very rapid so this method is limited to nonturbid waters close to the light source
- *Metal ions*—Silver, copper, mercury
- *Alkalis and acids*
- *pH adjustment*—To under 3.0 or over 11.0
- *Oxidizing agents*—Bromine, ozone, potassium permanganate, and chlorine

The vast majority of drinking water systems in the United States use chlorine for disinfection. Along with meeting the desired characteristics listed above, chlorine has the added advantage of a long history of use and is fairly well understood. Although some small water systems may use other disinfectants, we concentrate on chlorine here.

Chlorination

The addition of chlorine or chlorine compounds to water is called *chlorination*. Chlorination is considered to be the single most important process for preventing the spread of waterborne disease. Chlorine has many attractive features that contribute to its wide use in industry. Five key attributes of chlorine are

1. It damages the cell wall.
2. It alters the permeability of the cell (the ability to pass water in and out through the cell wall).
3. It alters the cell protoplasm.
4. It inhibits the enzyme activity of the cell so it is unable to use its food to produce energy.
5. It inhibits cell reproduction.

Chlorine is available in a number of different forms:

- As pure elemental gaseous chlorine, a greenish-yellow gas possessing a pungent and irritating odor that is heavier than air, nonflammable, and nonexplosive; when released to the atmosphere, this form is toxic and corrosive
- As solid calcium hypochlorite (in tablets or granules)
- As a liquid sodium hypochlorite solution (in various strengths)

The selection of one form of chlorine over the others for a given water system depends on the amount of water to be treated, configuration of the water system, local availability of the chemicals, and skill of the operator.

One of the major advantages of using chlorine is the effective residual that it produces. A residual indicates that disinfection is completed and the system has an acceptable bacteriological quality. Maintaining a residual in the distribution system provides another line of defense against pathogenic organisms that could enter the distribution system and helps to prevent regrowth of those microorganisms that were injured but not killed during the initial disinfection stage.

Common chlorination terms include the following:

- *Chlorine reaction*—Regardless of the form of chlorine used for disinfection, the reaction in water is basically the same. The same amount of disinfection can be expected, provided the same amount of available chlorine is added to the water. The standard units used to express the concentration of chlorine in water are milligrams per liter (mg/L) and parts per million (ppm); these terms indicate the same quantity.
- *Chlorine dose*—The amount of chlorine added to the system. It can be determined by adding the desired residual for the finished water to the chlorine demand of the untreated water. Dosage can be either milligrams per liter (mg/L) or pounds per day. The most common is mg/L.
- *Chlorine demand*—The amount of chlorine used by iron, manganese, turbidity, algae, and microorganisms in the water. Because the reaction between chlorine and microorganisms is not instantaneous, demand is relative to time. For example, the demand 5 minutes after applying chlorine will be less than the demand after 20 minutes. Demand, like dosage, is expressed in mg/L. The chlorine demand is determined as follows:

$$Cl_2 \text{ demand} = Cl_2 \text{ dose} - Cl_2 \text{ residual} \quad (11.7)$$

- *Chlorine residual*—The amount of chlorine (determined by testing) that remains after the demand is satisfied. Residual, like demand, is based on time. The longer the time after dosage, the lower the residual will be, until all of the demand has been satisfied. Residual, like dosage and demand, is expressed in mg/L. The presence of a *free residual* of at least 0.2 to 0.4 ppm usually provides a high degree of assurance that the disinfection of the water is complete. *Combined residual* is the result of combining free chlorine with nitrogen compounds. Combined residuals are also called *chloramines*. The *total chlorine residual* is the mathematical combination of free and combined residuals. Total residual can be determined directly with standard chlorine residual test kits.

- *Chorine contact time*—A key item in predicting the effectiveness of chlorine on microorganisms. It is the interval (usually only a few minutes) between the time when chlorine is added to the water and the time the water passes by the sampling point. Contact time is the "T" in CT. CT is calculated based on the free chlorine residual prior to the first consumer times the contact time in minutes:

$$CT = \text{Concentration} \times \text{Contact time} = \text{mg/L} \times \text{Minutes} \tag{11.8}$$

A certain minimum time period is required for the disinfecting action to be completed. The contact time is usually a fixed condition determined by the rate of flow of the water and the distance from the chlorination point to the first consumer connection. Ideally, the contact time should not be less than 30 minutes, but even more time is needed at lower chlorine doses, in cold weather, or under other conditions.

Pilot studies have shown that specific CT values are necessary for the inactivation of viruses and *Giardia*. The required CT value will vary depending on pH, temperature, and the organisms to be killed. Charts and formulas are available to make this determination. The USEPA has set a CT value of 3-log ($CT_{99.9}$) inactivation to ensure that the water is free of *Giardia*. State drinking water regulations provide charts containing CT values for various pH and temperature combinations. Filtration, in combination with disinfection, must provide 3-log removal or inactivation of *Giardia*. Charts in the USEPA Surface Water Treatment Rule guidance manual list the required CT values for various filter systems.

Under the 1996 Interim Enhanced Surface Water Treatment Rule, the USEPA requires systems that filter to remove 99% (2 log) of *Cryptosporidium* oocysts. To be sure that the water is free of viruses, a combination of filtration and disinfection that provides 4-log (99.99%) removal of viruses has been judged the best for drinking water safety. Viruses are inactivated more easily than cysts or oocysts.

Chlorine Chemistry

The reactions of chlorine with water and the impurities that might be in the water are quite complex, but a basic understanding of these reactions can aid the operator in keeping the disinfection process operating at its highest efficiency. When dissolved in pure water, chlorine reacts with H^+ ions and OH^- radicals in the water. Two of the products of this reaction (the actual disinfecting agents) are *hypochlorous acid* (HOCl) and the *hypochlorite radical* (OCl^-). If microorganisms are present in the water, the HOCl and the OCl^- penetrate the microbe cells and react with certain enzymes. This reaction disrupts the metabolism of the organisms and kills them. The chemical equation for hypochlorous acid is as follows:

$$Cl_2 \text{ (chlorine)} + H_2O \text{ (water)} \leftrightarrow HOCl \text{ (hypochlorous acid)} + HCl \text{ (hydrochloric acid)} \tag{11.9}$$

Note: The symbol ↔ indicates that the reactions are reversible.

Hypochlorous acid (HOCl) is a weak acid, meaning that it dissociates slightly into hydrogen and hypochlorite ions, but it is a strong oxidizing and germicidal agent. Hydrochloric acid (HCl) in the above equation is a strong acid and retains more of the properties of chlorine. HCl tends to lower the pH of the water, especially in swimming pools where the water is recirculated and continually chlorinated. The total hypochlorous acid and hypochlorite ions in water constitute the *free available chlorine*. Hypochlorites act in a manner similar to HCl when added to water, because hypochloric acid is formed.

When chlorine is first added to water containing some impurities, the chlorine immediately reacts with the dissolved inorganic or organic substances and is then unavailable for disinfection. The amount of chlorine used in this initial reaction is the *chlorine demand* of the water.

If dissolved ammonia (NH_3) is present in the water, the chlorine will react with it to form compounds called *chloramines*. Only after the chlorine demand is satisfied and the reaction with all the dissolved ammonia is complete is the chlorine actually available in the form of HOCl and OCl^-. The equation for the reaction of hypochlorous acid (HOCl) and ammonia (NH_3) is as follows:

$$\text{HOCl (hypochlorous acid)} + \text{NH}_3 \text{ (ammonia)} \leftrightarrow \text{NH}_2\text{Cl (monochloramine)} + \text{H}_2\text{O (water)} \tag{11.10}$$

Note: The chlorine as hypochlorous acid and hypochlorite ions remaining in the water after the above reactions are complete is known as *free available chlorine,* and it is a very active disinfectant.

Breakpoint Chlorination

To produce a free chlorine residual, enough chlorine must be added to the water to produce what is referred to as *breakpoint chlorination*, which is the point at which near complete oxidation of nitrogen compounds is reached; any residual beyond breakpoint is mostly free chlorine (see Figure 11.9). When chlorine is added to natural waters, the chlorine begins combining with and oxidizing the chemicals in the water before it begins disinfecting. Although residual chlorine will be detectable in the water, the chlorine will be in the combined form with a weak disinfecting power. As we see in Figure 11.9, adding more chlorine to the water at this point actually decreases the chlorine residual as the additional chlorine destroys the combined chlorine compounds. At this stage, water may have a strong swimming pool or medicinal taste and odor. To avoid such taste and odor issues, add still more chlorine to produce a free residual chlorine. Free chlorine has the highest disinfecting power. The point at which most of the combined chlorine compounds have been destroyed and the free chlorine starts to form is the *breakpoint.*

The chlorine breakpoint of water can only be determined by experimentation. This simple experiment requires 20 1000-mL breakers and a solution of chlorine. Place the raw water in the beakers and dose with progressively larger amounts of chlorine; for example, we might start with 0 in the first beaker, then 0.5 mg/L, then 1.0 mg/L, and so on. After a period of time, say 20 minutes, test each beaker for total chlorine residual and plot the results.

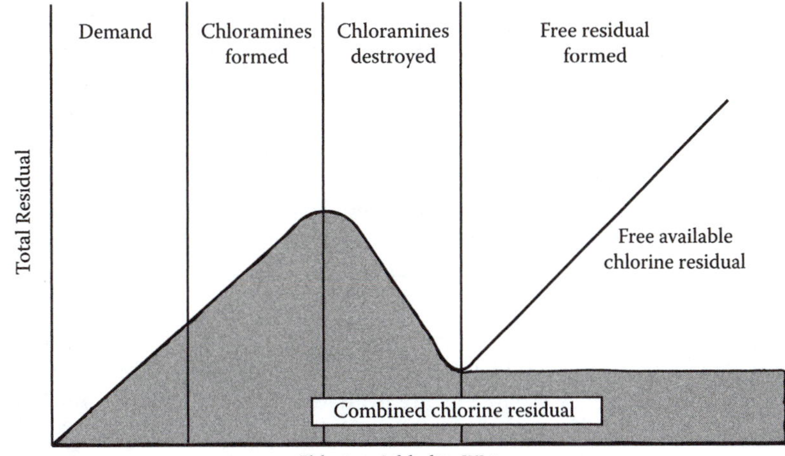

FIGURE 11.9 Breakpoint chlorination curve.

Breakpoint Chlorination Curve

When the curve starts, no residual exists, even though a dosage was applied. This is the *initial demand*, when microorganisms and interfering agents are using the result of the chlorine. After the initial demand, the curve slopes upward. Chlorine combining to form chloramines produces this part of the curve. All of the residual measured on this part of the curve is combined residual. At some point, the curve begins to drop back toward zero. This portion of the curve results from a reduction in combined residual, which occurs because enough chlorine has been added to destroy (oxidize) the nitrogen compounds used to form combined residuals. The breakpoint is the point where the downward slope of the curve breaks upward. At this point, all of the nitrogen compounds that could be destroyed have been destroyed. After breakpoint, the curve begins to move upward again, usually at a 45° angle. Only on this part of the curve can free residuals be found. Notice that the breakpoint is not zero. The distance that the breakpoint is above zero is a measure of the remaining combined residual in the water. This combined residual exists because some of the nitrogen compound will not have been oxidized by chlorine. If irreducible combined residual is more than 15% of the total residual, chlorine odor and taste complaints will be high.

Gas Chlorination

Gas chlorine is provided in 100-lb to 1-ton containers. Chlorine is placed in the container as a liquid. The liquid boils at room temperature and is reduced to a gas that builds pressure in the cylinder. At room temperature (70°F), a chlorine cylinder will have a pressure of 85 psi; 100- to 150-lb cylinders should be maintained in an upright position and chained to the wall. To prevent a chlorine cylinder from rupturing in a fire, the cylinder valves are equipped with special fusible plugs that melt between 158 and 164°F.

Chlorine gas is 99.9% chlorine. A gas chlorinator meters the gas flow and mixes it with water, which is then injected as a water solution of pure chlorine. As the compressed liquid chlorine is withdrawn from the cylinder, it expands as a gas, withdrawing heat from the cylinder. Care must be taken not to withdraw the chlorine at too fast a rate; if the operator attempts to withdraw more than about 40 lb of chlorine per day from a 150-lb cylinder, it will freeze up.

> ***Note:*** All chlorine gas feed equipment sold today is vacuum operated. This safety feature ensures that, if a break occurs in one of the components in the chlorinator, the vacuum will be lost, and the chlorinator will shut down without allowing gas to escape.

Chlorine gas is a highly toxic lung irritant, and special facilities are required for storing and housing it. Chlorine gas will expand to 500 times its original compressed liquid volume at room temperature (1 gallon of liquid chlorine will expand to about 67 ft^3). Its advantage as a drinking water disinfectant is the convenience afforded by a relatively large quantity of chlorine available for continuous operation for several days or weeks without the need for mixing chemicals. Where water flow rates are highly variable, the chlorination rate can be synchronized with the flow.

Chlorine gas has a very strong, characteristic odor that can be detected by most people at concentrations as low as 3.5 ppm. Highly corrosive in moist air, it is extremely toxic and irritating in concentrated form. Its toxicity ranges from being a throat irritant at 15 ppm to causing rapid death at 1000 ppm. Although chlorine does not burn, it supports combustion, so open flames should never be used around chlorination equipment.

When changing chlorine cylinders, an accidental release of chlorine may occasionally occur. To handle this type of release, a NIOSH-approved, self-contained breathing apparatus (SCBA) must be worn. Special emergency repair kits are available from the Chlorine Institute for use by emergency response teams to deal with chlorine leaks. Because chlorine gas is 2.5 times heavier than air, exhaust and inlet air ducts should be installed at floor level. A leak of chlorine gas can be found with a strong ammonia mist solution, as a white cloud develops when ammonia mist and chlorine combine.

Hypochlorination

Combining chlorine with calcium or sodium produces hypochlorites. Calcium hypochlorites are sold in powder or tablet forms and can contain chlorine concentrations up to 67%. Sodium hypochlorite is a liquid (bleach, for example) and is found in concentrations up to 16%. Chlorine concentrations of household bleach range from 4.75 to 5.25%. Most small system operators find using these liquid or dry chlorine compounds more convenient and safer than chlorine gas.

The compounds are mixed with water and fed into the water with inexpensive solution feed pumps. These pumps are designed to operate against high system pressures but can also be used to inject chlorine solutions into tanks, although injecting chlorine into the suction side of a pump is not recommended as the chlorine may corrode the pump impeller.

Calcium hypochlorite can be purchased as tablets or granules, with approximately 65% available chlorine (10 lb of calcium hypochlorite granules contain only 6.5 lb of chlorine). Normally, 6.5 lb of calcium hypochlorite will produce a concentration of 50 mg/L chlorine in 10,000 gal of water. Calcium hypochlorite can burn (at 350°F) if combined with oil or grease. When mixing calcium hypochlorite, operators must wear chemical safety goggles, a cartridge breathing apparatus, and rubberized gloves. Always place the powder in the water. Placing the water into the dry powder could cause an explosion.

Sodium hypochlorite is supplied as a clear, greenish-yellow liquid in strengths from 5.25 to 16% available chlorine. Often referred to as "bleach," it is, in fact, used for bleaching. Common household bleach is a solution of sodium hypochlorite containing 4.75 to 5.25% available chlorine. The amount of sodium hypochlorite required to produce a 50-mg/L chlorine concentration in 10,000 gal of water can be calculated using the solutions equation:

$$C_1 \times V_1 = C_2 \times V_2 \tag{11.11}$$

where

C = Solution concentration (mg/L or %).
V = Solution volume (liters, gallons, quarts, etc.).

In this example, C_1 and V_1 are associated with the sodium hypochlorite, and C_2 and V_2 are associated with the 10,000 gallons of water with a 50-mg/L chlorine concentration (and 10,000 mg/L = 1%). Therefore:

$$C_1 = \frac{5.25\% \times 10{,}000 \text{ mg/L}}{1.0\%} = 52{,}500 \text{ mg/L}$$

V_1 = Unknown volume of sodium hypochlorite.
C_2 = 50 mg/L.
V_2 = 10,000 gallons.

Thus,

$$C_1 \times V_1 = C_2 \times V_2$$

$$52{,}500 \text{ mg/L} \times V_1 = 50 \text{ mg/L} \times 10{,}000 \text{ gal}$$

$$V_1 = \frac{50 \text{ mg/L} \times 10{,}000 \text{ gal}}{52{,}500 \text{ mg/L}}$$

$$= 9.52 \text{ gal sodium hypochlorite}$$

Sodium hypochlorite solutions are introduced to the water in the same manner as calcium hypochlorite solutions. The purchased stock bleach is usually diluted with water to produce a feed solution that is pumped into the water system.

Hypochlorites must be stored properly to maintain their strengths. Calcium hypochlorite must be stored in airtight containers in cool, dry, dark locations. Sodium hypochlorite degrades relatively quickly even when properly stored; it can lose more than half of its strength in 3 to 6 months. Operators should purchase hypochlorites in small quantities to be sure they are used while still strong. Old chemicals should be discarded safely.

The pumping rate of a chemical metering pump is usually manually adjusted by varying the length of the piston or diaphragm stroke. Once the stroke is set, the hypochlorinator feeds accurately at that rate; however, chlorine measurements must be made occasionally at the beginning and end of the well pump cycle to ensure correct dosage. A metering device may be used to vary the hypochlorinator feed rate, synchronized with the water flow rate. Where a well pump is used, the hypochlorinator is connected electrically with the on/off controls of the pump so the chlorine solution is not fed into the pipe when the well is not pumping.

Determining Chlorine Dosage

Proper disinfection requires calculation of the amount of chlorine that must be added to the water to produce the required dosage. The type of calculation used depends on the form of chlorine being used. The basic chlorination calculation used is the same one used for all chemical addition calculations—the *pounds formula*:

$$\text{Pounds} = \text{mg/L} \times 8.34 \text{ lb/gal} \times \text{MG} \tag{11.12}$$

where

Pounds = Pounds of available chlorine required.
mg/L = Desired concentration in milligrams per liter.
8.34 lb/gal = Conversion factor.
MG = Millions of gallons of water to be treated.

■ Example 11.3

Problem: Calculate the number of pounds of gaseous chlorine needed to treat 250,000 gal of water with 1.2 mg/L of chlorine.

Solution:

$$\text{Pounds} = 1.2 \text{ mg/L} \times 8.34 \text{ lb/gal} \times 0.25 \text{ MG} = 2.5 \text{ lb}$$

Note: Hypochlorites contain less than 100% available chlorine; thus, we must use more hypochlorite to get the same number of pounds of chlorine into the water.

If we substitute calcium hypochlorite with 65% available chlorine in our example, 2.5 lb of available chlorine is still needed, but more than 2.5 lb of calcium hypochlorite is required to provide that much chlorine. Determine how much of the chemical is needed by dividing the pounds of chlorine required by the decimal form of the percent available chlorine. Because 65% is the same as 0.65, we need to add

$$\frac{2.5 \text{ lb}}{0.65 \text{ available chlorine}} = 3.85 \text{ lb Ca (OCl)} \tag{11.13}$$

to get that much chlorine.

In practice, because most hypochlorites are fed as solutions, we often need to know how much chlorine solution we should feed. In addition, the practical problems faced in day-to-day operation are never so clearly stated as the practice problems we work; for example, small water systems do not usually deal with water flow in million gallons per day. Real-world problems usually require a lot of intermediate calculations to get everything ready to plug into the pounds formula.

THOUGHT-PROVOKING QUESTIONS

11.1 Is chlorination the best way in which to disinfect drinking water? Explain.

11.2 If you had a choice to replace chlorine with some other disinfectant, which replacement would you choose?

REFERENCES AND RECOMMENDED READING

Craun, G.F. (1981). Outbreaks of waterborne disease in the United States. *Journal AWWA*, 73(7), 360.

Culp, G.L. and Culp, R.L. (1974). Outbreaks of waterborne disease in the United States. *Journal AWWA*, 73(7), 360.

Harr, J. (1995). *A Civil Action*. New York: Vintage Books.

Montgomery, J.M. (1985). *Water Treatment Principles and Design*. New York: John Wiley & Sons.

Singer, P.C. and Chang, S.D. (1989). Correlations between trihalomethanes and total organic halides formed during water treatment. *Journal AWWA*, 81(8), 61–65.

Snead, M.C. et al. (1980). *Benefits of Maintaining a Chlorine Residual in Water Supply Systems*, EPA600/2-80-010. Washington, DC: U.S. Environmental Protection Agency.

Spellman, F.R. (1999). *Choosing Disinfection Alternatives for Water/Wastewater Treatment*. Boca Raton: FL: CRC Press.

TWUA. (1988). *Manual of Water Utility Operations*, 8th ed. Austin: Texas Water Utilities Association.

USEPA. (1997). *Community Water System Survey*, Vols. I and II, EPA-815-R-97-001a. Washington, DC: U.S. Environmental Protection Agency.

USEPA. (1998). *National Primary Drinking Water Regulations: Interim Enhanced Surface Water Treatment Final Rule*. Washington, DC: U.S. Environmental Protection Agency.

12 Water Treatment Calculations

Because of huge volume and flow conditions, the quality of natural water cannot be modified significantly within the body of water (Gupta, 1997). Consequently, the quality control approach is directed toward water withdrawn from a source for a specific use. The drawn water is treated prior to its use.

INTRODUCTION

In the following sections, we present the basic, often used, daily operational calculations along with engineering calculations used for solving more complex computations. This presentation method is in contrast to the presentation methods used in typical water mathematics texts. The author deviates from the norm based on and because of practical real-world experience. That is, experience demonstrates that the environmental engineer tasked with managing a water or wastewater treatment plant not only is responsible for computation of many complex math operations (engineering calculations) but is also responsible for overseeing proper plant operation (including math operations at the operator level). Obviously, engineers are well versed in basic math operations; however, they often need to refer to example plant operation calculations in a variety of texts. In this text, the format used, though unconventional, is designed to provide both basic operations math and more complex engineering math in one ready format.

WATER SOURCE AND STORAGE CALCULATIONS

WATER SOURCES

Approximately 40 million cubic miles of water cover or reside within the Earth. The oceans contain about 97% of all water on Earth. The other 3% is freshwater: (1) snow and ice on the surface of the Earth contain about 2.25% of the water, (2) usable ground water is approximately 0.3%, and (3) surface freshwater is less than 0.5%. In the United States, for example, average rainfall is approximately 2.6 ft (a volume of 5900 km^3). Of this amount, approximately 71% evaporates (about 4200 km^3), and 29% goes to stream flow (about 1700 km^3).

Beneficial freshwater uses include manufacturing, food production, domestic and public needs, recreation, hydroelectric power production, and flood control. Stream flow withdrawn annually is about 7.5% (440 km^3). Irrigation and industry use almost half of this amount (3.4%, or 200 km^3/yr). Municipalities use only about 0.6% (35 km^3/yr) of this amount. Historically, in the United States, water usage has been increasing (as might be expected); for example, in 1900, 40 billion gallons of freshwater were used. In 1975, usage increased to 455 billion gallons. Projected use in 2000 was about 720 billion gallons.

The primary sources of freshwater include the following:

- Captured and stored rainfall in cisterns and water jars
- Groundwater from springs, artesian wells, and drilled or dug wells
- Surface water from lakes, rivers, and streams
- Desalinized seawater or brackish groundwater
- Reclaimed wastewater

Water Source Calculations

Water source calculations covered in this section apply to wells and pond or lake storage capacity. Specific well calculations discussed include well drawdown, well yield, specific yield, well casing disinfection, and deep-well turbine pump capacity.

Well Drawdown

Drawdown is the drop in the level of water in a well when water is being pumped. Drawdown is usually measured in feet or meters. One of the most important reasons for measuring drawdown is to make sure that the source water is adequate and not being depleted. The data collected to calculate drawdown can indicate if the water supply is slowly declining. Early detection can give the system time to explore alternative sources, establish conservation measures, or obtain any special funding that may be needed to get a new water source. Well drawdown is the difference between the pumping water level and the static water level:

$$\text{Drawdown (ft)} = \text{Pumping water level (ft)} - \text{Static water level (ft)} \tag{12.1}$$

■ Example 12.1

Problem: The static water level for a well is 70 ft. If the pumping water level is 90 ft, what is the drawdown?

Solution:

$$\text{Drawdown} = \text{Pumping water level (ft)} - \text{Static water level (ft)} = 90\text{ ft} - 70\text{ ft} = 20\text{ ft}$$

■ Example 12.2

Problem: The static water level of a well is 122 ft. The pumping water level is determined using the sounding line. The air pressure applied to the sounding line is 4.0 psi, and the length of the sounding line is 180 ft. What is the drawdown?

Solution: First calculate the water depth in the sounding line and the pumping water level:

1. Water depth in sounding line = 4.0 psi × 2.31 ft/psi = 9.2 ft
2. Pumping water level = 180 ft – 9.2 ft = 170.8 ft

Then calculate drawdown as usual:

$$\text{Drawdown} = \text{Pumping water level (ft)} - \text{Static water level (ft)} = 170.8\text{ ft} - 122\text{ ft} = 48.8\text{ ft}$$

Well Yield

Well yield is the volume of water per unit of time that is produced from the well pumping. Usually, well yield is measured in terms of gallons per minute (gpm) or gallons per hour (gph). Sometimes, large flows are measured in cubic feet per second (cfs). Well yield is determined by using the following equation:

$$\text{Well yield (gpm)} = \text{Gallons produced} \div \text{Duration of test (min)} \tag{12.2}$$

■ Example 12.3

Problem: When the drawdown level of a well was stabilized, it was determined that the well produced 400 gal during a 5-min test. What was the well yield?

Solution:

Well yield = Gallons produced ÷ Duration of test (min) = 400 gal ÷ 5 min = 80 gpm

■ **Example 12.4**

Problem: During a 5-min test for well yield, a total of 780 gal was removed from the well. What was the well yield in gpm? In gph?

Solution:

Well yield = Gallons produced ÷ Duration of test (min) = 780 gal ÷ 5 min = 156 gpm

Then convert gpm flow to gph flow:

$$156 \text{ gpm} \times 60 \text{ min/hr} = 9360 \text{ gph}$$

Specific Yield

Specific yield is the discharge capacity of the well per foot of drawdown. The specific yield may range from 1 gpm/ft drawdown to more than 100 gpm/ft drawdown for a properly developed well. Specific yield is calculated using Equation 12.3:

$$\text{Specific yield (gpm/ft)} = \text{Well yield (gpm)} \div \text{Drawdown (ft)} \tag{12.3}$$

■ **Example 12.5**

Problem: A well produces 260 gpm. If the drawdown for the well is 22 ft, what is the specific yield in gpm/ft?

Solution:

Specific yield = Well yield (gpm) ÷ Drawdown (ft) = 260 gpm ÷ 22 ft = 11.8 gpm/ft

■ **Example 12.6**

Problem: The yield for a particular well is 310 gpm. If the drawdown for this well is 30 ft, what is the specific yield in gpm/ft?

Solution:

Specific yield = Well yield (gpm) ÷ Drawdown (ft) = 310 gpm ÷ 30 ft = 10.3 gpm/ft

Well Casing Disinfection

A new, cleaned, or repaired well normally contains contamination that may remain for weeks unless the well is thoroughly disinfected. This may be accomplished by using ordinary bleach at a concentration of 100 parts per million (ppm) of chlorine. The amount of disinfectant required is determined by the amount of water in the well. The following equation is used to calculate the pounds of chlorine required for disinfection:

$$\text{Chlorine (lb)} = \text{Chlorine (mg/L)} \times \text{Casing volume (MG)} \times 8.34 \text{ lb/gal} \tag{12.4}$$

■ **Example 12.7**

Problem: A new well is to be disinfected with chlorine at a dosage of 50 mg/L. If the well casing diameter is 8 in. and the length of the water-filled casing is 110 ft, how many pounds of chlorine will be required?

Solution: First calculate the volume of the water-filled casing:

$$0.785 \times 0.67 \times 67 \times 110 \text{ ft} \times 7.48 \text{ gal/ft}^3 = 290 \text{ gal}$$

Then determine the pounds of chlorine required using the mg/L to lb equation:

$$\text{Chlorine (lb)} = \text{Chlorine (mg/L)} \times \text{volume (MG)} \times 8.34 \text{ lb/gal}$$

$$= 50 \text{ mg/L} \times 0.000290 \text{ MG} \times 8.34 \text{ lb/gal} = 0.12 \text{ lb}$$

Deep-Well Turbine Pumps

The deep-well turbine pump is used for high-capacity deep wells. The pump, usually consisting of more than one stage of centrifugal pump, is fastened to a pipe called the *pump column*; the pump is located in the water. The pump is driven from surface through a shaft running inside the pump column. The water is discharged from the pump up through the pump column to the surface. The pump may be driven by a vertical shaft, electric motor at the top of the well, or some other power source, usually through a right-angle gear drive located at the top of the well. A modern version of the deep-well turbine pump is the submersible type of pump, where the pump (as well as a close-coupled electric motor built as a single unit) is located below water level in the well. The motor is built to operate submerged in water.

Vertical Turbine Pump Calculations

The calculations pertaining to well pumps include head, horsepower, and efficiency calculations. *Discharge head* is measured to the pressure gauge located close to the pump discharge flange. The pressure (psi) can be converted to feet of head using the following equation:

$$\text{Discharge head (ft)} = \text{Pressure (psi)} \times 2.31 \text{ ft/psi} \tag{12.5}$$

Total pumping head (*field head*) is a measure of the lift below the discharge head pumping water level (*discharge head*). Total pumping head is calculated as follows:

$$\text{Pumping head (ft)} = \text{Pumping water level (ft)} + \text{Discharge head (ft)} \tag{12.6}$$

■ Example 12.8

Problem: The pressure gauge reading at a pump discharge head is 4.1 psi. What is this discharge head expressed in feet?

Solution:

$$4.1 \text{ psi} \times 2.31 \text{ ft/psi} = 9.5 \text{ ft}$$

■ Example 12.9

Problem: The static water level of a pump is 100 ft. The well drawdown is 26 ft. If the gauge reading at the pump discharge head is 3.7 psi, what is the total pumping head?

Solution:

$$\text{Total pumping head (ft)} = \text{Pumping water level (ft)} + \text{Discharge head (ft)} \tag{12.7}$$

$$= (100 \text{ ft} + 26 \text{ ft}) + (3.7 \text{ psi} \times 2.31 \text{ ft/psi})$$

$$= 126 \text{ ft} + 8.5 \text{ ft}$$

$$= 134.5 \text{ ft}$$

Five types of *horsepower* calculations are used for vertical turbine pumps; it is important to have a general understanding of these five horsepower types:

- *Motor horsepower* refers to the horsepower supplied to the motor. The following equation is used to calculate motor horsepower:

$$\text{Motor horsepower (input horsepower)} = \frac{\text{Field brake horsepower}}{\text{Motor efficiency}/100} \tag{12.8}$$

- *Total brake horsepower* refers to the horsepower output of the motor. The following equation is used to calculate total brake horsepower:

$$\text{Total brake horsepower} = \text{Field horsepower} + \text{Thrust bearing loss (hp)} \tag{12.9}$$

- *Field horsepower* refers to the horsepower required at the top of the pump shaft. The following equation is used to calculate field horsepower:

$$\text{Field horsepower} = \text{Bowl horsepower} + \text{Shaft loss (hp)} \tag{12.10}$$

- *Bowl or laboratory horsepower* refers the horsepower at the entry to the pump bowls. The following equation is used to calculate bowl horsepower:

$$\text{Bowl horsepower (lab horsepower)} = \frac{\text{Bowl head (ft)} \times \text{Capacity (gpm)}}{3960 \times (\text{Bowl efficiency}/100)} \tag{12.11}$$

- *Water horsepower* refers to the horsepower at the pump discharge. The following equation is used to calculate water horsepower:

$$\text{Water horsepower} = \frac{\text{Field head (ft)} \times \text{Capacity (gpm)}}{3960} \tag{12.12}$$

  or the equivalent equation

$$\text{Water horsepower} = \frac{\text{Field head (ft)} \times \text{Capacity (gpm)}}{33{,}000 \text{ ft-lb/min}}$$

■ Example 12.10

Problem: The pumping water level for a well pump is 150 ft and the discharge pressure measured at the pump discharge centerline is 3.5 psi. If the flow rate from the pump is 700 gpm, what is the water horsepower? (Use Equation 12.12.)

Solution: First calculate the field head. The discharge head must be converted from psi to ft:

$$3.5 \text{ psi} \times 2.31 \text{ ft/psi} = 8.1 \text{ ft}$$

The water horsepower is, therefore,

$$150 \text{ ft} + 8.1 \text{ ft} = 158.1 \text{ ft}$$

The water horsepower can now be determined:

$$\text{Water horsepower} = \frac{158.1 \text{ ft} \times 700 \text{ gpm} \times 8.34 \text{ lb/gal}}{33{,}000 \text{ ft-lb/min}} = 28$$

■ Example 12.11

Problem: The pumping water level for a pump is 170 ft. The discharge pressure measured at the pump discharge head is 4.2 psi. If the pump flow rate is 800 gpm, what is the water horsepower? (Use Equation 12.12.)

Solution: The field head must first be determined. In order to determine field head, the discharge head must be converted from psi to ft:

$$4.2 \text{ psi} \times 2.31 \text{ ft/psi} = 9.7 \text{ ft}$$

The field head can now be calculated:

$$170 \text{ ft} + 9.7 \text{ ft} = 179.7 \text{ ft}$$

And then the water horsepower can be calculated:

$$\text{Water horsepower} = \frac{179.7 \text{ ft} \times 800 \text{ gpm} \times 8.34 \text{ lb/gal}}{33{,}000 \text{ ft-lb/min}} = 36$$

■ Example 12.12

Problem: A deep-well vertical turbine pump delivers 600 gpm. If the lab head is 185 ft and the bowl efficiency is 84%. What is the bowl horsepower? (Use Equation 12.11.)

Solution:

$$\begin{aligned}\text{Bowl horsepower} &= \frac{\text{Bowl head (ft)} \times \text{Capacity (gpm)}}{3960 \times (\text{Bowl efficiency}/100)} \\ &= \frac{185 \text{ ft} \times 600 \text{ gpm}}{3960 \times (84.0/100)} \\ &= \frac{185 \text{ ft} \times 600 \text{ gpm}}{3960 \times 0.84} \\ &= 33.4\end{aligned}$$

■ Example 12.13

Problem: The bowl bhp is 51.8 bhp. If the 1-inch diameter shaft is 170 ft long and is rotating at 960 rpm with a shaft fiction loss of 0.29 hp loss per 100 ft, what is the field bhp?

Solution: Before field bhp can be calculated, the shaft loss must be factored in:

$$\frac{(0.29 \text{ hp loss}) \times (170 \text{ ft})}{100} = 0.5$$

Now determine the field horsepower:

$$\text{Field horsepower} = \text{Bowl horsepower} + \text{Shaft loss (hp)} = 51.8\text{ hp} + 0.5\text{ hp} = 52.3\text{ bhp}$$

■ Example 12.14

Problem: The field horsepower for a deep-well turbine pump is 62 bhp. If the thrust bearing loss is 0.5 hp and the motor efficiency is 88%, what is the motor input horsepower? (Use Equation 12.8.)

Solution:

$$\text{Motor input horsepower} = \frac{\text{Field (total) bhp}}{\text{Motor efficiency}/100} = \frac{62\text{ bhp} + 0.5\text{ hp}}{0.88} = 71\text{ mhp}$$

When we speak of the *efficiency* of any machine, we are speaking primarily of a comparison of what is put out by the machine (e.g., energy output) compared to its input (e.g., energy input). Horsepower efficiency, for example, is a comparison of horsepower output of the unit or system with horsepower input to that unit or system—the unit's efficiency. With regard to vertical turbine pumps, there are four types efficiencies considered with vertical turbine pumps:

- Bowl efficiency
- Field efficiency
- Motor efficiency
- Overall efficiency

The general equation used in calculating percent efficiency is shown below:

$$\% = \frac{\text{Part}}{\text{Whole}} \times 100 \tag{12.13}$$

Vertical turbine pump *bowl efficiency* is easily determined using a pump performance curve chart provided by the pump manufacturer. Field efficiency is determined using Equation 12.14:

$$\text{Field efficiency (\%)} = \frac{\text{Field head (ft)} \times \text{Capacity (gpm)}}{3960 \times \text{Total bhp}} \times 100 \tag{12.14}$$

■ Example 12.15

Problem: Given the data below, calculate the field efficiency of the deep-well turbine pump:

Field head—180 ft
Capacity—850 gpm
Total bhp—61.3 bhp

Solution:

$$\text{Field efficiency} = \frac{\text{Field head (ft)} \times \text{Capacity (gpm)}}{3960 \times \text{Total bhp}} \times 100 = \frac{180\text{ ft} \times 850\text{ gpm}}{3960 \times 61.3} \times 100 = 63\%$$

Overall efficiency is a comparison of the horsepower output of the system with that entering the system. Equation 12.15 is used to calculate overall efficiency:

$$\text{Overall efficiency (\%)} = \frac{\text{Field efficiency (\%)} \times \text{Motor efficiency (\%)}}{100} \quad (12.15)$$

■ Example 12.16

Problem: The efficiency of a motor is 90%. If the field efficiency is 83%, what is the overall efficiency of the unit?

Solution:

$$\text{Overall efficiency} = \frac{\text{Field efficiency (\%)} \times \text{Motor efficiency (\%)}}{100} = \frac{83\% \times 90\%}{100} = 74.7\%$$

Water Storage

Water storage facilities for water distribution systems are required primarily to provide for fluctuating demands of water usage (to provide a sufficient amount of water to average or equalize daily demands on the water supply system). In addition, other functions of water storage facilities include increasing operating convenience, leveling pumping requirements (to keep pumps from running 24 hours a day), decreasing power costs, providing water during power source or pump failure, providing large quantities of water to meet fire demands, providing surge relief (to reduce the surge associated with stopping and starting pumps), increasing detention time (to provide chlorine contact time and satisfy the desired contact time value requirements), and blending water sources.

Water Storage Calculations

The storage capacity, in gallons, of a reservoir, pond, or small lake can be estimated as follows:

$$\text{Capacity (gal)} = \text{Average length (ft)} \times \text{Average width (ft)} \times \text{Average depth (ft)} \times 7.48\ \text{gal/ft}^3 \quad (12.16)$$

■ Example 12.17

Problem: A pond has an average length of 250 ft, an average width of 110 ft, and an estimated average depth of 15 ft. What is the estimated volume of the pond in gallons?

Solution:

$$\text{Volume} = \text{Average length (ft)} \times \text{Average width (ft)} \times \text{Average depth (ft)} \times 7.48\ \text{gal/ft}^3$$

$$= 250\ \text{ft} \times 110\ \text{ft} \times 15\ \text{ft} \times 7.48\ \text{gal/ft}^3$$

$$= 3{,}085{,}500\ \text{gal}$$

■ Example 12.18

Problem: A small lake has an average length of 300 ft and an average width of 95 ft. If the maximum depth of the lake is 22 ft, what is the estimated gallons volume of the lake?

Note: For small ponds and lakes, the average depth is generally about 0.4 times the greatest depth; therefore, to estimate the average depth, measure the greatest depth and multiply that number by 0.4.

Solution: First, the average depth of the lake must be estimated:

$$\text{Estimated average depth} = 22 \text{ ft} \times 0.4 \text{ ft} = 8.8 \text{ ft}$$

Then, the lake volume can be determined:

$$\text{Volume} = \text{Average length (ft)} \times \text{Average width (ft)} \times \text{Average depth (ft)} \times 7.48 \text{ gal/ft}^3$$

$$= 300 \text{ ft} \times 95 \text{ ft} \times 8.8 \text{ ft} \times 7.48 \text{ gal/ft}^3$$

$$= 1{,}875{,}984 \text{ gal}$$

Copper Sulfate Dosing

Algae control by applying copper sulfate is perhaps the most common *in situ* treatment of lakes, ponds, and reservoirs; the copper ions in the water kill the algae. Copper sulfate application methods and dosages will vary depending on the specific surface water body being treated. The desired copper sulfate dosage may be expressed in mg/L copper, lb copper sulfate per ac-ft, or lb copper sulfate per acre.

For a dose expressed as mg/L copper, the following equation is used to calculate lb copper sulfate required:

$$\text{Copper sulfate (lb)} = \frac{\text{Copper (mg/L)} \times \text{Volume (MG)} \times 8.34 \text{ lb/gal}}{\text{\% Available copper}/100} \quad (12.17)$$

■ **Example 12.19**

Problem: For algae control in a small pond, a dosage of 0.5 mg/L copper is desired. The pond has a volume of 15 MG. How many pounds of copper sulfate will be required? (Copper sulfate contains 25% available copper.)

Solution:

$$\text{Copper sulfate} = \frac{\text{Copper (mg/L)} \times \text{Volume (MG)} \times 8.34 \text{ lb/gal}}{\text{\% Available copper}/100}$$

$$= \frac{0.5 \text{ mg/L} \times 15 \text{ MG} \times 8.34 \text{ lb/gal}}{25/100}$$

$$= 250 \text{ lb}$$

For calculating lb copper sulfate per ac-ft, use the following equation (assuming a desired copper sulfate dose of 0.9 lb/ac-ft):

$$\text{Copper sulfate (lb)} = \frac{0.9 \text{ lb Copper sulfate} \times \text{ac-ft}}{1 \text{ ac-ft}} \quad (12.18)$$

■ **Example 12.20**

Problem: A pond has a volume of 35 ac-ft. If the desired copper sulfate dose is 0.9 lb/ac-ft, how many lb of copper sulfate will be required?

Solution:

$$\text{Copper sulfate (lb)} = \frac{0.9 \text{ lb Copper sulfate} \times \text{ac-ft}}{1 \text{ ac-ft}}$$

$$\frac{0.9 \text{ lb Copper sulfate}}{1 \text{ ac-ft}} = \frac{x \text{ lb Copper sulfate}}{35 \text{ ac-ft}}$$

Then solve for x:

$$x = 0.9 \times 35 = 31.5 \text{ lb copper sulfate}$$

The desired copper sulfate dosage may also be expressed in terms of lb copper sulfate per acre. The following equation is used to determine lb copper sulfate (assuming a desired copper sulfate dose of 5.2 lb/ac):

$$\text{Copper sulfate (lb)} = (5.2 \text{ lb copper sulfate} \times \text{acres}) \div 1 \text{ ac} \tag{12.19}$$

■ **Example 12.21**

Problem: A small lake has a surface area of 6.0 ac. If the desired copper sulfate dose is 5.2 lb/ac, how many pounds of copper sulfate are required?

Solution:

$$\text{Copper sulfate (lb)} = (5.2 \text{ lb copper sulfate} \times 6.0 \text{ ac})/1 \text{ ac} = 31.2 \text{ lb copper sulfate}$$

COAGULATION, MIXING, AND FLOCCULATION CALCULATIONS

Coagulation

Following screening and the other pretreatment processes, the next unit process in a conventional water treatment system is a mixer where the first chemicals are added in what is known as *coagulation.* The exception to this situation occurs in small systems using groundwater, when chlorine or other taste and odor control measures are introduced at the intake and are the extent of treatment. The term *coagulation* refers to the series of chemical and mechanical operations by which coagulants are applied and made effective. These operations are comprised of two distinct phases: (1) rapid mixing to disperse coagulant chemicals by violent agitation into the water being treated, and (2) flocculation to agglomerate small particles into well-defined floc by gentle agitation for a much longer time. The coagulant must be added to the raw water and perfectly distributed into the liquid; such uniformity of chemical treatment is reached through rapid agitation or mixing. Coagulation is a reaction caused by adding salts or iron or aluminum to the water. Common coagulants (salts) include the following:

- Alum (aluminum sulfate)
- Sodium aluminate
- Ferric sulfate
- Ferrous sulfate
- Ferric chloride
- Polymers

Mixing

To ensure maximum contact between the reagent and suspended particles, coagulants and coagulant aids must be rapidly dispersed (mixed) throughout the water; otherwise, the coagulant will react with water, dissipating some of its coagulating power. To ensure complete mixing and optimum plug-flow reactor operation, proper detention time in the basin is required. Detention time can be calculated using the following procedures:

For complete mixing:

$$t = \frac{V}{Q} = \left(\frac{1}{K}\right)\left(\frac{C_i - C_e}{C_e}\right) \quad (12.20)$$

For plug flow:

$$t = \frac{V}{Q} = \frac{L}{v} = \left(\frac{1}{K}\right)\left(\ln\frac{C_i}{C_e}\right) \quad (12.21)$$

where

t = Detention time of the basin (min).
V = Volume of basin (m^3 or ft^3).
Q = Flow rate (m^3/s or cfs).
K = Rate constant.
C_i = Influent reactant concentration (mg/L).
C_e = Effluent reactant concentration (mg/L).
L = Length of rectangular basin (m or ft).
v = Horizontal velocity of flow (m/s or ft/s).

■ **Example 12.22**

Problem: Alum dosage is 40 mg/L and K = 90 per day based on lab tests. Compute the detention times for complete mixing and plug flow reactor for 90% reduction.

Solution: First find C_e:

$$C_e = (1 - 0.9) \times C_i = 0.1 \times C_i = 0.1 \times 40 \text{ mg/L} = 4 \text{ mg/L}$$

Now calculate t for complete mixing (Equation 12.20):

$$t = \frac{V}{Q} = \left(\frac{1}{K}\right)\left(\frac{C_i - C_e}{C_e}\right) = \left(\frac{1}{90/\text{day}}\right)\left(\frac{40 \text{ mg/L} - 4 \text{ mg/L}}{4 \text{ mg/L}}\right) = \left(\frac{1d}{90}\right) \times \left(\frac{1440 \text{ min}}{1 \text{ day}}\right) = 144 \text{ min}$$

Finally, calculate t for plug flow using the following formula:

$$t = \left(\frac{1}{K}\right)\left(\ln\frac{C_i}{C_e}\right) = \left(\frac{1440}{90}\right)\left(\ln\frac{40}{4}\right) = 36.8 \text{ min}$$

Flocculation

Flocculation follows coagulation in the conventional water treatment process. *Flocculation* is the physical process of slowly mixing the coagulated water to increase the probability of particle collision. Through experience, we see that effective mixing reduces the required amount of chemicals and greatly improves the sedimentation process, which results in longer filter runs and higher quality finished water. The goal of flocculation is to form a uniform, feather-like material similar to snowflakes—a dense, tenacious floc that traps the fine, suspended, and colloidal particles and carries them down rapidly in the settling basin. To increase the speed of floc formation and the strength and weight of the floc, polymers are often added.

Coagulation and Flocculation General Calculations

Proper operation of the coagulation and flocculation unit processes requires calculations to determine chamber or basin volume, chemical feed calibration, chemical feeder settings, and detention time.

Chamber and Basin Volume Calculations

To determine the volume of a square or rectangular chamber or basin, we use Equation 12.22 or Equation 12.23:

$$\text{Volume (ft}^3\text{)} = \text{Length (ft)} \times \text{Width (ft)} \times \text{Depth (ft)} \tag{12.22}$$

$$\text{Volume (gal)} = \text{Length (ft)} \times \text{Width (ft)} \times \text{Depth (ft)} \times 7.48 \text{ gal/ft}^3 \tag{12.23}$$

■ **Example 12.23**

Problem: A flash mix chamber is 4 ft square with water to a depth of 3 ft. What is the volume of water (in gallons) in the chamber?

Solution:

$$\begin{aligned}\text{Volume} &= \text{Length (ft)} \times \text{Width (ft)} \times \text{Depth (ft)} \times 7.48 \text{ gal/ft}^3\\ &= 4 \text{ ft} \times 4 \text{ ft} \times 3 \text{ ft} \times 7.48 \text{ gal/ft}^3\\ &= 359 \text{ gal}\end{aligned}$$

■ **Example 12.24**

Problem: A flocculation basin is 40 ft long by 12 ft wide with water to a depth of 9 ft. What is the volume of water (in gallons) in the basin?

Solution:

$$\begin{aligned}\text{Volume} &= \text{Length (ft)} \times \text{Width (ft)} \times \text{Depth (ft)} \times 7.48 \text{ gal/ft}^3\\ &= 40 \text{ ft} \times 12 \text{ ft} \times 9 \text{ ft} \times 7.48 \text{ gal/ft}^3\\ &= 32{,}314 \text{ gal}\end{aligned}$$

■ **Example 12.25**

Problem: A flocculation basin is 50 ft long by 22 ft wide and contains water to a depth of 11 ft, 6 in. How many gallons of water are in the tank?

Solution: First convert the 6-in. portion of the depth measurement to feet:

$$(6 \text{ in.}) \div (12 \text{ in./ft}) = 0.5 \text{ ft}$$

Then calculate basin volume:

$$\begin{aligned}\text{Volume} &= \text{Length (ft)} \times \text{Width (ft)} \times \text{Depth (ft)} \times 7.48 \text{ gal/ft}^3\\ &= 50 \text{ ft} \times 22 \text{ ft} \times 11.5 \text{ ft} \times 7.48 \text{ gal/ft}^3\\ &= 94{,}622 \text{ gal}\end{aligned}$$

Detention Time

Because coagulation reactions are rapid, detention time for flash mixers is measured in seconds, whereas the detention time for flocculation basins is generally between 5 and 30 min. The equation used to calculate detention time is shown below:

$$\text{Detention time (min)} = \text{Volume of tank (gal)} \div \text{Flow rate (gpm)} \quad (12.24)$$

■ Example 12.26

Problem: The flow to a flocculation basin that is 50 ft long by 12 ft wide by 10 ft deep is 2100 gpm. What is the detention time in the tank (in minutes)?

Solution:

$$\text{Tank volume (gal)} = 50 \text{ ft} \times 12 \text{ ft} \times 10 \text{ ft} \times 7.48 \text{ gal/ft}^3 = 44{,}880 \text{ gal}$$

$$\text{Detention time} = \text{Volume of tank (gal)} \div \text{Flow rate (gpm)} = 44{,}880 \text{ gal} \div 2100 \text{ gpm} = 21.4 \text{ min}$$

■ Example 12.27

Problem: A flash mix chamber is 6 ft long by 4 ft with water to a depth of 3 ft. If the flow to the flash mix chamber is 6 MGD, what is the chamber detention time in seconds (assuming that the flow is steady and continuous)?

Solution: First, convert the flow rate from gpd to gps so the time units will match:

$$6{,}000{,}000 \div (1440 \text{ min/day} \times 60 \text{ sec/min}) = 69 \text{ gps}$$

Then calculate detention time:

$$\begin{aligned}\text{Detention time} &= \text{Volume of tank (gal)} \div \text{Flow rate (gpm)}\\ &= (6 \text{ ft} \times 4 \text{ ft} \times 3 \text{ ft} \times 7.48 \text{ gal/ft}^3) \div 69 \text{ gps}\\ &= 7.8 \text{ sec}\end{aligned}$$

Determining Dry Chemical Feeder Setting (lb/day)

When adding (dosing) chemicals to the water flow, a measured amount of chemical is called for. The amount of chemical required depends on such factors as the type of chemical used, the reason for dosing, and the flow rate being treated. To convert from mg/L to lb/day, the following equation is used:

$$\text{Chemical added (lb/day)} = \text{Chemical (mg/L)} \times \text{Flow (MGD)} \times 8.34 \text{ lb/gal} \quad (12.25)$$

■ Example 12.28

Problem: Jar tests indicate that the best alum dose for water is 8 mg/L. If the flow to be treated is 2,100,000 gpd, what should the lb/day settling be on the dry alum feeder?

Solution:

$$\begin{aligned}\text{Chemical added (lb/day)} &= \text{Chemical (mg/L)} \times \text{Flow (MGD)} \times 8.34 \text{ lb/gal}\\ &= 8 \text{ mg/L} \times 2.10 \text{ MGD} \times 8.34 \text{ lb/gal}\\ &= 140 \text{ lb/day}\end{aligned}$$

■ **Example 12.29**

Problem: Determine the desired lb/day setting on a dry chemical feeder if jar tests indicate an optimum polymer dose of 12 mg/L and the flow to be treated is 4.15 MGD.

Solution:

$$\text{Polymer (lb/day)} = 12 \text{ mg/L} \times 4.15 \text{ MGD} \times 8.34 \text{ lb/gal} = 415 \text{ lb/day}$$

Determining Chemical Solution Feeder Setting (gpd)

When solution concentration is expressed as pound chemical per gallon solution, the required feed rate can be determined as follows:

$$\text{Chemical (lb/day)} = \text{Chemical (mg/L)} \times \text{Flow (MGD)} \times 8.34 \text{ lb/gal} \tag{12.26}$$

Then convert the lb/day dry chemical to gpd solution:

$$\text{Solution (gpd)} = \text{Chemical (lb/day)} \div \text{lb Chemical per gal solution} \tag{12.27}$$

■ **Example 12.30**

Problem: Jar tests indicate that the best alum dose for water is 7 mg/L. The flow to be treated is 1.52 MGD. Determine the gpd setting for the alum solution feeder if the liquid alum contains 5.36 lb of alum per gallon of solution.

Solution: First calculate the lb/day of dry alum required, using the mg/L to lb/day equation:

$$\begin{aligned}\text{Dry alum (lb/day)} &= \text{Chemical (mg/L)} \times \text{Flow (MGD)} \times 8.34 \text{ lb/gal}\\ &= 7 \text{ mg/L} \times 1.52 \text{ MGD} \times 8.34 \text{ lb/gal}\\ &= 89 \text{ lb/day}\end{aligned}$$

Then calculate gpd solution required:

$$\text{Alum solution (gpd)} = 89 \text{ lb/day} \div 5.36 \text{ lb alum per gal solution} = 16.6 \text{ gpd}$$

Determining Chemical Solution Feeder Setting (mL/min)

Some solution chemical feeders dispense chemical as milliliters per minute (mL/min). To calculate the mL/min solution required, use the following equation:

$$\text{Feed rate (mL/min)} = (\text{gpd} \times 3785 \text{ mL/gal}) \div (1440 \text{ min/day}) \tag{12.28}$$

■ **Example 12.31**

Problem: The desired solution feed rate was calculated to be 9 gpd. What is this feed rate expressed as mL/min?

Solution:

$$\begin{aligned}\text{Feed rate} &= (\text{gpd} \times 3785 \text{ mL/gal}) \div (1440 \text{ min/day})\\ &= (9 \text{ gpd} \times 3785 \text{ mL/gal}) \div (1440 \text{ min/day})\\ &= 24 \text{ mL/min}\end{aligned}$$

■ **Example 12.32**

Problem: The desired solution feed rate has been calculated to be 25 gpd. What is this feed rate expressed as mL/min?

Solution:

$$\begin{aligned}\text{Solution} &= (\text{gpd} \times 3785\ \text{mL/gal}) \div (1440\ \text{min/day})\\ &= (25\ \text{gpd} \times 3785\ \text{mL/gal}) \div (1440\ \text{min/day})\\ &= 65.7\ \text{mL/min feed rate}\end{aligned}$$

Sometimes we will need to know the mL/min solution feed rate but we do not know the gpd solution feed rate. In such cases, calculate the gpd solution feed rate first, using the following the equation:

$$\text{gpd} = \frac{\text{Chemical (mg/L)} \times \text{Flow (MGD)} \times 8.34\ \text{lb/gal}}{\text{Chemical (lb)/Solution (gal)}} \tag{12.29}$$

DETERMINING PERCENT STRENGTH OF SOLUTIONS

The strength of a solution is a measure of the amount of chemical solute dissolved in the solution. We use the following equation to determine the percent strength of a solution:

$$\%\ \text{Strength} = \frac{\text{Chemical (lb)}}{\text{Water (lb)} + \text{Chemical (lb)}} \times 100 \tag{12.30}$$

■ **Example 12.33**

Problem: If a total of 10 oz. of dry polymer is added to 15 gal of water, what is the percent strength (by weight) of the polymer solution?

Solution: Before calculating percent strength, the ounces of chemical must be converted to pounds of chemical:

$$(10\ \text{oz.}) \div (16\ \text{oz./lb}) = 0.625\ \text{lb chemical}$$

Now calculate percent strength:

$$\begin{aligned}\%\ \text{Strength} &= \frac{\text{Chemical (lb)}}{\text{Water (lb)} + \text{Chemical (lb)}} \times 100\\ &= \frac{0.625\ \text{lb chemical}}{(15\ \text{gal} \times 8.34\ \text{lb/gal}) + 0.625\ \text{lb}} \times 100\\ &= \frac{0.625\ \text{lb chemical}}{125.7\ \text{lb solution}} \times 100\\ &= 0.5\%\end{aligned}$$

■ **Example 12.34**

Problem: If 90 g (1 g = 0.0022 lb) of dry polymer is dissolved in 6 gal of water, what percent strength is the solution?

Solution: First, convert grams of chemical to pounds of chemical:

$$90 \text{ g polymer} \times 0.0022 \text{ lb/g} = 0.198 \text{ lb polymer}$$

Now calculate percent strength of the solution:

$$\begin{aligned}\% \text{ Strength} &= \frac{\text{lb Polymer}}{\text{lb Water} + \text{lb Polymer}} \times 100 \\ &= \frac{0.198 \text{ lb Polymer}}{(6 \text{ gal } \times 8.34 \text{ lb/gal}) + 0.198 \text{ lb Polymer}} \times 100 \\ &= 4\%\end{aligned}$$

Determining Percent Strength of Liquid Solutions

When using liquid chemicals to make up solutions (e.g., liquid polymer), a different calculation is required, as shown below:

$$\frac{\text{Liquid polymer (lb)} \times \text{Liquid polymer (\% strength)}}{100} = \frac{\text{Polymer solution (lb)} \times \text{Polymer solution (\% strength)}}{100} \quad (12.31)$$

■ **Example 12.35**

Problem: A 12% liquid polymer is to be used in making up a polymer solution. How many pounds of liquid polymer should be mixed with water to produce 120 lb of a 0.5% polymer solution?

Solution:

$$\frac{\text{Liq. polymer (lb)} \times \text{Liq. polymer (\% strength)}}{100} = \frac{\text{Polymer sol. (lb)} \times \text{Polymer sol. (\% strength)}}{100}$$

$$\frac{x \text{ lb} \times 12}{100} = \frac{120 \text{ lb} \times 0.5}{100}$$

$$x = \frac{120 \times 0.005}{0.12}$$

$$x = 5 \text{ lb}$$

Determining Percent Strength of Mixed Solutions

The percent strength of solution mixture is determined using the following equation:

$$\% \text{ Strength} = \frac{\left(\frac{\text{Sol. 1 (lb)} \times \text{Sol. 1 (\% strength)}}{100}\right) + \left(\frac{\text{Sol. 2 (lb)} \times \text{Sol. 2 (\% strength)}}{100}\right)}{\text{Sol. 1 (lb)} + \text{Sol. 2 (lb)}} \times 100 \quad (12.32)$$

■ Example 12.36

Problem: If 12 lb of a 10% strength solution is mixed with 40 lb of a 1% strength solution, what is the percent strength of the solution mixture?

Solution:

$$\%\text{ Strength of mix} = \frac{\left(\dfrac{\text{Sol. 1 (lb)} \times \text{Sol. 1 (\% strength)}}{100}\right) + \left(\dfrac{\text{Sol. 2 (lb)} \times \text{Sol. 2 (\% strength)}}{100}\right)}{\text{Sol. 1 (lb)} + \text{Sol. 2 (lb)}} \times 100$$

$$= \frac{(12\text{ lb} \times 0.1) + (40\text{ lb} \times 0.1)}{12\text{ lb} + 40\text{ lb}} \times 100 = \frac{1.2\text{ lb} + 0.40\text{ lb}}{52\text{ lb}} \times 100 = 3.1\%$$

Dry Chemical Feeder Calibration

Occasionally, we need to perform a calibration calculation to compare the actual chemical feed rate with the feed rate indicated by the instrumentation. To calculate the actual feed rate for a dry chemical feeder, place a container under the feeder, weigh the container when empty, then weigh the container again after a specified length of time (e.g., 30 min). The actual chemical feed rate can be calculated using the following equation:

$$\text{Chemical feed rate (lb/min)} = \frac{\text{Chemical applied (lb)}}{\text{Length of application (min)}} \tag{12.33}$$

If desired, the chemical feed rate can be converted to lb/day:

$$\text{Chemical feed rate (lb/day)} = \text{Feed rate (lb/min)} \times 1440\text{ min/day} \tag{12.34}$$

■ Example 12.37

Problem: Calculate the actual chemical feed rate (lb/day) if a container is placed under a chemical feeder and a total of 2 lb is collected during a 30-min period.

Solution: First calculate the lb/min feed rate:

$$\text{Chemical feed rate} = \frac{\text{Chemical applied (lb)}}{\text{Length of application (min)}} = \frac{2\text{ lb}}{30\text{ min}} = 0.06\text{ lb/min}$$

Then calculate the lb/day feed rate:

$$\text{Chemical feed rate} = 0.06\text{ lb/min} \times 1440\text{ min/day} = 86.4\text{ lb/day}$$

■ Example 10.38

Problem: Calculate the actual chemical feed rate (lb/day) if a container is placed under a chemical feeder and a total of 1.6 lb is collected during a 20-min period.

Solution: First calculate the lb/min feed rate:

$$\text{Chemical feed rate} = \frac{\text{Chemical applied (lb)}}{\text{Length of application (min)}} = \frac{1.6 \text{ lb}}{20 \text{ min}} = 0.08 \text{ lb/min}$$

Then calculate the lb/day feed rate:

$$\text{Chemical feed rate} = 0.08 \text{ lb/min} \times 1440 \text{ min/day} = 115 \text{ lb/day}$$

Solution Chemical Feeder Calibration

As with other calibration calculations, the actual solution chemical feed rate is determined and then compared with the feed rate indicated by the instrumentation. To calculate the actual solution chemical feed rate, first express the solution feed rate in MGD. When the MGD solution flow rate has been calculated, use the mg/L equation to determine chemical dosage in lb/day. If solution feed is expressed as mL/min, first convert mL/min flow rate to gpd flow rate:

$$\text{gpd} = \frac{(\text{mL/min}) \times 1440 \text{ min/day}}{3785 \text{ mL/gal}} \tag{12.35}$$

Then calculate chemical dosage:

$$\text{Chemical dosage (lb/day)} = \text{Chemical (mg/L)} \times \text{Flow (MGD)} \times 8.34 \text{ lb/day} \tag{12.36}$$

■ Example 12.39

Problem: A calibration test is conducted for a solution for a solution chemical feeder. During a 5-min test, the pump delivered 940 mg/L of the 1.20% polymer solution. (Assume that the polymer solution weighs 8.34 lb/gal.) What is the polymer dosage rate in lb/day?

Solution: The flow rate must be expressed as MGD; therefore, the mL/min solution flow rate must first be converted to gpd and then MGD. The mL/min flow rate is calculated as:

$$(940 \text{ mL}) \div (5 \text{ min}) = 188 \text{ mL/min}$$

Next convert the mL/min flow rate to gpd flow rate:

$$\frac{188 \text{ mL/min} \times 1440 \text{ min/day}}{3785 \text{ mL/gal}} = 72 \text{ gpd}$$

Then calculate the polymer feed rate:

$$\text{Polymer feed rate} = 12{,}000 \text{ mg/L} \times 0.000072 \text{ MGD} \times 8.34 \text{ lb/day} = 7.2 \text{ lb/day}$$

■ Example 12.40

Problem: A calibration test is conducted for a solution chemical feeder. During a 24-hr period, the solution feeder delivers a total of 100 gal of solution. The polymer solution is a 1.2% solution. What is the lb/day feed rate? (Assume that the polymer solution weighs 8.34 lb/gal.)

Solution: The solution feed rate is 100 gal per day, or 100 gpd. Expressed as MGD, this is 0.000100 MGD. Use the mg/L to lb/day equation to calculate the actual feed rate:

$$\begin{aligned}\text{Chemical feed rate} &= \text{Chemical (mg/L)} \times \text{Flow (MGD)} \times 8.34\ \text{lb/day}\\ &= 12{,}000\ \text{mg/L} \times 0.000100\ \text{MGD} \times 8.34\ \text{lb/day}\\ &= 10\ \text{lb/day}\end{aligned}$$

The actual pumping rates can be determined by calculating the volume pumped during a specified time frame; for example, if 60 gal are pumped during a 10-min test, the average pumping rate during the test is 6 gpm. Actual volume pumped is indicated by the drop in tank level. By using the following equation, we can determine the flow rate in gpm:

$$\text{Flow rate (gpm)} = \frac{0.785 \times D^2 \times \text{Drop in level (ft)} \times 7.48\ \text{gal/ft}^3}{\text{Duration of test (min)}} \tag{12.37}$$

■ Example 12.41

Problem: A pumping rate calibration test is conducted for a 15-min period. The liquid level in the 4-ft-diameter solution tank is measured before and after the test. If the level drops 0.5 ft during the 15-min test, what is the pumping rate in gpm?

Solution:

$$\begin{aligned}\text{Flow rate} &= \frac{0.785 \times D^2 \times \text{Drop in level (ft)} \times 7.48\ \text{gal/ft}^3}{\text{Duration of test (min)}}\\ &= \frac{0.785 \times (4\ \text{ft} \times 4\ \text{ft}) \times 0.5\ \text{ft} \times 7.48\ \text{gal/ft}^3}{15\ \text{min}}\\ &= 3.1\ \text{gpm}\end{aligned}$$

Determining Chemical Usage

$$\text{Average use (lb/day)} = \frac{\text{Total chemical used (lb)}}{\text{Number of days}} \tag{12.38}$$

$$\text{Average use (gpd)} = \frac{\text{Total chemical used (gal)}}{\text{Number of days}} \tag{12.39}$$

Then we can calculate days supply in inventory:

$$\text{Days supply in inventory} = \frac{\text{Total chemical in inventory (lb)}}{\text{Average use (lb/day)}} \tag{12.40}$$

$$\text{Days supply in inventory} = \frac{\text{Total chemical in inventory (gal)}}{\text{Average use (gpd)}} \tag{12.41}$$

■ Example 12.42

Problem: The chemical used for each day during a week is given below. Based on these data, what was the average lb/day chemical use during the week?

Monday	88 lb/day
Tuesday	93 lb/day
Wednesday	91 lb/day
Thursday	88 lb/day
Friday	96 lb/day
Saturday	92 lb/day
Sunday	86 lb/day

Solution:

$$\text{Average use} = \frac{\text{Total chemical used (lb)}}{\text{Number of days}} = \frac{634 \text{ lb}}{7 \text{ days}} = 90.6 \text{ lb/day}$$

■ Example 12.43

Problem: The average chemical use at a plant is 77 lb/day. If the chemical inventory is 2800 lb, how many days supply is this?

Solution:

$$\text{Days supply in inventory} = \frac{\text{Total chemical in inventory (lb)}}{\text{Average use (lb/day)}} = \frac{2800 \text{ lb}}{77 \text{ lb/day}} = 36.4 \text{ days}$$

Paddle Flocculator Calculations

The gentle mixing required for flocculation is accomplished by a variety of devices. Probably the most common device in use is the basin equipped with mechanically driven paddles. Paddle flocculators have individual compartments for each set of paddles. The useful power input imparted by a paddle to the water depends on the drag force and the relative velocity of the water with respect to the paddle (Droste, 1997). For paddle flocculator design and operation, environmental engineers are mainly interested in determining the velocity of a paddle at a set distance, the drag force of the paddle on the water, and the power input imparted to the water by the paddle. Because of slip, factor k, the velocity of the water will be less than the velocity of the paddle. If baffles are placed along the walls in a direction perpendicular to the water movement, the value of k decreases because the baffles obstruct the movement of the water (Droste, 1997). The frictional dissipation of energy depends on the relative velocity, v. The relative velocity can be determined using Equation 12.42:

$$v = v_p - v_t = v_p - kv_p = v_p(1 - k) \tag{12.42}$$

where

v_t = Water velocity.
v_p = Paddle velocity.

To determine the velocity of the paddle at a distance r from the shaft, we use Equation 12.43:

$$v_p = \frac{2\pi N}{60}(r) \tag{12.43}$$

where N is the rate of revolution of the shaft (rpm). To determine the drag force of the paddle on the water we use Equation 12.44:

$$F_D = 1/2pC_DAv^2 \tag{12.44}$$

where
F_D = Drag force.
C_D = Drag coefficient.
A = Area of the paddle.

To determine the power input imparted to the water by an elemental area of the paddle the usual equation used is Equation 12.45:

$$dP = dF_Dv = 1/2pC_Dv^3dA \tag{12.45}$$

SEDIMENTATION CALCULATIONS

Sedimentation, the solid–liquid separation by gravity, is one of the most basic processes of water and wastewater treatment. In water treatment, plain sedimentation, such as the use of a presedimentation basin for grit removal and a sedimentation basin following coagulation–flocculation, is the most commonly approach used.

Tank Volume Calculations

The two common tank shapes of sedimentation tanks are rectangular and cylindrical. The equations for calculating the volume for each type tank are shown below.

Calculating Tank Volume

For rectangular sedimentation basins, we use Equation 12.46:

$$\text{Volume (gal)} = \text{Length (ft)} \times \text{Width (ft)} \times \text{Depth (ft)} \times 7.48\ \text{gal/ft}^3 \tag{12.46}$$

For circular clarifiers, we use Equation 12.47:

$$\text{Volume (gal)} = 0.785 \times (\text{Diameter})^2 \times \text{Depth (ft)} \times 7.48\ \text{gal/ft}^3 \tag{12.47}$$

■ **Example 12.44**

Problem: A sedimentation basin is 25 ft wide by 80 ft long and contains water to a depth of 14 ft. What is the volume of water in the basin, in gallons?

Solution:

$$\begin{aligned}\text{Volume} &= \text{Length (ft)} \times \text{Width (ft)} \times \text{Depth (ft)} \times 7.48\ \text{gal/ft}^3\\ &= 80\ \text{ft} \times 25\ \text{ft} \times 14\ \text{ft} \times 7.48\ \text{gal/ft}^3\\ &= 209{,}440\ \text{gal}\end{aligned}$$

■ **Example 12.45**

Problem: A sedimentation basin is 24 ft wide by 75 ft long. When the basin contains 140,000 gal, what would the water depth be?

Solution:

$$\text{Volume} = \text{Length (ft)} \times \text{Width (ft)} \times \text{Depth (ft)} \times 7.48\ \text{gal/ft}^3$$

$$140{,}000\ \text{gal} = 75\ \text{ft} \times 24\ \text{ft} \times x\ \text{ft} \times 7.48\ \text{gal/ft}^3$$

$$x\ \text{ft} = \frac{140{,}000}{75 \times 24 \times 7.48} = 10.4\ \text{ft}$$

Detention Time

Detention time for clarifiers varies from 1 to 3 hr. The equations used to calculate detention time are shown below.

Basic detention time equation:

$$\text{Detention time (hr)} = \text{Volume of tank (gal)} \div \text{Flow rate (gph)} \tag{12.48}$$

Rectangular sedimentation basin equation:

$$\text{Detention time (hr)} = \frac{\text{Length (ft)} \times \text{Width (ft)} \times \text{Depth (ft)} \times 7.48\ \text{gal/ft}^3}{\text{Flow rate (gph)}} \tag{12.49}$$

Circular basin equation:

$$\text{Detention time (hr)} = \frac{0.785 \times (\text{Diameter, ft})^2 \times \text{Depth (ft)} \times 7.48\ \text{gal/ft}^3}{\text{Flow rate (gph)}} \tag{12.50}$$

■ Example 12.46

Problem: A sedimentation tank has a volume of 137,000 gal. If the flow to the tank is 121,000 gph, what is the detention time in the tank (in hours)?

Solution:

$$\text{Detention time} = \text{Volume of tank (gal)} \div \text{Flow rate (gph)} = 137{,}000\ \text{gal} \div 121{,}000\ \text{gph} = 1.1\ \text{hr}$$

■ Example 12.47

Problem: A sedimentation basin is 60 ft long by 22 ft wide and has water to a depth of 10 ft. If the flow to the basin is 1,500,000 gpd, what is the sedimentation basin detention time?

Solution: First, convert the flow rate from gpd to gph so the times units will match:

$$(1{,}500{,}000\ \text{gpd}) \div (24\ \text{hr/day}) = 62{,}500\ \text{gph}$$

Then calculate detention time:

$$\text{Detention time (hr)} = \frac{0.785 \times D^2 \times \text{Depth (ft)} \times 7.48\ \text{gal/ft}^3}{\text{Flow rate (gph)}}$$

Surface Overflow Rate

The surface overflow rate—similar to the hydraulic loading rate (flow per unit area)—is used to determine loading on sedimentation basins and circular clarifiers. Hydraulic loading rate, however, measures the total water entering the process, whereas surface overflow rate measures only the water overflowing the process (plant flow only).

Note: Surface overflow rate calculations do not include recirculated flows. Other terms used synonymously with surface overflow rate are *surface loading rate* and *surface settling rate*.

Surface overflow rate is determined using the following equation:

$$\text{Surface overflow rate} = \text{Flow (gpm)} \div \text{Area (ft}^2\text{)} \quad (12.51)$$

■ Example 12.48

Problem: A circular clarifier has a diameter of 80 ft. If the flow to the clarifier is 1800 gpm, what is the surface overflow rate in gpm/ft^2?

Solution:

Surface overflow rate = Flow (gpm) ÷ Area (ft^2) = (1800 gpm) ÷ (0.785 × 80 ft × 80 ft) = 0.36 gpm/ft^2

■ Example 12.49

Problem: A sedimentation basin 70 ft by 25 ft receives a flow of 1000 gpm. What is the surface overflow rate in gpm/ft^2?

Solution:

Surface overflow rate = Flow (gpm) ÷ Area (ft^2) = (1000 gpm) ÷ (70 ft × 25 ft) = 0.6 gpm/ft^2

Mean Flow Velocity

The measure of average velocity of the water as it travels through a rectangular sedimentation basin is known as mean flow velocity. Mean flow velocity is calculated using Equation 12.52:

$$\text{Flow } (Q)\ (\text{ft}^3/\text{min}) = \text{Cross-sectional area } (A)\ (\text{ft}^2) \times \text{Volume } (V)\ (\text{ft/min}) \quad (12.52)$$

$$Q = A \times V$$

■ Example 12.50

Problem: A sedimentation basin is 60 ft long by 18 ft wide and has water to a depth of 12 ft. When the flow through the basin is 900,000 gpd, what is the mean flow velocity in the basin in ft/min?

Solution: Because velocity is desired in ft/min, the flow rate in the $Q = A \times V$ equation must be expressed in ft^3/min (cfm):

$$\frac{900{,}000 \text{ gpd}}{1440 \text{ min/day} \times 7.48 \text{ gal/ft}^3} = 84 \text{ cfm}$$

Then, use $Q = A \times V$ to calculate velocity:

$$Q = A \times V$$

$$84 \text{ cfm} = (18 \text{ ft} \times 12 \text{ ft}) \times x \text{ fpm}$$

$$x = 84 \text{ cfm} \div (18 \text{ ft} \times 12 \text{ ft}) = 0.4 \text{ fpm}$$

■ Example 12.51

Problem: A rectangular sedimentation basin 50 ft long by 20 ft wide has a water depth of 9 ft. If the flow to the basin is 1,880,000 gpd, what is the mean flow velocity in ft/min?

Solution: Because velocity is desired in ft/min, the flow rate in the $Q = A \times V$ equation must be expressed in ft^3/min (cfm):

$$1{,}880{,}000 \text{ gpd} \div (1440 \text{ min/day} \times 7.48 \text{ gal/ft}^3) = 175 \text{ cfm}$$

Then, use the $Q = A \times V$ equation to calculate velocity:

$$Q = A \times V$$

$$175 \text{ cfm} = (20 \text{ ft} \times 9 \text{ ft}) \times x \text{ fpm}$$

$$x = 175 \text{ cfm} \div (20 \text{ ft} \times 9 \text{ ft}) = 0.97 \text{ fpm}$$

Weir Loading Rate (Weir Overflow Rate)

Weir loading rate (weir overflow rate) is the amount of water leaving the settling tank per linear foot of weir. The result of this calculation can be compared with design. Normally, weir overflow rates of 10,000 to 20,000 gal/day/ft are used in the design of a settling tank. Typically, the weir loading rate is a measure of the flow in gallons per minute (gpm) over each foot of weir. The weir loading rate is determined using the following equation:

$$\text{Weir loading rate (gpm/ft)} = \text{Flow (gpm)} \div \text{Weir length (ft)} \tag{12.53}$$

■ Example 12.52

Problem: A rectangular sedimentation basin has a total of 115 ft of weir. What is the weir loading rate in gpm/ft when the flow of 1,110,000 gpd?

Solution:

$$\text{Flow} = (1{,}110{,}000 \text{ gpd}) \div (1440 \text{ min/day} = 771 \text{ gpm}$$

$$\text{Weir loading rate} = \text{Flow (gpm)} \div \text{Weir length (ft)} = (771 \text{ gpm}) \div (115 \text{ ft}) = 6.7 \text{ gpm/ft}$$

■ Example 12.53

Problem: A circular clarifier receives a flow of 3.55 MGD. If the diameter of the weir is 90 ft, what is the weir loading rate in gpm/ft?

Solution:

$$\text{Flow} = (3{,}550{,}000 \text{ gpd}) \div (1440 \text{ min/day}) = 2465 \text{ gpm}$$

$$\text{Weir length} = 3.14 \times 90 \text{ ft} = 283 \text{ ft}$$

$$\text{Weir loading rate} = \text{Flow (gpm)} \div \text{Weir length (ft)} = (2465 \text{ gpm}) \div (283 \text{ ft}) = 8.7 \text{ gpm/ft}$$

Percent Settled Biosolids

The percent settled biosolids test (*volume over volume test*, or V/V test) is conducted by collecting a 100-mL slurry sample from the solids contact unit and allowing it to settle for 10 min. After 10 min, the volume of settled biosolids at the bottom of the 100-mL graduated cylinder is measured and recorded. The equation used to calculate percent settled biosolids is shown below:

$$\%\ \text{Settled biosolids} = \frac{\text{Settled biosolids volume (mL)}}{\text{Total sample volume (mL)}} \times 100 \tag{12.54}$$

■ Example 12.54

Problem: A 100-mL sample of slurry from a solids contact unit is placed in a graduated cylinder and allowed to set for 10 min. The settled biosolids at the bottom of the graduated cylinder after 10 min is 22 mL. What is the percent of settled biosolids of the sample?

Solution:

$$\%\ \text{Settled biosolids} = \frac{\text{Settled biosolids volume (mL)}}{\text{Total sample volume (mL)}} \times 100 = \frac{22\text{ mL}}{100\text{ mL}} \times 100 = 22\%$$

■ Example 12.55

Problem: A 100-mL sample of slurry from a solids contact unit is placed in a graduated cylinder. After 10 min, a total of 21 mL of biosolids settled to the bottom of the cylinder. What is the percent settled biosolids of the sample?

Solution:

$$\%\ \text{Settled biosolids} = \frac{\text{Settled biosolids volume (mL)}}{\text{Total sample volume (mL)}} \times 100 = \frac{21\text{ mL}}{100\text{ mL}} \times 100 = 21\%$$

Determining Lime Dosage (mg/L)

During the alum dosage process, lime is sometimes added to provide adequate alkalinity (HCO_3^-) in the solids contact clarification process for the coagulation and precipitation of the solids. To determine the lime dose required, in mg/L, three steps are required. In Step 1, the total alkalinity required to react with the alum to be added and provide proper precipitation is determined using the following equation:

$$\text{Total alkalinity required} = \text{Alkalinity reacting with alum} + \text{Alkalinity in the water} \tag{12.55}$$

↑

(1 mg/L alum reacts with 0.45 mg/L alkalinity)

■ Example 12.56

Problem: Raw water requires an alum dose of 45 mg/L, as determined by jar testing. If a residual 30-mg/L alkalinity must be present in the water to ensure complete precipitation of alum added, what is the total alkalinity required (in mg/L)?

Solution: First calculate the alkalinity that will react with 45 mg/L alum:

$$\frac{0.45 \text{ mg/L alkalinity}}{1 \text{ mg/L alum}} = \frac{x \text{ mg/L alkalinity}}{45 \text{ mg/L alum}}$$

$$0.45 \times 45 = x$$

$$x = 20.25 \text{ mg/L alkalinity}$$

Then calculate the total alkalinity required:

$$\text{Total alkalinity required} = \text{Alkalinity reacting with alum} + \text{Alkalinity in the water}$$

$$= 20.25 \text{ mg/L} + 30 \text{ mg/L} = 50.25 \text{ mg/L}$$

■ Example 12.57

Problem: Jar tests indicate that 36 mg/L alum is optimum for a particular raw water. If a residual 30-mg/L alkalinity must be present to promote complete precipitation of the alum added, what is the total alkalinity required (in mg/L)?

Solution: First calculate the alkalinity that will react with 36-mg/L alum:

$$\frac{0.45 \text{ mg/L alkalinity}}{1 \text{ mg/L alum}} = \frac{x \text{ mg/L alkalinity}}{36 \text{ mg/L alum}}$$

$$0.45 \times 36 = x$$

$$x = 16.2 \text{ mg/L alkalinity}$$

Then calculate the total alkalinity required:

$$\text{Total alkalinity required} = 16.2 \text{ mg/L} + 30 \text{ mg/L} = 46.2 \text{ mg/L}$$

In Step 2, we make a comparison between required alkalinity and alkalinity already in the raw water to determine how many mg/L alkalinity should be added to the water. The equation used to make this calculation is shown below:

$$\text{Alkalinity to be added} = \text{Total alkalinity required} - \text{Alkalinity present in water} \quad (12.56)$$

■ Example 12.58

Problem: A total of 44-mg/L alkalinity is required to react with alum and ensure proper precipitation. If the raw water has an alkalinity of 30 mg/L as bicarbonate, how much mg/L alkalinity should be added to the water?

Solution:

$$\text{Alkalinity to be added} = \text{Total alkalinity required} - \text{Alkalinity present in water}$$

$$= 44 \text{ mg/L} - 30 \text{ mg/L}$$

$$= 14 \text{ mg/L}$$

In Step 3, after determining the amount of alkalinity to be added to the water, we determine how much lime (the source of alkalinity) must be added. We accomplish this by using the ratio shown in Example 12.59.

■ Example 12.59

Problem: It has been calculated that 16 mg/L alkalinity must be added to a raw water. How much mg/L lime will be required to provide this amount of alkalinity? (1 mg/L alum reacts with 0.45 mg/L and 1 mg/L alum reacts with 0.35 mg/L lime.)

Solution: First determine the mg/L lime required by using a proportion that relates bicarbonate alkalinity to lime:

$$\frac{0.45 \text{ mg/L alkalinity}}{0.35 \text{ mg/L lime}} = \frac{16 \text{ mg/L alkalinity}}{x \text{ mg/L lime}}$$

Then cross-multiply:

$$0.45x = 16 \times 0.35$$

$$x = \frac{16 \times 0.35}{0.45}$$

$$x = 12.4 \text{ mg/L lime}$$

In Example 12.60, we use all three steps to determine the lime dosage (mg/L) required.

■ Example 12.60

Problem: Given the following data, calculate the lime dose required, in mg/L:

Alum dose required (determined by jar tests) = 52 mg/L
Residual alkalinity required for precipitation = 30 mg/L
1 mg/L alum reacts with 0.35 mg/L lime
1 mg/L alum reacts with 0.45 mg/L alkalinity
Raw water alkalinity = 36 mg/L

Solution: To calculate the total alkalinity required, we must first calculate the alkalinity that will react with 52 mg/L alum:

$$\frac{0.45 \text{ mg/L alkalinity}}{1 \text{ mg/L alum}} = \frac{x \text{ mg/L alkalinity}}{52 \text{ mg/L alum}}$$

$$0.45 \times 52 = x$$

$$23.4 \text{ mg/L alkalinity} = x$$

The total alkalinity requirement can now be determined:

$$\begin{aligned} \text{Total alkalinity required} &= \text{Alkalinity reacting with alum} + \text{Residual alkalinity} \\ &= 23.4 \text{ mg/L} + 30 \text{ mg/L} \\ &= 53.4 \text{ mg/L} \end{aligned}$$

Next calculate how much alkalinity must be added to the water:

$$\text{Alkalinity to be added} = \text{Total alkalinity required} - \text{Alkalinity present in water}$$
$$= 53.4\text{ mg/L} - 36\text{ mg/L}$$
$$= 17.4\text{ mg/L}$$

Finally, calculate the lime required to provide this additional alkalinity:

$$\frac{0.45\text{ mg/L alkalinity}}{0.35\text{ mg/L lime}} = \frac{17.4\text{ mg/L alkalinity}}{x\text{ mg/L alum}}$$
$$0.45x = 17.4 \times 0.35$$
$$x = \frac{17.4 \times 0.35}{0.45}$$
$$x = 13.5\text{ mg/L lime}$$

Determining Lime Dosage (lb/day)

After the lime dose has been determined in terms of mg/L, it is a fairly simple matter to calculate the lime dose in lb/day, which is one of the most common calculations in water and wastewater treatment. To convert from mg/L to lb/day lime dose, we use the following equation:

$$\text{Lime (lb/day)} = \text{Lime (mg/L)} \times \text{Flow (MGD)} \times 8.34\text{ lb/gal} \tag{12.57}$$

■ Example 12.61

Problem: The lime dose for a raw water has been calculated to be 15.2 mg/L. If the flow to be treated is 2.4 MGD, how many lb/day lime will be required?

Solution:

$$\text{Lime (lb/day)} = \text{Lime (mg/L)} \times \text{Flow (MGD)} \times 8.34\text{ lb/gal}$$
$$= 15.2\text{ mg/L} \times 2.4\text{ MGD} \times 8.34\text{ lb/gal}$$
$$= 304\text{ lb/day lime}$$

■ Example 12.62

Problem: The flow to a solids contact clarifier is 2,650,000 gpd. If the lime dose required is determined to be 12.6 mg/L, how many lb/day lime will be required?

Solution:

$$\text{Lime (lb/day)} = \text{Lime (mg/L)} \times \text{Flow (MGD)} \times 8.34\text{ lb/gal}$$
$$= 12.6\text{ mg/L} \times 2.65\text{ MGD} \times 8.34\text{ lb/gal}$$
$$= 278\text{ lb/day lime}$$

Determining Lime Dosage (g/min)

To convert from lb/day lime to g/min lime, use Equation 12.82:

Note: 1 lb = 453.6 g.

$$\text{Lime (g/min)} = \frac{\text{Lime (lb/day)} \times 453.6 \text{ g/lb}}{1440 \text{ min/day}} \tag{12.58}$$

■ Example 12.63

Problem: A total of 275 lb/day lime will be required to raise the alkalinity of the water passing through a solids-contact clarification process. How many g/min lime does this represent?

Solution:

$$\text{Lime (g/min)} = \frac{\text{Lime (lb/day)} \times 453.6 \text{ g/lb}}{1440 \text{ min/day}} = \frac{275 \text{ lb/day} \times 453.6 \text{ g/lb}}{1440 \text{ min/day}} = 86.6 \text{ g/min}$$

■ Example 12.64

Problem: A lime dose of 150 lb/day is required for a solids-contact clarification process. How many g/min lime does this represent?

Solution:

$$\text{Lime (g/min)} = \frac{\text{Lime (lb/day)} \times 453.6 \text{ g/lb}}{1440 \text{ min/day}} = \frac{150 \text{ lb/day} \times 453.6 \text{ g/lb}}{1440 \text{ min/day}} = 47.3 \text{ g/min}$$

Particle Settling (Sedimentation)*

Particle settling (sedimentation) may be described for a singular particle by the Newton equation (Equation 12.64) for terminal settling velocity of a spherical particle. For the engineer, knowledge of this velocity is basic in the design and performance of a sedimentation basin. The rate at which discrete particles will settle in a fluid of constant temperature is given by the following equation:

$$u = \left(\frac{4g(p_p - p)d}{3C_D p} \right)^{1/2} \tag{12.59}$$

where

u = Settling velocity of particles (m/s, ft/s).
g = Gravitational acceleration (m/s^2, ft/s^2).
p_p = Density of particles (kg/m^3, lb/ft^3).
p = Density of water (kg/m^3, lb/ft^3).
d = Diameter of particles (m, ft).
C_D = Coefficient of drag.

* Much of the information presented in this section is based on USEPA, *Guidance Manual for Compliance with the Interim Enhanced Surface Water Treatment Rule: Turbidity Provisions*, EPA 815-R-99-012, U.S. Environmental Protection Agency, Washington, DC, 1999.

The terminal settling velocity is derived by equating the drag, buoyant, and gravitational forces acting on the particle. At low settling velocities, the equation is not dependent on the shape of the particle and most sedimentation processes are designed so as to remove small particles, ranging from 1.0 to 0.5 μm, which settle slowly. Larger particles settle at higher velocity and will be removed whether or not they follow Newton's law or Stokes' law—the governing equation when the drag coefficient is sufficiently small (0.5 or less) as is the case for colloidal products (McGhee, 1991).

Typically, a large range of particle sizes will exist in the raw water supply. There are four types of sedimentation (Gregory and Zabel, 1990):

Type 1—Discrete particle settling (particles of various sizes, in a dilute suspension, which settle without flocculating)
Type 2—Flocculant settling (heavier particles coalesced with smaller and lighter particles)
Type 3—Hindered settling (high densities of particles in suspension resulting in an interaction of particles)
Type 4—Compression settling

The values of the drag coefficient depend on the density of water (p), relative velocity (u), particle diameter (d), and viscosity of water (μ), which gives the Reynolds number (Re) as

$$\text{Re} = \frac{p \times u \times d}{\mu} \tag{12.60}$$

As the Reynolds number increases, the value of C_D increases. For Re less than 2, C_D is related to Re by the linear expression as follows:

$$C_D = \frac{24}{\text{Re}} \tag{12.61}$$

At low levels of Re, the Stokes equation for laminar flow conditions is used (Equations 12.60 and 12.61 substituted into Equation 12.59):

$$u = \frac{G(p_p - p)d^2}{18\mu} \tag{12.62}$$

In the region of higher Reynolds numbers (2 < Re < 500–1000), C_D becomes (Fair et al. 1968):

$$C_D = \frac{24}{\text{Re}} + \frac{3}{\sqrt{\text{Re}}} + 0.34 \tag{12.63}$$

Note: In the region of turbulent flow (500–1000 < Re < 200,000), C_D remains approximately constant at 0.44.

The velocity of settling particles results in Newton's equation (AWWA and ASCE, 1990):

$$u = 1.74\left(\frac{(p_p - p)gd}{p}\right)^{1/2} \tag{12.64}$$

Note: When the Reynolds number is greater than 200,000, the drag force decreases substantially and C_D becomes 0.10. No settling occurs at this condition.

■ **EXAMPLE 12.65**

Problem: Estimate the terminal settling velocity in water at a temperature of 21°C of spherical particles with specific gravity 2.40 and average diameter of (a) 0.006 mm and (b) 1.0 mm.

Solution for part (a): Use Equation 12.62.

Given:

Temperature (T) = 21°C
p = 998 kg/m^3
μ = 0.00098 N s/m^2
d = 0.06 mm = 6 × 10^{-5} m
g = 9.81 m/s^2

$$u = \frac{g(p_p - p)d^2}{18\mu} = \frac{9.81 \text{ m/s}^2 \times (2400 \text{ kg/m}^3 - 998 \text{ kg/m}^3) \times (6 \times 10^{-5} \text{ m})^2}{18\mu} = 0.00281 \text{ m/s}$$

Use Equation 12.60 to check the Reynolds number:

$$\text{Re} = \frac{p \times u \times d}{\mu} = \frac{998 \times 0.00281 \times (6 \times 10^{-5})}{0.00098} = 0.172$$

Stokes' law applies, because Re < 2.

Solution for part (b): Use Equation 12.62.

$$u = \frac{9.81 \times (2400 - 998) \times (0.001)^2}{18 \times 0.00098} = \frac{0.137536}{0.01764} = 0.779 \text{ m/s}$$

Use Equation 12.60 to check the Reynolds number; assume that the irregularities of the particles $\Phi = 0.80$:

$$\text{Re} = \frac{\Phi \times p \times u \times d}{\mu} = \frac{0.80 \times 998 \times 0.779 \times 0.001}{0.00098} = 635$$

Because Re > 2, Stokes' law does not apply. Use Equation 12.59 to calculate u:

$$C_D = \frac{24}{635} + \frac{3}{\sqrt{635}} + 0.34 = 0.50$$

$$u^2 = \frac{4g(p_p - p)d}{3C_D p} = \frac{4 \times 9.81 \times (2400 - 998) \times 0.001}{3 \times 0.50 \times 998}$$

$$u = 0.192 \text{ m/s}$$

Recheck Re:

$$\text{Re} = \frac{\Phi \times p \times u \times d}{\mu} = \frac{0.80 \times 998 \times 0.192 \times 0.001}{0.00098} = 156$$

Calculate u using the new Re value:

$$C_D = \frac{24}{156} + \frac{3}{\sqrt{156}} + 0.34 = 0.73$$

$$u^2 = \frac{4 \times 9.81 \times 1402 \times 0.001}{3 \times 0.73 \times 998}$$

$$u = 159 \text{ m/s}$$

Recheck Re:

$$\text{Re} = \frac{0.80 \times 998 \times 0.159 \times 0.001}{0.00098} = 130$$

Calculate u using the new Re value:

$$C_D = \frac{24}{130} + \frac{3}{\sqrt{130}} + 0.34 = 0.79$$

$$u^2 = \frac{4 \times 9.81 \times 1402 \times 0.001}{3 \times 0.79 \times 998}$$

$$u = 0.152 \text{ m/s}$$

Thus, the estimated velocity is approximately 0.15 m/s.

Overflow Rate (Sedimentation)

Overflow rate, detention time, horizontal velocity, and weir loading rate are the parameters typically used for sizing sedimentation basin. The *theoretical detention time* (plug flow theory) is computed from the volume of the basin divided by average daily flow:

$$t = \frac{24V}{Q} \tag{12.65}$$

where

t = Detention time (hr).
24 = 24 hr per day.
V = Volume of basin (m^3, gal).
Q = Average daily flow (m^3/d, MGD).

The overflow rate is a standard design parameter that can be determined from discrete particle settling analysis. The overflow rate or surface loading rate is calculated by dividing the average daily flow by the total area of the sedimentation basin.

$$u = \frac{Q}{A} = \frac{Q}{lw} \tag{12.66}$$

where

u = Overflow rate ($m^3/m^2 \cdot d$, gpd/ft^2).
Q = Average daily flow (m^3/d, gpd).

A = Total surface area of basin (m^2, ft^2).
l = Length of basin (m, ft).
w = Width of basin (m, ft).

Note: All particles having a settling velocity greater than the overflow rate will settle and will be removed.

Rapid particle density changes due to temperature, solids concentration, or salinity can induce density current which can cause severe short-circuiting in horizontal tanks (Hudson, 1989).

■ EXAMPLE 12.66

Problem: A water treatment plant has two clarifiers treating 2.0 MGD of water. Each clarifier is 14 ft wide, 80 ft long, and 17 ft deep. Determine: (a) detention time, (b) overflow, (c) horizontal velocity, and (d) weir loading rate, assuming the weir length is 2.5 times the basin width.

Solution:

(a) Compute detention time (t) for each clarifier:

$$Q = \frac{2\text{ MGD}}{2} = \frac{1{,}000{,}000\text{ gal}}{\text{day}} \times \frac{1\text{ ft}}{7.48\text{ gal}} \times \frac{1\text{ day}}{24\text{ hr}} = 5570\text{ ft}^3/\text{hr} = 92.8\text{ ft}^3/\text{min}$$

$$t = \frac{V}{Q} = \frac{14\text{ ft} \times 80\text{ ft} \times 17\text{ ft}}{5570\text{ ft}^3/\text{hr}} = 3.42\text{ hr}$$

(b) Compute overflow rate u:

$$u = \frac{Q}{\text{Length} \times \text{Width}} = \frac{1{,}000{,}000\text{ gpd}}{14\text{ ft} \times 80\text{ ft}} = 893\text{ gpd/ft}$$

(c) Compute horizontal velocity V:

$$V = \frac{Q}{\text{Width} \times \text{Depth}} = \frac{92.8\text{ ft}^3/\text{min}}{14\text{ ft} \times 17\text{ ft}} = 0.39\text{ ft/min}$$

(d) Compute weir loading rate u_w:

$$u_w = \frac{Q}{2.5 \times \text{Width}} = \frac{1{,}000{,}000\text{ gpd}}{2.5 \times 14\text{ ft}} = 28{,}571\text{ gpd/ft}$$

WATER FILTRATION CALCULATIONS

Water filtration is a physical process of separating suspended and colloidal particles from waste by passing the water through a granular material. The process of filtration involves straining, settling, and adsorption. As floc passes into the filter, the spaces between the filter grains become clogged, reducing this opening and increasing removal. Some material is removed merely because it settles on a media grain. One of the most important processes is adsorption of the floc onto the surface of individual filter grains. In addition to removing silt and sediment, floc, algae, insect larvae, and any other large elements, filtration also contributes to the removal of bacteria and protozoa such as *Giardia lamblia* and *Cryptosporidium*. Some filtration processes are also used for iron and manganese removal.

The Surface Water Treatment Rule (SWTR) specifies four filtration technologies, although SWTR also allows the use of alternative filtration technologies (e.g., cartridge filters). These include slow sand filtration/rapid sand filtration, pressure filtration, diatomaceous earth filtration, and direct filtration. Of these, all but rapid sand filtration are commonly employed in small water systems that use filtration. Each type of filtration system has advantages and disadvantages. Regardless of the type of filter, however, filtration involves the processes of *straining* (where particles are captured in the small spaces between filter media grains), *sedimentation* (where the particles land on top of the grains and stay there), and *adsorption* (where a chemical attraction occurs between the particles and the surface of the media grains).

Flow Rate through a Filter (gpm)

Flow rate in gpm through a filter can be determined by simply converting the gpd flow rate, as indicated on the flow meter. The gpm flow rate can be calculated by taking the meter flow rate (gpd) and dividing by 1440 min/day as shown below:

$$\text{Flow rate (gpm)} = \text{Flow rate (gpd)} \div 1440 \text{ min/day} \tag{12.67}$$

■ Example 12.67

Problem: The flow rate through a filter is 4.25 MGD. What is this flow rate expressed as gpm?

Solution:

$$\text{Flow rate} = 4.25 \text{ MGD} \div 1440 \text{ min/day} = 4{,}250{,}000 \text{ gpd} \div 1440 \text{ min/day} = 2951 \text{ gpm}$$

■ Example 12.68

Problem: During a 70-hr filter run, a total of 22.4 million gal of water are filtered. What is the average flow rate through the filter in gpm during this filter run?

Solution:

$$\begin{aligned}\text{Flow rate} &= \text{Total gallons produced} \div \text{Filter run (min)}\\ &= 22{,}400{,}000 \text{ gal} \div (70 \text{ hr} \times 60 \text{ min/hr})\\ &= 5333 \text{ gpm}\end{aligned}$$

■ Example 12.69

Problem: At an average flow rate of 4000 gpm, how long of a filter run (in hours) would be required to produce 25 MG of filtered water?

Solution: Write the equation as usual, filling in known data:

$$\text{Flow rate (gpm)} = \text{Total gallons produced} \div \text{Filter run (min)}$$

$$4000 \text{ gpm} = 25{,}000{,}000 \text{ gal} \div (x \text{ hr} \times 60 \text{ min/hr})$$

Then solve for x:

$$x = 25{,}000{,}000 \text{ gal} \div (4000 \text{ gpm} \times 60 \text{ min/hr}) = 104 \text{ hr}$$

■ Example 12.70

Problem: A filter box is 20 ft by 30 ft (including the sand area). If the influent valve is shut, the water drops 3 in./min. What is the rate of filtration in MGD?

Solution:

Given:
Filter box = 20 ft × 30 ft
Water drops = 3 in./min

Step 1. Find the volume of water passing through the filter:

$$\text{Volume} = \text{Area} \times \text{Height}$$

$$\text{Area} = \text{Width} \times \text{Length}$$

Note: The best approach to performing calculations of this type is a step-by-step one, breaking down the problem into what is given and what is to be found.

Area = 20 ft × 30 ft = 600 ft^2
Convert 3.0 in. into feet: 3/12 = 0.25 ft
Volume = 600 ft^2 × 0.25 ft = 150 ft^3 of water passing through the filter in one minute

Step 2. Convert cubic feet to gallons:

$$150 \text{ ft}^3 \times 7.48 \text{ gal/ft}^3 = 1122 \text{ gpm}$$

Step 3. The problem asks for the rate of filtration in MGD. To find MGD, multiply the number of gallons per minute by the number of minutes per day:

$$1122 \text{ gpm} \times 1440 \text{ min/day} = 1.62 \text{ MGD}$$

■ Example 12.71

Problem: The influent valve to a filter is closed for 5 min. During this time, the water level in the filter drops 0.8 ft (10 in.). If the filter is 45 ft long and 15 ft wide, what is the gpm flow rate through the filter? Water drop equals 0.16 ft/min.

Solution: First calculate cfm flow rate using the $Q = A \times V$ equation:

$$Q = \text{Length (ft)} \times \text{Width (ft)} \times \text{Drop velocity (ft/min)} = 45 \text{ ft} \times 15 \text{ ft} \times 0.16 \text{ ft/min} = 108 \text{ cfm}$$

Then convert cfm flow rate to gpm flow rate:

$$108 \text{ cfm} \times 7.48 \text{ gal/ft}^3 = 808 \text{ gpm}$$

Filtration Rate

One measure of filter production is filtration rate (generally ranging from 2 to 10 gpm/ft^2). Along with filter run time, it provides valuable information for operation of filters. It is the gallons of water filtered per minute through each square foot of filter area. Filtration rate is determined using Equation 12.68:

$$\text{Filtration rate (gpm/ft}^2\text{)} = \text{Flow rate (gpm)} \div \text{Filter surface area (ft}^2\text{)} \tag{12.68}$$

■ Example 12.72

Problem: A filter 18 ft by 22 ft receives a flow of 1750 gpm. What is the filtration rate in gpm/ft²?

Solution:

$$\begin{aligned} \text{Filtration rate} &= \text{Flow rate (gpm)} \div \text{Filter surface area (ft}^2\text{)} \\ &= 1750 \text{ gpm} \div (18 \text{ ft} \times 22 \text{ ft}) \\ &= 4.4 \text{ gpm/ft}^2 \end{aligned}$$

■ Example 12.73

Problem: A filter 28 ft long by 18 ft wide treats a flow of 3.5 MGD. What is the filtration rate in gpm/ft²?

Solution:

$$\text{Flow rate} = 3{,}500{,}000 \text{ gpd} \div (1440 \text{ min/day}) = 2431 \text{ gpm}$$

$$\text{Filtration rate} = \text{Flow rate (gpm)} \div \text{Filter surface area (ft}^2\text{)} = 2431 \text{ gpm} \div (28 \text{ ft} \times 18 \text{ ft}) = 4.8 \text{ gpm/ft}^2$$

■ Example 12.74

Problem: A filter 45 ft long by 20 ft wide produces a total of 18 MG during a 76-hr filter run. What is the average filtration rate in gpm/ft² for this filter run?

Solution: First calculate the gpm flow rate through the filter:

$$\begin{aligned} \text{Flow rate} &= \text{Total gallons produced} \div \text{Filter run (min)} \\ &= 18{,}000{,}000 \text{ gal} \div (76 \text{ hr} \times 60 \text{ min/hr}) \\ &= 3947 \text{ gpm} \end{aligned}$$

Then calculate the filtration rate:

$$\text{Filtration rate} = \text{Flow rate (gpm)} \div \text{Filter area (ft}^2\text{)} = 3947 \text{ gpm} \div (45 \text{ ft} \times 20 \text{ ft}) = 4.4 \text{ gpm/ft}^2$$

■ Example 12.75

Problem: A filter is 40 ft long by 20 ft wide. During a test of flow rate, the influent valve to the filter is closed for 6 min. The water level drop during this period is 16 inches. What is the filtration rate for the filter in gpm/ft²?

Solution: First calculate the gpm flow rate using the $Q = A \times V$ equation:

$$\begin{aligned} Q &= \text{Length (ft)} \times \text{Width (ft)} \times \text{Drop velocity (ft/min)} \times 7.48 \text{ gal/ft}^3 \\ &= (40 \text{ ft} \times 20 \text{ ft} \times 1.33 \text{ ft} \times 7.48 \text{ gal/ft}^3) \div 6 \text{ min} \\ &= 1326 \text{ gpm} \end{aligned}$$

Then calculate the filtration rate:

$$\text{Filtration rate} = \text{Flow rate (gpm)} \div \text{Filter area (ft}^2\text{)} = 1326 \text{ gpm} \div (40 \text{ ft} \times 20 \text{ ft}) = 1.6 \text{ g/ft}^2$$

Unit Filter Run Volume

The unit filter run volume (UFRV) calculation indicates the total gallons passing through each square foot of filter surface area during an entire filter run. This calculation is used to compare and evaluate filter runs. UFRVs are usually at least 5000 gal/ft^2 and generally in the range of 10,000 gpd/ft^2. The UFRV value will begin to decline as the performance of the filter begins to deteriorate. The equation to be used in these calculations is shown below:

$$\text{Unit filter run volume} = \text{Total gallons filtered} \div \text{Filter surface area (ft}^2\text{)} \tag{12.69}$$

■ Example 12.76

Problem: The total water filtered during a filter run (between backwashes) is 2,220,000 gal. If the filter is 18 ft by 18 ft, what is the unit filter run volume in gal/ft^2?

Solution:

$$\begin{aligned}\text{Unit filter run volume} &= \text{Total gallons filtered} \div \text{Filter surface area (ft}^2\text{)}\\ &= 2{,}220{,}000 \text{ gal} \div (18 \text{ ft} \times 18 \text{ ft})\\ &= 6852 \text{ gal/ft}^2\end{aligned}$$

■ Example 12.77

Problem: The total water filtered during a filter run is 4,850,000 gal. If the filter is 28 ft by 18 ft, what is the unit filter run volume in gal/ft^2?

Solution:

$$\begin{aligned}\text{Unit filter run volume} &= \text{Total gallons filtered} \div \text{Filter surface area (ft}^2\text{)}\\ &= 4{,}850{,}000 \text{ gal} \div (28 \text{ ft} \times 18 \text{ ft})\\ &= 9623 \text{ gal/ft}^2\end{aligned}$$

Equation 12.69 can be modified as shown in Equation 12.70 to calculate the unit filter run volume given filtration rate and filter run data:

$$\text{Unit filter run volume} = \text{Filtration rate (gpm/ft}^2\text{)} \times \text{Filter run time (min)} \tag{12.70}$$

■ Example 12.78

Problem: The average filtration rate for a filter was determined to be 2 gpm/ft^2. If the filter run time was 4250 minutes, what was the unit filter run volume in gal/ft^2?

Solution:

$$\begin{aligned}\text{Unit filter run volume} &= \text{Filtration rate (gpm/ft}^2\text{)} \times \text{Filter run time (min)}\\ &= 2 \text{ gpm/ft}^2 \times 4250 = 8500 \text{ gal/ft}^2\end{aligned}$$

The problem indicates that, at an average filtration rate of 2 gal entering each square foot of filter each minute, the total gallons entering during the total filter run is 4250 times that amount.

■ Example 12.79

Problem: The average filtration rate during a particular filter run was determined to be 3.2 gpm/ft^2. If the filter run time was 61.0 hr, what was the UFRV in gal/ft^2 for the filter run?

Solution:

$$\text{UFRV} = \text{Filtration rate (gpm/ft}^2) \times \text{Filter run (hr)} \times 60 \text{ min/hr}$$
$$= 3.2 \text{ gpm/ft}^2 \times 61.0 \text{ hr} \times 60 \text{ min/hr} = 11{,}712 \text{ gal/ft}^2$$

Backwash Rate

In filter backwashing, one of the most important operational parameters to be determined is the amount of water in gallons required for each backwash. This amount depends on the design of the filter and the quality of the water being filtered. The actual washing typically lasts 5 to 10 min and usually amounts to 1 to 5% of the flow produced.

■ Example 12.80

Problem: A filter has the following dimensions:

Length = 30 ft
Width = 20 ft
Depth of filter media = 24 in.

Assuming that a backwash rate of 15 gal/ft²/min is recommended and 10 min of backwash is required, calculate the amount of water in gallons required for each backwash.

Solution: Given the above data, find the amount of water in gallons required:

1. Area of filter = 30 ft × 20 ft = 600 ft^2
2. Gallons of water used per square foot of filter = 15 gal/ft^2/min × 10 min = 150 gal/ft^2
3. Gallons required for backwash = 150 gal/ft^2 × 600 ft^2 = 90,000 gal

Typically, backwash rates will range from 10 to 25 gpm/ft^2. The backwash rate is determined by using Equation 12.71:

$$\text{Backwash rate} = \text{Flow rate (gpm)} \div \text{Filter area (ft}^2) \tag{12.71}$$

■ Example 12.81

Problem: A filter 30 ft by 10 ft has a backwash rate of 3120 gpm. What is the backwash rate in gpm/ft^2?

Solution:

$$\text{Backwash rate} = \text{Flow rate (gpm)} \div \text{Filter area (ft}^2) = 3120 \text{ gpm} \div (30 \text{ ft} \times 10 \text{ ft}) = 10.4 \text{ gpm/ft}^2$$

■ Example 12.82

Problem: A filter 20 ft by 20 ft has a backwash rate of 4.85 MGD. What is the filter backwash rate in gpm/ft^2?

Solution:

$$\text{Flow rate} = 4{,}850{,}000 \text{ gpd} \div (1440 \text{ min/day}) = 3368 \text{ gpm}$$

$$\text{Backwash rate} = \text{Flow rate (gpm)} \div \text{Filter area (ft}^2) = 3368 \text{ gpm} \div (20 \text{ ft} \times 20 \text{ ft}) = 8.42 \text{ gpm/ft}^2$$

Backwash Rise Rate

Backwash rate is occasionally measured as the upward velocity of the water during backwashing, expressed as in./min rise. To convert from gpm/ft^2 backwash rate to an in./min rise rate, use either Equation 12.72 or Equation 12.73:

$$\text{Backwash rate (in./min)} = \frac{\text{Backwash rate (gpm/ft}^2) \times \text{(12 in./ft)}}{7.48\ \text{gal/ft}^3} \quad (12.72)$$

$$\text{Backwash rate (in./min)} = \text{Backwash rate (gpm/ft}^2) \times 1.6 \quad (12.73)$$

■ Example 12.83

Problem: A filter has a backwash rate of 16 gpm/ft^2. What is this backwash rate expressed as an in./min rise rate?

Solution:

$$\text{Backwash rate} = \frac{\text{Backwash rate (gpm/ft}^2) \times \text{(12 in./ft)}}{7.48\ \text{gal/ft}^3} = \frac{16\ \text{gpm/ft}^2 \times \text{(12 in./ft)}}{7.48\ \text{gal/ft}^3} = 25.7\ \text{in./min}$$

■ Example 12.84

Problem: A filter 22 ft long by 12 ft wide has a backwash rate of 3260 gpm. What is this backwash rate expressed as an in./min rise?

Solution: First calculate the backwash rate as gpm/ft^2:

$$\text{Backwash rate} = \text{Flow rate (gpm)} \div \text{Filter area (ft}^2) = 3260\ \text{gpm} \div (22\ \text{ft} \times 12\ \text{ft}) = 12.3\ \text{gpm/ft}^2$$

Then convert gpm/ft^2 to the in./min rise rate:

$$\text{Rise rate} = (12.3\ \text{gpm/ft}^2 \times 12\ \text{in./ft}) \div 7.48\ \text{gal/ft}^3 = 19.7\ \text{in./min}$$

Volume of Backwash Water Required (gal)

To determine the volume of water required for backwashing, we must know both the desired backwash flow rate (gpm) and the duration of backwash (min):

$$\text{Backwash water volume (gal)} = \text{Backwash (gpm)} \times \text{Duration of backwash (min)} \quad (12.74)$$

■ Example 12.85

Problem: For a backwash flow rate of 9000 gpm and a total backwash time of 8 min, how many gallons of water will be required for backwashing?

Solution:

$$\begin{aligned}\text{Backwash water volume} &= \text{Backwash (gpm)} \times \text{Duration of backwash (min)} \\ &= 9000\ \text{gpm} \times 8\ \text{min} = 72{,}000\ \text{gal}\end{aligned}$$

■ Example 12.86

Problem: How many gallons of water would be required to provide a backwash flow rate of 4850 gpm for a total of 5 min?

Solution:

$$\text{Backwash water volume} = \text{Backwash (gpm)} \times \text{Duration of backwash (min)}$$

$$= 4850 \text{ gpm} \times 7 \text{ min} = 33{,}950 \text{ gal}$$

Required Depth of Backwash Water Tank (ft)

The required depth of water in the backwash water tank is determined from the volume of water required for backwashing. To make this calculation, simply use Equation 12.75:

$$\text{Volume (gal)} = 0.785 \times (\text{Diameter})^2 \times \text{Depth (ft)} \times 7.48 \text{ gal/ft}^3 \tag{12.75}$$

■ Example 12.87

Problem: The volume of water required for backwashing has been calculated to be 85,000 gal. What is the required depth of water in the backwash water tank to provide this amount of water if the diameter of the tank is 60 ft?

Solution: Use the volume equation for a cylindrical tank, filling in known data, then solve for x:

$$\text{Volume (gal)} = 0.785 \times (\text{Diameter})^2 \times \text{Depth (ft)} \times 7.48 \text{ gal/ft}^3$$

$$85{,}000 \text{ gal} = 0.785 \times (60 \text{ ft})^2 \times x \text{ ft} \times 7.48 \text{ gal/ft}^3$$

$$x = 85{,}000 \div (0.785 \times 60 \times 60 \times 7.48)$$

$$x = 4 \text{ ft}$$

■ Example 12.88

Problem: A total of 66,000 gal of water will be required for backwashing a filter at a rate of 8000 gpm for a 9-min period. What depth of water is required if the backwash tank has a diameter of 50 ft?

Solution: Use the volume equation for cylindrical tanks:

$$\text{Volume (gal)} = 0.785 \times (\text{Diameter})^2 \times \text{Depth (ft)} \times 7.48 \text{ gal/ft}^3$$

$$66{,}000 \text{ gal} = 0.785 \times (50 \text{ ft})^2 \times x \text{ ft} \times 7.48 \text{ gal/ft}^3$$

$$x = 66{,}000/(0.785 \times 50 \times 50 \times 7.48)$$

$$x = 4.5 \text{ ft}$$

Backwash Pumping Rate (gpm)

The desired backwash pumping rate (gpm) for a filter depends on the desired backwash rate in gpm/ft^2 and the ft^2 area of the filter. The backwash pumping rate (gpm) can be determined by using Equation 12.76:

$$\text{Backwash pumping rate (gpm)} = \text{Desired backwash rate (gpm/ft}^2) \times \text{Filter area (ft}^2) \tag{12.76}$$

■ **EXAMPLE 12.89**

Problem: A filter is 25 ft long by 20 ft wide. If the desired backwash rate is 22 gpm/ft^2, what backwash pumping rate (gpm) will be required?

Solution: The desired backwash flow through each square foot of filter area is 20 gpm. The total gpm flow through the filter is therefore 20 gpm times the entire square foot area of the filter:

$$\begin{aligned}\text{Backwash pumping rate} &= \text{Desired backwash rate (gpm/ft}^2) \times \text{Filter area (ft}^2)\\ &= 20\ \text{gpm/ft}^2 \times (25\ \text{ft} \times 20\ \text{ft})\\ &= 10{,}000\ \text{gpm}\end{aligned}$$

■ **EXAMPLE 12.90**

Problem: The desired backwash pumping rate for a filter is 12 gpm/ft^2. If the filter is 20 ft long by 20 ft wide, what backwash pumping rate (gpm) will be required?

Solution:

$$\begin{aligned}\text{Backwash pumping rate} &= \text{Desired backwash rate (gpm/ft}^2) \times \text{Filter area (ft}^2)\\ &= 12\ \text{gpm/ft}^2 \times (20\ \text{ft} \times 20\ \text{ft})\\ &= 4800\ \text{gpm}\end{aligned}$$

PERCENT PRODUCT WATER USED FOR BACKWASHING

Along with measuring filtration rate and filter run time, another aspect of filter operation that is monitored for filter performance is the percent of product water used for backwashing. The equation for percent of product water used for backwashing calculations used is shown below:

$$\text{Backwash water (\%)} = \frac{\text{Backwash water (gal)}}{\text{Water filtered (gal)}} \times 100 \tag{12.77}$$

■ **EXAMPLE 12.91**

Problem: A total of 18,100,000 gal of water was filtered during a filter run. If backwashing used 74,000 gal of this product water, what percent of the product water was used for backwashing?

Solution:

$$\text{Backwash water} = \frac{\text{Backwash water (gal)}}{\text{Water filtered (gal)}} \times 100 = \frac{74{,}000\ \text{gal}}{18{,}100{,}000\ \text{gal}} \times 100 = 0.4\%$$

■ **EXAMPLE 12.92**

Problem: A total of 11,400,000 gal of water was filtered during a filter run. If backwashing used 48,500 gal of this product water, what percent of the product water was used for backwashing?

Solution:

$$\text{Backwash water} = \frac{\text{Backwash water (gal)}}{\text{Water filtered (gal)}} \times 100 = \frac{48{,}500\ \text{gal}}{11{,}400{,}000\ \text{gal}} \times 100 = 0.43\%$$

Percent Mudball Volume

Mudballs are heavier deposits of solids near the top surface of the medium that break into pieces during backwash, resulting in spherical accretions of floc and sand (usually less than 12 in. in diameter). The presence of mudballs in the filter media is checked periodically. The principal objection to mudballs is that they diminish the effective filter area. To calculate the percent mudball volume we use Equation 12.78:

$$\text{\% Mudball volume} = \frac{\text{Mudball volume (mL)}}{\text{Total sample volume (mL)}} \times 100 \tag{12.78}$$

■ Example 12.93

Problem: A 3350-mL sample of filter media was taken for mudball evaluation. The volume of water in the graduated cylinder rose from 500 mL to 525 mL when mudballs were placed in the cylinder. What is the percent mudball volume of the sample?

Solution: First determine the volume of mudballs in the sample:

$$525 \text{ mL} - 500 \text{ mL} = 25 \text{ mL}$$

Then calculate the percent mudball volume:

$$\text{\% Mudball volume} = \frac{\text{Mudball volume (mL)}}{\text{Total sample volume (mL)}} \times 100 = \frac{25 \text{ mL}}{3350 \text{ mL}} \times 100 = 0.75\%$$

■ Example 12.94

Problem: A filter is tested for the presence of mudballs. The mudball sample has a total sample volume of 680 mL. Five samples were taken from the filter. When the mudballs were placed in 500 mL of water, the water level rose to 565 mL. What is the percent mudball volume of the sample?

Solution: The mudball volume is the volume that the water rose:

$$565 \text{ mL} - 500 \text{ mL} = 65 \text{ mL}$$

Because five samples of media were taken, the total sample volume is 5 times the sample volume:

$$5 \times 680 \text{ mL} = 3400 \text{ mL}$$

$$\text{\% Mudball volume} = \frac{\text{Mudball volume (mL)}}{\text{Total sample volume (mL)}} \times 100 = \frac{65 \text{ mL}}{3400 \text{ mL}} \times 100 = 1.9\%$$

Filter Bed Expansion

In addition to backwash rate, it is also important to expand the filter media during the wash to maximize the removal of particles held in the filter or by the media; that is, the efficiency of the filter wash operation depends on the expansion of the sand bed. Bed expansion is determined by measuring the distance from the top of the unexpanded media to a reference point (e.g., top of the filter wall) and from the top of the expanded media to the same reference. A proper backwash rate should expand the filter 20 to 25%. Percent bed expansion is given by dividing the bed expansion by the total depth of expandable media (i.e., media depth less support gravels) and multiplied by 100, as follows:

Expanded measurement = Depth to top of media during backwash (in.)

Unexpanded measurement = Depth to top of media before backwash (in.)

Bed expansion = Unexpanded measurement (inches) – Expanded measurement (in.)

$$\text{Bed expansion (\%)} = \frac{\text{Bed expansion measurement (in.)}}{\text{Total depth of expandable media (in.)}} \times 100 \tag{12.79}$$

■ Example 12.95

Problem: The backwashing practices for a filter with 30 in. of anthracite and sand are being evaluated. While at rest, the distance from the top of the media to the concrete floor surrounding the top of filter is measured to be 41 in. After the backwash has begun and the maximum backwash rate is achieved, a probe containing a white disk is slowly lowered into the filter bed until anthracite is observed on the disk. The distance from the expanded media to the concrete floor is measured to be 34.5 in. What is the percent bed expansion?

Solution:

Given:
Unexpanded measurement = 41 in.
Expanded measurement = 34.5 in.

$$\text{Bed expansion} = 41 \text{ in.} - 34.5 \text{ in.} = 6.5 \text{ in.}$$

$$\text{\% Bed expansion} = \frac{\text{Bed expansion measurement (in.)}}{\text{Total depth of expandable media (in.)}} \times 100 = \frac{6.5 \text{ in.}}{30 \text{ in.}} \times 100 = 22\%$$

Filter Loading Rate

Filter loading rate is the flow rate of water applied to the unit area of the filter. It is the same value as the flow velocity approaching the filter surface and can be determined by using Equation 12.80:

$$u = Q/A \tag{12.80}$$

where
u = Loading rate ($m^3/(m^2 \cdot d$, gpm/ft^2).
Q = Flow rate (m^3/d, ft^3/d, gpm).
A = Surface area of filter (m^2, ft^2).

Filters are classified as slow sand filters, rapid sand filters, and high-rate sand filters on the basis of loading rate. Typically, the loading rate for rapid sand filters is 120 $m^3/m^2 \cdot d$ (83 $L/m^2 \cdot min$ or 2 gal/min/ft^2). The loading rate may be up to five times this rate for high-rate filters.

■ Example 12.96

Problem: A sanitation district is to install rapid sand filters downstream of the clarifiers. The design-loading rate is selected to be 150 m^3/m^2. The design capacity of the waterworks is 0.30 m^3/s (6.8 MGD). The maximum surface per filter is limited to 45 m^2. Design the number and size of filters and calculate the normal filtration rate.

Solution: Determine the total surface area required:

$$A = \frac{Q}{u} = \frac{0.30 \text{ m}^3/\text{sec } (85{,}400 \text{ sec/day})}{150 \text{ m}^3/\text{m}^2 \cdot \text{day}} = \frac{25{,}920}{150} = 173 \text{ m}^2$$

Determine the number of filters:

$$\text{No. of filters} = \frac{173 \text{ m}^2}{45 \text{ m}} = 3.8$$

Select 4 m. The surface area (*A*) for each filter is

$$A = 173 \text{ m}^2 \div 4 = 43.25 \text{ m}^2$$

We can use 6 m × 7 m or 6.4 m × 7 m or 6.42 m × 7 m. If a 6-m × 7-m filter is installed, the normal filtration rate is

$$u = \frac{Q}{A} = \frac{0.30 \text{ m}^3/\text{sec} \times 86{,}400 \text{ sec/day}}{4 \text{ m} \times 6 \text{ m} \times 7 \text{ m}} = 154.3 \text{ m}^3/\text{m}^2 \cdot \text{day}$$

Filter Medium Size

Filter medium grain size has an important effect on the filtration efficiency and on backwashing requirements for the medium. The actual medium selected is typically determined by performing a grain size distribution analysis—sieve size and percentage passing by weight relationships are plotted on logarithmic-probability paper. The most common parameters used in the United States to characterize a filter medium are effective size (ES) and uniformity coefficient (UC) of medium size distribution. The ES is that grain size for which 10% of the grains are smaller by weight; it is often abbreviated by d_{10}. The UC is the ratio of the 60th percentile (d_{60}) to the 10th percentile. The 90th percentile (d_{90}) is the size for which 90% of the grains are smaller by weight. The d_{90} size is used for computing the required filter backwash rate for a filter medium. Values of d_{10}, d_{60}, and d_{90} can be read from an actual sieve analysis curve. If such a curve is not available and if a linear log–probability plot is assumed, the values can be interrelated by Equation 12.81 (Cleasby, 1990):

$$d_{90} = d_{10}(10^{1.67 \log \text{UC}}) \tag{12.81}$$

■ Example 12.97

Problem: A sieve analysis curve of a typical filter sand gives $d_{10} = 0.52$ mm and $d_{60} = 0.70$ mm. What are its uniformity coefficient and d_{90}?

Solution:

$$\text{UC} = d_{60}/d_{10} = 0.70 \text{ mm}/0.52 \text{ mm} = 1.35$$

$$d_{90} = d_{10}(10^{1.67 \log \text{UC}}) = 0.52 \text{ mm} \times (10^{1.67 \log 1.35}) = 0.52 \text{ mm} \times (10^{0.218}) = 0.86 \text{ mm}$$

Mixed Media

An innovation in filtering systems has offered a significant improvement and economic advantage to rapid rate filtration: the mixed media filter bed. Mixed media filter beds offer specific advantages in specific circumstances, but will give excellent operating results at a filtering rate of 5 gal/ft²/min.

Moreover, the mixed media filtering unit is more tolerant of higher turbidities in the settled water. For improved process performance, activated carbon or anthracite is added on the top of the sand bed. The approximate specific gravities of ilmenite (<60% TiO_2), silica sand, anthracite, and water are 4.2, 2.6, 1.5, and 1.0, respectively. The economic advantage of the mixed bed media filter is based on filter area; it will safely produce 2-1/2 times as much filtered water as a rapid sand filter. When settling velocities are equal, the particle sizes for media of different specific gravity can be computed by using Equation 12.82:

$$\frac{d_1}{d_2} = \left(\frac{s_2 - s}{s_1 - s}\right)^{2/3} \tag{12.82}$$

where

d_1, d_2 = Diameters of particles 1 and 2, respectively.

s_1, s, s_2 = Specific gravity of particle 1, water, and particle 2, respectively.

■ Example 12.98

Problem: Estimate the particle size of ilmenite sand (specific gravity = 4.2) that has the same settling velocity as silica sand that is 0.60 mm in diameter (specific gravity = 2.6).

Solution: Find the diameter of ilmenite sand using Equation 12.82:

$$d = 0.6 \text{ mm} \times \left(\frac{2.6-1}{4.2-1}\right)^{2/3} = 0.38 \text{ mm}$$

Head Loss for Fixed Bed Flow

When water is pumped upward through a bed of fine particles at a very low flow rate the water percolates through the pores (void spaces) without disturbing the bed. This is a fixed bed process. The head loss (pressure drop) through a clean granular-media filter is generally less than 0.9 m (3 ft). With the accumulation of impurities, head loss gradually increases until the filter is backwashed. The Kozeny equation, shown below, is typically used for calculating head loss through a clean fixed bed flow filter.

$$\frac{h}{L} = \left(\frac{k\mu(1-\varepsilon)^2}{gp\varepsilon^3}\right)\left(\frac{A}{V}\right)^2 u \tag{12.83}$$

where

h = Head loss in filter depth L (m, ft).

k = Dimensionless Kozeny constant (5 for sieve openings, 6 for size of separation).

μ = Absolute viscosity of water (N·s/m², lb·s/ft²).

ε = Porosity (dimensionless).

g = Acceleration of gravity (9.81 m/s or 32.2 ft/s).

p = Density of water (kg/m³, lb/ft³).

A/V = Grain surface area per unit volume of grain.

= Specific surface S (shape factor = 6.0–7.7).

= $6/d$ for spheres.

= $6/\psi d_{eq}$ for irregular grains.

ψ = Grain sphericity or shape factor.

d_{eq} = Grain diameter of spheres of equal volume.

u = Filtration (superficial) velocity (m/s, fps).

■ **Example 12.99**

Problem: A dual-medium filter is composed of 0.3 m of anthracite (mean size 2.0 mm) that is placed over a 0.6-m layer of sand (mean size 0.7 mm) with a filtration rate of 9.78 m/hr. Assume that the grain sphericity is ψ = 0.75 and the porosity for both is 0.42. Although normally taken from the appropriate table at 15°C, we provide the head loss data of the filter at 1.131×10^{-6} m²·s.

Solution: Determine head loss through the anthracite layer using the Kozeny equation, Equation 12.83:

$$\frac{h}{L} = \left(\frac{k\mu(1-\varepsilon)^2}{g\rho\varepsilon^3}\right)\left(\frac{A}{V}\right)^2 u$$

where

$k = 6$
$g = 9.81$ m/s²
$\mu_p = v = 1.131 \times 10^{-6}$ m²·s (from the appropriate table)
$\varepsilon = 0.40$
$A/V = 6/0.75d = 8/d = 8/0.002$
$u = 9.78$ m/hr = 0.00272 m/s
$L = 0.3$ m

Then,

$$h = 6 \times \frac{1.131 \times 10^{-6}}{9.81} \times \frac{(1-0.42)^2}{0.42^3} \times \left(\frac{8}{0.002}\right)^2 \times 0.00272 \times 0.3 = 0.0410 \text{ m}$$

Compute the head loss passing through the sand. Use the same data but insert:

$k = 5$
$d = 0.0007$ m
$L = 0.6$ m

$$h = 5 \times \frac{1.131 \times 10^{-6}}{9.81} \times \frac{(0.58)^2}{0.42^3} \times \left(\frac{8}{0.0007}\right)^2 \times 0.00272 \times 0.6 = 0.5579 \text{ m}$$

Compute total head loss (h):

$$h = 0.0410 \text{ m} + 0.5579 \text{ m} = 0.599 \text{ m}$$

Head Loss through a Fluidized Bed

If the upward water flow rate through a filter bed is very large, then the bed mobilizes pneumatically and may be swept out of the process vessel. At an intermediate flow rate the bed expands and is in what we call an *expanded* state. In the fixed bed the particles are in direct contact with each other, supporting each other's weight. In the expanded bed, the particles have a mean free distance between particles and the drag force of the water supports the particles. The expanded bed has some of the properties of the water (i.e., of a fluid) and is called a *fluidized bed* (Chase, 2002). Simply, *fluidization* is defined as upward flow through a granular filter bed at sufficient velocity to suspend the grains in the water. Minimum fluidizing velocity (U_{mf}) is the superficial fluid velocity required to start fluidization; it is important in determining the required minimum backwashing flow rate. Wen and Yu (1966) proposed that the U_{mf} equation include the constants (over a wide range of particles) 33.7 and 0.0408 but exclude the porosity of fluidization and the shape factor (Wen and Yu, 1966):

$$U_{mf} = \frac{\mu}{pd_{eq}} \times (1135.69 + 0.0408G_n)^{0.5} - \frac{33.7\mu}{pd_{eq}} \quad (12.84)$$

where

μ = Absolute viscosity of water (N·s/m², lb·s/ft²).
p = Density of water (kg/m³, lb/ft³).
$d_{eq} = d_{90}$ sieve size (used instead of d_{eq}).
G_n = Galileo number.

$$d_{eq}^3 p(p_s - p) g / \mu^2 \quad (12.85)$$

Other variables used are as expressed in Equation 12.83.

Note: Based on the studies of Cleasby and Fan (1981), we use a safety factor of 1.3 to ensure adequate movement of the grains.

■ EXAMPLE 12.100

Problem: Estimate the minimum fluidized velocity and backwash rate for a sand filter. The d_{90} size of the sand is 0.90 mm. The density of the sand is 2.68 g/cm³.

Solution: Compute the Galileo number. From the data given and the applicable table, at 15°C:

$p = 0.999$ g/cm³
$\mu = 0.0113$ N s/m² = 0.00113 kg/ms = 0.0113 g/cm·s
$\mu p = 0.0113$ cm²/s
$g = 981$ cm/s²
$d = 0.90$ cm
$p_s = 2.68$ g/cm³

Using Equation 12.85:

$$G_n = d_{eq}^3 p(p_s - p) g / \mu^2 = (0.090)^3 \times 0.999 \times (2.68 - 0.999) \times 981 / (0.0113)^2 = 9405$$

Compute U_{mf} using Equation 12.84:

$$U_{mf} = \frac{\mu}{pd_{eq}} \times (1135.69 + 0.0408G_n)^{0.5} - \frac{33.7\mu}{pd_{eq}}$$

$$= \frac{0.0113}{0.999 \times 0.090} \times (1135.69 + 0.0408 \times 9405)^{0.5} - \frac{33.7 \times 0.0113}{0.999 \times 0.090}$$

$$= 0.660 \text{ cm/s}$$

Compute backwash rate. Apply a safety factor of 1.3 to U_{mf} as backwash rate:

$$\text{Backwash rate} = 1.3 \times 0.660 \text{ cm/s} = 0.858 \text{ cm/s}$$

$$0.858 \times \frac{\text{cm}^3}{\text{cm}^2 \cdot \text{s}} \times \frac{\text{L}}{1000\ \text{cm}^3} \times \frac{1}{3.785} \times \frac{\text{gal}}{\text{L}} \times 929 \times \frac{\text{cm}^2}{\text{ft}^2} \times \frac{60\ \text{s}}{\text{min}} = 12.6\ \text{pgm/ft}^2$$

Horizontal Washwater Troughs

Washwater troughs are used to collect backwash water as well as to distribute influent water during the initial stages of filtration. Washwater troughs are normally placed above the filter media in the United States. Proper placement of these troughs is very important to ensure that the filter medium is not carried into the troughs during the backwash and removed from the filter. These backwash troughs are constructed from concrete, plastic, fiberglass, or other corrosion-resistant materials. The total rate of discharge in a rectangular trough with free flow can be calculated by using Equation 12.86:

$$Q = C \times w \times h^{1.5} \tag{12.86}$$

where

Q = Flow rate (cfs).
C = Constant (2.49).
w = Trough width (ft).
h = Maximum water depth in trough (ft).

■ Example 12.101

Problem: Troughs are 18 ft long, 18 in. wide, and 8 ft to the center with a horizontal flat bottom. The backwash rate is 24 in./min. Estimate (a) the water depth of the troughs with free flow into the gullet, and (b) the distance between the top of the troughs and the 30-in. sand bed. Assuming 40% expansion and 6 in. of freeboard in the troughs and 6 in. of thickness.

Solution: Estimate the maximum water depth (h) in the trough:

$$\text{Velocity } (V) = 24 \text{ in./min} = 2 \text{ ft/60 s} = 1/30 \text{ fps}$$

$$A = 18 \text{ ft} \times 8 \text{ ft} = 144 \text{ ft}^2$$

$$Q = V \times A = 1/30 \text{ fps} \times 144 \text{ cfs} = 4.8 \text{ cfs}$$

Using Equation (12.86):

$$w = 1.5 \text{ ft}$$

$$Q = Cwh^{1.5} = 2.49wh^{1.5}$$

$$h = \left(Q/2.49w\right)^{2/3} = \left[4.8(2.49 \times 1.5)\right]^{2/3} = 1.18 \text{ ft (approx. 14 in.} = 1.17 \text{ ft)}$$

Determine the distance (y) between the sand bed surface and the top troughs:

$$\text{Freeboard} = 6 \text{ in.} = 0.5 \text{ ft}$$

$$\text{Thickness} = 8 \text{ in.} = 0.67 \text{ ft (bottom of trough)}$$

$$y = 2.5 \text{ ft} \times 0.4 + 1.17 \text{ ft} + 0.5 \text{ ft} + 0.5 \text{ ft} = 3.2 \text{ ft}$$

Filter Efficiency

Water treatment filter efficiency is defined as the effective filter rate divided by the operating filtration rate as shown in Equation 12.87 (AWWA and ASCE, 1998):

$$E = \frac{R_e}{R_o} = \frac{UFRV - UBWV}{UFRV} \tag{12.87}$$

where

E = Filter efficiency (%).
R_e = Effective filtration rate (gpm/ft^2).
R_o = Operating filtration rate (gpm/ft^2).
$UFRV$ = Unit filter run volume (gal/ft^2).
$UBWV$ = Unit backwash volume (gal/ft^2).

■ Example 12.102

Problem: A rapid sand filter operates at 3.9 gpm/ft^2 for 48 hours. Upon completion of the filter run, 300 gal/ft^2 of backwash water is used. Find the filter efficiency.

Solution: Calculate the operating filtration rate (R_o):

$$R_o = 3.9 \text{ gpm/ft}^2 \times 60 \text{ min/hr} \times 48 \text{ hr} = 11{,}232 \text{ gal/ft}^2$$

Calculate the effective filtration rate (R_e):

$$R_e = 11{,}232 \text{ gal/ft}^2 - 300 \text{ gal/ft}^2 = 10{,}932 \text{ gal/ft}^2$$

Calculate the filter efficiency (E) using Equation 12.87:

$$E = 10{,}932 \text{ gal/ft}^2 \div 11{,}232 \text{ gal/ft}^2 = 97.3\%$$

WATER CHLORINATION CALCULATIONS

Chlorine is the most commonly used substance for disinfection of water in the United States. The addition of chlorine or chlorine compounds to water is called *chlorination*. Chlorination is considered to be the single most important process for preventing the spread of waterborne disease.

Chlorine Disinfection

Chlorine can destroy most biological contaminants by various mechanisms, including the following:

- Damaging the cell wall
- Altering the permeability of the cell (the ability to pass water in and out through the cell wall)
- Altering the cell protoplasm
- Inhibiting the enzyme activity of the cell so it is unable to use its food to produce energy
- Inhibiting cell reproduction

Chlorine is available in a number of different forms: (1) as pure elemental gaseous chlorine (a greenish-yellow gas possessing a pungent and irritating odor that is heavier than air, nonflammable, and nonexplosive), which, when released to the atmosphere, is toxic and corrosive; (2) as solid calcium hypochlorite (in tablets or granules); or (3) as a liquid sodium hypochlorite solution (in various strengths). The choice of one form of chlorine over another for a given water system depends on the amount of water to be treated, configuration of the water system, the local availability of the chemicals, and the skill of the operator. One of the major advantages of using chlorine is the effective residual that it produces. A residual indicates that disinfection is completed and the system has an acceptable bacteriological quality. Maintaining a residual in the distribution system helps to prevent regrowth of those microorganisms that were injured but not killed during the initial disinfection stage.

Determining Chlorine Dosage (Feed Rate)

The expressions milligrams per liter (mg/L) and pounds per day (lb/day) are most often used to describe the amount of chlorine added or required. Equation 12.88 can be used to calculate either mg/L or lb/day chlorine dosage:

$$\text{Chlorine feed rate (lb/day)} = \text{Chlorine (mg/L)} \times \text{Flow (MGD)} \times 8.34 \text{ lb/gal} \quad (12.88)$$

■ Example 12.103

Problem: Determine the chlorinator setting (lb/day) required to treat a flow of 4 MGD with a chlorine dose of 5 mg/L.

Solution:

$$\begin{aligned}\text{Chlorine feed rate} &= \text{Chlorine (mg/L)} \times \text{Flow (MGD)} \times 8.34 \text{ lb/gal}\\ &= 5 \text{ mg/L} \times 4 \text{ MGD} \times 8.34 \text{ lb/gal}\\ &= 167 \text{ lb/day}\end{aligned}$$

■ Example 12.104

Problem: A pipeline that is 12 in. in diameter and 1400 ft long is to be treated with a chlorine dose of 48 mg/L. How many pounds of chlorine will this require?

Solution: First determine the gallon volume of the pipeline:

$$\begin{aligned}\text{Volume} &= 0.785 \times (\text{Diameter})^2 \times \text{Length (ft)} \times 7.48 \text{ gal/ft}^3\\ &= 0.785 \times (1 \text{ ft})^2 \times 1400 \text{ ft} \times 7.48 \text{ gal/ft}^3\\ &= 8221 \text{ gal}\end{aligned}$$

Now calculate the pounds chlorine required:

$$\begin{aligned}\text{Chlorine} &= \text{Chlorine (mg/L)} \times \text{Volume (MG)} \times 8.34 \text{ lb/gal}\\ &= 48 \text{ mg/L} \times 0.008221 \text{ MG} \times 8.34 \text{ lb/gal}\\ &= 3.3 \text{ lb}\end{aligned}$$

■ Example 12.105

Problem: A chlorinator setting is 30 lb per 24 hr. If the flow being chlorinated is 1.25 MGD, what is the chlorine dosage expressed as mg/L?

Solution:

$$\text{Chlorine (lb/day)} = \text{Chlorine (mg/L)} \times \text{Flow (MGD)} \times 8.34 \text{ lb/gal}$$

$$30 \text{ lb/day} = x \text{ mg/L} \times 1.25 \text{ MGD} \times 8.34 \text{ lb/gal}$$

$$x = 30 \div (1.25 \times 8.34)$$

$$x = 2.9 \text{ mg/L}$$

■ Example 12.106

Problem: A flow of 1600 gpm is to be chlorinated. At a chlorinator setting of 48 lb per 24 hr, what would be the chlorine dosage in mg/L?

Solution: Convert the gpm flow rate to MGD flow rate:

$$1600 \text{ gpm} \times 1440 \text{ min/day} = 2{,}304{,}000 \text{ gpd} = 2.304 \text{ MGD}$$

Now calculate the chlorine dosage in mg/L:

$$\text{Chlorine (lb/day)} = \text{Chlorine (mg/L)} \times \text{Flow (MGD)}$$

$$48 \text{ lb/day} = x \text{ mg/L} \times 2.304 \text{ MGD} \times 8.34 \text{ lb/gal}$$

$$x = 48 \div (2.304 \times 8.34)$$

$$x = 2.5 \text{ mg/L}$$

Calculating Chlorine Dose, Demand, and Residual

Common terms used in chlorination include the following:

- *Chlorine dose*—The amount of chlorine added to the system. It can be determined by adding the desired residual for the finished water to the chlorine demand of the untreated water. Dosage can be either milligrams per liter (mg/L) or pounds per day (lb/day). The most common is mg/L:

$$\text{Chlorine dose (mg/L)} = \text{Chlorine demand (mg/L)} + \text{Chlorine (mg/L)}$$

- *Chlorine demand*—The amount of chlorine used by iron, manganese, turbidity, algae, and microorganisms in the water. Because the reaction between chlorine and microorganisms is not instantaneous, demand is relative to time. For instance, the demand 5 minutes after applying chlorine will be less than the demand after 20 minutes. Demand, like dosage is expressed in mg/L. The chlorine demand is as follows:

$$\text{Chlorine demand (mg/L)} = \text{Chlorine dose (mg/L)} - \text{Chlorine residual (mg/L)}$$

- *Chlorine residual*—The amount of chlorine (determined by testing) remaining after the demand is satisfied. Residual, like demand, is based on time. The longer the time after dosage, the lower the residual will be, until all of the demand has been satisfied. Residual, like dosage and demand, is expressed in mg/L. The presence of a *free residual* of at least 0.2 to 0.4 ppm usually provides a high degree of assurance that the disinfection of the water is complete. *Combined residual* is the result of combining free chlorine with nitrogen compounds. Combined residuals are also called chloramines. *Total chlorine residual* is the mathematical combination of free and combined residuals. Total residual can be determined directly with standard chlorine residual test kits.

The following examples illustrate the calculation of chlorine dose, demand, and residual using Equation 12.89:

$$\text{Chlorine dose (mg/L)} = \text{Chlorine demand (mg/L)} + \text{Chlorine residual (mg/L)} \tag{12.89}$$

■ Example 12.107

Problem: A water sample is tested and found to have a chlorine demand of 1.7 mg/L. If the desired chlorine residual is 0.9 mg/L, what is the desired chlorine dose in mg/L?

Solution:

$$\begin{aligned}\text{Chlorine dose} &= \text{Chlorine demand (mg/L)} + \text{Chlorine residual (mg/L)}\\ &= 1.7\text{ mg/L} + 0.9\text{ mg/L}\\ &= 2.6\text{ mg/L}\end{aligned}$$

■ Example 12.108

Problem: The chlorine dosage for water is 2.7 mg/L. If the chlorine residual after 30 min of contact time is found to be 0.7 mg/L, what is the chlorine demand expressed in mg/L?

Solution:

$$\text{Chlorine dose (mg/L)} = \text{Chlorine demand (mg/L)} + \text{Chlorine residual (mg/L)}$$

$$2.7\text{ mg/L} = x\text{ mg/L} + 0.6\text{ mg/L}$$

$$x\text{ mg/L} = 2.7\text{ mg/L} - 0.7\text{ mg/L}$$

$$x = 2.0\text{ mg/L}$$

■ Example 12.109

Problem: What should the chlorinator setting (lb/day) be to treat a flow of 2.35 MGD if the chlorine demand is 3.2 mg/L and a chlorine residual of 0.9 mg/L is desired?

Solution: Determine the chlorine dosage in mg/L:

$$\begin{aligned}\text{Chlorine dose} &= \text{Chlorine demand (mg/L)} + \text{Chlorine residual (mg/L)}\\ &= 3.2\text{ mg/L} + 0.9\text{ mg/L}\\ &= 4.1\text{ mg/L}\end{aligned}$$

Calculate the chlorine dosage (feed rate) in lb/day:

$$\begin{aligned}\text{Chlorine (lb/day)} &= \text{Chlorine (mg/L)} \times \text{Flow (MGD)} \times 8.34\text{ lb/gal}\\ &= 4.1\text{ mg/L} \times 2.35\text{ MGD} \times 8.34\text{ lb/gal}\\ &= 80.4\text{ lb/day}\end{aligned}$$

Breakpoint Chlorination Calculations

To produce a free chlorine residual, enough chlorine must be added to the water to produce what is referred to as *breakpoint chlorination*, the point at which near complete oxidation of nitrogen compounds is reached. Any residual beyond breakpoint is mostly free chlorine. When chlorine is added to natural waters, the chlorine begins combining with and oxidizing the chemicals in the water before it begins disinfecting. Although residual chlorine will be detectable in the water, the chlorine will be in the combined form with a weak disinfecting power. Adding more chlorine to the water at this point actually decreases the chlorine residual as the additional chlorine destroys the combined chlorine compounds. At this stage, water may have a strong swimming pool or medicinal taste and odor. To avoid this taste and odor, add still more chlorine to produce a free chlorine residual. Free chlorine has the highest disinfecting power. The point at which most of the combined chlorine compounds have been destroyed and the free chlorine starts to form is the *breakpoint.*

Note: The actual chlorine breakpoint of water can only be determined by experimentation.

To calculate the actual increase in chlorine residual that would result from an increase in chlorine dose, we use the mg/L to lb/day equation as shown below:

$$\text{Increase in chlorine (lb/day)} = \text{Expected increase (mg/L)} \times \text{Flow (MGD)} \times 8.34 \text{ lb/gal} \quad (12.90)$$

Note: The actual increase in residual is simply a comparison of new and old residual data.

■ Example 12.110

Problem: A chlorinator setting is increased by 2 lb/day. The chlorine residual before the increased dosage was 0.2 mg/L. After the increased chlorine dose, the chlorine residual was 0.5 mg/L. The average flow rate being chlorinated is 1.25 MGD. Is the water being chlorinated beyond the breakpoint?

Solution: Calculate the expected increase in chlorine residual using the mg/L to lb/day equation:

$$\text{Increase in chlorine (lb/day)} = \text{Expected increase (mg/L)} \times \text{Flow (MGD)} \times 8.34 \text{ lb/gal}$$

$$2 \text{ lb/day} = x \text{ mg/L} \times 1.25 \text{ MGD} \times 8.34 \text{ lb/gal}$$

$$x = (2 \text{ lb/day}) \div (1.25 \text{ MGD} \times 8.34 \text{ lb/gal})$$

$$x = 0.19 \text{ mg/L}$$

The actual increase in residual chlorine is

$$0.5 \text{ mg/L} - 0.19 \text{ mg/L} = 0.31 \text{ mg/L}$$

■ Example 12.111

Problem: A chlorinator setting of 18 lb chlorine per 24 hr results in a chlorine residual of 0.3 mg/L. The chlorinator setting is increased to 22 lb per 24 hr. The chlorine residual increased to 0.4 mg/L at this new dosage rate. The average flow being treated is 1.4 MGD. On the basis of these data, is the water being chlorinated past the breakpoint?

Solution: Calculate the expected increase in chlorine residual:

$$\text{Increase in chlorine (lb/day)} = \text{Expected increase (mg/L)} \times \text{Flow (MGD)} \times 8.34\ \text{lb/gal}$$

$$4\ \text{lb/day} = x\ \text{mg/L} \times 1.4\ \text{MGD} \times 8.34\ \text{lb/gal}$$

$$x = (4\ \text{lb/day}) \div (1.4\ \text{MGD} \times 8.34\ \text{lb/gal})$$

$$x = 0.34\ \text{mg/L}$$

The actual increase in residual chlorine is

$$0.4\ \text{mg/L} - 0.3\ \text{mg/L} = 0.1\ \text{mg/L}$$

Calculating Dry Hypochlorite Feed Rate

The most commonly used dry hypochlorite, calcium hypochlorite contains about 65 to 70% available chlorine, depending on the brand. Because hypochlorites are not 100% pure chorine, more pounds per day must be fed into the system to obtain the same amount of chlorine for disinfection. The equation used to calculate the lb/day hypochlorite needed is Equation 12.91:

$$\text{Hypochlorite (lb/day)} = \frac{\text{Chlorine (lb/day)}}{\%\ \text{Available chlorine}/100} \tag{12.91}$$

■ Example 12.112

Problem: A chlorine dosage of 110 lb/day is required to disinfect a flow of 1,550,000 gpd. If the calcium hypochlorite to be used contains 65% available chlorine, how many lb/day hypochlorite will be required for disinfection?

Solution: Because only 65% of the hypochlorite is chlorine, more than 110 lb of hypochlorite will be required:

$$\text{Hypochlorite} = \frac{\text{Chlorine (lb/day)}}{\%\ \text{Available chlorine}/100} = \frac{110}{65/100} = \frac{110}{0.65} = 169\ \text{lb/day}$$

■ Example 12.113

Problem: A water flow of 900,000 gpd requires a chlorine dose of 3.1 mg/L. If calcium hypochlorite (65% available chlorine) is to be used, how many lb/day of hypochlorite are required?

Solution: Calculate the lb/day chlorine required:

$$\text{Chlorine} = \text{Chlorine (mg/L)} \times \text{Flow (MGD)} \times 8.34\ \text{lb/gal}$$

$$= 3.1\ \text{mg/L} \times 0.90\ \text{MGD} \times 8.34\ \text{lb/gal}$$

$$= 23\ \text{lb/day}$$

Calculate the lb/day hypochlorite:

$$\text{Hypochlorite} = \frac{\text{Chlorine (lb/day)}}{\%\ \text{Available chlorine}/100} = \frac{23}{65/100} = \frac{23}{0.65} = 35\ \text{lb/day}$$

■ Example 12.114

Problem: A tank contains 550,000 gal of water and is to receive a chlorine dose of 2.0 mg/L. How many pounds of calcium hypochlorite (65% available chlorine) will be required?

Solution:

$$\text{Hypochlorite} = \frac{\text{Chlorine (mg/L)} \times \text{Volume (MG)} \times 8.34 \text{ lb/gal}}{\text{\% Available chlorine}/100}$$

$$= \frac{2.0 \text{ mg/L} \times 0.550 \text{ MG} \times 8.34}{65/100} = \frac{9.2}{0.65} = 14.2 \text{ lb}$$

■ Example 12.115

Problem: A total of 40 lb of calcium hypochlorite (65% available chlorine) is used in a day. If the flow rate treated is 1,100,000 gpd, what is the chlorine dosage in mg/L?

Solution: Calculate the lb/day chlorine dosage:

$$\text{Hypochlorite} = \frac{\text{Chlorine (lb/day)}}{\text{\% Available chlorine}/100}$$

$$40 \text{ lb/day} = \frac{x \text{ lb/day}}{0.65}$$

$$0.65 \times 40 = x$$

$$26 \text{ lb/day} = x$$

Then calculate mg/L chlorine using the mg/L to lb/day equation and filling in the known information:

$$26 \text{ lb/day chlorine} = x \text{ mg/L chlorine} \times 1.10 \text{ MGD} \times 8.34 \text{ lb/gal}$$

$$x = 26 \text{ lb/day}/(1.10 \text{ MGD} \times 8.34 \text{ lb/gal})$$

$$x = 2.8 \text{ mg/L}$$

■ Example 12.116

Problem: A flow of 2,550,000 gpd is disinfected with calcium hypochlorite (65% available chlorine). If 50 lb of hypochlorite are used in a 24-hr period, what is the mg/L chlorine dosage?

Solution: Calculate the lb/day chlorine dosage:

$$50 \text{ lb/day hypochlorite} = (x \text{ lb/day chlorine}) \div 0.65$$

$$x = 32.5 \text{ lb/day chlorine}$$

Calculate mg/L chlorine:

$$x \text{ mg/L chlorine} \times 2.55 \text{ MGD} \times 8.34 \text{ lb/gal} = 32.5 \text{ lb/day}$$

$$x = 1.5 \text{ mg/L chlorine}$$

Calculating Hypochlorite Solution Feed Rate

Liquid hypochlorite (i.e., sodium hypochlorite) is supplied as a clear, greenish-yellow liquid in strengths varying from 5.25 to 16% available chlorine. Often referred to as *bleach*, it is, in fact, used for bleaching. Common household bleach is a solution of sodium hypochlorite containing 5.25% available chlorine. When calculating gallons per day (gpd) liquid hypochlorite, the lb/day hypochlorite required must be converted to gpd hypochlorite. This conversion is accomplished using Equation 12.92:

$$\text{Hypochlorite (gpd)} = \frac{\text{Hypochlorite (lb/day)}}{8.34\ \text{lb/gal}} \tag{12.92}$$

■ Example 12.117

Problem: A total of 50 lb/day sodium hypochlorite is required for disinfection of a 1.5-MGD flow. How many gallons per day hypochlorite is this?

Solution: Because lb/day hypochlorite has already has been calculated, we simply convert lb/day to gpd hypochlorite required:

$$\text{Hypochlorite (gpd)} = \frac{\text{Hypochlorite (lb/day)}}{8.34\ \text{lb/gal}} = \frac{50\ \text{lb/day}}{8.34\ \text{lb/gal}} = 6.0\ \text{gpd}$$

■ Example 12.118

Problem: A hypochlorinator is used to disinfect the water pumped form a well. The hypochlorite solution contains 3% available chlorine. A chlorine dose of 1.3 mg/L is required for adequate disinfection throughout the system. If the flow being treated is 0.5 MGD, how many gpd of the hypochlorite solution will be required?

Solution: Calculate the lb/day chlorine required:

$$\text{Chlorine (lb/day)} = 1.3\ \text{mg/L} \times 0.5\ \text{MGD} \times 8.34\ \text{lb/gal} = 5.4\ \text{lb/day}$$

Calculate the lb/day hypochlorite solution required:

$$\text{Hypochlorite (lb/day)} = 5.4\ \text{lb/day chlorine}/0.03 = 180\ \text{lb/day}$$

Calculate the gpd hypochlorite solution required:

$$\text{Hypochlorite (gpd)} = 180\ \text{lb/day}/8.34\ \text{lb/gal} = 21.6\ \text{gpd}$$

Calculating Percent Strength of Solutions

If a teaspoon of salt is dropped into a glass of water it gradually disappears. The salt dissolves in the water, but a microscopic examination of the water would not show the salt. Only examination at the molecular level, which is not easily done, would show the salt and water molecules intimately mixed. If we taste the liquid, we would know that the salt is there. We can recover the salt by evaporating the water. In a solution, the molecules of the salt, the *solute,* are homogeneously dispersed among the molecules of water, the *solvent.* This mixture of salt and water is homogeneous on a molecular level. Such a homogeneous mixture is called a *solution*. The composition of a solution can be varied within certain limits. The three common states of matter are gas, liquid, and solid. In this discussion, of course, we are only concerned, at the moment, with the solid (calcium hypochlorite) and liquid (sodium hypochlorite) states.

Calculating Percent Strength Using Dry Hypochlorite

To calculate the percent strength of a chlorine solution, we use Equation 12.93:

$$\%\text{ Chlorine strength} = \frac{\text{Hypochlorite (lb)}\left(\dfrac{\%\text{ Available chlorine}}{100}\right)}{\text{Water (lb)} + \text{Hypochlorite (lb)}\left(\dfrac{\%\text{ Available chlorine}}{100}\right)} \times 100 \tag{12.93}$$

■ Example 12.119

Problem: If a total of 72 oz. of calcium hypochlorite (65% available chlorine) is added to 15 gal of water, what is the percent chlorine strength (by weight) of the solution?

Solution: Convert the ounces of hypochlorite to pounds of hypochlorite:

$$(72\text{ oz.}) \div (16\text{ oz./lb}) = 4.5\text{ lb chemical}$$

$$\%\text{ Chlorine strength} = \frac{\text{Hypochlorite (lb)}\left(\dfrac{\%\text{ Available chlorine}}{100}\right)}{\text{Water (lb)} + \text{Hypochlorite (lb)}\left(\dfrac{\%\text{ Available chlorine}}{100}\right)} \times 100$$

$$= \frac{4.5\text{ lb} \times 0.65}{(15\text{ gal} \times 8.34\text{ lb/gal}) + (4\text{ lb} \times 0.65)} \times 100$$

$$= \frac{2.9\text{ lb}}{125.1\text{ lb} + 2.9\text{ lb}} \times 100$$

$$= \frac{2.9\text{ lb}}{126} \times 100$$

$$= 2.3\%$$

Calculating Percent Strength Using Liquid Hypochlorite

To calculate the percent strength of liquid solutions, such as sodium hypochlorite, a different equation is required:

$$\begin{aligned}&\text{Liquid hypochlorite (gal)} \times 8.34\text{ lb/gal} \times \left(\frac{\%\text{ Strength of hypochlorite}}{100}\right)\\&= \text{Hypochlorite solution (gal)} \times 8.34\text{ lb/gal} \times \left(\frac{\%\text{ Strength of hypochlorite}}{100}\right)\end{aligned} \tag{12.94}$$

■ **Example 12.120**

Problem: A 12% liquid hypochlorite solution is to be used in making up a hypochlorite solution. If 3.3 gal of liquid hypochlorite are mixed with water to produce 25 gal of hypochlorite solution, what is the percent strength of the solution?

Solution:

$$\text{Liquid hypochlorite (gal)} \times 8.34\ \text{lb/gal} \times \left(\frac{\%\ \text{Strength of hypochlorite}}{100}\right)$$

$$= \text{Hypochlorite solution (gal)} \times 8.34\ \text{lb/gal} \times \left(\frac{\%\ \text{Strength of hypochlorite}}{100}\right)$$

$$3.3\ \text{gal} \times 8.34\ \text{lb/gal} \times \left(\frac{12}{100}\right) = 25\ \text{gal} \times 8.34\ \text{lb/gal} \times \left(\frac{x}{100}\right)$$

$$x = \frac{\cancel{100} \times 3.3 \times \cancel{8.34} \times 12}{25 \times \cancel{8.34} \times \cancel{100}}$$

$$x = \frac{3.3 \times 12}{25} = 1.6\%$$

CHEMICAL USAGE CALCULATIONS

In a typical plant operation, chemical usage is recorded each day. Such data provide a record of daily use from which the average daily use of the chemical or solution can be calculated. To calculate average use in pounds per day (lb/day), we use Equation 12.95. To calculate average use in gallons per day (gpd), we use Equation 12.96:

$$\text{Average use (lb/day)} = \text{Total chemical used (lb)} \div \text{Number of days} \tag{12.95}$$

$$\text{Average use (gpd)} = \text{Total chemical used (gal)} \div \text{Number of days} \tag{12.96}$$

To calculate the days supply in inventory in, we use Equation 12.97 or Equation 12.98.

$$\text{Days supply in inventory} = \text{Total chemical in inventory (lb)} \div \text{Average use (lb/day)} \tag{12.97}$$

$$\text{Days supply in inventory} = \text{Total chemical in inventory (gal)} \div \text{Average use (gpd)} \tag{12.98}$$

■ **Example 12.121**

Problem: Calcium hypochlorite usage for each day during a week is given below. Based on these data, what was the average lb/day hypochlorite chemical use during the week?

Monday	50 lb/day
Tuesday	55 lb/day
Wednesday	51 lb/day
Thursday	46 lb/day
Friday	56 lb/day
Saturday	51 lb/day
Sunday	48 lb/day

Solution:

$$\text{Average use} = \text{Total chemical used (lb)} \div \text{Number of days} = 357\ \text{lb} \div 7\ \text{days} = 51\ \text{lb/day}$$

■ **Example 12.122**

Problem: The average calcium hypochlorite use at a plant is 40 lb/day. If the chemical inventory in stock is 1100 lb, how many days supply is this?

Solution:

$$\text{Days supply in inventory} = \text{Total chemical in inventory (lb)} \div \text{Average use (lb/day)}$$
$$= (1100 \text{ lb in inventory}) \div (40 \text{ lb/day average use})$$
$$= 27.5 \text{ days}$$

CHLORINATION CHEMISTRY

Chlorine is used in the form of free elemental chlorine or as hypochlorites. Temperature, pH, and organic content in the water influence its chemical form in water. When chlorine gas is dissolved in water, it rapidly hydrolyzes to hydrochloric acid (HCl) and hypochlorous acid (HOCl):

$$Cl_2 + H_2O \leftrightarrow H^+ + Cl^- + HOCl \tag{12.99}$$

The equilibrium constant is (White, 1972):

$$K_H = \frac{[H^+][Cl^-][HOCl]}{\left[Cl_{2(aq)}\right]} = 4.48 \times 10^4 \text{ at } 25°C \tag{12.100}$$

Henry's law is used to explain the dissolution of gaseous chlorine, $Cl_{2(aq)}$. Henry's law describes the effect of the pressure on the solubility of the gases: There is a linear relationship between the partial pressure of gas above a liquid and the mole fraction of the gas dissolved in the liquid (Fetter, 1999). The Henry's law constant, K_H (as shown in Equation 12.100), is a measure of the compound transfer between the gaseous and aqueous phases. K_H is presented as a ratio of the compound's concentration in the gaseous phase to that in the aqueous phase at equilibrium:

$$K_H = \frac{P}{C_{water}} \tag{12.101}$$

where

K_H = Henry's law constant.
P = Compound's partial pressure in the gaseous phase.
C_{water} = Compound's concentration in the aqueous solution.

Note: The unit of the Henry's law constant is dependent on the choice of measure; however, it can also be dimensionless.

For our purposes, Henry's law can be expressed as (Downs and Adams, 1973):

$$Cl_{2(g)} = \frac{Cl_{2(aq)}}{H \text{ (mol/L} \cdot \text{atm)}} = \frac{\left[Cl_{2(aq)}\right]}{P_{Cl_2}} \tag{12.102}$$

where

$[Cl_{2(aq)}]$ = Molar concentration of Cl_2.
P_{Cl2} = Partial pressure of chlorine in the atmosphere.

The disinfection capabilities of hypochlorous acid (HOCl) are generally higher than those of hypochlorite ions (OCl^-) (Water, 1978):

$$\text{Henry's law constant } (H) \text{ (mol/L} \cdot \text{atm)} = 4.805 \times 10^{-6} \exp\left(\frac{2818.48}{T}\right) \tag{12.103}$$

Hypochlorous acid is a weak acid and subject to further dissociation to hypochlorite ions (OCl^-) and hydrogen ions:

$$HOCl \leftrightarrow OCl^- + H^+ \tag{12.104}$$

Its acid dissociation constant, K_a, is

$$K_H = \frac{[OCl^-][H^+]}{[HOCl]} \tag{12.105}$$

$$= 3.7 \times 10^{-8} \text{ at } 25°C$$

$$= 2.61 \times 10^{-8} \text{ at } 20°C$$

The value of K_a for hypochlorous acid is a function of temperature in Kelvin (K) as follows (Morris, 1966):

$$\ln K_a = 23.184 - 0.058T - 6908/T \tag{12.106}$$

REFERENCES AND RECOMMENDED READING

AWWA and ASCE. (1990). *Water Treatment Plant Design*, 2nd ed. New York: McGraw-Hill.
AWWA and ASCE. (1998). *Water Treatment Plant Design*, 3rd ed. New York: McGraw-Hill.
Chase, G.L. (2002). *Solids Notes: Fluidization*. Akron, OH: University of Akron.
Cleasby, J.L. (1990). Filtration. In: *Water Quality and Treatment: A Handbook of Community Water Supplies*, 4th ed., Pontius, F.W., Ed. New York: McGraw-Hill.
Cleasby, J.L. and Fan, K.S. (1981). Predicting fluidization and expansion of filter media. *Journal of the Environmental Engineering Division*, 107(EE3), 355–471.
Downs, A.J. and Adams, C.J. (1973). *The Chemistry of Chlorine, Bromine, Iodine, and Astatine*. Oxford: Pergamon.
Droste, R.L. (1997). *Theory and Practice of Water and Wastewater Treatment*. New York: John Wiley & Sons.
Fair, G.M., Geyer, J.C., and Okun, D.A. (1968). *Water and Wastewater Engineering*. Vol. 2. *Water Purification and Wastewater Treatment and Disposal*. New York: John Wiley & Sons.
Fetter, C.W. (1998). *Handbook of Chlorination*. New York: Litton Educational.
Gregory, R. and Zabel, T.R. (1990). Sedimentation and flotation. In: *Water Quality and Treatment: A Handbook of Community Water Supplies*, 4th ed., Pontius, F.W., Ed. New York: McGraw-Hill.
Gupta, R.S. (1997). *Environmental Engineering and Science: An Introduction*. Rockville, MD: Government Institutes.
Hudson, Jr., H.E. (1989). Density considerations in sedimentation. *Journal of the American Water Works Association*, 64(6), 382–386.
McGhee, T.J. (1991). *Water Resources and Environmental Engineering*, 6th ed. New York: McGraw-Hill.
Morris, J.C. (1966). The acid ionization constant of HOCl from 5°C to 35°C. *Journal of Physical Chemistry*, 70(12), 3789.
USEPA. (1999). Individual filter self assessment. In: *EPA Guidance Manual: Turbidity Provisions*. Washington, DC: U.S. Environmental Protection Agency, pp. 5.1–5.17.
Water, G.C. (1978). *Disinfection of Wastewater and Water for Reuse*. New York: Van Nostrand Reinhold.
Wen, C.Y. and Yu, Y.H. (1966). Minimum fluidization velocity. *AIChE Journal*, 12(3), 610–612.
White, G.C. (1972). *Handbook of Chlorination*. New York: Litton Education.

Glossary

A

Absorption: Any process by which one substance penetrates the interior of another substance.

Acid: Refers to water or other liquid that has a pH less than 5.5.

Acid rain: Precipitation with higher than normal acidity, caused primarily by sulfur and nitrogen dioxide air pollution.

Acidic deposition: The transfer of acidic or acidifying substances from the atmosphere to the surface of the Earth or to objects on its surface. Transfer can be either by wet deposition (rain, snow, dew, fog, frost, hail) or by dry deposition (gases, aerosols, or fine to coarse particles).

Acre-foot (acre-ft.): The volume of water required to cover an acre of land to a depth of 1 foot; equivalent to 43,560 cubic feet or 32,851 gallons.

Activated carbon: A very porous material that after being subjected to intense heat to drive off impurities can be used to adsorb pollutants from water.

Adsorption: The process by which one substance is attracted to and adheres to the surface of another substance, without actually penetrating its internal structure.

Aeration: A physical treatment method that promotes biological degradation of organic matter. The process may be passive (when waste is exposed to air), or active (when a mixing or bubbling device introduces the air).

Aerobic bacteria: A type of bacteria that requires free oxygen to carry out metabolic function.

Algae: Chlorophyll-bearing nonvascular, primarily aquatic species that have no true roots, stems, or leaves; most algae are microscopic, but some species can be as large as vascular plants.

Algal bloom: The rapid proliferation of passively floating, simple plant life, such as blue–green algae, in and on a body of water.

Alkaline: Refers to water or other liquid that has a pH greater than 7.

Alluvial aquifer: A water-bearing deposit of unconsolidated material (sand and gravel) left behind by a river or other flowing water.

Alluvium: General term for sediments of gravel, sand, silt clay, or other particulate rock material deposited by flowing water, usually in the beds of rivers and streams, on a flood plain, on a delta, or at the base of a mountain.

Alpine snow glade: A marshy clearing between slopes above the timberline in mountains.

Amalgamation: The dissolving or blending of a metal (commonly gold and silver) in mercury to separate it from its parent material.

Ammonia: A compound of nitrogen and hydrogen (NH_3) that is a common byproduct of animal waste. Ammonia readily converts to nitrate in soils and streams.

Anadromous fish: Migratory species that are born in freshwater, live mostly in estuaries and ocean water, and then return to freshwater to spawn.

Anaerobic: Pertaining to, taking place in, or caused by the absence of oxygen.

Anomalies: As related to fish, externally visible skin or subcutaneous disorders, including deformities, eroded fins, lesions, and tumors.

Anthropogenic: Having to do with or caused by humans.

Anticline: A fold in the Earth's crust, convex upward, whose core contains stratigraphically older rocks.

Aquaculture: The science of faming organisms that live in water, such as fish, shellfish, and algae.

Aquatic: Living or growing in or on water.

Aquatic guidelines: Specific levels of water quality which, if reached, may adversely affect aquatic life. These are nonenforceable guidelines issued by a governmental agency or other institution.

Aquifer: A geologic formation, group of formations, or part of a formation that contains sufficient saturated permeable material to yield significant quantities of water to springs and wells.

Arroyo: A small, deep, flat-floored channel or gully of an ephemeral or intermittent stream, usually with nearly vertical banks cut into unconsolidated material.

Artificial recharge: Augmentation of natural replenishment of groundwater storage by some method of construction, by spreading the water, or by pumping water directly into an aquifer.

Atmospheric deposition: The transfer of substances from the air to the surface of the Earth, either in wet form (rain, fog, snow, dew, frost, hail) or in dry form (gases, aerosols, particles).

Atmospheric pressure: The pressure exerted by the atmosphere on any surface beneath or within it; equal to 14.7 pounds per square inch at sea level.

Average discharge: As used by the U.S. Geological Survey, the arithmetic average of all complete water years of record of surface water discharge whether consecutive or not. The term "average" generally is reserved for average of record and "mean" is used for averages of shorter periods—namely, daily, monthly, or annual mean discharges.

B

Background concentration: Concentration of a substance in a particular environment that is indicative of minimal influence by human (anthropogenic) sources.

Backwater: A body of water in which the flow is slowed or turned back by an obstruction such as a bridge or dam, an opposing current, or the movement of the tide.

Bacteria: Single-celled microscopic organisms.

Bank: The sloping ground that borders a stream and confines the water in the natural channel when the water level, or flow, is normal.

Bank storage: The change in the amount of water stored in an aquifer adjacent to a surface water body resulting from a change in stage of the surface water body.

Barrier bar: An elongate offshore ridge, submerged at least at high tide and built up by the action of waves or currents.

Base flow: The sustained low flow of a stream, usually groundwater inflow to the stream channel.

Basic: The opposite of acidic; refers to water that has a pH greater than 7.

Basin and range physiography: A region characterized by a series of generally north-trending mountain ranges separated by alluvial valleys.

Bed material: Sediment comprising the streambed.

Bed sediment: The material that temporarily is stationary in the bottom of a stream or other watercourse.

Bedload: Sediment that moves on or near the streambed and is in almost continuous contact with the bed.

Bedrock: A general term used for solid rock that underlies soils or other unconsolidated material.

Benthic invertebrates: Insects, mollusk, crustaceans, worms, and other organisms without a backbone that live in, on, or near the bottom of lakes, streams, or oceans.

Benthic organism: A form of aquatic life that lives on or near the bottom of stream, lakes, or oceans.

Bioaccumulation: The biological sequestering of a substance at a higher concentration than that at which it occurs in the surrounding environment or medium. Also, the process whereby a substance enters organisms through gills, epithelial tissues, diet, or other means.

Bioavailability: The capacity of a chemical constituent to be taken up by living organisms either through physical contact or by ingestion.

Biochemical: Refers to chemical processes that occur inside or are mediated by living organisms.

Biochemical oxygen demand (BOD): The amount of oxygen required by bacteria to stabilize decomposable organic matter under aerobic conditions.

Biochemical process: A process characterized by, produced by, or involving chemical reactions in living organism.

Biodegradation: Transformation of a substance into new compounds through biochemical reactions or the actions of microorganisms such as bacteria.

Biological treatment: A process that uses living organisms to bring about chemical changes.

Biomass: The amount of living matter, in the form of organisms, present in a particular habitat, usually expressed as weight per unit area.

Biota: All living organisms of an area.

Bog: A nutrient-poor, acidic wetland dominated by a waterlogged, spongy mat of sphagnum moss that ultimately forms a thick layer of acidic peat; generally has no inflow or outflow and is fed primarily by rain water.

Brackish water: Water with a salinity intermediate between seawater and freshwater (containing from 1000 to 10,000 mg/L of dissolved solids).

Breakdown product: A compound derived by chemical, biological, or physical action upon a pesticide. The breakdown is a natural process that may result in a more toxic or a less toxic compound and a more persistent or less persistent compound.

Breakpoint chlorination: The addition of chlorine to water until the chlorine demand has been satisfied and free chlorine residual is available for disinfection.

C

$C \times T$ value: The product of the residual disinfectant concentration (C), in milligrams per liter, and the corresponding disinfectant contact time (T), in minutes. Minimum $C \times T$ values are specified by the Surface Water Treatment Rule as a means of ensuring adequate killing or inactivation of pathogenic microorganisms in water.

Calcareous: Refers to substance formed of calcium carbonate or magnesium carbonate by biological deposition or inorganic precipitation, or containing those minerals in sufficient quantities to effervesce when treated with cold hydrochloric acid.

Capillary fringe: The zone above the water table in which water is held by surface tension. Water in the capillary fringe is under a pressure less than atmospheric.

Carbonate rocks: Rocks (such as limestone or dolostone) that are composed primarily of minerals (such as calcite and dolomite) containing a carbonate ion.

Center pivot irrigation: An automated sprinkler system with a rotating pipe or boom that supplies water to a circular area of an agricultural field through sprinkler heads or nozzles.

Channel scour: Erosion by flowing water and sediment on a stream channel; results in removal of mud, silt, and sand on the outside curve of a stream bend and the bed material of a stream channel.

Channelization: The straightening and deepening of a stream channel to permit the water to move faster or to drain a wet area for farming.

Chemical treatment: A process that results in the formation of a new substance or substances. The most common chemical water treatment processes include coagulation, disinfection, water softening, and filtration.

Chlordane: Octachlor-4,7-methanotetrahydroindane; an organochlorine insecticide no longer registered for use in the United States. Technical chlordane is a mixture in which the primary components are *cis*- and *trans*-chlordane, *cis*- and *trans*-nonachlor, and heptachlor.

Chlorinated solvent: A volatile organic compound containing chlorine; some common solvents are trichloroethylene, tetrachloroethylene, and carbon tetrachloride.

Chlorination: The process of adding chlorine to water to kill disease-causing organisms or to act as an oxidizing agent.

Chlorine demand: A measure of the amount of chlorine that will combine with impurities and is therefore unavailable to act as a disinfectant.

Chlorofluorocarbons: A class of volatile compounds consisting of carbon, chlorine, and fluorine; commonly called *freons*, which have been used in refrigeration mechanisms, as blowing agents in the fabrication of flexible and rigid foams, and, until banned from use several years ago, as propellants in spray cans.

Cienaga: A marshy area where the ground is wet due to the presence of seepage of springs.

Clean Water Act (CWA): Federal law passed in 1972 (with subsequent amendments) intended to restore and maintain the chemical, physical, and biological integrity of the nation's waters. Its long-range goal is to eliminate the discharge of pollutants into navigable waters and to make national waters fishable and swimmable.

Climate: The sum total of the meteorological elements that characterize the average and extreme conditions of the atmosphere over a long period of time at any one place or region of the Earth's surface.

Coagulants: Chemicals that cause small particles to stick together to form larger particles.

Coagulation: A chemical water treatment method that causes very small suspended particles to attract one another and form larger particles. This is accomplished by the addition of a coagulant, which neutralizes the electrostatic charges causing the particles to repel each other.

Coliform bacteria: A group of bacteria predominantly inhabiting the intestines of humans or animals but also occasionally found elsewhere. The presence of these bacteria in water is used as an indicator of fecal contamination (contamination by animal or human wastes).

Color: A physical characteristic of water. Color is most commonly tan or brown from oxidized iron, but contaminants may cause other colors, such as green or blue. Color differs from turbidity, which is the cloudiness of the water.

Combined sewer overflow: The discharge of untreated sewage and stormwater to a stream when the capacity of a combined storm/sanitary sewer system is exceeded by storm runoff.

Communicable diseases: Usually caused by *microbes*—microscopic organisms including bacteria, protozoa, and viruses. Most microbes are essential components of our environment and do not cause disease. Those that do are called *pathogenic organisms*, or simply *pathogens*.

Community: In ecology, the species that interact in a common area.

Community water system: A public water system that serves at least 15 service connections used by year-round residents or regularly serves at least 25 year-round residents.

Composite sample: A series of individual or grab samples taken at different times from the same sampling point and mixed together.

Concentration: The ratio of the quantity of any substance present in a sample of a given volume or a given weight compared to the volume or weight of the sample.

Cone of depression: The depression of heads around a pumping well caused by withdrawal of water.

Confined aquifer (artesian aquifer): An aquifer that is completely filled with water under pressure and that is overlain by material that restricts the movement of water.

Confining layer: A body of impermeable or distinctly less permeable material located stratigraphically adjacent to one or more aquifers; this layer restricts the movement of water into and out of the aquifers.

Confluence: The flowing together of two or more streams; the place where a tributary joins the main stream.

Conglomerate: A coarse-grained sedimentary rock composed of fragments larger than 2 millimeters in diameter.

Constituent: A chemical or biological substance in water, sediment, or biota that can be measured by an analytical method.

Consumptive use: The quantity of water that is not available for immediate rescue because it has been evaporated, transpired, or incorporated into products, plant tissue, or animal tissue.

Contact recreation: Recreational activities, such as swimming and kayaking, in which contact with water is prolonged or intimate and in which there is a likelihood of ingesting water.

Contaminant: A toxic material found as an unwanted residue in or on a substance.

Contamination: Degradation of water quality compared to original or natural conditions due to human activity.

Contributing area: The area in a drainage basin that contributes water to streamflow or recharge to an aquifer.

Core sample: A sample of rock, soil, or other material obtained by driving a hollow tube into the undisturbed medium and withdrawing it with its contained sample.

Criterion: A standard rule or test on which a judgment or decision can be based.

Cross connection: Any connection between safe drinking water and a nonpotable water or fluid.

D

Datum plane: A horizontal plane to which ground elevations or water surface elevations are referenced.

Deepwater habitat: Permanently flooded lands lying below the deepwater boundary of wetlands.

Degradation products: Compounds resulting from transformation of an organic substance through chemical, photochemical, and/or biochemical reactions.

Denitrification: A process by which oxidized forms of nitrogen such as nitrate are reduced to form nitrites, nitrogen oxides, ammonia, or free nitrogen; commonly brought about by the action of denitrifying bacteria and usually resulting in the escape of nitrogen to the air.

Detection limit: The concentration of a constituent or analyte below which a particular analytical method cannot determine, with a high degree of certainty, the concentration.

Diatoms: Single-celled, colonial, or filamentous algae with siliceous cell walls constructed of two overlapping parts.

Direct runoff: The runoff entering stream channels promptly after rainfall or snowmelt.

Discharge area (groundwater): Area where subsurface water is discharged to the land surface, to surface water, or to the atmosphere.

Discharge: The volume of fluid passing a point per unit of time, commonly expressed in cubic feet per second, million gallons per day, gallons per minute, or seconds per minute per day.

Disinfectants and disinfection byproducts (DBPs): A term used in connection with state and federal regulations designed to protect public health by limiting the concentration of either disinfectants or the byproducts formed by the reaction of disinfectants with other substances in the water (such as trihalomethanes, or THMs).

Disinfection: A chemical treatment method involving the addition of a substance (e.g., chlorine, ozone, hydrogen peroxide) that destroys or inactivates harmful microorganisms or inhibits their activity.

Dispersion: The extent to which a liquid substance introduced into a groundwater system spreads as it moves through the system.

Dissociate: The process of ion separation that occurs when an ionic solid is dissolved in water.

Dissolved constituent: Operationally defined as a constituent that passes through a 0.45-micrometer filter.

Dissolved oxygen (DO): The oxygen dissolved in water, usually expressed in milligrams per liter, parts per million, or percent of saturation.

Dissolved solids: Any material that can dissolve in water and be recovered by evaporating the water after filtering the suspended material.

Diversion: A turning aside or alteration of the natural course of a flow of water, normally considered to be the water physically leaving the natural channel. In some areas, this can be consumptive use direct from another stream, such as for livestock watering. In other areas, a diversion consists of such actions as taking water through a canal, pipe, or conduit.

Dolomite: A sedimentary rock consisting chiefly of magnesium carbonate.

Domestic withdrawals: Water used for normal household purposes, such as drinking, food preparation, bathing, washing clothes and dishes, flushing toilets, and watering lawns and gardens. The water may be obtained from a public supplier or may be self-supplied. Also called *residential water use.*

Drainage area (of a stream): At a specified location, the area, measured in a horizontal plane, enclosed by a drainage divide.

Drainage basin: The land area drained by a river or stream.

Drainage divide: Boundary between adjoining drainage basins.

Drawdown: The difference between the water level in a well before pumping and the water level in the well during pumping. Also, for flowing wells, the reduction of the pressure head as a result of the discharge of water.

Drinking water standards: Water quality standards addressing, for example, suspended solids, unpleasant taste, and microbes harmful to human health. Drinking water standards are included in state water quality rules.

Drinking water supply: Any raw or finished water source that is or may be used as a public water system or as drinking water by one or more individuals.

Drip irrigation: An irrigation system in which water is applied directly to the root zone of plants by means of applicators (orifices, emitters, porous tubing, or perforate pipe) operated under low pressure. The applicators can be placed on or below the surface of the ground or can be suspended from supports.

Drought: A prolonged period of less-than-normal precipitation such that the lack of water causes a serious hydrologic imbalance.

E

Ecoregion: An area of similar climate, landform, soil, potential natural vegetation, hydrology, or other ecologically relevant variables.

Ecosystem: Community of organisms considered together with the nonliving factors of its environment.

Effluent: Outflow from a particular source, such as a stream that flows from a lake or liquid waste that flows from a factory or sewage-treatment plant.

Effluent limitations: Standards developed by the U.S. Environmental Protection Agency to define the levels of pollutants that could be discharged into surface waters.

Electrodialysis: The process of separating substances in a solution by dialysis, using an electric field as the driving force.

Electronegativity: The tendency for atoms that do not have a complete octet of electrons in their outer shell to become negatively charged.

Emergent plants: Erect, rooted, herbaceous plants that may be temporarily or permanently flooded at the base but do not tolerate prolonged inundation of the entire plant.

Enhanced Surface Water Treatment Rule (ESWTR): A revision of the original Surface Water Treatment Rule that includes new technology and requirements to deal with newly identified problems.

Environment: The sum of all conditions and influences affecting the life of organisms.

Environmental sample: A water sample collected from an aquifer or stream for the purpose of chemical, physical, or biological characterization of the sampled resource.

Environmental setting: Land area characterized by a unique combination of natural and human-related factors, such as row-crop cultivation or glacial-till soils.

Ephemeral stream: A stream or part of a stream that flows only in direct response to precipitation; it receives little or no water from springs, melting snow, or other sources, and its channel is at all times above the water table.

EPT richness index: An index based on the sum of the number of taxa in three insect orders, Ephemeroptera (mayflies), Plecoptera (stoneflies), and Trichoptera (caddisflies), which are composed primarily of species considered to be relatively intolerant to environmental alterations.

Erosion: The process whereby materials of the Earth's crust are loosened, dissolved, or worn away and simultaneously moved from one place to another.

Eutrophication: The process by which water becomes enriched with plant nutrients, most commonly phosphorus and nitrogen.

Evaporite minerals (deposits): Minerals or deposits of minerals formed by evaporation of water containing salts. These deposits are common in arid climates.

Evaporites: A class of sedimentary rocks composed primarily of minerals precipitated from a saline solution as a result of extensive or total evaporation of water.

Evapotranspiration: The process by which water is discharged to the atmosphere as a result of evaporation from the soil and surface-water bodies and transpiration by plants.

F

Facultative bacteria: A type of anaerobic bacteria that can metabolize its food either aerobically or anaerobically.

Fall line: Imaginary line marking the boundary between the ancient, resistant crystalline rocks of the Piedmont province of the Appalachian Mountains and the younger, softer sediments of the Atlantic Coastal Plain province in the Eastern United States. Along rivers, this line commonly is reflected by waterfalls.

Fecal bacteria: Microscopic single-celled organisms (primarily fecal coliforms and fecal streptococci) found in the wastes of warm-blooded animals. Their presence in water is used to assess the sanitary quality of water for body-contact recreation or for consumption. Their presence indicates contamination by the wastes of warm-blooded animals and the possible presence of pathogenic (disease-producing) organisms.

Federal Water Pollution Control Act (1972): The objective of the Act is "to restore and maintain the chemical, physical, and biological integrity of the nation's waters." This Act and subsequent Clean Water Act amendments represent the most far-reaching water pollution control legislation ever enacted. They provide for comprehensive programs for water pollution control, uniform laws, and interstate cooperation, as well as grants for research, investigations, and training on national programs on surveillance, the effects of pollutants, pollution control, and the identification and measurement of pollutants. Additionally, they allot grants and loans for the construction of treatment works. The Act established national discharge standards with enforcement provisions, as well as several milestone achievement dates. It required secondary treatment of domestic waste by publicly owned treatment works and the application of "best practicable" water pollution control technology by industry by 1977. Virtually all industrial sources have achieved compliance. (Because of economic difficulties and cumbersome federal requirements, certain publicly owned treatment works obtained an extension to 1988 for compliance.) The Act mandates a strong pretreatment program to control toxic pollutants discharged by industry into publicly owned treatment works and called for new levels of technology to be imposed during the 1980s and 1990s, particularly for controlling toxic pollutants. The 1987 amendments require regulation of stormwater from industrial activity.

Fertilizer: Any of a large number of natural or synthetic materials, including manure and nitrogen, phosphorus, and potassium compounds, spread on or worked into soil to increase its fertility.

Filtrate: Liquid that has been passed through a filter.

Filtration: A physical treatment method for removing solid (particulate) matter from water by passing the water through porous media such as sand or a manmade filter.

Flocculation: Water treatment process that follows coagulation; it uses gentle stirring to bring suspended particles together so they will form larger, more settleable clumps called *floc*.

Flood: Any relatively high streamflow that overflows the natural or artificial banks of a stream.

Flood attenuation: A weakening or reduction in the force or intensity of a flood.

Flood irrigation: The application of irrigation water whereby the entire surface of the soil is covered by ponded water.

Flood plain: A strip of relatively flat land bordering a stream channel that is inundated at times of high water.

Flow line: The idealized path followed by particles of water.

Flowpath: An underground route for groundwater movement, extending from a recharge (intake) zone to a discharge (output) zone such as a shallow stream.

Fluvial: Pertaining to a river or stream.

Freshwater: Water that contains less than 1000 mg/L dissolved solids.

Freshwater chronic criteria: The highest concentration of a contaminant that freshwater aquatic organisms can be exposed to for an extended period of time (4 days) without adverse effects.

G

Grab sample: A single water sample collected at one time from a single point.

Groundwater: Freshwater found under the surface of the Earth, usually in aquifers. Groundwater is a major source of drinking water and a source of growing concern in areas where leaching agricultural or industrial pollutants or substances from leaking underground storage tanks are contaminating it.

H

Habitat: The part of the physical environment in which a plant or animal lives.

Hardness: A characteristic of water caused primarily by the salts of calcium and magnesium. It causes deposition of scale in boilers, damage in some industrial processes, and sometimes an objectionable taste. It can also decrease the effectiveness of soap.

Headwaters: The source and upper part of a stream.

Hydraulic conductivity: The capacity of a rock to transmit water. It is expressed as the volume of water at the existing kinematic viscosity that will move in unit time under a unit hydraulic gradient through a unit area measured at right angles to the direction of flow.

Hydraulic gradient: The change of hydraulic head per unit of distance in a given direction.

Hydrogen bonding: The term used to describe the weak but effective attraction that occurs between polar covalent molecules.

Hydrograph: Graph showing variation of water elevation, velocity, streamflow, or other property of water with respect to time.

Hydrologic cycle: Literally, the water–earth cycle; the movement of water in all three physical forms through the various environmental media (air, water, biota, soil).

Hydrology: The science that deals with water as it occurs in the atmosphere, on the surface of the ground, and underground.

Hydrostatic pressure: The pressure exerted by water at any given point in a body of water at rest.

Hygroscopic: Refers to a substance that readily absorbs moisture.

I

Impermeability: The incapacity of a rock to transmit a fluid.

Index of Biotic Integrity (IBI): An aggregated number, or index, based on several attributes or metrics of a fish community that provides an assessment of biological conditions.

Indicator sites: Stream sampling sites located at outlets of drainage basins with relatively homogeneous land use and physiographic conditions; most indicator-site basins have drainage areas ranging form 20 to 200 square miles.

Infiltration: The downward movement of water from the atmosphere into soil or porous rock.

Influent: Water flowing into a reservoir, basin, or treatment plant.

Inorganic: Containing no carbon; matter other than plant or animal.

Inorganic chemical: A chemical substance of mineral origin not having carbon in its molecular structure.

Inorganic soil: Soil with less than 20% organic matter in the upper 16 inches.

Instantaneous discharge: The volume of water that passes a point at a particular instant of time.

Instream use: Water use taking place within the stream channel for such purposes as hydroelectric power generation, navigation, water quality improvement, fish propagation, and recreation. Sometimes called *nonwithdrawal use* or *in-channel use.*

Intermittent stream: A stream that flows only when it receives water from rainfall runoff or springs, or from some surface source such as melting snow.

Internal drainage: Surface drainage whereby the water does not reach the ocean, such as drainage toward the lowermost or central part of an interior basin or closed depression.

Intertidal: Alternately flooded and exposed by tides.

Intolerant organisms: Organisms that are not adaptable to human alterations to the environment and thus decline in numbers where alterations occur.

Invertebrate: An animal having no backbone or spinal column.

Ion: A positively or negatively charged atom or group of atoms.

Ionic bond: The attractive forces between oppositely charged ions, such as the forces between the sodium and chloride ions in a sodium chloride crystal.

Irrigation: Controlled application of water to arable land to supply requirements of crops not satisfied by rainfall.

Irrigation return flow: The part of irrigation applied to the surface that is not consumed by evapotranspiration or taken up by plants and that migrates to an aquifer or surface water body.

Irrigation withdrawals: Withdrawals of water for application on land to assist in the growing of crops and pastures or to maintain recreational lands.

K

Kill: Dutch term for stream or creek.

L

Lacustrine: Pertaining to, produced by, or formed in a lake.

Leachate: A liquid that has percolated through soil containing soluble substances and that contains certain amounts of these substances in solution.

Leaching: The removal of materials in solution from soil or rock; also refers to movement of pesticides or nutrients from land surfaces to groundwater.

Limnetic: The deepwater zone (greater than 2 meters deep).

Littoral: The shallow-water zone (less than 2 meters deep).

Load: Material that is moved or carried by streams, reported as weight of material transported during a specified time period, such as tons per year.

M

Main stem: The principal trunk of a river or a stream.

Marsh: A water-saturated, poorly drained area, intermittently or permanently water covered, having aquatic and grasslike vegetation.

Maturity (stream): The stage in the development of a stream at which it has reached its maximum efficiency, when velocity is just sufficient to carry the sediment delivered to it by tributaries; characterized by a broad, open, flat-floored valley having a moderate gradient and gentle slope.

Maximum contaminant level (MCL): The maximum allowable concentration of a contaminant in drinking water, as established by state and/or federal regulations. Primary MCLs are health related and mandatory. Secondary MCLs are related to the aesthetics of the water and are highly recommended but not required.

Mean discharge: The arithmetic mean of individual daily mean discharges of a stream during a specific period, usually daily, monthly, or annually.

Membrane filter method: A laboratory method used for coliform testing. The procedure uses an ultrathin filter with a uniform pore size smaller than bacteria (less than a micron). After water is forced through the filter, the filter is incubated in a medium that promotes the growth of coliform bacteria. Bacterial colonies with a green–gold sheen indicate the presence of coliform bacteria.

Method detection limit: The minimum concentration of a substance that can be accurately identified and measured with current lab technologies.

Midge: A small fly in the family Chironomidae. The larval (juvenile) life stages are aquatic.

Minimum reporting level (MRL): The smallest measured concentration of a constituent that may be reliably reported using a given analytical method. In many cases, the MRL is used when documentation for the method detection limit is not available.

Mitigation: Actions taken to avoid, reduce, or compensate for the effects of human-induced environmental damage.

Modes of transmission of disease: The ways in which diseases spread from one person to another.

Monitoring: Repeated observation, measurement, or sampling at a site on a scheduled or event basis for a particular purpose.

Monitoring well: A well designed for measuring water levels and testing groundwater quality.

Multiple-tube fermentation method: A laboratory method used for coliform testing, which uses a nutrient broth placed in a culture tubes. Gas production indicates the presence of coliform bacteria.

N

National Pollutant Discharge Elimination System (NPDES): Permit program authorized by the Clean Water Act that controls water pollution by regulating discharges to any water body. It sets the highest permissible effluent limits prior to making any discharge.

National Primary Drinking Water Regulations (NPDWRs): Regulations developed under the Safe Drinking Water Act that establish maximum contaminant levels, monitoring requirements, and reporting procedures for contaminants in drinking water that endanger human health.

Near Coastal Water Initiative: Initiative developed in 1985 to provide for management of specific problems in waters near coastlines that are not dealt with in other programs.

Nitrate: An ion consisting of nitrogen and oxygen (NO_3). Nitrate is a plant nutrient that is very mobile in soils.
Nonbiodegradable: Substances that do not break down easily in the environment.
Non-point source contaminant: A substance that pollutes or degrades water that comes from lawn or cropland runoff, the atmosphere, roadways, and other diffuse sources.
Non-point source water pollution: Water contamination that cannot be traced to a discrete source (such as a discharge pipe); instead, it originates from a broad area (such as leaching of agricultural chemicals from cropland) and enters the water resource diffusely over a large area.
Nonpolar covalently bonded: Refers to a molecule composed of atoms that share their electrons equally, resulting in a molecule that does not have polarity.
Nutrient: Any inorganic or organic compound needed to sustain plant life.

O

Organic: Containing carbon, but possibly also containing hydrogen, oxygen, chlorine, nitrogen, and other elements.
Organic chemical: A chemical substance of animal or vegetable origin having carbon in its molecular structure.
Organic detritus: Any loose organic material in streams (such as leaves, bark, or twigs) removed and transported by mechanical means, such as disintegration or abrasion.
Organic soil: Soil that contains more than 20% organic matter in the upper 16 inches.
Organochlorine compound: Synthetic organic compounds containing chlorine. As generally used, term refers to compounds containing mostly or exclusively carbon, hydrogen, and chlorine.
Outwash: Soil material washed down a hillside by rainwater and deposited upon more gently sloping land.
Overland flow: The flow of rainwater or snowmelt over the land surface toward stream channels.
Oxidation: A chemical treatment method where a substance either gains oxygen or loses hydrogen or electrons through a chemical reaction.
Oxidizer: A substance that oxidizes another substance.

P

Parts per million (ppm): The number of weight or volume units of a constituent present with each 1 million units of a solution or mixture. Formerly used to express the results of most water and wastewater analyses, ppm is being replaced by milligrams per liter (mg/L). For drinking water analyses, concentrations expressed in parts per million and milligrams per liter are equivalent. A single ppm can be compared to a shot glass full of water inside a swimming pool.
Pathogens: Types of microorganisms that can cause disease.
Perched groundwater: Unconfined groundwater separated from an underlying main body of groundwater by an unsaturated zone.
Percolation: The movement, under hydrostatic pressure, of water through interstices of a rock or soil (except the movement through large openings such as caves).
Perennial stream: A stream that normally has water in its channel at all times.
Periphyton: Microorganisms that coat rocks, plants, and other surfaces on lake bottoms.
Permeability: The capacity of a rock for transmitting a fluid; a measure of the relative ease with which a porous medium can transmit a liquid.
pH: A measure of the acidity (less than 7) or alkalinity (greater than 7) of a solution; a pH of 7 is considered neutral.
Phosphorus: A nutrient essential for growth that can play a key role in stimulating aquatic growth in lakes and streams.

Photosynthesis: The synthesis of compounds with the aid of light.

Physical treatment: Any water treatment process that does not produce a new substance (e.g., screening, adsorption, aeration, sedimentation, filtration).

Point source: A discrete source of water pollutants.

Polar covalent bond: Occurs when the shared pair of electrons between two atoms are not equally held; thus, one of the atoms becomes slightly positively charged and the other atom becomes slightly negatively charged.

Polar covalent molecule: Molecule containing one or more polar covalent bonds; exhibits partial positive and negative poles, causing them to behave like tiny magnets. Water is the most common polar covalent substance.

Pollutant: Any substance introduced into the environment that adversely affects the usefulness of the resource.

Pollution: The presence of matter or energy whose nature, location, or quantity produces undesired environmental effects. Under the Clean Water Act, the term is defined as a manmade or human-induced alteration of the physical, biological, or radiological integrity of water.

Polychlorinated biphenyls (PCBs): A mixture of chlorinated derivatives of biphenyl, marketed under the trade name Aroclor, with a number designating the chlorine content (e.g., Aroclor 1260). PCBs were used in transformers and capacitors for insulating purposes and in gas pipeline systems as a lubricant. Further sales or new uses were banned by law in 1979.

Polycyclic aromatic hydrocarbons (PAHs): A class of organic compounds with a fused-ring aromatic structure. PAHs result from incomplete combustion of organic carbon (including wood), municipal solid waste, and fossil fuels, as well as from natural or anthropogenic introduction of uncombusted coal and oil. PAHs included benzo(*a*)pyrene, fluoranthene, and pyrene.

Population: A collection of individuals of one species or mixed species making up the residents of a prescribed area.

Porosity: The ratio of the volume of voids in a rock or soil to the total volume.

Potable water: Water that is safe and palatable for human consumption.

Precipitation: Any or all forms of water particles that fall from the atmosphere, such as rain, snow, hail, and sleet. The act or process of producing a solid phase within a liquid medium.

Pretreatment: Any physical, chemical, or mechanical process used before the main water treatment processes. It can include screening, presedimentation, and chemical addition.

Primary Drinking Water Standards: Regulations on drinking water quality (under the Safe Drinking Water Act) considered essential for preservation of public health.

Primary treatment: The first step of treatment at a municipal wastewater treatment plant; typically involves screening and sedimentation to remove materials that float or settle.

Public supply withdrawals: Water withdrawn by public and private water suppliers for use within a general community. The water is used for a variety of purposes, such as domestic, commercial, industrial, and public water use.

Public water system: As defined by the Safe Drinking Water Act, any system, publicly or privately owned, that serves at least 15 service connections 60 days out of the year or serves an average of 25 people at least 60 days out of the year.

Publicly owned treatment works (POTW): A waste treatment works owned by a state, local government unit, or Indian tribe, usually designed to treat domestic wastewaters.

R

Rain shadow: A dry region on the lee side of a topographic obstacle, usually a mountain range, where rainfall is noticeably less than on the windward side.

Reach: A continuous part of a stream between two specified points.

Reaeration: The replenishment of oxygen in water from which oxygen has been removed.

Receiving waters: A river, lake, ocean, stream, or other water source into which wastewater or treated effluent is discharged.

Recharge: The process by which water is added to a zone of saturation, usually by percolation from the soil surface.

Recharge area (groundwater): An area within which water infiltrates the ground and reaches the zone of saturation.

Reference dose (RfD): An estimate of the amount of a chemical that a person can be exposed to on a daily basis that is not anticipated to cause adverse systemic health effects over the person's lifetime.

Representative sample: A sample containing all of the constituents present in the water from which it was taken.

Return flow: That part of irrigation water that is not consumed by evapotranspiration and that returns to its source or another body of water.

Reverse osmosis (RO): Solutions of differing ion concentration are separated by a semipermeable membrane. Typically, water flows from the chamber with lesser ion concentration into the chamber with the greater ion concentration, resulting in hydrostatic or osmotic pressure. In RO, enough external pressure is applied to overcome this hydrostatic pressure, thus reversing the flow of water. This results in the water on the other side of the membrane becoming depleted in ions and demineralized.

Riffle: A shallow part of the stream where water flows swiftly over completely or partially submerged obstructions to produce surface agitation.

Riparian: Pertaining to or situated on the bank of a natural body of flowing water.

Riparian rights: A concept of water law under which authorization to use water in a stream is based on ownership of the land adjacent to the stream.

Riparian zone: Pertaining to or located on the bank of a body of water, especially a stream.

Runoff: That part of precipitation or snowmelt that appears in streams or surface water bodies.

Rural withdrawals: Water used in suburban or farm areas for domestic and livestock needs. The water generally is self-supplied and includes domestic use, drinking water for livestock, and other uses such as dairy sanitation, evaporation from stock-watering ponds, and cleaning and waste disposal.

S

Safe Drinking Water Act (SDWA): A federal law passed in 1974 with the goal of establishing federal standards for drinking water quality, protecting underground sources of water, and setting up a system of state and federal cooperation to ensure compliance with the law.

Saline water: Water that is considered unsuitable for human consumption or for irrigation because of its high content of dissolved solids (generally expressed as mg/L of dissolved solids); seawater is generally considered to contain more than 35,000 mg/L of dissolved solids. A general salinity scale is

	Concentration of dissolved solids (mg/L)
Slightly saline	1000–3,000
Moderately saline	3000–10,000
Very saline	10,000–35,000
Brine	More than 35,000

Saturated zone: A subsurface zone in which all the interstices or voids are filled with water under pressure greater than that of the atmosphere.

Screening: A pretreatment method that uses coarse screens to remove large debris from the water to prevent clogging of pipes or channels to the treatment plant.

Secondary Drinking Water Standards: Regulations developed under the Safe Drinking Water Act that established maximum levels of substances affecting the aesthetic characteristics (taste, color, or odor) of drinking water.

Secondary maximum contaminant level (SMCL): The maximum level of a contaminant or undesirable constituent in public water systems that, in the judgment of the U.S. Environmental Protection Agency, is required to protect the public welfare. SMCLs are secondary (nonenforceable) drinking water regulations established by the USEPA for contaminants that may adversely affect the odor or appearance of such water.

Secondary treatment: The second step of treatment at a municipal wastewater treatment plant. This step uses growing numbers of microorganisms to digest organic matter and reduce the amount of organic waste. Water leaving this process is chlorinated to destroy any disease-causing microorganisms before its release.

Sedimentation: A physical treatment method that involves reducing the velocity of water in basins so the suspended material can settle out by gravity.

Seep: A small area where water percolates slowly to the land surface.

Seiche: A sudden oscillation, caused by the wind, of a moderate-size body of water.

Sinuosity: The ratio of the channel length between two points on a channel to the straight-line distance between the same two points; a measure of meandering.

Soil horizon: A layer of soil that is distinguishable from adjacent layers by characteristic physical and chemical properties.

Soil moisture: Water occurring in the pore spaces between the soil particles in the unsaturated zone from which water is discharged by the transpiration of plants or by evaporation from the soil.

Solution: Formed when a solid, gas, or another liquid in contact with a liquid becomes dispersed homogeneously throughout the liquid. The substance that dissolves is called a *solute*, and the liquid in which it dissolves is called the *solvent.*

Solvated: Refers to when either a positive or negative ion becomes completely surrounded by polar solvent molecules.

Sorb: To take up and hold, by either absorption or adsorption.

Sorption: General term for the interaction (binding or association) of a solute ion or molecule with a solid.

Specific yield: The ratio of the volume of water that will drain under the influence of gravity to the volume of saturated rock.

Spring: Place where a concentrated discharge of groundwater flows at the ground surface.

Surface runoff: Runoff that travels over the land surface to the nearest stream channel.

Surface tension: The molecules at the surface of water hold onto each other tightly because there are no molecules pulling on them from the air above. As the molecules on the surface stick together, they form an invisible "skin."

Surface water: All water naturally open to the atmosphere, and all springs, wells, or other collectors that are directly influenced by surface water.

Surface Water Treatment Rule (SWTR): A federal regulation established by the U.S. Environmental Protection Agency under the Safe Drinking Water Act that imposes specific monitoring and treatment requirements on all public drinking water systems that draw water from a surface water source.

Suspended sediment: Sediment that is transported in suspension by a stream.

Suspended solids: Different from suspended sediment only in the way in which the sample is collected and analyzed.

Synthetic organic chemicals (SOCs): Generally applies to manufactured chemicals (e.g., herbicides, pesticides, chemicals widely used in industry) that are not as volatile as volatile organic compounds.

T

Total head: The height above a datum plane of a column of water. In a groundwater system, it is composed of elevation head and pressure head.

Total suspended solids (TSS): Solids present in wastewater.

Transpiration: The process by which water passes through living organisms, primarily plants, into the atmosphere.

Trihalomethanes (THMs): A group of compounds formed when natural organic compounds from decaying vegetation and soil (such as humic and fulvic acids) react with chlorine.

Turbidity: A measure of the cloudiness of water caused by the presence of suspended matter, which shelters harmful microorganisms and reduces the effectiveness of disinfecting compounds.

U

Unconfined aquifer: An aquifer whose upper surface is a water table free to fluctuate under atmospheric pressure.

Unsaturated zone: A subsurface zone above the water table in which the pore spaces may contain a combination of air and water.

V

Vehicle of disease transmission: Any nonliving object or substance contaminated with pathogens.

Vernal pool: A small lake or pond that is filled with water for only a short time during the spring.

W

Wastewater: The spent or used water from individual homes, a community, a farm, or an industry that contains dissolved or suspended matter.

Water budget: An accounting of the inflow to, outflow from, and storage changes of water in a hydrologic unit.

Water column: An imaginary column extending through a water body from its floor to its surface.

Water demand: Water requirements for a particular purpose, such as irrigation, power, municipal supply, plant transpiration, or storage.

Water softening: A chemical treatment method that uses either chemicals to precipitate or a zeolite to remove those metal ions (typically Ca^{2+}, Mg^{2+}, Fe^{3+}) responsible for hard water.

Water table: The top water surface of an unconfined aquifer at atmospheric pressure.

Waterborne disease: Water is a potential vehicle of disease transmission, and waterborne disease is possibly one of the most preventable types of communicable illness. The application of basic sanitary principles and technology has virtually eliminated serious outbreaks of waterborne diseases in developed countries. The most prevalent waterborne diseases include typhoid fever, dysentery, cholera, infectious hepatitis, and gastroenteritis.

Watershed: The land area that drains into a river, river system, or other body of water.

Wellhead protection: The protection of the surface and subsurface areas surrounding a water well or well field supplying a public water system from contamination by human activity.

Y

Yield: The mass of material or constituent transported by a river in a specified period of time divided by the drainage area of the river basin.

Index

A

B

C

D

E

F

I

J

K

L

N

O

P

Q

R

S

T

U

V

W

Z